Geological Society of America
Memoir 194

Paleozoic and Mesozoic tectonic evolution of central Asia: From continental assembly to intracontinental deformation

Edited by

Marc S. Hendrix
Department of Geology
University of Montana
Missoula, Montana 59812
USA

and

Gregory A. Davis
Department of Earth Sciences
University of Southern California
Los Angeles, California 90089
USA

2001

Published by The Geological Society of America, Inc.
3300 Penrose Place, P.O. Box 9140, Boulder, Colorado 80301
www.geosociety.org

Printed in U.S.A.

GSA Books Science Editor Abhijit Basu
GSA Books Editor Rebecca Herr
Cover design by Margo Good

Library of Congress Cataloging-in-Publication Data

Paleozoic and Mesozoic tectonic evolution of central Asia : from continental assembly to intracontinental deformation / edited by Marc S. Hendrix and Gregory A. Davis.
    p. cm. —(Memoir ; 194)
    Includes bibliographical references and index.
    ISBN 0-8137-1194-0
    1. Geology, Stratigraphic—Paleozoic. 2. Geology, Stratigraphic—Mesozoic.
3. Geology, Structural—Asia, Central. I. Hendrix, Marc S., 1963- II. Davis, Gregory A. (Gregory Arlen), 1935- III. Memoir (Geological Society of America) ; 194.

QE654.P2412   2000
551.7′2′0958—dc21

                                      00-064048

Cover: Fenster and klippe northeast of Kashgar (Wenguri), in the western Tarim basin, northwest China. The décollement places Lower Permian marine turbidites and chert on Cretaceous nonmarine red beds (see Carroll et al., Chapter 2, this volume).

10 9 8 7 6 5 4 3 2 1

Contents

Geological Society of America
Memoir 194
2001

Preface

Asia is the largest mosaic of assembled tectonic elements on Earth. Although there are several large cratonic blocks in the mosaic, much of this tectonic amalgam consists of a variety of smaller entities of variable character and origin, including volcano-plutonic arc assemblages, collapsed ocean basins and their sedimentary cover, and small blocks of transitional crustal nature. The assembly history of these tectonic elements and the evolution of the Asian continent to its present form is long and complicated. Most of the geologic evidence suggests that individual tectonic entities accreted piecemeal to a growing Asian continental nucleus, mainly by addition to the southern and eastern sides of the Asian continent.

Further complicating the tectonic history of Asia is its protracted record of intracontinental deformation. Asia is commonly regarded as the quintessential natural laboratory for the study of intracontinental deformation. Much of this intracontinental deformation has been associated with continental margin growth and propagation of related deformation into the continental interior. For example, peaks with elevation >7400 m occur in the Tian Shan (Heavenly Mountains) 1200 km north of the Indus-Tsangpo suture, the nominal boundary between the currently converging Indian and Asian plates. In recent years, the vast majority of tectonic studies in Asia have focused on the Cenozoic history of collision and intracontinental deformation associated with the ongoing India-Asia collision. However, an equally impressive and relatively overlooked history of pre-Cenozoic intracontinental deformation exists in the orogenic belts of central and eastern Asia and in the fill of associated sedimentary basins.

Appreciating this history of ancient intracontinental deformation is crucial to understanding the nature and kinematics of modern intracontinental deformation in Asia, because considerable evidence suggests that tectonic inheritance has played a significant role in localizing major orogenic belts within the continent. For example, a fundamental tectonic boundary exists between a collection of tectonically amalgamated magmatic arcs in southern Mongolia and adjacent portions of China and rocks of continental affinity to the south. This tectonic boundary, alternatively called the Suolon suture or the Junggar-Hegen suture, extends across much of northern China, from the Sino-Kazakhstan border in the west to the longitude of the Korean Peninsula to the east (see Hendrix et al., this volume, Fig. 1A). Although the ocean basin(s) that formerly separated the two sides of this suture zone are generally thought to have closed by late Paleozoic time, several episodes of major contractile deformation have occurred since then in close association with this plate boundary. As documented in several of the papers that follow, large-scale contractional Mesozoic structures are present in the Junggar, Tarim, and Turpan basins of western China, the Beishan and Daqing Shan along tectonic strike to the east, and the Yinshan belt in northeastern China, even further east along tectonic strike. Significant modern shortening in response to the India-Asia collision now appears to be localized in this ancient perisutural region, as demonstrated by significant deformation and elevation of the Chinese Tian Shan in contrast to the little-deformed Tarim basin, located much closer to the Indus-Tsangpo suture.

Despite the long-recognized importance of Asia as the world's largest mosaic of assembled tectonic elements, and its profound record of ancient and modern intracontinental deformation, much of the continent has been made accessible to the international earth science community only within the past 10–15 years. For example, Mongolia is situated in the middle of central Asia and is key to understanding the tectonic history of the continent, but this country was closed to non-Soviet citizens prior to 1990. Likewise, since the late 1980s, a relaxation of travel restrictions has occurred in much of China, allowing onground examination of many key localities that formerly could be studied only by remote sensing.

This volume presents ground-based work conducted during the past decade by a number of different scientists, laboratories, and institutes on the Paleozoic and Mesozoic history of assembly and intracontinental deformation of central and eastern Asia. Many of the chapters in this volume, particularly those focusing on Mongolian geology, are among the first ground-truth reports available in the English language for these areas. In presenting these chapters, we elected to group them thematically. The first group of chapters focuses on the structural, thermochronologic, and sedimentary records of the history of Paleozoic assembly of central and eastern Asia, both in Mongolia and central and western China. The second group of chapters focuses on the record of Mesozoic deformation in orogenic belts of central and eastern Asia, mainly through structural, geochronologic, and thermochronologic documentation and analysis. The final group of chapters in the volume focuses on the record of intracontinental deformation in Asia as preserved in the fill of Asia's vast sedimentary basins. These papers examine the record of changing paleoenvironments, thermal histories, provenance, and paleoclimates and document the sedimentary record of ancient contractile, extensional, and strike-slip deformation in central and eastern Asia.

This volume would not have been possible without the generous and continuous support of a large number of individuals, organizations, and institutions. Much of central and eastern Asia remains a frontier, and working conditions can be very difficult. Many sites described in this volume can be accessed only after several days of off-road driving, and it is commonly necessary for expeditions to carry all food, water, and motor-vehicle fuel en route. Obviously, such field work would not be possible without the support of Asian colleagues. Individual agencies, institutions, and individuals that provided critical field and logistical support are acknowledged in each chapter. Similarly, a large number of governmental, private, and industrial institutions provided critical financial backing for many of the studies presented in this volume, and those agencies and companies are acknowledged in each chapter within the volume.

Marc S. Hendrix
Missoula, Montana

Gregory A. Davis
Los Angeles, California

March 2000

Geological Society of America
Memoir 194
2001

Assembly of central Asia during the middle and late Paleozoic

Christoph Heubeck*
BP Amoco, P.O. Box 3092, Houston, Texas 77253, USA

ABSTRACT

The core of Asia was assembled during the late Paleozoic by numerous collisions between small continental blocks and by accretionary growth along convergent margins. Although the contributions of the major cratons to the overall growth of Asia through time are relatively well understood, the integration of the kinematics, sedimentary environments, and paleogeography of the large number of smaller elements into the picture is in its infancy. The set of paleogeographic maps in this study is a result of superpositioning fixist paleoenvironmental data sets from China and the former Soviet Union on a digital model of tectonic elements. This merger showed only moderate agreement across the common border through time, but fit known plate tectonic boundaries. Structural, paleontological, paleomagnetic, and paleoclimatic data from other sources appear to be less ambiguous indicators than facies correlation in determining the relationship of individual tectonic elements to their neighbors.

Prior to final closure in the Permian, the western segment of the Turkestan ocean separated a northern domain composed of Siberia, Baltica, and the Altaids from an east-west elongate group of elements that extended from North China through Tarim and the Tadjik depression westward to the Scythian-Turanian platform. This group shows indications of geographic continuity beginning in the Early Carboniferous. The southern margin of this group of largely continental blocks formed an east-west-oriented subduction zone along which the Paleotethys was subducted northward. One or several marginal basins existed between the Hindu Kush and the Kunlun during the Early Carboniferous. During the late Paleozoic, Yili, the North Tian Shan arc, and the volcanic arcs of southern Mongolia, all of which belonged to the Turkestan ocean realm, accreted to the North China–Tarim–Turan assemblage, typically along diachronously closing basins. Diachronous, west to east closure of the Turkestan ocean throughout the late Paleozoic and collision of the Altaids with the eastern European margin contributed to the construction of the nuclear Eurasian landmass.

Large-scale kinematic problems in central Asia include, among others, the eastward and westward extent of the Tarim block during Paleozoic time, the timing of closure of the Turkestan ocean, the tectonic framework of the Aral Sea–pre-Caspian region, and the magnitude of Permian strike-slip tectonics.

INTRODUCTION

The geographic core of Asia is a mosaic of continental blocks and arc fragments separated by accretionary complexes (Figs. 1 and 2). This core was constructed by numerous collisions among tectonic blocks that were small compared to the cratons (North China, South China, India, Arabia, Baltica, Siberia) surrounding them. Tectonic assembly took place largely in the Paleozoic and early Mesozoic. The growth in available paleomagnetic, structural, geochronologic, and sedimentary data on central Asia, in particular over the past decade,

---

*Present address: Freie Universität Berlin, Department of Geosciences, 12249 Berlin, Germany.

Heubeck, C., 2001, Assembly of central Asia during the middle and late Paleozoic, *in* Hendrix, M.S., and Davis, G.A., eds., Paleozoic and Mesozoic tectonic evolution of central Asia: From continental assembly to intracontinental deformation: Boulder, Colorado, Geological Society of America Memoir 194, p. 1–22.

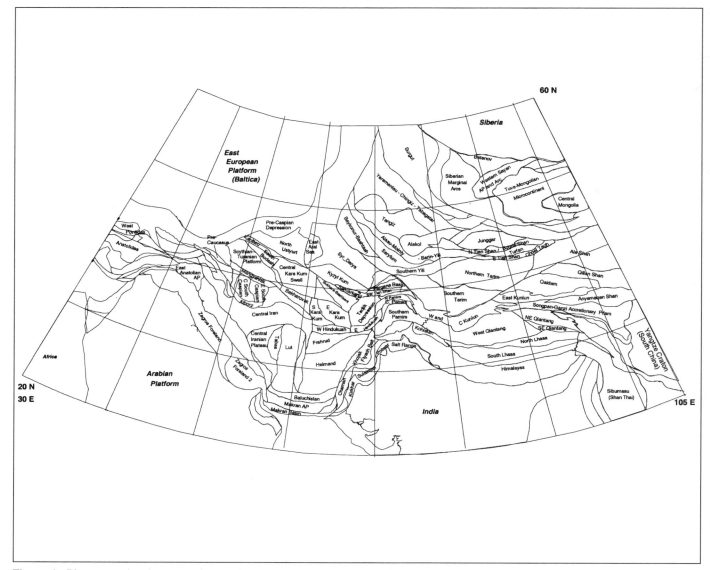

Figure 1. Plate tectonic elements of central Asia. See text for definition of individual polygons and discussion of nature of their margins.

has allowed the reconstruction of its tectonic evolution in increasing detail (e.g., Burrett, 1974; Zonenshain and Gorodnitsky, 1977; Zonenshain et al., 1987a, 1990; Şengör, 1984; Zhang et al., 1984; Watson et al., 1987; Boulin, 1991; Chen et al., 1993; Mossakovsky et al., 1993; Şengör and Natal'in, 1996).

One of the tools commonly used to better understand continental dynamics is a digital plate kinematic model that is consistent through time and in space (e.g., Scotese and Golonka, 1992). Construction of such a model generates rigorous, quantifiable interactions among tectonic elements that make apparent problems of rate, scale, or position. This approach is best applied to problems of a scale in which the motion of the elements is governed by the rigid-body principles of classical plate tectonics. In multiply or plastically deformed regions such as

the regions of central Asia discussed here, however, the rigid-body approach to continental dynamics reaches its limits, and simplifying assumptions must be used. These assumptions are defined and discussed in the following.

Plate kinematic maps and their logical continuation, paleogeographic maps, have found a number of applications. In the academic environment, maps showing the spatial and temporal distribution of land and sea, the positions of shorelines, major depositional environments, and an estimate of topography contribute to the understanding of biogeography and paleoclimate (e.g., Rowley et al., 1985; A.M. Ziegler et al., 1979; P.A. Ziegler, 1990; Parrish, 1982; Golonka et al., 1994; Kutzbach and Ziegler, 1994). Paleogeographic maps also help construct hypotheses to be tested and modified by new data. In the petro-

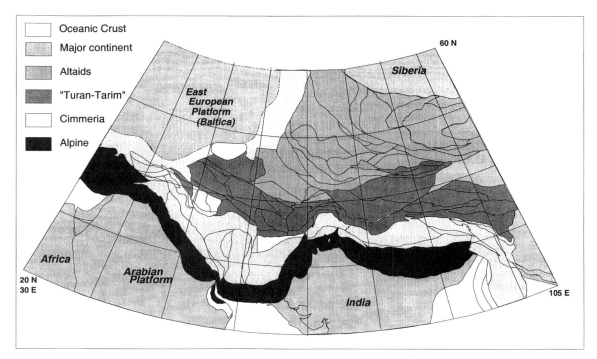

Figure 2. Major tectonic orogenic belts of Asia, as discussed in this chapter.

leum industry, paleogeographic base maps are used to predict expected type of source rocks, regional variations of expected depositional systems, trap types and timing of formation, and thermal basin history. Properly used, these tools are helpful to explorationists searching for productive trends and to those ranking petroliferous basins on a regional or worldwide basis (e.g., Parrish, 1982; Moore and Barron, 1994).

This study attempts to illustrate the Devonian–Permian assembly of central Asia using a series of time-slice paleogeographic maps. The digital model was excerpted from a study of the Phanerozoic assembly of Asia undertaken by Amoco in the mid-1990s. Previous published models can be assigned to three different groups: (1) sketch maps principally dealing with the territory of the former Soviet Union (e.g., Zonenshain et al., 1990; Şengör et al., 1993a; Mossakovsky et al., 1993; Didenko et al., 1994; Şengör and Natal'in, 1996); (2) sketch maps dealing principally with the territory of China (e.g., Yin and Nie, 1996); and (3) digital models, some of them global. Models of this latter group bridge the national borders but usually show limited spatial resolution (e.g., Rowley et al., 1985; Ronov et al., 1989; Scotese and McKerrow, 1990; Scotese and Langford, 1995; Ziegler et al., 1997). The model proposed here has four objectives. First, it attempts to integrate data from both China and the former Soviet Union and to improve the level of detail of previous digital models. A second objective is to quantify the sketches shown by Yin and Nie (1996) for the tectonic assembly of China and to examine the consequences of their reconstructions for the assembly of Asia outside China. Third, I attempted to examine the compatibility of the detailed paleo-

facies and paleogeographic information available (in a fixist framework) on both sides of the China–former Soviet Union border. A particular emphasis was placed on compiling geologic constraints on the assembly of the little known region between the Caucasus and Tarim.

The study area extends from the Caspian Sea to western Mongolia and from the Songpan-Ganzi complex of central China to the southern Urals (Figs. 1, 2, and 3). The southern limit of the study area is outlined by the paleo-Asian suture, which closed in the Late Triassic and Early Jurassic and attached the Tethysides to Asia (Şengör and Natal'in, 1996). I intentionally excluded the bulk of Mongolia from the area of interest because it is studied in more detail in this volume; consequently, I placed stronger emphasis on the blocks between Tarim and the Caspian Sea.

METHOD

I digitized tectonically significant, mappable fault traces of central Asia, such as ophiolitic belts, suture zones, and selected major faults from published maps (Commission for the Geologic Map of the World, 1976; Chen et al., 1985; Cheng et al., 1986) and constructed closed polygons denoting the present-day geologic outline of individual tectonic elements (Fig. 1). A modified version of the PALEOMAP software (Scotese, 1998) was used to interactively move individual or groups of tectonic elements on a virtual sphere into their positions relative to each other. Poles of rotation were derived from summaries of

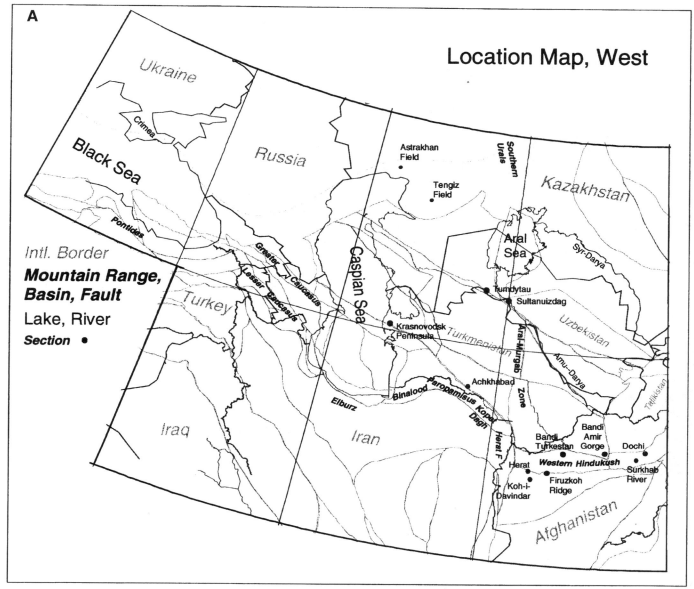

Figure 3. A (this page), B (facing page): Location maps to localities mentioned in text. Line weights represent tectonic boundaries (light gray, from Fig. 1, for orientation), thin black lines represent coastal boundaries, thick black lines represent international political boundaries.

published paleomagnetic data (e.g., Zonenshain et al., 1987a, 1987b, 1990; Sharps et al., 1989; Zhao et al., 1990, 1993; Besse and Courtillot, 1991; Nie, 199; Enkin et al., 1992; Chen et al., 1993; Gilder et al., 1996). Latitudinal constraints were derived from paleoclimatic data (e.g., A.M. Ziegler et al., 1979; A.M. Ziegler, 1990; Nie et al., 1990; Nie, 1991). Detailed paleogeographic maps of Asia exist in atlas form both of the former Soviet Union and China (Vinogradov et al., 1968; Wang, 1985; Lai and Wang, 1988). These map series represent compilations of enormous quantities of stratigraphic data and served as my single largest source of information of stratigraphic and paleoenvironmental information for central Asia (Fig. 3). However,

these maps were assembled in a fixist framework, not taking into account horizontal motions of tectonic elements relative to each other, and end at national boundaries. In this study, geochronologic and sedimentologic data defined the relative position of tectonic elements to each other and their sequence and geometry of collision. This model consists of ~75 elements for central Asia (Fig. 1). The Decade of North American Geology time scale (Palmer, 1983) was followed where possible.

The polyorogenic character of the central Asian mountain belts makes the unambiguous identification, location, and dating of former plate-bounding zones impossible. Commonly, former transform boundaries have been reactivated as thrusts;

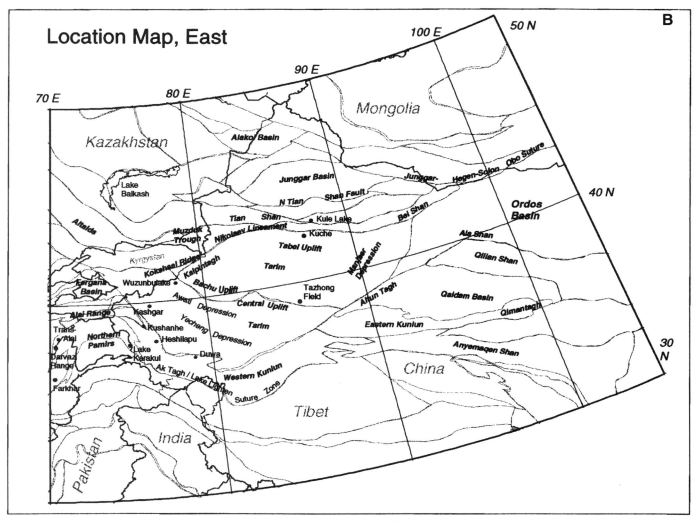

Figure 3. (*continued*)

oroclinal bending has deformed, dismembered, and obscured critical structures; and major faults have reorganized to accommodate deformation along more efficient splays. Choices in location and nature of former plate boundaries must be made. These are discussed in the following.

In using a detailed rigid-body approach in a region of complex continental deformation, it is crucial to define the intent and limitations of the tools. None of the digitized polygons should be considered terranes ("fault-bounded geological entities of regional extent that characterized by geological histories different from those of neighbouring terranes"; Schermer et al., 1984). Their boundaries should not necessarily be considered suture zones in the sense of Şengör and Natal'in (1996, p. 487) ("a line along which two preexisting nonsubductable pieces of lithosphere have been opposed following the demise of intervening oceanic lithosphere"). Rather, many polygon boundaries are merely proxies to accommodate distributed zones of inter-

nal deformation. I deemed them necessary to simplify the actual tectonic pattern. For example, Mesozoic intraplate deformation in the lowlands of central Asia was taken into account by individually digitizing clearly intraplate polygons (e.g., central Karakum swell, Kopet Dagh, central Turan; Fig. 1) along major fault zones and slightly separating them from each other throughout the time period under consideration. This enlarged the Turanian block into a Paleozoic precursor.

Individual blocks should not be considered to be necessarily underlain by buoyant continental lithosphere. In many cases, polygons are placeholders for objects constructed principally of arc-related material or subduction-accretion complexes (e.g., Altaids, Kunlun, Qaidam, North Tian Shan arc; Fig. 1). Rigid-body mechanics do not apply to these elements of such composition and size, and any present-day geometry is only a very crude approximation or former size and shape. Their usefulness, however, is in reminding us that these elements existed as tectonic entities approximately at that time and approximately at the locations shown.

The greatest uncertainties in defining individual tectonic blocks exist in the desert lowlands of central Asia between the Caucasus and Tarim (Fig. 3). There, the outlines of the individual blocks were taken from published Russian literature (e.g., Vinogradov et al., 1968; Popkov, 1995; Zonenshain et al., 1987a, 1987b; Clarke, 1994). Where the geometrical definition of tectonic elements was ambiguous, I chose to err on the side of detail. The margins of Tarim and adjacent blocks were digitized from Chen et al. (1985) and the *Geological Atlas of the World* (Commission for the Geologic Map of the World, 1976).

Geodynamic reconstructions followed a hierarchical model and an iterative process. First, the locations of the major cratons (East European Platform, Siberia, North China, South China, Tarim) were determined at specific times using the data sources discussed above. Second, their movement was interpolated through time. Third, smaller elements were placed into their proper positions relative to these cratons and moved with respect to them. Within the limits of the study area, the position of Siberia, almost undefined by paleomagnetic data for the late Paleozoic (Smethurst et al., 1998), posed the largest uncertainties in the geodynamic reconstructions.

As a consequence of the incomplete tectonic understanding of this region, the disparate nature of the data, and the limitation of the rigid-body approach to continental geodynamics, the model presented here is necessarily simplistic and nonunique. The results should be seen as an attempt to provide a framework for more detailed future studies, to identify problems between the existing data sets, and to predict timing and geometry of deformation. Data integration in the form of paleogeographic maps or geodynamic models, especially between geologic and paleomagnetic data, is far from complete and occasionally controversial (e.g., Enkin et al., 1992; Nie and Rowley, 1994; Courtillot et al., 1994; Şengör and Natal'in, 1996). On a global scale, A.M. Ziegler et al. (1997) accomplished this integration admirably in their set of detailed global paleogeographical maps for the Permian. In the area of interest, I generally attempted to give equal weight to geochronologic, sedimentary, geomagnetic, biostratigraphic, and paleoclimatic data. Choices had to be made, which are documented in the discussion of the individual time slices.

In the following, paleogeographic constraints on the individual tectonic elements of central Asia are sequentially discussed in a clockwise sense around the Paleozoic Turkestan ocean: from North China westward to the Caspian Sea, then north and east through Kazakhstan toward Siberia (Fig. 3).

DISCUSSION

Geodynamic evolution of central Asia

Pre-Devonian rocks of central Asia are commonly either poorly exposed (Scythian-Turanian platform, Tarim, Ala Shan) or severely disrupted and poorly preserved (e.g., Kunlun, Pamirs,

Hindu Kush, Qaidam, Junggar) (Figs. 1 and 2). The best evidence comes from the exposed margins of large, relatively coherent block such as Tarim, North China, Siberia, and South China. The pre-Devonian locations of smaller entities, with their paucity of biostratigraphic, paleomagnetic, and paleoclimatic indicators, can be estimated only with great uncertainty; Şengör and Natal'in (1996) showed the limits of this technique for the Altaids.

The majority of maps showing pre-Devonian geodynamic patterns show North China, South China, and in some cases, Tarim as part of Cambrian-Ordovician Gondwana (e.g., Burrett, 1974; Burrett et al., 1987; Metcalfe, 1996). Not enough is known on the location of North China, South China, and Tarim and their relationship with Gondwana. Nie et al. (1990) and Nie (1991) related the supraregional Ordovician-Carboniferous unconformity of North China to rift-related uplift and separation of this block from Gondwana. Burrett et al. (1987) also argued, largely on the basis of paleoclimatic data, for an association of North and South China with Gondwana, most likely Australia. However, considering the desired level of detail and the method chosen, the Devonian appeared to be the earliest time period for which a semiquantitative paleogeographic reconstruction could be made. Reconstructions for earlier time periods would either be reduced in their level of detail or would portray an unwarranted accuracy of position.

Middle Devonian, 380 Ma

Regrettably, there are few constraints on the paleogeographic position of North China during the middle Paleozoic because the uppermost Ordovician to lowermost Carboniferous is represented by a major supraregional unconformity spanning ~150 m.y. (Figs. 4 and 5). Nie (1991), interpolating paleocli-

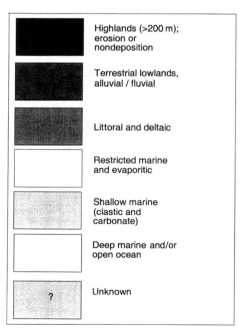

Figure 4. Legend to paleogeographic maps.

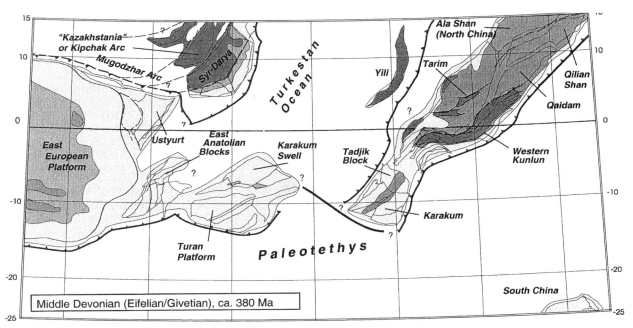

Figure 5. Middle Devonian paleogeography of central Asia. General notes to following reconstructions (see Figs. 4–9): Middle Devonian, Early Carboniferous, and Late Carboniferous maps are Lambert conic conformal projections (1st parallel: 20°N, 2nd parallel, 10°S). Early Permian and Late Permian maps are Lambert conic conformal projections (1st parallel: 60°N, 2nd parallel, 20°N). Devonian paleogeography: north China paleogeography is modified from Liu (1990) and Wang (1985). Western China paleogeography is from Petroleum Industry Press (1987). Southwest Asian, Uralian, and Kazakhstanian paleogeography is modified from Vinogradov et al. (1968), Leven (1997), Vachard (1980), and others (see text). Strike-slip separation between Turan and Tadjik is conjectural and is only invoked to avoid excessively southerly position of Turan. North China–Tarim connection follows reasoning of Zhou and Graham (1996).

matic data from the strata bracketing this unconformity, suggested a location in the dry subtropical belt (10°–15° latitude for Upper Ordovician strata, 5°–20° latitude for Lower Carboniferous strata). Some sedimentary record of this vast span of time may be preserved in locations marginal to the platform, such as in the Ala Shan (west of the Ordos basin, Fig. 3), the Qilian Shan, and Korea. Paleomagnetic data for the Silurian and Devonian from these locations indicate a near-equatorial latitude of ~7°N (Zhao et al., 1993).

In addressing the question of structural basement continuity between North China and Tarim, I follow the reasoning of Zhou and Graham (1996) and Yue et al. (this volume). Zhou and Graham argued, on the basis of the Late Silurian collision of the combined Qilian Shan–Qaidam block with the Bei Shan (which apparently was part of North China) and the apparent structural continuity of Qaidam basement with Tarim basement, that Tarim and North China had established a continental link by the Devonian (Petroleum Industry Press, 1987). Yue et al. (this volume) extend this argument by demonstrating a close correlation between the continental margins of both blocks, modified by post-Paleozoic strike-slip faulting. Yin and Nie (1994) suggested that terminal collision between Qaidam and the North China block may have contributed to widespread uplift of the North China block, resulting in the regional unconformity between Ordovician and Carboniferous strata. Paleomagnetic

data, however, show that the two blocks did not begin moving as a single entity until the Cretaceous. Whereas Mesozoic discordances are small enough to be accounted for by adjustments within the Asian collage, the Paleozoic differences are substantial and have given rise to protracted discussions, summarized in Yue et al. (this volume). In the map series (Figs. 5–9), I have therefore shown North China and Tarim loosely connected but sufficiently separated to allow adjustments in orientation with respect to each other.

The question of structural contiguity between Tarim and Qaidam was discussed by Sobel and Arnaud (1999), who concluded that the Altyn Tagh Range between these two basins contains an early Paleozoic oceanic basin of unknown extent that closed by southward subduction prior to the Middle Devonian and formed the Lapeiquan suture. To the north of this suture are Precambrian crystalline rocks best interpreted as Tarim basement; to the south of it, such rocks are absent but replaced by Proterozoic and Paleozoic volcanic and metasedimentary rocks that resemble rocks known from the other margins of the Qaidam basin. Overall, outcrop data around the margins of Qaidam and seismic records indicate that the Qaidam basement is unlike the Tarim basement. Pre-Devonian strata probably representative of Qaidam basement consist of deformed and metamorphosed fragments of Proterozoic and early Paleozoic magmatic arc assemblages (Petroleum Industry

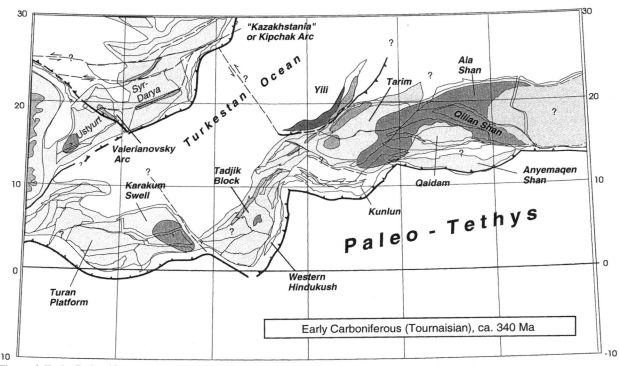

Figure 6. Early Carboniferous paleogeography of central Asia. Strike-slip movement along Aral-Murgab zone is conjectural. Pale-ozoic stratigraphy of interior of Scythian-Turanian-Tadjik blocks is largely unknown and inferred from ir margins; existing strati-graphic information is largely taken from Vinogradov et al. (1968). Likewise, Qaidam interior is not exposed, largely undrilled, and is therefore shown only schematically. No record is preserved for this time period in north China. Position of Junggar basin is poorly defined; see text for discussion. Note opening of marginal basins along Hindu Kush–Kunlun margin facing Paleotethys.

Press, 1987). In contrast, the widespread stable shelf assem-blages of the early Paleozoic, characteristic for the Tarim block, have not been found in Qaidam (Wang and Coward, 1990). Post-Silurian strata on both Tarim and Qaidam, however, are sufficiently similar to warrant the treatment of the two blocks as contiguous for the Devonian to Permian time period (Fig. 5). For example, Upper Devonian strata in the Qaidam basin are widespread along all margins and generally overlie the de-formed lower Paleozoic basement with angular unconformity (Petroleum Industry Press, 1987; Wang and Coward, 1990). They are largely continental redbeds that were initially de-posited in fault-bounded depressions; they exhibit rapid lateral changes in thickness and lithology, are associated with inter-mediate volcanics in the Qimantag Mountains to the south, and grade upward into marine clastics and carbonates. They resem-ble the thick continental redbeds of similar age in Tarim (Petro-leum Industry Press, 1987).

Northward subduction of oceanic crust beneath the western Kunlun had probably begun in the early Paleozoic (Fig. 5). There, a southward-younging accretionary complex developed throughout the Paleozoic (for a synthesis see Şengör and Na-tal'in, 1996). The pre-Devonian basement complex in the west-ern Kunlun is represented by, in part, isoclinally folded and probably structurally duplicated schist and phyllite of great ap-parent thickness, metagraywacke, and local limestone. Defor-

mation was accompanied by widespread arc-related plutonism and volcanism (Şengör and Okurogullari, 1991; Yao and Hsü, 1992; Mattern et al., 1996). This accretionary complex is poorly known. The early Paleozoic arc complex was metamorphosed and uplifted in the Middle Devonian. The cause may have been a change in plate motion or subduction of a spreading ridge (Şengör and Okurogullari, 1991). A pre-Carboniferous marginal basin was probably involved in the northwestern Kunlun (Matte et al., 1996; Sobel and Arnaud, 1999), which may have closed by southward subduction. Unmetamorphosed Upper Devonian terrestrial deposits and Carboniferous clastics unconformably overlie the metasedimentary and metavolcanic basement rocks.

In the Middle and Late Devonian, the Tarim block under-went its most significant Paleozoic deformation. The hitherto apparently stable basement fragmented internally through block faulting and flexing, thereby segmenting Tarim into a number of subbasins and uplifts (Hu et al., 1969; Tian et al., 1989; Ministry of Geology and Mineral Resources [MGMR], 1993; McKnight, 1993; Fig. 5). In contrast to the thermal events in the Kunlun at its southern margin, a major magmatic or metamorphic pulse did not accompany this deformation within the Tarim block. This deformation (early Hercynian of the Chinese literature) el-evated, tilted, and segmented the Tabei and Central uplifts. As a result, the Manjaer, Awati, and Yecheng depressions of Tarim (Fig. 3) deepened and received detritus from the rising ranges

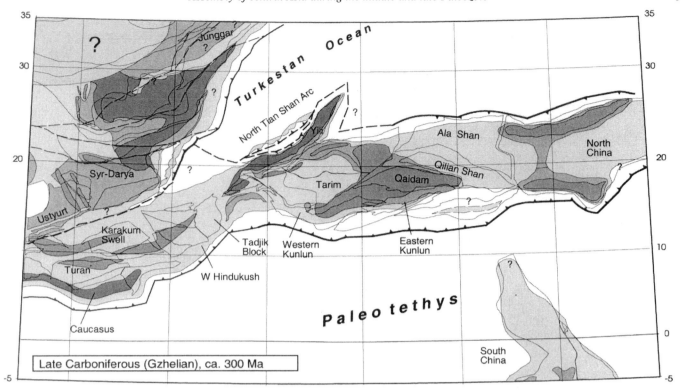

Figure 7. Late Carboniferous paleogeography of central Asia. Great majority of paleoenvironmental information is from Vinogradov et al. (1968) and Wang (1985). Closure along Aral-Murgab zone is conjectural. Note that paleogeography of Qaidam, Kunlun, and Anyemaqen Shan is poorly known and shown schematically. Width of remaining Turkestan ocean in Late Carboniferous is unknown. Location of Junggar basin on either side of Turkestan ocean depends on nature of its (unexposed) basement; see text for discussion. Note collapse of Early Carboniferous backarc basins along south-facing margin of paleo-Asia.

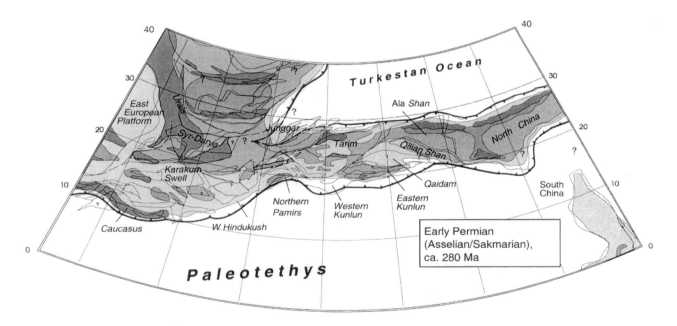

Figure 8. Early Permian paleogeography of central Asia. Position of Junggar basin is poorly defined. Paleogeography of East European platform, Turan plate, Ural region, and Altaid blocks is modified from Vinogradov et al. (1968) and other sources (see text). Paleogeography of Qaidam and Kunlun is largely schematic. Position of Mongolian arcs within Turkestan ocean is unknown and is not displayed. Change in map projection was made to avoid excessive distortion in northern latitudes by Mercator projection of previous figures.

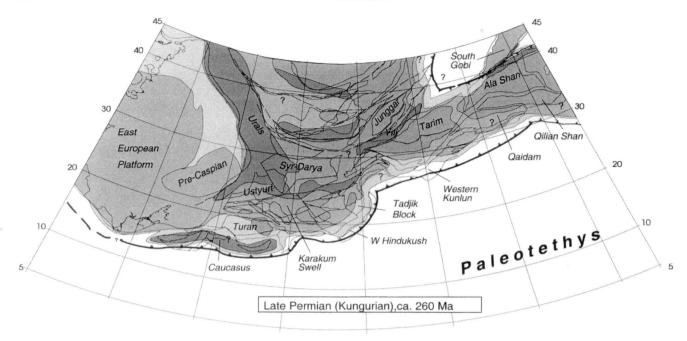

Figure 9. Late Permian paleogeography of central Asia. Great majority of paleoenvironmental boundaries is from Vinogradov et al. (1968) and Wang (1985). Note extensive occurrence of continental clastic facies and development of evaporitic pre-Caspian depression.

within and outside Tarim (Carroll et al., this volume). Mapping along the southern margin of Tarim (MGMR, 1993) indicates that large quantities of siliciclastic detritus that had been generated during uplift of the Kunlun arc complex to the south were carried northward. Alluvial and fluvial deposits, shed under arid conditions and showing rapid facies and thickness changes in outcrop and in seismic sections, filled the depressions (Liao et al., 1992; McKnight, 1993). A renewed period of tectonic stability was reached only in the latest Devonian or earliest Carboniferous and is indicated by the widespread occurrence of the shoreline facies Donghe sandstone and correlative facies over the eroded pre-Carboniferous land surface. This event forms a distinct angular unconformity on regional seismic lines. A more detailed knowledge of the timing, sediment provenance, and distribution of depositional systems for this event, which is crucial for the understanding of Paleozoic petroleum systems in Tarim, is lacking.

In the far western Chinese Tian Shan, northwest of the Kalpintagh, the Middle to Upper Devonian strata consist of coastal or marine facies and include lagoonal or carbonate-dominated shallow shelf facies rocks (Anonymous, 1991). These strata may correlate along strike with the Devonian marine sequences of the northern margin of the Tarim plate exposed in the South Tian Shan. For example, the Kule Lake section in the south-central Tian Shan, north of Kuche (Fig. 3), exposes Lower Devonian, thinly bedded, cherty limestone and minor shale at the base, grading upward into pure radiolarian chert. This sequence is overlain abruptly by fine-grained siliciclastic turbidites (Wang et al., 1996; Zhou et al., this volume).

Compared with the underlying Upper Silurian carbonate shelf strata, the Devonian represents a gradual deepening of the north-facing platform.

Whether Tarim ever extended farther to the west is unknown. The evidence is circumstantial because Paleozoic strata of western Tarim dip south and west beneath the Tertiary Yecheng depression (Fig. 3). In the Tadjik depression of northeastern Afghanistan, Jurassic evaporites form a regional decollement that prevents exposures of Paleozoic strata despite intensive folding of the suprasalt sedimentary sequence (e.g., Leith, 1982; Thomas et al., 1994). The predominant view in the Russian literature is that Tarim was contiguous to the west with a number of loosely coherent, semiconsolidated pieces of continental crust, collectively termed the Tarim-Karakum microcontinent of Zonenshain et al. (1987a, 1987b, 1990) or the Ustyurt-Tarim continent of Mukhin et al. (1989). The Chinese literature, to my knowledge, does not address the question of a western extension of the Paleozoic Tarim block.

Middle Paleozoic strata representing the northern margin of the Tadjik depression and the Bukhara-Beshkent block (part of the poorly defined Karakum block, Fig. 1) are exposed in the foothills of the southern Tian Shan, just south of the Ghissar suture (Burtman, 1975; Fig. 3). Partly metamorphosed Silurian to Devonian rocks there consist of thick, mixed siliciclastic-carbonate marine strata (Burtman, 1975; Mukhin et al., 1989; Fig. 5) and represent a gradually subsiding, passive, north-facing continental margin. Evidence for tectonic loading of this margin, and of carbonate sedimentation being increasingly replaced by clastic sedimentation, becomes stronger in the Late

Devonian (Mukhin et al., 1989). Farther south, regionally metamorphosed mixed carbonate-siliciclastic shelf sequences of ?Silurian–Devonian age are exposed in the eastern Hindu Kush of northeastern Afghanistan (Vachard, 1980; Wensink, 1991; Fig. 1). Along the southern margin of the Tadjik block in the Hindu Kush, thick Devonian limestones suggest the existence of a shallow-water carbonate platform (Vachard, 1980). Metavolcanics are rare (Wensink, 1991). Upper Devonian–Lower Carboniferous rocks of the southern Turanian plate are exposed in the Kopet Dagh of northeastern Iran (Fig. 3; Baud and Stampfli, 1989; Alavi et al., 1997). They consist of slightly metamorphosed volcaniclastic and pyroclastic rocks interlayered with limestone. In my reconstruction (Fig. 5), I indicate that the South and East Karakum blocks, together with the Tadjik block, may have been tectonically separated from the blocks farther to the west (central Karakum swell, Bakharosk, Turan platform) along the north-south–oriented Aral-Murgab zone, the site of Permian-Triassic rifting (Clarke, 1994). This separation, however, is conjectural because both the basal rift fill and the basement geology of northern Turkmenistan and southwestern Kazakhstan in general are known largely from geophysical data only (e.g., Avdeyev, 1984). The Devonian reconstruction for this region is poorly defined and awaits better paleomagnetic or stratigraphic data.

The reconstruction of the Altaids in this map sequence largely follows that of Şengör et al. (1993a) and Şengör and Natal'in (1996). The kinematic sequence is simplified and the number of blocks greatly reduced because the Altaids are a prime example of accretionary tectonics in which rigid-body tectonics is clearly not applicable. West-directed subduction of the Uralian paleo-ocean below the Mugodzhar arc, part of the Uralian margin of the East European platform, began in the Devonian (Fig. 5; Zonenshain et al., 1984; Brown et al., 1998; Puchkov, 1997). The Mugodzhar arc, in turn, was separated from the stable shallow-water East European platform by an internally complex, north-south elongate marginal basin (the Magnitogorsk oceanic basin of Puchkov, 1997, or Sakmara-Magnitogorsk marginal sea of Şengör and Natal'in, 1996; see also A.M. Ziegler, 1990; Matte et al., 1993). Starting in the Ordovician, parts of the magmatic arc between Baltica and Siberia (the Kipchak arc of Şengör et al., 1993a) were stacked and shortened by left-lateral strike-slip faulting. Their western elements had begun to interact with the Mugodzhar arc by the Late Devonian (Puchkov, 1997). Rifting and spreading west of the Mugodzhar arc created the oceanic seafloor of the north Caspian depression (Burke, 1977; Şengör and Natal'in, 1996).

The major ocean separating Siberia sensu lato from the Cathaysian blocks (e.g., North China, South China, Tarim) continued to be subducted northward beneath the Kipchak arc (Şengör et al., 1993a). For its western part, I use Burtman's (1975) term Turkestan ocean. The Mongol-Okhotsk ocean (e.g., Kosygin and Parfenov, 1981; Parfenov and Natal'in, 1985; Rowley et al., 1985; Zonenshain et al., 1990; Nie et al., 1990, 1993), in contrast, separated Siberia proper from microcontinents and arc

complexes summarily known as Amuria in the Russian literature (e.g., Zonenshain et al., 1990).

Early Carboniferous (Tournaisian), 340 Ma

Marine transgression from the north over an eroded peneplain restored sedimentation to the North China platform in the Namurian, and widespread coal deposits began to form in the Middle Carboniferous (Liu, 1990) (Fig. 6). Limited occurrence of evaporites of Early Carboniferous age suggest a paleo-latitude of 5°–20° (Nie, 1991).

Outcrops of Lower Carboniferous strata in Qaidam are limited to the basin margin. They are generally separated from the underlying Devonian redbeds by an unconformity and consist of thick, in part evaporitic, platformal carbonate, interbedded with calcareous sand and shale representing littoral and neritic environments (Wang and Coward, 1990). In the Qimantagh (Fig. 3), southwest of the Qaidam basin, the platform carbonates are interbedded with acid to intermediate volcanics and thick sequences of partially metamorphosed, fine- to coarse-grained clastics. This transition may indicate proximity to the active margin to the south. Overall, the lithology and sequence of Lower Carboniferous strata strongly resemble those of Tarim described in the following.

Paleomagnetic data suggest that a major clockwise rotation of Tarim took place between Late Devonian and Late Carboniferous (Li, 1990; Fang et al., 1990; Gilder et al., 1996). This is consistent with a diachronous east-to-west closure of the ocean to the north of Tarim. Shallow-marine limestone in southwest Tarim and evaporites to the east suggest that Tarim occupied a position in the subtropical dry belt of ~15°–30° latitude (Nie, 1991). The end-Devonian–earliest Carboniferous marine transgression from the west and the northwest over the eroded Devonian topography of Tarim represented the beginning of a period of marine and restricted marine deposition (Hu et al., 1969). This transgression reached as far as east-central Tarim where evaporitic tidal platform and lagoonal facies dominate. The lagoonal facies, in turn, bordered the subaerially exposed east Tarim platform. However, in many places, the relief created by the Devonian Hercynian orogeny in Tarim was not eroded completely, and the transgression left uncovered parts of the Tabei uplift, the Tazhong uplift, the Bachu uplift, and its western along-strike extension, in the present-day Kalpintagh. These uplifts remained as large islands in the shallow Carboniferous sea (Hu et al., 1969; McKnight, 1993; P. Brenckle, 1995, personal commun.).

In northern Tarim, the collision between Tarim and Yili (also known as Central Tian Shan) is recorded in the southern Tian Shan suture (the Qinbulak-Qawabulak fault, part of the Nikolaev lineament; Figs. 1 and 3). This suture is commonly interpreted as representing the northward subduction of oceanic crust and accretion of the north-facing passive margin of Tarim to an active continental margin on the south side of Yili. Collision occurred probably in the Late Devonian–Early Carboniferous

(Allen et al., 1992; Gao et al., 1998) and was complete by the Tournaisian. A well-defined suture zone in northern Tarim marks the collision. Discontinuous belts of ultramafic rock, fault-bound slivers exposing blueschist facies metapelite, and Early Carboniferous plutons occur along the Nikolaev lineament (Liou et al., 1989; Allen et al., 1992; Shi et al., 1994; Gao et al., 1998). There is also good evidence for an elongate, east-west–oriented flysch-dominated trough along the southern margin of the Tian Shan at the time of collision (Carroll et al., 1995, this volume; Zhou et al., this volume; Fig. 6). This basin probably represents a thrust-loaded precollisional foreland basin (Zhou et al., this volume). Most of its original width is buried beneath the south-directed thrust sheets of the southern Tian Shan. Across the border in Kyrgystan to the west, an identical sedimentation style occurred in rocks of similar age in the Kokshaal Ridge and the Muzduk trough (Fig. 3; Sinitsyn, 1957; Burtman, 1975; Zonenshain et al., 1990). This may be the along-strike continuation of the Yili flysch trough. Better characterization of this trough through mapping and dating could better define the history of western Yili.

In the far-western Kalpintagh, close to the border to Kyrgystan, a carbonate ramp margin developed in the Carboniferous, reportedly with reefal mounds (Wuzunbulake section; Anonymous, 1991; Fig. 3). I speculate that this western margin of Tarim may have faced the Turkestan ocean and was not yet affected by the collision of Yili with northern Tarim.

Several previous reconstructions have asserted that the Yili microcontinent was part of Kazakhstania (or the Altaid assemblage) and that with the Yili-Tarim collision, the Altaids were colliding with Tarim. Figures 6–9 show an alternative and readily testable interpretation in which Yili, after collision with Tarim, is inserted in the Altaid stack by subsequent strike-slip faulting. This allows the Turkestan ocean to remain open longer. It also allows (1) the North Tian Shan arc to accrete to the free northern face of Yili, and (2) the principal suture to occur either in or north of Junggar basin from which it can readily be extended eastward toward the Hegen suture (Figs. 3 and 6).

Subduction-related sedimentation and volcanism along the southern margin of paleo-Asia for this time can be traced from the Anyemaqen Shan (Fig. 6) westward along the southern margin of the eastern Kunlun, offset left-laterally along the Altun Tagh fault, and continuing westward through the western Kunlun and along the Ak Tagh–Lake Lighten suture zone (Fig. 3; Şengör et al., 1993b). The Anyemaqen Shan appears to be largely a product of Carboniferous arc magmatism and associated accretionary tectonics; it may have occupied a southerly position relative to Qaidam and possibly constituted a linking segment to the subduction zone dipping below the (present-day) southern margin of North China.

A short-lived backarc basin opened and closed in the northern Pamirs and the western Kunlun in the Early Carboniferous (Fig. 6). In the Chinese literature, its remnants are described as a >600-km-long string of dismembered and discontinuous ophiolitic fragments in the Chinese territory of the Kunlun. It has become known as the northern ophiolite zone of west Kunlun (Yang et al., 1996), the western part of the Kudi mélange (Yao and Hsü, 1992), the north Kuda tectonic belt (Li et al., 1996), and the Oytag-Kudi suture (Mattern et al., 1996, which also contains a good field description of the Kudi ophiolite). The easternmost ophiolites of this belt can be documented at the southern end of the Altun Tagh fault (Yang et al., 1996), from where they can be traced westward through the western Kunlun and, after right-lateral offset on the Karakorum fault, into the north Pamir suture. Across the border, the Akzhilga zone in Tadjik territory, west of Lake Karakul, has virtually identical characteristics (Fig. 6; Kravchenko, 1979; Bazhenov and Burtman, 1982; Ruzhentsev and Shvolman, 1983; Burtman and Molnar, 1993; Novikov et al., 1994) and may be the along-strike continuation of the Oytag-Kudi suture. Ophiolitic sections associated with calc-alkaline volcanism of Early Carboniferous age extend farther into the Darvaz and Trans-Alai Mountains of northern Afghanistan (Fig. 3), where the orogenic strike turns south into northwestern Afghanistan (Bazhenov and Burtman, 1982).

Age comparisons appear to confirm the continuation of a single backarc basin across the China-Tadjikistan border. Pillow basalts of the Kudi ophiolite are ca. 360 Ma (Famennian-Tournaisian) in age (Jiang et al., 1992, in Yang et al., 1996). Goniatites in limestone stringers interbedded with pillow basalts in the Trans-Alai Mountains of Tadjikistan have been dated as early Namurian in age (Bazhenov and Burtman, 1982), and limestone in the upper part of the section contains Visean foraminifera (Budanov and Pashkov, 1988 [cited in Burtman and Molnar, 1993]). South of the backarc suture, the Kurgovat microcontinent of the northern Pamirs may represent a piece of continental crust rifted off as the backarc basin opened (Bazhenov and Burtman, 1982; Ruzhentsev and Shyolman, 1982, in Burtman and Molnar, 1993). The northern margin of the backarc basin fill is apparent in Carboniferous strata in extreme southwestern Tarim (e.g., Duwa section; Fig. 3; Anonymous, 1991) which record thick deep-water limestones, siliciclastic turbidites, and subordinate calc-alkaline volcanics. The general opinion in the literature is that the backarc basin collapsed in the Carboniferous, probably by southward subduction of its oceanic crust beneath the Kunlun arc (Yang et al., 1986). In contrast, Mattauer et al. (1985) favored a Cambrian to Middle Devonian southward subduction of a major ocean basin, the proto-Tethys, beneath the southern Kunlun, closure of this ocean by the Middle Devonian, and subsequent uplift, erosion, and blanketing by Late Devonian molasse deposits. Further field work and age dating defining Carboniferous events along this margin need to be conducted to confidently choose one model over the other.

The available geologic evidence suggests that the marginal arc along the Kunlun-Pamir margin of the Tarim-Tadjik block may have continued southwest into the eastern and western Hindu Kush (Fig. 1). Tournaisian volcanic rocks, interbedded with limestone and siliciclastic strata, are exposed in the Koh-i-Davindar area near Herat in western Afghanistan (Fig. 3;

Bazhenov and Burtman, 1982; Wensink, 1991) and along the Firuzkoh Ridge (Fig. 3; Vachard, 1980). They are unconformably overlain by thick deep-water clastics that grade into Upper Carboniferous strata. Farther east, in the Dochi and Surkhab River region of the western Hindu Kush, Early Carboniferous (likely Tournaisian) calc-alkaline volcanic rocks with interbedded pillow basalts were built on the Devonian shelf carbonates (Vachard, 1980; Şengör, 1984; Wensink, 1991; Burtman and Molnar, 1993; Fig. 3). These, in turn, are unconformably overlain by Visean limestone and sandstone (Boulin and Bouyx, 1977). Boulin (1991) and Burtman and Molnar (1993) also described Early Carboniferous syntectonic metamorphism and deformation from the Hindu Kush proper. Therefore, the available stratigraphic and structural evidence suggests the collapse of a ?pre-Early Carboniferous marginal basin along the southern margin of the Tadjik and Karakum blocks (Vachard, 1980; Boulin, 1988, 1991; Burtman and Molnar, 1993). The interior of these blocks was dominated by shallow-water carbonate environments characteristic of an epicontinental platform in a warm, subtropical climate (Fig. 6; Boulin, 1988).

The feature dominating the Early Carboniferous reconstruction of central Asia, however, is the immensely long, overall east-west–oriented, south-facing arc-trench complex created by the north-dipping subduction of Paleotethyan oceanic crust (Fig. 6). This margin is readily traced from the eastern Kunlun westward as far as western Turkey (Khain, 1979; Boulin, 1988, 1991; Şengör, 1979, 1984, 1990; Şengör et al., 1993b). Its least known segment is between the western Hindu Kush and the Greater Caucasus. There, apparently two Paleotethys-related subduction-collision belts exist: a northern belt can be traced from the western Hindu Kush by strong, northwest-trending magnetic anomalies along the northern margin of the Kopet Dagh to the Paleozoic metavolcanics of the Krasnovodsk Peninsula on the eastern shore of the Caspian Sea (Figs. 1 and 3). From there, the suture is inferred to trend across the Caspian Sea into the Greater Caucasus and possibly to the Crimea Peninsula (Bazhenov and Burtman, 1982; Boulin, 1991; Okay and Sahinturk, 1997; Fig. 3). All along this trend, middle Paleozoic basement appears to be overlain by late Paleozoic to Early Triassic shelf and deep-water strata, and intruded by late Paleozoic calc-alkaline plutonic rocks. Ophiolitic rocks are unconformably overlain by Jurassic molasse facies (Boulin, 1991). Alternatively, a Paleozoic orogenic belt can also be traced along a more southerly trend through ophiolitic fragments, partially metamorphosed deep-water sedimentary rocks, and calc-alkaline igneous rocks from the Herat fault of the western Hindu Kush into the Paropamisus, the Binalood, the central Elburz (Alborz) Mountains, the Lesser Caucasus, and the Pontides of northern Turkey (Fig. 3; e.g., Stöcklin, 1974; Ruttner, 1984; Alavi et al., 1997; Boulin, 1991). Along the Iranian segment of this line, the metamorphism and magmatism is late Paleozoic, and ophiolites are unconformably overlain by Lower Jurassic strata (Boulin, 1991). Boulin (1991) suggested that this along-strike duplication of the late Paleozoic active margin of southwest Asia may

be an artifact of Cenozoic indentation of the Arabian block into Eurasia and duplication of a segment of the Paleozoic active margin by dextral strike slip along the remarkably straight, present-day active Greater Caucasus–Achkhabad fault system. I have chosen the southern trend as the margin of paleo-Asia because of the lack of evidence for significant strike-slip displacement along the mentioned fault system. Evidence for and against such displacement, and its implications, should be tested and could be readily modeled.

East of the Hindu Kush, the main Paleotethys subduction zone likely passed south of the north Pamir–west Kunlun backarc basin along the Tanymas-Akbaytal thrust (Boulin, 1991; Bazhenov and Burtman, 1982, and references therein) and the western Kunlun fault (Cheng, 1986; Gaetani et al., 1990). This fault is equivalent to the northern main suture zone of Huang and Chen (1987) and passes into the Lighten Lake cryptic suture zone of Baud (1989). In the western Kunlun, this suture (the Taaxi-Qiaoertianshan-Hongshanhu suture; Mattauer et al., 1985) or the Ak Tagh–Qizil Yilga–Lake Lighten suture of Şengör and Okurogullari (1991) also forms the northern border of the Gondwana-related Karakorum (Gaetani et al., 1990).

The location and arrangement of continental blocks bordering the western Turkestan ocean on both sides are poorly known (Fig. 6). The location of the former oceanic basin can be constrained by lining up ophiolitic remnants along its suture zone, by describing the location of the magmatic arc along its subducting northern margin, and by mapping land to shelf to slope facies belts along its southern, passive margin. Ophiolitic remnants are discontinuously exposed along the Nikolaev lineament of the Tian Shan and southern Yili (Burtman, 1975; Carroll et al., 1995). They can be traced westward across the Fergana fault, along the southern margin of the Fergana basin (Leith, 1982; Khaimov, 1986), where the best ophiolite of the Turkestan ocean described so far is documented from the Alay Range (Peive, 1973, *in* Zonenshain et al., 1990, p. 69), and through the Alai Range of the southwest Tian Shan (Fig. 3; Burtman, 1975). From there, the suture continues westward along the northern margin of the Ghissar Range and the Samarkand and Kyzyl Kum blocks (Nuratau zone, Fig. 1) to the ophiolitic fragments exposed in the Sultanuizdag Ridges south of the Aral Sea (Figs. 1 and 3). There, the major volcanic activity is Late Silurian and mainly Devonian (Lutts and Feldman, 1992). Tectonic maps of the Sultanuizdag region (e.g., Lutts and Feldman, 1992) generally show the ophiolitic trend to turn sharply north and become the Kazakhstan-Uralian suture (Avdeyev, 1984; Zonenshain et al., 1984, 1990; Şengör et al., 1993a); however, the continuation of the north-south–trending part of the Turkestan suture into the Transuralian-Kazakhstanian suture (the Urkash fault of the southern Urals) is poorly understood (Popkov, 1995; Puchkov, 1997). The second means of constraining the position of the Early Carboniferous Turkestan ocean, calc-alkaline magmatism along its subducting margin, is readily found in the Beltau-Kurama volcanic belt of Bashkirian age, which extends from the eastern Fergana basin to the Aral

Sea (Zonenshain et al., 1990; part of the Valerianov-Chatkal unit of Şengör et al., 1993a). Due to poor outcrops, its details are poorly known. It is likely, however, that the northern margin of the Turkestan ocean during the Early Carboniferous was a complex tectonic zone of oblique subduction and left-lateral stacking of arc material, carbonate platforms, and small continental nuclei cored by Precambrian rocks. Abidov et al. (1997) provided a good example of the tectonic and stratigraphic complexity of a limited region south of Lake Aral. This apparent complexity contrasts with the absence of active-margin tectonics along the southern margin of the Turkestan ocean. In the Mangyshlak Mountains east of the Caspian Sea (Fig. 1), for example, mixed carbonate-clastic Paleozoic sequences, some deposited as turbidites, lack volcanic strata and suggest a northward-facing passive margin.

Along the southwestern shelf of the stable East European Platform, in the north Caspian region, massive reef complexes and shallow-water carbonate platforms, some of which form giant oil reservoirs today (Tengiz, Astrakhan, Karachaganak), developed throughout the Early Carboniferous along the margins of the pre-Caspian depression (e.g., Pavlov et al., 1988; Zonenshain et al., 1990; Cook et al., 1991; Murzagaliyev, 1996). This basin had opened in the Middle Devonian, possibly as a pull-apart basin, by the rifting of the Ustyurt platform from Baltica and closed incompletely in the Permian; it is widely believed to be underlain by trapped oceanic crust (Fig. 6; Zonenshain et al., 1990).

The leading western margin of the stacked Altaid arcs, the Valerianovskaya magmatic arc (Fig. 6), overrode the Uralian ocean by eastward subduction (Zonenshain et al., 1984). This arc was active only for ~30 m.y. (Visean–Middle Carboniferous). Oceanic crust was subducted until the end of the Middle Carboniferous, after which continental collision between the Valerianovskaya magmatic arc (with its heterogeneous Kazakhstanian hinterland) and the Magnitogorsk arc began (ca. 320 Ma). A marginal basin with doubly convergent subduction zones separated this arc from the East European Platform. Rapid collapse of this basin (the Magnitogorsk marginal basin) in the Tournaisian and Visean resulted in westward thrusting, loading of the East European carbonate shelf, development of a deep flysch trough, and in the first terrigenous influx onto the shelf from the east (Zilair Formation) (Zonenshain et al., 1984; Puchkov, 1997).

Along the southern margin of the Kipchak arc, at least one marginal basin of Middle Carboniferous age existed. Remnants of this basin are exposed in the Ghissar zone of the southwest Tian Shan (Figs. 1 and 3) (Kravchenko, 1979; Leith, 1982; Zonenshain et al., 1992; Mukhin et al., 1989). There, oceanic pillow basalts and sheeted dike complexes are overlain by Middle Carboniferous calc-alkaline volcanics and Late Carboniferous flysch (Kravchenko, 1979). Farther east, Early Carboniferous strata along the northeastern margins of the Junggar basin include thick volcaniclastic fluvial to alluvial conglomerate with interbedded coal (Feng et al., 1989; Carroll et al., 1990, their Junggar platform). In contrast, thick andesitic volcaniclas-tics exposed along the southern margin of the Junggar basin were deposited under deep-marine conditions. Carroll et al. (1990) interpreted this contrast as one line of evidence supporting major separation of these areas by the Junggar ocean, which I interpret simplistically in the Late Carboniferous reconstruction (Fig. 7) as the northern margin of the closing Turkestan ocean or one of its northern embayments (Carroll et al., 1990; their Fig. 8, A and B).

Late Carboniferous (Gzhelian), 300 Ma

South-directed subduction of oceanic crust beneath the northern margin of the North China block began in the Late Carboniferous at the latest (Wang, 1985, 1986; Şengör et al., 1991; Nie et al., 1990) (Fig. 7). There, structurally disrupted Upper Carboniferous volcanic rocks are interbedded with mixed carbonate-siliciclastic strata; the fossil content suggests a warm climate and an open marine environment (Mueller et al., 1991). The onset of subduction apparently reversed the regional northward-sloping relief of the North China block and lifted up a narrow, east-west elongate range, possibly the hinterland of a magmatic arc. The majority of North China, however, was dominated by paralic and marginal marine settings, in which fine-grained clastic, restricted-marine limestone, and thick coal deposits formed under a hot and wet climate (Stephanian–Sakmarian Taiyuan Formation; Liu, 1990). Phytogeographic studies on the plant fossils (Yao, 1983, *in* Nie, 1991; Smith, 1988) suggest that North China was in a tropical climatic zone (Liu, 1990; Nie et al., 1993; Fig. 7).

Late Carboniferous strata exposed along the margins of the Qaidam basin are largely paralic clastics and carbonates, some including coal seams. Upper Carboniferous strata transgress over Devonian or basement rocks in the Altyn Tagh (Petroleum Industry Press, 1987).

On the Tarim block, Late Carboniferous paleogeography was similar to that of the Early Carboniferous, but characterized by greater tectonic stability in its interior and by overall regression. Groves and Brenckle (1997) demonstrated biostratigraphically that the Late Carboniferous rock record in western Tarim is extremely discontinuous over a time span of 30–40 m.y. Most time is represented by regional hiatuses, and deposition in restricted marine, lagoonal, and evaporitic facies occurred only occasionally. In eastern Tarim, a tidal- and braided-stream-dominated delta prograded westward across the Tazhong area at the end of the Carboniferous. These deposits are interbedded with some marine carbonates representing tidal marsh-lagoonal systems.

In the Late Carboniferous–Early Permian, the northern margin of the combined Yili-Tarim block collided with a magmatic arc of Devonian–Late Carboniferous age along the North Tian Shan fault, closing the North Tian Shan sea (Allen et al., 1992, 1993; Bazhenov et al., 1993; Carroll et al., 1995; Gao et al., 1998; Fig. 7). Production of tholeiites, presumably representing arc magmatism, terminated in the Middle or Late Carboniferous (Wang, 1986; Wang et al., 1990; Windley et al., 1990; Allen et al., 1992). A foreland basin developed over and north of the extinct arc by the Late Carboniferous–Early Permian. The polarity of subduction zones north of the central Tian Shan is not well constrained, but Allen et al. (1992) suggested a dual divergent subduction beneath the central Tian Shan and beneath the North Tian Shan arc. What existed north of this arc is debatable. Allen et al. (1992) argued that the aforementioned collision destroyed the last oceanic lithosphere in the region, linking the Tarim region to the Siberia region. Carroll et al. (1995), in contrast, favored the continued existence of the Junggar ocean north of the North Tian Shan arc, to be trapped as Junggar basin floor and filled mainly during the Early Permian. The latter scenario poses the question of the location of the terminal suture of the Turkestan ocean. Feng et al. (1989) showed that virtually all strata exposed along the western margin of the Junggar basin have structural and stratigraphic affinities with the rocks to the north but are dissimilar from those to the south. Consequently, they placed the Junggar basin along the northern, probably highly irregular and obliquely closing margin of the Turkestan basin. The suture representing the main branch of the Turkestan ocean would then be near the southern basin edge but north of the Tian Shan arc. In this reconstruction (Fig. 7), I opted to display the scenario suggested by Feng et al. (1989), but the regional geometry and paleogeography of the terminal closure of the Turkestan ocean in the Junggar area has not yet been fully resolved.

A major difference of this reconstruction (Fig. 7) to Şengör and Natal'in's (1996, their Fig. 21.37) map is the enormous separation these authors show for the Late Carboniferous to exist between Tarim and North China. The south-facing active margin along this part of paleo-Asia was in their view taken up by arc segments, which were subsequently stacked by strike-slip faulting and now form the Qilian Shan and the basement to Qaidam basin. Regional stratigraphic correlation and first-order structural analysis of the widespread Devonian redbeds in this part of China would provide a test to verify such intense reshuffling. In the absence of solid evidence for major Carboniferous strike-slip movements in the Qilian Shan and the Qaidam basin, and because of the widespread low-relief, low-energy depositional environments present in this region (Fig. 7), I display these blocks as being relatively rigidly attached to North China and Tarim.

Available paleomagnetic data place Tarim at the end of the Carboniferous in the Northern Hemisphere at $29.0° ± 3.5°N$ (Fang et al., 1990). The floral assemblage of western Tarim, in contrast, suggests that Tarim occupied a tropical climatic zone (Laveine et al., 1996). In order to reconcile these two data sets, I placed Tarim midway near 20°N, similar to Şengör and Natal'in's (1996) reconstruction. Sharps et al. (1989) and Li (1990) calculated an 18° northward movement and ~46° clockwise rotation of Tarim between the Late Devonian and the Late Carboniferous from their paleomagnetic data. This rotation is consistent with a diachronous, east to west closure of the ocean

between Yili and Tarim, which is also indicated by the record of gradual east to west regression at the end of the Carboniferous.

Cluster analysis of phytogeographic data (Rowley et al., 1985) and occurrence of the pteridosperm *Paripteris,* an index fossil of the northern margin of Paleotethys, in coals of western Tarim and of central Europe (Laveine et al., 1992, 1996) suggest that a land connection existed at least intermittently between western Tarim and central Europe by late Namurian–early Westphalian time (Bashkirian, ca. 315 Ma). Nie et al. (1990) used analogous reasoning to explain the widespread distribution of the seed-bearing late Paleozoic *Gigantopteris* flora in the Cathaysian continental blocks at times when they were apparently tectonically independent (Devonian–Permian) by noting that island arcs and other smaller tectonic entities may have provided temporary land bridges that may have disappeared without a trace. Given the complex south-facing margin of paleo-Asia, land bridges provided by magmatic arcs or their remnants after collapse against a continental margin seem to be a reasonable mechanism to explain the observed floral distribution. Indirectly supporting this hypothesis are the warm-water fusulinids and corals on the Tadjik block which are in places mixed with populations known from the East European Platform (Leven, 1971; Fontaine et al., 1977; Boulin, 1988). They indicate that a biotic exchange mechanism in shallow-marine environments between central Asia and Europe existed.

The interior of the Turan platform, the Karakum swell, and the Hindu Kush–Tadjik block (Fig. 1) were characterized by carbonate shelf and siliciclastic marginal marine sedimentation (Fig. 7). The southern margin of this elongate group of continental blocks, extending from the Caucasus to the Qilian Shan, continued to be dominated by north-dipping subduction. Thick flysch-type sequences (e.g., Doshi section, Farkhar section; Fig. 4; Boulin, 1988; Leven, 1997; Duwa section, Anonymous, 1991) developed along this margin. The Early Carboniferous north Pamir backarc basin collapsed by the end of the Early Carboniferous, and Middle Carboniferous to Permian deep-water sediments overlie the suture (Burtman and Molnar, 1993).

By the end of the Carboniferous, the collapse of the Kipchak arc was nearly complete, and transpressional shortening of the arc material, wedged between the Siberian and Baltica cratons, caused widespread emergence (Zonenshain et al., 1984; Şengör et al., 1993a; Şengör and Natal'in, 1996). Late Carboniferous oblique collision of the leading edge of the Altaids with the partially collapsed Mugodzhar arc produced major deformation in the Urals. Much of the deformation occurred by westward thrusting and north-south–oriented sinistral strike-slip faulting. Flysch- and molasse-type sediments were shed westward from the rising thrust sheets into a gradually subsiding foreland basin (Zonenshain et al., 1984; Puchkov, 1997).

Nie et al. (1993) recognized that the Junggar basin and adjacent ranges between the Karamay–Kelameili Shan ophiolites to the north and the Tian Shan to the south were fundamentally an eastern extension of the Altaids into northern China. He concluded that this region represented part of the late Paleozoic continental margin north of the Turkestan ocean (Fig. 5). The basement of the Junggar basin is conjectural (see Allen et al., 1991, 1992, 1995 for discussion); it is probably similar in nature to the Paleozoic arc volcanics, forearc-related sedimentary rocks, and ophiolitic fragments exposed in the ranges surrounding the basin.

Diachronous closure of the Turkestan ocean accelerated. Moscovian flysch in great thickness, often including large blocks of Devonian limestone, characterized the margins of the closing Turkestan ocean along a 1000-km-long segment from the Tumdytau Ridge of Turkmenistan (Kyzyl Kum block; Fig. 1) through the southwest Tian Shan and along the southern margin of the Fergana basin (Burtman, 1975; Kravchenko, 1979; Leith, 1982; Zonenshain et al., 1990; Mukhin et al., 1989). In contrast, Şengör and Natal'in (1996) disavow the existence and the scissor-like closing of the Turkestan ocean. Rather, they invoke extensive left-lateral strike-slip movements among the Altaid and Manchurian elements within a very wide shear zone between Siberia and North China, virtually without the involvement of oceanic crust, to juxtapose North China–affiliated elements with the Altaids. In their view, arc-related magmatism, which continued in the Junggar basin to the end of the Carboniferous, was related to northward subduction of oceanic crust. The Tarim (and Yili) blocks, currently located directly south of Junggar, therefore had to be located elsewhere, and moved into their present place by subsequent strike-slip displacement along some of the major east-west–oriented faults. Although no data contradict this scenario, my view is that the liberal use of very large strike-slip displacements is not supported by existing data, and that a simpler scenario is that shown in Figure 7.

Early Permian (Asselian-Sakmarian), ca. 280 Ma

North China terrestrial and marginal marine environments began to prograde south during the Late Carboniferous as a result of the gradual uplift of its active northern margin and regional regression (cf. Figs. 7 and 8). Progradation continued throughout the Permian, and facies belts, including the economically important coal deposits, gradually shifted south (Liu, 1990; A.M. Ziegler, 1990). Paleomagnetic and paleoclimatic data indicate an overall northward migration of the North China block from equatorial latitudes in the Late Carboniferous to a low-latitude position during the Early Permian (Liu, 1990; Nie et al., 1990; Nie, 1991).

The nature of Permian strata from the Qaidam basin is generally poorly known. Thin marine clastic and carbonate sediments of Early Permian age, conformable with the underlying Late Carboniferous strata, have been reported from the eastern and southern parts of Qaidam (Petroleum Industry Press, 1987; Wang and Coward, 1990). Farther south, Lower Permian strata thicken as they grade into the mixed carbonate-clastic, volcanic-bearing active-margin assemblages of the eastern Kunlun (Fig. 8).

Early Permian paleomagnetic data for Tarim, compiled by Nie (1991), place the mean position of central Tarim (40°N, 83°E) at about 30° ± 4°N. The reconstruction (Fig. 8) shows Tarim at the lower limit of this latitudinal band in order to preserve continuity with North China. Lower Permian strata in Tarim are characterized by a gradual east-to-west regression in which shallow-marine limestone is replaced by shoreline and nonmarine siliciclastic strata (Hu et al., 1969; Carroll et al., 1995; also see Boguzidaliya section, Duwa section, Anonymous, 1991). As in the Late Carboniferous, siliciclastic and carbonate deposition was intermittent (Groves and Brenckle, 1997). The stratigraphically youngest marine strata are found in the Kashgar area (Liao et al., 1992). In northwest Tarim, these strata are interbedded with Early Permian basalt flows and cut by dikes of Early Permian age (Hu et al., 1969; Li, 1995; McWilliams and Hacker, personal commun., in Carroll et al., 1995). These dikes are considered part of a short-lived (284–275 Ma) intracontinental extensional event following collision with Yili.

The composite assemblage of (from south to north) Kunlun, Tarim, Qaidam, Yili, and the North Tian Shan magmatic arc collided with the collapsed Altaid arcs toward the end of the Early Permian, thereby effectively closing this segment of the Turkestan ocean (Fig. 8). Collision is indicated by the termination of andesitic volcanism in the Tian Shan, development of unconformities between late Early Permian and underlying sequences, intrusion of widespread granites on both sides of the suture, and phytogeographic data (Rowley et al., 1985; Smith, 1988; Coleman, 1989; Nie et al., 1990; Allen et al., 1992). The associated shortening and reactivation of older structures may have deformed the Tarim block internally (late Hercynian orogeny) sufficiently to cause a widespread regional block faulting (normal and reverse) at the end of the Early Permian.

Thick turbiditic siliciclastics were shed from the North Tian Shan arc and the Bogda Shan arc northward into the Junggar basin and southward into the intramontane Turfan basin, beginning in the Late Carboniferous and continuing into the Early Permian (Carroll et al., 1990; Allen et al., 1992, 1993). In marked contrast, Late Carboniferous redbeds in terrestrial environments dominate the northern margin of the Junggar basin (Peng and Zhang, 1989; Carroll et al., 1990). The transition to nonmarine environments along the southern margin of Junggar and in the adjacent Turfan basin occurred at the end of the Early Permian (Carroll et al., 1990; Fig. 8).

In the southern Kunlun, deep-marine siliciclastic strata, interbedded with thin limestones and tuffites (Bazar Dara Slates), continued to accumulate in a forearc subduction-accretion complex (Mattauer et al., 1996; Şengör and Okurogullari, 1991).

Northward-directed subduction continued along the Caucasus and the Hindu Kush margins. Stratigraphic sections throughout northern Afghanistan are similar to those in Tarim, i.e., older, predominantly marginal marine environments gradually replaced by terrestrial redbeds (Bandi Turkestan, Bandi

Amir Gorge sections; Lys and Lapparent, 1971; Auden, 1974; Vachard, 1980; Leven, 1997). Calc-alkaline volcanics along the southern margin of the Tadjik and Karakum blocks record continued subduction of the Paleotethys. There, Lower Permian fluvial and alluvial strata generally overlie the Carboniferous deep-water clastics with angular unconformity (Boulin, 1988; Burtman and Molnar, 1993). The sequence is apparently similar in stratigraphy, fossil content, and deformational record to those in the northern Pamirs (Leven, 1997), and suggests a collisional event at the end of the Permian; however, there is no tectonic evidence of it along the active margin to the south.

Along the Uralian margin, intensive deformation, associated with the magmatism, metamorphism, and foreland basin fill continued from the Late Carboniferous (Fig. 8). The foreland basin was markedly asymmetrical. A molasse facies developed between and immediately in front of the thrust sheets but graded westward rapidly into a flysch-filled basin-axis trough. Artinskian flysch is as much as 1000 m thick (Puchkov, 1997). Farther west, a chain of Asselian to early Artinskian barrier reefs marked the western margin of the foredeep (Chuvashov and Nairn, 1993, *in* Puchkov, 1997; Fig. 8). The large majority of the Altaid collage collided with the East European margin and became emergent. Only along their southern margin, where northward subduction of the Turkestan ocean continued, a complex magmatic arc with a series of backarc basins still existed (Zonenshain et al., 1984).

Late Permian (Kungurian), ca. 260 Ma

In the Late Permian, vast tracts of central Asia were dominated by terrestrial deposition, owing to global eustatic sea-level lowstands (Ross and Ross, 1994) and widespread erosion. Extensive fluvial and alluvial deposits extended from the North China platform through the Ala Shan, Qilian Shan, Qaidam, and Tarim into central Asia (Fig. 9).

The Kungurian-Kazanian lower Shihezi Group of northern China consists of thick, southward-shed alluvial-fluvial sandstone and conglomerate (Liu, 1990), interspersed with lacustrine mudstone. These strata mark the rapid uplift and erosion of the northern margin of North China as a result of diachronous, west to east collision with the Amuria collage along the Junggar-Hegen-Solon Obo suture (Wang and Liu, 1986; Şengör, 1987; Zhao et al., 1990, 1993; Mueller et al., 1991; Pruner, 1992; Amory et al., 1994; Hendrix et al., 1996). This region is described in greater detail by Lamb and Badarch (this volume) and Johnson et al. (this volume). Floral evidence and some evaporite locations show that the climate of North China changed to arid at the beginning of the Late Permian (Ziegler et al., 1997). This may be due to a rain-shadow effect from the rising mountains to the south or due to continued northward drift documented by a number of paleomagnetic studies (listed in Nie, 1991; Zhang, 1997). Coal deposits continued to form only near the south-facing margin under deltaic-marginal marine influence (Fig. 8).

The Late Permian of the Qaidam region has been documented only in rocks from its southern margin. There, littoral clastic strata are interbedded with limestone, generally grading from terrestrial into marine environments (Petroleum Industry Press, 1987).

Paleopoles for the Late Permian from North China are very different from those of Tarim, suggesting that these two blocks were probably not joined in Permian time (Bai et al., 1987; McFadden et al., 1988; Sharps et al., 1989); this is in disagreement with geologic data sets. Possibly, this contradiction can be explained in the light of Late Permian–Triassic block rotations that are postulated for the origin of the Junggar, Turfan, and Alakol basins (Allen et al., 1995). Interpretation of Late Permian paleomagnetic data for Tarim, compiled by Nie (1991), place the mean position of central Tarim (40°N, 83°E) at about $31° \pm 5°$N, compatible with the reconstructed latitude of the southern Altaids. The Late Permian of Tarim is characterized by widespread fluvial and alluvial deposition (Carroll et al., 1995, this volume). Major Permian magmatism occurred in southwest Tarim, in the Altun Tagh, and in the Anyemaqen Shan, south of the Qaidam block. Şengör et al. (1993b) interpreted the magmatism as reflecting Late Permian backarc opening in the Anyemaqen Shan and subsequent closure. In the western Kunlun, the thick marine clastics of the Bazar Dara Slates record the continuation of active margin tectonics (Gaetani et al., 1990; Mattern et al., 1996; Fig. 9). At that time, the southern Kunlun basement was intruded by numerous late Paleozoic calc-alkaline plutons distributed over a distance of at least 1000 km lateral extent (Mattern et al., 1996).

By the Late Permian, the Tadjik, Karakum, central Karakum swell, Kyzylkum, Ustyurt, and Syr-Darya blocks (Fig. 1) were largely covered by Late Permian fluvial, alluvial, and lacustrine deposits, which compose voluminous redbeds (Fig. 9; Alavi, 1991; Alavi et al., 1997). They may express Late Permian brittle adjustments to Late Carboniferous and Early Permian deformation and are locally very thick. For example, in the Koh-i-Davindar (Fig. 3) of western Afghanistan, ~3800 m of Permian redbeds overlie Carboniferous flysch with angular unconformity and are in turn overlain with angular unconformity by Triassic strata (Wensink, 1991). In the Achkhabad area of northeastern Iran, the thick conglomerate is related to uplift and erosion of the active margin (Alavi et al., 1997). Along the southern margin of the Turan plate, a Permian accretionary complex that formed in a trench-forearc setting is preserved in the Binalood Mountains of northeastern Iran (Fig. 3; Alavi, 1991). Pyroclastic strata and carbonate olistostromes interbedded with the siliciclastic turbidites attest to the proximity of active volcanism and shallow-water carbonate environments, respectively. Clearly, this assemblage represents the active southern margin of the Turan block (Stocklin, 1974; Alavi, 1991).

Deformation in the Ural mountain belt continued through the Late Permian. A major, in part evaporitic, postcollisional foreland basin developed in the Kungurian in the pre-Caspian depression (Puchkov, 1997; Zholtayev, 1998), covering the Permian siliciclastic and Carboniferous carbonate strata and filling the north Caspian depression (Fig. 9). Later Permian environments in the Uralian foreland included marine, lacustrine, and continental settings. During the Tatarian (latest Permian), the foreland basin extended as far west as Moscow (Vinogradov et al., 1968; P.A. Ziegler, 1990; Maslyayev, 1989). Simultaneously, incipient backarc spreading and/or orogenic collapse caused subsidence of the west Siberian platform east of the Urals. The Uralian foreland basin and the rapidly and diachronously closing Turkestan ocean constitute the only remaining marine environments in central Asia at that time (Vinogradov et al., 1968).

One comparatively well studied segment of the closing Turkestan ocean is the Junggar basin area. There, the transition from marine to terrestrial sedimentation occurred at the end of the Early Permian, and Upper Permian strata in the Junggar basin are dominantly lacustrine in character (Allen et al., 1993; Carroll et al., 1995). This is accompanied by block rotation and extension within a regional, overall sinistral shear zone (Bazhenov et al., 1993; Allen et al., 1995).

Paleomagnetic data suggest that several thousand kilometers of latitudinal separation remained between the amalgamated North China–Amurian continent and stable Siberia in the Permian and Triassic (Opdyke et al., 1986; Zhao and Coe, 1989; Zhao et al., 1990; Pruner, 1992; Courtillot et al., 1994; Ziegler et al., 1997). Geologic data indicate that the far-eastern part of this Mongol-Okhotsk ocean did not close until Late Jurassic to Early Cretaceous time (Kosygin and Parfenov, 1981; Klimetz, 1983; Berzin et al., 1994; Şengör and Natal'in, 1996), and that the Uda magmatic arc, bordering stable Siberia to the south, remained active at least through the Middle Jurassic (Kosygin and Parfenov, 1981; Zonenshain et al., 1990). Only Late Cretaceous paleopoles show North China block and Siberia in their present-day orientation with respect to each other.

SUMMARY

The late Paleozoic marks the time in which the core of present-day Asia assembled from numerous smaller fragments. Understanding this process is crucial toward our approach of recognizing similar continent-building processes elsewhere. This study attempted to integrate available tectonic and paleoenvironmental data from both sides of the former Soviet Union–Chinese border into a modern, internally consistent geodynamic framework and compare its results with several previous studies. This study found the following.

1. The level of detail required for a semirealistic modeling of central Asia surpasses the limits of rigid-body kinematics. Accretionary tectonics (Şengör et al., 1993a; Şengör and Natal'in, 1996), intraplate deformation, common backarc basin formation, and destruction and modifications of major faults by reactivation render the proper recognition of true former plate tectonic boundaries difficult. The model ac-

counted for these factors by reducing the number of actual blocks considered and by treating internally deformed entities as several subblocks.

2. Paleoenvironmental maps from the former Soviet Union (Vinogradov et al., 1968) and China (Wang, 1985) show poor to moderate agreement across the border, principally due to different levels of detail, lack of mapped structures, and different stratigraphic nomenclature. Few suspected former plate tectonic boundaries are recognized in either atlas.

3. The model shows the largest uncertainties associated with the position of the Scythian-Turanian-Afghan elements. Paleobotanical evidence suggests an (intermittent?) land bridge between Tarim and Europe as early as the Bashkirian (ca. 315 Ma), for which a plausible geometry is shown. However, there is no independent tectonic, stratigraphic, or paleomagnetic information to validate this particular scenario.

4. Several other major problems of middle to late Paleozoic plate kinematics and paleogeography remain to be addressed. These include: (1) the continuity of Tarim to the east (Qaidam–Qilian Shan–North China); but see Yue et al., this volume; (2) the potential duplication of the southwestern margin of paleo-Asia in the South Caspian region; (3) the tectonic framework of the Aral Sea region; (4) the Paleozoic stratigraphy and tectonic framework of the southwest Paleo-Asian blocks (Tadjik-Karakum-Turan); (5) the magnitude of Permian strike-slip displacements in central Asia; and (6) the geometry of the closure of the Turkestan ocean west of the Junggar basin.

ACKNOWLEDGMENTS

Chris Scotese and Malcolm Ross wrote the original PALEO-MAP software. Shawn Stephens, Everett Rutherford, and Shangyou Nie advised me in the early stages of this project on data management, plate tectonic analysis, and Chinese literature, respectively. The Paleogeographic Mapping Project at the University of Texas, Arlington, and the paleoclimatic research group at the University of Chicago provided data and advice. Discussions with Ed Sobel improved my thoughts on central Asian tectonics. Critical reviews by Kevin Burke, An Yin, Shangyou Nie, Brad Hacker, Chris Scotese, and Marc Hendrix significantly improved the manuscript. BP Amoco management permitted publication of this study, which was performed while I was with Amoco in Houston, Texas.

REFERENCES CITED

Abidov, A.A., Abetov, A.Y., Kirshin, A.V., and Avazhkodzhayev, K.K., 1997, Geodynamic evolution of Kuanysh-Koskalin tectonic zone in the Paleozoic (East Ustyurt, Karakalpakstan): Petroleum Geology, v. 31, p. 388–391.

Alavi, M., 1991, Sedimentary and structural characteristics of the Paleo-Tethys remnants in northeastern Iran: Geological Society of America Bulletin, v. 103, p. 983–992.

Alavi, M., Vaziri, H., Seyed-Emanmi, K., and Lasemi, Y., 1997, The Triassic and associated rocks of the Nakhlak and Aghdarband areas in central and northeastern Iran as remnants of the southern Turanian active continental margin: Geological Society of America Bulletin, v. 109, p. 1563–1575.

Allen, M.B., Windley, B.F., and Zhang C., 1992, Paleozoic collisional tectonics and magmatism of the Chinese Tien Shan, central Asia: Tectonophysics, v. 220, p. 89–115.

Allen, M.B., Windley, B.F., Zhang Chi, and Guo Jinghui., 1993, Evolution of the Turfan Basin, Chinese central Asia: Tectonics, v. 12, p. 889–896.

Allen, M.B., Şengör, A.M.C., and Natalin, B.A., 1995, Junggar, Turfan, and Alakol basins as Late Permian to ?Early Triassic extensional structures in a sinistral shear zone in the Altaid orogenic collage, Central Asia: Geological Society of London Journal, v. 152, p. 327–338.

Amory, J.Y., Hendrix, M.S., Lamb, M., Keller, A.M., Badarch, G., and Tomurtoggo, O., 1994, Permian sedimentation and tectonics of southern Mongolia: Implications for a time-transgressive collision with north China: Geological Society America Abstracts with Programs, v. 26, no. 7, p. A-242.

Anonymous, 1991, Structural analysis, *in* Petroleum geology of the Tarim Basin: Beijing, Science Press, 182 p.

Auden, J.B., 1974, Afghanistan-Pakistan, *in* Spencer, A.M., ed., Mesozoic-Cenozoic orogenic belts: Data for orogenic studies: Edinburgh, Scottish Academic Press, p. 235–253.

Avdeyev, A.V., 1984, Ophiolite zones and the geologic history of Kazakhstan from the mobilist standpoint: International Geology Review, v. 26, p. 995–1005.

Bai, Y., Chen, G., Sun, Q., Sun, Y., Li, Y., Dong, Y., and Sun, D., 1987, Late Paleozoic polar wander path for the Tarim Platform and its tectonic significance: Tectonophysics, 139, p. 145–153.

Baud, A., 1989, The western end of the Tibetan Plateau, *in* Şengör, A.M.C., et al., Tectonic evolution of the Tethyan region: Dordrecht, Boston, Reidel Publishing Company, p. 505–506.

Baud, A., and Stampfli, G.M.S., 1989, Tectonogenesis and evolution of a segment of the Cimmerides: The volcano-sedimentary Triassic of Aghdarband (Kopet-Dagh, North-east Iran), *in* Şengör, A.M.C., et al., eds., Tectonic evolution of the Tethyan region: NATO ASI Series, Series C, 259, p. 265–275.

Bazhenov, M.L., and Burtman, V.S., 1982, The kinematics of the Pamir Arc: Geotectonics, v. 16, p. 288–301.

Bazhenov, M.L., Chauvin, A., Audibert, M., and Levashova, N.M., 1993, Permian and Triassic paleomagnetism of the southwestern Tian Shan; timing and mode of tectonic rotation: Earth and Planetary Science Letters, v. 118, p. 195–212.

Berzin, N.A., Coleman, R.G., Dobretsov, N.L., Zonenshain, L.P., Xiao Xuchan, and Chang, E.Z., 1994, Geodynamic map of the western Paleoasian Ocean: Geologiya i Geofizika, v. 35, p. 8–29 (in Russian).

Besse, J., and Courtillot, V., 1991, Revised and synthetic apparent polar wander paths for the African, Eurasian, North American, and Indian plates, and true polar wander since 200 Ma: Journal for Geophysical Research, v. 96, p. 4029–4050.

Boulin, J., 1988, Hercynian and Eocimmerian events in Afghanistan and adjoining regions: Tectonophysics, v. 148, p. 253–278.

Boulin, J., 1991, Structures in southwest Asia and evolution of the eastern Tethys: Tectonophysics, v. 196, p. 211–268.

Boulin, J., and Bouyx, E., 1977, Introduction à la géologie de l'Hindou Kouch occidental, *in* Livre à la mémoire de Albert F. de Lapparent (1905–1975) consacré aux Recherches géologiques dans les chaines alpines de l'Asie du Sud-Ouest: Mémoire Hors Série, Société Géologique de France, v. 8, p. 87–105.

Brown, D., Juhlin, C., Alvarez-Marron, J., Perez-Estaun, A., and Oslianski, A., 1998, Crustal-scale structure and evolution of an arc-continent collision zone in the southern Urals, Russia: Tectonics, v. 17, p. 158–171.

Burke, K., 1977, Aulacogens and continental breakup: Annual Review of Earth and Planetary Sciences, v. 5, p. 371–396.

Burrett, C.F., 1974, Plate tectonics and the fusion of Asia: Earth and Planetary Science Letters, v. 21, p.181–189.

Burrett, C.F., Long, J., and Stait, B., 1987, Early-middle Paleozoic biogeography of Asian terranes derived from Gondwana, *in* McKerrow, W.S., and Scotese, C.R., eds., Paleozoic paleogeography and biogeography: Geological Society [London] Memoir 12, p. 163–174.

Burtman, V.S., 1975, Structural geology of the Variscan Tien Shan, USSR: American Journal of Science, v. 275-A, p. 157–186.

Burtman, V.S., and Molnar, P., 1993, Geological and geophysical evidence for deep subduction of continental crust beneath the Pamir: Geological Society of America Special Paper 281, 76 p.

Carroll, A.R., Liang, Y., Graham, S.A., Xiao, X., Hendrix, M.S., Chu, J., and McKnight, C.L., 1990, Junggar basin, northwest China: Trapped late Paleozoic ocean: Tectonophysics, v. 181, p. 1–14.

Carroll, A.R., Graham, S.A., Hendrix, M.S., Don Ying, and Da Zhou, 1995, Late Paleozoic tectonic amalgamation of northwestern China: Sedimentary record of the northern Tarim, northwestern Turpan, and southern Junggar basins: Geological Society of America Bulletin, v. 107, p. 571–594.

Chen, Y., Courtillot, V., Cogne, J.-P., Besse, J., Yang, Z., and Enkin, R., 1993, The configuration of Asia prior to the collision of India: Cretaceous paleomagnetic constraints: Journal of Geophysical Research, v. 98, p. 21927–21941.

Chen, Z., Wu, N., Zhang, D., Hu, J., Shen, G., Wu, G., Tang, H., and Hu, Y., 1985, Geologic map of Xinjiang Uygur autonomous region: Beijing, Geologic Publishing House, scale 1: 2 000 000.

Cheng, Y.C., ed., 1986, Metamorphic map of China: Beijing, Geologic Publishing House.

Clarke, J.W., 1994, Petroleum potential of the Amu Dar'ya Province, western Uzbekistan and eastern Turkmenistan: International Geology Review, v. 36, p. 407–415.

Coleman, R.G., 1989, Continental growth of northwest China: Tectonics, v. 8, p. 621–636.

Commission for the Geologic Map of the World, 1976, Geological world atlas: Paris, United Nations Educational, Scientific, and Cultural Organization, scale 1:10 000 000.

Cook, H.E., Taylor, M.E., Zhemchuzhnikov, S.V., Apollonov, M.K., Ergaliev, G.K., Sargaskaev, Z.S., Dubinina, S.V., and Melnikova, L., 1991, Comparison of two early Paleozoic carbonate submarine fans—Western United States and southern Kazakhstan, Soviet Union, *in* Cooper, J.D., and Stevens, C.H., eds., Paleozoic paleogeography of the western United States—II: Pacific Section, SEPM (Society for Sedimentary Geology) Book 67, p. 847–872.

Courtillot, V., Enkin, R., Yang, Z., Chen, Y., Bazhenov, M., Besse, J., Cogne, J.-P., Coe, R., Zhao, X., and Gilder, S., 1994, Reply (to The configuration of Asia prior to the collision of India: Cretaceous paleomagnetic constraints): Journal of Geophysical Research, v. 99, p. 18043–18048.

Didenko, A.N., Mossakovskii, A.A., Pecherski, D.M., Ruzhentsev, S.V., Samygin, S.G., and Kheraskova, T.N., 1994, Geodynamics of the central Asian Paleozoic Ocean: Geologiya i Geofizika, v. 35, p. 59–75.

Enkin, R.J., Yang, Z., Chen, Y., and Courtillot, V., 1992, Paleomagnetic constraints on the geodynamic history of the major blocks of China from the Permian to the present: Journal of Geophysical Research, v. 97, p. 13953–13989.

Fang, D., Chen, H., Jin, G., Guo, Y., Wang, Z., Tan, X., and Yin, S., 1990, Late Paleozoic and Mesozoic paleomagnetism and tectonic evolution of the Tarim terrane, *in* Wiley, T.J., et al., eds., Terrane analysis of China and the Pacific rim: Houston, Texas, Circum-Pacific Council for Energy and Mineral Resources Earth Science Series, v. 13, p. 251–256.

Feng, Y., Coleman, R.G., Tilton, G., and Xiao, X., 1989, Tectonic evolution of the west Junggar region, Xinkiang, China: Tectonics, v. 8, p. 729–752.

Fontaine, H., and Semenoff-Tian-Chansky, P., 1977, Apercu sur les coraux du Carbonifere de l'Hazarajat et des autres regions de l'Afghanistan: Bulletin de la Société Géologique de France, v. 19, p. 235–237.

Gaetani, M., Gosso, G., and Pognante, U., 1990, A geological transect from Kun Lun to Karakorum (Sinking, China): The western termination of the Tibetan Plateau: Preliminary note: Terra Nova, v. 2, p. 23–30.

Gao Jun, Li Maosong, Xiao Xuchang, Tang Yaoqing, and He Guoqi, 1998, Paleozoic tectonic evolution of the Tianshan orogen, northwestern China: Tectonophysics, v. 287, p. 213–231.

Gilder, S.A., Zhao, X., Coe, R., Meng, Z., Courtillot, V., and Besse, J., 1996, Paleomagnetism and tectonics of the southern Tarim basin, northwestern China: Journal of Geophysical Research, v. 101, p. 22015–22031.

Golonka, J., Ross, M.I., and Scotese, C.R., 1994, Phanerozoic paleogeographic and paleoclimatic modeling maps, *in* Embry, A.F., et al., eds., Pangea: Global environments and resources: Canadian Society of Petroleum Geologists Memoir 17, p. 1–48.

Groves, J.R., and Brenckle, P.L., 1997, Graphic correlation in frontier petroleum provinces: Application to upper Paleozoic sections in the Tarim basin, western China: American Association of Petroleum Geologists Bulletin, v. 81, p. 1259–1266.

Hendrix, M.S., Graham, S.A., Amory, J.Y., and Badarch, G., 1996, Noyon Uul syncline, southern Mongolia: Lower Mesozoic sedimentary record of the tectonic amalgamation of central Asia: Geological Society of America Bulletin, v. 108, p. 1256–1274.

Hu Bing, Wang Jing-Bin, Gao Zhen-Jia, and Fang Xiao-Di, 1969, Problems of the Paleozoics of the Tarim Platform: International Geological Review, v. 11, p. 650–666.

Izvestiya Akademii Nauk Kazakhskoy SSR, 1987, 1:1,500,000 geologic map of Kazakhstan and Central Asia: Alma-Ata, USSR.

Khaimov, R.N., 1986, The Paleozoic sediments of the Fergana Basin: Their potential for oil and gas: International Geology Review, v. 28, p. 75–79.

Khain, V.Y., 1979, The Late Triassic volcanic/plutonic belt in the North Caucasus–Turkmenia–North Afghanistan and the opening of the northern part of the Tethys: Akademiya Nauk SSSR Doklady, v. 249, p. 1190–1192 (in Russian).

Klimetz, M.P., 1983, Speculations on the Mesozoic plate tectonic evolution of eastern China: Tectonics, v. 2, p. 139–166.

Kosygin, Y.A., and Parfenov, L.M., 1981, Structural evolution of eastern Siberia and adjacent areas: American Journal of Science, v. 275A, p. 187–208.

Kravchenko, K.N., 1979, Tectonic evolution of the Tien Shan, Pamir, and Karakorum, *in* Farah, A., and DeJong, K.A., eds., Geodynamics of Pakistan: Quetta, Geological Survey of Pakistan, p. 25–40.

Kutzbach, J.E., and Ziegler, A.M., 1994, Simulation of Late Permian climate and biomes with an atmosphere-ocean model: Comparisons and observations, *in* Allen, J.R.L., et al., eds., Palaeoclimates and their modelling: With special reference to the Mesozoic era: London, Chapman & Hall, p. 119–132.

Lai, J., and Wang, J., 1988, Atlas of the paleogeography of Xinjiang: Urumqi, China, Xinjiang People's Publishing House, 92 p. (in Chinese with English abstract).

Laveine, J.-P., Lemoigne, Y., and Zhang, S., 1992, Pangaea, Paleotethys, and *Paripteris*: Paris, Académie des Sciences Comptes Rendus, v. 314, ser. II, p. 1103–1110.

Laveine, J.-P., Zhang Shenzhen, Lemoigne, Y., Desheng An, Qishi Zheng, and Jingyuan Cao, 1996, The upper Paleozoic floras of Hotan area (Xinjiang Province, Northwest China), and their paleogeographical significance: Paris, Académie des Sciences Comptes Rendus, v. 322, ser. IIa, p. 781–790.

Leith, W., 1982, Rock assemblages in central Asia and the evolution of the southern Asian margin: Tectonics, v. 1, p. 303–318.

Leven, E.J., 1971, Les gisements permiens et les Fusulinides de l'Afghanistan du Nord: Museum National d'Histoire Naturelle Paris, Notes et Memoires du Moyen Orient, v. 12, 35 p.

Leven, E.J., 1997, Permian stratigraphy and fusulinida of Afghanistan with their paleogeographic and paleotectonic implications: Geological Society of America Special Paper 316, 134 p.

Li, Y., 1990, An apparent polar wander path from the Tarim block, China: Tectonophysics, v. 181, p. 31–41.

Li Yong'an, Cao Yundong, and Sun Dongjiang, 1996, Structural geology along the Sino-Pakistan Highway in the western Kunlun Mountains of China:

30th International Geological Congress Field Trip Guide T113/T363: Beijing, Geological Publishing House (Beijing), 21 p.

Liao, Z., Wang, Y., Wang, K., Jiang, N., Xia, F., Zhou, Y., Sun, F., and Li, S., 1992, Carboniferous of Tarim and its adjacents, *in* Zhou, Z., and Chen, P., eds., Biostratigraphy and geological evolution of Tarim: Beijing, Science Press, p. 202–242.

Liou, J.G., Wang, X., Coleman, R.G., Zhang, Z.M., and Maruyama, S., 1989, Blueschists in major suture zones of China: Tectonics, v. 8, p. 609–619.

Liu Guanghua, 1990, Permo-Carboniferous paleogeography and coal accumulation and their tectonic control in the North and south China continental plates: International Journal of Coal Geology, v. 16, p. 73–117.

Lutts, B.G., and Feldman, M.N., 1992, Geodynamic interpretation of Paleozoic magmatism in the Sultanuizdag Ridge, Uzbekistan: Geotectonics, v. 26, p. 307–314.

Lys, M., and de Lapparent, A.F., 1971, Foraminiferes et microfacies de Permien de l'Afghanistan central: Museum National d'Histoire Naturelle Paris, Notes et Memoires du Moyen Orient, v. 12, 98 p.

Maslyayev, G.A., 1989, Paleogeologic reconstructions of the Russian Platform: International Geology Review, v. 31, p. 217–227.

Mattauer, M., Matte, P., Malavieille, J., Tapponier, P., Maluski, H., Xu, Z., Lu, Y., and Tang, Y., 1985, Tectonics of the Qinling belts, build-up and evolution of eastern Asia: Nature, v. 327, p. 496–500.

Matte, P., Maluski, H., Caby, R., Kepeshinskas, A., and Sobolev, S., 1993, Geodynamic model and Ar/Ar dating for the generation and emplacement of the high pressure (HP) metamorphic rocks in SW Urals: Paris, Académie des Sciences Comptes Rendus, v. 317, p. 1667–1674.

Matte, P., Tapponier, P., Arnaud, N., Bourjot, L., Avouac, J.P., Vidal, P., Liu Qiang, Pan Yusheng, and Wang Yi, 1996, Tectonics of western Tibet, between the Tarim and the Indus: Earth and Planetary Science Letters, v. 142, p. 311–330.

Mattern, F., Schneider, W., Li, Y., and Li, X., 1996, A traverse through the western Kunlun (Xinjiang, China): Tentative geodynamic implications for the Paleozoic and Mesozoic: Geologische Rundschau, v. 85, p. 705–722.

McFadden, P.L., Ma, X.H., McElhinny, M.W., and Zhang, Z.K., 1988, Permo-Triassic magnetostratigraphy in China, northern Tarim: Earth and Planetary Science Letters, v. 87, p. 152–160.

McKnight, C.L., 1993, Structural styles and tectonic significance of Tian Shan foothill fold-thrust belts, northwest China [Ph.D. thesis]: Stanford, California, Stanford University, 207 p.

Metcalfe, I., 1996, Pre-Cretaceous evolution of SE Asian terranes, *in* Hall, R., and Blundell, D., eds., Tectonic evolution of Southeast Asia: Geological Society [London] Special Publication 106, p. 97–122.

Ministry of Geology and Mineral Resources, 1993, Regional geology of Xinjiang Uygur Autonomous Region: Beijing, Geological Publishing House, Geological Memoir Series, v. 1, p. 783–841.

Moore, G.T., and Barron, E.J., 1994, Source rock, reservoir, and evaporitic seal predictions from general circulation models: Late Jurassic to mid-Cretaceous case studies: Geological Society of America Abstracts with Programs, v. 26, no. 7, p. A-372.

Mossakovsky, A.A., Ruzhentsev, S.V., Samygin, S.G., and Kheratovska, T.N., 1993, The central-Asian fold belt: Geodynamic evolution and formation history: Geotektonika, no. 6, p. 3–32 (in Russian).

Mueller, F., Rogers, J.W., Jin Yu-Gan, Wang Huayu, Li Wenguo, Chronic, J., and Mueller, J.F., 1991, Late Carboniferous to Permian sedimentation in Inner Mongolia, China, and tectonic relationships between north China and Siberia: Journal of Geology, v. 99, p. 251–263.

Mukhin, P.A., Abdullayev, K.A., Minayev, V.Y., Khristov, S.Y., and Egamberdyyev, S.A., 1989, The Paleozoic geodynamics of central Asia: International Geology Review, v. 30, p. 1073–1083.

Murzagaliyev, D.M., 1996, Sub-salt carbonate reservoirs on the shelf of the North Caspian and their oil-gas prospects: Petroleum Geology, v. 30, p. 1–5.

Nie, S.Y., 1991, Paleoclimatic and paleomagnetic constraints on the Paleozoic reconstructions of south China, north China and Tarim: Tectonophysics, v. 196, p. 279–308.

Nie, S.Y., and Rowley, D.R., 1994, Paleomagnetic constraints on the geodynamic history of the major blocks of China from the Permian to the present: Comment: Journal of Geophysical Research, v. 99, p. 18035–18042.

Nie, S.Y., Rowley, D.B., and Ziegler, A.M., 1990, Constraints on the location of Asian microcontinents in Paleo-Tethys during the late Palaeozoic, *in* McKerrow, W.S., and Scotese, C.R., eds., Paleozoic paleogeography and biogeography: Geological Society [London] Memoir 12, p. 397–409.

Nie, S.Y., Rowley, D.B., van der Voo, R., and Li, M., 1993, Paleomagnetism of late Paleozoic rocks in the Tianshan, northwestern China: Tectonics, v. 12, p. 568–579.

Novikov, V.L., Romanko, A.Y., Salikov, F.S., Suprychev, V.V., Yefremova, L.B., and Savichev, A.T., 1994, Late Paleozoic volcanism and geodynamics of the North Pamir region: Transactions of the U.S.S.R. Academy of Sciences, Earth Science Section, v. 327A, p. 106–110.

Okay, A.I., and Sahinturk, O., 1997, Tectonics and petroleum geology of the Black Sea and surrounding regions, *in* Robinson, A.G., ed., Regional and petroleum geology of the Black Sea and surrounding region: American Association of Petroleum Geologists Memoir 68, p. 291–311.

Opdyke, N., Huang, K., Xu, G., Zhang, W.Y., and Kent, D.V., 1986, Paleomagnetic results from the Triassic of the Yangtze Platform: Journal of Geophysical Research, v. B91, p. 9553–9568.

Palmer, A.R., compiler, 1983, The Decade of North American Geology geologic time scale: Geology, v. 11, p. 503–504.

Parrish, J.T., 1982, Upwelling and petroleum source beds, with reference to the Paleozoic: American Association of Petroleum Geologists Bulletin, v. 66, p. 750–774.

Pavlov, N.D., Salov, Y.A., Gogonenkov, G.N., Akopov, Y.I., Tolstykh, A.A., and Denisyuk, R.S., 1988, Geological-geophysical model of the Tengiz paleo-atoll based on seismostratigraphy: International Geology Review, v. 30, p. 1057–1069.

Peng, X., and Zhang, G., 1989, Tectonic features of the Junggar Basin and their relationship with oil and gas distribution *in* Hsü, K.J., and Zhu X., eds., Chinese sedimentary basins: Amsterdam, Elsevier, p. 17–31.

Popkov, V.I., 1995, Tectonics west of the Turan Platform: Petroleum Geology, v. 29, p. 71–102.

Pruner, P., 1992, Paleomagnetism and paleogeography of Mongolia from the Carboniferous to the Cretaceous—Final report: Physics of the Earth and Planetary Interiors, 70, p. 169–177.

Puchkov, V.N., 1997, Structure and geodynamics of the Uralian orogen, *in* Burg, J.-P., and Ford, M., eds., Orogeny through time: Geological Society [London] Special Publication 121, p. 201–236.

Ronov, A., Khain, V., and Balukhovsky, A., 1989, Atlas of lithological-paleogeographical maps of the world: Moscow, Academy of Sciences, 79 p.

Ross, C.A., and Ross, J.R.P., 1994, Permian sequence stratigraphy and fossil zonation, *in* Embry, A.F., et al., eds., Pangea: Global environments and resources: Canadian Society of Petroleum Geologists Memoir 17, p. 219–231.

Rowley, D.B., Raymond, A., Parrish, J.T., Lottes, A.L., Scotese, C.R., and Ziegler, A.M., 1985, Carboniferous paleogeographic, phytogeographic, and paleoclimatic reconstructions: International Journal of Coal Geology, v. 5, p. 7–42.

Ruttner, A.W., 1984, The pre-Liassic basement of the eastern Kopet Dagh range: Neues Jahrbuch für Paläontologie Abhandlungen, v. 168, p. 256–268.

Ruzhentsev, S.V., and Shvolman, V.A., 1983, Tectonics of the northwestern Himalayas and Kohistan: Geotectonics, v. 17, p. 138–148.

Schermer, E.R., Howell, D.G., Jones, D.L., 1984, The origin of allochthonous terranes: Perspectives on the growth and shaping of continents: Annual Review of Earth and Planetary Sciences, v. 12, p. 107–131.

Scotese, C.R., 1998, Computer software to produce plate-tectonic reconstructions, *in* Almond, J., et al., eds., Gondwana 10: Event stratigraphy of Gondwana, Special Abstracts Issue: Journal of African Earth Sciences, v. 27, p. 171–172.

Scotese, C.R., and Golonka, J., 1992, Paleogeographic atlas: PALEOMAP progress report 20-0692: Department of Geology, University of Texas at Arlington, 34 p.

Scotese, C.R., and Langford, R.P., 1995, Pangea and the palegeography of the Permian, *in* Scholle, P.A., et al., eds., The Permian of northern Pangea, Volume 1, Paleogeography, paleoclimates, and stratigraphy: Berlin, Springer-Verlag, p. 3–19.

Scotese, C.R., and McKerrow, W.S., 1990, Revised world maps and introduction, *in* McKerrow, W.S., and Scotese, C.R., eds., Paleozoic paleogeography and biogeography: Geological Society [London] Memoir 12, p. 1–21.

Şengör, A.M.C., 1979, Mid-Mesozoic closure of Permo-Triassic Tethys and its implications: Nature, v. 279, p. 590–593.

Şengör, A.M.C., 1984, The Cimmeride orogenic system and the tectonics of Eurasia: Geological Society of America Special Paper 195, 82 p.

Şengör, A.M.C., 1987, Tectonic subdivisions and evolution of Asia: Technical University of Istanbul Bulletin, v. 40, p. 355–435.

Şengör, A.M.C., 1990, Tethyan evolution of central Asia: American Association of Petroleum Geologists Bulletin, v. 74, p. 761.

Şengör, A.M.C., 1991, The Paleo-Tethyan suture: A line of demarcation between two fundamentally different architectural styles in the structure of Asia: The Island Arc, v. 1, p. 78–91.

Şengör, A.M.C., and Natal'in, B., 1996, Paleotectonics of Asia: Fragments of a synthesis, *in* Yin, A., and Harrison, M., eds., The tectonic evolution of Asia: New York, Cambridge University Press, p. 486–640.

Şengör, A.M.C., and Okurogullari, A.H., 1991, The role of accretionary wedges in the growth of continents: Asiatic examples from Argand to plate tectonics: Eclogae Geologicae Helvetiae, v. 84, p. 535–597.

Şengör, A.M.C., Natal'in, B.A., and Burtman, V.S., 1993a, Evolution of the Altaid tectonic collage and Paleozoic crustal growth in Eurasia: Nature, v. 364, p. 299–307.

Şengör, A.M.C., Cin, A., Rowley, D.B., and Nie, S.Y., 1993b, Space-time patterns of magmatism along the Tethysides: A preliminary study: Journal of Geology, v. 101, p. 51–84.

Sharps, R., McWilliams, M., Li, Y.P., Cox, A., Zhang, Z., Zhai, Z., Gao, Z., Li, Y.A., and Li, Q., 1989, Lower Permian paleomagnetism of the Tarim block, northwestern China: Earth and Planetary Science Letters, v. 92, p. 275–291.

Shi, Y., Lu H., Jia, D., Cai, D., Wu, S., and Chen, C., 1994, Paleozoic plate-tectonic evolution of the Tarim and western Tianshan regions, western China: International Geological Review, v. 36, p. 1058–1066.

Sinitsyn, V.M., 1957, Northwestern part of the Tarim basin: Moscow, Academiya Nauk SSSR, 257 p.

Smethurst, M.A., Khramov, A.N., and Torsvik, T.H., 1998, The Neoproterozoic and Paleozoic paleomagnetic data for the Siberian Platform: From Rodinia to Pangea: Earth Science Reviews, v. 43, p. 1–24.

Smith, A.B., 1988, Late Paleozoic biogeography of East Asia and paleontological constraints on plate tectonic reconstructions: Royal Society of London Philosophical Transactions, ser. A, v. 326, p. 189–227.

Sobel, E., and Arnaud, N., 1999, A possible middle Paleozoic suture in the Altyn Tagh, NW China: Tectonics, v. 18, p. 64–74.

Stöcklin, J., 1974, Possible ancient continental margins in Iran, *in* Burk, C.A., and Drake, C.L., eds., The geology of continental margins: Berlin, Springer Verlag, p. 873–887.

Stöcklin , J., 1977, Structural correlation of the alpine rangès between Iran and Central Asia, *in* Livre à la mémoire de Albert F. de Lapparent (1905–1975) consacré aux Recherches géologiques dans les chaines alpines de l'Asie du Sud-Ouest: Mémoire Hors Série, Société Géologique de France, v. 8, p. 333–353.

Thomas, J.C., Chauvin, A., Gapais, D., Bazhenov, M.L., Perroud, H., Cobbold, P.R., and Burtman, V.S., 1994, Paleomagnetic evidence for Cenozoic block rotations in the Tadjik depression (Central Asia): Journal of Geophysical Research, v. 99, p. 15141–15160.

Tian Zaiyi, Chai Guilin, and Kang Yuzhu, 1989, Tectonic evolution of the Tarim Basin, *in* Hsü, K.J., and Zhu, X., eds., Chinese sedimentary basins: Amsterdam, Elsevier, p. 33–43.

UNESCO, 1991, World geological atlas: Paris, scale 1:2 000 000.

Vachard, D., 1980, Tethys et Gondwana au Paléozoique superieur: Les donnees Afghanes: Paris, Documents et Travaux de l'Institut Geologique Albert de Lapparent (IGAL), no. 2, 463 p.

Vinogradov, A.P., et al., eds., 1968, Atlas of the lithological-paleogeographical maps of the USSR: Min. Geol.-Akademiya Nauk SSSR, Moscow.

Wang Baoyu, Li Qiang, and Liu Jianbing, 1996, Geology and tectonics of the west Tianshan mountains along Dushanzi-Kuqa highway in Xinkiang: 30th International Geological Congress, Field Trip Guidebook T364: Beijing, Geological Publishing House, 24 p.

Wang, H., compiler, 1985, Atlas of the paleogeography of China: Beijing, Cartographic Publishing House, 143 p.

Wang, H., 1986, Geotectonic development of China, *in* Yang, Z., et al., eds., The geology of China: Oxford, Clarendon Press, p. 235–276.

Wang Qinmin, and Coward, M.P., 1990, The Chaidam Basin (NW China): Formation and hydrocarbon potential: Journal of Petroleum Geology, v. 13, p. 93–112.

Wang, Q., and Liu, X., 1986, Paleoplate tectonics between Cathaysia and Angaraland in Inner Mongolia of China: Tectonics, v. 5, p. 1073–1088.

Watson, M.P., Hayward, A.B., Parkinson, D.N., and Zhang Z.M., 1987, Plate tectonic history, basin development and petroleum source rock deposition onshore China: Marine and Petroleum Geology, v. 4, p. 205–225.

Wensink, H., 1991, Late Precambrian and Paleozoic rocks of Iran and Afghanistan, *in* Moullade, M., and Nairn, A.E.M., eds., The Phanerozoic geology of the world, I: The Paleozoic: Amsterdam, Elsevier, p. 147–218.

Windley, B.F., Allen, M.B., Zhang, C., Zhao, Z.Y., and Wang, G.R., 1990, Paleozoic accretion and Cenozoic redeformation of the Chinese Tien Shan Range, Central Asia: Geology, v. 18, p. 128–131.

Yang, Z., Cheng, Y.Q., and Wang, H.W., eds., 1986, The geology of China: Oxford, Clarendon Press, 303 p.

Yao, Y., and Hsu, K.J., 1992, Origin of the Kunlun Mountains by arc-arc and arc-continent collisions: The Island Arc, v. 3, p. 75–89.

Yin, A., and Nie, S.Y., 1994, Palinspastic reconstruction of western and central China from the middle Paleozoic to the early Mesozoic: Eos (Transactions, American Geophysical Union), v. 75, supplement, p. 629.

Yin, A., and Nie, S.Y., 1996, A Phanerozoic palinspastic reconstruction of China and its neighboring regions, *in* Yin, A., and Harrison, T.M., eds., The tectonic evolution of Asia: Cambridge, Cambridge University Press, p. 442–485.

Zhang, K.J., 1997, North and south China collision along the eastern and southern north China margin: Tectonophysics, v. 270, p. 145–156.

Zhang, Z.M., Liou, J.G., and Coleman, R.G., 1984, An outline of the plate tectonics of China: Geological Society of America Bulletin, v. 95, p. 295–312.

Zhao, X., and Coe, R.S., 1989, Tectonic implications of Permo-Triassic paleomagnetic results from North and South China, *in* Hillhouse, J.W., ed., Deep structure and past kinematics of accreted terranes: American Geophysical Union Geophysical Monograph, v. 50, p. 267–283.

Zhao, X., Coe, R.S., Zhou, Y., Wu, H., and Wang, J., 1990, New paleomagnetic results from northern China: Collision and suturing with Siberia and Kazakhstan: Tectonophysics, v. 181, p. 43–81.

Zhao, X., Coe, R., Wu, H., and Zhao, Z., 1993, Silurian and Devonian paleomagnetic poles from north China and implications for Gondwana: Earth and Planetary Science Letters, v. 117, p. 497–506.

Zholtayev, G.Z., 1998, Geodynamic model of north Caspian syneclize in the Paleozoic: Petroleum Geology, v. 32, p. 250–261.

Zhou Da, and Graham, S.A., 1996, Extrusion of the Altyn Tagh wedge: A kinematic model for the Altyn Tagh fault and palinspastic resconstruction of northern China: Geology, v. 24, p. 427–430.

Ziegler, A.M., 1990, Phytogeographic patterns and continental configurations during the Permian period, *in* McKerrow, W.S., and Scotese, C.R., eds., Paleozoic paleogeography and biogeography: Geological Society [London] Memoir 12, p. 363–377.

Ziegler, A.M., Scotese, C.R., McKerrow, W.S., Johnson, M.E., and Bambach, R.K., 1979, Paleozoic paleogeography: Annual Reviews of Earth and Planetary Science, v. 7, p. 473–502.

Ziegler, A.M., Hulver, M.L., and Rowley, D.B., 1997, Permian world topography and climate, *in* Martini, P., ed., Late glacial and postglacial environmental changes: New York, Oxford, Oxford University Press, p. 111–146.

Ziegler, P.A., 1990, Geological atlas of Western and Central Europe (second edition): Avon, Geological Society [London] Publishing House, 239 p.

Zonenshain, L.P., and Gorodnitsky, A.M., 1977, Paleo-Mesozoic and Mesozoic reconstructions of the continents and oceans: Geotectonics, v. 11, p. 159–172.

Zonenshain, L.P., Korinevsky, V.G., Kazmin, V.G., Pechersky, D.M., Khain, V.V., and Matveenkov, V.V., 1984, Plate tectonic model of the south Urals development: Tectonophysics, v. 109, p. 95–135.

Zonenshain, L.P., Kuzmin, M.I., and Natapov, L.M., 1987a, Phanerozoic palinspastic reconstructions for the USSR: Geotectonics, v. 21, p. 487–502.

Zonenshain, L.P., Kuzmin, M.I., and Kononov, M.V., 1987b, Absolute reconstructions of the position of the continents during Paleozoic and early Mesozoic time: Geotectonics, v. 21, p. 199–212.

Zonenshain, L.P., Kuzmin, M.I., and Natapov, L.M., 1990, Geology of the USSR: A plate-tectonic synthesis, *in* Page, B.M., ed., American Geophysical Union Geodynamics Series, v. 21, 242 p.

MANUSCRIPT ACCEPTED BY THE SOCIETY JUNE 5, 2000

Geological Society of America
Memoir 194
2001

Paleozoic tectonic amalgamation of the Chinese Tian Shan: Evidence from a transect along the Dushanzi-Kuqa Highway

Da Zhou
ExxonMobil Exploration Company, P.O. Box 4778, Houston, Texas 77060, USA
Stephan A. Graham*
Department of Geological and Environmental Sciences, Stanford University, Stanford, California 94305, USA
Edmund Z. Chang
Department of Geological and Environmental Sciences, Stanford University, Stanford, California 94305, USA
Baoyu Wang
Institute of Geology and Mineral Resources of Xinjiang, Urumqi, Xinjiang Uygur Autonomous Region, China
Bradley Hacker
Department of Geological Sciences, University of California, Santa Barbara, California 93106, USA

ABSTRACT

The Du-Ku Highway traverses the Tian Shan from the cities of Dushanzi to Kuqa in western China and provides a section across the full width of this Paleozoic orogen. The north Tian Shan consists of an accretionary complex of deformed ophiolite, arc volcanic assemblages, and siliciclastic sedimentary rocks derived mainly from volcanic arcs. The central Tian Shan is underlain by Precambrian basement covered mainly by Carboniferous andesitic tuff, Paleozoic marine carbonate and siliciclastic strata, and abundant Carboniferous–Early Permian granitoids. The south Tian Shan features continuous sequences of Silurian to Lower Carboniferous carbonate, chert, and siliciclastic strata of the deformed Paleozoic passive continental margin of the Tarim block.

A high pressure-temperature (*P-T*) blueschist belt, parallel to and south of the Nikolaev line, which bounds the central and south Tian Shan, is exposed to the west of the Du-Ku Highway. Previously published radiometric ages bracket the blueschist metamorphism of the blueschist belt as 415–315 Ma. A paired coeval magmatic belt to the north was active from at least 400 Ma to ca. 330 Ma.

Radiolarian fossils from chert associated with the north Tian Shan ophiolite, and Lower to Middle Carboniferous lithic turbidites structurally imbricated with ophiolite and tuff in the north Tian Shan, indicate that subduction-accretion along the north Tian Shan lasted at least through the end of the Early Carboniferous. Plutonism in the Borohoro arc along the northern margin of the central Tian Shan block culminated during Late Carboniferous–Early Permian time based on radiometric dating.

We infer that amalgamation of the Tian Shan proceeded in two steps: (1) accretion of the central Tian Shan block to the northwest margin of the Tarim block by the Middle Carboniferous, with attendant cessation of arc magmatism and high *P-T* subduction metamorphism; (2) accretion of the Junggar arc complex and the

*Corresponding author; E-mail: graham@pangea.stanford.edu

Zhou, D., et al., 2001, Paleozoic tectonic amalgamation of the Chinese Tian Shan: Evidence from a transect along the Dushanzi-Kuqa Highway, *in* Hendrix, M.S., and Davis, G.A., eds., Paleozoic and Mesozoic tectonic evolution of central Asia: From continental assembly to intracontinental deformation: Boulder, Colorado, Geological Society of America Memoir 194, p. 23–46.

Tarim–central Tian Shan composite block during Late Carboniferous–Early Permian time. Extensive preservation of upper Paleozoic tectonostratigraphic elements and thermo-chronologic and paleomagnetic considerations suggest that accretion occurred by soft, perhaps rotational or oblique, collision.

INTRODUCTION

The Tian Shan (Heavenly Mountains) of central Asia extends east-west for about 2500 km along the northern border of the Tarim basin and beyond (Fig. 1). Its present elevation, many peaks rising to above 4500 m and locally higher than 7000 m, reflects crustal shortening and thickening that resulted from the India-Asia collision. Late Cenozoic tectonics in the Tian Shan has been demonstrated by Landsat imagery and surface mapping of active deformation along the northern Chinese Tian Shan (Tapponnier and Molnar, 1979; Avouac et al., 1993), by surface mapping of folded and thrusted upper Cenozoic strata along the southern flank of the Tian Shan (Yin et al., 1998), by global positioning system (GPS) measurements of present-day crustal shortening of the Tian Shan in Kazakhstan (Abdrakhmatov et al., 1996), and by fault-plane solutions of earthquakes (Molnar and Deng, 1984; Molnar and Lyon-Caen, 1989). Neogene uplift has been detected by apatite fission-track records retrieved from folded and thrusted foreland basin sedimentary strata on both

sides of the Chinese Tian Shan (Hendrix et al., 1994; Sobel and Dumitru, 1997), and more recently along fault zones within the Tian Shan (Dumitru et al., this volume), although thermo-chronologic data also indicate that much of the westernmost Chinese Tian Shan has not been significantly unroofed since its tectonic amalgamation in the late Paleozoic (Yin et al., 1998; Dumitru et al., this volume).

Besides its history of Cenozoic intracontinental mountain-building, the Tian Shan has long been recognized as an orogenic belt associated with late Paleozoic tectonic amalgamation of central Asia (Zhang et al., 1984; Carroll et al., 1990, 1995; Wang et al., 1990; Windley et al., 1990; Allen et al., 1992; Gao et al., 1998; Chen et al., 1999), although many details remain to be resolved. Hendrix et al. (1992) and Hendrix (2000), on the basis of subsidence and provenance study of Mesozoic nonmarine sections of south Junggar and north Tarim basin, interpreted at least three periods of thrusting and uplifting of the Tian Shan during the Mesozoic. Their contention is supported by apatite fission-track data from the interior of the Tian Shan collected

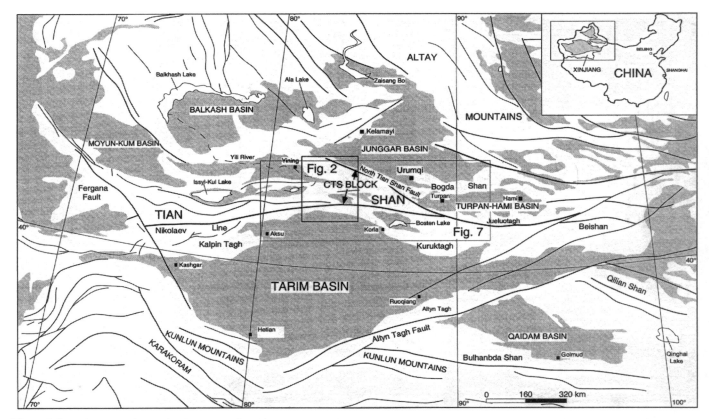

Figure 1. Location map of central Asia surrounding Tian Shan, showing Cenozoic sedimentary basins (gray shading), major faults (modified from Lee, 1982), and major tectonic blocks. CTS—central Tian Shan.

in our study and reported in a companion paper in this volume (Dumitru et al.). In sum, many interpretations of the history of the Chinese Tian Shan, particularly in the non-Chinese language literature, are based on inferences derived from field evidence from the flanks of the range, or from remotely sensed data, rather than from field observations in the interior of the range. Clearly, understanding of all facets of Tian Shan history, from Paleozoic accretion and amalgamation to Mesozoic–Cenozoic structural reactivation and uplift, will be optimized by more data from the core of the range.

This chapter presents a description of a geologic cross section across the western part of the Chinese Tian Shan, as exposed along the Dushanzi-Kuqa (Du-Ku) Highway (Figs. 1–3). This road provides rare, critical access to remote portions of the range at a longitude where the range is wide and fully expressed. Built for military purposes, the Du-Ku Highway was closed to non-Chinese citizens until the International Geological Congress in Beijing in 1996, so geologic studies along the highway route from the 1950s to mid-1990s were conducted by Chinese geoscientists (e.g., Wang et al., 1996). Before the opening of the Du-Ku Highway, non-Chinese scientists were limited to observations of the Chinese Tian Shan along the highway between Urumqi and Korla (e.g., Windley et al., 1990; Allen et al., 1992), 300 km to the east of the Du-Ku Highway (Fig. 1). Unfortunately, the Urumqi-Korla Highway was sited along a generally low-relief route where outcrops are relatively poor for great distances, so that the ability to observe many structures and petrotectonic elements of the Tian Shan is rather limited. Thus, a study of the Du-Ku Highway section is essential for reconstruction of an amalgamation history that involved continental blocks, ocean basins, and magmatic arcs. The Du-Ku section also provides an opportunity to study uplift history of the Tian Shan by thermochronologic study of plutonic and metamorphic rock samples from the interior of this orogen (Yin et al., 1998; Dumitru et al., this volume).

Although reconnaissance in nature, our study of the Du-Ku section incorporates all biostratigraphic and radiometric data available from published and unpublished reports (Wang Baoyu, 1994, unpublished data) and new ^{40}Ar/^{39}Ar ages obtained during this study. In this chapter, we first present our observations of petrotectonic elements exposed along the Du-Ku Highway, from north to south. Based on the Du-Ku transect, we examine previously published tectonic models and offer our view of the amalgamation history of the Tian Shan. In the companion paper by Dumitru et al. (this volume), we use the Du-Ku transect as a platform for discussing new apatite fission-track data from the interior Tian Shan that provide direct evidence of its uplift and exhumation history from the latest Paleozoic to Cenozoic.

GEOLOGY OF THE DU-KU HIGHWAY

Du-Ku cross section

The Du-Ku Highway begins in the north in the city of Dushanzi, which is along the boundary between the Junggar basin and the northern foothills of the Tian Shan (Fig. 2). From

there, the road trends south-southwesterly roughly perpendicular to the east-west structural trend of the range, and traverses the major, generally recognized petrotectonic elements that compose the late Paleozoic Tian Shan orogenic belt. These distinctive tectonic elements and their structural boundaries permit subdivision of the Chinese Tian Shan into three generally recognized geologic and physiographic components, the north, central (also known as the Yili terrane, e.g., Gao et al., 1998), and south Tian Shan (Wang et al., 1990). The complicated physiography of the Tian Shan, characterized by successive east-west–oriented ranges separated by intermontane lowlands (Figs. 2 and 3), owes its character to diverse origins and evolution of Paleozoic petrotectonic elements, overprinted by Mesozoic and Cenozoic deformation, as we demonstrate here (and in Dumitru et al., this volume). The Du-Ku Highway emerges from the south Tian Shan in the foothill topography of the fold and thrust belt along the northern Tarim basin, and tracks southward to the town of Kuqa (Figs. 2 and 3). The Kekesu section (Fig. 2) is 180 km west of where the Du-Ku Highway crosses the Bayanbulak Valley, and exposes a blueschist-bearing subduction complex. It is complementary to the Du-Ku section, because high pressure-temperature (*P-T*) metamorphic rocks apparently are structurally omitted from the tectonic boundary between the central Tian Shan and south Tian Shan along the Du-Ku Highway; accordingly, we incorporate in our discussion some information from unpublished Stanford University studies of the Tekes-Kekesu area (Jia, 1996).

Stanford University investigators made observations of the southern and northern portions of the Tian Shan during the 1987–1992 field seasons, followed in 1993 by a transect across the entire range on the Du-Ku Highway guided by an unpublished cross section created by the Xinjiang Bureau of Geology and Mineral Resources that follows the highway. Portions of this section were subsequently published in a guidebook for an International Geological Congress field trip (Wang et al., 1996). Our paper includes a modified version (Fig. 3) of the entire unpublished section. Note that the section follows the highway (although not exactly) and therefore locally transects structure obliquely; it perhaps is most aptly termed a graphic roadlog. Observations made along the road proper are supplemented by data from proprietary 1:200 000 geologic maps of the Tian Shan made by the Xinjiang Bureau of Geology and Mineral Resources. Despite some limitations of the graphic roadlog, it provides a first-order, ground-level cross section of the western Chinese Tian Shan.

North Tian Shan accretionary complex

Lower and Middle Carboniferous strata of the north Tian Shan rise abruptly from low-relief foothills underlain by strata of the Mesozoic–Cenozoic Junggar foreland basin (Figs. 2 and 3) (note that Carboniferous and Permian subdivisions of Chinese standard usage differ from international usage; see Carroll et al., 1995, for an alignment of Chinese stages with absolute time). Recent uplift (e.g., Avouac et al., 1993) is evinced by

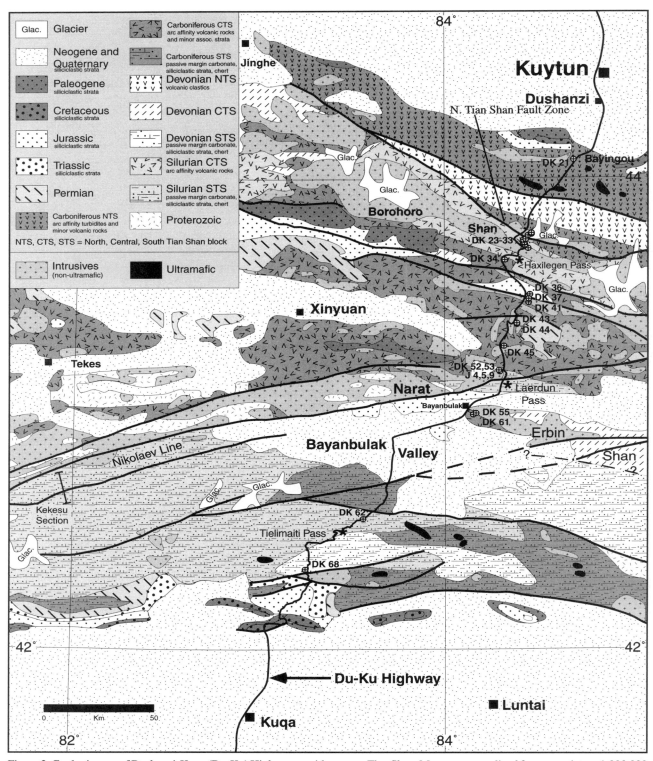

Figure 2. Geologic map of Dushanzi-Kuqa (Du-Ku) Highway corridor across Tian Shan. Map was generalized from proprietary 1:200 000 geologic mapping by Xinjiang Bureau of Geology and Mineral Resources. Cross section in Figure 3 follows Du-Ku Highway.

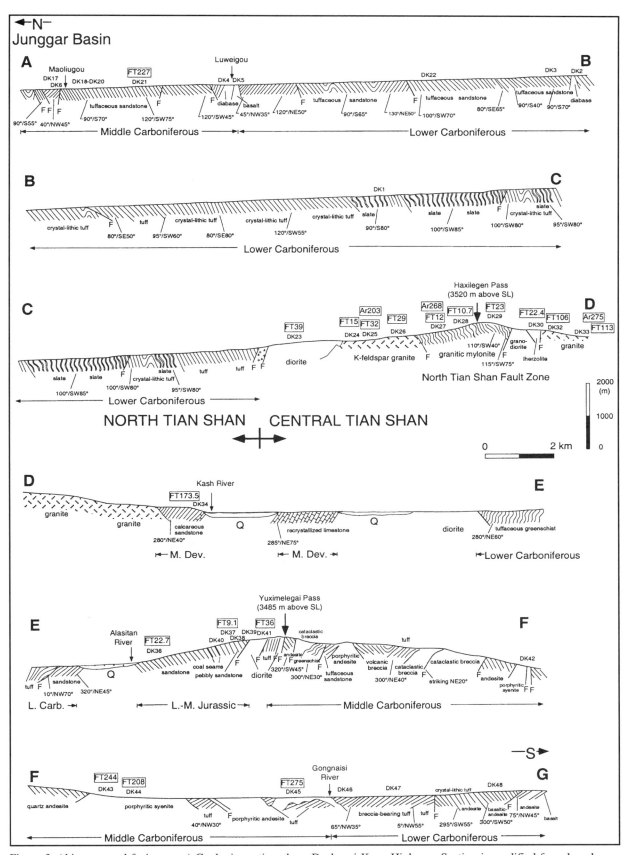

Figure 3. (this page and facing page) Geologic section along Dushanzi-Kuqa Highway. Section is modified from largely unpublished cross section and accompanying base map (Fig. 2) compiled by Xinjiang Bureau of Geology and Mineral Resources. Selected segments of section previously were published by Wang et al. (1996) and Gao et al. (1998). Note that section follows road and is not across-strike structure section. Lithology, available ages, attitudes of bedding or foliation, intrusive rocks, and observed faults are shown on section. Sample numbers (e.g., DK45) are posted at their locations. Fission-track ages (e.g., FT10.7) are discussed in Dumitru et al. (this volume). ^{40}Ar/^{39}Ar ages are shown above their sample numbers and are listed in Table 1. Single long cross section from north (A) to south (L) is formed by 14 segments of the section with two breaks or omissions due to Quaternary cover or lengthy strike exposure.

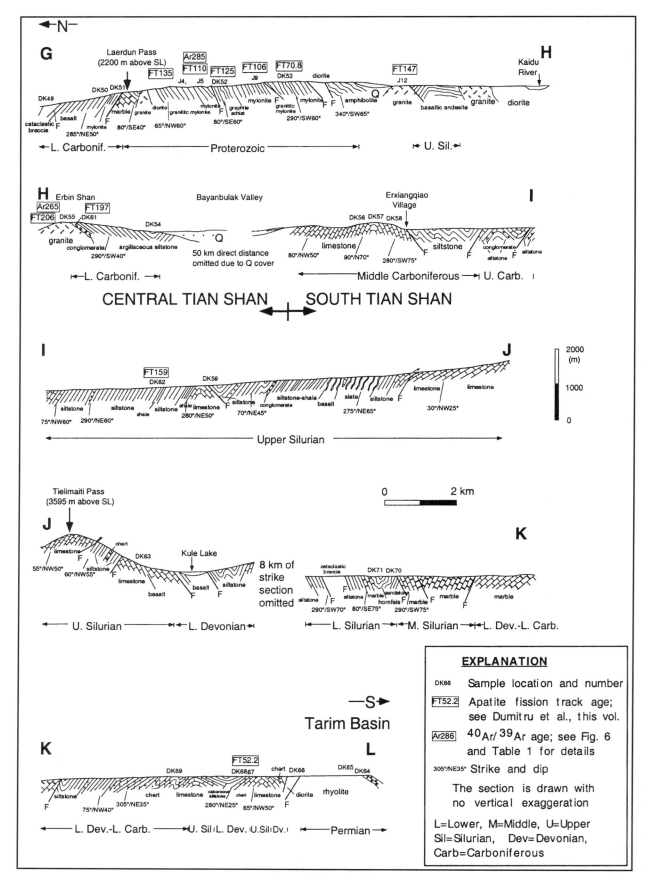

Figure 3. (*continued*)

elevated and dissected Cenozoic channel-fill and flood-plain deposits along the Kuitun River Valley, where the Dushanzi-Kuqa Highway enters the north Tian Shan, and by Carboniferous strata thrusted over Neogene deposits along the Du-Ku Highway (Gao et al., 1998). Carboniferous strata are intensely folded and faulted in a generally north-vergent direction, and can be described as broken formation where monolithologic, or as melange where ophiolite and other lithologies coexist. Rocks in this part of the north Tian Shan (Fig. 3, A-B) include crystal-lithic tuff, tuff, members of ophiolites, and lithic turbidites rich in andesitic and basaltic volcanic fragments and plagioclase feldspar (Fig. 4A). At Maoliugou (Fig. 3A) north of the Luweigou (Fig. 3, A-B)-Bayingou (Fig. 2) ophiolite melange zone, Middle Carboniferous crinoid and coral bearing (Wang et al., 1996; Fig. 5) volcaniclastic turbidites are well exposed. The sequence consists predominantly of monotonous greenish, massive, graded or planar laminated sandstone; crinoid fossils and feldspar crystals are displayed as distinctive white spots on weathered outcrop surfaces. One interval of high-density-turbidite sandstone in particular displays excellent normally graded bedding with gravel and pebbles at the bottom composed of coral and crinoid fossils (Fig. 4B), cemented fossil aggregates, and lithic sandstone rich in volcanic fragments.

Well exposed along the Du-Ku Highway at Luweigou (Fig. 3, A-B) and along strike several kilometers to the east at Bayingou (Fig. 2) are ophiolitic rocks incorporated within the folded and faulted volcanic and volcaniclastic rock of the north Tian Shan. The ophiolite at Luweigou is strongly dismembered and incompletely preserved. Pillow basalt and radiolarian chert are the most abundant rock types, but siliciclastic turbidites are also exposed. At Bayingou the ophiolite is better preserved and contains pillow basalt (Fig. 4C), gabbro, and radiolarian chert (Fig. 4D). We found serpentinite float but no outcrops, although ultramafic bodies were reported by Wang et al. (1990). Radiolaria from the red chert of the Bayingou ophiolite were identified as Early to early Late Carboniferous by Wang et al. (1990) (Fig. 5), whereas our group has collected Fammenian (Late Devonian) conodonts from these cherts (Carroll, 1991; Fig. 5).

The boundary between the north and central Tian Shan roughly follows the north Tian Shan fault (Figs. 1 and 2; Allen et al., 1992) or the Dzhungarian fault system of Tapponnier and Molnar (1979). In Figure 3 (C-D) we place the north-central Tian Shan boundary at the northern limit of Paleozoic plutons we believe to be of central Tian Shan tectonic affinity, rather than at the north Tian Shan fault 4–5 km to the south, because the fault here is a Cenozoic structural modification (see Dumitru et al., this volume) that locally diverges from the Paleozoic tectonic boundary.

Central Tian Shan magmatic arc assemblage

Calc-alkalic magmatic arc assemblages (Wang et al., 1990) are exposed (Fig. 3, C-D–H-I) between Haxilegen Pass and Erbin Shan along the Du-Ku Highway cross section. The outcrops along the central portion of this part of the highway are composed of tuff, crystal-lithic tuff, porphyritic tuff of andesitic composition, granite, K-feldspar granite, diorite, granodiorite, and dacite (Figs. 3 and 4E). Carbonate strata and volcanic sandstone as old as Devonian (Wang et al., 1996; Fig. 5) occur locally in the central Tian Shan (Fig. 3, D-E). Lower to Middle Jurassic lacustrine siltstone, shale, coaly siltstone, sandstone, and conglomerate of alluvial-fluvial facies are exposed in the Alasitan River Valley and the northern flank of Yuximelegai Pass (Fig. 3). Although the Jurassic sequence is in fault contact with Lower to Middle Carboniferous volcanic rocks along the highway, coarse-grained volcanic and metamorphic fragments (Fig. 4F) within Jurassic strata indicate local derivation from underlying Paleozoic rocks, and elsewhere in the range Jurassic strata are mapped in depositional contact with the Paleozoic rocks (Xinjiang Bureau of Geology and Mineral Resources [XBGMR], 1993).

Granitic plutons are exposed along the northern and southern margins of the central Tian Shan, at Haxilegen Pass and Erbin Shan, respectively (Figs. 2 and 3, C-D, H-I). The Haxilegen Pass area (Fig. 3, C-D) exposes granite, diorite, granodiorite, and K-feldspar granite. The northern slope of Haxilegen Pass consists of highly faulted and fragmented tuff and crystal-lithic tuff intruded by a granitoid pluton (Fig. 4E; contact is faulted along line of section C-D in Fig. 3). Fossils in the interbedded volcaniclastic strata indicate an Early Carboniferous age (XBGMR, 1993; Fig. 5). Three samples from the Haxilegen Pass area were dated by the ^{40}Ar/^{39}Ar method (Fig. 6; Table 1): biotite from DK27, a mylonitized granite, gave a weighted mean age of 267.7 Ma; K-feldspar from DK25, the K-feldspar granite (Fig. 4E) intruded into the volcaniclastic assemblage, shows that the pluton is older than 202.9 Ma; and biotite from lherzolite DK33 on the southern slope of the pass yielded a plateau age of 275.3 Ma (Fig. 3). A whole-rock Rb-Sr isochron age of 292 ± 15 Ma was reported by Wang et al. (1990) from biotite granite of the Haxilegen belt, significantly older than the 267.7 Ma age reported here. The enigmatic lherzolite rock body exposed on the southern slope of Haxilegen Pass is apparently a fault sliver within granite (Fig. 3, C-D), and may be a piece of north Tian Shan ophiolite translated along the north Tian Shan fault zone that extends through the pass. Apatite fission-track data from the lherzolite (Dumitru et al., this volume) indicate early Miocene exhumation, possibly related to strike-slip motion of the bounding faults.

Laerdun Pass, along strike of the Nikolaev tectonic line (Wang et al., 1990) farther to the west, marks the southern margin of the central Tian Shan, and consists of granitic mylonite, actinolite schist, graphite schist, and chlorite schist (Fig. 3, G-H). These rocks were interpreted previously as Precambrian gneiss, schist, migmatite, and amphibolite. However, field observation and examination of thin sections in this study indicated that the so-called gneiss, migmatite, and amphibolite are in fact mylonitized, greenschist facies granitoid. A Rb-Sr whole-rock isochron age of 335 ± 28 Ma obtained from seven mylonitic granites by Zhu and Sun (1986) reflects the crystallization age of the metagranites. We obtained two ^{40}Ar/^{39}Ar ages

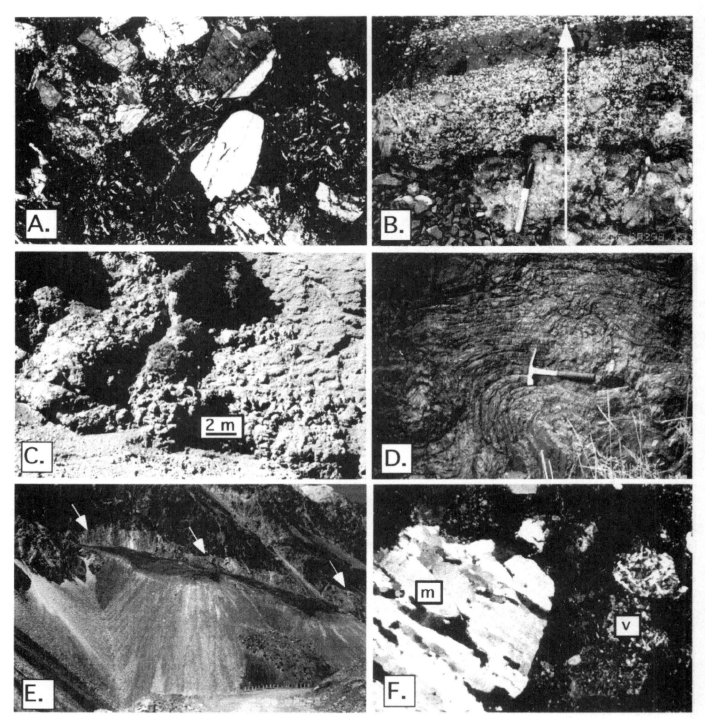

Figure 4. Outcrop photos and photomicrographs from north Tian Shan and central Tian Shan. A: Photomicrograph, crossed polars, of typical siliciclastic sandstone (DK6) from north Tian Shan accretionary complex. Framework clasts are composed almost exclusively of volcanic lithic fragments and plagioclase grains. Width of photo is ~4mm. B: Outcrop photo of normally graded turbidite (arrow highlights upward-fining trend) exposed in Maoliugou, north Tian Shan. Gravel clasts at bottom of bed are either volcanic or volcaniclastic rocks similar in composition to underlying and overlying sandstone layers. Intervals containing abundant crinoid fossils (white rounded objects) alternate with those rich in volcanic lithic fragments. C: Pillow basalt at Bayingou, where type section of north Tian Shan ophiolite was documented by Wang et al. (1990). D: Tightly folded and thin-bedded red radiolarian chert at Bayingou, north Tian Shan. E: Exposure of light colored K-feldspar granite (lower hillside) intruding deformed and metamorphosed dark colored Carboniferous volcanigenic slates on upper part of north slope of Haxilegen Pass. Arrows point to intrusive contact. $^{40}Ar/^{39}Ar$ plateau age of granite is 203 Ma. For scale, note meter-high guard-rail posts along road. F: Photomicrograph of Jurassic pebbly sandstone (sample DK 38, Fig. 3, E-F) from north of Yuximelegai Pass consisting exclusively of volcanic (v) and metamorphic lithic fragments (m). Width of photo is ~5 mm.

Fossils reported from the Dushanzi-Kuqa Road corridor across the Tian Shan

North Tian Shan

Upper Carboniferous
- Corals: *Lophophyllidium* sp., *Meniscophyllum* sp.
- Cephalopods: *Branneroceras* sp., *Goniatites* sp.
- Plants: *Mesocalamites* sp., *Calamites* sp.

Middle Carboniferous
- Radiolaria: *Ceratoikiscum* sp., *Astroentactinia crassata*, *Bissylentactinia* sp., *Entatinosphaera chstactorata*

Lower Carboniferous
- Corals: *Lophophyllidium* sp.
- Bryozoa: *Fistulipora* sp., *Septepora* sp., *Fenestella* sp., *Meekopora* sp., *Polypora* sp.
- Trilobites: *Philipsia?* sp.

---Fault contact---

Devonian
- Corals: *Pachycanaclicula* cf. *karcevae*, *Crassialveolites* sp., *Thamnopora* sp., *Lithophyllum* sp. *Stromatopora*: *Actinostroma* sp.
- Conodonts (from cherts correlative to those shown in Fig. 4D; cited by Carroll, 1991): *Delotaxis* sp., *Palmatolepis* sp., *Polygnathus* cf. *perplexus* or *P. nodocostatus*, *Polygnathus* sp.
- Radiolaria (from cherts correlative to those shown in Fig. 4D; cited by Carroll, 1991): *Ceratoikiscum* sp.

Central Tian Shan

Upper Carboniferous
- Fusulinids: *Ozawainella umbonala*, *Profusulinella prisce*, *Palaeestapfella* cf. *moelleri*
- Brachiopods: *Chorestites* sp.
- Bivalves: *Edmodia* sp., *Myrlina tianshanensis*, *Schizodus* sp.

_____Unconformity_____

Middle Carboniferous
- Corals: *Campophyllum* aff. *klaeri*, *Yuanophyllum kansuense*, *Multithecopora regutaris*, *Acachuolasma sinense*, *Chaetetes luindhoi*, *C.* ex gr. *longtanensis*, *C. pinnatus*, *Lithostrotionella styleris*
- Brachiopods: *Neospirifer derjawini*, *Avonia stepanovi*, *Christites pseudobisulcatus*, *C. crassicustatus*, *C. nilekensis*, *C. jiguleensis*, *C. javorskii*, *C. mansuyi*, *Orthotetes radiata*, *Buxtonia cabricula*, *B. juresaensis*, *Marginifera janischewskiani*, *M. praegratiosus*, *M. schartimiensis*, *M. orientalis*, *Pustula pustulosa*, *Productus productus*, *P. concinus*, *P. chaoi*, *Brachythyrina pingusiformis*, *B. swallrviformis*, *B. borochoroensis*, *Phricodothyris ovata*, *Linoproductus corrugatus*, *L. cora*, *L. leneatus*, *L. undatus*, *Cancrinella undata*, *C. cancriniformis*, *C. truacatus*, *C. undifera*, *Balakhonia kekdscharensis*, *Paechelmaunia borochoroensis*, *Chonetes sinkiangensis*, *C. carboniferus*, *C. asiatica*, *C. pygmeus*, *Krotovia karpinskiana*, *K. aculeata*, *Stiatfera kokdscharensi*, *Echinoconchus punctatus*, *E. elegans*, *Schizophoria resupinata*, *Pughax swalloritformis*, *Dielasma itaitubense*, *Dictyoclostus taldybnlakensis*, *Wellerella osagensi*, *Husdia grandicosta*
- Fusulinids: *Pseudostaffella* cf. *irinovkensis*, *P. umbilicata*, *P. larionovae* var. *polasnensis*, *Fusulina aminodiscus*, *Schubertella obscura*
- Ammonoids: *Goniatites choetawensis*, *Gastriceras perovnatum*, *G. suborientale*, *Lyrgoniatites cloudi*
- Plants: *Calamites undulates*, *Angaropteridium cardiopteroides*

Lower Carboniferous
- Corals: *Rotiphyllum rushiahum*, *R. zhaosuense*, *Arachnolasma cylindricum*, *A. sinense*, *Disphyllum platiformi*, *D. muhicystatum*, *Gangamophyllum xinjiangense*, *G. kumpanivasljuk*, *G. kumpan*, *Hexaphyllia tenuis*, *H. lyelli*, *H. zhaosuensis*, *Kizilia concavitabulata*, *Lithostration irregulari*, *L. mccoyanum*, *Neoclisiophyllum zhaosuense*, *Palaeosmilia murchisoni*, *P. stutchburgi*, *P. zhaosuensis*, *P. regia*, *Siphonophyllia magma*, *S. sinjiangensis*
- Brachiopods: *Martinia semigiobos*, *Echinoconchus elegans*, *E. punctatus*, *Chonetipustula ferganensis*, *Chonetes sinkiangensis*, *Gigantoproductus striatosulcatus*, *G. protveformis*, *G. rectestrius*, *G.*

Figure 5. Fossils reported from the Dushanzi-Kuqa Road corridor across the Tian Shan. Compiled from Carroll (1991), Cheng et al. (1985), Institute of Geology and Mineral Resources (1991), Li (1993), Lin and Wang (1984), Qiao (1989), Wang (1983), and Wang et al. (1996).

akeshakensis, G. praemoderatus, Megachonetes zimmermanni, Schuchertella radialis,
Semiphanus semiphanus, S. miknailovensis, Striatifera striata, S. angusta, Tubaria
akeshakensis
_____Disconformity_____

Upper Devonian

   Plants: *Leptophloeum rhombicum, Sublepidodendron mirebile, Lepidosigillaria* sp.
---Fault contact---

Middle Devonian

   Corals: *Coenitis* sp., *Thamnopora* sp., *Striatopora* sp., *Placocoenites* sp.,*Keriophyllum* sp.,
         Crassialveolites sp., *Tyrganolites* sp., *T.* cf. *eugeni, Spongophyllum* sp.

Upper Silurian

   Corals: *Palaeofavosites xinertaiensis, Taxopora huochengensis, Cladopora* sp., *Heliolites* sp.
   Brachiopods: *Protochonetes* sp., *Plectatrypa* sp., *Camarotoechia* sp.

Middle Silurian

   Corals: *Angopora bolhinurensis, Palaeofavosites jifukensis, Mesofavosites jingheensis, Syringopora*
         nanshanensis, Fletcheriella jucta, Catenipora spirordensis, Halysites nikaensis, H. abnormis, H.
         tianshanensis, Propora nilkaensis, Zelophyllum borohoroshanense, Yassia nilkaense,
         Palaeophyllum jingheense, Amplexoides bolohoroshanense, Microplasma sp.
   Brachiopods: *Pentamerus* cf. *oblongus*
   Cephalopods: *Sichuanoceras* sp., *Parahelenites* cf. *rarus*

Lower Silurian

   Graptolites: *Streptograptus* sp., *Pristiograptus* sp, *Monograptus* cf. *sedgwichii*

Middle Ordovician–Upper Ordovician

   Corals: *Leolasma abnorma, Sinkiangolasma simplex, S. primitica, Catenipora* sp., *Rhabdotetradium*
         quadradium, Eofletcheria sp., *Paratetradium xinertaiensis, Eofletcheria xinertaiensis,*
         Agetolites asiaticus, A. insuetus, Agetolitella vera, Pentamerus cf. *oblongus*
   Graptolites: *Dicellograptus sextans, D. szechuangensis, D. tenuiculus, D. minutus, Climacograptus*
         putillus, Nemagraptus gracillis, Pseudoclimacograptus scharenbergi, Glyptograptus
         teretiusculus
   Trilobites: *Huochengia xinertaiensis, H. convexus*

Lower Ordovician

   Trilobites: *Hysterolenus oblongus, Neoagnostus* sp., *Neohedina spisconica, N. xinjiangensis,*
         Paraboligella lata, P. sayramensis, Rhadinopleura irifmis, Inkouia brevica, I. cavernosa,
         Clavatellus sinicus, Geragnostus sp., *Pseodokainella jingheensis, Shumardia keguginensis,*
         Shumardops longifrons, Asaphus lepisurus
   Graptolites: *Adelograptus antiquus, A. simples, A.* cf. *hunne bergensis, A.* cf. *lapworthi, Clonograptus*
         cf. *lenellus, Didymograptus deflexus, Tetragraptus fruticosus, T. approximatus, Oncograptus*
         upsilon, Cardiograptus morsus, Isograptus maximodivergens

Cambrian

   Trilobites: *Agnostascus* sp., *Glyptagnostus* sp., *Lotaganostus* sp., *Xystridura* sp., *Peronopsis* sp.,
         Hypagnostus sp., *Goniagnostus* sp., *Ptychagnostus* sp., *Redlichia* sp., *Colodiscus* sp.,
         Kootenia sp.

South Tian Shan

Upper Carboniferous

   Fusulinids: *Triticites pseudosimplex, T. pcosillas, T. parvulus, T. simplex, T. notus, T. tianshanensis,*
         T. variabilis, Montiparus minutus, M. thombiconicus, Schubertella paramelonica var. *minor,*
         S. latat var. elliptica, S. kingi, Pseudoschwagerina laohutaiensis, P. cinjiangensi, P. nakazawai,
         P. glomerosa, P. sphaerica, Schwagerin levicula, Rugosofusulina stabilis, R. hutiensis,
         Pseudofusulina kankarinensis, P. fusulinoides, P. subashiensi, P. pusilla, Schwagerina
         elkoensis, S. angusta, S. levicula, S. pailensis, Boultonia willsi, Quasifusulina baichengensis,
         Q. tenuissima, Paraschwagerina gigantea, Hemifusulina shengi, Eoparafusulina laohutaiensis,
         E. ovata compacta, E. bella, Zallia xinjiangensis, Paraschwagerina mukhamediarovica rauser

Figure 5. (*continued*)

Brachiopods: *Cancrinella koninchana, Linoproductus cora, L. lineatus, Dictyoclostus* cf.
    taiyuanfuensis, D. uralicus, Schellwienella cf. *kueichowensis, Maritinia glabra, M. simiplana,*
    M. shansiensis, Phricodothyris asiatica, Dielasma aff. *plica, Notothyris* aff. *mediterreanea*
Corals: *Tachylasma elongatum, Carinthiaphyllum laohutaiensis*
_____Disconformity_____

Middle Carboniferous
Corals: *Kionophyllum* cf. *dibunum, Diphyphyllum* sp., *Caninia* sp., *Caninophyllum* sp.
Brachiopods:, *Choristites* cf. *monsuyi, C. trautscholdi, Neospirifer* sp., *Martinia semiconverxa,*
    Productus sp., *Dictyoclostus* sp.
Fusulinids: *Fusulina quasicylinderica, F. cylindrica, F. bocki, F. pseudobocki, F. pseudocolahina, F.*
    schwagerinoides, Fusulinella laxc, Pseudostaffella sphaeroidec, P. latispiralis, P.
    paracompressa, P. miner, Eostafella postmasquensis, E. pseudostruvei var. *angusta, Millerella*
    marblensis, Ozawainella pseudorhomboidalis, O. vozhgalica, Profusulinella praelibrovich, P.
    rhombides, P. bella, P. parva, Protriticites sp., *Schubertella quasiobscura, Rugosofusulina* sp.,
    Schwagerina sp.
_____Disconformity or unconformity_____

Lower Carboniferous
Corals: *Kueichouphyllum* sp., *Striatifera strata, Aulina carinata*
Brachiopods: *Syringopora* cf. *weiningensis, Fluctuaria yeyuengouensis, F. undata, Linoproductus*
    simenensis, Striatifera striata, Productus cf. *productus, Gigantoproductus latissimus*
Fusulinids: *Eostaffella paropuolvae, E. mosquensis, E. ozawainellaeformis, E. kashirica, E. ikensis* var
    ikensis, E. xinjiangensis
_____Unconformity_____

Upper Devonian
Corals: *Temnophyllum* sp.
Brachiopods: *Ambocoelia* cf. *sinensis, Ilmeria* cf. *maslovi, Yunnanellina* sp., *Tenticospirifer* aff
    gortani, T. cf. *tenticulum, Camarotoechia sikuangshanensis, Productella lachrymosa asiatica*

Middle Devonian
Corals: *Tabulophyllum* sp., *Zaphrentis* sp., *Syringopora* cf. *supergigantea, Thamnopora* cf. *major,*
    Cladopora sp., *Diplochone* sp., *Ilmeria* sp. *Fasciphyllum* sp., *Syringopora obesus, S.* cf.
    grandis, Disphllum breviseptatu, Chaetetes sp., *Heliolites* cf. *insolens* var. *dunbeiense,*
    Pachyfavosites sp., *Emmonsia* cf. *radiformis, Xystriphyllum kumuxense, Cyathophyllum* sp.,
    Alveolites sp., *Squameiofavosites* sp., *Favosites* aff. *polymorphus* var. *sibirica, Keriophyllum*
    sp., *Tryplasma* sp., *Heliolites insolens* subsp. *dunbeiensis, Natalophyllum* sp., *Coenites* sp.,
    Digonophyllum sp., *Pseudomicroplasma* cf. *flabelliforme, Temnophyllum* cf. *liunmaensis*
Brachiopods: *Indospirifer* sp., *Atrypa desquamata, A. aspera, Cryptonella* sp., *Athyris gurdoni* var.
    transversalis, Schizophoria cf. *kutsingensis, Gyridula* sp., *Stringocephalus grandobesus, S.*
    obesus var grandis, S. burtnini, Euryspirifer sp.

Upper Silurian–Lower Devonian
Corals: *Pachypora cristata, Favosites intricatus, F. forbesi* var. *eifeliensis, F. aphragma,*
    Sinkiangophyllum sinkiangensis, Rhizophyllum sp., *Tryplasma* cf. *maximum,*
    Leptoinophyllum sp., *Squameofavosits* sp.
Brachiopods: *Condhidium ex* gr. *khighti, Spirifer koegeler, Stropheodonta nobilis, Camarotoechia* cf.
    boloniensis, Atryps. cf. *desquamata, A. ex* gr. *camata, Karpinskia* sp., *Grunewaldtia* sp.,
    Uncinulus sp., *Schuchertella* sp.

Middle Silurian–Upper Silurian
Corals: *Mesofavosites baichengensis, Pachyfavosites* cf. *gesamiaoensis, Stelliporella* cf. *keyiensis, S.*
    baichengensis, Dictyofavosites sp., *Squameofavosites* sp., *Heliolites* sp., *Fletcheriella* cf.
    sichuanensis, Tryplasma sp., *Amplexoides* sp., *Songophyllum* sp.

Figure 5. (*continued*)

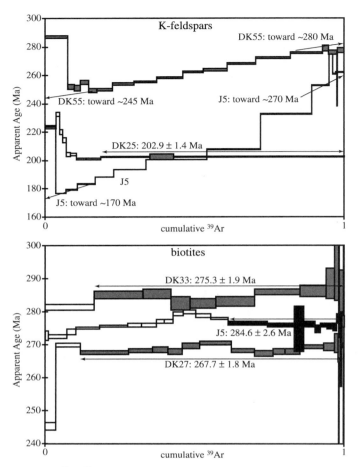

Figure 6. ^{40}Ar/^{39}Ar ages of Tian Shan samples obtained in this study, following laboratory procedures outlined in Hacker et al. (1996). Sample locations are shown in Figure 3 and results are tabulated in Table 1; data are interpreted in text.

on granitic mylonite (sample J5 in Figs. 3 and 6). Biotite gave a plateau age of 284.6 Ma, and our K-feldspar spectrum shows compatible high-temperature step ages of ca. 270 Ma decreasing serially into low-temperature step ages indicating either thermal or deformational disturbance ca. 170 Ma.

The Erbin Shan consists of granite intruded in siliciclastic and carbonate strata bearing Devonian coral and brachiopod fossils (XBGMR, 1993; Fig. 5) exposed along the southern margin of the central Tian Shan block (Fig. 2). A U-Pb zircon age of 378 Ma from the Erbin Shan granite was reported by Hu et al. (1986). Lower Carboniferous siliciclastic and carbonate strata bearing brachiopod, coral, and fusulinid fossils of the Visean Stage (XBGMR, 1991) unconformably overlie the Erbin Shan granite, with a basal conglomerate consisting exclusively of granitic clasts (Figs. 3 and 7A). K-feldspar separated from the Erbin Shan granite (sample DK55 in Fig. 6) gave a ^{40}Ar/^{39}Ar spectrum with maximum step ages of about 278 Ma and minimum ages of about 245 Ma. This age is younger than the Visean limestone (Fig. 7B) overlying the granite, and may reflect thermal disturbance by Late Carboniferous–Early Permian magma-

tism that resulted from south-dipping subduction along the north Tian Shan, and/or exhumation cooling. The latter explanation is preferred by Dumitru et al. (this volume), who interpreted cooling of sample DK55 from 350°C to <80°C ca. 270–250 Ma to be the consequence of 9 km of exhumation. However, Carboniferous strata of the Erbin Shan and Bayanbulak area are little deformed or metamorphosed and are <3 km thick in the area (proprietary 1:200 000 geologic mapping by Xinjiang Bureau of Geology and Mineral Resources). Considering that Cenozoic exhumation in the central Tian Shan was probably 2–3 km (see Dumitru et al., this volume), preservation of Carboniferous carbonate strata would be extremely unlikely if exhumation was the sole cause of the Early Permian cooling from 350°C to <80°C. Thus, reheating by Early Permian Borohoro arc magmatism, followed by rapid cooling without significant exhumation, is a viable interpretation of thermochronologic relations in the Erbin Shan.

South Tian Shan passive continental margin sequence

The Nikolaev line (Fig. 2) apparently marks a significant tectonic boundary in the western Chinese Tian Shan (Wang et al., 1990): north of the line, Devonian–Carboniferous strata are volcanogenic in the central Tian Shan terrane, whereas coeval strata south of the line in the south Tian Shan are basically nonvolcanic.

Folded and thrusted sequences of a Paleozoic passive margin constitute the south Tian Shan belt from the intermontane lowland called the Bayanbulak Valley south across Tielimaiti Pass to the outcrop of Permian rhyolite and overlying Mesozoic foreland basin strata north of Kuqa in the Tarim basin (Figs. 2 and 3; Hendrix et al., 1992). Upper Silurian strata north of Tielimaiti Pass consist of siliciclastic sandstone, shale, and calcareous conglomerate of turbidite and olistostrome origin (Fig. 3, H-I, I-J). In contrast, Lower and Upper Silurian strata exposed at and south of Tielimaiti Pass in the south Tian Shan consist of thick massive shelfal limestone beds containing abundant coral and brachiopod fossils (Figs. 5 and 7B). Stratigraphically and structurally above the Silurian massive limestone, in and south of the Kule Lake area, is a succession of Lower Devonian thin-bedded cherty limestone overlain by radiolarian chert (Fig. 7D) and siliciclastic siltstone and fine-grained sandy turbidite deposits rich in volcanic lithic fragments. This sequence was designated as Lower Devonian based on corals (Wang et al., 1996), or as Middle Devonian to Early Carboniferous based on radiolaria and conodont fossils in cherts (Gao, 1993) (Fig. 5).

PALEOZOIC TECTONIC EVOLUTION OF THE TIAN SHAN

Paleozoic magmatic arcs of the Tian Shan

Calc-alkalic magmatism is closely associated with subduction of oceanic crust, and its temporal and spatial distributions are important for paleotectonic reconstruction. Figure 8 delineates plutons exposed in the Chinese Tian Shan, and Tables 1–3

Figure 7. Outcrop photos from central (CTS) and south Tian Shan (STS). A: Lower Carboniferous basal conglomerate unconformably overlying Erbin Shan granitoid of CTS. Conglomerate is composed mostly of granitic gravel and few metamorphic clasts. B: Massive fossiliferous limestone of Early Carboniferous age exposed on north slope of STS, south of Bayanbulak Valley. Lens cap is 5 cm in diameter. C: Silurian massive limestone (upper cliffs) and dark colored calcareous shale (lower slopes) units exposed on Tielimaiti Pass, STS. D: Thin-bedded radiolarian chert and cherty limestone of possible Middle Devonian–Early Carboniferous age exposed at Kule Lake area, south slope of Tielimaiti Pass.

list the radiometric ages available in literature and obtained during this study along the Du-Ku Highway transect. Most ages were obtained by either K-Ar or $^{40}Ar/^{39}Ar$ methods and therefore are comparable.

It appears that two plutonic belts exist in the western Chinese Tian Shan: a northeast-east–trending belt (here termed the Narat arc after the Narat Shan) along the southern margin of the central Tian Shan block; and another northwest-west–trending belt (Borohoro arc) along the northern margin of the central Tian Shan terrane and Tarim block (Fig. 8). Plutons of the Narat arc were intruded during the Early Carboniferous with the exception of a few in Early Devonian (Table 2). On the trend of the Narat arc westward into Kazakhstan, alleged early Paleozoic plutons exist in the Lake Issyl–Kul area (Sobolev et al., 1982). In contrast, limited available ages of the Borohoro arc cluster around Late Carboniferous to Early Permian (Table 3), although more dating is needed to discriminate between magma-emplacement and uplift-cooling ages.

The relationships between intrusions and fossiliferous sedimentary rocks also define the age of the Narat arc. East of the

Du-Ku Highway, fossiliferous Lower Carboniferous strata with a basal conglomerate unconformably overlie porphyritic biotite granite (Liu, 1985). About 0.5 km off the highway at Bayanbulak, the basal conglomerate of Visean Stage (Early Carboniferous) consists almost exclusively of granitic gravel (Figs. 3 and 7A), and unconformably overlies Erbin Shan granite. In Narat Shan, volcanic rocks of the Lower Carboniferous Dahalajunshan Formation (Zhang, 1986) unconformably overlie granite.

Thus, radiometric ages and unconformity relationships indicate that the Narat arc was active from Devonian to Early Carboniferous, whereas the Borohoro arc was most active during Late Carboniferous and Early Permian time.

South Tian Shan suture and associated high P/T metamorphism

Although ophiolite and high-pressure metamorphic rocks have not been identified at the boundary between the central Tian Shan and Tarim–south Tian Shan along the Du-Ku Highway,

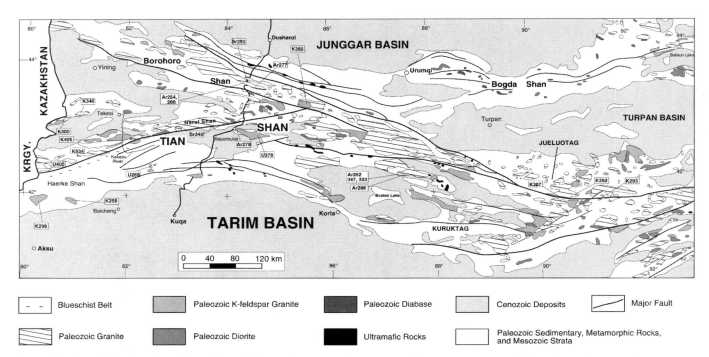

Figure 8. Paleozoic igneous intrusive rocks and ultramafic rocks in Chinese Tian Shan region with available radiometric ages shown in boxes (K, Ar, Sr, U indicating anaytical method, followed by age in m.y.). See Tables 2 and 3 for tabulated data and sources.

they are exposed in high ranges south of the Nikolaev line and west of the Bayanbulak Valley (Fig. 2). Because petrologic and geochronologic studies of blueschist belts are crucial in identifying fossil subduction complexes and reconstructing the tectonic evolution of orogens, the south Tian Shan blueschist belt has been studied by many geologists (Wang et al., 1990; Gao, 1993; Gao et al., 1995; Jia, 1996; Chen et al., 1999). We briefly review observations of field relations made by one of us (Da Zhou) and available geochronologic data from the south Tian Shan suture complex.

Outcrops of the south Tian Shan blueschist occur in the western Chinese Tian Shan and extend westward into Kazahkstan (Dobretsov et al., 1987). A blueschist-greenschist metamorphic complex is relatively well exposed in the Kekesu River gorge west of the Bayanbulak Valley and 50 km south of the village of Tekes (Kekesu Section in Fig. 2). The complex consists of metasiliciclastic rocks, metabasite, and marble (Gao et al., 1995, 1998; Jia, 1996). Protoliths of mafic blueschist and mafic greenschist were identified as mid-ocean ridge basalt (MORB) tholeiite, protoliths of metasediment blueschist as graywacke containing MORB tholeiite clasts, and protoliths of metasediment greenschist as pelite, pelitic limestone, arenite, and graywacke (Gao, 1993). Gao et al. (1995) estimated that peak metamorphism of metabasite blueschist occurred at P-T condition of 9–10 kbar and 450°C.

Table 4 lists radiometric ages obtained from the south Tian Shan blueschist belt. Glaucophane and phengite separated from the Kekesu blueschist yielded $^{40}Ar/^{39}Ar$ plateau ages of 315 ± 2 Ma and 345.4 ± 6.5 Ma, respectively (Gao et al., 1995). Py-

roxene from ophiolite of the south Tian Shan at Chang'awuzi, 80 km west of the Kekesu River, yielded a $^{40}Ar/^{39}Ar$ plateau age of 439 ± 27 Ma (Hao and Liu, 1993). This age was interpreted as the time of formation of oceanic crust (Hao and Liu, 1993). However, pyroxene is not a good mineral for $^{40}Ar/^{39}Ar$ dating because it contains little potassium. An $^{40}Ar/^{39}Ar$ age of 350 Ma was also obtained by Xiao et al. (1990) from Chang'awuzi sodic amphibole, which represents the minimum age of the blueschist metamorphism. Jia (1996) obtained another $^{40}Ar/^{39}Ar$ total fusion age (without a plateau) of ca. 315 Ma from phengite separated from the Chang'awuzi blueschists. Phengite in garnet-muscovite-blueschist from Qongkushitai, 40 km east of the Kekesu section, yielded a $^{40}Ar/^{39}Ar$ plateau age of 415 ± 2 Ma and an isochron age of 420 ± 4 Ma, interpreted as the minimum ages for peak metamorphism of the blueschist complex (Gao, 1993). Across the border in Kazahkstan, the metamorphic belt contains glaucophanized eclogite, glaucophane schist, garnet-biotite schist, and greenschist (Sobolev et al., 1982), with K/Ar ages of 520–550 Ma (Sobolev et al., 1982), and 410 ± 15 Ma (Dobretsov et al., 1987), although details of lab procedures were not reported. Silurian coral fossils were reported to have been found in the marbles within the blueschist metamorphic complex (Gao et al., 1994). Rapid exhumation is required for preservation of blueschist (Maruyama et al., 1996), and because high P-T blueschist metamorphism conditions in subduction zones typically are close to the $^{40}Ar/^{39}Ar$ blocking temperature (350°C), the $^{40}Ar/^{39}Ar$ ages of the south Tian Shan blueschist samples likely approximate time of metamorphism.

TABLE 1. $^{40}Ar/^{39}Ar$ AGES OF SAMPLES FROM DU-KU TRANSECT, TIAN SHAN, NORTHWEST CHINA

Sample No.	Location	Lithology	Mineral Analyzed	TFA (Ma)	WMPA (Ma)	IA (Ma)	MSWD	$^{40}Ar/^{36}Ar$	Steps	$\%^{39}Ar$
DK-25	North Tian Shan	K-spar granite	K-feldspar	204.3 ± 1.4	**202.9 ± 1.4**	Insufficient spread in isotopic ratios			9–18/18	81
DK-27	Haxilegen Pass	Mylonitized granite	Biotite	267.1 ± 1.8	267.7 ± 1.8	267.2 ± 1.8	1.72	461 ± 128	3–18/18	88
DK-33	Haxilegen Pass	Lherzolite	Biotite	275.3 ± 1.9	**275.3 ± 1.9**	275.3 ± 1.9	0.5	256 ± 72	14–28/28	39
DK-55	Erbin Shan	Granite	K-feldspar	265.3 ± 2.4	See text for interpretation of age spectrum.				n/a	n/a
831J5	Laerdun Pass	Mylonitized granite	Biotite	284.0 ± 2.6	**284.6 ± 2.6**	285.6 ± 2.7	0.78	210 ± 47	2–13/13	84
831J5	Laerdun Pass	Mylonitized granite	K-feldspar	210.3 ± 2.0	See text for interpretation of age spectrum.				n/a	n/a

Note: TFA, total fusion age; WMPA, weighted mean plateau age (WMA, weighted mean ages in italic); IA, isochron age; preferred age in bold. $^{40}Ar/^{36}Ar$, ratio of trapped ^{40}Ar.

MSWD, mean square weighted deviation (Wendt and Carl, 1991), which expresses the goodness of fit of the isochron (Roddick, 1978).

WMPA and IA are based on temperature steps and fraction of ^{39}Ar listed in the last two columns.

TABLE 2. RADIOMETRIC AGES OF PLUTONS ALONG THE SOUTHERN MARGIN OF THE CENTRAL TIAN SHAN BLOCK

Location	Material dated	Method	Age (Ma)	Reference
Nalati shan	Biotite	K-Ar	327	Gao (1993)
Nalati shan	Biotite	^{40}Ar/^{39}Ar plateau	355.1 ± 10.7	Hao and Lui (1993)
Nalati shan	Whole rock	Rb-Sr isochron	339	Wang et al. (1990)
Laerdun pass	Whole rock	Rb-Sr isochron	334.9 ± 28	Zhu and Sun (1986)
Erbin shan	Zircon	U-Pb	378	Wang et al. (1996)
	K-feldspar	Ar/Ar plateau	265.3 ± 2.4	This study
Buheda	Sphene	U-Pb	344	Wang et al. (1990)
Chaihengbulake	Sphene	U-Pb	330	Wang et al. (1996)
Zhaosu	Biotite	K-Ar	334.4	Gao (1993)
Zhaosu	Muscovite	K-Ar	404	Gao (1993)
Zhaosu	Biotite	K-Ar	345.6	Gao (1993)
Zhaosu	Granite	K-Ar	300	XBGMR (1993)*
Chang'awuzi	Granite	K-Ar	534	XBGMR (1993)*
Chang'awuzi	Granite	U-Pb	400	XBGMR (1993)*
Chang'awuzi	Granite	K-Ar	405	XBGMR (1993)*

*Xinjiang Bureau of Geology and Mineral Resources.

The radiometric ages listed in Table 4 apparently testify to subduction of oceanic crust from at least Late Silurian through Early Carboniferous. Because the southwestern Tian Shan south of the blueschist belt lacks Paleozoic calc-alkaline magmatic rocks, and because the central Tian Shan terrane north of the blueschist belt has abundant volcanic and intrusive rocks of Silurian and Carboniferous ages (Table 3), we infer here that subduction was north dipping, as have many other workers in recent years (e.g., Allen et al., 1992; Gao et al., 1998; Chen et al., 1999). Some of the alkalic granites and rhyolites exposed farther south along the north margin of Tarim (Fig. 8) yield Early Permian ages (Hendrix, 1992; Yin et al., 1998), and possibly are

late to postcollisional granites (Coleman, 1989) or part of the Early Permian bimodal magmatism that was widespread in the Tarim basin and Tian Shan (Carroll et al., 1995; this volume).

Early Carboniferous unconformity

In the central Tian Shan and northern Tarim basin, outcrop and subsurface data document an angular unconformity between Lower Carboniferous and older rocks (Zhang and Wu, 1985; Kang et al., 1992; Carroll et al., 1995). In the northwestern Tarim basin, the stratigraphic sequence above the unconformity starts with locally derived nonmarine conglomerate, and

TABLE 3. RADIOMETRIC AGES OF PLUTONS ALONG THE NORTHERN MARGIN OF THE CENTRAL TIAN SHAN-TARIM BLOCK

Location	Material dated	Method	Age (Ma)	Reference
Borohoro Shan	Whole rock	Rb-Sr	292 ± 15	Wang et al.(1990)
Haxilegen	Biotite	Ar/Ar plateau	267.7 ± 1.8	This Study
	Biotite	Ar/Ar plateau	275.3 ± 1.9	This Study
Manas	Granite	K-Ar	285	XBGMR (1993)*
Jueluotage	Granite	K-Ar	292	Wang et al. (1990)
	Granite	K-Ar	300	Wang et al. (1990)
	Granite	K-Ar	307	XBGMR (1993)*
	Granite	K-Ar	293	XBGMR (1993)*
	Granite	K-Ar	293	XBGMR (1993*)
Heaven Lake	Biotite	K-Ar	313	Hu et al. (1986)
	Biotite	K-Ar	265.5	Hu et al. (1986)
	Biotite	K-Ar	256.3	Hu et al. (1986)
Borohoro Shan	Granitoid-biotite	Ar/Ar	254.5 ± 4.8, 260.1 ± 3.4	Yin et al. (1998)
Bosten Lake	Gneiss-biotite	Ar/Ar	286	Yin et al. (1998)
Bosten Lake	Diorite-hornblende	Ar/Ar	262	Yin et al. (1998)
Bosten Lake	Diorite-biotite	Ar/Ar	347	Yin et al. (1998)
Bosten Lake	Diorite-K-spar	Ar/Ar	333	Yin et al. (1998)

*Xinjiang Bureau of Geology and Mineral Resources.

TABLE 4. RADIOMETRIC AGES OF SOUTH TIAN SHAN HIGH-PRESSURE METAMORPHIC ROCKS

Location	Mineral (rock) dated	Method	Age (Ma)	Laboratory	Reference
Chang'awuzi	Glaucophane from blueschist	$^{40}Ar/^{39}Ar$ plateau	350.89 ± 1.96	Institute of Geology, CAGS*	Xiao et al. (1990)
Qiongkushitai	Phengite from blueschist	$^{40}Ar/^{39}Ar$ plateau	415.37 ± 2.27	Institute of Geology, CAGS	Gao (1993)
Qiongkushitai	Phengite from blueschist	$^{40}Ar/^{39}Ar$ isochron	419.62 ± 3.92	Institute of Geology, CAGS	Gao (1993)
Chang'awuzi	Pyroxene from gabbro	$^{40}Ar/^{39}Ar$ plateau	439.4 ± 26.7	Institute of Geology, Academia Sinica	Hao and Liu (1993)
Kekesu	Phengite from blueschist	$^{40}Ar/^{39}Ar$ plateau	345.39 ± 6.51	Institute of Geology, CAGS	Gao et al. (1995)
Chang'awuzi	Phengite from blueschist	$^{40}Ar/^{39}Ar$ total fusion	ca. 315	Stanford University	Jia (1996)
Kazakhstan	Blueschist-greenschist	unknown	410 ± 15	unknown	Dobretsov et al. (1987)

*CAGS: Chinese Academy of Geological Sciences, Beijing, China

changes upward within 30 stratigraphic meters to paralic and shelfal siliciclastic and carbonate deposits (Carroll et al., 1995). In the Wushi area of the northwesternmost Tarim basin, carbonate olistostromes were deposited in a relatively deep water environment, and paleocurrent indicators suggest southerly provenance (i.e., from the Tarim basin) (Carroll et al., 1995). The Carboniferous transgressive deposits above the unconformity were interpreted by Carroll et al. (1995) as initial fill of a foreland basin developed after the amalgamation of the central Tian Shan and Tarim blocks. According to their reconstruction, loading by an inferred south-vergent overthrust of the central Tian Shan terrane onto the Tarim block flexed the north Tarim basin and formed a transgressive depositional sequence above the folded and erosionally truncated precollision strata.

The discovery of a similar unconformity and transgressive sequence in the central Tian Shan terrane (Figs. 3, H-I, and 7A), and the fact that the basal conglomerate units at all localities we have visited in northern Tarim and the central Tian Shan are dominated by local provenance (carbonate in Tarim, Carroll et al., 1995; and granite in Erbin Shan, Fig. 7A) are significant to models of basin evolution, as we discuss in our tectonic synthesis. For this discussion, however, these stratigraphic and deposystem relations (i.e., widespread Lower Carboniferous marine carbonate deposits in the south Tian Shan and northern Tarim basin) seemingly exclude the possibility of prominent mountain ranges in the Tian Shan region during the Early Carboniferous, except perhaps for locally emergent active volcanoes. A northerly derived (i.e., central Tian Shan derived) siliciclastic wedge, as predicted in the Early Carboniferous foreland basin model outlined by Carroll et al. (1995), is absent or not preserved, and only in the Late Permian is a Tian Shan-sourced nonmarine foreland sedimentary wedge evident in northern Tarim basin (Carroll et al., 1995, this volume). Thus, the lack of a major, emergent collisional mountain belt in the Carboniferous is suggestive of an accretion event which best might be described as a soft col-

lision. The unroofing of Devonian granite prior to Early Carboniferous transgressive deposition along the southern margin of the central Tian Shan, as in the Erbin Shan, requires an explanation. One possibility, discussed in our tectonic synthesis, is tectonism associated with rotational closing of the ocean basin that existed between the central Tian Shan and Tarim blocks.

Precambrian basement

Precambrian rocks occur in some, but not all, of the Tian Shan terranes, and the occurrence or absence of Precambrian basement is very important for interpretations of the tectonic evolution of the orogenic belt. Precambrian outcrops in or adjacent to the Chinese Tian Shan are mainly distributed in three areas: the Kuruktag in northeast Tarim, the Kalpin uplift in the northwest Tarim, and the central Tian Shan terrane (Fig. 1; XBGMR, 1993). The lithology, metamorphism, stratigraphy, and major stratigraphic boundaries of those areas are summarized in Figure 9. Middle to Late Proterozoics ages were assigned to various strata and metamorphic complexes in all of these areas on the basis of microflora, stromatolites, and radiometric data derived from the literature. Evaluation of radiometric ages from these metamorphic complexes is difficult because few original data are published. It is obvious that more systematic and meaningful isotopic dating is crucial for better documentation of these Precambrian rocks. Even though the absolute stratigraphic position of the Precambrian rocks in each area is at least somewhat problematic, it is possible to compare the stratigraphy of the three areas, assuming that there is consistency in age assignment and stratigraphic division based on microflora and stromatolites published in the Chinese literature.

From studies by Chinese geologists and our own limited observations of Precambrian rocks of the region, we draw the following conclusions from comparisons of the Precambrian stratigraphy of the three areas (Fig. 9).

	Central Tian Shan Block		Tarim Block	
			Aksu Region (NW Tarim)	Kuruk Tagh Region (NE Tarim)
Post-Sinian Cover — Age System	Dahalajun Shan Fm. (C1)	Volcaniclastic, lava flow, and shallow marine siliciclastic and carbonate rocks	Xiaoerbulake Fm. (Lower Cambrian) — Dolomites, asphaltic dolomites, cherty phosphate rocks. Bearing trilobites, brachiopods, hyolithes, ostracods, and sponge spicules.	Xishanbulake (Lower Cambrian) — Gray, gray-black thin-bedded limestone, and chert. Local andesitic porphyrite. Cherty phosphorite at the base.
		—Angular Unconformity—		
600 Ma — Sinian	Talishayi Fm.	Pebbly ss., varved mudstones, reddish conglomerates, and calcirudites 50 m	Upper Sinian — Qigebulake Group: shelfal dolomite and siliciclastic rocks. ~ 150 m; Sugaitebulake Group: nonmarine to paralic siliciclastic arenite, and shelfal carbonate and siliciclastic arenite, with top bearing glauconite. ~200 - 900 m	Kuluketage Group — Upper part: predominant siliciclastic rocks, minor chert and carbonate. Many are thought to be glacial deposits. Lower part: intermediate to acid volcaniclastics and lava flow. several thousands of meters
800		Angular Unconformity	Youermeinake Group: red and green conglomerates and pebbly sandstone and arenite. ~ 50 m (probably of glacial origin)	—Angular Unconformity—
	Qingbaikou — Kushitai Group	Stromatolithic limestones, cherty limestones, dolomitic limestones, and silt- and sandstones. 547 - 1480 m	Qiaoenbulake Group: gray-green colored coarse-medium lithic arkoses, fine sandstone, and muddy and calcareous siltstones. ~ 500 - 1500 m	Paergangtage Group — Upper part (614 m): stromatolithic carbonates (marbles and dolomite). Lower part (>140 m): schist and quartzite.
1000		—Disconformity—	—Angular Unconformity—	—contact not clear ?—
	Jixian — Kekesu Group	Stromatolithic dolomites, dolomitic limestones, limestones, cherty limestones, sandstones, and comglomerates at the base. > 780 m	Akesu Group metamorphosed — ~ 700 Ma metaphorphic age (whole rock-phengitic mica Rb-Sr isochron, and phengitic mica K-Ar age) (Nakajima et al., 1990). Psammitic, pelitic, and mafic schist and minor quartzite, meta-ironstone, and metachert of low-grade green-schist facies and greenschist-blueschist transitional facies (Liou et al., 1989).	Erijgan Group metamorphosed — marble, schist, quartzite, iron bearing quartzite, and minor conglomerate.
1400		—Angular Unconformity—		—Angular Unconformity?—
— Changcheng — Tekesi Group metam'd		Metamorphic arenites, phyllites, meta-mafic volcanics, quartzites, and minor stromatolithic marble. ~ 3000 m	lower age limit unknown ?	Xingditage Group — Metamorphosed clastic and carbonate rocks.
1800				

Figure 9. Summary of Middle-Upper Proterozoic stratigraphy of central Tian Shan block and north Tarim (compiled from Gao et al., 1985a, 1985b; Gao and Peng, 1985; Liou et al., 1989; Zhang and Liu, 1991; Zhou et al., 1991).

1. The age and/or stratigraphic position of an unconformity between the unmetamorphosed cover strata and metamorphosed basement rocks differs between the central Tian Shan and the Tarim block. In the central Tian Shan block, the unconformity between the Jixian system and Changcheng system developed at ca. 1400 Ma (unpublished 1:200 000 mapping of XBGMR; Xinjiang Stratigraphic Table Compiling Group, 1981). In the Aksu and Kuruktag regions of the Tarim block, the ages of unconformities are much younger, although in detail age estimates vary in the literature. Gao et al. (1985a, 1985b) and Gao and Peng (1985) assigned strata directly above the unconformity to the lower Sinian for both Aksu and Kuruktag regions based on stromatolites and microflora. Zhang and Liu (1991) concluded, however, that the Qiaoenbulak Group (lower Sinian, according to Gao et al., 1985a, 1985b; Gao and Peng, 1985) actually belongs to the Qingbaikou system (1000–800 Ma), based on microflora from Qiaoenbulak Group in the Kalpin uplift. Rb-Sr whole-rock-phengite isochron and K-Ar phengite ages of the blueschist-greenschist transitional facies basement (Aksu

Group) are about 700 Ma (Liou et al., 1989; Nakajima et al., 1990), which implies that the Qiaoenbulak above the unconformity must be in the upper Sinian. In either scenario, it seems apparent that the central Tian Shan terrane was consolidated at least 400–700 m.y. earlier than the Precambrian basement of the northern part of the Tarim block.

2. The lithology and depositional environments of the cover strata of the two blocks are also distinct from each other. On the central Tian Shan block, Sinian strata are thin (or absent) in the Tekes area (Fig. 2), and they are interpreted to be glacial deposits (Gao and Peng, 1985). Unconformably below Sinian strata are more than 2000 m of platformal stromatolitic carbonate. In contrast, thick siliciclastic nonmarine, shallow-marine to deep-marine sedimentary rocks of Sinian age were deposited along the northern margin of the Tarim block (Zhou et al., 1991). Many intervals of volcaniclastic strata and lava flows occur in the Kuruktag area (Gao and Peng, 1985).

3. If 700 Ma represents the time of the blueschist-greenschist metamorphism of the Aksu Group in northwest Tarim, this

part of the block probably was subjected to rifting during the late Sinian and formed a north-facing (in present orientation) passive continental margin (see Carroll et al., this volume). Because no radiometric ages are available from the Kuruktag Group, whether the Kuruktag margin was coevally developed with the Kalpin margin will remain unclear until tested by radiometric dating. Extensive volcaniclastic rocks and lava flows of variable compositions occur throughout the Kuruktag Group, dated as lower and upper Sinian by microflora (Gao et al., 1985b; Gao and Peng, 1985). The lower upper Sinian in the Aksu region also contains basalt flows and volcaniclastic intervals to 90 m thick (Gao et al., 1985a; Carroll et al., this volume). These volcanic rocks are so extensive that systematic and high-quality isotopic dating would considerably improve the Sinian chronostratigraphy of the Tarim block.

Despite the uncertainty in the ages of Sinian strata of north Tarim, we can roughly divide the evolution of the northern margin of the Tarim block into pre-Sinian and Sinian stages. During the pre-Sinian stage, metamorphism and crystallization formed the basement of the Tarim block. Rifting occurred along the northern Tarim margin during the Sinian and a passive continental margin developed and persisted through the early Paleozoic (Carroll et al., this volume).

Although currently adjacent to the Aksu region, the central Tian Shan block had a different Precambrian history, hence is inferred here to have originated from a different parent continent with respect to the Tarim block. If the stromatolite and microflora stratigraphy (Gao et al, 1985a, 1985b; Gao and Peng, 1985) is correct, the central Tian Shan terrane consolidated ca. 1400 Ma, about 400–700 m.y. earlier than the Tarim block. Therefore, it is appropriate to regard the central Tian Shan terrane as a separate entity from Precambrian to the latest Paleozoic.

Tectonic amalgamation of the Chinese Tian Shan

On the basis of our study of the Du-Ku transect, previous studies by our research group in north Tarim and south Junggar basins (Carroll et al., 1990, 1995; Graham et al., 1990), and our synthesis of the literature, we postulate an amalgamation history of the western Chinese Tian Shan region that follows Carroll et al. (1995) in general form, but that offers some elaborations from the perspective of the interior of the Tian Shan.

Three tectonostratigraphic elements—the Junggar arc terranes (substrate of southern Junggar basin and some rocks of the north Tian Shan), the central Tian Shan terrane, and the Tarim block (and its deformed distal cover strata of the south Tian Shan)—are defined for reconstruction of the amalgamation history of the Tian Shan as delineated in Figure 1. The Tarim and central Tian Shan blocks consist of crystalline basement of at least Proterozoic age with Sinian and younger cover strata, whereas the Junggar terrane consists of rocks of oceanic affinity no older than Devonian (Carroll et al., 1990; XBGMR, 1993).

The south Tian Shan blueschist belt along the Nicolaev line records subduction from possibly Late Silurian–Early Devonian to Early Carboniferous time in the western Chinese Tian Shan (Table 4). A parallel coeval magmatic arc (Table 2), the Narat arc, developed along the southern margin of the central Tian Shan terrane. We infer that subduction and arc activity continued until the end of Early Carboniferous in the western Tian Shan, as depicted in Figure 10, and as posited by Gao et al. (1998) and Chen et al. (1999). The northwestern margin of the Tarim block persisted as a carbonate-dominated marine environment until the Late Carboniferous, or even the late Early Permian farther west in the Wenguri area, north of Kashgar (Carroll et al., 1995). These lines of evidence, when combined with the chronology of magmatism discussed in the following, seem to suggest that final central Tian Shan–Tarim accretion in the western Tian Shan did not occur until the Middle Carboniferous, ~40 m.y. later than Carroll et al.'s (1995) inference of a Late Devonian–Early Carboniferous collision based on the pre-Early Carboniferous angular unconformity in north Tarim.

These seemingly conflicting data and interpretations (i.e., a Devonian–Carboniferous collision versus continued subuction capped by a Middle Carboniferous central Tian Shan–Tarim collision) can be reconciled when the amalgamation of Tarim and the central Tian Shan is seen as an event of finite duration, rather than instantaneous. Collisional events in their full expression, from initial encounter through mountain building and foreland development, have durations that range from a few million years in the case of colliding arcs to more than 100 m.y. when large continents collide (Ingersoll et al., 1995). Available paleomagnetic data, summarized in Chen et al. (1999), suggest an initial encounter of the present-day eastern tip of the Tarim block with the central Tian Shan near the Devonian-Carboniferous boundary, followed by clockwise rotational motion of Tarim to close a remnant ocean basin through the Early Carboniferous and complete Tarim–central Tian Shan suturing (as illustrated in Fig. 16 of Carroll et al., this volume). This diachronous suturing, common in collisions (Graham et al., 1975; Ingersoll et al., 1995), assures that tectonism associated with the Tarim–central Tian Shan collision likely was important throughout at least the 40 m.y. of the Early Carboniferous.

As the central Tian Shan and Tarim blocks became conjoined, the west-northwest–trending Borohoro arc extended continuously from the central Tian Shan block to the northeast margin of the Tarim block, creating a north-facing active margin on a connected central Tian Shan–Tarim composite block (Fig. 11). Radiometric ages (Table 3) from plutons of this north-facing arc suggest a Late Carboniferous–Early Permian age, whereas probable forearc basin and accretionary wedge volcaniclastic turbidite deposits exposed in the north Tian Shan along the Du-Ku Highway indicate arc activity as early as the Early Carboniferous or even the Devonian (Figs. 1 and 3). The timing of initiation of arc magmatism along the northeast margin of the Tarim block (Kuruktag region) is uncertain in detail. However, the existence of a single continuous Late

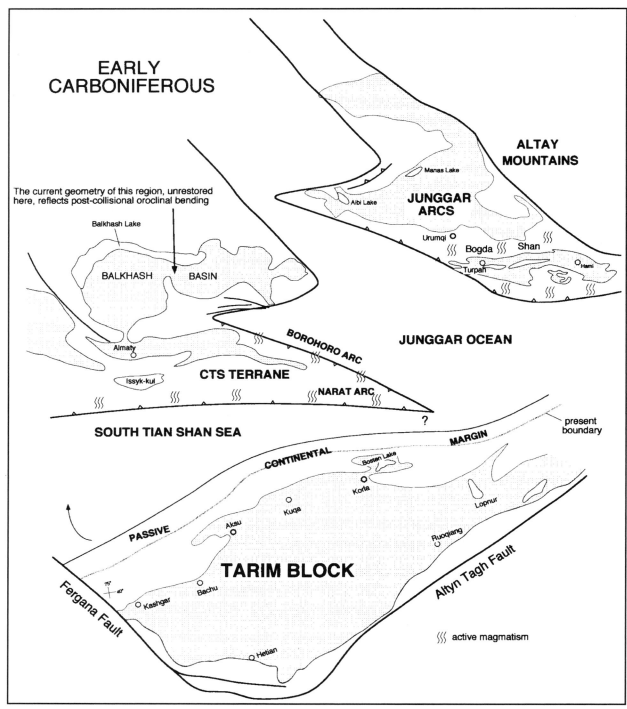

Figure 10. Schematic reconstruction of tectonics and paleogeography of Tarim block, central Tian Shan (CTS) block, and Junggar arc complex during Early Carboniferous time. Figure is nonpalinspastic: shapes of Cenozoic basins (Fig. 1) are retained in order to clearly show tectonic affinities of current outcrops. Orientation and internal details of CTS block and dimensions of Junggar ocean are unknown. Whether eastern end of CTS block was attached to Tarim block in this time frame is unclear due to Cenozoic deformation and lack of data.

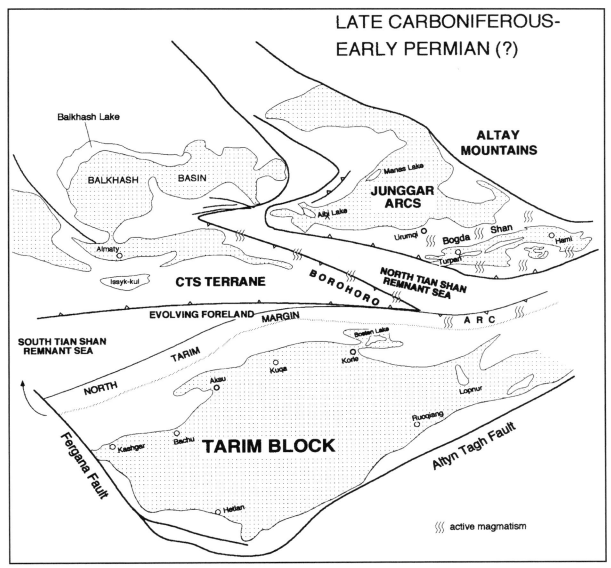

Figure 11. Schematic reconstruction of tectonics and paleogeography of Tarim-central Tian Shan (CTS) block and Junggar arc complex during Late Carboniferous–Early Permian time. Figure is nonpalinspastic: shapes of Cenozoic basins (Fig. 1) are retained in order to clearly show tectonic affinities of current outcrops. Arc magmatism ceased along southern CTS margin, although south Tian Shan sea was not completely closed. Arc magmatism was active along north CTS margin and possibly northeast Tarim margin during Late Carboniferous and possibly Early Permian time.

Carboniferous–Early Permian magmatic belt from the central Tian Shan to northeast Tarim indicates that the two blocks were connected by that time.

Both the Narat and Borohoro (excluding northeast Tarim) arcs were active during Devonian–Early Carboniferous time (Fig. 10). Subduction in the south Tian Shan and magmatism of the Narat arc ceased by Middle Carboniferous time (Tables 2 and 4), which may indicate complete consumption of oceanic crust between the central Tian Shan and Tarim terranes in the south Tian Shan region. The westernmost south Tian Shan remained a marine carbonate environment until Late Carbonif-

erous or even Early Permian time (Carroll et al., 1995). South-dipping subduction probably commenced in Late Carboniferous in Kurukag area along the northeast margin of Tarim, when the central Tian Shan terrane was connected with the Tarim block north and east of Korla (Fig. 11). Continuous subduction along this northern margin during Late Carboniferous–Early Permian time finally brought together the amalgamated Tarim–central Tian Shan terrane and the Junggar island arc complex (cf. Carroll et al., 1990; Allen et al., 1992).

Carroll et al. (1995) regarded the Lower Carboniferous transgressive sedimentary sequences of northwestern Tarim basin as deposits of a foreland basin developed by tectonic loading related to accretion of the south-verging central Tian Shan terrane, whereas we emphasize continued subduction during rotational closure of the south Tian Shan remnant ocean basin in the Early Carboniferous. We regard the absence of emergent mountain ranges in the Tian Shan region during the Carboniferous, and the persistence of Late Carboniferous or Early Permian marine carbonate environments, as indications of soft collision between the central Tian Shan and Tarim blocks. As the collisional process continued, emergence of a Tian Shan collisional mountain range in the Permian was recorded by the northerly-derived nonmarine foreland basin Permian sequence of northern Tarim (Carroll et al., 1995, this volume).

The late Cenozoic collision of Australia with various small blocks to the north that elevated Irian Jaya–Papua New Guinea may offer some instructive insights about the late Paleozoic evolution of the Chinese Tian Shan; these may help to reconcile the differences between the interpretation of Carroll et al. (1995), with its emphasis on foreland development and early inferred collision, and the interpretation presented here. As emphasized by Pigram and Davies (1987), Cenozoic collision along the northern margin of Australia has been complex, but for our purposes the history can be simplified to the following critical events and features: oceanic crust north of the northern passive margin of Australia was consumed beneath an opposing convergent plate boundary until initial collision with various island arcs and associated blocks, some with cratonal underpinnings, by the end of the Oligocene (Pigram and Davies, 1987); as the continental underpinnings of northern Australia became involved in collision, New Guinea became emergent and a foreland basin evolved; this basin is clearly a collisional flexural basin (Haddad and Watts, 1999), and although it is now diachronously filling with siliciclastics derived from the uplifted New Guinea collisional belt, for most of the late Cenozoic and locally to the present, the foreland has been dominated by marine carbonate deposystems (Pigram et al., 1989) due to arid climate and the insular nature of the suture zone. Continued accretion is manifest in the ongoing collision of the now-amalgamated Australia–southern New Guinea composite block with the Bismark arc off northeast Papua New Guinea, which involves the diachronous closure of the Solomon Sea remnant ocean basin (Ingersoll et al., 1995; Galewsky and Silver, 1997). Thus, the amalgamation of Australia with terranes to the north has been in progress for at least 25–30 m.y., and is not yet complete.

Applying these observations to the late Paleozoic Tian Shan, we envision an initial encounter between Tarim and the central Tian Shan occurring near the Devonian-Carboniferous boundary (perhaps reflected in the Devonian-Carboniferous unconformity observed in northern Tarim and the central Tian Shan), with subsequent rotational closure of the south Tian Shan remnant ocean basin throughout the Early Carboniferous accounting for blueschist-metamorphic and arc-igneous radiometric ages in the paired belts of the southern central Tian Shan (Fig. 10) The rotational closure probably implies as yet un-

documented late Paleozoic strike-slip faulting within the Tian Shan orogen. Closure of this remnant ocean basin was followed by uplift and emergence of the southern Tian Shan and evolution of the northern Tarim foreland basin, which persisted as a low-latitude, marine carbonate–dominated foreland in the later Carboniferous, until supplanted by Tian Shan–derived nonmarine foreland siliciclastics in the Permian. As in Cenozoic New Guinea, the initial collisional event was followed by accretion of an oceanic arc terrane to the north. The amalgamated Tarim–central Tian Shan terrane closed with the Junggar arc terranes in the late Paleozoic in a manner partially analogous to the ongoing collision between amalgamated Australia–New Guinea and the Bismark arc; alternatively, an analog scenario employing the closing Moluccan Sea bounded by facing arcs was offered by Allen et al. (1992). The full process of sequential collision recorded in the Tian Shan, as interpeted here, occurred over at least 75 m.y., whereas the Cenozoic example afforded by New Guinea has been in process for 25–30 m.y. The north Tarim, south Junggar, and Tian Shan region became a nonmarine environment in Late Permian (Carroll et al., 1992, 1995; XBGMR, 1993) after island arcs in Junggar region were accreted to the Tarim–central Tian Shan terrane to the south and Altay orogen–Mongolia to the north, marking final consolidation of this entire region of central Asia.

The soft-collisional terrane accretion we infer for the late Paleozoic Tian Shan is an end-member type of orogeny that involves small blocks (e.g., island arcs) instead of two large continents (e.g., India and Eurasia). Central Asia was characterized during the Paleozoic by widespread arc accretion (Şengör et al., 1993; Şengör and Natal'in, 1996), similar to the tectonic regime of modern southeast Asia. The mobility of island arcs (e.g., bending and slicing; e.g., Şengör and Natal'in, 1996), unconstrained boundaries (e.g., new subduction initiated outboard after accretion), and thinner and denser immature arcs make soft-collisional accretion possible.

CONCLUSIONS

1. Paleozoic amalgamation of the Tian Shan orogenic belt involved three major tectonic units: the Tarim block, the central Tian Shan block, and the Junggar arc complex. The first two have different Precambrian crystalline basements, whereas the Junggar arc complex is composed of volcanic and sedimentary formations of oceanic affinity no older than Devonian.

2. The central Tian Shan terrane converged with the Tarim block through ocean basin closure from at least Late Silurian to Early Carboniferous time, generating coeval magmatic and blueschist belts along its southern margin. Polarity of the paired belts indicates north-dipping subduction.

3. Subduction of oceanic crust of the south Tian Shan ocean basin in the western Tian Shan apparently ceased by Middle Carboniferous time. Marine carbonate environments per-

sisted on Tarim's northwest margin until Late Carboniferous–Early Permian time, and marine facies prevailed in the central Tian Shan terrane during Late Carboniferous and perhaps Early Permian time (XBGMR, 1993), which suggests that no emergent ancestral Tian Shan existed until at least Early Permian time.

4. Although active volcanism occurred along the northern margin of the central Tian Shan terrane from at least Devonian to Late Carboniferous time, available ages of the voluminous northwest-west–trending granitoid plutons that extend from the central Tian Shan terrane to the northeast margin of the Tarim block cluster from the Late Carboniferous to the Early Permian, thereby constraining the minimum age of the amalgamation between the Tarim–central Tian Shan block and Junggar arc complex.

ACKNOWLEDGMENTS

Our ideas about the Tian Shan have been greatly improved by discussions and/or field work with Alan Carroll, Marc Hendrix, Robert Coleman, Trevor Dumitru, Edward Sobel, Cleavy McKnight, Clifford Hopson, Don Ying, and Wang Zouxun. We thank Mingming Jia for additional rock samples, Melissa Lamb for assistance in sample preparation, and D.L. Miller for drafting assistance. The manuscript was significantly improved by reviews from Mark Allen, Marc Hendrix, and Shangyou Nie. This study was supported by the Stanford-China Geosciences Industrial Affiliates Program, a consortium of 26 petroleum companies over the period 1988–1999.

REFERENCES CITED

Abdrakhmatov, K.Y., Aldazhanov, S.A., Hager, B.H., Hamburger, M.W., Herring, T.A., Kalabaev, K.B., Makarov, V.I., Molnar, P., Panasyuk, S.V., Prilepin, M.T., Reilinger, R.E., Sadybakasov, I.S., Souter, B.J., Trapeznikov, Y.A., Tsurkov, V.Y., and Zubovich, A.V., 1996, Relatively recent construction of the Tien Shan inferred from GPS measurements of present-day crustal deformation rates: Nature, v. 384, p. 450–453.

Allen, M.B., Windley, B.F., and Zhang, C., 1992, Paleozoic collisional tectonics and magmatism of the Chinese Tien Shan, central Asia: Tectonophysics, v. 220, p. 89–115.

Avouac, J.P., Tapponnier, P., Bai, M., You, H., and Wang, G., 1993, Active thrusting and folding along the northern Tien Shan and late Cenozoic rotation of the Tarim relative to Dzungaria and Kazakhstan: Journal of Geophysical Research, v. 98, p. 6755–6804.

Carroll, A.R., 1991, Late Paleozoic tectonics, sedimentation, and petroleum potential of the Junggar and Tarim basins, northwest China [Ph.D. thesis]: Stanford, California, Stanford University, 405 p.

Carroll, A.R., Liang, Y., Graham, S.A., Xiao, X., Hendrix, M.S., Chu, J., and McKnight, C.L., 1990, Junggar basin, northwest China: Trapped late Paleozoic ocean: Tectonophysics, v. 181, p. 1–14.

Carroll, A.R., Brassell, S.C., and Graham, S.A., 1992, Upper Permian lacustrine oil shale of the southern Junggar basin, northwest China: American Association of Petroleum Geologists Bulletin, v. 76, p. 1874–1902.

Carroll, A.R., Graham, S.A., Hendrix, M.S., Ying, X., and Zhou, D., 1995, Late Paleozoic tectonic amalgamation of northwestern China: Sedimentary record of the northern Tarim, northwestern Turpan, and southern Junggar Basins: Geological Society of America Bulletin, v. 107, p. 571–594.

Chen, C., Lu, H., Jia, D., Cai, D., and Wu, S., 1999, Closing history of the southern Tianshan oceanic basin, western China: An oblique collisional orogeny: Tectonophysics, v. 302, p. 23–40.

Cheng, S., Xiao, B., and Wang, W., 1985, The lower Paleozoic of Tianshan, Xinjiang: Xinjiang Geology, v. 3, no. 2, p. 26–87.

Coleman, R.G., 1989, Continental growth of northwest China: Tectonics, v. 8, p. 621–635.

Dobretsov, N.L., Coleman, R.G., Liou, J.G., and Maruyama, S., 1987, Blueschist belts in Asia and possible periodicity of blueschist facies metamorphism: Ofioliti, v. 12, p. 445–456.

Galewsky, J., and Silver, E.A., 1997, Tectonic controls on facies transitions in an oblique collision: The western Soloman Sea, Papua New Guinea: Geological Society of America Bulletin, v. 109, p. 1266–1278.

Gao, J., 1993, Plate tectonics and geodynamics of the orogenesis of the southwest Tian Shan [Ph.D. thesis]: Beijing, Chinese Academy of Geological Sciences, 90 p.

Gao, J., Xiao, X., Tang, Y., Zhao, M., and Wang, J., 1994, New observation of the Upper Silurian Akeyazi and Qiongkushitai formations on the northern slope of the Halik mountains: Regional Geology of China, v. 1994, p. 240–245.

Gao, J., He, G., Li, M., Xiao, X., Tang, Y., Wang, J., and Zhao, M., 1995, The mineralogy, petrology, metamorphic PTDt trajectory and exhumation mechanism of blueschists, south Tian Shan, northwestern China: Tectonophysics, v. 250, p. 151–168.

Gao J., Li, M., Xiao, X., Tang, Y., and He, G., 1998, Paleozoic tectonic evolution of the Tianshan orogen, northwest China: Tectonophysics, v. 287, p. 213–231.

Gao, Z., and Peng, C., 1985, The Precambrian of Tianshan, Xinjiang: Xinjiang Geology, v. 3, p. 14–25.

Gao, Z., Wang, W., Peng, C., Li, Y., and Xiao, B., 1985a, The Sinian in Aksu-Wushi Region, Xinjiang, China: Urumqi, Xinjiang People's Publishing House, 184 p.

Gao, Z., Wang, W., Peng, C., Li, Y., and Xiao, B., 1985b, The Sinian in Xinjiang: Urumqi, China, Xinjiang People's Publishing House, 173 p.

Graham, S.A., Dickinson, W.R., and Ingersoll, R.V., 1975, Himalayan-Bengal model for flysch dispersal in the Appalachian-Ouachita system: Geological Society of America Bulletin, v. 86, p. 273–286.

Graham, S.A., Brassell, S., Carroll, A.R., Xiao, X., Demaison, G., McKnight, C.L., Liang, Y., Chu, J., and Hendrix, M.S., 1990, Characteristics of selected petroleum source rocks, Xinjiang Uygur Autonomous Region, northwest China: American Association of Petroleum Geologists Bulletin, v. 74, p. 493–512.

Hacker, B.R., Mosenfelder, J.L., and Gnos, E., 1996, Rapid emplacement of the Oman ophiolite: Thermal and geochronologic constraints: Tectonics, v. 15, p. 1230–1247.

Haddad, D., and Watts, A.B., 1999, Subsidence history, gravity anomalies, and flexure of the northeast Australian margin in Papua New Guinea: Tectonics, v. 18, p. 827–842.

Hao, J., and Liu, X., 1993, The age of ophiolite melange and tectonic evolutional model in south Tianshan area: Scientia Geologica Sinica, v. 28, p. 93–95.

Hendrix, M.S., 1992, Sedimentary basin analysis and petroleum potential of Mesozoic strata, northwest China [Ph.D. thesis]: Stanford, California, Stanford University, 65 p.

Hendrix, M.S., 2000, Evolution of Mesozoic sandstone compositions, southern Junggar, northern Tarim, and western Turpan basins, northwest China: A detrital record of the ancestral Tian Shan: Journal of Sedimentary Research, v. 70, p. 520–532.

Hendrix, M.S., Graham, S.A., Carroll, A.R., Sobel, E.R., McKnight, C.L., Schulein, B.J., and Wang, Z., 1992, Sedimentary record and climatic implications of recurrent deformation in the Tian Shan: Evidence from Mesozoic strata of the north Tarim, south Junggar, and Turpan basins, northwest China: Geological Society of America Bulletin, v. 104, p. 53–79.

Hendrix, M.S., Dumitru, T.A., and Graham, S.A., 1994, Late Oligocene–early Miocene unroofing in the Chinese Tian Shan: An early effect of the India-Asia collision: Geology, v. 22, p. 487–490.

Hu, A., Zhang, Z., Liu, J., Peng, J., Zhang, J., Zhao, D., Yang, S., and Zhou, W., 1986, U-Pb age and evolution of Precambrian metamorphic rocks of the middle Tianshan uplift zone, eastern Tianshan, China: Geochimica, v. 1986, p. 23–35.

Ingersoll, R.V., Graham, S.A., and Dickinson, W.R., 1995, Remnant ocean basins, in Busby, C.J., and Ingersoll, R.V., eds., Tectonics of sedimentary basins: Oxford, Blackwell Science, p. 363–392.

Institute of Geology and Mineral Resources, 1991, The Palaeozoic eratherm of Xinjiang, (Volume 2), Stratigraphic summary of Xinjiang: Urumqi, China, Xinjinag People's Publishing House, 482 p.

Jia, M., 1996, Petrology, geochronology, and tectonic significance of the south Tian Shan blueschist belt, NW China [M.S. thesis]: Stanford, California, Stanford University, 50 p.

Kang, Y., Huang, Y., Li, B., Zhang, Z., Yan, H., Lin, Z., and He, X., eds., 1992, Oil and gas field in marine Paleozoic of Tarim Basin: Wuhan, China University of Geoscience Press, 176 p.

Lee, C.Y., 1982, Tectonic map of Asia: Beijing, Geological Publishing House, scale 1:8 000 000.

Li, Y., 1993, New discovery of Cephalopada and discussion on the age of Kuruer Formation of Silurian in the Bolhinur Mts., Xinjiang: Xinjiang Geology, v. 11, p. 204–206.

Lin, B., and Wang, B., 1984, Middle and Late Silurian Tabulata and Helioliteda from the northern side of Bolhinur Mountain in Xianjiang: Xinjiang Geology, v. 2, p. 34–46.

Liou, J.G., Graham, S.A., Maruyama, S., Wang, X., Xiao, X., Carroll, A.R., Chu, J., Feng, Y., Hendrix, M.S., Liang, Y.H., McKnight, C.L., Tang, Y., Wang, Z.X., Zhao, M., and Zhu, B, 1989, Proterozoic blueschist belt in western China: Best documented Precambrian blueschists in the world: Geology, v. 17, p. 1127–1131.

Liu, J., 1985, The genetic type, evolution and significance for ore deposits of Hercynian granites in Tianshan: Xinjiang Geology, v. 3, p. 66–75.

Maruyama, S., Liou, J.G., and Terabayashi, M., 1996, Blueschists and eclogites of the world and their exhumation: International Geology Review, v. 38, p. 485–594.

Molnar, P., and Deng, Q.D., 1984, Faulting associated with large earthquakes and the average rate of deformation in central and eastern Asia: Journal of Geophysical Research, v. 89, p. 6203–6228.

Molnar, P., and Lyon-Caen, H., 1989, Fault plane solutions of earthquakes and active tectonics of the Tibetan Plateau and its margins: Geophysical Journal International, v. 99, p. 123–153.

Nakajima, T., Maruyama, S., Uchiumi, S., Liou, J.G., Wang, X., Xiao, X., and Graham, S.A., 1990, Evidence for Late Proterozoic subduction from 700-Myr-old blueschists in China: Nature, v. 346, p. 263–265.

Pigram, C.J., and Davies, H.L., 1987, Terranes and accretion history of the New Guinea orogen: Bureau of Mineral Resources Journal of Australian Geology and Geophysics, v. 10, p. 193–212.

Pigram, C.J., Davies, H.L., Feary, D.A., and Symonds, P.A., 1989, Tectonic controls on carbonate platform evolution in southern Papua New Guinea: Passive margin to foreland basin: Geology, v. 17, p. 199–202.

Qiao, X., 1989, On graptolite-bearing strata and the graptolite fauna division of Ordovician in the western North Tianshan: Xinjiang Geology, v. 7, p. 81–90.

Roddick, J.C., 1978, The application of isochron diagrams in $^{40}Ar/^{39}Ar$ dating: A discussion: Earth and Planetary Science Letters, v. 41, p. 233–244.

Şengör, A.M., and Natal'in, B.A., 1996, Turkic-type orogeny and its role in the making of the continental crust: Annual Review of Earth and Planetary Sciences, v. 24, p. 263–337.

Şengör, A.M.C., Natal'in, B.A., and Burtman, V.S., 1993, Evolution of the Altaid tectonic collage and Palaeozoic crustal growth in Eurasia: Nature, v. 364, p. 299–307.

Sobel, E.R., and Dumitru, T.A., 1997, Thrusting and exhumation around the margins of the western Tarim basin during the India-Asia collision: Journal of Geophysical Research, v. 102, p. 5043–5063.

Sobolev, V.S., Lepezin, G.G., and Dobretsov, N.L., eds., 1982, Metamorphic complexes of Asia: Moscow, Nauka Publishers, 322 p.

Tapponnier, P., and Molnar, P., 1979, Active faulting and Cenozoic tectonics of the Tien Shan, Mongolia, and Baykal regions: Journal of Geophysical Research, v. 84, p. 3425–3459.

Wang, B., Li, Q., and Liu, J., 1996, Geology and tectonics of the west Tianshan Mountains along Dushanzi-Kuqa Highway in Xinjiang, in Hou, H., and Liu, L., eds., 30th International Geological Congress Field Trip Guide, Volume 5: Beijing, Geological Publishing House, p. T364.1–T364.26.

Wang, Z., Wu, J., Lu, X., Zhang, J., and Liu, C., 1990, Polycyclic tectonic evolution and metallogeny of the Tianshan mountains: Beijing, Science Press, 217 p.

Wendt, I., and Carl, C., 1991, The statistical distribution of the mean squared weighted deviation: Chemical Geology, v. 86, p. 275–285.

Windley, B.F., Allen, M.B., Zhang, C., Zhao, Z.Y., and Wang, G.R., 1990, Paleozoic accretion and Cenozoic redeformation of the Chinese Tien Shan Range, central Asia: Geology, v. 18, p. 128–131.

Xiao, X., Tang, Y., Li, J., Zhao, M., Feng, Y., and Zhu, B., 1990, Geotectonic evolution of northern Xinjiang: Xinjiang Geological Sciences, v. 1, p. 47–48.

Xinjiang Bureau of Geology and Mineral Resources, 1991, The Paleozoic of Xinjiang: Urumqi, China, Xinjiang People's Publishing House, 482 p.

Xinjiang Bureau of Geology and Mineral Resources, 1993, Regional geology of Xinjiang Uygur Autonomous Region: Geological Memoirs, ser. 1, 841 p.

Xinjiang Stratigraphic Table Compiling Group, 1981, Xinjiang stratigraphic tables: Beijing, Geological Publishing House, 496 p.

Yin, A., Nie, S., Craig, P., Harrison, T.M., Ryerson, F.J., Xianglin, Q., and Geng, Y., 1998, Late Cenozoic tectonic evolution of the southern Chinese Tian Shan: Tectonics, v. 17, p. 1–27.

Zhang, L., and Wu, N., 1985, The geotectonics and evolution of the Tian Shan: Xinjiang Geology, v. 3, p. 1–14.

Zhang, Z., 1986, Discussion on the age of the Dahalajunshan Formation: Xinjiang Geology, v. 4, p. 67–73.

Zhang, Z., and Liu, S., 1991, Microflora of Qiaoenbulake Group in Kalpin region and its geological age, in Jia, R., ed., Research of petroleum geology of northern Tarim basin in China: Beijing, China University of Geoscience Press, p. 29–35.

Zhang, Z., Liou, J.G., and Coleman, R.G., 1984, An outline of the plate tectonics of China: Geological Society of America Bulletin, v. 95, p. 295–312.

Zhou, J., Zhou, D., Zhao, M., and Qin, T., 1991, The Sinian stratigraphy and depositional characteristics of Kuluketage region, Xinjiang, China, in Jia, R., ed., Research of petroleum geology of northern Tarim basin in China: Beijing, China University of Geoscience Press, p. 118–125.

Zhu, J., and Sun, W., 1986, Approach of diagenetic age and evolution of metamorphic rocks of middle Tianshan: Xinjiang Geology, v. 4, p. 47–52.

MANUSCRIPT ACCEPTED BY THE SOCIETY JUNE 5, 2000

Geological Society of America
Memoir 194
2001

Sinian through Permian tectonostratigraphic evolution of the northwestern Tarim basin, China

Alan R. Carroll
Department of Geology and Geophysics, 1215 West Dayton Street, Madison, Wisconsin 53706 USA
Stephan A. Graham
Edmund Z. Chang
Department of Geological and Environmental Sciences, Stanford University, Stanford, California 94305 USA
Cleavy McKnight
Department of Geology, Baylor University, P.O. Box 97354, Waco, Texas 76798 USA

ABSTRACT

Sinian through Permian sedimentary rocks of the Kalpin and Bachu uplifts, northwest Tarim basin, record three major periods of basin evolution, as represented by stratigraphic megasequences divided by major unconformities. Each megasequence is marked by distinctive sedimentary facies, sediment dispersal patterns, sandstone provenance, subsidence history, and in two cases coeval magmatism. The same megasequences are recognized in both the surface Kalpin and largely subsurface Bachu uplifts, indicating that these areas shared an essentially identical history at least through the end of the Paleozoic. The Sinian–Ordovician megasequence overlies an angular basal unconformity with older metamorphic rocks. Siliciclastic facies directly above the unconformity are coarse grained and contain interbedded basalt flows. These facies grade upward into shallow-marine limestone and dolomite and interbedded deeper marine graptolitic shale, apparently as a result of thermal subsidence following a period of extension. Silurian and Devonian facies unconformably overlie Middle Ordovician strata, and are exclusively siliciclastic. They grade upward from green shelfal siltstone and sandstone into red fluvial sandstone and mudstone; paleocurrent indicators within the fluvial facies indicate derivation from the east. These deposits correspond in age with a proposed suture in the Altyn Tagh range adjacent to the eastern Tarim basin, suggesting that they may have been shed from a rising collisional orogen in that area. A pronounced angular unconformity separates Devonian strata from Carboniferous to Lower Permian fluvial and marine facies, which contain quartz-rich sandstone derived from recycling of underlying strata. Carboniferous–Permian rocks include relatively deep marine Carboniferous facies that are preserved in the most northwestern outcrop exposures of the Kalpin uplift. These progressively lap out to the southeast, where only thin Lower Permian fluvial and shallow-marine facies are preserved. These facies and thickness relationships suggest deposition in a flexural foreland basin, brought about by an ongoing collision between the Tarim and the central Tian Shan blocks. Lower Permian fluvial facies interbedded with basalt flows sharply overlie the marine facies in the Kalpin uplift. The basalts are closely tied in age with northwest-southeast–trending dikes, sills, and plutons in the Bachu uplift. The significance of this magmatism is unclear, but it may relate to limited extension normal to the collisional front.

Carroll, A.R., et al., 2001, Sinian through Permian tectonostratigraphic evolution of the northwestern Tarim basin, China, *in* Hendrix, M.S., and Davis, G.A., eds., Paleozoic and Mesozoic tectonic evolution of central Asia: From continental assembly to intracontinental deformation: Boulder, Colorado, Geological Society of America Memoir 194, p. 47–69.

INTRODUCTION

The Tarim basin of northwest China (Fig. 1) contains as much as 13–14 km of Sinian through Quaternary sedimentary fill deposited over several major episodes of basin subsidence, and represents one of the most important tectonic elements of central Asia. It provides one of the longest and most complete

records of regional tectonics and paleogeography available on the continent. The Tarim basin also contains significant petroleum reserves, which recently have been the subject of exploration interest (e.g., Gao and Ye, 1997). Several authors have written general summaries of the basin as a whole (e.g., Tian et al., 1989; Wang et al., 1992; Li et al., 1996), focusing on its large-scale tectonic evolution and petroleum potential. The ef-

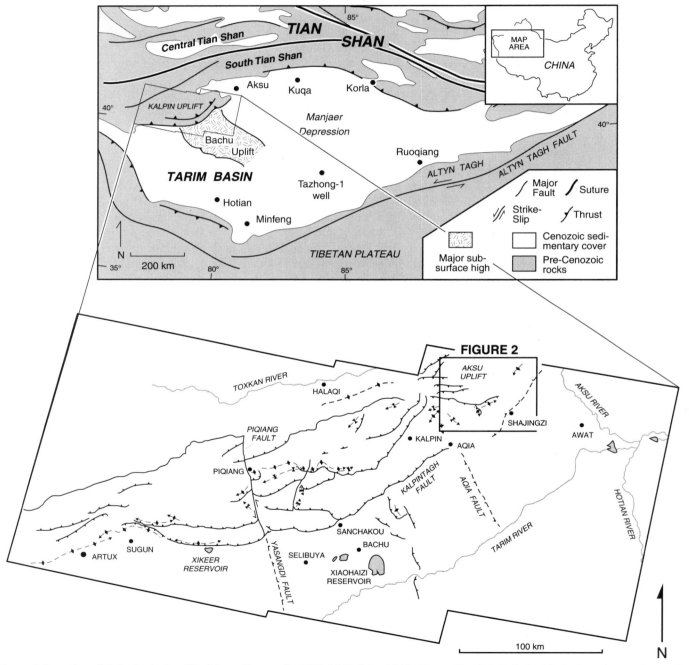

Figure 1. Location of Tarim basin (modified from Chen et al., 1985; McKnight, 1993). Yasangdi and Aqia faults (dashed) are known from subsurface seismic studies and have little surface expression.

fects of Cenozoic deformation during the Indian collision are particularly striking, hence a number of studies have considered the role of these structures in accommodating regional strain (e.g., Tapponnier and Molnar, 1979; McKnight et al., 1989; Nishidai and Berry, 1990; McKnight, 1993; Dong et al., 1998; Yin et al., 1998; Allen et al., 1999). In contrast, the sedimentary record of earlier deformations within and adjacent to the Tarim block has received far less attention (cf. Hendrix et al., 1992; Carroll et al., 1995; Allen et al., 1999)

Precambrian rocks crop out in ranges surrounding all sides of the Tarim basin (Chen et al., 1985). In contrast to the Paleozoic oceanic or accretionary substrate that underlies the Junggar basin to the north (Hopson et al., 1989; Kwon et al., 1989; Carroll et al., 1990; Şengör et al., 1993; Allen et al., 1995), the Tarim basin is most likely entirely underlain by Precambrian basement (Zhang et al., 1984; Li et al., 1996), although the nature of this basement beneath the cryptic Manjaer depression remains open to speculation (Şengör et al., 1996; Fig. 1). Cenozoic thin-skinned folding and thrusting of the Kalpin uplift, which probably occurred above a basal decollement in Upper Cambrian evaporites, is well documented at a reconnaissance scale (Nishidai and Berry, 1990; McKnight, 1993; Yin et al., 1998; Allen et al., 1999). Estimates of shortening derived from these studies range from 20% to 50%. Additional Paleozoic outcrop exposures are available within the Bachu uplift (Fig. 1), a largely subsurface structural high that intersects the Kalpin uplift.

Sinian and Paleozoic strata of the northwest Tarim basin have been previously studied by various Chinese researchers, but unfortunately this literature is difficult to access and evaluate by workers outside China (due to difficulties in aquiring and translating Chinese publications, incomplete reporting of spe-

cific field and laboratory data sets, and problems in verifying the accuracy of the data that are reported). The purpose of this paper is to examine the evolution of major sedimentary megasequences exposed adjacent to the northwestern Tarim basin, and to interpret the tectonic record they provide. This study is based on investigations we conducted over a number of field seasons (1987, 1988, 1991, and 1992) in the Kalpin and neighboring Bachu uplifts (Fig. 1). Our studies have focused in two principal geographic areas: the region between Aksu and village of Yingan in the northeastern Kalpin uplift, and near the Xiaohaizi reservoir in the northwestern Bachu uplift.

STRATIGRAPHY AND SEDIMENTARY FACIES

Sedimentary rocks of the northwestern Tarim basin may be subdivided into four distinct tectono-stratigraphic packages, based on their internal characteristics and on the position of major angular unconformities. Because each of these packages represent major, discrete phases of basin evolution, we refer to them as megasequences (see Hubbard et al., 1985, and Hubbard, 1988, for further discussion of the megasequence concept). Very similar facies are present in both the Kalpin and Bachu uplifts, but surface exposures are stratigraphically less complete and of poorer quality within the Bachu uplift. The following descriptions refer principally to Paleozoic rocks in the northeastern Kalpin uplift near Aksu, Sishichang, Wushi, and Yingan, and in the northwestern Bachu uplift near the Xiaohaizi reservoir (Figs. 1 and 2). Age assignments are supported by a variety of previously reported fossil evidence, summarized in Table 1, and by limited radiometric dating of volcanic units, as described in the following.

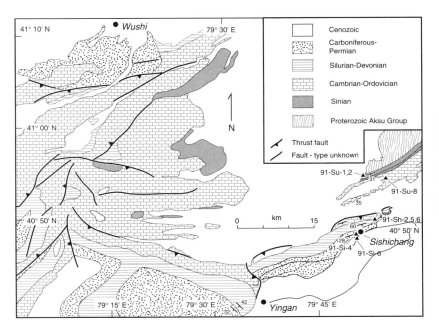

Figure 2. Geology of Aksu-Yingan area, northeastern Kalpin uplift (modified from unpublished 1:200 000 geologic mapping of Xinjiang Bureau of Geology and Mineral Resources). Triangles indicate position of apatite fission-track samples (see Dumitru et al., this volume).

TABLE 1. PARTIAL LIST OF SINIAN THROUGH PERMIAN INDEX FOSSILS, KALPIN UPLIFT

Age Assignment	Formation	Group	Fauna and/or Flora
Shajingzi	Upper Permian	Ostracod	*Darwinula jatskovae, Darwinuloides bugurulanica*
Kaipaizileke	Lower Permian	Plant	*Sphenophyllum, Calamites*
Kupukuziman	Lower Permian	Palynomorph	*Kraueselisporites, Latosporites, Deltoidospora*
Kangkelin	Lower Permian	Fusulinid, conodont	*Robustoschwagerina* sp., *Sweetognathus whitei*
Sishichang	Upper Carboniferous	Fusulinid, conodont	*Pseudoschwagerina leei, Streptognathus gracilis*
Bijingtawu	Lower Carboniferous	Fusulinid	*Fusulinella cacki, Fusulina chemovi*
Kongtaiaiken	Lower Carboniferous	Bivalve	*Delepinea subcarinata, Vitiliproductus tulensis, Gigantoproductus chaoi, Productus* cf. *asiaticus*
Yimungantawu	Lower Devonian	Bivalve, gastropod	*Eurymella* cf. *xitunensis, Grammatodon* sp., *Modiomorpha yunnanensis, Phestia* sp., *Edmontia* sp., *Letodesma* sp., *Pyncnomphalus* sp., *Planitrochus* sp.
Tataaiertage	Middle-Upper Silurian	Gastropod, chitin	*Horologium* sp., *Cingulochitina wronal*
Kalpintage	Lower Silurian	Graptolite	*Glyptograptus tamariscus linearis, Climacograptus tarascoides*
Yingan	Upper Ordovician	Graptolite	*Climacograptus spiniferous*
Qilang	Middle Ordovician	Graptolite	*Crynoides americanus*
Kanling	Middle Ordovician	Graptolite, conodont, cephalopod	*Pygodus anserinus, Prioniodus lingulatus, Lituites ningkiangense, Trilacinoceras discors*
Saergan	Middle Ordovician	Graptolite	*Glytograptus teretiusculus, Didymograptus jiangxiensis, Pterograptus elegans*
Qiulitage	Lower Ordovician	Conodont	*Eoplacognathus suecicus, Amorphognathus variabilis, Baltoniodus* aff. *Navis, Paroistodus proteus, Paltodus deltifer*
Shayilike	Upper-Middle Cambrian	Trilobite	*Kunmingaspis kalpinensis, Chittidilla nanjiangensis*
Wusongger	Lower Cambrian	Trilobite	*Paokannia* sp.
Xiaoerbulake	Lower Cambrian	Trilobite	*Kepinaspis kepingensis, Tianshanocephalus tianshanensis, Metaredlichioides kalpinensis, Shizhudiscus sugaitensis*
Yurtusi	Lower Cambrian	(Uncertain)	*Lapworthella* sp. nov. indet., *Paragloborilus spinatus, Anabarites* sp., *Protohertzina anabarica*
Qegebulake (Sinian)	Upper Sinian	Stromatolite	*Paniscollenia, Nucleela, Colleniella, Jurusania, Statifera, Gongylina, Linella, Tungussia, Conophyton*

Note: Derived from fauna and floral occurrences reported by Wang and Rui (1987), Chen et al. (1991), Liu and Xiong (1991), Wang (1991), Xiong (1991), Xinjiang Stratigraphic Table Compiling Group (1981); Fan and Ma (1991); and Carroll et al. (1995).

Sinan to Ordovician megasequence

The oldest rocks of the Kalpin uplift are Upper Proterozoic blueschist-greenschist facies metamorphic rocks of the Aksu Group, exposed locally near Aksu (Figs. 1 and 2). These rocks record high pressure-temperature ($P-T$) conditions associated with subduction-accretion or collision in a poorly known tectonic setting, and have yielded metamorphic ages ranging from 698 to 754 Ma (K-Ar and Rb-Sr ages of 698–728 Ma from phengite, and a ^{40}Ar/^{39}Ar age of 754 Ma from crossite; Liou et al., 1989, 1996; Nakajima et al., 1990). These rocks compose crystalline basement for the overlying unmetamorphosed sedimentary sequences. They are intruded by a series of northwest-southeast–trending diabase dikes. Liou et al. (1989, 1996) interpreted these dikes to be pre-Sinian, but the dikes have not been directly dated.

Proterozoic sedimentary rocks have been subdivided into a lower clastic section assigned to the lower Sinian (the Qiaoenbulak and Yulmenack Formations) and a lithologically diverse upper Sinian section (Sugaitebulake and Qegebulake Formations; Gao et al., 1985; Fig. 3). We have not observed the lower Sinian section, but Gao et al. (1985) reported that it locally reaches ~2000 m in thickness and includes glaciogenic turbidite facies of the Qiaoenbulake Formation that were deformed prior to the deposition of tillite facies of the Yulmenack Formation.

The upper Sinian rocks are exposed only at the east end of the Kalpin uplift (Fig. 2) and include a basal conglomeratic phase (Sugaitebulake Formation) that laps unconformably onto the Aksu Group. The character of these deposits is locally variable, ranging from boulder conglomerate to pebbly, coarse sandstone. The conglomerate contains clasts of underlying lithologies, including diabase similar to the Aksu Group dikes. The clasts are moderately to well rounded, with a maximum diameter 1.3 m. The conglomerate grades upward into interbedded red mudstone and fine- to coarse-grained sandstone of the Sugaitebulake Formation (Fig. 4A). The sandstone beds are typically 20 cm to 1 m thick, lenticular, and contain meter-scale planar and trough cross-beds (Fig. 4B), amalgamated beds, and ripples. We interpret these facies to represent braided stream deposits. This interval also contains at least three stratiform basaltic units (Fig. 4A), which we interpret to be flows on the basis of their vesicular character and possible columnar jointing. However, it is also possible that some units could be sills (R. Ressetar, 1999, personal commun.).

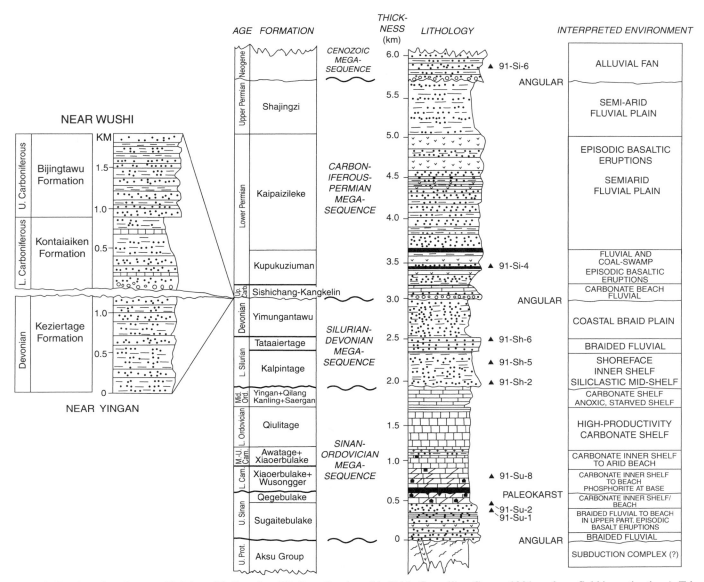

Figure 3. Stratigraphy of eastern Kalpin uplift (based on Xinjiang Stratigraphic Table Compiling Group, 1981, and our field investigations). Triangles indicate positions of apatite fission-track samples (see Dumitru et al., this volume).

The upper Sugaitebulake Formation contains interbedded red mudstone and sandstone with planar-parallel lamination to low-angle cross-stratification, and limestone containing flat-pebble intraclast conglomerate and stromatolite. We interpret this succession as marking a transition from nonmarine to shallow-marine depositional conditions. The uppermost Sinian Qegebulake Formation consists of interbedded limestone, dolomite, and mudstone. The carbonate facies are characterized by abundant stromatolite (Fig. 4C) and flat-pebble conglomerate, and are interpreted as intertidal to supratidal. The Qegebulake Formation is truncated by an unconformity that Gao et al. (1985) inferred to represent only a brief period of time, but that is noteworthy for the paleokarst features associated with it (Fig. 4, D and E).

Cambrian and Lower Ordovician strata exposed in the eastern Kalpin uplift constitute a 1–1.5-km-thick section of predominantly carbonate facies (Figs. 2 and 3). Except for a very thin, oolitic phosphorite-shale interval at its base, Lower Cambrian facies largely are dominated by stromatolite and flat-pebble conglomerate facies, interpreted as representing intertidal to supratidal environments. Dolomitic Middle to Upper Cambrian strata are also stromatolitic (Fig. 5A) and apparently record periods of restriction and emergence, based on reported interbedded evaporite (Fan and Ma, 1991) and observed red mudstone facies.

Topographically prominent massive limestone beds of the overlying Lower Ordovician Qiulitage Formation consist of thin-bedded micrite to calcarenite with diverse, normal marine fauna (Fan and Ma, 1991) indicating shallow shelf to supratidal

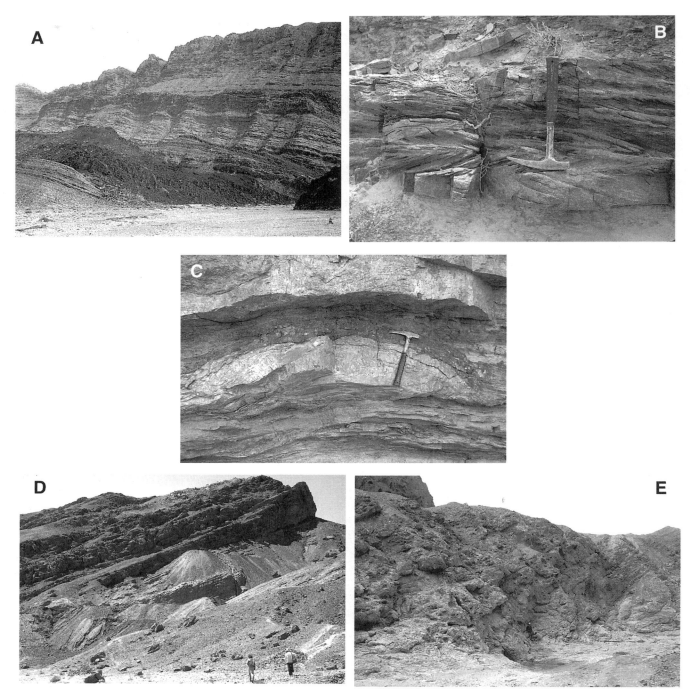

Figure 4. Outcrop photographs of Sinian sedimentary facies southwest of Aksu. A. Siliciclastic strata of Sugaitebulake Formation, interbedded with basalt flow in lower third of view. Thickness of exposed section is approximately 200 m. B: Cross-beds in basal upper Sinian Yurmeinake Formation sandstone, interpreted to represent braided fluvial environment. Note that sense of flow in this view is predominantly left to right (to southeast) C: Stromatolite in lowermost part of uppermost Sinian Qegebulake Formation, documenting shallow-marine depositional conditions. D: Sinian-Cambrian boundary (marked by change to more resistant-weathering massive Cambrian carbonate facies in upper half of exposure). Thickness of exposed section is approximately 50 m. E: Brecciated Sinian carbonates in karst zone immediately below contact with Cambrian.

Figure 5. Outcrop photographs of Cambrian and Ordovician facies between Sishichang and Sanchakou (see Figs. 1 and 2 for locations). A: Stromatolite hemispheroids exposed on bedding-plane surface in Middle–Upper Cambrian Shayilike Formation dolomite, southwest of Aksu. B: Lower Ordovician Qiulitage Formation limestone facies of Bachu uplift, exposed at Sanchakou. C: Nodular shelf limestone of Middle Ordovician Yingan Formation. D: Storm-deposited bed in Middle Ordovician Yingan Formation (carbonate intraclasts concentrated in layer marked by compass; also see Fig. 7).

environments. Similar Qiulitage Formation facies represent the oldest strata exposed within the Bachu uplift at several localities, including Sanchakou (Figs. 1 and 5B).

The Middle Ordovician generally reflects deeper water, but probably still shelfal environments, in the Kalpin uplift. The Saergan Formation is a graptolitic, pyritic, carbonaceous, laminated black shale interpreted to represent anoxic to suboxic depositional conditions (Graham et al., 1990). The Saergan Formation has been reported to occur in the Bachu uplift (Xinjiang Stratigraphic Table Compiling Group, 1981), but was not investigated during this study. Overlying Ordovician formations represent a return to a well-oxygenated carbonate shelf environment (Fig. 6). These rocks are extensively bioturbated and occasionally punctuated by conglomeratic and fossiliferous beds 5–10 cm thick, interpreted as storm deposits (Figs. 5, C and

D, and 7). Ordovician units above the Saergan Formation have not been reported from Bachu uplift outcrops.

Silurian to Devonian megasequence

Lithology changes sharply and markedly across an unconformity that separates the Middle Ordovician carbonate sequence from the overlying Lower Silurian green siliciclastic strata of the Kalpintage Formation (Figs. 6, 7, and 8, A and B). This unconformity locally removes all or part of the Ordovician Yingan and Qilang Formations (Fan and Ma, 1991), indicating erosional relief of tens to hundreds of meters. At Sishichang, a discontinuous, fossiliferous conglomeratic lag <1 m thick is on the unconformity surface (Fig. 7). The lower Kalpintage Formation consists of interbedded fissile green siliciclastic

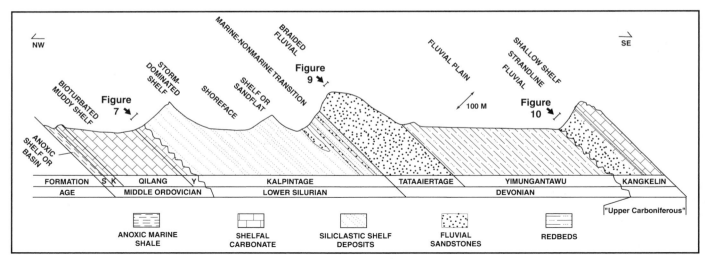

Figure 6. Outcrop transect through Paleozoic sedimentary rocks near Sishichang.

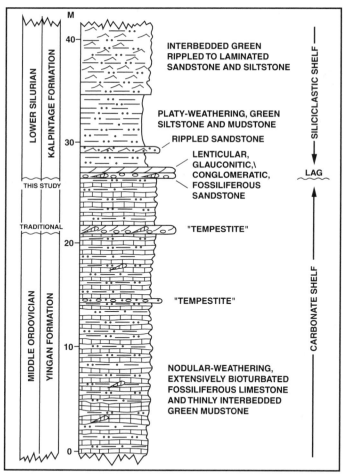

Figure 7. Measured outcrop section though Ordovician–Silurian contact near Sishichang.

mudstone, siltstone, and sandstone, interpreted to be deposited on a storm-dominated shelf. The siltstone is calcareous, planar-parallel to wavy laminated, and locally nodular and bioturbated. Thin fine- to medium-grained sandstone beds are planar-parallel laminated to rippled, and locally contain coarse-grained glauconitic layers, bedding-plane trace fossils, and centimeter-scale grooves on bed soles. This facies association is interbedded with medium- to coarse-grained sandstone with pervasive trough cross-beds to 1 m in amplitude, interpreted as shoreface deposits. The upper Kalpintage Formation includes a lenticular, upward-fining succession of white sandstone beds deposited above a basal scour, with meter-scale trough cross-beds and 15–20 cm mud balls at the base, interpreted as a tidal channel deposit. Possible tidal deposits have also been reported near the village of Yingan (Fig. 2; S.J. Vincent, 1999, personal commun.).

Redbeds with local mudcracks and paleosols increase in frequency upward near the top of the Kalpintage Formation; the rocks are mapped as the Tataaiertage Formation at the point where redbeds dominate. The Tataaiertage Formation was formerly assigned to the Devonian (Xinjiang Stratigraphic Tables Compiling Group, 1981), but the transitional nature of the contact with the underlying green, marine Silurian rocks suggests that it is fully conformable with the underlying Kalpintage Formation. Fan and Ma (1991) assigned a Silurian age to the Tataaiertage Formation based on the gradational nature of the contact and the local occurrence of Silurian marine fossils (Table 1). They inferred that previously reported plant fossils (*Lepidodendropsis*) were actually not found in place. The Tataaiertage section contains mostly unfossiliferous, trough cross-bedded sandstone interpreted to represent braided fluvial deposits (Figs. 8C and 9). Finer-grained overlying facies of the Yimungantawu Formation may represent deposits on a broad fluvial plain. The Yimungantawu Formation has been assigned to the Lower Devonian on the basis of nonmarine bivalves (Fan and Ma, 1991; Table 1). The overlying Keziertage Formation,

Figure 8. Outcrop photographs of Silurian and Devonian facies near Sishichang. A: Unconformable contact between Middle Ordovician (light colored rocks on right) and Lower Silurian (dark colored rocks on left). Apparent angularity is artifact of angle of view at this location. B: Overview of siliciclastic facies of Kalpintage and Tataaiertage Formations, looking south (standing near base of section; also refer to Fig. 6). Total thickness of strata in this view is ~500 m. Foreground consists of Kalpintage Formation green mudstone and sandstone, interpreted as storm-dominated shelf deposits. Similar facies continue into partially covered strike valley, where they are interbedded with trough cross-bedded shoreface sandstone. Ridge in background contains lenticular, cross-bedded sandstone facies interpreted as shoreface to tidal deposits, overlain by tidal to coastal plain red mudstone mapped as base of Tataaiertage Formation. Resistant beds at ridge on skyline (left) consist of Tataaiertage Formation red sandstone facies interpreted as braided fluvial deposits. C: Cross-bedded sandstone of Tataaiertage Formation, interpreted as braided fluvial deposits. These cross-beds indicate predominant flow direction to west.

which locally was completely eroded away, contains trough cross-bedded sandstone facies interpreted as a return to braided fluvial deposition.

We observed a very similar Silurian–Devonian section in the Bachu uplift near the Xiaohaizi reservoir (Fig. 1). The Xiaohaizi section, however, is cut by numerous diabase dikes and sills, which obscure stratigraphic relationships. The Xiaohaizi section also includes several intervals of fluvial quartz-pebble

conglomerate beds with scoured bases. These are presumed to be Devonian, but are unfossiliferous.

Carboniferous to Permian megasequence

Kalpin uplift. Carboniferous and Permian facies of the northeast Kalpin uplift were described in detail by Carroll et al. (1995); these descriptions are therefore not repeated here.

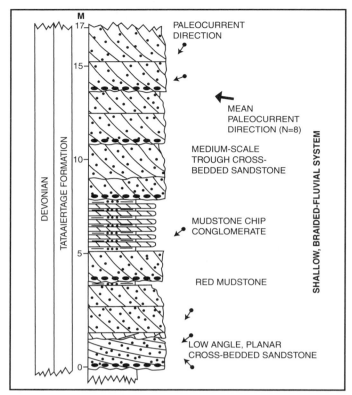

Figure 9. Measured section in Tataaiertage Formation 5 km northeast of Sishichang (see Fig. 6 for stratigraphic position).

Devonian (and possibly Silurian) strata are extensively truncated beneath an angular unconformity cut across the entire Kalpin uplift. The oldest strata above the unconformity are progressively younger to the southeast. Near Wushi (Fig. 2), at least 2000 m of carbonate and siliciclastic facies as old as Early Carboniferous unconformably overlie Cambrian through Silurian rocks. This succession deepens upward from fluvial gravels to siliciclastic turbidites, which then grade into shelf sandstone overlain by carbonate olistostrome. On the basis of regional relationships, the latter facies are interpreted to have been shed from northwest-facing carbonate platforms. At Sishichang (Fig. 2) the Lower Permian Sishichang Formation unconformably overlies Devonian redbeds (Carroll et al., 1995). Fluvial conglomerate and sandstone grade upward into tidal sandstone facies and shallow-marine algal and skeletal limestone of the Kangkelin Formation (Fig. 10; Carroll et al., 1995). Clasts in the conglomerates are mostly sedimentary, and have lithologies similar to underlying Devonian strata.

The Kangkelin Formation is sharply overlain by variegated red-green, nonmarine Permian strata of the Kupukuziman, Kaipaizileke, and Shajingzi Formations (Fig. 3). Carbonaceous strata, rooted fossil trees, fossil leaf horizons, and thin coal beds occur sporadically throughout the lower half of this section, supporting the inference of at least a seasonally humid environment. Two series of basaltic lavas, each of which consists of multiple thin flow horizons, occur within the Kupukuziman and

Kaipaizileke Formations and underlie a large area of the northwest Tarim basin (Chang, 1988; Liu and Li, 1991; Wang and Liu, 1991; Fig. 11). Each series totals ~150–200 m in thickness, but is interbedded within thicker intervals of nonmarine sedimentary rocks; marine interbeds occur to the west (M.B. Allen, 1999, personal commun.). Liu and Li (1991) and Wang and Liu (1991) presented geochemical evidence indicating that the flows range in composition from tholeiites to alkali basalts, consistent with within-plate magmatism. A variety of K-Ar whole-rock ages have been reported for these flows, ranging from 293.25 ± 7.70 to 285.24 ± 6.67 Ma for the lower series and 295.13 ± 7.1 to 228.83 ± 5.15 Ma for the upper series (Liu and Li, 1991; Wang and Liu, 1991). Carroll et al. (1995) reported an $^{40}Ar/^{39}Ar$ age of 277.53 ± 0.46 Ma for plagioclase separated from a flow in the lower series 20 km southwest of Sishichang, indicating an Early Permian age for these flows. Basalts of the northeast Kalpin uplift thicken from Sishichang to Yingan (Fig. 2), where dikes may indicate Permian vent areas. The Yingan area is the site of a major tear fault in the Kalpin thrust sheets (Figs. 1 and 2), possibly suggesting a long-lived zone of structural reactivation.

Age relationships within the nonmarine Permian exposures above the basalt flows are subject to controversy. Upper Permian rocks are most often depicted overlying an angular unconformity above the Lower Permian (Xinjiang Stratigraphic Table Compiling Group, 1981; unpublished 1:200 000 geologic mapping of the Xinjiang Bureau of Geology and Mineral Resources), but some workers maintain that this entire section may be Lower Permian (Li Wunfeng, 1992, personal commun.). The total original thickness of this interval is unknown; it is erosionally truncated and overlain by Cenozoic deposits.

Significant along-strike variations in Carboniferous–Permian stratigraphy occur within the Kalpin uplift. It generally appears that thicker Carboniferous strata are present to the southwest (Xinjiang Stratigraphic Table Compiling Group, 1981; unpublished 1:200 000 geologic mapping of the Xinjiang Bureau of Geology and Mineral Resources), although detailed field data have not been reported.

Bachu uplift. Carboniferous and Permian rocks of the Bachu uplift at Xiaohaizi are extensively intruded (Fig. 12, A and B), and generally are less well exposed than their Kalpin uplift equivalents. A mafic composite sill ~50–100 m thick (Fig. 12A) either coincides with the position of the Devonian-Carboniferous contact, or else is slightly above it. The angular unconformity seen in the Kalpin uplift is not clearly visible at Xiaohaizi. However, we measured ~10° of dip discordance between Devonian strata beneath the sills and Carboniferous strata above. The sills are cut by a composite syenitic to granitic pluton that radiates felsic dikes into the surrounding rocks (Fig. 12C). Lower Carboniferous to Lower Permian shallow-marine deposits overlie the sill complex and comprise interbedded mudstone, limestone, and poorly exposed gypsum (possibly nodular). We interpret these facies to represent back-barrier and lagoonal environments in which carbonate grains de-

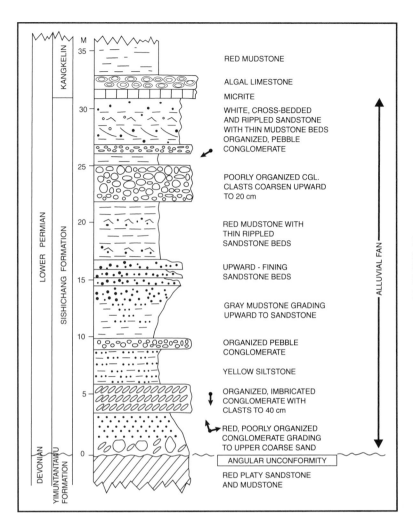

Figure 10. Measured section through angular contact between Yingan and Sishichang Formations 5 km northeast of Sishichang (see Fig. 6 for stratigraphic position). CGL.—conglomerate.

rived from storm washover alternated with mudstone and gypsum deposition under restricted conditions. The upper part of the succession also contains brecciated limestone, algal laminites, and flat-pebble conglomerate indicating supratidal environments. All of these facies are extensively cut by northwest-southeast–trending mafic dikes, which locally compose as much as 20% of the total outcrop width. These dikes, and gabbroic dikes associated with an alkali igneous complex to the south of Xiaohaizi, have $^{40}Ar/^{39}Ar$ ages that are essentially identical to the lower series of basalt flows discussed here (Carroll et al., 1995). However, the Xiaohaizi exposures do not include flows.

Groves and Brenckle (1997) used graphical correlation techniques to infer that the Carboniferous–Permian succession at Xiaohaizi is actually far less stratigraphically complete than it appears in outcrop. They argued that the Xiaohaizi section and several sections within the Kalpin uplift contain hiatuses that represent at least as much geologic time as the preserved sedimentary rocks. Furthermore, these hiatuses are generally longer than can be explained by third-order sea-level changes, and suggest instead control by local geologic processes. We did not observe any obvious unconformities within this succession.

PALEOCURRENTS AND SANDSTONE PROVENANCE

Paleocurrent measurements were collected at several locations from Sinian through Permian facies in the Aksu-Sishichang area (Fig. 13). Note that although the numbers of measurements reported here are not statistically significant, they are representative of larger, visually identified populations of similar features. Sandstone point-counts were conducted using a modified Gazzi-Dickinson method (Table 2; see Graham et al., 1993, and Ingersoll et al., 1984, for complete discussions of techniques). The raw data were recalculated into detrital modes and plotted on standard ternary diagrams (Fig. 14) to permit comparison with previously recognized provenance types (cf. Dickinson, 1985).

Sinian paleocurrent measurements generally indicate transport directions to the south or southeast (Fig. 13A). These measurements represent mostly decimeter-scale planar cross-beds within coarse-grained, fluvial sandstone facies (e.g., Fig. 4B). Sinian sandstone samples vary widely in composition and reflect mixed provenance (Fig. 14A), related both to metamorphic basement lithologies and to interstratified Sinian volcanic rocks.

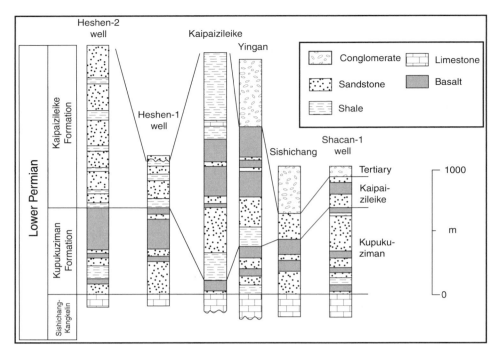

Figure 11. Occurrence of Lower Permian basalt in northwest Tarim basin (modified from Xinjiang Stratigraphic Table Compiling Group, 1981; Chang, 1988; Liu and Li, 1991; Wang and Liu, 1991; J. Liu, 1995, personal commun.). Basalt intervals represented in upper part of figure are actually composite intervals of interbedded basalt and siliclastic sedimentary rocks; total thickness of basalt flows is therefore less than represented.

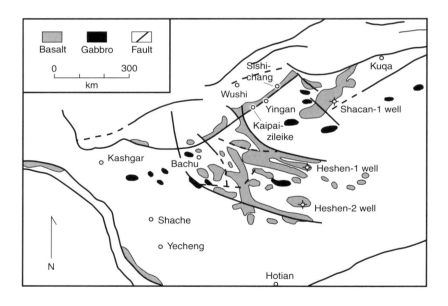

The relatively high content of polycrystalline quartz (Qp) most likely reflects input from metamorphic rocks of the Aksu Group.

Cross-beds, ripples, and groove casts in Silurian sandstone facies display mixed transport directions (Fig. 13B). In contrast, fluvial Devonian sandstone facies show strong west to southwest transport directions based on trough cross-beds, parting lineation, and grooves (Figs. 8C and 13C). Silurian and Devonian sandstone contains mostly quartz and lithic rock fragments (Fig. 14B), suggesting a recycled orogenic provenance (cf. Dickinson, 1985). Together these observations circumstantially support the presence of a Devonian orogenic sediment source area east of the study area. Uniformly high plagioclase/total feldspar (P/F) and locally high lithic volcanic/total lithic (Lv/L) ratios in Silurian–Devonian sandstones likely record a partly volcanic provenance, probably south of the Altyn Tagh fault (Fig. 1).

Net northwest sediment transport in the Kalpin uplift area during the Carboniferous is expected from regional considerations, but our paleocurrent data (Fig. 13D) are too sparse to lend further support to this interpretation. Allen et al. (1999) reported northwest-directed paleocurrent directions in turbidites of the Upper Carboniferous Sasikebulake Formation (located west

Figure 12. Outcrop photographs of intrusions at Xiaohaizi reservoir (see Fig. 1 for location). A: South-dipping diabasic sills intruding Devonian and Carboniferous strata (utility poles in foreground are ~10 m high). B: Small dike cutting cross-bedded Devonian fluvial sandstone, and feeding sill that has intruded cross-laminae. C: Sills shown in A (right), cut by pluton (left). Height of ridge at right is approximately 50 m above foreground.

of Halaqi; Fig. 1). Carboniferous to Lower Permian sandstone older than the Kupukuziman Formation is relatively quartzose (Fig. 14C), reflecting derivation from low-lying areas within the Tarim craton. Carboniferous subcrop maps indicate that Precambrian basement was nowhere exposed within the interior of Tarim (Lu and Qi, 1994, personal commun.); pre-Kupukuziman Formation sandstone therefore was almost certainly derived from weathering and reworking of underlying Silurian–Devonian sandstone and conglomerate.

In contrast, Lower Permian sandstone of the Kupukuziman Formation marks a dramatic shift in both paleocurrents and provenance (Figs. 13E and 14D). They are dominated by felsic volcanic material derived from the northwest. Felsic-volcanic rock fragments with pyroclastic textures, granitic rock fragments, angular monocrystalline quartz, and potassium-feldspar grains are all common constituents. Carroll et al. (1995) inter-preted these sandstones to be derived from rhyolitic volcanics and related granites in the southern Tian Shan. However, northwest-directed paleocurrents have also been noted in time-equivalent sandstone facies with similar modal compositions located to the southwest of our study area (S.J. Vincent, 1999, personal commun.). This suggests the presence of another rhy-olitic or granitic detrital source to the southeast, most likely within the buried Bachu uplift.

SUBSIDENCE HISTORY

The subsidence history of the Aksu-Sishichang and Xiao-haizi areas was reconstructed using the backstripping methods of Bond and Kominz (1984; also see Steckler and Watts, 1978; Van Hinte, 1978; Schlater and Christie, 1980, for further details of the method). Absolute ages are based on the time scale of

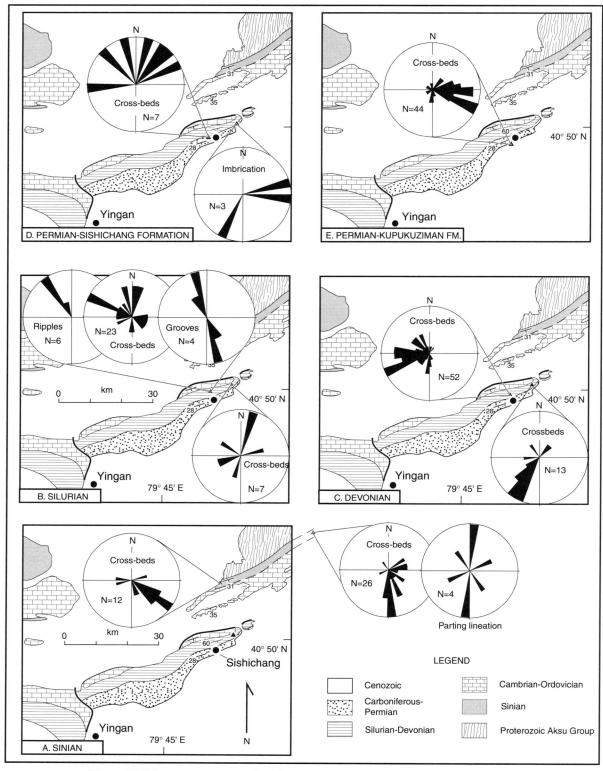

Figure 13. Paleocurrent summary for Aksu-Yingan area (see Figs. 1 and 2 for location).

TABLE 2. SANDSTONE POINT-COUNT DATA

Sample number	Age	N	Qm	Qp	Chert	K	P	Lv	CO$_3$	Ls	Lm	Micas	other	PMC
Kupukuziman Formation														
87-A-23	P1	501	77	9	17	79	4	207	0	24	0	0	7	77
87-A-27	P1	500	131	9	1	70	32	169	0	4	6	3	15	60
87-A-28	P1	506	98	26	12	40	23	48	0	120	0	2	39	98
87-A-29	P1	482	131	11	0	41	77	97	0	63	0	1	15	46
88-A-28A	P1	566	73	1	1	61	24	279	0	0	0	0	15	112
88-A-28B	P1	578	77	2	5	124	29	186	0	57	0	0	20	78
88-A-28C	P1	502	69	3	2	87	21	185	0	30	0	2	28	75
88-A-28D	P1	500	63	0	1	68	26	238	0	32	0	0	12	60
88-A-28E	P1	500	69	2	1	43	35	249	0	10	0	0	21	70
88-A-41A	P1	499	58	1	0	85	16	255	0	23	0	0	7	54
88-A-41B	P1	604	73	1	2	128	30	203	0	50	0	0	13	104
Kongtaiaiken-Sishichang Formations														
87-A-18	P1	500	295	25	23	11	6	12	0	23	8	0	40	57
87-A-20	P1	500	323	31	9	2	0	7	0	10	0	0	16	102
87-A-21	P1	613	277	38	8	9	25	58	0	46	7	2	83	60
88-A-21	P1	500	301	18	9	0	0	1	13	0	1	0	0	157
88-A-30	P1	677	428	32	13	4	0	0	0	17	3	0	3	177
88-A-31	P1	500	343	14	22	2	6	2	0	25	7	0	2	77
88-A-40F	P1	669	455	25	5	3	2	0	0	4	0	0	6	169
92-Ba-65	P1	500	188	8	2	52	84	58	0	15	0	0	12	81
87-W-15	C2	500	258	54	14	2	14	3	11	21	5	0	19	99
87-W-5	C2	500	236	20	18	3	11	0	76	6	2	0	33	95
92-Ba-60	C2	500	402	1	0	1	0	5	0	0	0	0	8	83
92-Ba-61	C2	500	266	0	0	1	2	1	0	3	0	0	4	223
92-Y-1	C2	500	336	8	1	0	0	12	1	10	0	1	6	125
92-W-6	C	500	369	15	6	2	1	14	0	3	4	0	7	79
Kalpintage-Yimungantawu Formations														
87-A-16	D	499	236	20	10	10	89	11	0	0	0	5	0	118
87-A-14	D	500	166	6	3	0	1	10	0	0	3	0	4	307
87-A-12	S	490	191	20	15	5	15	85	0	28	0	7	6	118
87-A-13	S	499	207	15	35	20	24	0	0	26	0	14	1	157
91-SH-5	S	500	327	12	3	2	2	35	0	4	1	0	18	96
92-AQ-1	S	500	255	2	2	3	9	22	0	29	38	1	15	124
92-AQ-2	S	500	306	6	5	2	11	7	0	37	20	0	9	97
92-AQ-3	S	500	126	11	1	1	5	63	0	39	76	2	55	121
92-Ba-66	S	500	381	3	1	0	3	13	1	1	0	0	5	92
92-Ba-67	S	500	334	9	4	0	1	28	5	27	0	0	7	85
92-Ba-68	S	500	316	4	7	0	2	42	14	12	0	0	2	101
Yurminake-Qegebulake Formations														
87-A-33	Z	524	129	7	0	41	183	55	0	18	17	20	31	23
87-A-34	Z	508	140	29	4	0	143	102	0	11	3	6	35	35
88-A-2	Z	500	397	0	1	10	3	0	0	0	0	0	34	55
88-A-3	Z	500	180	10	4	37	9	39	7	1	2	0	1	210
88-A-39C	Z	500	132	47	2	42	31	43	0	0	0	1	14	188
88-A-39D	Z	500	271	23	9	5	8	10	0	3	0	0	24	147
88-A-39E	Z	500	152	37	9	58	5	31	0	0	0	5	8	195
88-A-39F	Z	500	127	42	6	50	25	24	0	0	0	4	17	205
88-A-39G	Z	500	334	10	4	0	0	14	0	0	0	0	22	116
91-SU-2	Z	500	278	0	1	65	3	0	2	2	0	0	14	135

Note: The following abbreviations are used: Q—Quaternary, P1—Lower Permian, C—Carboniferous, C2—Upper Carboniferous, D—Devonian, S—Silurian, Z—Sinian, Qm—monocrystalline quartz, Qp—polycrystalline quartz, K—potassium feldspar, P—plagioclase, Lv—lithic volcanic, CO$_3$—carbonate grains, Ls—lithic sedimentary, Lm—lithic metamorphic, PMC—porosity, matrix, and cement.

A.R. Carroll et al.

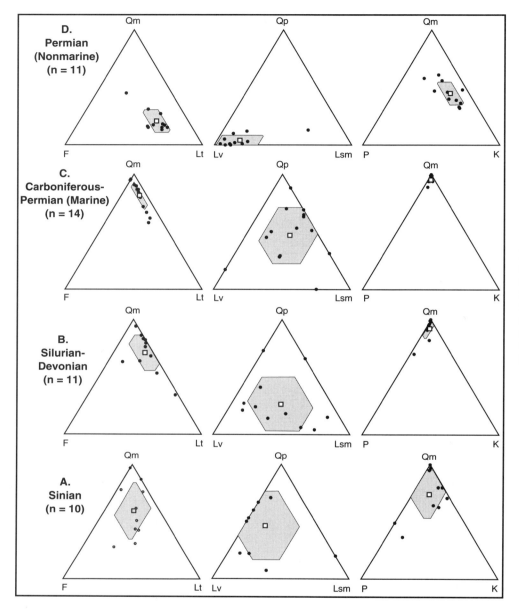

Figure 14. Modal compositions of sandstone samples (see Table 2 for raw data and text for further explanation). Qm—monocrystalline quartz, F—total feldspar, Lt—total lithics + polycrystalline quartz, Qp—polycrystalline quartz, Lv—volcanic lithics, Lsm—sedimentary + metamorphic lithics, P—plagioclase, K—potassium feldspar.

Harland et al. (1990), except for the Permian. Permian ages are based on Ross et al. (1994) because this time scale more closely fits known Lower Permian paleontological and radiometric constraints at Sishichang. The lower age limit of Sinan sedimentary rocks near Aksu is very well defined by radiometric dating of micas contained in pelitic schist underlying the basal unconformity (Nakajima et al., 1990). A late Sinian (Vendian) age is further supported by reported acritarch and stromatolite occurrences (Gao et al., 1985). The amount of missing time represented by the basal unconformity, however, is unknown. Stratigraphic thicknesses for all units are based on appropriate reference sections (Xinjiang Stratigraphic Table Compiling Group, 1981), modified on the basis of our field investigations.

Because of the incompleteness of individual outcrop sections, the reference sections each represent local composites of more than one actual measured section. Some errors (on the order of 10%–20%) may therefore have been introduced due to stratigraphic thickness variations, but generally the backstripped sections are representative of the true vertical thickness of strata present at each location.

Paleobathymetries are based on our interpretation of marine depositional environments, and approximate error ranges are estimated. Paleoelevations during deposition of nonmarine units represent best guesses, based on the occurrence of similar depositional environments elsewhere and on the present elevation of northwest Tarim; estimated error ranges are correspondingly large. Unconformities are based on missing strata as indicated by the Xinjiang Stratigraphic Tables, paleontologic and radiometric data of Carroll et al. (1995), graphic correlation of Carboniferous and Permian marine fauna by Groves and Brenckle (1997), and our field investigations.

The total and tectonic subsidence histories at both locations were very similar at both localities during lower to middle Paleozoic time (Fig. 15), both in terms of timing and magnitude. Sinian through Cambrian rocks do not crop out within the Bachu uplift; direct comparison of subsidence during this interval therefore is not possible. We know of no reported well penetrations of these units in the Bachu uplift, but their presence has been inferred in the subsurface (e.g., Fan and Ma, 1991). Tectonic subsidence rates in the Aksu-Sishichang area gradually decreased from Sinian through Cambrian time, and appear to have remained relatively constant and slow during the Ordovician. Greater Early Ordovician subsidence occurred at the Bachu uplift, resulting in preservation of strata that are ~50% thicker than in the Aksu-Kalpin area. Upper Ordovician strata are not reported from the Xiaohaizi area, due either to erosional removal by the base-Silurian unconformity or simply to lack of exposure. We were not able to locate an exposed Ordovician-Silurian contact in this area.

Subsidence rates increased in both areas during the Silurian following development of the widespread basal-Silurian unconformity, and increased subsidence continued into the Devonian. The thickness of preserved Devonian strata varies greatly within the Kalpin uplift, depending on the magnitude of erosion at the basal Carboniferous unconformity. For example, the Upper Devonian Keziertage Formation is absent at Sishichang, but locally reaches nearly 1200 m in thickness above the Yimungantawu Formation (Xinjiang Stratigraphic Table Compiling Group, 1981). At its maximum the Silurian–Devonian megasequence in the Kalpin uplift reaches more than 2200 m in thickness in the Kalpin uplift, and therefore records a significant period of tectonic subsidence.

The record of Carboniferous to earliest Permian subsidence varies greatly depending on location. Most or all of this interval is represented by an unconformity in the Aksu-Sishichang area, whereas sedimentation at Xiaohaizi initially appears to have been more continuous. Deposition of the Kupukuziman Formation initiated a major new episode of basin formation, characterized by rapid subsidence.

Fission-track analysis of Sinian through Permian samples from the Aksu-Sishichang area indicates that this entire section cooled below annealing temperature simultaneously during the latest Permian to Early Triassic (details of these analyses and estimates of the timing and magnitude of unroofing are presented in Dumitru et al., this volume; these results are therefore not be repeated here). We infer that substantial erosion of Permian rocks occurred at this time (Fig. 15). There is no stratigraphic or fission-track evidence that large thicknesses of Mesozoic strata ever covered the Kalpin uplift. Fission-track shortening of ~15%

in these samples could have occurred either due to relatively minor Mesozoic burial, or due to burial beneath now-eroded Cenozoic sediments. About 1700 m of Neogene-Quaternary nonmarine sedimentary rocks cover the Bachu uplift near Selibuya (Fig. 1); involvement of these strata in Cenozoic thrusting (Allen et al., 1999) indicates that they may have originally also covered Paleozoic rocks in the Kalpin uplift.

DISCUSSION AND CONCLUSIONS

Sinian–Ordovician megasequence

The basal unconformity that marks the onset of Sinian deposition most likely corresponds with the initiation of rifting of a preexisting Proterozoic continent. Shi et al. (1995) suggested that Sinian extension was followed by a period of thermal subsidence. The Sinian paleogeography of this region is very poorly known, but local variations in the thickness of lower Sinian strata of ~2000 m and the presence of basal upper Sinian boulder conglomerate indicates substantial relief on the basal unconformity. South-directed paleocurrents indicate either that Tarim Proterozoic rocks continue to the north of the study area, or that the original source area for these sediments subsequently rifted away. There is no evidence for renewed tectonic activity during the latest Sinian through Ordovician; the observed changes in sedimentary environments were apparently driven mostly by changing relative sea level. The Tarim block occupied low paleolatitudes by the early Paleozoic (Fig. 16A), latitudes that were conducive to extensive shallow-marine carbonate sedimentation. Graptolitic shale deposition probably records maximum sea-level highstands.

Silurian–Devonian megasequence

The base-Silurian unconformity corresponds to a basin-wide event also observed in the subsurface north of the Tazhong structure in the southeastern Tarim basin (site of the Tazhong-1 well, Fig. 1; Li et al., 1996; R. Ressetar, 1999, personal commun.), with relief of at least hundreds of meters. Increased basin subsidence and the reintroduction of siliciclastic detritus indicate that this unconformity records a new tectonic episode of basin evolution. The source area for Silurian to Devonian clastic sediments is unknown, but may be a middle Paleozoic orogenic belt developed to the east, within the Altyn Tagh range (Figs. 1 and 16B). Sobel and Arnaud (1999) proposed the existence of a middle Paleozoic suture (termed the Lapeiquan suture) between Archean rocks of the Tarim block and a Proterozoic block to the south. The oldest arc intrusions within the Proterozoic block were dated as 435 ± 20 Ma, and crosscutting postorogenic granite plutons were dated as 383 ± 7 Ma. Sobel and Arnaud (1999) therefore concluded that an intervening ocean basin closed after the Early Silurian and before the Middle Devonian. They also pointed out that sedimentary rocks of Silurian to Middle

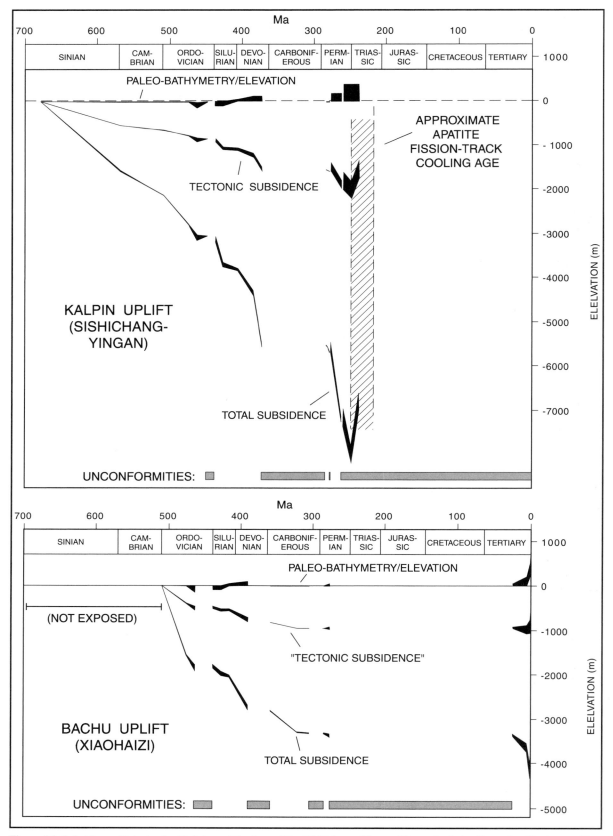

Figure 15. Subsidence histories for Kalpin and Bachu uplifts. Sishichang-Yingan history includes Devonian Keziertage Formation, which is present at Yingan but eroded at Sishichang. Time scale after Harland et al. (1990); Permian time scale is modified according to Ross et al. (1994). Apatite fission-track (AFTA) cooling age is from Dumitru et al. (this volume).

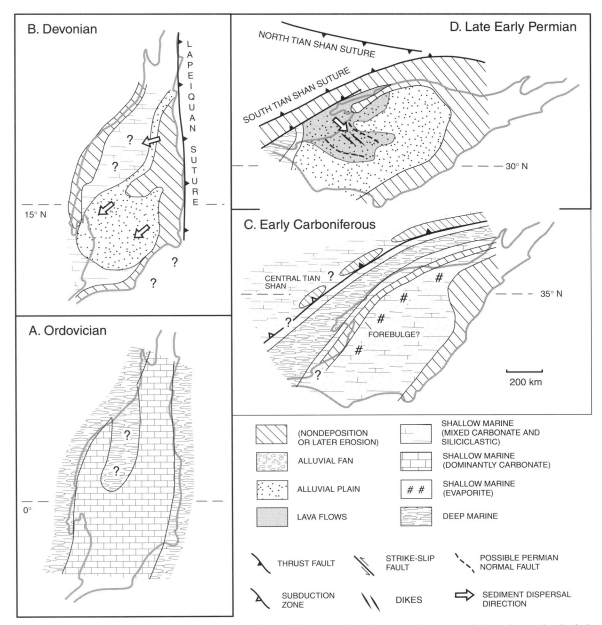

Figure 16. Schematic paleogeographic reconstructions of Tarim basin and surrounding areas. Orientation and paleolatitude of Tarim block compiled from paleomagnetic studies and summaries by Bai et al. (1987), Li et al. (1988a, 1988b), Li (1990), Sharps et al. (1989), Fang et al. (1990), and Zhao et al. (1996). Paleogeography from this study and from previous work by Hu et al. (1965), Zhang et al. (1983), Lai and Wang (1988), Allen et al. (1991, 1993), Fan and Ma (1991), Wang et al. (1991), Zhou et al. (1991), Carroll et al. (1995), Chen et al. (1999), and Zhou et al. (this volume).

Devonian age are not known from this area. We noted the absence of Silurian or Devonian sedimentary rocks along the southeastern margin of the Tarim basin (between Ruoqiang and Minfeng), although Devonian quartz-pebble conglomerates are present in the southwestern Tarim near Hotian (Fig. 1). Together, these observations suggest that an orogenic belt roughly parallel to the present southeast margin of the Tarim basin may

have shed Silurian–Devonian clastic sediments to the northwest (Fig. 16B). The compositions of Silurian–Devonian sandstone samples from the Kalpin uplift (Table 2; Figure 14B) are consistent with derivation from the preorogenic lithologies described by Sobel and Arnaud (1999). However, it is not clear what genetic relationship, if any, existed between the Lapeiquan suture and basin subsidence in the Kalpin and Xiaohaizi areas.

Apparently increasing rates of tectonic subsidence during the Late Devonian (Fig. 15) are suggestive of flexural subsidence (cf. Dickinson, 1976; Jordan, 1981), but this interpretation is speculative due to the poor age data on these nonmarine facies.

Carboniferous–Permian megasequence

The base-Carboniferous angular unconformity is present virtually throughout the Tarim basin, including the Kalpin uplift and the southwest Tarim basin near Hotian (our field observations), in the subsurface of the southeastern Tarim basin (Li et al., 1996), and in the Altyn Tagh range (Sobel and Arnaud, 1999). Carroll et al. (1995) and Allen et al. (1999) proposed that a flexural foredeep formed along the northern Tarim margin in response to a Late Devonian to Early Carboniferous collision between Tarim and the Yili block (Wang et al., 1990; Allen et al., 1991, 1993; Hsü et al., 1994; Gao et al., 1998; Fig. 16C). We further propose that flexural loading of the northern Tarim block by the collisional orogen may have created a flexural forebulge with in the northern Tarim basin, coincident with the Sishichang-Yingan area of the Kaplin uplift (Fig. 16C). A forebulge would explain the absence of Lower Carboniferous strata in the Sishichang-Yingan area (they are present at both Wushi and Xiaohaizi), and would help restrict marine circulation and promote the deposition of evaporite facies.

Observations by Zhou et al. (this volume) north of Kuqa (Fig. 1) support at least limited Late Devonian to Early Carboniferous uplift of the southern margin of the central Tian Shan. They report that the Erbin Shan granite, which intrudes Devonian strata and has a reported U-Pb age of 378 Ma (Hu et al., 1986), is unconformably overlain by Carboniferous conglomerate and carbonate facies bearing Visean marine fossils. This timing of this nonconformity corresponds with the development of the Devonian–Carboniferous angular unconformity in the Tarim basin, strongly suggesting a causal relationship. The presence of Lower Carboniferous marine facies within the Tian Shan does not support continued widespread orogenic uplift, however, as implied by Carroll et al (1995) and Allen et al. (1999).

Paleomagnetic, stratigraphic, and petrologic evidence suggests that this collision was diachronous, occurring later to the west (Gao et al., 1998; Fang et al., 1990; Li, 1990; Windley et al., 1990; Biske, 1995; Carroll et al., 1995; Chen et al., 1999; Zhou et al., this volume). Zhou et al. (this volume) note the presence of blueschist facies metamorphic rocks as young as Early Carboniferous in the western Tian Shan along the suture between the central and southern Tian Shan, and a coeval magmatic arc along the southern edge of the central Tian Shan terrane. They infer that subduction continued in this area through the Early Carboniferous (Fig. 16C), as proposed by Gao et al. (1998) and Chen et al. (1999). This conclusion requires that the foreland basin model proposed for the northern Tarim basin by Carroll et al. (1995) and subsequently adapted by Allen et al. (1999) be modified in recognition that the actual collision between the northwestern Tarim block and the central Tian Shan

terrane may not have occurred until the Middle Carboniferous. The origin of Early Carboniferous Tarim basin tectonic subsidence therefore becomes somewhat enigmatic. One hypothesis is that vertical movements of the northwest Tarim passive margin occurred in response to flexure of the downgoing remnant oceanic lithosphere as the Tarim block approached the trench, but prior to the collision. Further detailed field mapping coupled with mechanical modeling will be required to evaluate this hypothesis. Alternatively, oblique convergence between Tarim and the central Tian Shan during or after collision may have resulted in strike-slip faulting along their boundary (as yet undocumented), thereby altering the original spatial relationships between the Carboniferous arc and the northwestern Tarim basin. It is conceivable that the entire range of depositional environments present at Wushi resulted from changing eustatic sea level, and that the absence of most or all of the Carboniferous at Sishichang reflects coastal onlap. However, global sea level generally appears to have fallen during the Early to Middle Carboniferous (Ross and Ross, 1988), which seems to conflict with the overall deepening trend we observed at Wushi.

The Early Permian magmatic event represented by mafic intrusions and flows in central and northwest Tarim and by rhyolite in the southern Tian Shan–northern Tarim coincides with dramatic shifts in sediment dispersal patterns and provenance, and a marked increase in basin subsidence (Fig. 16D). A prominent northwest-southeast aeromagnetic anomaly associated with the Bachu uplift is most likely related to Lower Permian intrusive rocks, rather than a middle Paleozoic suture as proposed by Yin and Nie (1994). Lower Permian dikes record continental extension in a direction approximately normal to the Paleozoic northwest passive margin of Tarim; their orientation suggests that northwest-southeast compression continued during emplacement. Continued uplift in the south Tian Shan is also indicated by southeast-directed sediment transport, and by felsic-volcanic lithic grains and granitic rock fragments in Lower Permian sandstone. The sandstone grains appear to have been derived principally from Lower Permian rhyolite that crops out in the south Tian Shan; there were additional contributions from uplifted and eroded granitic plutons.

Carroll et al. (1995) suggested that Early Permian extension may have resulted from collision along an irregular continental margin, similar to the scenario proposed by Şengör et al. (1978) for the upper Rhine graben. Alternatively, extension may have resulted from mantle flow out of the collisional zone in the least principal stress direction (cf. Flower et al., 1998). An alternate hypothesis may be that the Permian magmatism is unrelated to regional tectonics, but rather reflects the influence of a mantle plume (M.B. Allen, 1999, personal commun.).

ACKNOWLEDGMENTS

We are grateful to many individuals for their assistance with various aspects of this investigation or for helpful discussions, including M.B. Allen, J. Amory, J. Chu, B. Hacker, M.

Hendrix, L. Lamb, J.G. Liou, J. Liu, X. Liu, M. McWilliams, E.R. Sobel, S.J. Vincent, X. Wang, D. Ying, D. Zhou, and X. Xiao. Financial assistance was provided by the Stanford-China Industrial Affiliates, a group of companies that has included Agip, Amoco, Anadarko, Anschutz, BHP Petroleum, British Petroleum, Canadian Hunter, Chevron, Conoco, Elf-Aquitaine, Enterprise Oil, Exxon, Fletcher Challenge, Japanese National Oil Corporation, Mobil, Occidental, Pecten, Phillips, Statoil, Sun, Texaco, Transworld Energy International, Triton, Union Texas, and Unocal. We thank M.B. Allen, S.J. Vincent, R. Ressetar, and M.S. Hendrix for careful and helpful reviews.

REFERENCES CITED

Allen, M.B., Windley, B.F., Zhang, C., Zhao, Z.Y., and Wang, G.R., 1991, Basin evolution within and adjacent to the Tien Shan range, NW China: Geological Society of London Journal, v. 148, p. 369–378.

Allen, M.B., Windley, B.F., and Zhang, C., 1993, Paleozoic collisional tectonics and magmatism of the Chinese Tien Shan, central Asia: Tectonophysics, v. 220, p. 89–115.

Allen, M.B., Şengör, A.M.Ç., and Nalal'in, B.A., 1995, Junggar, Turfan, and Alakol basins as Late Permian to ?Early Triassic extensional structures in a sinistral shear zone in the Altaid orogenic collage, central Asia: Geological Society of London Journal, v. 152, p. 327–338.

Allen, M.B., Vincent, S.J., and Wheeler, P.J., 1999, Late Cenozoic tectonics of the Kepingtage thrust zone: Interactions of the Tien Shan and Tarim basin, northwest China: Tectonics, v. 18, p. 639–654.

Bai, Y., Chen, G., Sun, Q., Li, Y.A., Dong, Y., and Sun, D., 1987, Late Paleozoic polar wander path for the Tarim platform and its tectonic significance: Tectonophysics, v. 139, p. 145–153.

Biske, Y.S., 1995, Late Paleozoic collision of the Tarimskiy and Kirghiz-Kazakh palaeocontinents: Geotectonics, v. 29, p. 26–34.

Bond, G.C., and Kominz, M.A., 1984, Construction of tectonic subsidence curves for the early Paleozoic miogeocline, southern Canadian Rocky Mountains: Implications for subsidence mechanisms, age of breakup, and crustal thinning: Geological Society of America Bulletin, v. 95, p. 155–173.

Carroll, A.R., Liang, Y., Graham, S.A., Xiao, X., Hendrix, M.S., Chu, J., and McKnight, C.L., 1990, Junggar basin, northwest China: Trapped late Paleozoic ocean: Tectonophysics, v. 181, p. 1–14.

Carroll, A.R., Graham, S.A., Hendrix, M.S., Ying, D., and Zhou, D., 1995, Late Paleozoic tectonic amalgamation of northwestern China: Sedimentary record of the northern Tarim, northwestern Turpan, and southern Junggar basins: Geological Society of America Bulletin, v. 107, p. 571–594.

Chang, J., 1988, Correlation of Lower Permian series of terrestrial facies in west part of Tarim platform: Xinjiang Geology, v. 6, p. 2–14 (in Chinese with English abstract).

Chen, C., Lu, H., Jia, D., Cai, D., and Wu, S., 1999, Closing history of the southern Tianshan oceanic basin, western China: An oblique collisional history: Tectonophysics, v. 302, p. 23–40.

Chen, M., Wang, J., Wang, X., and Ma, Y., 1991, Biostratigraphic boundary between Carboniferous and Permian in Kalpin region, Xinjiang, *in* Jia, R., ed., Research of petroleum geology of the northern Tarim basin in China, Volume 1: Wuhan, China University of Geoscience Press, p. 86–93 (in Chinese with English abstract).

Chen, Z., Wu, N., Zhang, D., Hu, J., Huang, H., Shen, G., Wu, G., Tang, H., and Hu, Y., 1985, Geologic map of Xinjiang Uygur Autonomous Region: Beijing, Geologic Publishing House, scale 1:2 000 000.

Dickinson, W.R., 1976, Plate tectonic evaluation of sedimentary basins: American Association of Petroleum Geologists Continuing Education Course Note Series, no. 1, 62 p.

Dickinson, W.R., 1985, Interpreting provenance relations from detrital modes of sandstone, *in* Zuffa, G.G., ed., Provenance of arenites: Hingham, D. Reidel Publishing Company, p. 333–361.

Dong, J., Lu, H., Cai, D., Wu, S., Shi, Y., and Chen, C., 1998, Structural features of northern Tarim basin: Implications for regional tectonics and petroleum traps: American Association of Petroleum Geologists Bulletin, v. 82, p. 147–159.

Fan, J., and Ma, B., 1991, Oil-and-gas geology of Tarim, *in* Biostratigraphy and geologic evolution of Tarim, Volume 4: Beijing, Science Press, 466 p.

Fang, D., Chen, H., Jin, G., Guo, Y., Wang, Z., Tan, X., and Yin, S., 1990, Late Paleozoic and Mesozoic paleomagnetism and tectonic evolution of the Tarim terrane, *in* Wiley, T.J., et al., eds., Terrane analysis of China and the Pacific Rim: Circum-Pacific Council for Energy and Mineral Resources Earth Science Series, v. 13, p. 251–255.

Flower, M.F.J., Mocanu, V., and Russo, R.M., 1998, Eurasian mantle provinciality: a record of collision-induced mantle flow?: Eos (Transactions, American Geophysical Union), v. 79, p. F848.

Gao, C., and Ye, D., 1997, Petroleum geology of the Tarim basin, NW China: Recent advances: Journal of Petroleum Geology, v. 20, p. 239–244.

Gao, J., Li, M., Xiao, X., Tang, Y., and He, G., 1998, Paleozoic tectonic evolution of the Tianshan orogen, northwest China: Tectonophysics, v. 287, p. 213–231.

Gao, Z., Wang, W., Peng., C., and Xiao, B., 1985, The Sinian in Aksu-Wushi area, Xinjiang, China: Urumqi, Xinjiang People's Publishing House, 184 p.

Graham, S.A., Brassell, S., Carroll, A.R., Xiao, X., Demaison, G., McKnight, C.L., Liang, Y., Chu, J., and Hendrix, M.S., 1990, Characteristics of selected petroleum-source rocks, Xinjiang Uygur Autonomous Region, northwest China: American Association of Petroleum Geologists Bulletin, v. 74, p. 493–512.

Graham, S.A., Hendrix, M.S., Wang, L.B., and Carroll, A.R., 1993, Collisional successor basins of western China: Impact of tectonic inheritance on sand composition: Geological Society of America Bulletin, v. 105, p. 323–344.

Groves, J.R., and Brenckle, P.L., 1997, Graphic correlation in frontier petroleum provinces: Application to upper Paleozoic sections in the Tarim basin, western China: American Association of Petroleum Geologists Bulletin, v. 81, p. 1259–1266.

Harland, W.B., Armstrong, R.L., Cox, A.V., Craig, L.E., Smith, A.G., and Smith, D.G., 1990, A geologic time scale 1989: Cambridge, Cambridge University Press, 263 p.

Hendrix, M.S., Graham, S.A., Carroll, A.R., Sobel, E.R., McKnight, C.L., Schulein, B.J., and Wang, Z., 1992, Sedimentary record and climatic implications of recurrent deformation in the Tian Shan: Evidence from Mesozoic strata of the north Tarim, south Junggar, and Turpan basins, northwest China: Geological Society of America Bulletin, v. 104, p. 53–79.

Hopson, C., Wen, J., Tilton, G., Tang, Y., Zhu, B., and Zhao, M., 1989, Paleozoic plutonism in east Junggar, Bogdashan, and eastern Tianshan, northwest China: Eos (Transactions, American Geophysical Union), v. 70, p. 1403–1404.

Hu, A., Zhang, Z., Liu, J., Peng, J., Zhang, J., Zhao, D., Yang, S., and Zhou, W., 1986, U-Pb age and evolution of Precambrian metamorphic rocks of the middle Tianshan uplift zone, eastern Tianshan, China: Geochimica et Cosmochimica Acta, v. 1986, p. 23–35.

Hu, B., Wang, J.B., Gao, Z., and Fang, X.D., 1965, Problems of the Paleozoics of Tarim platform: International Geology Review, v. 11, p. 650–665.

Hsü, K.J., Yao, Y., Hao, J., Hsü, P., Li, J., and Wang, Q., 1994, Origin of the Chinese Tianshan by arc-arc collisions: Eclogae Geologicae Helvetiae, v. 87, p. 265–292.

Hubbard, R.J., 1988, Age and significance of sequence boundaries on Jurassic and Early Cretaceous rifted continental margins: American Association of Petroleum Geologists Bulletin, v. 72, p. 49–72.

Hubbard, R.J., Pape, J., and Roberts, D.G., 1985, Depositional sequence mapping to illustrate the evolution of a passive continental margin, *in* Berg, O.R., and Woolverton, D., eds., Seismic stratigraphy II: An integrated approach to hydrocarbon exploration: American Association of Petroleum Geologists Memoir 39, p. 93–115.

Ingersoll, R.V., Bullard, T.F., Ford, R.L., Grimm, J.P., Pickle, J.D., and Sares, S., 1984, The effect of grain size on detrital modes: A test of the Gazzi-Dickinson point-counting method: Journal of Sedimentary Petrology, v. 54, p. 103–116.

Jordan, T.E., 1981, Thrust loads and foreland basin evolution, western United States: American Association of Petroleum Geologists Bulletin, v. 65, p. 2506–2520.

Kwon, S.T., Tilton, G.R., Coleman, R.G., and Feng, Y., 1989, Isotopic studies bearing on the tectonics of the west Junggar region, Xinjiang, China: Tectonics, v. 8, p. 797–727.

Lai, J., and Wang, J., 1988, Atlas of the paleogeography of Xinjiang: Urumqi, Xinjiang People's Publishing House, 92 p. (in Chinese with English abstract).

Li, D., Liang, D., Jia, C., Wang, G., Wu, Q., and He, D., 1996, Hydrocarbon accumulations in the Tarim basin, China: American Association of Petroleum Geologists Bulletin, v. 80, p. 1587–1603.

Li, Y., 1990, An apparent polar wander path from the Tarim Block, China: Tectonophysics, v. 181, p. 31–41.

Li, Y., McWilliams, M., Cox, A., Sharps, R., Li, Y.A., Gao, Z., Zhang, Z., and Zhai, Y., 1988a, Late Permian paleomagnetic poles from dikes of the Tarim craton, China: Geology, v. 16, p. 275–278.

Li, Y., Li, Y.A., McWilliams, M., Sharps, R., Zhai, Y., Zhang, Z., Li, Q., Cox, A., and Gao, Z., 1988b, Paleomagnetic results of the Devonian for the Tarim craton and its tectonic implications: Journal of Changchung University Earth Sciences, v. 18, p. 447–484.

Liou, J.G., Graham, S.A., Maruyama, S., Wang, X., Xiao, X., Carroll, A.R., Chu, J., Feng, Y., Hendrix, M.S., Liang, Y.H., McKnight, C.L., Tang, Y., Wang, Z.X., Zhao, M., and Zhu, B., 1989, Proterozoic blueschist belt in western China: Best documented Precambrian blueschists in the world: Geology, v. 17, p. 1127–1131.

Liou, J.G., Graham, S.A., Maruyama, S., and Zhang, R.Y., 1996, Characteristics and tectonic significance of the late Proterozoic Aksu blueschists and diabasic dikes, northwest Xinjiang, China: International Geology Review, v. 38, p. 228–244.

Liu, C., and Xiong, J., 1991, New understanding of Carboniferous and Permian stratigraphy in the northern region of Tarim basin, in Jia, R., ed., Research of petroleum geology of the northern Tarim basin in China, Volume 1: Wuhan, China University of Geoscience Press, p. 64–73 (in Chinese with English abstract).

Liu, J., and Li, W., 1991, Petrologic characteristics and ages of basalt in north Tarim, in Jia, R., ed., Research of petroleum geology of northern Tarim basin in China, Volume 1: Wuhan, China University of Geoscience Press, p. 194–201 (in Chinese with English abstract).

McKnight, C.L., 1993, Structural styles and tectonic significance of Tian Shan foothill fold and thrust belts, northwest China [Ph.D. thesis]: Stanford, California, Stanford University, 207 p.

McKnight, C.L., Carroll, A.R., Chu, J., Hendrix, M.S., Graham, S.A., and Lyon, R.J.P., 1989, Stratigraphy and structure of the Kalpin uplift, Tarim basin, northwest China: Proceedings of the Seventh Thematic Conference on Remote Sensing for Exploration Geology: Ann Arbor, Michigan, Environmental Research Institute of Michigan, p. 1085–1096.

Nakajima, T., Maruyama, S., Uchiumi, S., Liou, J.G., Wang, X., Xiao, X., and Graham, S.A., 1990, Evidence for late Proterozoic subduction from 700-Myr-old blueschists in China: Nature, v. 346, p. 263–265.

Nishidai, T., and Berry, J.L., 1990, Structure and hydrocarbon potential of the Tarim basin (NW China) from satellite imagery: Journal of Petroleum Technology, v. 13, p. 35–58.

Ross, C.A., and Ross, J.R., 1988, Late Paleozoic transgressive-regressive deposition, in Wilgus, C.K., et al., eds., Sea-level changes: An integrated approach: Society of Economic Paleontologists and Mineralogists Special Publication 42, p. 227–247.

Ross, C.A., Baud, A., and Menning, M., 1994, A time scale for Project Pangea, in Embry, A.F., et al., eds., Pangea: Global environments and resources: Canadian Society of Petroleum Geologists Memoir 17, p. 81–83.

Schlater, J.G., and Christie, P.A.F., 1980, Continental stretching: An explanation of the post-mid-Cretaceous subsidence of the central North Sea basin: Journal of Geophysical Research, v. 85, p. 3711–3739.

Şengör, A.M.Ç., Burke, K., and Dewey, J.F., 1978, Rifts at high angles to orogenic belts: Tests for their origin and the upper Rhine graben as an example: American Journal of Science, v. 278, p. 24–40.

Şengör, A.M.Ç., Natal'in, B.A., and Burtman, V.S., 1993, Evolution of the Altaid tectonic collage and Paleozoic crustal growth in Eurasia: Nature, v. 364, p. 299–307.

Şengör, A.M.C., Graham, S.A., and Biddle, K.T., 1996, Is the Tarim basin underlain by a Neoproterozoic oceanic plateau?: Geological Society of America Abstracts with Programs, v. 28, no. 7, p. A-67.

Sharps, R., McWilliams, M., Li, Y., Cox, A., Zhang, Z., Zhai, Y., Gao, Z., Li, Y., and Li, Q., 1989, Lower Permian paleomagnetism of the Tarim block, northwestern China: Earth and Planetary Science Letters, v. 92, p. 275–291.

Shi, Y., Lu, H., Jia, D., Cai, D., Wu, S., Chen, C., Howell, D.G., and Valin, Z.C., 1995, Paleozoic plate-tectonic evolution of the Tarim and western Tianshan regions, western China: International Geology Review, v. 36, p. 1058–1066.

Sobel, E.R., and Arnaud, N., 1999, A possible middle Paleozoic suture in the Altyn Tagh, NW China: Tectonics, v. 18, p. 64–74.

Steckler, M.S., and Watts, A.B., 1978, Subsidence of the Atlantic-type continental margin off New York: Earth and Planetary Science Letters, v. 41, p. 1–13.

Tapponnier, P., and Molnar, P., 1979, Active faulting and Cenozoic tectonics of the Tien Shan, Mongolia, and Baykal regions: Journal of Geophysical Research, v. 84, p. 3425–3459.

Tian, Z., Chai, G., and Kang. Y., 1989, Tectonic evolution of the Tarim basin, in Hsü, K.J., and Zhu, X., eds., Chinese sedimentary basins: Amsterdam, Elsevier, p. 33–43.

Van Hinte, J.E., 1978, Geohistory—Application of micropaleontology in exploration geology: American Association of Petroleum Geologists Bulletin, v. 62, p. 201–222.

Wang, G., Liu, C., and Xiong, J., 1991, Permo-Carboniferous sedimentary facies in northeastern Tarim basin, in Jia, R., ed., Research of petroleum geology of northern Tarim basin in China, Volume 1: Wuhan, China University of Geoscience Press, p. 160–168 (in Chinese with English abstract).

Wang, Q., Nishidai, T., and Coward, M.P., 1992, The Tarim Basin, NW China; formation and aspects of petroleum geology: Journal of Petroleum Geology, v. 15, p. 5–34.

Wang, S., 1991, Sporopollen assemblage of Lower Permian Kaipaizileike formation, north Tarim, in Jia, R., ed., Research of petroleum geology of northern Tarim basin in China, Volume 1: Wuhan, China University of Geoscience Press, p. 94–99 (in Chinese with English abstract).

Wang, T., and Liu, J., 1991, A preliminary investigation on formative phase and rifting of Tarim basin, in Jia, R., ed., Research of petroleum geology of northern Tarim basin in China, Volume 2: Wuhan, China University of Geoscience Press, p. 115–124 (in Chinese with English abstract).

Wang, Z., and Rui, L., 1987, Conodont sequence across the Carboniferous-Permian boundary, in Wang, C., ed., Carboniferous boundaries in China: Beijing, Science Press, p. 151–159.

Wang, Z., Wu, J., Lu, X., Liu, C., and Zhang, J., 1990, Polycyclic tectonic evolution and metallogeny of the Tian Shan mountains: Beijing, Science Press, 217 p.

Windley, B.F., Allen, M.B., Zhang, C., Zhao, Z.Y., and Wang, G.R., 1990, Paleozoic accretion and Cenozoic redeformation of the Chinese Tien Shan range, central Asia: Geology, v. 18, p. 128–131.

Xinjiang Stratigraphic Table Compiling Group, 1981, Xinjiang stratigraphic tables: Beijing, Geological Publishing House, 496 p.

Xiong, J., 1991, Permo-Carboniferous conodont sequence in northeastern Tarim basin, *in* Jia, R., ed., Research of petroleum geology of northern Tarim basin in China, Volume 1: Wuhan, China University of Geoscience Press, p. 74–85 (in Chinese with English abstract).

Yin, A., and Nie, S., 1994, Palinspastic reconstruction of western and central China from the middle Paleozoic to the early Mesozoic: Eos (Transactions, American Geophysical Union) v. 75, p. 629.

Yin, A., Nie, S., Craig, P., and Harrison, T.M., 1998, Late Cenozoic tectonic evolution of the southern Chinese Tian Shan: Tectonics, v. 17, p. 1–27.

Zhang, Z., Wu, S., Gao, Z., and Ziao, S.B., 1983, Research on sedimentary model from Late Carboniferous to Early Permian Epoch in Kalpin region, Xinjiang: Xinjiang Geology, v. 1, p. 9–20 (in Chinese with English abstract).

Zhang, Z., Liou, J.G., and Coleman, R.G., 1984, An outline of the plate tectonics of China: Geological Society of America Bulletin, v. 95, p. 295–312.

Zhao, X., Core, R.S., Gilder, S.A., and Frost, G.M., 1996, Paleomagnetic constraints on the palaeogeography of China: Implications for Gondwanaland: Australian Journal of Earth Sciences, v. 43, p. 643–672.

Zhou, D., Zhou, J., Zhao, M., and Qing, T., 1991, Sedimentation and hydrocarbon potential of Cambrian and Ordovician continental margins in northeastern Tarim basin, *in* Jia, R., ed., Research of petroleum geology of northern Tarim basin in China, Volume 1: Wuhan, China University of Geoscience Press, p. 126–137 (in Chinese with English abstract).

MANUSCRIPT ACCEPTED BY THE SOCIETY JUNE 5, 2000

Geological Society of America
Memoir 194
2001

Uplift, exhumation, and deformation in the Chinese Tian Shan

Trevor A. Dumitru, Da Zhou*, Edmund Z. Chang, and Stephan A. Graham
Department of Geological and Environmental Sciences, Stanford University, Stanford, California 94305, USA
Marc S. Hendrix
Department of Geology, University of Montana, Missoula, Montana 59812, USA
Edward R. Sobel
Institut für Geowissenschaften, Universität Potsdam, D-14415 Potsdam, Germany
Alan R. Carroll
Department of Geology and Geophysics, University of Wisconsin, Madison, Wisconsin 53706, USA

ABSTRACT

The terranes composing the basement of the Tian Shan were originally sutured together during two collisions in Late Devonian–Early Carboniferous and Late Carboniferous–Early Permian time. Since then, the range has repeatedly been uplifted and structurally reactivated, apparently as a result of the collision of island arcs and continental blocks with the southern margin of Asia far to the south of the range. Evidence for these deformational episodes is recorded in the sedimentary histories of the Junggar and Tarim foreland basins to the north and south of the range and by the cooling and exhumation histories of rocks in the interior of the range. Reconnaissance apatite fission-track cooling ages from the Chinese part of the range cluster in three general time periods, latest Paleozoic, late Mesozoic, and late Cenozoic. Latest Paleozoic cooling is recorded at Aksu (east of Kalpin) on the southern flank of the range, at two areas in the central Tian Shan block along the Dushanzi-Kuqa Highway, and by detrital apatites at Kuqa that retain fission-track ages of their sediment source areas. Available $^{40}Ar/^{39}Ar$ cooling ages from the range also cluster within this time interval, with very few younger ages. These cooling ages may record exhumation and deformation caused by the second basement suturing collision between the Tarim–central Tian Shan composite block and the north Tian Shan.

Apatite data from three areas record late Mesozoic cooling, at Kuqa on the southern flank of the range and at two areas in the central Tian Shan block. Sedimentary sections in the Junggar and Tarim foreland basins contain major unconformities, thick intervals of alluvial conglomerate, and increased subsidence rates between about 140 and 100 Ma. These data may reflect deformation and uplift induced by collision of the Lhasa block with the southern margin of Asia in latest Jurassic–Early Cretaceous time. Large Jurassic intermontane basins are preserved within the interior of the Tian Shan and in conjunction with the fission-track data suggest that the late Mesozoic Tian Shan was subdivided into a complex of generally east-west–trending, structurally controlled subranges and basins.

Apatite data from five areas record major late Cenozoic cooling, at sites in the basin-vergent thrust belts on the northern and southern margins of the range, and along the north Tian Shan fault system in the interior of the range. The thrust belts

*Now at ExxonMobile Exploration Company, P.O. Box 4778, Houston, Texas 77060, USA

Dumitru, T.A., et al., 2001, Uplift, exhumation, and deformation in the Chinese Tian Shan, *in* Hendrix, M.S., and Davis, G.A., eds., Paleozoic and Mesozoic tectonic evolution of central Asia: From continental assembly to intracontinental deformation: Boulder, Colorado, Geological Society of America Memoir 194, p. 71–99.

and fault system have been sites of active shortening and exhumation since at least ca. 25 Ma, apparently induced by the collision of the Indian subcontinent with the southern margin of Asia. On the basis of regional relations, the north Tian Shan fault system is likely an important active right-lateral transpressional structure that has reactivated the north Tian Shan–central Tian Shan suture zone.

In general, most of the Chinese Tian Shan appears to have been exhumed only limited amounts through Mesozoic and Cenozoic time. Within our sampling areas, only limited areas along the north Tian Shan fault zone and in parts of the range-margin thrust belts were exhumed more than ~3 km during the Cenozoic India-Asia collision. Modern intermontane basins are present within the Tian Shan and help divide it into a number of subranges, much like the late Mesozoic Tian Shan. This modern physiography likely reflects in part reactivation of pre-Cenozoic structural trends.

INTRODUCTION

The Tian Shan (Heavenly Mountains) extends east-west for ~2500 km through western China, Kazakstan, and Kirghizstan and reaches elevations as high as 7400 m (Figs. 1 and 2). Much

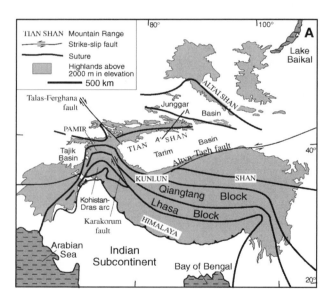

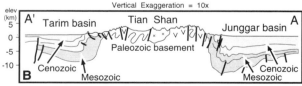

Figure 1. A: Schematic map of central Asia showing location of Tian Shan within context of basins and highlands formed by Cenozoic collision between India and southern margin of Asia. Also shown are Qiangtang block, Lhasa block, and Kohistan-Dras arc, which collided with southern margin of Asia in Mesozoic time. Modified from Watson et al. (1987). B: Cross section showing deformed Paleozoic and older basement in core of Tian Shan and Mesozoic-Cenozoic foreland basins on flanks of range. Foreland basin strata have been thrust outward from range core during Cenozoic shortening driven by India-Asia collision.

of the modern relief of the range is a result of contraction and uplift driven by the collision of the India subcontinent with the southern margin of Asia, which began in early Tertiary time and continues today (e.g., Patriat and Achache, 1984; Tapponnier et al., 1986; Klootwijk et al., 1992; Avouac et al., 1993; Sobel and Dumitru, 1997). However, the basement terranes composing the Tian Shan were originally sutured together by Late Carboniferous–Early Permian time (e.g., Allen et al., 1992; Carroll et al., 1995; Zhou et al., this volume) and since then have been subjected to repeated structural reactivations. These reactivations may have been caused by earlier collisions of smaller continental fragments and volcanic arcs with the southern margin of Asia, such as the collision of the Qiangtang block in Late Triassic time and the Lhasa block in latest Jurassic–Early Cretaceous time (Fig. 1) (e.g., Hendrix et al., 1992).

Very little is known about the postassembly deformational history of the Tian Shan, particularly in pre-Cenozoic time. In conjunction with stratigraphic and sedimentological studies in the Chinese part of the range, we have collected about 65 rock samples for apatite fission-track analysis. The apatite fission-track method may be used to date the cooling of rock units through the subsurface temperature window of about 125 °C to 60 °C. Assuming a nominal geothermal gradient of about 22 °C/km (discussed further in the following), this is equivalent to exhumation up through a depth window of about 5–2 km beneath the Earth's surface. Such cooling is commonly a result of uplift and exhumation of rock units during large-scale deformation in the upper crust and so may be used to constrain the timing, magnitude, and location of such deformation.

The 65 fission-track samples make an extremely sparse sampling set for a mountain range as large and complex as the Tian Shan, so results of this study are necessarily reconnaissance in nature. Different areas of the range retain fission-track records of a variety of cooling events ranging in age from late Paleozoic to late Cenozoic. The available data show a tendency for these events to cluster in three general time periods, referred to here as the latest Paleozoic, late Mesozoic, and late Cenozoic cooling episodes, which appear to correlate in time with significant plate margin collisional events.

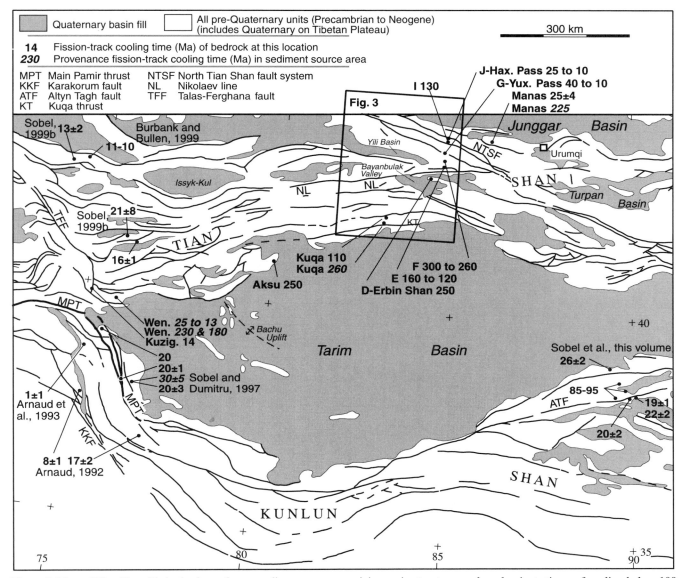

Figure 2. Map of Tian Shan, Tarim basin, and surrounding areas summarizing main structures and predominate times of cooling below 100 °C derived from fission-track data. Map modified from Academy of Geological Sciences of China (1975), Xinjiang Bureau of Geology and Mineral Resources (1985), Chen et al. (1985), Jiao et al. (1988), Pan (1992), and Brunel et al. (1994).

BACKGROUND GEOLOGY
AND TECTONIC HISTORY

The basement terranes currently exposed in the Tian Shan amalgamated during two poorly understood Paleozoic collisional events, a Late Devonian to Early Carboniferous collision between the Tarim and central Tian Shan blocks, followed by a Late Carboniferous–Early Permian collision between the Tarim–central Tian Shan composite block and island arcs currently exposed in the north Tian Shan and Bogda Shan (e.g., Fig. 3) (Wang et al., 1990; Allen et al., 1992; Carroll et al., 1995, this volume; Zhou et al., this volume). Widespread Carboniferous granitic intrusions and Early Permian rhyolitic and basaltic

eruptions represent the last significant igneous activity in the range (Allen et al., 1992; Carroll et al., 1995, this volume).

Landsat imagery, Chinese regional maps, seismicity, and local field studies suggest that latest Cenozoic strain in the Chinese part of the range is concentrated in basin-vergent thrust systems on the northern and southern margins of the range (Fig. 3) (e.g., Tapponnier and Molnar, 1979; Xinjiang Bureau of Geology and Mineral Resources [BGMR], 1985, 1993; Ma, 1986; Avouac et al., 1993; Roecker et al., 1993; Yin et al., 1998). Very few data have been published about the structural history of the interior of the range. Tapponnier and Molnar (1979) used Landsat imagery to identify a network of very large active east-northeast– and west-northwest–striking strike-slip fault systems

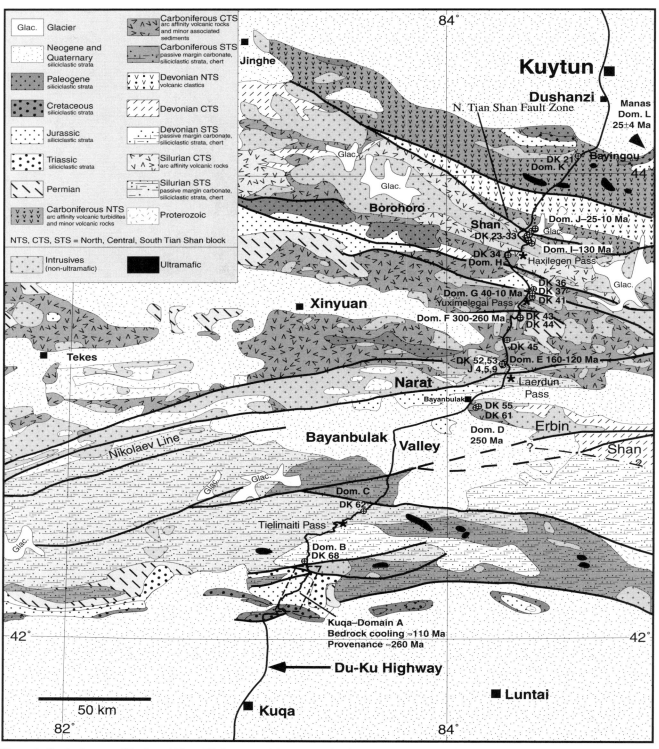

Figure 3. Geologic map of Dushanzi-Kuqa Highway corridor across Tian Shan, and fission-track data. Range is composed of three strongly deformed Paleozoic assemblages: (1) south Tian Shan, characterized by passive margin succession overlying Tarim block basement; (2) central Tian Shan block characterized by abundant arc affinity volcanic rocks; and (3) north Tian Shan, also characterized by abundant arc affinity volcanic rocks. Weakly deformed Jurassic strata in interior of range are remnants of Mesozoic intermontane basins. East-west–trending belts of Mesozoic and Cenozoic siliciclastic strata on northern and southern margins of range mark locations of Cenozoic basin-vergent thrust belts. North Tian Shan fault zone appears to be Cenozoic strike-slip or transpressional structure that has reactivated Paleozoic suture between central Tian Shan and north Tian Shan. Map from Zhou et al. (this volume), based on proprietary mapping by Xinjiang Bureau of Geology and Mineral Resources. Note that many contacts on map are faults; only most important faults are shown. Fission-track sample locations and data domains (A–L) are indicated. Ages indicated for most domains are main times of bedrock cooling below ~100 °C indicated by fission-track modeling.

in central Asia, which they inferred to have accommodated east-ward lateral escape of crustal blocks induced by the collision of India with Asia. One such structure, the Talas-Ferghana fault, cuts across the Tian Shan in Kazakstan and Kirghizstan (Figs. 1 and 2) (e.g., Burtman et al., 1996; Sobel, 1999a). Many recent compilation maps show a major active west-northwest–striking, right-lateral, strike-slip or transpressional structure, sometimes named the north Tian Shan fault system, crossing the core of the range in China (Figs. 2 and 3) (e.g., Xinjiang BGMR, 1985, 1993; Ma, 1986; Hendrix et al., 1992; Yin et al., 1998; Zhou et al., this volume). However, this structure was not identified by Tapponnier and Molnar (1979), and data documenting Cenozoic offset along it do not appear to have been published. Almost all published K-Ar and $^{40}Ar/^{39}Ar$ ages from the Tian Shan are older than about 250 Ma (Yin et al., 1998; Zhou et al., this volume), suggesting that all metamorphism and ductile fabrics in the sampled areas are Paleozoic or older, and that post-Paleozoic exhumation has totaled <~15 km.

Substantial work has been done on the sedimentological and structural histories of the Paleozoic to Holocene Tarim and Junggar basins, which flank the Tian Shan on the south and north (Figs. 1–3) (Hendrix et al., 1992; Carroll et al., 1995, this volume; Li et al., 1996; Sobel and Dumitru, 1997; Yin et al., 1998; Sobel, 1999a). These basins resided in foreland settings during Mesozoic and Cenozoic time, and the basin margins have been uplifted and exhumed by late Cenozoic folding and thrusting and thus are exposed for study (e.g., Figs. 1B and 3). This folding and thrusting generally verges out from the range and into the basins and presumably reflects compressive stresses transmitted to the Tian Shan from the India-Asia collision zone far to the south. In this chapter we refer to these exposed sedimentary rocks as the Tarim and Junggar thrust belts and use the term Tian Shan core for the rocks exposed in the interior of the range. Paleocurrent and provenance data from the two thrust belts indicate that the core of the Tian Shan generally formed a persistent topographic high from late Paleozoic to Holocene time, shedding detritus into the basins to the north and south (Carroll et al., 1990, 1995, this volume; Hendrix et al., 1992; Graham et al., 1993; Hendrix, 2000). Geographically widespread episodes of particularly coarse grained sedimentation and erosional unconformities within the basin sections record renewed deformation in the core of the range at several times, suggesting that Mesozoic accretion of smaller terranes onto the south Asian continental margin uplifted the Tian Shan, presumably by reactivating Carboniferous–Permian structures, analogous to the major deformation induced by the Cenozoic collision of the Indian subcontinent (Hendrix et al., 1992). However, there is as yet essentially no direct evidence of deformation within the core of the range. In particular, slip during the Mesozoic does not appear to have been definitively documented on any faults within the Chinese part of the Tian Shan.

The Cenozoic strata exposed in the foreland thrust belts potentially retain important information on the deformation and uplift of the Tian Shan during the India-Asia collision. However, control on the depositional ages of these clastic strata has generally proven problematic. Sobel and Dumitru (1997) concluded that published data on Cenozoic sediments exposed around the western Tarim basin did not reliably constrain the timing of initiation of Cenozoic deformation in the Tian Shan. Yin et al. (1998) suggested that initial significant thrusting in the northern Tarim basin at Kuqa (Figs. 2 and 3) may have begun at 24–21 Ma, based on a tentative magnetostratigraphic age assignment for a facies change from lacustrine to braided-fluvial deposition.

There are several especially noteworthy aspects of Cenozoic deformation in the Tian Shan. First, the Tian Shan lies more than 1000 km north of the Indus-Tsangpo suture, which marks the collision between India and Asia (Fig. 1). Thus, stresses from the collision have been transmitted remarkably long distances to deform the Tian Shan, while the intervening Tarim basin has remained relatively undeformed (e.g., Li et al., 1996). Second, the India-Asia collision began in Paleocene–Eocene time, but major deformation in the Tian Shan appears to have initiated ca. 25 Ma, on the basis of the very sparse data currently available (Hendrix et al., 1994; Sobel and Dumitru, 1997; Yin et al., 1998). Thus, there was a substantial time lag between collision at the southern margin of Asia and the migration of deformation into the Tian Shan. Numerous studies of the Himalaya and Tibet suggest that a major shift from extrusion-dominated to crustal thickening-dominated tectonics occurred in latest Oligocene–early Miocene time, approximately coincident with the start of unroofing in the Tian Shan (e.g., Harrison et al., 1992). This suggests that unroofing in the Tian Shan may have been a distant effect of that shift in tectonic style. Third, Cenozoic deformation has apparently been partitioned into contrasting domains dominated by contraction and strike-slip faulting. Some of this partitioning may reflect the influence and reactivation of preexisting older structures.

The history of deformation in the Tian Shan during the Cenozoic provides a loose analog for the possible effects of earlier Mesozoic collisions at the southern margin of Asia. Specifically, (1) transmission of stresses into the Tian Shan from far distant collisions appears quite plausible, (2) collisions are protracted events (~55–45 m.y. thus far for the India-Asia collision) and so might cause extended episodes of deformation; and (3) there may be significant time lags between the initiation of collision and the initiation of deformation in the Tian Shan. The suture zones from these earlier collisions are preserved within the Himalaya and Tibet (Fig. 1) and are closer to the Tian Shan than the Indus-Tsangpo suture (although still on the opposite side of the Tarim basin) so these collisions might be expected to have had stronger effects on the Tian Shan. On the other hand, the blocks that collided in the Mesozoic were much smaller and less rigid than the Indian subcontinental indenter, which consists primarily of a very large, strong, cold Precambrian craton (Fig. 1).

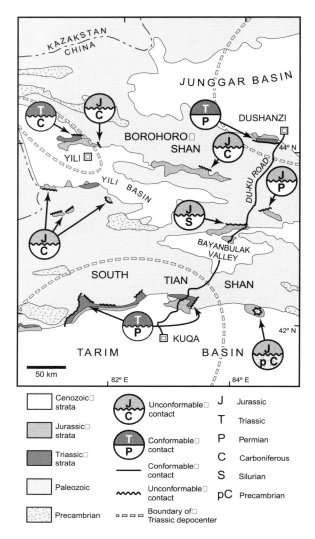

Figure 4. Summary map of depositional contact relations between Jurassic and Triassic strata and underlying Paleozoic and Precambrian rocks in central Tian Shan (Graham et al., 1994), based on Chinese maps (e.g., Chen et al., 1985; Xinjiang Bureau of Geology and Mineral Resources, 1993). Deposition of Jurassic clastic deposits in intermontane basins within several areas on Tian Shan indicates unroofing and exposure of underlying Paleozoic units by Jurassic time. Triassic strata are preserved within core of Yili basin and are overlapped by Jurassic strata. Jurassic strata are unmetamorphosed and generally much less deformed than underlying pre-Mesozoic units. Preservation of Jurassic units is loose indicator that post-Jurassic exhumation has been limited to no more than several kilometers in much of interior of Tian Shan. Yili basin and Bayanbulak Valley are modern intermontane basins within Tian Shan that serve as possible analogs for Mesozoic intermontane basins.

MESOZOIC INTERMONTANE BASINS WITHIN THE TIAN SHAN CORE

Strata in the Tarim and Junggar basins have provided most of the data on the history of the Tian Shan during the Mesozoic. In addition to these strata north and south of the range, large tracts of poorly known Jurassic strata are preserved in intermontane basins within the interior of the range (Figs. 3 and 4) (Xinjiang BGMR, 1985, 1993; Graham et al., 1994; Zhou et al., this volume). These units unconformably overlie strongly deformed Paleozoic rocks (Fig. 4). The most extensive of the intermontane sequences occur in the modern intermontane Turpan basin (Fig. 2) which, on the basis of facies, paleocurrent, and provenance data, was an intermontane basin in Jurassic time, partitioned from the Junggar basin to the north by an ancestral Bogda Shan (Huang et al. 1991; Hendrix et al., 1992; Greene et al., this volume). Within the Tian Shan itself, several smaller areas of Jurassic strata are exposed (e.g., Fig. 3) (Chen et al., 1985; Xinjiang BGMR, 1993; Graham et al., 1994). In one accessible area, strata are tilted ~45° but are otherwise not strongly deformed (Zhou et al., this volume). Preservation of these relatively much less deformed sequences suggests that total post-Jurassic exhumation in these areas has been limited. The modern intermontane basins within the Tian Shan, such as the Yili basin and Bayanbulak Valley (Figs. 2–4), may be localized topographically by Cenozoic reactivation of older structures, and so may provide a rough analog for the older Jurassic basins.

FISSION-TRACK METHODS

General principles

The use of apatite fission-track methods for reconstructing the time-temperature histories of rock samples relies on the facts that new tracks accumulate at an essentially constant rate from the fission decay of trace ^{238}U present within apatite crystals, allowing the calculation of fission-track ages, while at the same time, tracks are partially or totally erased by thermal annealing at elevated subsurface temperatures. Annealing is slight at temperatures $<\sim 60\,°C$, progressively more severe between about 60 and 110–125 °C, and total at >110–125 °C. Total annealing resets the fission-track age to zero, whereas partial annealing reduces the fission-track age and reduces the lengths of individual tracks by amounts directly dependent on the specific temperature (e.g., Naeser, 1979; Gleadow et al., 1986; Green et al., 1989a, 1989b; Dumitru, 2000).

In this study, both primary apatite from plutonic and metamorphic rocks and detrital apatite from sandstones were analyzed (e.g., Dumitru, 2000). The fission-track system records the low-temperature cooling histories of rocks, so ages from plutonic and metamorphic rocks are generally related to final uplift and exhumation of the rocks rather than their earlier igneous crystallization or metamorphism. Data from detrital apatite may yield two different types of information, depending on the maximum paleoburial temperature (T_{max}) undergone by the sandstone sample in question. Where sedimentary horizons were buried to sufficient depth that T_{max} exceeded ~110–135 °C, the fission-track clock is reset to zero age and fission-track data record information on the time-temperature (t-T) cooling path of the sandstone horizon as it passed through the apatite fission-track partial annealing temperature window of ~125–60 °C during exhumation (e.g., Green et al., 1989a, 1989b). Assuming a nominal geothermal gradient of 22 °C/km (discussed herein), this is equivalent to unroofing through depths of ~5–2 km below the Earth's surface. With shallower burial at $T_{max} <$~50 °C, detrital apatites in sandstones instead retain fission-track ages of their sediment source areas. Such ages generally date cooling of rocks in the source area, such as following igneous activity or exhumation (e.g., Cerveny et al., 1988), and different apatite grains within a given sample may yield different ages if they were derived from different source areas. In the current study, shallowly buried samples yielded late Paleozoic, Mesozoic, and Oligocene–Miocene provenance ages and thus constrain cooling in sediment source areas during those times. Where T_{max} was between about 50 and 125 °C, the fission-track clock is partially reset, and interpretation of data patterns permits less precise estimates of provenance ages and cooling ages.

Interpretation of fission-track data involves the analysis of several fission-track parameters in addition to the sample fission-track age. Track length data are useful because lengths of individual tracks are shortened during partial annealing by an amount directly dependent on the temperature. Because new tracks are continuously formed over geologic time, each track formed at a different time and therefore existed to respond and shorten during a different portion of the total thermal history of the sample. A moderate heating event (e.g., 90 °C) occurring 75% of the way through a sample's history, for example, will shorten the 75% of the tracks formed before the event but will not affect the 25% of the tracks formed after the heating. The distribution of track lengths, generally measured as a histogram of the lengths of 100–150 individual tracks, therefore records details of the *t-T* path. In the case just mentioned, 75% of the tracks in the histogram will be short and 25% will be long (ignoring certain biasing factors). This *t-T* information may be recovered by modeling of the time-constant track production and temperature-dependent track annealing processes to determine the spectrum of *t-T* paths consistent with the observed age and track length data (e.g., Green et al., 1989a; Corrigan, 1991; Gallagher, 1995).

Single-grain age data are useful because different apatite grains in a sample may anneal at somewhat different temperatures, a kinetic effect related to the differing chemical composition of the grains. Thus exposure to certain temperatures (~95–125 °C in most cases) may totally anneal some grains and reset their ages to zero, while other grains are only partially annealed. In this case, ages of the youngest, totally reset grains may date exhumation and cooling (e.g., Green et al., 1989a; Sobel and Dumitru, 1997).

Figure 5 illustrates expected idealized fission-track data patterns for *t-T* histories where sediments are eroded from a source terrane, deposited, buried, and heated in a sedimentary basin, then exhumed back to the surface during an unroofing event. In path A, the sample was only shallowly buried in the basin at a maximum burial temperature (T_{max}) of 30 °C. At these low burial temperatures, annealing is minimal and all track lengths remain long and the sample retains the fission-track age of a cooling event in the sediment source area. Figure 5 assumes that all grains were derived from a single source area of uniform apatite age, and the single-grain ages for path A cluster within a $\pm 2\sigma$ swath on a radial plot (see Fig. 5) that records this provenance age. If grains were derived from multiple sources with different ages, there would be a spread in ages reflecting this, but all age clusters would be at least as old as the depositional age of the sediment. As burial depths and T_{max} increase (Fig. 5, paths B and C, 60–85 °C), track lengths shorten and the apparent sample age becomes younger. In path D (100 °C), the track length distribution becomes bimodal. The bimodality is due to combining a component of early formed tracks that have been strongly shortened by burial heating with a component of long tracks that formed after the sample cooled. Single-grain ages are widely spread in a distinctive mixed age wedge. The youngest age component approximates the cooling age because it includes grains that were totally annealed before cooling (generally, the more F-rich grains), whereas the oldest age component (Cl-rich grains) approaches the original provenance age. With even deeper burial (path E, 115 °C), length distributions are dominated by long tracks formed after maximum temperatures because most of the older, shortened tracks have been totally erased. Single-grain ages still show a broad spread, but the young cluster is dominant. Finally, when T_{max} exceeds ~125 °C (path F), all tracks formed before cooling are totally erased. Length distributions are long again, because all preexhumation tracks have been completely erased. If cooling is fairly rapid, all single-grain ages will cluster within a $\pm 2\sigma$ swath and date the time of cooling.

Analytical methods

Laboratory procedures used in this study were essentially identical to those used by Dumitru et al. (1995, their Table 2) and are summarized in the footnote to Table 1. For track length analyses, 100–150 horizontal confined tracks (Laslett et al., 1982) were measured in each sample, provided that many were present. For age determinations, 20–40 good-quality grains per sample were dated, again assuming that sufficient grains were present. Following convention, all statistical uncertainties on ages and mean track lengths are quoted at the $\pm 1\sigma$ level, but $\pm 2\sigma$ uncertainties are taken into account for geologic interpretation. Maximum burial paleotemperatures of sandstone samples were estimated using the methods of Dumitru (1988). For samples that yielded good quality track length data and relatively tightly clustered single-grain age distributions, the 1998 version of the Monte Trax fission-track modeling program of Gallagher (1995) was used to determine the spectrum of time-temperature histories consistent with the observed age and track length data. The footnote of Table 1 lists the specific modeling parameters use.

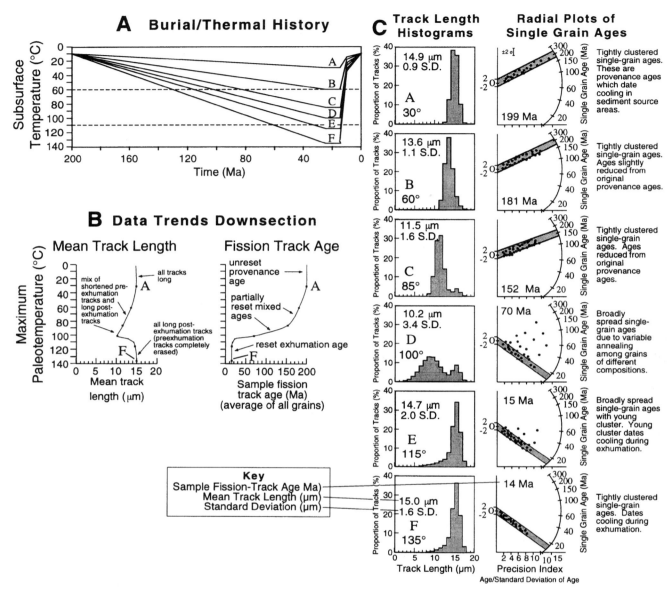

Figure 5. A: Hypothetical burial, exhumation, and thermal histories for sedimentary basin. B: Idealized fission-track age and mean track length trends versus depth in section for thermal histories in A. C: Idealized track length histograms and single-grain age distributions for these thermal histories (Sobel and Dumitru, 1997; see also Dumitru, 2000). Single-grain age distributions are displayed on radial plots, special plots developed to accommodate large and widely varying statistical uncertainties on single-grain ages (Galbraith, 1990; Galbraith and Laslett, 1993; see also Dumitru, 2000). In these plots, individual grain ages are read by projecting line from plot origin (0) through each data point onto radial age scale. Key features of radial plots are that all grain ages have error bars of equal length (one example is shown in plot for path A) and more precise single-grain ages plot farther to right. If all grains have statistically concordant ages, data points cluster within ±2σ swath (e.g., A, B, C, F). If there are significant differences between ages of individual grains, points scatter outside of single swath (e.g., D, E). See text for description of systematic data trends.

TABLE 1. FISSION TRACK SAMPLE LOCALITY, COUNTING, AND AGE DATA

Sample number	Irradiation number	Latitude (°N)	Longitude (°E)	Elevation (m)	No xls	Spontaneous Rho-S	Spontaneous NS	Induced Rho-I	Induced NI	P(χ²) (%)	Dosimeter Rho-D	Dosimeter ND	Age ± 1σ (Ma)
Manas Samples*													
M1 (89-M-52)	SU002-17	42°58'30"	85°49'20"	1160†	22	1.2380	632	2.2440	1145	12.0	1.7920	7838	186.9 ± 10.0
M2 (89-M-44)	SU002-16	43°56'40"	85°51'50"	1160†	20	0.7480	355	2.5910	1229	0.7	1.7920	7838	97.6 ± 9.9
M3 (89-M-30)	SU002-15	43°55'10"	85°52'40"	1160†	39	0.4270	452	5.2180	5591	<0.1	1.7920	7838	32.5 ± 5.3
M4 (89-M-14)	SU002-14	43°51'50"	85°49'40"	1160†	37	0.3910	977	3.2320	8080	<0.1	1.7920	7838	38.1 ± 4.0
M5 (89-M-7)	SU002-12	43°51'20"	85°49'00"	1160†	16	0.1800	92	2.2790	1170	0.3	1.7920	7838	29.7 ± 5.7
Aksu Samples													
91-SI6	SU016-04	40°48.3'†	79°50.2'†	1370†	41	0.4140	586	0.5207	737	58.0	1.4990	4116	226 ± 13
91-SU1	SU016-05	40°48.3'†	79°50.0'†	1370†	21	1.3960	913	1.7100	1119	<0.1	1.4920	4116	240 ± 21
91-SU2	SU016-06	40°50.4'†	79°52.7'†	1370†	16	2.2630	803	2.9650	1052	<0.1	1.4920	4116	211 ± 16
91-SH5	SU016-08	40°50.4'†	79°52.7'†	1370†	16	1.5240	1000	2.3120	1517	0.2	1.4850	4116	185 ± 12
91-SH6	SU016-09	40°50.4'†	79°52.7'†	1370†	20	1.1200	1090	1.3450	1309	<0.1	1.4780	4116	238 ± 17
91-SI4	SU006-09	40°54.9'†	79°54.1'†	1370†	33	0.8721	1476	1.1980	2027	<0.1	1.5830	4534	222 ± 12
91-SH2	SU006-08	40°55.1'†	79°50.8'†	1370†	14	0.6619	379	0.8505	487	1.8	1.5870	4534	219 ± 23
91-SU8	SU006-14	40°55.1'†	79°50.8'†	1370†	12	2.1170	986	3.6980	1722	0.4	1.5620	4534	168 ± 11
Kuqa Samples													
K-1(89-K-115)	SU002-22	42°06'30"	83°06'00"	1980†	20	0.7999	535	3.1340	2096	<0.1	1.7920	7838	103.3 ± 16.0
K-2(89-K-34-38)	SU002-20	42°08'30"	83°06'50"	1980†	10	1.4420	272	2.4920	470	44.0	1.7920	7838	197.1 ± 15.0
K-3(89-K-113)	SU002-21	42°09'30"	83°06'30"	1980†	20	0.5195	202	3.5260	1371	5.0	1.7920	7838	50.0 ± 4.4
K-4(89-K-17B)	SU002-19	42°14'20"	83°14'00"	1980†	13	1.3480	224	5.0290	836	1.0	1.7920	7838	96.7 ± 12.0
K-5(89-K-15)	SU002-18	42°14'30"	83°14'30"	1980†	20	0.4523	368	1.8210	1482	0.2	1.7920	7838	96.7 ± 9.3
K-6(89-K-118)	SU002-23	42°15'30"	83°15'20"	1980†	20	0.7312	499	3.7540	2562	0.1	1.7920	7838	67.1 ± 5.5
K-7(89-K-1)	SU014-35	42°17'20"	83°16'30"	1980†	20	0.5300	574	1.5370	1664	4.4	1.3050	4064	86.3 ± 5.4
Du-Ku Samples													
DK21 (Dom. K)	SU023-04	44°05.0'	84°42.1'	1220	19	0.1705	115	0.2431	164	8.7	1.7090	5073	227.2 ± 28.0
DK23 (Dom. J)	SU023-06	43°46.0'	84°28.1'	2800	35	0.2750	212	2.3650	1823	2.0	1.7300	5073	38.7 ± 3.5
DK24 (Dom. J)	SU023-07	43°46.0'	84°28.1'	2800	48	0.0361	97	0.7999	2147	0.2	1.7510	5073	15.3 ± 2.1
DK25 (Dom. J)	SU023-08	43°45.7'	84°27.0'	3360	38	0.0782	137	0.8233	1443	3.5	1.7510	5073	32.0 ± 3.5
DK26 (Dom. J)	SU023-09	43°44.6'	84°26.2'	3200	36	0.1461	323	1.7440	3855	0.8	1.7720	5073	28.6 ± 2.2
DK27 (Dom. J)	SU023-10	43°44.4'	84°25.5'	3440	25	0.3967	563	11.2800	16013	94.0	1.7720	5073	12.0 ± 0.5
DK28 (Dom. J)	SU023-11	43°44.4'	84°25.5'	3440	31	0.0479	58	1.5490	1875	31.0	1.7930	5073	10.7 ± 1.4
DK29 (Dom. J)	SU023-12	43°43.8'	84°25.8'	3760	29	0.3168	481	4.6860	7116	35.0	1.7930	5073	23.4 ± 1.1
DK30 (Dom. J)	SU023-14	43°43.7'	84°25.3'	3480	24	0.2402	241	3.7640	3776	47.0	1.8250	5073	22.4 ± 1.5
DK32 (Dom. I)	SU023-15	43°43.2'	84°25.9'	2920	29	0.7804	811	2.5680	2669	34.0	1.8250	5073	106.1 ± 4.5
DK33 (Dom. I)	SU023-16	43°42.4'	84°27.0'	3080	36	0.8339	2042	2.6030	6373	0.1	1.8460	5073	113.1 ± 4.4
DK34 (Dom. H)	SU023-17	43°39.4'	84°19.4'	2560	38	0.3542	198	0.7174	401	0.5	1.8460	5073	173.5 ± 21.0
DK36 (Dom. G)	SU023-18	43°30.6'	84°27.4'	3010	15	0.0686	12	1.0860	190	61.0	1.8670	5073	22.7 ± 6.8
DK37 (Dom. G)	SU023-19	43°29.5'	84°27.1'	3280	5	0.0234	2	0.9235	79	8.3	1.8670	5073	9.1 ± 6.5

TABLE 1. FISSION TRACK SAMPLE LOCALITY, COUNTING, AND AGE DATA (continued)

Sample number	Irradiation number	Latitude (°N)	Longitude (°E)	Elevation (m)	No xls	Spontaneous Rho-S	Spontaneous NS	Induced Rho-I	Induced NI	$P(\chi^2)$ (%)	Dosimeter Rho-D	Dosimeter ND	Age ± 1σ (Ma)
DK41 (Dom. G)	SU023-20	43°28.6'	84°27.0'	3360	9	0.9919	162	10.0800	1646	38.0	1.8880	5073	35.8 ± 3.0
DK43 (Dom. F)	SU023-21	43°23.2'	84°23.2'	2400	24	2.1250	2467	3.1070	3606	43.0	1.8880	5073	244.5 ± 7.3
DK44 (Dom. F)	SU023-22	43°23.0'	84°23.2'	2280	11	2.1420	610	3.7250	1061	85.0	1.9090	5073	208.3 ± 11.0
DK45 (Dom. F)	SU023-23	43°17.4'	84°18.9'	1760	25	1.4210	2835	1.8600	3710	84.0	1.9090	5073	275.4 ± 7.9
DK52 (Dom. E)	SU034-13	43°11.3'	84°17.3'	2560	28	0.3795	839	0.9857	2179	75.0	1.7030	4808	125.3 ± 5.4
J4 (Dom. E)	SU034-12	43°11.3'†	84°17.3'†	2560†	11	1.5650	566	3.7500	1356	2.6	1.7030	4808	135.7 ± 9.7
J5 (Dom. E)	SU034-15	43°11.3'†	84°17.3'†	2560†	14	0.3879	399	1.1660	1199	6.4	1.7330	4808	110.3 ± 6.6
J9 (Dom. E)	SU034-16	43°11.1'†	84°17.4'†	2600†	10	1.4500	719	4.5320	2247	44.0	1.7330	4808	106.1 ± 4.8
DK53 (Dom. E)	SU034-11	43°10.9'	84°17.6'	2640	5	0.3858	57	1.7600	260	60.0	1.6830	4808	70.8 ± 10.0
J12 (Dom. E)	SU034-10	#	#	#	5	0.1923	43	0.4204	94	0.6	1.6830	4808	146.9 ± 52.0
DK55 (Dom. D)	SU034-09	43°00.1'	84°09.8'	2710	25	1.5440	2108	2.3800	3248	1.3	1.6630	4808	204.9 ± 8.4
DK61 (Dom. D)	SU023-24	42°59.9'	84°08.9'	2660	21	1.0660	1128	1.9890	2104	34.0	1.9300	5073	196.6 ± 7.8
DK62 (Dom. C)	SU023-25	42°32.7'	83°33.6'	2960	2	0.5675	16	1.3120	37	16.0	1.9300	5073	159.0 ± 48.0
DK68 (Dom. B)	SU023-26	42°19.7'	83°11.7'	2000	15	0.3415	84	2.4400	600	30.0	1.9400	5073	52.2 ± 6.1

Note: Abbreviations are: No xls—number of individual crystals (grains) dated; Rho-S—spontaneous track density ($\times 10^6$ tracks per square centimeter); NS—number of spontaneous tracks counted; Rho-I—induced track density in external detector (muscovite) ($\times 10^6$ tracks per square centimeter); NI—number of induced tracks counted; $P(\chi^2)$ – χ^2 probability (Galbraith, 1981; Green, 1981); Rho-D—induced track density in external detector adjacent to dosimetry glass ($\times 10^6$ tracks per square centimeter); ND, number of tracks counted in determining Rho-D. Age is the sample central fission-track age (Galbraith and Laslett, 1993), calculated using zeta calibration method (Hurford and Green, 1983). Analyst: T.A. Dumitru.

The following is a summary of key laboratory procedures. Apatites were etched for 20 s in 5N nitric acid at room temperature. Grains were dated by external detector method with muscovite detectors. Samples were irradiated in well thermalized positions of Texas A&M University (SU002, SU006) or Oregon State University (SU014, SU016, SU023, SU034) reactor. CN5 dosimetry glasses with muscovite external detectors were used as neutron flux monitors. External detectors were etched in 48% HF. Tracks counted with Zeiss Axioskop microscope with 100× air objective, 1.25× tube factor, 10× eyepieces, transmitted light with supplementary reflected light as needed; external detector prints were located with Kinetek automated scanning stage (Dumitru, 1993). Only grains with c axes subparallel to slide plane were dated. Ages calculated using zeta calibration factor of 389.5. Confined tracks lengths were measured only in apatite grains with c axes subparallel to slide plane; only horizontal tracks measured (within ±~5°–10°), following protocols of Laslett et al. (1982). Lengths were measured with computer digitizing tablet and drawing tube, calibrated against stage micrometer (Dumitru, 1993). Data reduction was done with program by D. Coyle.

Summary of thermal history modeling methods: modeling done with February 11, 1998 version of "code_trax" program, an updated version of the Monte Trax program of Gallagher (1995). Modeling parameters used: (1) used raw track length data (actual lengths of each track) and actual track counts (NS and NI for each grain), (2) used ±10% uncertainty on observed age, ±0.35 micron uncertainty on mean length, and ±0.5 micron uncertainty on standard deviation of track length distribution, (3) used least likelihood evaluation method, (4) used initial track length of 16.3 microns, (5) used Durango apatite annealing model of Laslett et al. (1987), (6) modeled with 100 simulated tracks and 500 to 2000 Monte Carlo runs (genetic algorithm not used), (7) output plots show all runs which pass least likelihood test for both age and track length data.

*Previously published in Data Repository form by Hendrix et al. (1994).

†Approximate location and/or elevation.

#Location unknown. Approximate location shown in cross section of Zhou et al. (this volume).

Geothermal gradients

Fission-track data constrain time-temperature histories and it is desirable to convert these into time-depth histories. If the geothermal gradient is constant, this is straightforward. However, gradients vary over time and in response to rapid burial and exhumation events. In this chapter a paucity of constraints on the thermal history over the past 300 m.y. forces us to simply assume a nominal constant geothermal gradient of 22 °C/km and a surface temperature of 10 °C, based on limited available data on Cenozoic thermal gradients in basins in the region (Zhang, 1989; Fan et al., 1990; see also Sobel and Dumitru, 1997). These are fairly typical values based on worldwide geothermal gradient and heat-flow data sets (e.g., Turcotte and Schubert, 1982).

FISSION-TRACK DATA

Figure 2 shows the locations of the fission-track sampling areas. We first summarize previously published data from the Manas section on the northern flank of the Tian Shan. The interpretation of this area is especially straightforward and serves as a model for the interpretation of results from other areas. We then present results from other areas, proceeding generally from southwest to northeast.

It is useful to keep several points in mind in reading these descriptions. First, the apatite fission-track system is sensitive to cooling only within a temperature window of about 60–125 °C, so cooling events outside this range are not recorded. Second, samples from different burial depths within a single section may cool through the fission-track temperature sensitivity window at different times, so apparent fission-track ages of samples from a single area may be expected to vary substantially. Third, later exhumation and cooling events tend to overprint and erase the fission-track evidence of earlier events. Thus earlier events certainly affected broader areas than revealed by the fission-track data, but fission-track evidence is only retained in areas not subjected to large amounts of younger overprinting deformation and exhumation.

Manas section, southern margin of Junggar basin

The Junggar basin is the major Mesozoic–Cenozoic foreland basin north of the Tian Shan (Figs. 1–3). Its main depocenter lies immediately adjacent to the Tian Shan, where it is ~11 km thick and has been upturned and exposed along the northern flank of the range (Hendrix et al., 1992). Hendrix et al. (1994) reported fission-track data from five Junggar sandstone samples exposed in a north-dipping homoclinal section in the Manas River valley (Fig. 6; Table 1). The data display the overall pattern of younger sample fission-track age with increasing paleodepth and paleotemperature that is expected from increased annealing structurally downsection (Fig. 6B).

In the shallowest sample (M1), single grain ages are tightly clustered ca. 186 Ma and are much older than the sandstone's

depositional age (ca. 80 Ma) (Fig. 6C). Individual grains in sandstones commonly have a range of annealing susceptibilities (correlated with apatite composition), and the lack of any grains significantly younger than the depositional age indicates only slight thermal annealing at burial temperatures <~85–90 °C (Green et al., 1989a). M1 has a tight track-length distribution with a mean track length of 12.4 μm, shorter than the 14.5–15.5 μm lengths in samples exposed only to near-surface temperatures (Gleadow et al., 1986; Green et al., 1989b). A tight distribution with a mean length of ~12.4 μm and few tracks longer than 14 μm suggests exposure to burial temperatures of ~60–90 °C late in the sample's history, after most of the tracks had formed (Green et al., 1989a, 1989b; Corrigan, 1993). These observations suggest that M1 was buried ~2.3–3.6 km (assuming a 22 °C/km gradient) sometime in the middle or late Cenozoic, before being exhumed to the surface.

In sample M2, about one-half of the single-grain ages are distinctly younger than the depositional age, whereas the remainder are about as old as the depositional age. The mean track length is 11.9 μm. These observations suggest partial annealing at maximum burial temperatures of ~80–100 °C. The track-length distribution is tight, and few tracks are longer than 13 μm, so cooling occurred sometime in the middle or late Cenozoic, after most of the tracks had formed.

The three deepest samples (M3, M4, M5) have much younger, middle Tertiary sample apparent fission-track ages. The radial plots of single-grain ages indicate that these samples have been strongly but not totally annealed. The plots show pronounced clusters of young grains ca. ~25 Ma, with no significantly younger grains, but with small proportions (~10%–25%) of much older grains. This single-grain age pattern indicates exposure to temperatures of ~95–120 °C (paleodepths of ~4–5 km) ca. 25 Ma, sufficient to anneal fully all but the most retentive grains (e.g., Green et al., 1989a, 1989b; Dumitru, 2000). The broad track-length distributions with numerous short tracks are further evidence that annealing was not total. Cooling of the samples to temperatures below ~80–95 °C ca. 25 Ma is indicated by the resetting of most of the single-grain ages at that time.

An age of 24.7 ± 3.8 Ma (±1σ) for the young clustering of grains was calculated from the combined data from samples M3–M5 (Sobel and Dumitru, 1997, their Table 3), and is the best estimate of the time significant cooling and exhumation began. The amount of unroofing and cooling at this time is only roughly constrained. The samples probably cooled at least 10–20 °C to set the tight cluster of ca. 25 Ma grains, equivalent to at least 0.5–1 km of unroofing. The similarity in cooling age over the 12 km map distance between M3 and M5 suggests that the unroofing was a strong event and thus significantly greater than 0.5–1 km. The total amount of ca. 25 Ma to Holocene unroofing of M3–M5 was ~4–5 km, and the rugged modern topography and active seismicity of the Tian Shan suggest that a significant part of this total occurred in the latest Cenozoic, well after 25 Ma.

The Monte Trax program (Gallagher, 1995; see also footnote to Table 1) was used to model the fission-track data from

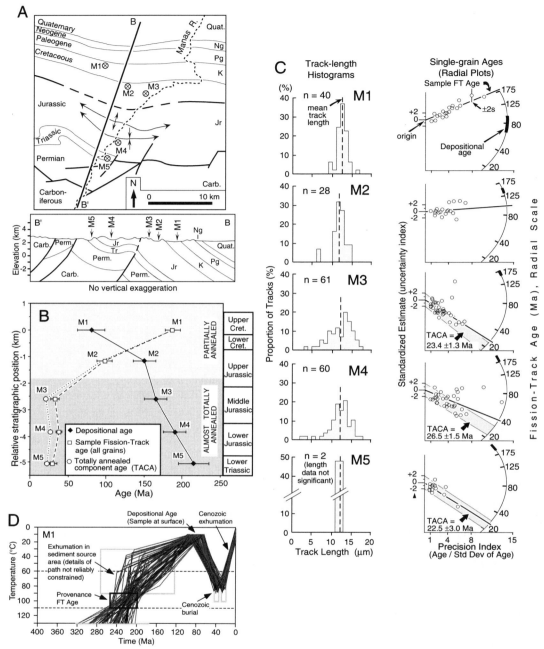

Figure 6. A: Map and cross section of Manas River section on northern flank of Tian Shan, based on unpublished Chinese maps and Hendrix et al. (1992). B: Plot of sample fission-track (FT) ages versus relative stratigraphic depth in section, showing systematic reduction in age with increasing depth. C: Fission-track length histograms and radial plots of single-grain fission-track ages. Note strong clustering of single grain ages ca. 25 Ma in three deepest samples, where shaded ±2σ swath denotes grains that were totally annealed before cooling ca. 25 Ma. These samples also contain small numbers of much older grains that had not been totally reset to zero age at 25 Ma. These data are interpreted to indicate that unroofing in this area initiated ca. 25 Ma, before which time deepest sample was buried at depth of ~5 km and temperature of ~100–125 °C. Figure is modified from Hendrix et al. (1994). Ages of totally annealed components calculated in Sobel and Dumitru (1997; Table 3). D: Modeling of fission-track data from Cretaceous sandstone sample M1. All paths shown are compatible with observed data. Modeling indicates cooling below ~100 °C in sediment source areas at 200–250 Ma. Note that although model plots extend to temperatures of 130 °C, histories hotter than about 110 °C cannot normally be constrained by fission-track data. Therefore specific cooling paths through 130–110 °C temperature interval should not be relied upon.

sample M1. Figure 6D shows the family of thermal histories that are compatible with the observed age and track length data. The depositional age of M1 is Cretaceous, and the modeling indicates cooling below ~100 °C at 200–250 Ma. This suggests that significant cooling occurred in the sediment source area(s) for the detrital apatite in M1 at that time, suggesting exhumation and deformation in source areas at about that time.

Wenguri section, southern Tian Shan thrust belt

The modern boundary between the Tian Shan and the Tarim basin is formed by the southern Tian Shan thrust belt, which extends east-west for about 1200 km (Fig. 2). Yin et al. (1998) divided the belt into four segments, based on differences in structural style. We have fission-track data from four locations along the belt.

Sobel and Dumitru (1997) reported data from sections near Wenguri and Kuzigongsu on the northwest margin of the Tarim basin (Fig. 2). The samples are Jurassic, Cretaceous, and Miocene sandstones of the Tarim basin section that have been uplifted and exhumed by erosion following late Cenozoic folding and thrusting in the area. This shortening is presumably a result of the India-Asia collision and has affected large areas around the northwestern and southwestern margins of the Tarim basin (Sobel and Dumitru, 1997).

At Wenguri, two samples of Cretaceous strata (W5 and W6 in Fig. 7A) show sample ages of 141 ± 13 and 155 ± 11 Ma. Almost all single-grain ages are tightly clustered and are older than the depositional age, indicative of a provenance age. Mean track lengths are 10.9 and 11.5 μm, suggesting burial at maximum temperatures of 80–90 °C; late Cenozoic cooling below ~60 °C is indicated by the low proportion of long tracks (e.g., Green et al., 1989b). These burial temperatures have reduced the fission-track ages. Modeling with the Monte Trax program of Gallagher (1995) corrects for this and yields a provenance age of ca. 160–250 Ma (Fig. 7B), suggesting cooling below ~100 °C in the sediment source area at that time.

In contrast to the Cretaceous samples, four samples of Miocene strata (W1–W4) show widely spread single-grain ages ranging from similar to the Miocene depositional ages to 100–200 Ma. The radial plots show that there are distinct clusters of young grains similar in age or only slightly older than the respective depositional ages (Fig. 7A). The ages of these young clusters decrease upsection. The older populations of grains are less well defined but are generally consistent with the provenance ages of the two Cretaceous samples discussed herein. All of these grains record provenance ages (Sobel and Dumitru, 1997).

These data indicate that unroofing in at least one sediment source area cooled rocks through ~100 °C (and subsequently exposed them at the surface) from at least Oligocene–Miocene time (23.5 ± 3.9, 25.0 ± 3.9 Ma) to middle Miocene time (16.9 ± 2.7, 13.1 ± 2.2 Ma) (Fig. 7A). The source areas may be areas to the north and east in the Tian Shan that are cut by south-

vergent Neogene thrust faults (Wang et al., 1992). The older grains may be derived from other source areas or from higher structural levels within the same source areas.

The modeled provenance age of the two Cretaceous samples is ca. 160–260 Ma and the Miocene samples also contain an old population of apatite generally compatible with this age. Paleocurrent indicators in the Cretaceous section suggest sediment transport from the east (Sobel and Dumitru, 1997). The Bachu uplift (Fig. 2) separated the north Tarim and southwest Tarim basins from the Triassic until the Miocene (e.g., Carroll et al., this volume). Therefore sediment was likely derived from parts of the Tian Shan north and west of Bachu. To the east, in north Tarim, the sedimentary record suggests an episode of Late Triassic–Early Jurassic deformation (Hendrix et al., 1992) that may have set these provenance ages.

Kuzigongsu section, southern Tian Shan thrust belt

Ten samples were analyzed from Jurassic to Paleogene strata near the Kuzigongsu River (Sobel and Dumitru, 1997). The deepest Jurassic sample yielded the most informative data (Fig. 7C). In this sample, almost all grains clustered at an age of 13.6 ± 2.2 Ma, with a few older grains. This indicates cooling of the sample from a T_{max} of 105–130 °C ca. 14 Ma. South-vergent thrusts south of the section may have accommodated the exhumation.

Kalpin (Aksu) uplift, southern Tian Shan thrust belt

The Kalpin uplift is a large belt of southeast-vergent folding and thrusting of Cenozoic age that has exposed large tracts of Paleozoic strata and lesser tracts of older units along the southern margin of the Tian Shan (Figs. 2 and 8) (McKnight, 1993, 1994; Carroll et al., 1995, this volume; Yin et al., 1998; Allen et al., 1999). The section in the general area consists of upper Proterozoic basement unconformably overlain by upper Proterozoic to Upper(?) Permian strata. The Permian strata are in turn overlain along a major angular unconformably by Neogene to Quaternary strata. The upper Proterozoic to Permian section constitutes a passive margin succession deposited on the northern margin of the Tarim block. In most areas the thrust style is thin skinned with a decollement within Cambrian strata. In the southern part of the belt, the stratigraphic thicknesses of the passive margin units in the thrust sheets are nearly constant and the thrusts are very regularly spaced at about 12–15 km, suggesting that folding instability controlled the thrust architecture (Yin et al., 1998).

We collected 11 samples from two eroded homoclines near Aksu that presumably are cored by buried Cenozoic thrust faults (Fig. 8A). One of these samples is from Neogene clastics, the rest are from Sinian (late Precambrian) to Permian units. The structural style near Aksu is somewhat different than the regularly spaced thrust imbricates farther to the west. There is apparently no decollement within the Cambrian strata, and thick

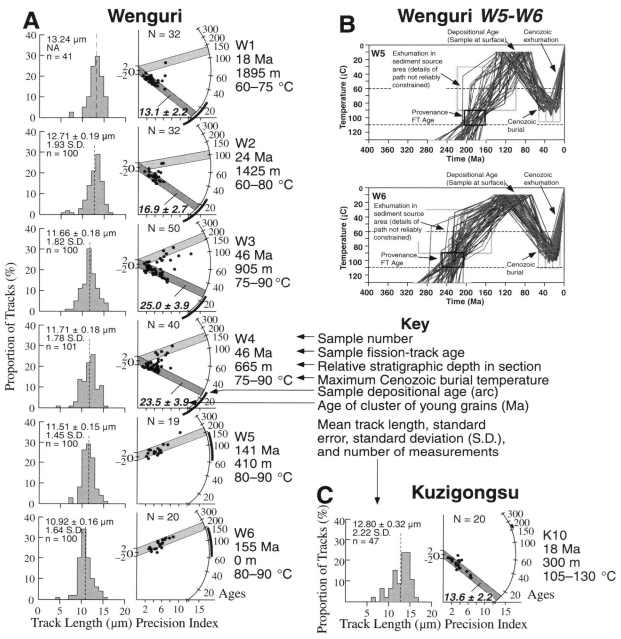

Figure 7. A: Fission-track data from Wenguri section on southern flank of Tian Shan. None of these samples have been buried deeply enough to strongly anneal fission tracks in apatite. Four shallowest samples are Miocene sandstones where most apatite grains yield single grain ages only slightly older than Miocene depositional ages, but there are also small numbers of much older grains. Note that young grains trend younger upsection, tracking depositional age trend. These samples record provenance fission-track information, indicating major unroofing in sediment source areas in Miocene time. Older grains were derived either from higher levels within these source areas or from other source areas with more limited Miocene unroofing. Two deepest samples are Cretaceous sandstones. All grains are older than depositional ages, indicating that these samples underwent only limited burial (estimated maximum burial temperatures of 80–90 °C). B: Modeling of two deepest samples, indicating provenance age of ca. 160–260 Ma. C: Data from stratigraphically deepest sample in Kuzigongsu section on southern flank of Tian Shan. This sample has Jurassic depositional age. Only this sample was sufficiently buried to reset most of grains. Note strong clustering of grains at 14 Ma (with a few older grains), indicating major cooling ca. 14 Ma. See Sobel and Dumitru (1997) for maps, cross sections, and data from additional partially reset samples from shallower in Kuzigongsu section.

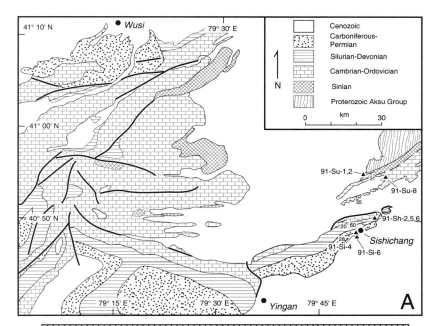

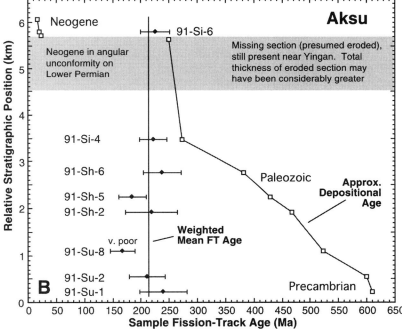

Figure 8. A: Geologic map of Aksu area on southern flank of Tian Shan, showing sampling localities. Map modified from unpublished 1:200 000 mapping by Xinjiang Bureau of Geology and Mineral Resources, Gao et al. (1985), and Carroll et al. (1995, this volume). B: Plot of fission-track ages vs. relative stratigraphic position in section. Note essentially constant ages through entire section. C (following page): Radial plots and track length histograms. Five samples labeled M yielded sufficient length data to permit modeling (see Fig. 9).

sections of Precambrian strata are involved in the folding (e.g., Liou et al., 1989).

The quality of the apatite in these samples proved to be poor and only eight samples yielded usable data. Sample ages are approximately constant through the section (ca. 214 Ma), and we infer that irregular variations in age within the section probably reflect minor errors induced by the poor sample quality rather than true variations in cooling histories (Fig. 8). In order to filter out some of this variation, we calculated a weighted mean age of 214 ± 6 Ma for six of the samples, excluding sample

SU-8, which had very poor quality apatite, and the Neogene sample. Useful track length data could be collected from five of the samples (Fig. 8C).

We used the Monte Trax program to model the histories of these five samples twice, first using the actual measured fission-track ages for each sample and second using the 214 Ma mean age, which is probably a better estimate of the true age. The modeling indicates major cooling of all samples through the 100 °C isotherm ca. 250 ± 10 Ma (Fig. 9), i.e., in latest Paleozoic time. The youngest Paleozoic strata in the area, which are

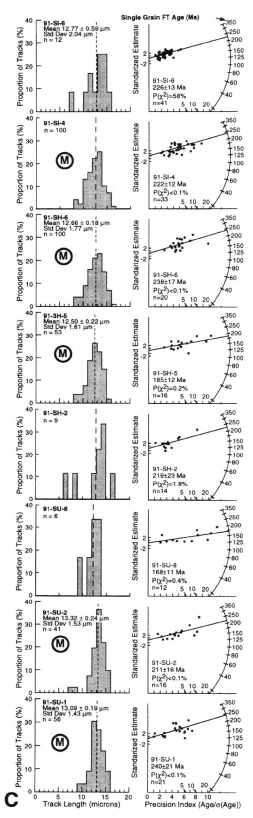

Figure 8. (*continued*)

nonmarine clastics, have generally been assigned an Upper Permian age (256–245 Ma on the time scale of Harlan et al., 1990) based on lithologic correlations, although no definitive data support this, and they could be Lower Permian (e.g., Carroll et al., this volume). In either case, cooling apparently began soon after the termination of deposition in the area.

The stratigraphically deepest sample SU-1 illustrates most aspects of the overall uplift history. Stratigraphic reconstructions suggest that this sample was buried about 7.5 km in middle Permian time (ca. 260 Ma) (Carroll et al., this volume). The fission-track data indicate major cooling beginning ca. 250 Ma (±10 m.y.) to temperatures of ~60–80 °C, corresponding to depths of about 1–3 km assuming a geothermal gradient between 22 and 50 °C/km (see following). Thus total unroofing of about 5 km in early Mesozoic time is indicated. All the other samples show similar unroofing but of lesser magnitude, because the initial preunroofing burial depths were less.

All samples (ignoring the Neogene sample) were at sufficient temperatures (>~110 °C) in the Late Permian to reset their fission-track ages to zero. Therefore all were buried at least about 2–4 km. This suggest that at least 2–4 km of Permian strata formerly buried the stratigraphically shallowest sample. Only about 1 km of such higher section is preserved in our sampling area, but about 2.2 km is preserved near Yingan (e.g., Carroll et al., this volume; Fig. 8A) and an even thicker overburden, subsequently partly removed by erosion in the Late Permian, is permissible.

All samples suggest slow cooling through middle Mesozoic to middle Cenozoic time (Fig. 9). In Cenozoic time, samples cooled from temperatures of ~50 °C to surface temperatures, indicating ~2 km of unroofing. The modeling permits this cooling to have occurred at any time within about the past 50 m.y. and cannot provide better definitions. Data from areas to the east and west suggest that Cenozoic folding and thrusting may have started ca. 21–24 Ma, and this is probably the best inference for the timing of major shortening in the Aksu area (Sobel and Dumitru, 1997; Yin et al., 1998).

The Kalpin fold and thrust belt exhibits a major angular unconformity between Permian and older strata and Neogene and younger foreland basin deposits. The fission-track data indicate that this major erosion initiated in middle Permian time, and about 5 km of erosion is indicated for the Sinian (Precambrian) age rocks in the Kalpin area. The Neogene fission-track sample yields an age essentially identical with the ca. 214 Ma age from the Sinian to Permian section. This suggests that the Neogene sample had a local source, consistent with the detrital mineralogy of the Neogene rocks in the area (Graham et al., 1993).

Mesozoic strata are entirely absent in the Kalpin region and stratigraphic relations suggest that the Kalpin area was relatively high throughout Mesozoic time (e.g., Sobel, 1999a; Carroll et al., this volume). The fission-track data are consistent with significant middle Permian deformation and exhumation followed by subsequent relative quiescence (slow erosion) during Mesozoic time.

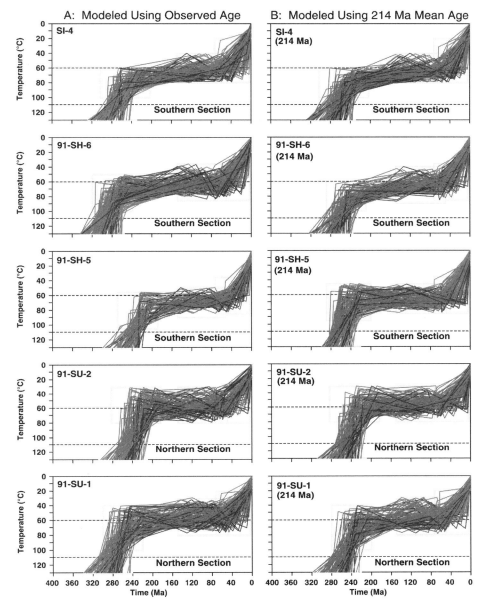

Figure 9. Modeling results from Aksu samples. Modeling was run twice, using individual observed ages (A): and using 214 Ma weighted mean age (B). 214 Ma runs probably yield most accurate results. For all samples, note strong cooling below 100 °C ca. 250 ± 10 Ma, indicating major exhumation at that time. Models then indicate relative quiescence until late Cenozoic time, when further cooling occurred. Specific time of late Cenozoic cooling is not well determined by data, because Cenozoic burial depths were shallower than main sensitivity range of fission-track system. As in Figure 6D, note that parts of histories hotter than ~110 °C are not significant.

A less likely alternate interpretation of the fission-track data from the Aksu area is that they reflect major heating of the section by Early Permian volcanism. As discussed by Carroll et al. (this volume), two series of basaltic lavas crop out in the sampling areas and underlie a large area of the northwest Tarim basin. In outcrop, each series totals ~150–200 m in thickness and consists of multiple thin flows. It appears unlikely that these volcanics reset the fission-track ages, because (1) the total flow thicknesses (and thus heat content) are quite small relative to the ~7.5 km total thickness of the section; (2) the individual flows are thin and so would rapidly lose virtually all of their heat to the atmosphere; and (3) available vitrinite reflectance data indicate minimal heating in the area (Graham

et al., 1990). However, this regional volcanism suggests that geothermal gradients in the area may have been higher in the Permian than the 22 °C/km we have assumed for other areas. We have therefore used assumed gradients between 22 and 50 °C/km for the Permian paleotemperature to depth conversions for Aksu discussed here.

Kuqa River transect, southern Tian Shan thrust belt

Yin et al. (1998) described the Baicheng-Kuqa thrust system as an ~400-km-long segment of the southern Tian Shan thrust system. The segment is characterized by a major south-vergent thrust, the Kuqa thrust. The thrust is exposed in the areas more than about 20 km east of the Kuqa River, where it juxtaposes Carboniferous strata over Neogene–Quaternary sediments (Fig. 10). To the west closer to the Kuqa River, it dies out into a broad anticline as the magnitude of slip decreases. Thrusting and folding in the Kuqa River area is complex; several north-vergent folds and thrusts are exposed on the forelimb of the south-vergent anticline. In general, the Cretaceous and younger strata are tightly folded, whereas folds in Permian to Jurassic strata are more open (Yin et al., 1998).

We collected seven samples in this area before Yin et al. (1998) completed their structural work. In Figure 10 we have projected these samples onto their map and cross section. The results from Kuqa are fairly complex, probably because several different fault slices were sampled (Fig. 10). The four strati-

graphically oldest samples yielded similar data; single-grain ages and sample ages cluster ca. 85 Ma and broad track length distributions are indicative of multistage cooling histories (Fig. 11). Modeling shows that the data are consistent with a two-stage cooling history, the first cooling starting from total annealing temperatures ($\geq$110–125 °C) with cooling through ~100 °C ca. 110 ± 20 Ma, followed by cooling from partial annealing temperatures of 60–90 °C in the late Cenozoic (after ca. 30 Ma) (Fig. 12B). The consistency of data over a 7 km map distance suggests that the ca. 110 Ma cooling was a strong unroofing event. Strong unroofing ca. 110 Ma is consistent with a major unconformity overlain by conglomerate in the Tarim basin section at about that time, as discussed in the following. Generally similar age cooling occurred along parts of the

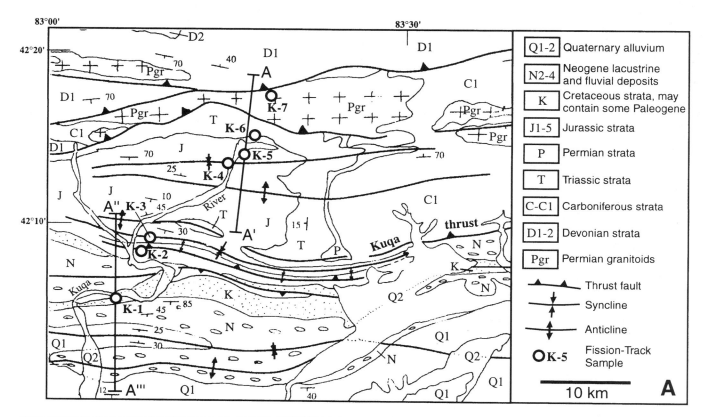

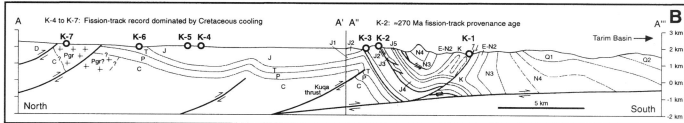

Figure 10. A: Geologic map of southern flank of Tian Shan north of town of Kuqa, with fission-track sample localities. Map reproduced directly from Yin et al. (1998), who compiled it primarily from mapping of Xinjiang Bureau of Geology and Mineral Resources (1966). B: Cross section A–A′′′. Southern part of section is directly from Yin et al. (1998), based on their new mapping in that area. Northern part of section is our tentative interpretation from map in A.

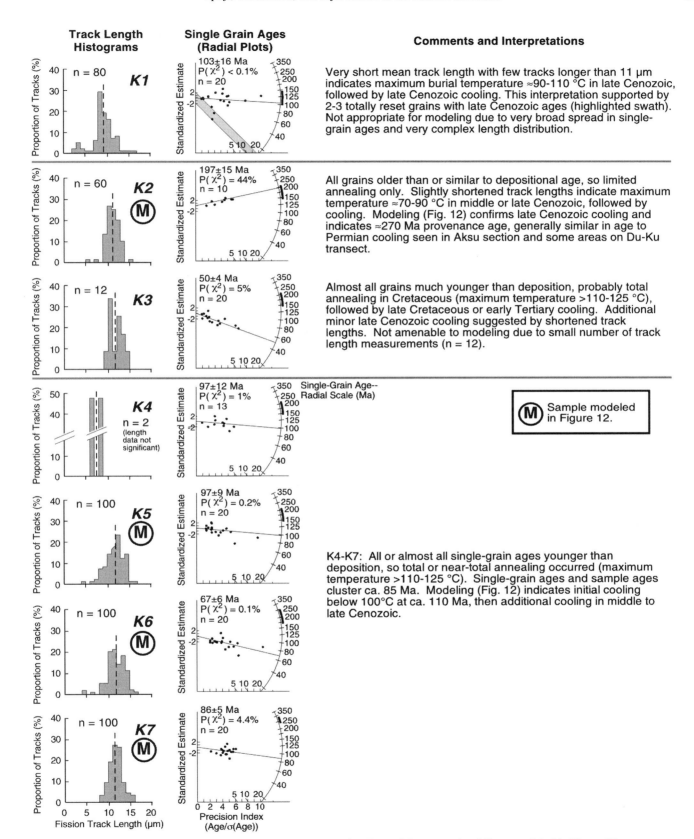

Figure 11. Fission-track data and interpretations from Kuqa samples. Four of these samples (M) are modeled in Figure 12.

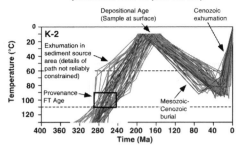

A: Paleozoic provenance age
(Maximum time on plot 400 Ma)

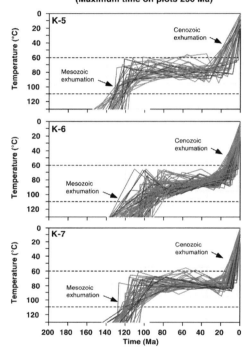

B: Mesozoic exhumation age
(Maximum time on plots 200 Ma)

Figure 12. Modeling results for four Kuqa samples. Northernmost three samples (K-5, K-6, K-7) indicate major cooling below 100 °C ca. 110 ± 20 Ma. Range in ages may be real and could represent somewhat different times of exhumation through 100 °C isotherm at different sample sites, which are ~7 km apart, or may reflect errors in data. Sample K-2, with Jurassic depositional age, records cooling below 100 °C in sediment provenance areas ca. 270 Ma. All samples also record late Cenozoic cooling, but modeling cannot constrain specific time. Parts of histories hotter than ~110 °C are not significant.

Dushanzi-Kuqa corridor, and provide additional evidence for a significant late Mesozoic cooling episode.

The stratigraphically youngest sample (K1) is from an area of complex Cenozoic deformation (Fig. 10). This sample has a very short mean track length of 9.5 μm with few tracks longer than 11 μm. This indicates exposure to burial temperatures of 90–110 °C (burial depth on the order of 4 km) in the late Cenozoic, after almost all of the tracks had formed, followed by un-

roofing to expose the sample at the surface (Fig. 11). Much of the overburden was probably tectonic burial during Cenozoic folding and thrusting, because the Cretaceous to Neogene stratigraphic thicknesses are too thin to bury the sample so deeply (Zhang, 1981).

Samples K2 and K3 are from Jurassic strata exposed on the limb of a syncline, where K3 is situated about 1 km stratigraphically downsection from K2. Modeling shows that K2 records cooling below ~100 °C ca. 260 Ma, much older than the depositional age. This age reflects cooling ages in the source areas for the Jurassic sediments and is similar to the latest Paleozoic cooling episode seen at Aksu and along parts of the Dushanzi-Kuqa corridor. Only sparse track length data could be collected for sample K3. Modeling (not shown) suggests cooling in early Tertiary time, but given the sparse length data this result is not very robust. Both samples record additional cooling in the late Cenozoic, from temperatures of ~80–90 °C, indicating about 3 km of unroofing.

The fission-track data unfortunately contribute little to precisely dating Cenozoic thrusting and exhumation history of the Kuqa area, because the samples have undergone too little Cenozoic exhumation for the fission-track system. Yin et al. (1998) suggested that initial significant thrusting may have begun at 24–21 Ma, based on the time of a major facies change from lacustrine to braided-fluvial sequences in the Tarim basin section. A generally similar timing for initiation of Cenozoic shortening has been reported from other areas around the Tarim and Junggar basins (Hendrix et al., 1994; Sobel and Dumitru, 1997).

The major ca. 110 Ma unroofing event recorded by the four northernmost samples is more interesting. A significant mid-Cretaceous unconformity is present in the north Tarim, south Junggar, and Turpan basins (Hendrix et al., 1992). Age control on the nonmarine Cretaceous strata in western China is problematic, but it appears that strata with ages from ca. 120 Ma to ca. 98 Ma are missing along the unconformity. The fission-track modeling indicates a minimum of about 1.5 km of unroofing at this time. This is apparently a major widespread event involving major erosion within parts of the Tian Shan core and adjacent basins.

Dushanzi-Kuqa corridor

The Dushanzi-Kuqa Highway (Du-Ku) crosses the core of the Tian Shan from Kuqa on the south to Dushanzi on the north (Fig. 3). Zhou et al. (this volume) summarize the geology along the road corridor, with emphasis on the Paleozoic and earlier history. Very little is known about the Mesozoic and Cenozoic deformational history of the area. Use of the road has been highly restricted and the time available for us to sample and make field observations was severely limited.

Figures 3, 13, and 14 and Table 1 summarize fission-track data from this area, and the fission-track samples are also indicated on the cross-sectional road log of Zhou et al. (this volume). The approach used to interpret these data is somewhat different from that used in the other sampling areas. In the other

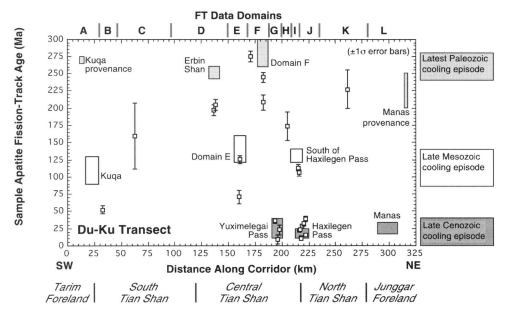

Figure 13. Plot of observed fission-track ages (points) versus location along Du-Ku corridor (Fig. 3). Ages are not shown for Kuqa and Manas (see Figs. 6 and 11). Plot shows domains A to L with consistent fission-track data discussed in text. Boxes indicate main times of cooling below ~100 °C indicated by modeling.

areas, suites of samples could be collected from individual areas with at least some degree of structural control, such as sample position within partially intact stratigraphic successions. An integrated interpretation could then be derived for the suite of samples. Along the Du-Ku transect, structural control is essentially unavailable and generally only a few samples were analyzed from each area. In this case, the track length and age data are then modeled to determine the range of permitted time-temperature histories. Broader geologic issues may then be addressed by interpreting the variations in time-temperature histories across the range.

The data from the transect can be tentatively divided into 12 domains, A–L, that yield reasonably consistent histories. Domains A and L are the Kuqa and Manas sections discussed previously, where the dominant times of cooling recorded by the fission-track system were ca. 110 and 25 Ma, respectively.

Domains B and C comprise two poor quality samples of lower Paleozoic sedimentary strata (Figs. 3, 13, and 14). These sample yielded ages of 52 ± 6 Ma and 159 ± 48 Ma. Useful track length data could not be collected and a detailed interpretation of these samples is not possible. One important conclusion is that late Cenozoic unroofing of these samples has been limited, <~5 km, because the samples have not been hot enough to totally reset fission-track ages to zero during late Cenozoic time.

Domain D comprises two samples in the Erbin Shan of the central Tian Shan block that have concordant ages of 197 ± 8 and 205 ± 8 Ma. Sample DK-55 is from a Devonian? granite (378 U-Pb age, Hu et al., 1986; Zhou et al., this volume) and DK-61 is from an overlying Lower Carboniferous conglomerate. DK-55 yielded a potassium feldspar $^{40}Ar/^{39}Ar$ standard step-heating age spectra that rises slowly from 252 to 277 Ma (Zhou et al., this volume). Modeling of the fission-track data from these samples indicates major cooling below 100 °C ca. 250 Ma, followed by slow cooling until a final episode of cool-

ing some time in the late Cenozoic (Fig. 15A). The early cooling ca. 250 Ma is essentially concordant with the cooling indicated by the $^{40}Ar/^{39}Ar$ data. This indicates strong cooling, probably from temperatures hotter than ~300–350 °C to cooler than ~80 °C at that time. This is very similar in timing to the latest Paleozoic cooling episode recorded at Aksu.

Domain E comprises six samples of metamorphic and granitic rocks (Fig. 3, 13, and 14). The ages of four of these samples, which were collected within about 2 km of each other, range from 106 to 135 Ma. The remaining two samples were of very poor quality and were not interpreted. Sample J5 yielded a biotite $^{40}Ar/^{39}Ar$ plateau age of 286.1 ± 2.6 Ma (Zhou et al., this volume), suggesting it cooled below ~350 °C in late Paleozoic time. Fission-track modeling indicates that these samples cooled below ~100 °C between ca. 160 and 120 Ma (Fig. 15B). Three of the samples exhibit relatively slow late Mesozoic cooling rates and the spread in cooling times may reflect the fact that different samples cooled through the ~100 °C isotherm at different times during a protracted period of relatively slow erosion in the area. This timing of cooling is generally similar to the late Mesozoic cooling recorded at Kuqa.

Domain F comprises three samples of igneous rocks with ages of 208 to 275 Ma. Samples DK-43 and DK-44 were collected about 1 km apart and yield compatible modeled histories (Fig. 15A) if DK-43 is assumed to have resided 0.5–1.0 km higher in the crust in late Paleozoic time. DK-43 cooled below 100 °C ca. 280 Ma, reaching ~65 °C thereafter. DK-44 cooled below 100 °C later, ca. 260 Ma, and then reached ~80 °C. These contrasting histories are compatible with slow cooling in late Paleozoic time. Sample DK-45 cooled below 100 °C ca. 300 Ma. All three samples then experienced further cooling in late Cenozoic time.

Domain G comprises three samples from near Yuximelegai Pass with ages of 9.1, 22.7, and 35.7 Ma. These young ages

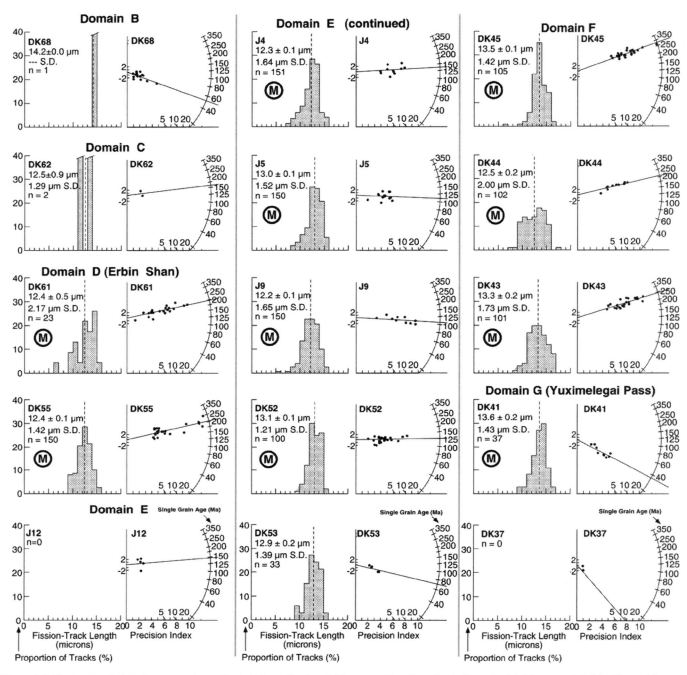

Figure 14. Fission-track data from samples collected along Dushanzi-Kuqa corridor. Samples indicated with M are modeled in Figure 15.

suggest that this area is a site of significantly more Cenozoic unroofing than areas to the south. Only DK-41 yielded sufficient track length data to allow modeling. This sample indicates a complex cooling history with major cooling below ~100 °C at about 40 Ma, and additional cooling in the past 15 m.y. (Fig. 15C). The other two samples that could not be modeled have younger sample ages and corroborate major cooling since ca. 25 Ma.

Domain H comprises a single poor sample with an age of 174 Ma. This sample was not interpreted. It is clear that it has been exhumed no more than about 5 km in Cenozoic time.

Domain I comprises two samples with ages of 106 and 113 Ma. These samples are just south across a possible major Cenozoic fault system from much younger samples in domain J. Modeling indicates cooling below 100 °C ca. 140–120 Ma (Fig. 15B), generally similar to the late Mesozoic cooling seen in Domain E.

Domain J comprises eight samples near Haxilegen Pass with ages of 11–39 Ma (Figs. 3, 13, and 14). Five of these samples yielded sufficient track length data to permit modeling. Sample DK-27 yields the youngest time of cooling, with cool-

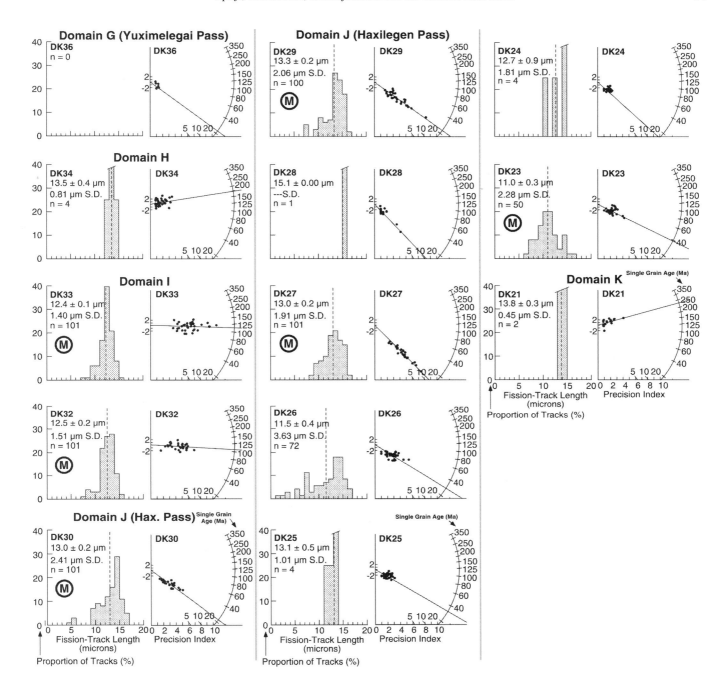

Figure 14. (*continued*)

ing below 100 °C ca. 11 Ma. The other four samples also underwent strong cooling within the past 20 m.y. Several of the samples record earlier cooling sometime between 70 and 40 Ma (Fig. 15C).

Haxilegen Pass is the approximate location of the north Tian Shan fault system, which forms the current boundary between the north Tian Shan and central Tian Shan blocks (Fig. 3) (Zhou et al., this volume). Essential no field data have been published on this fault system. Several recent regional maps (e.g., Ma, 1986; Hendrix et al., 1992; Yin et al., 1998) seem to show it as a major right-lateral strike-slip fault system (with various names), although it was not identified in the pioneering Landsat interpretation of Tapponnier and Molnar (1979). Assuming that such interpretations are generally correct, it is likely that a complex network of Cenozoic faults passes near Haxilegen Pass and that the fission-track samples are from various structural blocks. Thus, the fission-track data confirm that the fault system has been a site of active deformation and exhumation during the Cenozoic. Uplift and exhumation along major strike-slip fault systems is common and may be ascribed to either of two mechanisms. It

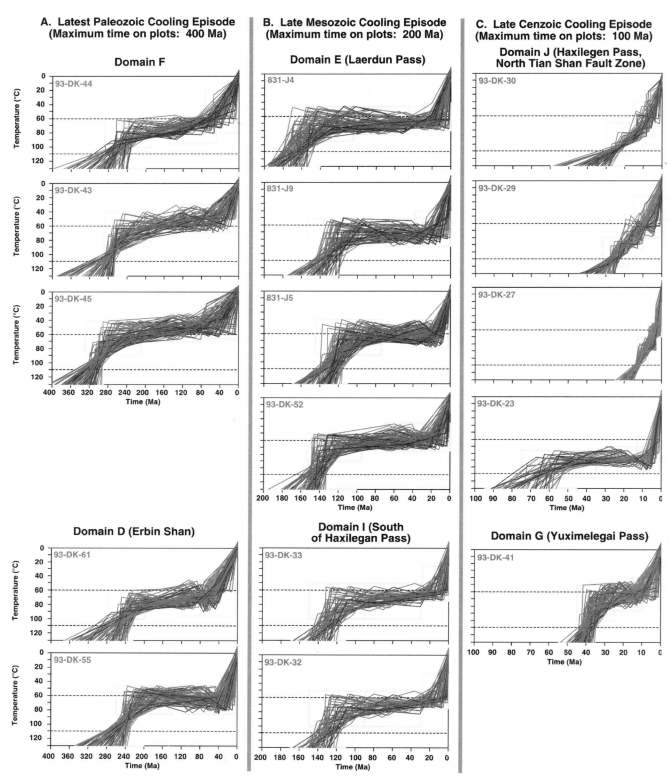

Figure 15. Modeling of samples from Dushanzi-Kuqa corridor. Various areas (domains) that record similar times of cooling are grouped together. See text for discussion. Note that model histories from Aksu (Fig. 9) and Kuqa sample K-2 (Fig. 12) are generally similar to latest Paleozoic histories shown here, and histories from Kuqa samples K-5 to K-7 are generally similar to late Mesozoic histories shown here. Parts of histories hotter than ~110 °C are not significant.

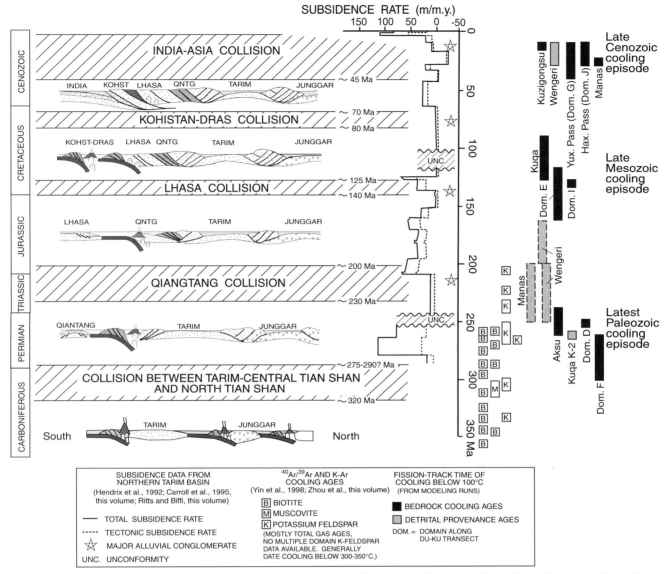

Figure 16. Summary diagram of indicators of timing of deformation in Chinese Tian Shan. Subsidence history in northern Tarim basin shows rapid subsidence during India-Asia collision and earlier periods of accelerated subsidence that appear to correlate with late Paleozoic and Mesozoic collisions at southern margin of Asia. Almost all available K-Ar and ^{40}Ar/^{39}Ar cooling ages are older than 250 Ma. Times of cooling below 100 °C appear to cluster in late Cenozoic, late Mesozoic, and latest Paleozoic time. See Figure 2 for locations. Cross sections modified from Watson et al. (1987). QNTG.—Qiantang.

may occur due to a component of regional transpression across the fault system, or may occur due to space problems along bends and steps in the fault system (e.g., Dumitru, 1991; Bürgmann et al., 1994). Predominately dip-slip secondary faults that root into the main strike-slip system are common in such settings.

Domain K comprises a single poor sample with an age of 227 Ma. This sample was not interpreted.

Domain L comprises the Manas section of the Junggar basin sequence discussed previously, where the fission-track data record major exhumation ca. 25 Ma (Fig. 6). The Manas section is not along the Du-Ku road, but is about 75 km to the east.

DISCUSSION AND CONCLUSIONS

Previous studies have indicated that the Tian Shan has undergone multiple episodes of deformation since initial suturing of the Tarim, central Tian Shan, and north Tian Shan blocks in Paleozoic time. The highly heterogeneous fission-track cooling record is certainly consistent with this. Figures 2 and 16 compile the times of cooling recorded by the fission-track system in the various areas and compare them with what is known about the timing of various collisional events on the southern margin of Asia. The available fission-track data cluster mainly in three time intervals, defining latest Paleozoic, late Mesozoic, and late Cenozoic cooling episodes (Fig. 16).

Latest Paleozoic cooling episode

Latest Paleozoic cooling is indicated at Aksu and in domains D and F along the Du-Ku corridor. Available ^{40}Ar/^{39}Ar cooling ages from the Tian Shan also tend to cluster at this time; there are very few younger ages (Fig. 16). At Aksu, additional timing constraints are provided by the age of the youngest essentially conformable nonmarine clastic strata preserved in the section, which must predate the time of exhumation. These units have generally been assigned an Upper Permian age, but are barren of fossils; a Lower Permian age is also possible (e.g., Carroll et al., this volume). In domain D in the Erbin Shan, potassium feldspar ^{40}Ar/^{39}Ar and apatite fission-track data indicate that sample DK-55 cooled from temperatures >300–350 °C to <80 °C ca. 270–250 Ma.

The collision of the Tarim–central Tian Shan composite block with the north Tian Shan apparently occurred in Late Carboniferous–Early Permian time. However, the data constraining this timing are very limited and it is plausible that the collision was diachronous east-west along the length of the Tian Shan (as was the India-Asia collision), that deformation within the Tian Shan was focused at different places at different times, and that exhumation continued for a period after the actual collision ended. The significant number of middle- to Late Permian ^{40}Ar/^{39}Ar cooling ages in the range (Fig. 16) suggests that strong deformation continued into Late Permian time. This is also consistent with interpretations from the Junggar basin, where Carroll et al. (1995) inferred a foreland basin setting linked to shortening and uplift in the Tian Shan, based on facies and subsidence patterns. Thus we tentatively ascribe the latest Paleozoic cooling episode to deformation and exhumation induced by the collision, but remain cautious because much of the cooling appears to significantly postdate the currently favored estimates for the age range for the collision.

Late Mesozoic cooling episode

Late Mesozoic cooling within the Tian Shan is recorded by fission-track data from Kuqa, as well as domains E and I within the Tian Shan core (Figs. 2, 3, and 16). This time of cooling coincides broadly with a significant unconformity within the stratigraphic section of the northern Tarim basin (Zhang, 1981) as well as a period of coarse conglomeratic sedimentation within the north Tarim and south Junggar foreland depocenters inferred to reflect episodic deformation and physiographic rejuvenation of the range (Hendrix et al., 1992). The general age of cooling in these areas is also consistent with deformation in the southeastern Junggar basin, where subsurface data show that Jurassic strata are cut by reverse faults and overlapped by undeformed Cretaceous strata (Li and Jiang, 1987). Deformation of this age is reasonably widespread across western China and

may reflect structural reactivation of the Tian Shan by the collision of the Lhasa block onto the south Asian continental margin during latest Jurassic–earliest Cretaceous time (e.g., Hendrix et al. 1992). The Lhasa block is located in the southern portion of the Tibetan Plateau, and extends from the Banggong suture south to the Indus suture (Fig. 1). Precollision flysch sequences associated with the Banggong suture are entirely Jurassic in age (Girardeau et al., 1984; see also Matte et al., 1996). These, along with Tithonian radiolaria preserved in deep-sea cherts that crop out in the suture, indicate that closing of the suture was some time during or after the latest Tithonian, perhaps continuing into the earliest Cretaceous (Smith, 1988). Uppermost Jurassic-lowermost Cretaceous nonmarine to shallow-marine sediments, which overlap the obducted Donqiao-Gyanco ophiolite associated with the Banggong suture, provide an upper bound for the age of accretion of the Lhasa block (Girardeau et al., 1984).

Uplift in the Mesozoic in the Tian Shan, however, was clearly heterogeneous across the core of the range, as indicated by fission-track records for late Mesozoic cooling in some areas, the preservation of records of latest Paleozoic cooling in other areas, and the deposition of Jurassic intermontane basin sections in still others areas (Figs. 2, 3, 4, and 16). The late Mesozoic Tian Shan was likely characterized by several uplifted ranges and intervening intermontane basins; a generally east-west trend in structural grain is indicated by the preserved trends of the intermontane basins. Thus the Mesozoic physiography of the Tian Shan somewhat resembled that of the modern Tian Shan, and Cenozoic reactivation of Mesozoic structural trends probably helped shape the modern range.

In addition to the Lhasa block collision, Hendrix et al. (1992) concluded that collision of the Qiangtang block and Kohistan-Dras arc system (Fig. 1) with the southern margin of Asia deformed the Chinese Tian Shan and resulted in pulses of coarse clastic sedimentation in the southern Junggar and northern Tarim basins. The available fission-track data do not provide evidence for these Mesozoic tectonic events, except perhaps the Qiangtang collision in provenance ages at Wenguri and Manas (Fig. 16). This suggests that the deformation they caused within the Tian Shan was too minor to be recorded by the fission-track system, or that any such deformation affected other unsampled areas of the range. In the Altyn Tagh, on the southeast side of the Tarim basin, Sobel et al. (this volume) document Early to Middle Jurassic cooling inferred to reflect the Qiangtang collision (Fig. 2). That region is much closer to the collision zone, so it is not surprising that a record of tectonism may be preserved there but not in the Tian Shan. Cooling that might be linked to the Lhasa block collision is not apparent in the transect of Sobel et al. (this volume), but may be recorded at a second transect farther east in the Altyn Tagh (Delville et al., this volume).

This preceding discussion assumes that collisions at Asia's southern margin were the tectonic drivers for Mesozoic deformation in the Tian Shan. However, there is no direct evidence for this in the cooling and sedimentation records. It is alternately possible that poorly understood interactions at other Asian margins drove Mesozoic deformation in the range, such as the Mesozoic closure of the Mongol-Okhosk seaway (e.g., Halim et al., 1998).

Late Cenozoic cooling episode

One of the most significant conclusions from the current data set is that much of the Chinese Tian Shan has undergone only modest total unroofing during Cenozoic time, despite the great relief and strong seismicity in the modern-day range. Most sampling areas underwent no more than about 3 km of late Cenozoic unroofing. The areas on the margins of the Tarim and Junggar basins that have undergone greater late Cenozoic unroofing include the base of the Kuzigongsu section (~4–5 km), the sediment source areas for the Miocene Wenguri samples (>5 km), sample K-1 at Kuqa (4–5km), and the base of the section at Manas (4–5 km) (Fig. 2). These Tarim and Junggar data may be somewhat biased toward areas of greatest unroofing because especially thick, intact sections were targeted for sampling. In addition, at least some of these areas were tectonically buried by thrusting and consequent thickening in late Cenozoic time, so part of the total unroofing reflects removal of this extra tectonic burial. Within the interior of the Chinese Tian Shan, only the Yuximelegai Pass and Haxilegen Pass areas reveal more than 5 km of late Cenozoic unroofing. Thus, these data lead to a general picture of late Cenozoic deformation of the Chinese Tian Shan where strong uplift, seismicity, and deformation are occurring, but the amount of erosion from the range has been limited (cf. Yin et al., 1998).

The available fission-track data have recorded major Cenozoic unroofing within the interior of the range in only two areas, near Yuximelegai and Haxilegen Passes (Fig. 3). Haxilegen Pass apparently is astride the north Tian Shan fault system (e.g., Ma, 1986; Zhou et al., this volume). The fission-track data demonstrate that this fault system has been strongly reactivated in Cenozoic time. General regional relations suggest that this may be a west-northwest–trending, right-lateral strike-slip or transpressional fault. As such, it may be an important member of the system of major strike-slip faults that is helping to accommodate deformation in central Asia induced by the India-Asia collision (e.g., Tapponnier and Molnar, 1979). However, the fission-track data cannot specifically document strike-slip motions on the system and so cannot rule out a predominately dip-slip sense of offset at Haxilegen Pass.

ACKNOWLEDGMENTS

This research was supported by the Stanford–China Geosciences Industrial Affiliates and the Stanford McGee and Shell Funds. We are grateful to Wang Baoyu, Xiao Xuchang, Fu Derong, Jia Mingming, Liu Xun, Wang Zuoxun, Don Ying, the Chinese Academy of Geological Sciences, and the Xinjiang Bureau of Geology and Mineral Resources for collaboration in the field. We thank Tina Bush-Rester and Dean Miller for help with sample preparation and drafting, David Coyle and Kerry Gallagher for providing software, and Nicholas Arnaud and Ann Blythe for reviewing the manuscript. The Texas A&M University Nuclear Science Center and University of Oregon Radiation Center provided sample irradiations.

REFERENCES CITED

Academy of Geological Sciences of China, 1975, Geological map of Asia: Beijing, Geologic Publishing House, scale 1:5 000 000.

Allen, M.B., Windley, B.F., and Zhang, C., 1992, Paleozoic collisional tectonics and magmatism of the Chinese Tien Shan, central Asia: Tectonophysics, v. 220, p. 89–115.

Allen, M.B., Vincent, S.J., and Wheeler, P.J., 1999, Late Cenozoic tectonics of the Kepingtage thrust zone: Interactions of the Tien Shan and Tarim basin, northwest China: Tectonics, v. 18, p. 639–654.

Arnaud, N.O., 1992, Apport de la thermochronologie ^{40}Ar/^{39}Ar sur feldspath potassique à la connaissance de la tectonique Cenozoique d'Asie [Ph.D. thesis]: Clermont-Ferrand, France, Université du Clermont-Ferrand 2, 161 p.

Arnaud, N., Brunel, M., Cantagrel, J.M., and Tapponnier, P., 1993, High cooling and denudation rates at Kongur Shan (Xinjiang, China) revealed by ^{40}Ar/^{39}Ar K-feldspar thermochronology: Tectonics, v. 12, p. 1335–1346.

Avouac, J.P., Tapponnier, P., Bai, M., You, H., and Wang, G., 1993, Active thrusting and folding along the northern Tien Shan and late Cenozoic rotation of the Tarim relative to Dzungaria and Kazakhstan: Journal of Geophysical Research, v. 98, p. 6755–6804.

Brunel, M., Arnaud, N., Tapponnier, P., Pan, Y.S., and Wang, Y., 1994, Kongur Shan normal fault: Type example of mountain building assisted by extension (Karakoram fault, eastern Pamir): Geology, v. 22, p. 707–710.

Burbank, D.W., and Bullen, M.E., 1999, Late Cenozoic rates of rock and surface uplift in the Central Kyrgyz Range, Northern Tien Shan [abs.]: Eos (Transactions, American Geophysical Union), v. 80, p. F1035.

Bürgmann, R., Arrowsmith, R., Dumitru, T., and McLaughlin, R., 1994, Rise and fall of the southern Santa Cruz Mountains, California, from fission-tracks, geomorphology, and geodesy: Journal of Geophysical Research, v. 99, p. 20181–20202.

Burtman, V.S., Skobelev, S.F., and Molnar, P., 1996, Late Cenozoic slip on the Talas-Ferghana fault, the Tien Shan, central Asia: Geological Society of America Bulletin, v. 108, p. 1004–1021.

Carroll, A.R., Liang, Y., Graham, S.A., Xiao, X., Hendrix, M.S., Chu, J., and McKnight, C.L., 1990, Junggar basin, northwest China: Trapped late Paleozoic ocean: Tectonophysics, v. 181, p. 1–14.

Carroll, A.R., Graham, S.A., Hendrix, M.S., Ying, X., and Zhou, D., 1995, Late Paleozoic tectonic amalgamation of northwestern China: Sedimentary record of the northern Tarim, northwestern Turpan, and southern Junggar basins: Geological Society of America Bulletin, v. 107, p. 571–594.

Cerveny, P.F., Naeser, N.D., Zeitler, P.K., Naeser, C.W., and Johnson, N.M., 1988, History of uplift and relief of the Himalaya during the past 18 million years: Evidence from fission-track ages of detrital zircons from sandstones of the Siwalik Group, *in* Kleinspehn, K.L., and Paola, C., eds., New perspectives in basin analysis: New York, Springer-Verlag, p. 43–61.

Chen, Z., Wu, N., Zhang, D., Hu, J., Huang, H., Shen, G., Wu, G., Tang, H., and Hu, Y., 1985, Geological map of Xinjiang Uygur Autonomous Region, China: Beijing, Geological Publishing House, scale 1:2 000 000.

Corrigan, J.D., 1991, Inversion of apatite fission-track data for thermal history information: Journal of Geophysical Research, v. 96, p. 10347–10360.

Corrigan, J.D., 1993, Apatite fission-track analysis of Oligocene strata in south Texas, U.S.A.: Testing annealing models: Chemical Geology, v. 104, p. 227–249.

Dumitru, T.A., 1988, Subnormal geothermal gradients in the Great Valley forearc basin, California, during Franciscan subduction: A fission-track study: Tectonics, v. 7, p. 1201–1221.

Dumitru, T.A., 1991, Major Quaternary uplift along the northernmost San Andreas fault, King Range, northwestern California: Geology, v. 19, p. 526–529.

Dumitru, T.A., 1993, A new computer-automated microscope stage system for fission-track analysis: Nuclear Tracks and Radiation Measurements, v. 21, p. 575–580.

Dumitru, T.A., 2000, Fission-track geochronology, in Knoller, J.S., et al., eds., Quaternary geochronology: Methods and applications: American Geophysical Union Reference Shelf, v. 4, p. 131–156.

Dumitru, T.A., Miller, E.L., O'Sullivan, P.B., Amato, J.M., Hannula, K.A., Calvert, A.C., and Gans, P.B., 1995, Cretaceous to recent extension in the Bering Strait region, Alaska: Tectonics, v. 14, p. 549–563.

Fan, S.F., Zhou, Z.Y., Pan, C.C., Han, L., and Zhu, Y.M., eds., 1990, Paleotemperature and oil and gas of Tarim: Oil and gas geology of Tarim basin: Beijing, Science Press, 77 p.

Galbraith, R.F., 1981, On statistical models for fission-track counts: Mathematical Geology, v. 13, p. 471–478.

Galbraith, R.F., 1990, The radial plot: Graphical display of spread in ages: Nuclear Tracks and Radiation Measurements, v. 17, p. 207–214.

Galbraith, R.F., and Laslett, G.M., 1993, Statistical models for mixed fission-track ages: Nuclear Tracks and Radiation Measurements, v. 21, p. 459–470.

Gallagher, K., 1995, Evolving temperature histories from apatite fission-track data: Earth and Planetary Science Letters, v. 136, p. 421–435.

Gao, Z., Wang, W., Peng, C., and Xiao, B., 1985, The Sinian in Aksu-Wushi area, Xinjiang, China: Urumqi, Xinjiang People's Publishing House, 184 p.

Girardeau, J., Marcoux, J., Allegre, C.J., Bassoullet, J.P., Tang, Y., Xiao, X., Zao, Y., Wang, X., 1984, Tectonic environment and geodynamic significance of the Neo-Cimmerian Donqiao ophiolite, Bangong-Nujiang suture zone, Tibet: Nature, v. 307, p. 27–31.

Gleadow, A.J.W., Duddy, I.R., Green, P.F., and Lovering, J.F., 1986, Confined fission-track lengths in apatite: A diagnostic tool for thermal history analysis: Contributions to Mineralogy and Petrology, v. 94, p. 405–415.

Graham, S.A., Brassell, S., Carroll, A.R., Xiao, X., Demaison, G., McKnight, C.L., Liang, Y., Chu, J., and Hendrix, M.S., 1990, Characteristics of selected petroleum source rocks, Xianjiang Uygur Autonomous Region, northwest China: American Association of Petroleum Geologists Bulletin, v. 74, p. 493–512.

Graham, S.A., Hendrix, M.S., Wang, L., and Carroll, A.R., 1993, Collisional successor basins of western China: Impact of tectonic inheritance on sand composition: Geological Society of America Bulletin, v. 105, p. 323–344.

Graham, S.A., Zhou, D., Chang, E.Z., and Hendrix, M.S., 1994, Origins of intermontane Jurassic strata of the Chinese Tian Shan: Geological Society of America Abstracts with Programs, v. 26, no. 7, p. A464.

Green, P.F., 1981, A new look at statistics in fission-track dating: Nuclear Tracks and Radiation Measurements, v. 5, p. 77–86.

Green, P.F., Duddy, I.R., Gleadow, A.J.W., and Lovering, J.F., 1989a, Apatite fission-track analysis as a paleotemperature indicator for hydrocarbon exploration, in Naeser, N.D., and McCulloh, T.H., eds., Thermal history of sedimentary basins: Methods and case histories: New York, Springer-Verlag, p. 181–195.

Green, P.F., Duddy, I.R., Laslett, G.M., Hegarty, K.A., Gleadow, A.J.W., and Lovering, J.F., 1989b, Thermal annealing of fission-tracks in apatite, 4. Quantitative modelling techniques and extension to geological timescales: Chemical Geology, v. 79, p. 155–182.

Halim, N., Kravchinsky, V., Gilder, S., Cogne, J.-P., Alexyutin, M., Sorokin, A., Courtillot, V., and Chen, Y., 1998, A palaeomagnetic study from the Mongol-Okhotsk region: Rotated Early Cretaceous volcanics and remagnetized Mesozoic sediments: Earth and Planetary Science Letters, v. 159, p. 133–145.

Harlan, W.B., Armstrong, R.L., Cox, A.V., Craig, L.E., Smith, A.G., and Smith, D.G., 1990, A geologic time scale 1989: Cambridge, Cambridge University Press, 263 p.

Harrison, T.M., Copeland, P., Kidd, W.S.F., and Yin, A., 1992, Raising Tibet: Science, v. 255, p. 1663–1670.

Hendrix, M.S., 2000, Evolution of Mesozoic sandstone compositions, southern Junggar, northern Tarim, and western Turpan basins, northwest China: A detrital record of the ancestral Tian Shan: Journal of Sedimentary Research, v. 70, p. 520–532.

Hendrix, M.S., Graham, S.A., Carroll, A.R., Sobel, E.R., McKnight, C.L., Schulein, B.J., and Wang, Z., 1992, Sedimentary record and climatic implications of recurrent deformation in the Tian Shan: Evidence from Mesozoic strata of the north Tarim, south Junggar, and Turpan basins, northwest China: Geological Society of America Bulletin, v. 104, p. 53–79.

Hendrix, M.S., Dumitru, T.A., and Graham, S.A., 1994, Late Oligocene–early Miocene unroofing in the Chinese Tian Shan: An early effect of the India-Asia collision: Geology, v. 22, p. 487–490.

Hu, A., Zhang, Z., Liu, J., Peng, J., Zhang, J., Zhao, D., Yang, S., and Zhou, W., 1986, U-Pb age and evolution of Precambrian metamorphic rocks of the middle Tianshan uplift zone, eastern Tianshan, China: Geochimica, v. 1986, p. 23–35.

Huang, D., Zhang, D., Li, J., and Huang, X., 1991, Hydrocarbon genesis of Jurassic coal measures in the Turpan basin, China: Organic Geochemistry, v. 10, p. 65–72.

Hurford, A.J., and Green, P.F., 1983, The zeta age calibration of fission-track dating: Chemical Geology, v. 41, p. 285–317.

Jiao, S.P., Zhang, Y.F., Yi, S.X., Ai, C.X., Zhao, Y.N., Wang, H.D., Xu, J.E., Hu, J.Q., and Guo, T.Y., 1988, Geological map of Qinghai-Xizang (Tibet) plateau and adjacent areas: Beijing, Geological Publishing House, scale 1:1 500 000.

Klootwijk, C.T., Gee, J.S., Peirce, J.W., Smith, G.M., and McFadden, P.L., 1992, An early India-Asia contact: Paleomagnetic constraints from Ninety East Ridge, ODP Leg 121: Geology, v. 20, p. 395–398.

Laslett, G.M., Kendall, W.S., Gleadow, A.J.W., and Duddy, I.R., 1982, Bias in the measurement of fission-track length distributions: Nuclear Tracks and Radiation Measurements, v. 6, p. 79–85.

Laslett, G.M., Green, P.F., Duddy, I.R., and Gleadow, A.J.W., 1987, Thermal annealing of fission-tracks in apatite, 2. A quantitative analysis: Chemical Geology, v. 65, p. 1–13.

Li, D., Liang, D., Jia, C., Wang, G., Wu, Q., and He, D., 1996, Hydrocarbon accumulations in the Tarim basin, China: American Association of Petroleum Geologists Bulletin, v. 80, p. 1587–1603.

Li, J., and Jiang, J., 1987, Survey of petroleum geology and the controlling factors for hydrocarbon distribution in the east part of the Junggar basin: Oil and Gas Geology, v. 8, p. 99–107.

Liou, J.G., Graham, S.A., Wang, X., Xiao, X., and Carroll, A., 1989, Proterozoic blueschist belt in western China: Best documented Precambrian blueschists in the World: Geology, v. 17, p. 1127–1131.

Ma, X., 1986, Lithospheric dynamics map of China and adjacent seas with explanatory notes: Beijing, Geological Publishing House, scale 1:4 000 000.

Matte, P., Tapponnier, P., Arnaud, N., Bourjot, L., Avouac, J.P., Vidal, P., Liu, Q., Pan Y., and Wang, Y., 1996, Tectonics of western Tibet, between the Tarim and the Indus: Earth and Planetary Science Letters, v. 142, p. 311–330.

McKnight, C.R., 1993, Structural styles and tectonic significance of Tian Shan foothill fold-thrust belts, northwest China [Ph.D. thesis]: Stanford, California, Stanford University, 207 p.

McKnight, C.R., 1994, Compressional episodes, structural styles, and shortening estimates, southern Tien Shan foreland, NW China: Geological Society of America Abstracts with Programs, v. 26, no. 7, p. 463.

Naeser, C.W., 1979, Fission track dating and geological annealing of fission-tracks, in Jäger, E., and Hunziker, J.C., eds., Lectures in isotope geology: New York, Springer-Verlag, p. 154–169.

Pan, Y.S., ed., 1992, Introduction to integrated scientific investigation on Karakorum and Kunlun Mountains, China: Beijing, Meterology Press, 92 p.

Patriat, P., and Achache, J., 1984, India-Eurasia collision chronology has implications for crustal shortening and driving mechanism of plates: Nature, v. 311, p. 615–621.

Roecker, S.W., Sabitova, T.M., Vinnik, L.P., Burmakov, Y.A., Golvanov, M.I., Mamatkanova, R., and Munirova, L., 1993, Three-dimensional elastic wave velocity structure of the western and central Tien Shan: Journal of Geophysical Research, v. 98, p. 15779–15795.

Smith, A.B., 1988, Late Palaeozoic biogeography of East Asia and palaeontological constraints on plate tectonic reconstructions: Royal Society of London Philosophical Transactions, sec. A, v. 326, p. 189–227.

Sobel, E.R., 1999a, Basin analysis of the Jurassic–Lower Cretaceous southwest Tarim basin, northwest China: Geological Society of America Bulletin, v. 111, p. 709–724.

Sobel, E., 1999b, Cenozoic exhumation of the Kyrgyz Tian Shan constrained by apatite fission-track thermochronology [abs.]: Eos (Transactions, American Geophysical Union), v. 80, p. F1037.

Sobel, E.R., and Dumitru, T.A., 1997, Thrusting and exhumation around the margins of the western Tarim basin during the India-Asia collision: Journal of Geophysical Research, v. 102, p. 5043–5063.

Tapponnier, P., and Molnar, P., 1979, Active faulting and Cenozoic tectonics of the Tien Shan, Mongolia, and Baykal regions: Journal of Geophysical Research, v. 84, p. 3425–3459.

Tapponnier, P., Peltzer, G., and Armijo, R., 1986, On the mechanics of the collision between India and Asia, *in* Coward, M.P., and Ries, A. C., eds., Collision tectonics: Geological Society [London] Special Publication 19, p. 115–157.

Turcotte, D.L., and Schubert, G., 1982, Geodynamics: Applications of continuum physics to geological problems: New York, John Wiley and Sons, 450 p.

Wang, Q.M., Nishidai, T., and Coward, M.P., 1992, The Tarim basin, NW China: Formation and aspects of petroleum geology: Journal of Petroleum Geology, v. 15, p. 5–34.

Wang, Z., Wu, J., Lu, X., Zhang, J., and Liu, C., 1990, Polycyclic tectonic evolution and metallogeny of the Tianshan mountains: Beijing, Science Press, 217 p.

Watson, M.P., Hayward, A.B., Parkinson, D.N., and Zhang, Z.M., 1987, Plate tectonic history, basin development, and petroleum source rock deposition, onshore China: Marine and Petroleum Geology, v. 4, p. 205–225.

Xinjiang Bureau of Geology and Mineral Resources, 1966, Geologic map of the Kuche region: Beijing, Ministry of Geology and Mineral Resources, scale 1:200 000.

Xinjiang Bureau of Geology and Mineral Resources, 1985, Geological map of southwestern Xinjiang: Beijing, Geological Publishing House, scale 1:500 000.

Xinjiang Bureau of Geology and Mineral Resources, 1993, Regional geology of Xinjiang Uygur Autonomous Region: Beijing, Geological Publishing House, Geological Memoirs, ser. 1, 841 p.

Yin, A., Nie, S., Craig, P., Harrison, T.M., Ryerson, F.J., Xianglin, Q., and Geng, Y., 1998, Late Cenozoic tectonic evolution of the southern Chinese Tian Shan: Tectonics, v. 17, p. 1–27.

Zhang, H.N., 1989, Geochemical characteristics and resource prediction of Meso-Cenozoic source rocks in Tarim, *in* Institute of Petroleum Geology of the Ministry of Geology and Mineral Resources, ed., Petroleum resources—Prospects and evaluations, Selected papers on petroleum and natural gas geology, Volume 2: Beijing, Geological Publishing House, p. 53–76.

Zhang, X., 1981, Regional stratigraphic chart of northwest China, Branch of Xingiang Uygar Autonomous Region: Beijing, Geological Publishing House, 496 p. (in Chinese).

MANUSCRIPT ACCEPTED BY THE SOCIETY JUNE 5, 2000

Geological Society of America
Memoir 194
2001

Tectonic correlation of Beishan and Inner Mongolia orogens and its implications for the palinspastic reconstruction of north China

Yongjun Yue, Juhn G. Liou, Stephan A. Graham
Department of Geological and Environmental Sciences, Stanford University, Stanford, California 94305-2115, USA

ABSTRACT

A correlation is established between the fault-truncated Beishan and Inner Mongolia orogens on the basis of their equivalent ophiolite zones and petrotectonic units. Two ophiolite zones within these two orogens exhibit similar lithologic sequences and identical age relationships. Each of the three fault-bounded tectonic units in the Beishan orogen has a corresponding unit in the Inner Mongolia orogen; they share identical Paleozoic continental margin assemblages. The proposed correlation between the Beishan and Inner Mongolia orogens suggests that the left-lateral offset along the Alxa–east Mongolia fault is at ~400 km and shows that at least the northeastern part of the Tarim block was connected with the north China craton in the Paleozoic, and the north Qilian suture extends westward to the north Altyn Tagh suture. This correlation also provides a mechanism for the development of the present-day narrow connection between the Tarim block and north China craton, and suggests that this corridor was formed by post-Paleozoic strike-slip faulting along the Altyn Tagh–Alxa–east Mongolia fault.

INTRODUCTION

Tectonic correlation between eastern and western north China across the Altyn Tagh and Alxa–east Mongolia faults (Fig. 1; Yue and Liou, 1999) is one of the most important issues of the tectonics of Asia. This issue has been intensely debated in terms of the relationship between the Tarim block and north China craton over the past 20 yr, and three hypotheses have been proposed.

Some geologists suggested that the north China craton has been contiguous with the Tarim block since Precambrian time on the basis of the observation that their Precambrian basements have similar lithology (e.g., Cheng, 1994) and are contiguous through a narrow connection, called the Anxi geowaist (Figs. 1 and 2; Huang et al., 1977; Li and Tang, 1983; Ren et al., 1987). However, other geologists argued that the Tarim block and north China craton were widely separated in the Paleozoic based on available paleomagnetic and paleontological data (e.g., Li et al., 1989; Wang et al., 1994; Zhou and Graham, 1996). Others argued that the Tarim block is composed of two different Pre-

cambrian massifs separated by an east-west–trending magmatic zone inferred on the basis of the central Tarim high areomagnetic anomaly (CTHM in Fig. 1; Yang, 1983; Liu et al., 1997).

In contrast with these debates, few discussions have been offered for the relationship between the Beishan and Inner Mongolia orogens, the northern bounding orogens of the northeastern Tarim block (i.e., Dunhuang block) and the north China craton, respectively (Fig. 1). These two orogens are located in the central part of a huge east-west–trending orogenic belt in north China, and have been depicted as continuous in many tectonic maps of China and Asia (e.g., Li et al., 1982; Ren et al., 1987). However, this tectonic scenario is not consistent with many recent geological observations, most notably that: (1) the east-west–trending Beishan orogen does not extend across the northeast-east–trending Alxa fault (Wu, 1990; Wang et al., 1993, 1994; Wu and He, 1993); and (2) the Inner Mongolia orogen is truncated by the left-lateral strike-slip east Mongolia fault and does not extend across that fault (Figs. 1 and 2; Suvorov, 1982; Yanshin, 1989; Lamb et al., 1999; Yue and Liou, 1999).

Yue, Y., Liou, J.G., and Graham, S.A., Tectonic correlation of Beishan and Inner Mongolia orogens and its implications for the palinspastic reconstruction of north China, *in* Hendrix, M.S., and Davis, G.A., eds., Paleozoic and Mesozoic tectonic evolution of central Asia: From continental assembly to intracontinental deformation: Boulder, Colorado, Geological Society of America Memoir 194, p. 101–116.

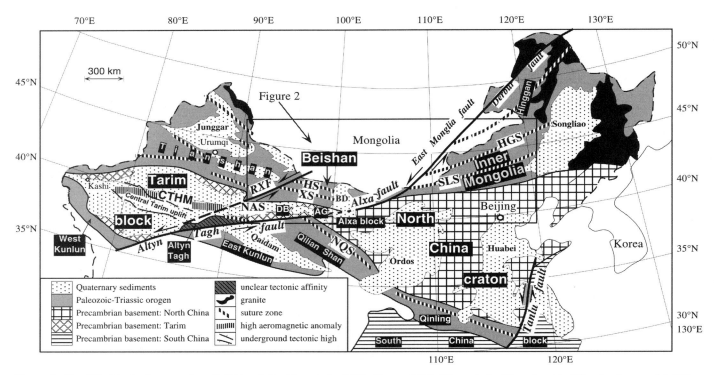

Figure 1. Simplified tectonic map of north China (after Ren et al., 1987; Cheng, 1994; Zhou et al., this volume). AG—Anxi geowaist, BD—Badanjilin desert, CTHM—central Tarim high areomagnetic anomaly, DB—Dunhuang block, HGS—Hegenshan suture, HS—Hongshishan suture, NAS—north Altyn Tagh suture, NQS—north Qilian suture, RXF—Ruoqiang-Xingxingxia fault, SLS—Solon-Linxi suture, XS—Xiaohuangshan suture.

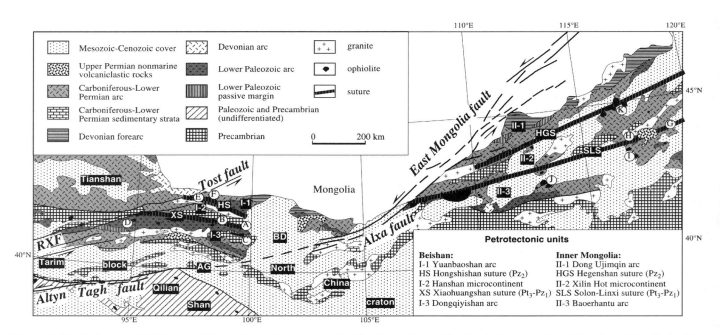

Figure 2. Tectonic map of Beishan and Inner Mongolia orogens. A—Xiaohuangshan, AG—Anxi Geowaist, B—Shibanjing, BD—Badanjilin desert, C—Xichangjing, D—Hongliuhe, E—Hongshishan, F—Qinghekou, G—Linxi, H—Wudaoshimen, I—Kedanshan, J—Ondor Sum, K—Hegenshan, RXF—Ruoqiang-Xingxingxia fault. Key for petrotectonic units in lower right. Other abbreviations as in Figure 1.

The purpose of this chapter is to investigate the relationship between the fault-truncated Beishan and Inner Mongolia orogens on the basis of their characteristic petrotectonic assemblages, and to explore the role of these orogens in palinspastic reconstruction of north China, especially the relationship between the Tarim block and the north China craton.

TECTONIC ZONATION OF THE BEISHAN AND INNER MONGOLIA OROGENS

According to Zuo and He (1990), the Beishan orogen is composed of three distinct tectonic units: the southern Dongqiyishan volcanic arc, the central Hanshan microcontinent and the northern Yuanbaoshan arc. The Dongqiyishan arc is characterized by extensive occurrence of Ordovician to Silurian and Carboniferous to Permian arc assemblages; the Hanshan microcontinent is characterized by Precambrian massifs and a Carboniferous to Permian arc complex; and the Yuanbaoshan arc is characterized by a continuous Ordovician to Permian arc-related sequence. The boundaries between these units are the Xiaohuangshan and Hongshishan ophiolitic sutures (Fig. 2).

Three similar tectonic units have also been recognized in the Inner Mongolia orogen (e.g., Cheng, 1994), the southern Baoerhantu volcanic arc, the central Xilin Hot microcontinent, and the northern Dong Ujimqin arc. Their major petrotectonic assemblages are the same as those occurring in the corresponding units of the Beishan orogen, and the tectonic boundaries between these units are Solon-Linxi and Hegenshan ophiolitic sutures (Fig. 2).

CORRELATION OF TWO OPHIOLITIC ZONES IN THE BEISHAN AND INNER MONGOLIA OROGENS

The Beishan orogen contains the southern Xiaohuangshan and the northern Hongshishan ophiolite zones (Zuo and He, 1990). The Xiaohuangshan ophiolite zone along the suture between the Dongqiyishan arc and the Hanshan microcontinent is ~50 km wide and ~400 km long. The ophiolites are structurally dismembered and composed of gabbro and severely serpentinized dunite and harzburgite. Mafic-ultramafic bodies of the ophiolite melange are generally <1 km wide (Zuo and He, 1990); some within the volcanic arc are less dismembered, much wider, and contain many more ophiolitic constituents. For example, the ophiolite at Xichangjing (location C in Fig. 2) includes a nearly complete sequence from dunite and harzburgite through gabbro, plagiogranite, and diabase dike complex upward to pillow lava and chert in one section. Ordovician microfossils occur in cherty rocks (Table 1), and the Rb-Sr whole-rock isochron of plagiogranite (from location C) gives a Late Proterozoic age of 729.4 ± 66.3 Ma. The former age is considered to be the age of sedimentary cover, and the latter age is considered to be the age when oceanic crust was formed at the mid-ocean ridge (Zuo and Li, 1996).

The northern Hongshishan ophiolite zone is ~30 km wide and extends roughly east-west for about 200 km. The ophiolitic rocks occur in Lower Carboniferous turbiditic strata as tectonic blocks of serpentinized dunite and harzburgite, massive gabbro, metamorphosed pillow lava, and chert. The age of the ophiolite was inferred by Zuo and He (1990) to be Carboniferous on the

TABLE 1. CORRELATION BETWEEN THE OPHIOLITES OF BEISHAN AND INNER MONGOLIA

Beishan orogen		Inner Mongolia orogen	
Ophiolite zone	Lithological assemblages	Ophiolite zone	Lithological assemblages
Pt₃ to O* Xiaohuangshan	Within suture: A[†]: dn[‡] + hz + gb B: dn + hz + gb Fossils: not reported	Pt₃ to O Linxi-Solon	Within suture: G: gb + diab + pb; H: gb + diab + pb + ch Fossils: *Ammodiscus* sp., *Acrotretidal, Sphaerallari, Panderodus* sp., *Chancelloria* (?) sp., *Variocymaatiosphaera* sp., *Gregalosphaera* cf. *Cyclopylora* Wang (MS), *Asteropylorus cruciporus* Wang (MS)
	Within arc: C: dn + px + gb + plgr + diab + pb + ch D: hz + prd + px + gb + an + plgr + diab + pb + ch Fossils: *Entactinia* sp., *Entactactinidae indet., Entactinidae aculeata* Naz.		Within arc: I: dn + hz + gb + pb + ch + lst J: dn + hz + gb + pb + ch Fossils: *Entactiniids, Actiommids, Liesphaerids, Plagoniids, Dityosphaerids, Albaillella, Asteropylorus cruciporus* Wang (MS), *Pylospaera* sp., *Ecfoprimitia* sp., etc.
Pz₂ Hongshishan	E: dn + hz + gb + pb + ch F: prd + px + gb + mtbs + ch	Pz₂ Hegenshan	K: dn + hz + prd + gb + diab + pb + ch Fossils: *Entactinia* sp., *Cenellipsis* (?) sp., *Tetrentactinia* sp.

* Abbreviation of ages: O—Ordovician; Pt₃—Late Proterozoic; Pz₂—Late Paleozoic.
[†] Abbreviation for locations: A—Xiaohuangshan, B—Shibanjing, C—Xichangjing, D—Hongliuhe, E—Hongshishan, F—Qinghekou, G—Linxi, H—Wudaoshimen, I—Kedanshan, J—Ondor Sum, K—Hegenshan (see Fig. 3).
[‡] Abbreviation of rock names: an—anorthite, ch—chert, diab—diabase, dn—dunite, gb—gabbro, lst—imestone, mtbs—metabasalt, pb—pillow basalt, plgr—plagiogranite, prd—peridotite, px = pyroxenite.
References: Cao, 1989; He and Shao, 1983; Liu, 1983; Wang et al., 1991; Zuo and He, 1990; Zuo and Li, 1996.

basis of its close relationship with Carboniferous arc assemblages, but has not yet been confirmed either by fossils or by radiometric dating.

Two similar ophiolite zones occur in the Inner Mongolia orogen: the southern Solon-Linxi ophiolite zone and the northern Hegenshan ophiolite zone. The Solon-Linxi ophiolite zone is similar to the Xiaohuangshan ophiolite zone in Beishan in terms of width and the mode of occurrences. The ophiolitic bodies along or close to the boundary between the Baoerhantu arc and the Xilin Hot microcontinent are generally small, and contain fewer ophiolitic constituents; the ophiolites within the volcanic arc are generally better preserved and contain more ophiolitic constituents. For example, the ophiolite at Kedanshan and Ondor Sum (locations I and J in Fig. 2, respectively) is composed of dunite, harzburgite, gabbro, pillow basalt, and chert $\pm$ limestone. The age of this ophiolite zone has been constrained to be late Proterozoic to Ordovician by a Rb-Sr isochron age of pillow basalt (630 Ma) and microfossils in chert (Table 1; He and Shao, 1983; Wang et al., 1991).

The northern Hegenshan ophiolite zone occurs as tectonic blocks of various ophiolitic components in upper Paleozoic volcaniclastic strata. The ophiolitic rocks are composed of serpentinized dunite, harzburgite, lherzolite, cumulate gabbro, diabase, pillow basalt, and chert. The radiolarian fossils in chert shown in Table 1 suggest a Middle to Late Devonian age (Liu, 1983)

Correlation of these ophiolite zones in the Inner Mongolia and Beishan orogens has not been made previously. However, on the basis of their similarities in field attributes, lithologies, and age relationship of these ophiolite zones and their adjacent petrotectonic units (described in the following), we suggest that the two ophiolite zones in the Inner Mongolia orogen are eastern extensions of those in the Beishan orogen.

CORRELATION OF THREE PETROTECTONIC UNITS IN THE BEISHAN AND INNER MONGOLIA OROGENS

The focus of this section is to correlate the three tectonic units in the Beishan orogen with the three tectonic units in the Inner Mongolia orogen (Fig. 2) based on the stratigraphic and geochemical characteristics of their petrotectonic assemblages. Figures 3 and 4 illustrate composite sections of Paleozoic strata in the three petrotectonic units in both the Beishan and Inner Mongolia orogens. These include the Dongqiyishan arc, Hanshan microcontinent, and Yuanbaoshan arc of the Beishan orogen, and the Baoerhantu arc, Xilin Hot microcontinent, and Dong Ujimqin arc of the Inner Mongolia orogen. These sections are compiled based on our observations from four field seasons in the Beishan and Inner Mongolia orogens and available published data (Gansu Province Bureau of Geology and Mineral Resources [BGMRG], 1989; Zuo and He, 1990; Inner Mongolia Autonomous Region BGMR, 1991; Q. Wang et al., 1991; Tang, 1992; T. Wang et al., 1994; Zuo and Li, 1996).

Stratigraphy

Dongqiyishan arc and Baoerhantu arc. The stratigraphic sections shown in Figure 3 suggest a strong similarity between the Dongqiyishan arc and the Baoerhantu arc. Both arcs are developed above Cambrian–Lower Ordovician continental marginal sequences and form a nearly continuous section upward through an Upper Permian nonmarine assemblage. Jurassic-Cretaceous fluvial-lacustrine assemblages unconformably overlie the Paleozoic continental margin strata; there are few Triassic strata (Inner Mongolia Autonomous Region BGMR, 1991). Among these assemblages, Cambrian–Lower Ordovician passive margin and Ordovician–Lower Permian arc strata occur as the most linear and narrow belts, hence are most appropriate for correlation purposes (Fig. 2).

Cambrian–Lower Ordovician passive margin strata. Passive margin assemblages are very rare in both arc terranes, probably due to poor preservation. The tectonic setting was inferred mainly on the basis of the absence of volcanics and the marine environment of the strata (Zuo and He, 1990; Wang et al., 1991).

In the Dongqiyishan arc terrane, a continuous Cambrian–Lower Ordovician section occurs in the Xichangjing area (location C in Fig. 2). Lower Cambrian strata primarily comprise a turbiditic suite composed of quartzofeldspathic sandstone and slate, and are considered to be formed on a continental slope; Middle and Upper Cambrian strata contain mainly chert, and a minor amount of turbiditic strata of quartzofeldspathic sandstone and silty slate. The Lower Ordovician is a suite of turbiditic quartzofeldspathic sandstone, silty slate, and chert.

In the Baoerhantu arc terrane, Lower Cambrian strata are composed of limestone with interbedded siltstone; no Middle to Upper Cambrian strata have been documented. Lower Ordovician strata consist of an ~400-m-thick turbidite sequence of sandstone, slate and chert, suggesting a continental slope environment.

The Cambrian strata in the Dongqiyishan arc contain abundant trilobites (Fig. 3), among which Lower Cambrian fossils, such as *Subeia beishanensis,* suggest endemism, whereas Middle to Late Cambrian fossils resemble those in southeast China (Gansu Province BGMR, 1989). The Lower Ordovician strata contain gastropods and cephalopods. The former include *Coreanoceras* sp. and *Manchuroceras* sp., which are identical to the Lower Ordovician gastropod fossils in the north China craton. In the Baoerhantu arc, Lower Cambrian strata contain no fossils except for some algae; no fossils have been reported for the Lower Ordovician turbidites.

Middle Ordovician to Silurian arc assemblage. Field observations suggest that the Middle Ordovician to Silurian volcanic arc assemblage conformably overlies the Cambrian to Lower Ordovician continental margin strata both in the Dongqiyishan and Baoerhantu arc terranes; no break of sedimentation has been reported between these two distinctive assemblages (Fig. 3; Xu, 1987; Zuo and He, 1990).

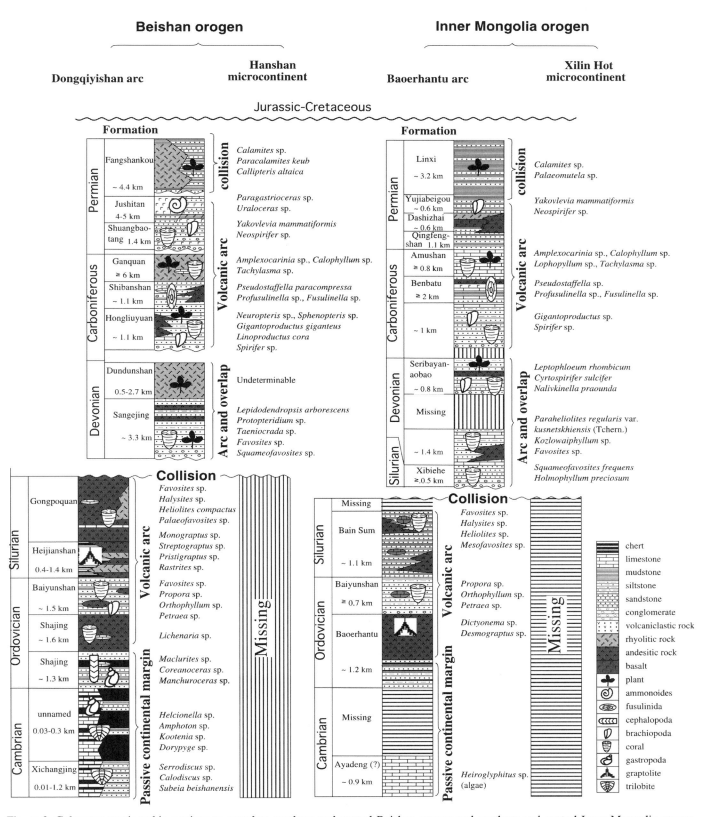

Figure 3. Columnar stratigraphic sections to correlate southern and central Beishan orogen and southern and central Inner Mongolia orogen (after Gansu Province Bureau of Geology and Mineral Resources [BGMR], 1989; Zuo and He, 1990; Inner Mongolia Autonomous Region BGMR, 1991; Wang et al., 1991; Tang, 1992; no vertical scale is applied).

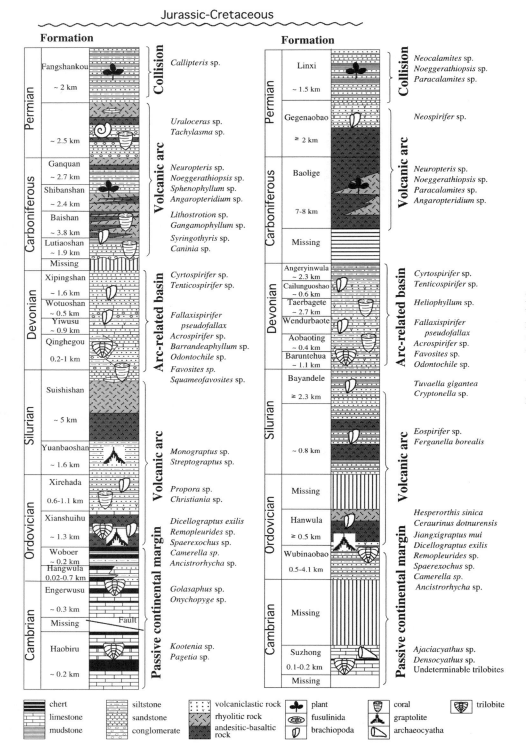

Figure 4. Columnar stratigraphic sections to correlate northern Beishan and northern Inner Mongolia orogens (after Gansu Province Bureau of Geology and Mineral Resources [BGMR], 1989; Zuo and He, 1990; Inner Mongolia Autonomous Region BGMR, 1991; Wang et al., 1991; Tang, 1992; no vertical scale is applied).

The Middle Ordovician strata in both arcs consist of basalt, andesite, dacite, and rhyolite with interbedded tuffaceous sandstone, siltstone, and chert; the Upper Ordovician strata in both arcs are mainly composed of turbiditic sandstone, siltstone, chert, limestone, and interbedded andesite (Fig. 3).

The Silurian strata in the Dongqiyishan arc consist mainly of basalt, andesite, dacite, and rhyolite. These volcanic rocks are interbedded with limestone and turbiditic sandstone and mudstone. The Silurian volcaniclastic assemblage in the Baoerhantu arc terrane consists mainly of turbidite and minor interbedded andesitic flows (Wang et al., 1991; Tang, 1992).

The Middle Ordovician strata in both arcs contain few fossils. The Upper Ordovician strata in both arcs contain similar coral fossils, including *Propora* sp., *Orthophyllum* sp., and *Petraea* sp. This is one of the two reasons why the Upper Ordovician strata in both arcs share a common formation name; the other reason is that these strata have a similar lithology (Fig. 3; Inner Mongolia Autonomous Region BGMR, 1991). The Silurian faunas in the Dongqiyishan and Baoerhantu arcs are considered to constitute one subdivision of fauna (Inner Mongolia Autonomous Region BGMR, 1991) and they contain many coral fossils, such as *Halysites* sp., *Favosites* sp., and *Heliolites* sp. (Fig. 3).

The Ordovician–Silurian arc assemblage is unconformably overlain by Lower Devonian shallow-marine coarse-grained sedimentary rocks in the Dongqiyishan arc and by Upper Silurian shallow-marine conglomerate and coarse sandstone in the Baoerhantu arc (Tang, 1990, 1992; Zuo and He, 1990).

Devonian volcanic arc and overlap assemblage. Only very limited Devonian strata crop out in the Dongqiyishan arc. Lower to Middle Devonian strata are in sharp lithologic contrast with the lower Paleozoic arc complex below the unconformity. They are composed of a basal conglomerate, overlying sandstone, and some interbeds of siltstone and andesite. Upper Devonian strata include volcaniclastic rocks, including rhyolitic, and dacitic agglomerate and tuff.

A basal, Upper Silurian conglomerate is present in the Baoerhantu arc terrane. Upper Silurian and Lower Devonian nonvolcanic strata consist of quartzofeldspathic sandstone, limestone, siltstone, and slate. At some localities, these strata are interbedded with basalt and andesite (Tang, 1992). No Middle Devonian section has been reported in this arc. Upper Devonian strata can be subdivided into three parts: the lower part is composed of conglomerate, sandstone, siltstone, and limestone; the middle part is andesitic lava and tuff; the upper part is quartz sandstone, siltstone, and intercalated limestone.

The Devonian strata in both arcs contain coral, brachiopod, and plant fossils (Fig. 3). These fossils, together with the coarseness and red and purple color of some clastic interbeds, suggest alternating nonmarine and marine environments.

The tectonic setting for the Devonian strata in both arcs is highly controversial. A Late Silurian–Devonian collisional environment was proposed based on upward-shoaling paleobathymetry in the Devonian, coarseness and rarity of the Devonian strata, sudden termination of the widespread early Paleozoic arc

volcanism, and the unconformity between the Upper Silurian–Lower Devonian conglomerate and the underlying arc assemblage (He and Shao, 1983; Zuo and He, 1990). However, a volcanic arc environment has been assigned for the Devonian strata based on the calc-alkaline nature and the arc affinity of Devonian volcanic rocks (e.g., Wang et al., 1991; Cheng, 1994). To reconcile this discrepancy, we propose that oceanic lithosphere may have resumed its southward subduction at a new location north of the microcontinent after the collision of the Hanshan and Xilin Hot microcontinents with the Dongqiyishan and Baoerhantu arcs. Much of the Devonian strata may have been formed in an arc and overlap setting. This suggestion is also consistent with the fact that the central microcontinent and the southern arc in both the Beishan and the Inner Mongolia orogens share a common arc and overlap sequence of upper Paleozoic rocks (Fig. 3).

Carboniferous–Lower Permian volcanic arc assemblage. In the Dongqiyishan arc, Lower Carboniferous and the lower part of Upper Carboniferous strata are composed of sandstone with some interbedded siltstone and limestone. The upper part of Upper Carboniferous strata, with recorded thickness exceeding 6 km, is composed mainly of rhyolite and rhyolitic tuff. Lower Permian strata are divided into two formations: the lower formation consists of conglomerate, sandstone, and basalt, and interbeds of shale and limestone; the upper formation is composed of turbidite at the base and pillow basalt at the top, and the total thickness is >3000 m (Fig. 3).

In the Baoerhantu arc, Lower Carboniferous rocks are composed of sandstone, siltstone, andesitic agglomerate, and tuff. Upper Carboniferous strata include two formations: the lower formation consists mainly of basalt, andesite, shale, and limestone and the upper formation is limestone and siltstone. Lower Permian strata contain three units. The lower unit is mainly a suite of turbidite composed of sandstone and shale; the middle unit contains mainly pillow basalt, andesite, and rhyolite; and the upper unit is composed of turbidite.

The Lower Carboniferous brachiopod fauna in both arcs is composed of spirifers and productids, and is correlatable with the brachiopod fauna of the north China craton (Fig. 3; Inner Mongolia Autonomous Region BGMR, 1991). In addition to containing abundant fusulinid, brachiopod, and coral fossils, the Upper Carboniferous strata in the Baoerhantu arc also contain a flora that has a strong Cathysian affinity (Fig. 3; Huang, 1983). Near Chifeng (Fig. 5), this flora comprises *Lepidodendron tachingshanensis, L. szeianum, Neuropteris ovata, N. plicata, Sphenophyllum oblongifolium, Annularia gracilesceus, Pecopteris hemitelioides, P. cyathea,* and *Asterophyllites aohanensis.* Near Ondor Sum, this flora includes *Lepidodendron tripunctatum, L. szeianum, L. posthumi, L. oculus-felis, Neuropteris ovata, N. plicata, Sphenophyllum oblongifolium, S. thonii, S. kawasakii,* and *Asterophyllites longifolius.* Similarly, the Lower Permian strata also contain plant fossils with a strong Cathysian affinity (Huang, 1983). In the Dongqiyishan arc, the Upper Carboniferous–Lower Permian strata contain mainly marine animal fossils, such as coral and brachiopods; Cathysian plant fossils mainly occur to the south of this arc (Fig. 5; Gansu Province BGMR, 1989).

Hanshan microcontinent and Xilin Hot microcontinent. In addition to sharing common upper Paleozoic arc assemblages (Fig. 3), the Hanshan and Xilin Hot microcontinents contain no or very few lower Paleozoic continental margin strata, and they are each cored by similar Precambrian massifs.

The Precambrian massif of the Hanshan microcontinent is composed mainly of the Beishan complex, which is a suite of migmatite, migmatitic gneiss, mica quartz schist, sericite quartz schist, phyllite, marble, and quartzite. Radiometric dating of migmatites yields a Rb-Sr isochron age of 649.14 ± 56 Ma and a Sm-Nd model age of 1774 ± 42 Ma (Zuo and Li, 1996); the latter age supports an Early Proterozoic age, previously inferred on the basis of metamorphic grade (Inner Mongolia Autonomous Region BGMR, 1991). The protoliths of the complex are thought to be of sedimentary origin, on the basis of the wide occurrence of peraluminous minerals (e.g., sericite, garnet, muscovite) and marble (Inner Mongolia Autonomous Region BGMR, 1991).

Precambrian metamorphic rocks inferred to be Early Proterozoic (Inner Mongolia Autonomous Region BGMR, 1991) are widespread in the Xilin Hot microcontinent and have been grouped as the Xilin Hot complex. This complex is composed of gneiss, quartzite, mica quartz schist, phyllite, slate, and marble, and is lithologically similar to the Beishan complex.

Yuanbaoshan arc and Dong Ujimqin arc. Figure 4 suggests that the Yuanbaoshan and Dong Ujimqin arcs have similar Paleozoic continental margin assemblages, including Cambrian–Lower Ordovician passive margin strata and Middle Ordovician–Lower Permian arc-related volcaniclastic assemblages. Like the southern Dongqiyishan and Baoerhantu arcs, these two arc terranes contain additional nonmarine Upper Permian strata (Fig. 4). Devonian sedimentary rocks are widespread in these two arc terranes, whereas only minor Devonian rocks are in the southern Dongqiyishan and Baoerhantu arcs (Fig. 2).

Cambrian–Lower Ordovician passive margin strata. The prearc Cambrian strata in both arcs include very similar thin-bedded limestone and chert, although more dolomite and chert crop out in the Yuanbaoshan arc. Lower Ordovician strata in the Yuanbaoshan arc are composed of two formations, both of which are composed predominantly of turbiditic sandstone, siltstone, and argillaceous and cherty mudstone. The Lower Ordovician strata in the Dong Ujimqin arc include turbiditic sequences to 4 km thick, composed mainly of interbedded sandstone, siltstone, and argillaceous and cherty slate.

Cambrian strata in the Yuanbaoshan arc contain abundant trilobite fossils (Fig. 4), which show a close relationship with the trilobites from southeast China. The Cambrian strata in the Dong Ujimqin arc have archaeocyatha fossils and indeterminate trilobites. The Lower to Middle Ordovician strata in the Yuanbaoshan and Dong Ujimqin arcs contain abundant trilobite, brachiopod, and graptolite fossils (Fig. 4). Like the Late Cambrian fossils, these fossils are more similar to the fossils from southeast China than from the north China craton (Inner Mongolia Autonomous Region BGMR, 1991).

Middle Ordovician to Silurian volcanic arc assemblage. Middle Ordovician strata in both arc terranes are composed mainly of andesite and andesitic tuff and interbedded basaltic and dacitic tuff. These strata conformably overlie the passive continental margin assemblage. Upper Ordovician strata in the Yuanbaoshan arc consist of a sequence of turbiditic sandstone, siltstone, and slate, but comparable strata have not been reported in the Dong Ujimqin arc. The Silurian strata in both arcs consist of basalt, andesite, and rhyolite with interdedded turbidites. In the Yuanbaoshan arc, Lower Silurian strata comprise a turbidite sequence of sandstone, siltstone, and chert, and regionally they may change to a suite of predominantly cherty and argillaceous shale. Middle-Upper Silurian strata consist of extremely thick sequences of andesitic, basaltic, dacitic, and rhyolitic volcanics and volcaniclastics in the lower part, and intercalated sandstone, siltstone, chert, and limestone in the upper part. In the Dong Ujimqin arc, Lower and Middle Silurian strata are composed of slate, sandstone, andesite, and limestone. Upper Silurian strata are composed of sandstone and limestone lenses in the lower part, and a suite of turbiditic sandstone and slate in the upper part.

The Upper Ordovician strata in the Yuanbaoshan arc contain coral and brachiopod fossils (Fig. 4); only the Lower and the uppermost Silurian strata contain fossils, such as graptolites and corals. The most abundant fossils in the Silurian strata in the Dong Ujimqin arc are brachiopods, which include *Tuvaella gigantea* and *Eospirifer* sp.

Devonian arc-related basin assemblage. Devonian strata in the Yuanbaoshan arc contain few lavas. Lower Devonian strata are composed of interbedded calcareous sandstone, conglomerate, and tuff, but is nontuffaceous in some areas; Middle Devonian strata consist of two parts: the lower part is a suite of interbedded calcareous sandstone, calcareous-argillaceous siltstone, and arenaceous and bioclastic limestone; the upper part is conglomerate, conglomeratic sandstone, quartzofeldspathic sandstone, tuffaceous sandstone, and limestone. The Upper Devonian is composed mainly of sandstone and interbeds of limestone. Hence, the Devonian in the Yuanbaoshan arc is mainly of sedimentary origin.

Devonian strata in the Dong Ujimqin arc are also predominantly sedimentary with few volcanic flows. The major rock types include siltstone, mudstone, tuff, and sandstone; the total thickness is >7 km.

The Devonian faunas in both the Yuanbaoshan and the Dong Ujimqin arcs are dominated by brachiopods, and share several common assemblages, such as the Lower Devonian *Fallaxispirifer-Paraspirifer* assemblage and the Upper Devonian *Cyrtospirifer-Tenticospirifer* assemblage (Fig. 4).

Carboniferous–Lower Permian volcanic arc assemblage. Two Lower Carboniferous formations occur in the Yuanbaoshan arc: the lower one consists of shallow-marine sandstone, slate, and limestone; the upper one is composed of mainly dacite, rhyolite, and andesitic lava and tuffs, interbedded with chert and limestone. The Upper Carboniferous also has two formations; the lower one is composed of basalt-andesite-dacite-rhyolite assemblages interbedded with siltstone, sandstone, chert, and limestone, and the upper one is composed of silty mudstone, sandstone, siltstone, limestone, andesite, dacite, and rhyolite. The Lower Permian is composed of andesitic and rhyolitic lava and tuffs, graywacke, conglomerate, and siltstone interlayers and lenses of limestone.

The Carboniferous volcaniclastic assemblage in the Dong Ujimqin arc contains andesitic basalt, andesite, dacite, and graywacke, and is >7 km thick. Similar to the Lower Permian volcanics in the Yuanbaoshan arc, minor Lower Permian andesitic lava and tuff are present in the Dong Ujimqin arc terrane.

The Lower Carboniferous strata in the Yuanbaoshan arc contain coral and brachiopod fossils (Fig. 3). The Upper Carboniferous strata in both arcs contain an Angara-type flora (Fig. 5; Huang, 1983; Inner Mongolia Autonomous Region BGMR, 1991). In the Yuanbaoshan arc, this flora includes *Angaropteridium?* sp., *Neoggerathiopsis?* sp., *?Sphenophyllum* sp., and *Nephropsis?* sp. to the northeast of Ejina Qi (Fig. 5; Wang et al., 1994, p. 62). To the northwest of Ejina Qi, this flora contains *Angaropteridium* sp., *Angaropteridium cardiopteroides,* and *Cardioneura* sp. (Inner Mongolia Autonomous Region BGMR, 1991, p. 169). In the Dong Ujimqin arc, this flora includes *Angaropteridium cardiopteroides, Angaridium potaninii, A. submongolicum, Neuropteris* cf. *izylensis, N.* cf. *paimbaensis, Neoggerathiopsis tschirkovae, N. longjiangensis, N.* cf. *derzavinii, Dicranophyllum* sp., *Paracalamites* sp. (Huang, 1983). The Angaran flora is very diagnostic, inasmuch as it is mainly distributed within the Siberian craton and its surrounding orogens before the Late Permian (Huang, 1983).

Petrochemistry

More than 200 petrochemical data sets from the Paleozoic volcanic rocks of the Beishan and Inner Mongolia orogens (Yue, 1985; Xu, 1987; Zuo and He, 1990; Wang et al., 1991; Tang, 1992; Yue and He, 1993) are used to refine the assignments of tectonic settings and stratigraphic correlations in the previous section. The samples were collected by various people and analyzed by many different laboratories over a long period of time, thus it is not possible to quantitatively estimate the errors carried in these data sets. However, we believe the errors to be insignificant inasmuch as these data sets provide consistent results.

A series of AFM and $\log\sigma$-$\log\tau$ diagrams ($\sigma = [K_2O + Na_2O]^2/[SiO_2 - 43]$, $\tau = [Al_2O_3 - Na_2O]/TiO_2$; Rittmann, 1973) are shown in Figures 6 and 7, respectively. The former diagram is very useful to differentiate tholeiitic series and calc-alkaline series, and the latter diagram can be used to discriminate non-orogenic and arc-related volcanics.

Ordovician to Silurian volcanics. On the AFM diagrams, most Ordovician–Silurian volcanic samples from the Dongqiyishan and Baoerhantu arcs show a clear calc-alkaline trend; some basaltic rocks from the latter arc are within tholeiitic field (Fig. 6, A and B). On the $\log\sigma$-$\log\tau$ diagram, 24 of the 25 samples from the Dongqiyishan arc plot as orogenic and island arc volcanics with one sample falling within the alkaline field (Fig. 7A); 41 of the 43 samples from the Baoerhantu arc plot as orogenic and island arc volcanics; one sample is within the alkaline field and one sample is near the orogenic–island arc–nonorogenic boundary but within the nonorogenic field (Fig. 7B).

Carboniferous volcanics. The Carboniferous volcanics from the Dongqiyishan arc and the Hanshan microcontinent are mainly rhyolitic, and have a calc-alkaline trend and strong arc affinity (Figs. 6C and 7C). Although 7 of the 14 samples from the Baoerhantu arc and Xilin Hot microcontinent have tholeiitic affinity, according to the AFM diagram (Fig. 6D), 10 of these 14 samples plot as orogenic and island arc volcanics, and the other 4 plot either on the orogenic–island arc–nonorogenic boundary or within the nonorogenic field (Fig. 7D). These results are consistent with the interpretation that these volcanics may have formed in a volcanic arc that may have undergone minor rifting. This suggestion is a reconciliation of the previous debate

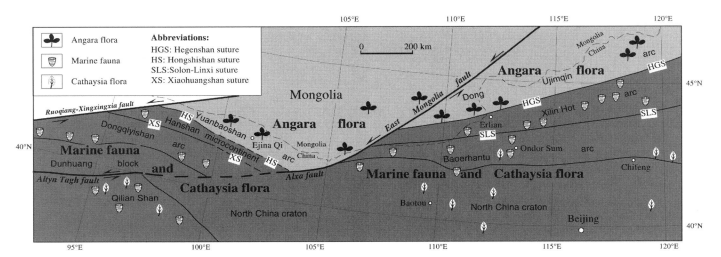

Figure 5. Distribution of Late Carboniferous faunas and floras in Beishan and Inner Mongolia orogens (after Huang, 1983; Gansu Province BGMR, 1989; Inner Mongolia Autonomous Region BGMR, 1991; Wang et al., 1994). Abbreviations as in Figure 1.

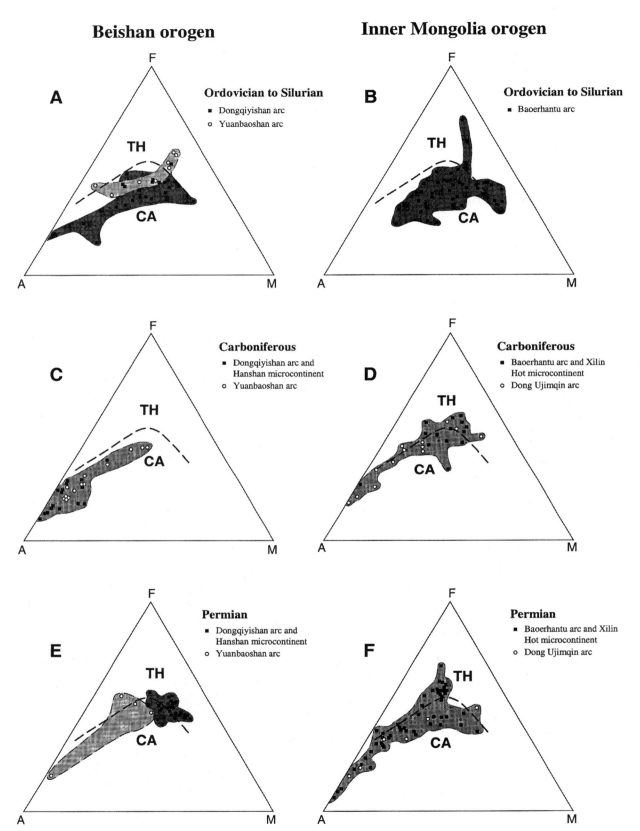

Figure 6. AFM diagrams for Paleozoic volcanic rocks of Beishan and Inner Mongolia orogens. (Boundary between tholeiitic [TH] and calc-alkaline [CA] fields is after Irvine and Baragar, 1971.)

Beishan orogen   **Inner Mongolia orogen**

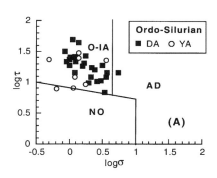
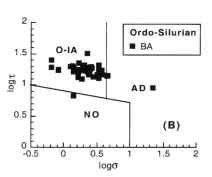

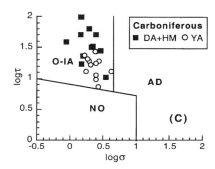
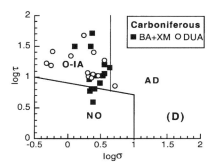

Figure 7. Log σ-log τ diagrams for Paleozoic volcanic rocks of Beishan and Inner Mongolia orogens. NO—nonorogenic volcanic lavas, O-IA—orogenic and island arc volcanic lavas, AD—alkaline derivatives of NO and O-IA (after Rittmann, 1973). BA—Baoerhantu arc, DA—Dongqiyishan arc, DUA—Dong Ujimqin arc, HM—Hanshan microcontinent, XM—Xilin Hot microcontinent, YA—Yuanbaoshan arc.

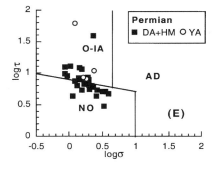
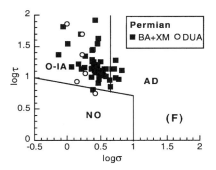

between a rift-setting origin (e.g., Shao, 1989; Tang, 1992) and an arc setting origin (e.g., Wang et al., 1991; Cheng, 1994). The samples from both the Yuanbaoshan and Dong Ujimqin arcs exhibit a very clear calc-alkaline trend in the AFM diagrams (Fig. 6, C and D) and fall within the orogenic–island arc field in the logσ-logτ diagram except for one sample that plots within the alkaline field (Fig. 7, C and D). The Carboniferous volcanics in the Beishan and the Inner Mongolia orogens are mainly calc-alkaline and have strong arc affinity.

Lower Permian volcanics. The Lower Permian volcanics from the Dongqiyishan arc and Hanshan microcontinent mainly plot at the boundary between tholeiitic and calc-alkaline fields, but do not show a clear trend on the AFM diagram (Fig. 6E). On the logσ-logτ diagram, 18 of the 31 samples plot as nonorogenic vol-

canics and 13 plot as orogenic–island arc volcanics (Fig. 7E). This feature also has been observed by previous studies, and three tectonic settings for these rocks have been proposed: (1) continental rift (e.g., Zuo and He, 1990), (2) volcanic arc (e.g., Cheng, 1994), and (3) rifting arc (e.g., Chen and Zhu, 1992). Based on the nature of the volcanics and their close relationship with Lower Permian turbidites, we believe that the proposed rifting arc setting is more plausible. The Lower Permian samples from the Baoerhantu arc and Xilin Hot microcontinent define a clear calc-alkaline trend on the AFM diagram (Fig. 6F; W.G. Ernst, 1999, personal commun.); on the logσ-logτ diagram, 49 of the 50 samples fall within the orogenic–island arc and alkaline fields (Fig. 7F).

Although the petrochemical data are not enough to define a trend for the volcanics from the Yuanbaoshan arc (Fig. 6E), all the four samples are within the orogenic and island arc field, suggesting an arc affinity (Fig. 7E). The samples of the Dong Ujimqin arc exhibit a clear calc-alkaline trend on the AFM diagram (Fig. 6F) and plot as orogenic and island arc volcanics, except for one sample that is near the boundary between the nonorogenic field and orogenic and island arc field but within the nonorogenic field (Fig. 7F).

Correlation of three tectonic units

The stratigraphic and petrochemical data presented here suggest a correlation between the three tectonic units in the Beishan orogen and the three tectonic units in the Inner Mongolia orogen. Both the Dongqiyishan arc in the Beishan orogen and the Baoerhantu arc in the Inner Mongolia orogen are characterized by Cambrian–Lower Ordovician passive continental margin, Ordovician-Silurian arc, Devonian arc and overlap, and Carboniferous–Lower Permian arc assemblages (Fig. 3). The Hanshan microcontinent in the Beishan orogen and the Xilin Hot microcontinent in the Inner Mongolia orogen are correlatable inasmuch as they contain similar Precambrian metamorphic assemblages, Devonian arc and overlap, and Carboniferous–Lower Permian arc assemblages. The Yuanbaoshan arc in the Beishan orogen and the Dong Ujimqin arc in the Inner Mongolia orogen share similar Cambrian–Lower Ordovician passive margin and Ordovician–Lower Permian arc or arc-related assemblages.

Moreover, the distribution of the Late Carboniferous floras and faunas also supports such a correlation. As can be seen in Figure 5, Angara-type plant fossils in the Inner Mongolia orogen occur only to the north of the Hegenshan suture zone; to the south of this suture zone marine fossils, such as corals, fusulinids, and brachiopods, occur with Cathysian-type plant fossils. In the Beishan orogen, similar Angara-type plant fossils occur in the Yuanbaoshan arc; this flora has been documented only to the north of the Hongshishan suture. To the south of this suture are coral, fusulinid, and brachiopod fossils, in addition to Cathysian plant fossils. Such a paleobiogeographic pattern is consistent with the proposed correlation between the three tectonic units in the Beishan orogen and those in the Inner Mongolia orogen.

CORRELATION OF BEISHAN AND INNER MONGOLIA OROGENS AND ITS IMPLICATIONS FOR TECTONICS OF NORTH CHINA

On the basis of the correlation of two ophiolite zones and three tectonic units, we conclude that each of the tectonostratigraphic units and ophiolite zones in the Beishan orogen has a corresponding element in the Inner Mongolia orogen, and that the Inner Mongolia orogen is the eastern extension of the Beishan orogen. It is thus apparent that each tectonic unit of the Beishan orogen and its counterpart in the Inner Mongolia orogen have been offset by the Alxa–east Mongolia fault. Matching all these corresponding tectonic units suggests a left-lateral offset of ~400 km on the Alxa–east Mongolia fault (Figs. 1 and 2).

Lamb et al. (1999) also noticed the large slip on the Alxa–east Mongolia fault and deduced ~200 km of left-lateral offset by matching rock units on either side of the fault. This difference in offset estimate is due to different considerations of the geology beneath the Badanjinlin desert (Fig. 2). We assume that the Beishan orogen basically continues its direction toward the Alxa fault beneath the desert, whereas Lamb et al. (1999) considered only outcrop distributions.

Implications for the tectonics of north China

The correlation between the Beishan and Inner Mongolia orogens has many implications for the palinspastic reconstruction of north China, especially for the north China craton–Tarim block relationship, and locality of the western extension of the north Qilian suture (Fig. 1).

Tarim block–north China craton relationship. As stated earlier, currently available data sets permit three competing hypotheses for the relationship between the Tarim block and the north China craton. These hypotheses are: (1) the Tarim block and north China craton have been contiguous since Precambrian time; (2) they were widely separated during Paleozoic time; and (3) the Tarim block is composed of at least two blocks; the northern block was contiguous with the north China craton in Paleozoic time. These differing hypotheses suggest that geological and geophysical data solely from the two blocks are not enough to constrain their relationship due to the extensive Cenozoic cover of the Tarim basin, the main body of the Tarim block (Fig. 1).

The correlation between the better exposed Beishan and Inner Mongolia orogens has provided a new insight into the Tarim block–north China craton relationship on the basis of the logic that two blocks are connected if they share a single continuous margin. The advantage of this approach is that it provides a directly testable hypothesis that rest on the analysis of a well exposed area (i.e., the Beishan and Inner Mongolia orogens) instead of relying on oversimplified broad comparisons of poorly exposed areas (e.g., the Tarim block). From the correlations described in the previous sections, the Beishan orogen to the north of the northeastern Tarim block and the Inner Mongolia orogen to the north of the north China craton record an almost identical Paleozoic evolutionary history and project along strike into each other after 400 km restoration along the truncating strike-slip fault. Therefore, at least the northeastern part of the Tarim block (i.e., Dunhuang block) apparently connected with the north China craton through Paleozoic time.

Restoration of the Inner Mongolia orogen to the Beishan orogen along the Altyn Tagh–Alxa–east Mongolia fault (Fig. 1) requires a similar amount of restoration of the north China craton. This enlarges the narrow connection between the north China craton and the northeastern Tarim block into much wider region (Fig. 8), suggesting that this strange narrow connection (i.e., the Anxi geowaist in Fig. 1) may have formed by post-Paleozoic strike-slip faulting.

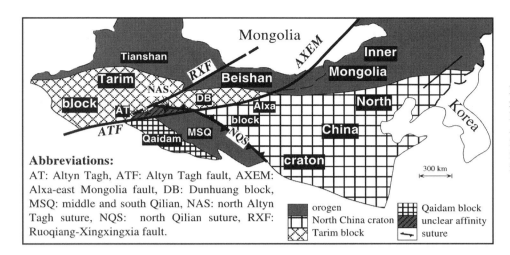

Figure 8. Tectonic pattern of north China reconstructed by restoring Inner Mongolia orogen to Beishan orogen along Altyn Tagh and Alxa–east Mongolia faults. Slip on Ruoqiang-Xingxingxia fault is not restored because of lack of information.

Abbreviations:
AT: Altyn Tagh, ATF: Altyn Tagh fault, AXEM: Alxa-east Mongolia fault, DB: Dunhuang block, MSQ: middle and south Qilian, NAS: north Altyn Tagh suture, NQS: north Qilian suture, RXF: Ruoqiang-Xingxingxia fault.

Western extension of the north Qilian suture. Three possibilities have been proposed for the western extension of the north Qilian suture (Fig. 1) beyond the Qilian Shan: (1) continuation into the Beishan orogen (e.g., Zhou and Graham, 1996); (2) equivalence with the north Altyn Tagh suture (e.g., Ge et al., 1991; Cui et al., 1996, 1999; Che et al., 1998; Xu et al., this volume); (3) continuation into the west Kunlun orogen (e.g., Zhang, 1985). Our correlation between the Beishan and Inner Mongolia orogens is consistent with and therefore supports the north Qilian–north Altyn Tagh suture connection. This connection is also consistent with the 400 ± 60 km post-Middle Jurassic left-lateral slip along the Altyn Tagh fault proposed by Ritts and Biffi (2000) on the basis of an offset Bajocian paleolake shoreline.

Both the Beishan and West Kunlun orogens contain a continuous sequence of upper Paleozoic volcaniclastic rocks, whereas the north Qilian suture in the Qilian Shan does not have equivalent upper Paleozoic rocks (e.g., Cheng, 1994). Instead, these two orogens have been suggested to be correlatable with the Inner Mongolia and east Kunlun orogens, respectively, in terms of their Paleozoic assemblages (Ge et al., 1991; Peltzer and Tapponnier, 1988). The north Qilian and north Altyn Tagh sutures contain similar early Paleozoic subduction complexes. The subduction complex along the north Qilian suture is composed mainly of ophiolite and ophiolitic melange, some of which has undergone blueschist to eclogite facies metamorphism. The age of the ophiolite was determined to be early Paleozoic based on consistent radiometric ages of 440–462 Ma for the blueschist eclogite (Xu et al., this volume). The north Altyn Tagh subduction complex is characterized by ophiolitic melanges of ultramafics, cumulate gabbro, pillow lava, and radiolarian chert; transitional blueschist greenschist with sodic amphiboles has been described in its eastern part (Qinghai Province BGMR, 1991). The $^{39}\mathrm{Ar}$-$^{40}\mathrm{Ar}$ dating of the sodium amphibole yields a plateau age of 457 ± 0.7 Ma, the same as the age of the north Qilian subduction complex (Sobel and Arnaud, 1999; Xu et al., this volume).

Tectonic zonation and evolution of north-central China

On the basis of the preceding correlations and discussion, we suggest the following tectonic zonation scheme for north-central China: from north to south, (1) the Beishan–Inner Mongolia orogen, (2) the north China–northeastern Tarim block, and (3) the north Qilian–north Altyn Tagh suture (Fig. 8). Xu et al. (this volume) suggest that south of the north Qilian–north Altyn Tagh suture is the middle-south Qilian–Altyn Tagh block (Fig. 8).

In Phanerozoic time, this region underwent two distinctive stages of evolution: Paleozoic assembly and post-Paleozoic strike-slip faulting. In the following we present our current tectonic model for the Phanerozoic history of this region.

Cambrian–Early Ordovician: Passive continental margin. The occurrence of Precambrian ophiolitic rocks (Wang et al., 1991; Zuo and Li, 1996), Early Cambrian microfossils in cherty rocks, and the occurrence of Lower Cambrian turbidites in the Dongqiyishan arc suggest that an ocean already existed to the north of the north China–northeastern Tarim block at the very beginning of the Phanerozoic (Fig. 9). Nonvolcanic Cambrian to Lower Ordovician strata further suggest that subduction did not begin until the Middle Ordovician. To the south of the north China–northeastern Tarim block, however, the oceanic lithosphere represented by the north Qilian–north Altyn Tagh ophiolite zone may have begun subducting during this period of time (Xu et al., this volume)

Middle Ordovician–Late Silurian: Volcanic arc. The initiation of calc-alkaline volcanism in both the Yuanbaoshan-Dong Ujimqin and the Dongqiyishan-Baoerhantu arcs suggests that the oceanic lithosphere began subducting southward beneath the north China–Dunhuang block and northward beneath a northern plate, the nature and extent of which are poorly understood (Webb et al., 1999). The continuous record of calc-alkaline arc volcanics indicates that this process continued to the Late Silurian (Fig. 9).

Latest Silurian–Devonian: Microcontinent-arc collision and subduction. The tectonic setting of the Yuanbaoshan–Dong

Ujimqin arc over this period is unknown. We infer an arc-related basin setting, inasmuch as Lamb (1998) recognized a Devonian arc north of this arc, and Devonian strata are widely distributed and mainly composed of sedimentary strata with a significant amount of tuffaceous rocks.

Along the northern margin of the north China–northeastern Tarim block, the Hanshan–Xilin Hot microcontinent collided with the Dongqiyishan–Baoerhantu arc, resulting in the deposition of Devonian coarse-grained sedimentary rocks and an unconformity between Upper Silurian or Lower Devonian and Ordovician–Silurian arc assemblages. During the collision process, oceanic lithosphere may have continued to subduct southward along a new subduction zone north of the Hanshan–Xilin Hot microcontinent, as suggested by the calc-alkaline nature and arc affinity of the Devonian volcanics.

Along the southern margin of the north China–northeastern Tarim block, the middle and south Qilian–Altyn Tagh block may have also collided with the north China–northeastern Tarim block approximately at the same time, resulting in the closure of the ocean manifested by the north Qilian–north Altyn Tagh ophiolite zone (Fig. 9; Xu et al., this volume).

Carboniferous–Early Permian: Volcanic arc. Widespread Carboniferous–Lower Permian arc volcanics suggest continued subduction during this period (Fig. 9). Some intraarc rifting may have occurred at some localities, resulting in the volcanic rocks with both rift and arc affinities.

Late Permian: Collision. The Upper Permian strata in both the Beishan and Inner Mongolia orogens are nonmarine, suggesting final closure of the ocean and collision between the northern Yuanbaoshan–Dong Ujimqin arc and the accreted north China–northeastern Tarim block. Mountain building processes induced by this collision may have also resulted in the lack of Triassic strata in both the Beishan and Inner Mongolia orogens (Fig. 9).

Post-Paleozoic strike-slip event. The tectonic pattern that resulted from the Paleozoic tectonic processes and events is

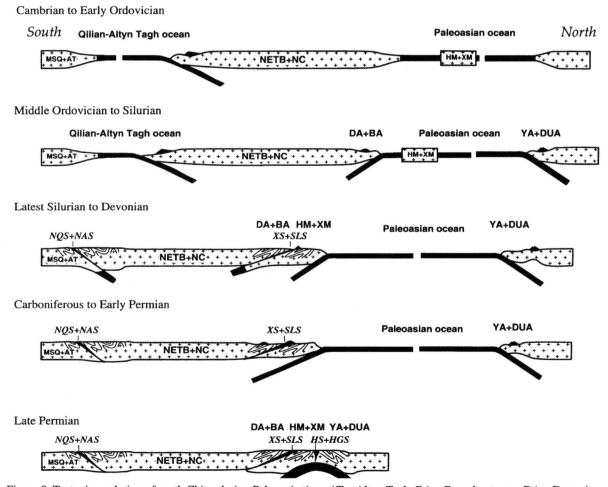

Figure 9. Tectonic evolution of north China during Paleozoic time. AT—Altyn Tagh, BA—Baoerhantu arc, DA—Dongqiyishan arc, DUA—Dong Ujimqin arc, HGS—Hegenshan suture, HM—Hanshan microcontinent, HS—Hongshishan suture, MSQ—middle and south Qilian, NAS—north Altyn Tagh suture, NC—north China craton, NETB—northeastern Tarim block, NQS—north Qilian suture, SLS—Solon-Linxi suture, XM—Xilin Hot microcontinent, XS—Xiaohuangshan suture, YA—Yuanbaoshan arc.

shown in Figure 8. However, this configuration has been significantly modified due to post-Paleozoic strike-slip faulting (Yue and Liou, 1999), and two stages of fault movement have been recognized. Stage I (inferred from Oligocene to ca. 13–16 Ma): 400 km of displacement separated the Inner Mongolia orogen from the originally contiguous Beishan orogen along a continuous Altyn Tagh–Alxa–east Mongolia fault (AXEM in Fig. 8), and offset the north China craton from the northeastern Tarim block and the north Qilian suture from the north Altyn Tagh suture. Stage II (inferred from 13–16 Ma to present): the Alxa–east Mongolia fault became inactive; additional offset along the Altyn Tagh fault has been mainly accommodated by shortening of the Qilian Shan and the Qaidam basin (for details see Yue and Liou, 1999).

It has been suggested that both the Beishan and Inner Mongolia orogens may also record a significant amount of shortening in the Jurassic and large-scale extension in the Cretaceous (Zheng and Zhang, 1993; Davis et al., 1998; Zheng et al, 1998; Webb et al., 1999). However, it is still difficult to assess the effects of these deformational events on the proposed correlation between the Beishan and Inner Mongolia orogens, inasmuch as this requires more detailed structural data.

CONCLUSIONS AND SUGGESTIONS FOR FUTURE STUDIES

Previous tectonic correlations between the eastern and western north China across the Altyn Tagh and Alxa–east Mongolia faults have been discussed predominantly in terms of the relationship between the Tarim block and north China craton. These studies have created many controversies inasmuch as the Tarim block is extensively covered with Quaternary sediments and detailed geophysical data are unavailable.

Stratigraphic and petrochemical data have revealed a strong similarity between the Beishan and Inner Mongolia orogens. Both of these two orogens are divided into three petrotectonic units by a southern early Paleozoic suture and a northern late Paleozoic suture. The Dongqiyishan arc of the Beishan orogen and the Baoerhantu arc of the Inner Mongolia orogen contain similar Cambrian–Lower Ordovician passive continental margin, Middle Ordovician–Silurian arc, Devonian arc and overlap, and Carboniferous–Lower Permian assemblages. The Hanshan microcontinent of the Beishan orogen and the Xilin Hot microcontinent of the Inner Mongolia orogen share similar Lower Proterozoic metamorphic, Devonian arc and overlap, and Carboniferous–Lower Permian arc assemblages. The Yuanbaoshan arc of the Beishan orogen and the Dong Ujimqin arc of the Inner Mongolia orogen share a common Cambrian–Lower Ordovician passive margin assemblage and a common Middle Ordovician–Lower Permian arc or arc-related assemblage.

The correlation between the Beishan and Inner Mongolia orogens suggests that at least the northeastern Tarim block and north China craton shared a common continental margin, and thus were connected during the Paleozoic. Moreover, this cor-

relation also suggests ~400 km of left-lateral strike slip on the northeast-east–northeast–trending Alxa–east Mongolia fault. Restoration of 400 km of left slip of Inner Mongolia orogen to the Beishan orogen along the Altyn Tagh and Alxa–east Mongolia fault (Yue and Liou, 1999) requires a similar amount of movement for the north China craton relative to the Tarim block. This restoration also suggests that the north Qilian suture extends westward to the north Altyn Tagh suture and suggests that the present narrow connection between the Tarim block and north China craton (i.e., the Anxi geowaist in Fig. 1) was formed by post-Paleozoic strike-slip faulting. The connection was smooth and wide before the faulting (Fig. 8).

We make two suggestions for future studies. First, previously some workers noted the possibility of large-scale left-lateral strike-slip faulting along the Alxa–east Mongolia fault (e.g., Suvorov, 1982). However, most proposed tectonic maps of Mongolia show some Paleozoic tectonic zones continuing across that fault (e.g., Tomurtogoo, 1996). If our proposed correlation is correct, all tectonic units in eastern Mongolia should have been offset by several hundred kilometers. Therefore, tectonic zonation of Mongolia will provide the most crucial test for our proposed correlation. Second, although our proposed correlation suggests a connection between the northeastern Tarim block and north China craton in the Paleozoic, it is still unclear if the rest of the Tarim block was contiguous with the north China craton because of the extensive Quaternary cover in the Tarim basin (Fig. 1). To address this issue, the relationship between the Tianshan and Beishan orogens needs to be investigated to determine if the middle and western parts of the Tarim block share a common Paleozoic continental margin with its northeastern part (i.e., Dunhuang block in Fig. 1). In addition, the western extension of the early Paleozoic north Qilian–north Altyn Tagh suture must be identified with certainty, and it must be verified if the northern and southern Tarim blocks were connected during early Paleozoic time.

ACKNOWLEDGMENTS

We thank Edmund Chang, Kim Hannula, Marc Hendrix, and Christoph Heubeck for constructive reviews. This project was supported by the Stanford–China Geosciences Industrial Affiliates Program, American Association of Petroleum Geologists Grant-in-Aid Fund, and McGee Fund of the School of Earth Sciences at Stanford University.

REFERENCES CITED

Cao, C., 1989, The ophiolite belts of northeastern China: Journal of Southeast Asian Sciences, v. 3, p. 233–236.

Che, Z., Liu, L., Liu, H., and Luo, J., 1998, The constituents of the Altun fault system and genetic characteristics of related Meso-Cenozoic petroleum-bearing basin: Regional Geology of China, v. 17, p. 377–384 (in Chinese).

Chen, S., and Zhu, Y., 1992, Rock chemistry and analysis of its structural environment for Carboniferous-Permian volcanics in Beishan, Xinjiang: Di Qiu Ke Xue, v. 17, p. 647–656 (in Chinese).

Cheng, Y., ed., 1994, An outline of the regional geology of China: Beijing, Geological Publishing House, 517 p. (in Chinese).

Cui, J., Tang, Z., Deng, J., Yue, Y., Li, J., and Lai, S., 1996, Early Paleozoic plate tectonic regime of the Altyn Tagh: Proceedings of the 30th Internation Geological Congress, v. 7, p. 59–74.

Cui, J., Tang, Z., Deng, J., Yue, Y., Meng, L., and Yu, Q., 1999, The Altyn Tagh fault system: Beijing, Geological Publishing House, 249 p. (in Chinese).

Davis, D.A., Wang, C., Zheng, Y., Zhang, J., Zhang, C., and Gehrels, G.E., 1998, The enigmatic Yinshan fold-and-thrust belt of northern China; new views on its intraplate contractional styles: Geology, v. 26, p. 43–46.

Gansu Province Bureau of Geology and Mineral Resources, 1989, Regional geology of Inner Mongolia Autonomous Region: Beijing, Geological Publishing House, 692 p. (in Chinese).

Ge, X., Duan, J., Li, C., Yang, H., and Tian, Y., 1991, A new recognition of the Altun fault zone and geotectonic pattern of northwest China: IGCP Project 321, Proceedings of First International Symposium on Gondwana Dispersion and Asian Accretion: Beijing, China University of Geosciences Press, p. 125–128.

He, G., and Shao, J., 1983, Determination of early Paleozoic ophiolites in southeastern Nei Mongol and their geotectonic significance: Contributions to the Project of Plate Tectonics of north China, no. 1, p. 241–250 (in Chinese).

Huang, B., 1983, On late Late Paleozoic palaeophyto-geographic regions of eastern Tianshan-Hingan fold belt and its geological significance: Contributions to the Project of Plate Tectonics of North China, no. 1, p. 138–152 (in Chinese).

Huang, T.K., Ren, J., Jiang, C., Zhang, Z., and Xu, Z., 1977, An outline of the tectonic characteristics of China: Acta Geologica Sinica, v. 59, p. 117–135.

Inner Mongolia Autonomous Region Bureau of Geology and Mineral Resources, 1991, Regional geology of Inner Mongolia Autonomous Region: Beijing, Geological Publishing House, 725 p. (in Chinese).

Irvine, T.N., and Baragar, W.R.A., 1971, A guide to the chemical classification of the common volcanic rocks: Canadian Journal of Earth Sciences, v. 8, p. 523–544.

Lamb, M.A., 1998, Paleozoic sedimentation, volcanism and tectonics of southern Mongolia [Ph.D. thesis]: Stanford, Stanford University, 207 p.

Lamb, M.A., Hanson, A.D., Graham, S.A., Badarch, G., and Webb, L.E., 1999, Left-lateral sense offset of Upper Proterozoic to Paleozoic features across the Gobi Onon, Tost, and Zuumbayan faults in southern Mongolia and implications for the central Asian faults: Earth and Planetary Science Letters, v. 173, p. 183–194.

Li, C., Wang, Q., Liu, X., and Tang, Y., 1982, Tectonic map of Asia: Beijing, Science Press, scale: 1:8 000 000.

Li, C., and Tang, Y., 1983, Some problems of subdivision of paleoplates in Asia: Acta Geologica Sinica, v. 57, p. 1–10 (in Chinese).

Li, Y.-P., Li, Y.-A., Zhang, Z., Zhai, Y., Li, Q., Gao, Z., Sharps, R., and McWilliams, M., 1989, Apparent polar wander path of the Tarim Massif in China: Acta Geologica Sinica, v. 63, p. 193–202 (in Chinese).

Liu, J., 1983, Study of the ophiolite from Hegenshan, Inner Mongolia, and its tectonic implications: Contributions to the Project of Plate Tectonics of north China, no. 1, p. 117–135 (in Chinese).

Liu, X., Wu, S., Du, D., Yao, J., Ding, X., and Wang, Y., 1997, The sedimentary-tectonic evolution of Tarim plate and its surrounding area: Urumqi, Xinjiang Science Technology and Hygiene Publishing House, 267 p. (in Chinese).

Peltzer, G., and Tapponnier, P., 1988, Formation and evolution of strike-slip faults, rifts, and basins during the India-Asia collision: An experimental approach: Journal of Geophysical Research, v. 93, p. 15085–15117.

Qinghai Province Bureau of Geology and Mineral Resources, 1991, Regional geology of Inner Mongolia Autonomous Region: Beijing, Geological Publishing House, 662 p. (in Chinese).

Ren, J., Jiang, C., Zhang, Z., and Qin, D., 1987, Geotectonic evolution of China: Beijing, Science Press, 203 p.

Ritts, B.D., and Biffi, U., 2000, Magnitude of post-Middle Jurassic (Bajocian) displacement on the Altyn Tagh fault, northwest China: Geological Society of America Bulletin, v. 112, p. 61–74.

Rittmann, A., 1973, Stable mineral assemblages of igneous rocks: New York, Springer-Verlag, 262 p.

Shao, J., 1989, Continental crust accretion and tectono-magmatic activity at the northern margin of the Sino-Korean Plate: Journal of Southeast Asian Earth Sciences, v. 3, p. 57–62.

Sobel, E., and Arnaud, N., 1999, A possible middle Paleozoic suture in the Altyn Tagh, NW China: Tectonics, v. 18, p. 64–74.

Suvorov, A.I., 1982, Structural map and faults of Mongolia: Izvestiya Akademii Nauk SSSR, Seriya Geologicheskaya, no. 6, p. 122–136 (in Russian).

Tang, K., 1990, Tectonic development of Paleozoic foldbelts at the north margin of the Sino-Korean craton: Tectonics, v. 9, p. 249–260.

Tang, K., 1992, Tectonic evolution and ore-forming regularity in the northern fold belt of the Sino-Korean plate: Beijing, Peking University Press, 277 p. (in Chinese).

Tomurtogoo, O., 1996, A new tectonic scheme of the Paleozoides in Mongolia: Proceedings of the 30th International Geological Congress, v. 7, p. 75–82.

Wang, Q., Liu, X., and Li, J., 1991, Plate tectonics between Cathaysia and Angaraland in China: Beijing, Peking University Press, 151 p. (in Chinese).

Wang, T., Wang, J., Liu, J., Wang, S., and Wu, J., 1993, Relationships between the north China and Tarim plates: Acta Geologica Sinica, v. 67, p. 287–300.

Wang, T., Wang, S., and Wang, J., 1994, The formation and evolution of Paleozoic continental crust in Alxa region: Lanzhou, Lanzhou University Press, 215 p. (in Chinese).

Webb, L.E., Graham, S.A., Johnson, C.L., Badarch, G., and Hendrix, M.S., 1999, Occurrence, age, and implications of the Yagan–Onch Hayrhan metamorphic core complex, southern Mongolia: Geology, v. 27, p. 143–147.

Wu, T., 1990, Lithospheric compositional and tectonic evolution of the northern Alxa block in the Paleozoic [Ph.D. thesis]: Beijing, Peking University, 119 p. (in Chinese).

Wu, T., and He, G., 1993, Tectonic units and their fundamental characteristics on the northern margin of the Alxa Block, Inner Mongolia: Acta Geologica Sinica, v. 67, p. 97–108.

Xu, D., 1987, Stratigraphic and petrochemical characteristics of a fossil volcanic-arc rock series—Baorhantu group with a discussion of its genesis: Contributions to the Project of Plate Tectonics of north China, no. 2, p. 101–112 (in Chinese).

Yang, H., 1983, Geomagnetic structure and oil-gas prospect of the Tarim basin, *in* Zhu, X., ed., The structure and evolution of the Mesozoic-Cenozoic basins of China: Beijing, Science Press, p. 212–219 (in Chinese).

Yanshin, A.L., 1989, Geological map of the Republic of Mongolia: Moscow, Academia Nauk USSR, scale 1:1 500 000 (in Russian).

Yue, Y., 1985, Hercynian tectonic evolution features of the northern Beishan [B.S. thesis]: Beijing, Peking University, 76 p. (in Chinese).

Yue, Y., and He, G., 1993, Early Permian bimodal volcanism in Linxi county and its tectonic significance: Geological Study, v. 26, p. 105–118 (in Chinese).

Yue, Y., and Liou, J.G., 1999, A two-stage evolution model for the Altyn Tagh fault, China: Geology, v. 27, p. 227–230.

Zhang, Z., 1985, Geology of the Altyn Tagh fault: Xian Institute of Geology Bulletin, v. 9, p. 20–32 (in Chinese).

Zheng, Y., and Zhang, Q., 1993, The Yagan metamorphic core complex and extensional detachment fault in Inner Mongolia: Acta Geologica Sinica, v. 67, p. 301–309.

Zheng, Y., Davis, G.A., Wang, C., Darby, B.J., and Hua, Y., 1998, Major thrust sheet in the Daqing Shan Mountains, Inner Mongolia, China: Science in China, ser. D, v. 41, p. 553–560.

Zhou, D., and Graham, S.A., 1996, Extrusion of the Altyn Tagh wedge: A kinematic model for the Altyn Tagh fault and palinspastic reconstruction of northern China: Geology, v. 24, p. 427–430.

Zuo, G., and He, G., 1990, Plate tectonics and ore-formation regularity of Beishan: Beijing, Peking University Press, 226 p. (in Chinese).

Zuo, G., and Li, M., 1996, Formation and evolution of the Early Paleozoic lithosphere in the Beishan area: Lanzhou, Gansu Science and Technology Press, 120 p. (in Chinese).

Manuscript Accepted by the Society June 5, 2000

Geological Society of America
Memoir 194
2001

Paleozoic sedimentary basins and volcanic arc systems of southern Mongolia: New geochemical and petrographic constraints

Melissa A. Lamb*
Department of Geological and Environmental Sciences, Stanford University, Stanford, California 94305-2115, USA
Gombosuren Badarch
Institute of Geology and Mineral Resources, Mongolian Academy of Sciences, 63 Peace Avenue,
Ulaan Baatar 210357, Mongolia

ABSTRACT

Paleozoic strata of southern Mongolia record significant portions of the geologic development of Central Asia but have only been studied by a handful of workers. In this chapter we present new data from these rocks and address key questions concerning the Paleozoic development of southern Mongolia and its relation to geologic events in China.

Major and trace element geochemical analyses of basalts from southern Mongolia suggest formation in volcanic arc, backarc basin, and within plate settings during the Devonian through Permian time periods. Sandstone provenance data from 12 localities document volcanic activity and record a shift from mature, quartzose Ordovician and Silurian sandstones to immature, volcanic lithic-rich Devonian and Carboniferous sandstones. Permian sandstone data point to volcanic activity and uplift of the Devonian and Carboniferous arc material.

These new data, combined with other sedimentologic and stratigraphic data, allow for better constrained tectonic models and paleogeographic reconstructions. Devonian strata of southern Mongolia represent components of two arc systems, including a trapped ocean basin that developed into a backarc basin. Carboniferous rocks record the development of a large arc system in southernmost Mongolia that may have extended westward into the Bogda Shan volcanic arc of China. The Carboniferous arc was probably a continuation of the Devonian arc system and was built in part upon Devonian strata. Permian strata record mainly nonmarine basins and the closing of the ocean basin between southern Mongolia and tectonic blocks of China.

INTRODUCTION

During the Paleozoic and early Mesozoic eras, a series of continental blocks, volcanic arc systems, accretionary wedges, and associated basins amalgamated to form central Asia (Fig. 1; Zhang et al., 1984; Coleman, 1989; Nie et al., 1990; Enkin et al., 1992; Şengör et al., 1993a, 1993b; Zorin et al., 1993; Carroll

et al., 1995; Chang et al., 1996). Paleozoic rocks of southern Mongolia record a significant part of this history (Fig. 2; Table 1). Most authors agree that this time period represents the growth of southern Mongolia through the development and accretion of volcanic arc material and, some authors suggest, fragments of continental crust (Voznesenskaya et al., 1989; Ruzhentsev and Pospelov, 1992; Şengör et al., 1993b; Zorin et al., 1993). Beyond this very general statement, however, there are markedly conflicting interpretations on the specifics of this history.

*Present address: Department of Geology, University of St. Thomas, Mail #OWS 153, 2115 Summit Avenue, St. Paul, Minnesota 55105.

Lamb, M.A., and Badarch, G., 2001, Paleozoic sedimentary basins and volcanic arc systems of southern Mongolia: New geochemical and petrographic constraints, *in* Hendrix, M.S., and Davis, G.A., eds., Paleozoic and Mesozoic tectonic evolution of central Asia: From continental assembly to intracontinental deformation: Boulder, Colorado, Geological Society of America Memoir 194, p. 117–149.

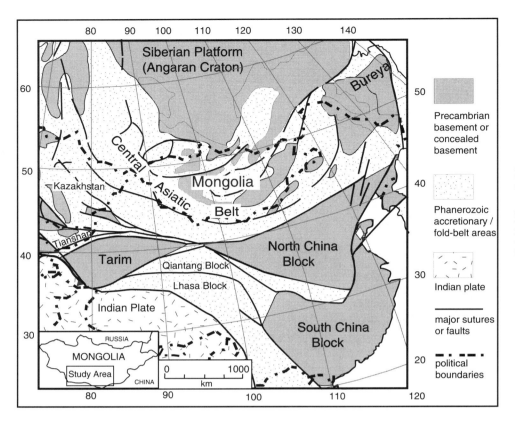

Figure 1. Tectonic and political boundaries of Asia (after Dobretsov et al., 1995). North China, south China, Lhasa, and Qiantang blocks represent mainly Mesozoic accretion, whereas much of Mongolia and central Asiatic belt represent Paleozoic accretion.

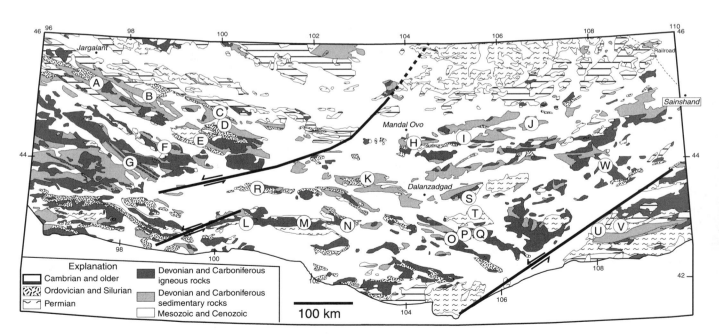

Figure 2. Simplified geologic map of southern Mongolia with selected faults, after Yanshin (1989). Letters denote localities examined as part of this study; see Table 1 for latitude and longitude of these localities. A—Tsahir Uul, B—Suj, C—Ulan Han Uul, D—Narin Sahir, E—Shin Jinst, F—Huvin Har, G—Edren, H—Mushgai Hudag, I—Borteg, J—Havtgai, K—Gurvan Sayhan, L—Tost Uul, M—Noyon Uul, N—Deng Nuru, O—Baga Nomgon, P—Nomgon, Q—Ih Uvgon, R—Nemegt, S—Tsogt Tsisi, T—Naftgar Uul, U—Lugin Gol, V—Hobsgol, W—Tsagan Suvarga.

TABLE 1. LATITUDE AND LONGITUDE FOR FIELD LOCALITIES

Locality	Map Letter	Latitude	Longitude	Age*
Tsahir Uul	A	45° 25.4'	97° 6.25'	D
Suj	B	45° 05.869'	98° 28.436'	D
Ulan Han Uul	C	44°53.255'	99° 49.582'	O-S
		44° 53.302'	99° 49.788'	D
Narin Sahir	D	44° 46.527'	99° 47.907'	D
Shin Jinst	E	44° 21.527'	99° 27.105'	O-S
		44° 22.105'	99° 27.0267'	D
		44° 23.243'	99° 26.278'	C
Huvin Har	F	44° 13.565'	98° 48.02'	D
Edren	G	43° 59.338'	97° 59.093'	D
		44° 1.912'	98° 10.11'	C
Mushgai Hudag	H	44° 22.67'	104° 18.167'	O-S
		44° 23.64'	104° 18.26'	D
		44° 24.228'	104° 18.13'	C
Borteg	I	44° 17.992'	105° 29.714'	D
Havtgai	J	44° 40.989'	106° 47.199'	O-S
		44° 41.330'	106° 48.916'	D
Gurvan Sayhan	K	43° 47.233'	103° 7.335'	D
Tost Uul	L	43° 9.7'	100° 54.1'	C
Noyon Uul	M	43° 07.819'	101° 53.821'	C
Deng Nuru	N	43° 04.935'	102° 43.253'	C
Baga Nomgon	O	42° 54.008'	104° 50.333'	C
Nomgon	P	42° 54.872'	105° 13.684'	C
Ih Uvgon	Q	42° 54.443'	105° 32.726'	C
Nemegt	R	43° 40.482'	100° 42.591'	C
Tsogt Tsisi	S	43° 36.258'	105° 28.419'	P
Naftgar Uul	T	43° 12.890'	105° 11.591'	P
Lugin Gol	U	42° 58'	108° 30'	P
Hobsgol	V	43° 36'	109° 45'	P
Tsagan Suvarga	W	43° 52.020'	108° 20.760'	D-C

*O = Ordovician, S = Silurian, D = Devonian, C = Carboniferous, P = Permian.

One of the principal difficulties in evaluating the contrasting interpretations has been a lack of data, due primarily to the remoteness of southern Mongolia and the small number of geoscientists who have worked in the area. Beginning in 1967, paleontologists, petrologists, and mapping geologists of the Soviet-Mongolian Scientific Research Expeditions produced geologic maps that we found to be lithologically accurate but often lacking in structural data. Few sedimentologists participated in these expeditions, and thus depositional environments and stratigraphic facing directions were not always recorded. Sample locations are often either not given or are indicated only on small-scale maps. As a result, much of the published work on southern Mongolia contains interpretations that are too detailed to be supported by their databases and thus are difficult to evaluate critically (e.g., Ruzhentsev et al., 1985; Voznesenskaya et al., 1989; Ruzhentsev and Pospelov, 1992).

Despite the lack of rigorous geologic constraints for this region, several authors have constructed complex tectonic models for Asia that include southern Mongolia. For example, Şengör et al. (1993b, 1996, personal commun.), Natal'in et al. (1996), and Şengör and Natal'in (1996) developed an intricate palinspastic reconstruction for the Altaids, the region bounded by the Siberian platform, north China block, and Tarim basin (central Asiatic belt in Fig. 1). This model contains more than 44 tectonostratigraphic units and delineates the development and history of movement for each unit throughout the Phanerozoic. Zonenshayn et al. (1990), Mossakovsky et al. (1993), and Scotese and McKerrow (1990; updated in Golonka et al., 1994) presented equally complex but different palinspastic models for Asia during the Phanerozoic. Nie et al. (1990, p. 405) presented a model of amalgamation for Asia during the late Paleozoic but admitted that "precise paleogeographic reconstructions . . . are not yet possible." Enkin et al. (1992) published a model for the amalgamation of China and surrounding blocks based on paleomagnetic data. They acknowledged that sutures within Mongolia are not well understood and decided, as a preliminary approach, to treat Mongolia as one block. This large number of complex and contrasting models reflects the present lack of data: with so little known about southern Mongolia, there are many degrees of freedom in constructing models. Expansion and improvement of the data set for this region therefore is crucial to define models and develop a tectonic framework for central Asia.

The goal of our work was to collect well-documented, reproducible data and begin to develop a well-defined geologic framework. We have focused on Ordovician through Carboniferous rocks but also examined Permian strata, partly within our own work, but also as part of two additional studies (Amory, 1996; Zinniker, 1997). This chapter builds upon our first paper in which we presented new stratigraphic and sedimentologic data and discussed possible paleogeographic and tectonic interpretations (Lamb and Badarch, 1997). In this chapter we report new geochemical and petrographic data, and incorporate data from both papers to update our paleogeographic models. Our models are necessarily simplistic due to the lack of a comprehensive geologic database for southern Mongolia, but are based on accurate and well-documented data and take into account one phase of post-Paleozoic structural deformation.

METHODS

We analyzed 21 whole-rock igneous samples from 4 localities for major, minor, and trace element abundances with X-ray fluorescence (XRF). Whole-rock igneous samples were trimmed, ground, and hammered into small chips, which were then examined and culled to minimize the amount of hydrothermal veins, lithic inclusions, and weathering. These were crushed in a tungsten carbide jawcrusher and ground to a powder in a tungsten carbide shatterbox. Major element analysis was conducted on glass discs made by fusing ignited sample powder with a lanthanum-bearing lithium borate flux (Norrish and Hutton, 1969). Samples for XRF trace element analysis were prepared with ~5 g of sample powder pressed into a boric acid backing using a hydraulic press (Norrish and Chappell, 1977). Major and trace element abundances (Tables 2 and 3) were analyzed at Stanford University using an automated Rigaku S-MAX wavelength-dispersive spectograph, equipped with an end-window, Rh-target X-ray tube. Major element matrix corrections were made following the method of Norrish and Hutton (1969). Trace element analyses were corrected for nonlinear backgrounds and spectral interferences after Norrish and Chappell (1977). Matrix absorption effects for all XRF trace elements were measured using a modified Compton scattering methods (Reynolds, 1967; Willis, 1989). All XRF analyses are presented in Table 3. Each analysis represents an average of separate analyses of two splits of each sample. Estimates of accuracy and precision (Table 3) were determined by analyzing well-characterized basalt, andesite, and granite standards (BVHO-1, AGV-1, and G-2) together with the unknowns. Detection limits for major elements range from 0.01 to 0.001 wt%; for Nb, Zr, Y, Sr, and Rb from 0.3 to 0.5 ppm; for Ga, Pb, Th, U, Zn, Cu, Ni, and Cr from 1.0 to 2.0 ppm; and for Ce, Ba, and La from 10 to 15 ppm. The detection limit for V is 4.0 ppm.

We point counted 93 sandstone samples from 12 localities in southern Mongolia using a modified Gazzi-Dickinson method (Dickinson, 1970; Ingersoll et al., 1984; Tables 4 and 5; Fig. 3). We counted 500 grains for each thin section to minimize uncertainty (VanderPlas and Tobi, 1965). Many of the more than 200 samples collected underwent minor alteration or low-grade

metamorphism that partially obscured original textures. This is most prevalent in the Ordovician and Silurian sandstones; in several localities, rocks of this age were mildly metamorphosed, folded, and deformed. In thin section, these rocks often display evidence of significant compaction and dissolution. Because these processes may preferentially remove grains of different mineralogies, thereby changing the rock from its original composition and rendering comparison with other samples invalid, these samples were not point counted. Visual inspection of these thin sections, however, suggests that they are similar in composition to rocks from these areas that were suitable for point counting. Devonian, Carboniferous, and Permian samples are less altered and better preserved than the older rocks. Within these samples, however, feldspar grains were commonly albitized and not well preserved, thus making it impossible to distinguish with confidence between plagioclase and potassium feldspar. These grains were labeled undifferentiated feldspar (Fu). In addition, volcanic lithic grains, which are fairly chemically labile, showed signs of postdepositional alteration. Although this did not impede their identification as volcanic rock fragments, it made it difficult to consistently quantify relative proportions of felsic, intermediate, and mafic grains, and so these data are not reported. These altered volcanic lithic grains were counted as Lv, not as Lm (Table 4). Raw point-count data (Table 4) were recalculated to detrital modes (Table 5) following the methods of Ingersoll et al. (1984).

GEOCHEMICAL RESULTS

In this chapter, we present and discuss new geochemical data from mafic volcanic units of southern Mongolia (Tables 2 and 3). Although we analyzed both mafic and felsic units within the study area, we focus here on the more mafic units, in part because they are more reliable when using trace and minor element data to suggest tectonic affiliations. Figure 4 contains two rock classification diagrams, one that uses major elements and one that uses less mobile, trace elements. The mafic samples plot primarily as basalts and andesite basalts on a Zr/TiO_2 versus Nb/Y diagram (Fig. 4A) and as basalts, basaltic andesites, and basaltic trachyandesites on a total alkali versus silica diagram (Fig. 4B). These samples similarly calculate as basalts, latite basalts, and latite andesites using the CIPW (Cross, Iddings, Pirrson, and Washington) norms and (International Union of Geodesy and Geophysics) volcanic rock classification. In addition, the major element data for these samples (Table 3) are similar to typical average values of basalts and andesites worldwide (Le Maitre, 1976; Hall, 1987; McBirney, 1993). Obtaining similar results from both major element and less-mobile trace element classification systems indicates that the samples from southern Mongolia did not undergo major alteration. Although there may have been some mobility of certain elements, these data primarily reflect original compositions.

We also compare minor and trace element abundances of these rocks to samples from known tectonic settings using

TABLE 2. SAMPLE INFORMATION FOR IGNEOUS SAMPLES USED IN GEOCHEMICAL ANALYSES

Sample	Lithology and formation name	Age* assignment	Quality of age assignment	Latitude (N)	Longitude (E)
94-SJ-119	Mafic to intermediate volcanic rock, Indertey Fm.	D3	Good	44° 21′ 54.8″	99° 28′ 54.2″
94-SJ-96	Mafic to intermediate volcanic rock, Indertey Fm.	D3	Good	44° 21′ 31.9″	99° 30′ 53.5″
94-SJ-203	Mafic dike that crosscuts the Indertey Fm.	P1	Fair	44° 23′ 18.4″	99° 24′ 42.1″
94-SJ-275	Diabase dike	P1	Fair	44° 22′ 4.2″	99° 28′ 35.0″
94-SJ-183	Basalt	P1	Good	44° 25′	99° 30′
94-SJ-184	Basalt	P1	Good	44° 25′	99° 30′
94-ED-33	Mafic to intermediate volcanic rock	D1	Fair	44° 2′ 4.2″	97° 58′ 15.3″
94-ED-34	Mafic to intermediate volcanic rock	D1	Fair	44° 2′ 4.2″	97° 58′ 15.3″
94-ED-46	Mafic to intermediate volcanic rock	P1	Fair	44° 5′ 27″	98° 0′ 29.4″
94-GS-9	Pillow basalts, Berke Uul Fm.	D	Good	43° 47′ 1.4″	103° 7′ 20.1″
94-GS-10	Pillow basalts, Berke Uul Fm.	D	Good	43° 47′ 1.4″	103° 7′ 20.1″
94-GS-74	Pillow basalts, Berke Uul Fm.	D	Good	43° 47′ 36.6″	103° 2′ 23.8″
94-GS-76	Mafic to intermediate volcanic rock	D3-C1	Fair	43° 47′ 43.6″	102° 49′ 5″
94-GS-87	Mafic to intermediate volcanic rock	C1	Fair	43° 46′ 26.8″	102° 52′ 29.2″
94-MH-30	Basalt, Batal Hudag Fm.	D3	Good	44° 24′	104° 18′ 10″
94-MH-34	Basalt, Batal Hudag Fm.	D3	Good	44° 24′	104° 18′ 10″
94-MH-42	Basalt, Batal Hudag Fm.	D3	Good	44° 24′	104° 18′ 10″
94-MH-70	Mafic to intermediate dike, crosscuts Mandal Ovoo Fm.	D3	Fair	44° 22′ 44″	104° 17′ 59.7″
94-MH-11	Mafic to intermediate dike, crosscuts Orgol Fm.	D3	Fair	44° 22′ 51.4″	104° 17′ 58.3″
94-MH-2	Mafic to intermediate dike, crosscuts Mandal Ovoo Fm.	D3	Fair	44° 22′ 44″	104° 17′ 59.7″
94-MH-214	Mafic to intermediate dike, crosscuts Mandal Ovoo Fm.	D3	Fair	44° 22′ 44″	104° 17′ 59.7″

Note: Fm.—Formation.
* D3—Upper Devonian, P1—Lower Permian, D—Devonian, C1—Lower Carboniferous.
Quality of age assignment is a subjective assessment based on authors' knowledge of the rocks, literature, and unpublished work.

tectono-magmatic discrimination diagrams and normalized multielement diagrams. Although these diagrams cannot unequivocally prove the tectonic setting in which a rock formed, they can suggest a tectonic affiliation (Rollinson, 1993). They are best used in conjunction with other geological data to determine the tectonic setting and geologic history of an area. Several different types of tectono-magmatic discrimination diagrams have been developed by plotting abundances of two or three elements or ratios of elements from rocks of known tectonic settings and defining fields for these tectonic settings (e.g., Pearce, 1982; Meschede, 1986). We plotted our samples on 14 of these diagrams and herein present four representative diagrams most appropriate for each locality (see Lamb, 1998, for additional diagrams).

Each set of samples represents a limited slice of geologic time at a specific locality and includes two to four samples, which do not constitute a comprehensive suite; however, these

results support interpretations based on other geological data presented here and in Lamb and Badarch (1997). It is important to note that our localities are 75–300 km apart in current coordinates and therefore we assume that they represent different magma sources.

Shin Jinst: Locality E

Devonian. The Upper Devonian sedimentary section near the town of Shin Jinst (Fig. 2, locality E) consists of marine sandstone, siltstone, and chert interbedded with volcanic flow units (Fig. 3). The volcanic units include both rhyolite and basalt to andesitic basalt. Two of the more mafic samples from this section, 94-SJ-119 and 94-SJ-96, were analyzed (Tables 2 and 3; Fig. 5). Sample 94-SJ-119 is from what appears to be a porphyritic tuff in the field; in thin section phenocrysts are recrystallized and there is secondary chlorite. Sample 94-SJ-96, a

M.A. Lamb and G. Badarch

TABLE 3. WHOLE-ROCK XRF ANALYSES FROM THE SHIN JINST, EDREN, GURVAN SAYHAN, AND MUSGHGAI HUDAG FIELD LOCALITIES

	94-SJ-119	94-SJ-96	94-SJ-203	94-SJ-275	94-SJ-183*	94-SJ-184	94-ED-33	94-ED-34	94-ED-46	94-GS-9	94-GS-10
wt%											
SiO_2	54.68	53.62	54.76	55.04	54.67	53.10	48.67	47.85	52.24	50.61	50.70
TiO_2	0.34	0.35	2.36	1.85	1.69	1.90	0.90	1.00	0.86	1.00	1.20
Al_2O_3	16.75	18.20	15.35	15.83	14.63	15.20	14.85	10.22	16.45	17.13	16.05
Fe_2O_3	10.96	11.33	10.39	9.38	9.53	9.48	13.78	15.11	10.48	12.90	14.31
MnO	0.14	0.14	0.17	0.17	0.14	0.15	0.27	0.21	0.18	0.21	0.20
MgO	5.26	5.84	3.50	4.33	6.00	5.89	5.34	8.01	6.60	5.00	4.24
CaO	7.36	4.63	6.79	6.40	7.03	7.48	9.84	14.79	8.66	6.71	6.96
Na_2O	4.07	6.09	3.50	3.82	3.63	3.10	3.59	1.61	1.74	4.83	5.06
K_2O	0.23	0.02	1.95	2.01	1.65	2.89	2.18	0.49	2.78	0.96	1.04
P_2O_5	0.04	0.06	1.21	0.93	0.71	0.67	0.54	0.53	0.35	0.17	0.33
Totals	99.82	100.28	99.97	99.76	*99.67	99.85	99.95	99.82	100.34	99.51	100.08
ppm											
Nb	0.5	0.6	34	27.3	12.8	12.6	2	1.5	3.8	2.3	3.4
Zr	12.7	17.3	474.5	439.9	300.5	280.4	43.2	38.4	67.2	67.2	121.5
Y	6.1	11.9	37.7	42.4	31.4	29.3	19.8	19.9	14.3	20.5	27.6
Sr	225	43	941	588	1424	1229	660	1053	749	571	370
U	1.1	1.2	2.4	2.5	2.6	2.6	2	1.5	0.9	1.5	2
Rb	3.1	**	23.3	52.8	35.9	66.2	63.5	1.9	54.3	12.2	15.3
Th	**	1.2	6.5	3.9	5.1	4.9	2	**	1.3	1.2	3.7
Pb	3.1	1.8	15.9	11.5	13.2	9.8	9.2	4.5	7.6	1.8	3.6
Ga	12.1	15.3	21.5	19.6	15.1	18.2	21.5	14.8	17.8	17.7	18.7
Zn	95	99	149	129	92	101	131	100	95	110	142
Cu	77	17	35	26	30	27	96	32	80	193	236
Ni	8.3	9.8	22.5	46.2	115.6	75.8	12.4	35.2	55.8	18.6	16.8
Cr	11	9	31	94	222	155	26	83	136	35	7
Ti	2050	2122	14118	11109	10108	11396	5384	5977	5150	5965	7206
V	283	285	159	158	168	185	508	491	268	347	414
Ce	13.9	6.9	179.6	109.7	119.2	101.3	37.4	29.9	15.2	24	49.2
Ba	91	18	916	607	956	1070	653	423	399	214	260
La	**	4	78	44	52	47	9	14	8	**	16

basalt to basaltic andesite from a possible pillow lava, is very fine grained in thin section and contains secondary chlorite and filled vesicles. These samples may be spillites and may have undergone some minor alteration. More samples are needed for a complete geochemical analysis.

Nevertheless, on discrimination diagrams, the Devonian samples plot consistently as plate margin basalts or volcanic arc basalts; Figure 5 shows four of these diagrams. A diagram by Pearce and Gale (1977; Fig. 5A) discriminates plate margin basalts from within-plate basalts using Zr and Ti, both of which increase in the within-plate setting while Y remains constant. On this diagram, the Devonian Shin Jinst samples plot as plate margin basalts. On the V-Ti diagram of Shervais (1982; Fig. 5B), the Devonian samples are within the arc tholeiite field and have a

V/Ti ratio most similar to mafic tholeiites of the Tonga and New Hebrides volcanic arcs. On a diagram not shown here, the TiO_2-MnO-P_2O_5 diagram of Mullen (1983; Lamb, 1998), these samples plot as calc-alkaline basalts. A diagram of Pearce (1982) uses Cr as a fractionation index and the element Y to discriminate between volcanic arc basalt and mid-ocean ridge basalt (MORB) or within-plate basalts (WPB; Fig. 5C). On this diagram, the Devonian samples plot outside of the defined fields but are closest to that of volcanic arc basalts. The Nb-Zr-Y diagram of Meschede (1986; Fig. 5D) contains overlapping fields and thus the Devonian samples plot as both volcanic arc basalts and normal MORB.

These diagrams support previous interpretations (Voznesenskaya and Badarch, 1991; Ruzhentsev and Pospelov, 1992; Lamb and Badarch, 1997) that volcanic arc activity occurred in the De-

TABLE 3. WHOLE-ROCK XRF ANALYSES FROM THE SHIN JINST, EDREN, GURVAN SAYHAN, AND MUSGHGAI HUDAG FIELD LOCALITIES (continued)

	94-GS-74	94-GS-76	94-GS-87	94-MH-30	94-MH-34	94-MH-42	94-MH-70	94-MH-11	94-MH-2	94-MH-214
wt%										
SiO_2	51.06	54.21	52.99	53.29	54.59	51.30	54.66	57.83	50.28	53.08
TiO_2	1.34	1.57	1.73	1.40	1.17	1.50	1.23	1.23	0.89	1.15
Al_2O_3	15.81	16.61	17.26	16.52	15.58	16.50	15.18	15.30	15.71	15.51
Fe_2O_3	13.25	9.47	9.69	13.27	10.24	11.15	10.52	12.52	10.75	12.72
MnO	0.20	0.15	0.16	0.18	0.13	0.23	0.21	0.14	0.18	0.19
MgO	4.24	4.71	5.19	5.06	5.18	6.95	6.02	6.88	8.10	5.26
CaO	8.01	6.03	7.69	7.01	4.40	5.99	7.65	2.75	10.92	8.45
Na_2O	4.30	3.72	3.50	3.70	3.82	5.70	3.40	2.80	2.13	3.09
K_2O	1.59	2.61	1.50	0.05	4.34	0.33	0.62	0.63	0.84	0.08
P_2O_5	0.38	0.50	0.67	0.23	0.12	0.27	0.19	0.15	0.07	0.10
Totals	100.18	99.58	*100.383	100.70	99.56	99.92	99.68	100.25	99.88	100.33
ppm										
Nb	5.6	9.2	12.7	2.3	1.4	8	2.8	1.8	0.6	0.8
Zr	128.2	210.2	241	104.3	57.7	118.6	102.9	82.7	35.3	46.2
Y	27.5	26.3	28	35.5	15.8	27.1	29.9	24	15.8	18.7
Sr	388	747	803	303	106	71	493	123	290	263
U	2	2.1	**	1.2	1.6	1.4	**	**	1.1	0.9
Rb	17.9	41.7	13.4	0.2	86.5	7.1	13.2	6.6	29.8	1.3
Th	4.2	1.8	2.4	1.7	**	**	1.3	1.8	**	1
Pb	5.2	5	4.2	2.4	**	1.3	20.4	**	1.7	2
Ga	19.8	18.1	18.3	19	10.2	15.3	16.5	18.2	15.7	15.5
Zn	124	101	96	126	94	101	114	95	76	102
Cu	198	66	66	31	120	54	44	55	68	88
Ni	36.8	55.2	71.7	6.6	48.3	100.2	47.2	22.6	51	13.2
Cr	42	58	54	7	177	222	140	69	202	54
Ti	8039	9418	10371	8393	7002	8998	7398	7386	5360	6918
V	388	196	186	326	265	231	226	344	286	318
Ce	55.7	47.2	59.8	15.7	6.7	17.2	15.9	8.6	9	6.3
Ba	481	558	403	28	101	22	92	71	179	62
La	27	23	25	2	**	5	**	**	**	**

* Results based on one analysis instead of two.
** = less than limit.
Analytical errors for these ranges of bulk composition (Lowe, 1995; G. Mahood, 1998, personal commun.) are approximately: SiO_2 ± 1.0%; TiO_2 ± 1.0%; Al_2O_3 ± 1.5%; Fe_2O_3 ± 1.0%; MnO ± 5.0%; MgO ± 1.0%; CaO ± 1.0%; Na_2O ± 3.0%; K_2O ± 5.0%; P2O5 ± 4.3%; Nb ± 5.0%; Zr ±2.0%; Y ± 12%; Sr ± 1.0%; U ±75%; Fe_2O_3 ± 1.0%; MnO ±5.0%; MgO ± 1.0%; CaO ± 1.0%; Na_2O ± 3.0%; K_2O ± 5.0%; P_2O_5 ± 4.3%; Nb ± 5.0%; Zr ±2.0%; Y ± 12%; Sr ± 1.0%; U ±75%; Rb ±5.0%; Th ± 20.0 to 50.0%; Pb ± 20%; Ga ± 3.8%; Zn ± 2.0%; Cu ± 5.0%; Ni ±5.0 to 10.0%; Cr ± 5.0 to 20.0; V ± 3.0%; Ce ±10.0; Ba ± 5.0%; La ± 25.0 to 75.0%; Rb ±5.0%; Th ± 20.0 to 50.0%; Pb ± 20%; Ga ± 3.8%; Zn ± 2.0%; Cu ± 5.0%; Ni ±5.0 to 10.0%; Cr ± 5.0 to 20.0; V ± 3.0%; Ce ± 10.0; Ba ± 5.0%; La ± 25.0 to 75.0%.

vonian. The sequence in which the Devonian basalts are located also contains several felsic volcanic units and marine sedimentary strata. This combination of volcanic and sedimentary attributes could be produced within a mature island arc; an arc forming on the outermost edge of a continent, perhaps within the sedimentary prism of a continental margin; or an arc that changes along strike from continental to oceanic. One possible analog is the New Zealand–Kermadec–Tonga arc system. Shin Jinst strata have a geochemical signature similar to that of the Tonga volcanics, yet may be near a continental source (as suggested by sandstone data discussed herein) like that of New Zealand (Wilson, 1989). Another possible analog may be the central Aleutian arc, just west of the oceanic to continental transition. This part of the Aleutian arc contains the widest range of volcanic compositions, including the most mafic and most felsic rocks documented along the entire arc (Fournelle et al., 1994; Vallier et al., 1994).

TABLE 4. RAW SANDSTONE POINT-COUNT DATA

Sample	Qm	Qp	Chert	K	P	Fu	Lv	Ls	CO$_3$	Lm	Lu	bt	ms	chl	mu	opq	hv	cmt	mtx	por	U	N
Ordovician and Silurian																						
94-SJ-19	188	35	17	2	2	24	0	6	0	2	0	0	0	2	0	6	0	0	213	0	3	500
94-SJ-111a	134	22	23	0	29	35	7	1	0	0	1	3	5	2	0	2	0	0	236	0	0	500
94-SJ-111b	131	16	16	0	63	0	0	2	0	0	0	1	4	0	0	2	0	0	264	0	1	500
94-SJ-112a	232	39	62	0	1	27	15	2	0	5	0	0	1	0	0	6	0	32	74	0	4	500
94-SJ-112b	203	44	45	0	3	34	3	13	0	2	0	0	2	0	0	5	0	0	142	0	4	500
94-UH-6	292	12	28	0	83	10	10	1	0	1	0	0	1	1	0	1	0	18	29	0	13	500
94-UH-8	275	15	26	1	53	14	9	9	0	5	3	0	0	1	0	1	1	20	49	0	18	500
94-UH-10	195	20	40	15	153	0	32	0	0	0	3	0	0	0	0	0	5	26	7	0	4	500
94-UH-31	75	28	37	16	105	0	165	0	0	5	0	0	0	0	0	0	7	1	30	0	31	500
94-UH-32	130	32	83	0	13	141	9	0	2	4	0	0	2	0	0	1	0	8	69	0	6	500
94-UH-36	176	35	131	0	84	0	4	0	0	2	0	0	0	0	0	0	7	27	32	0	2	500
94-UH-36	174	59	103	3	68	0	26	0	0	3	2	0	0	0	0	0	3	24	18	0	17	500
94-UH-37	188	24	90	0	0	48	3	1	0	3	0	0	3	0	0	3	0	15	98	0	24	500
94-UH-38	240	53	71	0	0	0	26	5	0	15	1	0	0	0	0	0	0	39	35	0	15	500
94-UH-46	309	38	57	19	0	0	3	0	0	7	1	0	0	0	0	0	1	23	30	0	12	500
94-UH-51	181	82	105	0	0	14	16	0	0	17	0	0	0	0	0	0	0	0	20	0	65	500
94-UH-59	117	34	105	1	100	34	29	4	0	3	1	0	0	0	0	1	0	21	50	0	0	500
94-UH-58	103	68	78	0	32	41	110	1	0	4	0	0	0	0	0	0	3	18	34	0	8	500
94-UH-17	136	37	54	0	96	85	34	5	0	1	0	0	0	1	0	0	0	12	33	0	6	500
95-HG-6	288	20	2	0	0	34	0	0	0	1	0	0	0	0	10	1	0	9	119	0	16	500
95-HG-8	315	21	12	0	0	26	0	0	0	4	0	0	3	0	0	0	0	5	99	0	15	500
Devonian and Carboniferous																						
94-KK-1	33	27	13	11	111	0	204	30	0	9	10	0	0	0	0	0	0	21	31	0	0	500
94-KK-3	32	13	7	9	147	0	131	8	0	10	11	0	0	0	0	0	3	119	10	0	0	500
94-KK-8	28	11	2	33	323	0	12	2	0	16	5	0	0	1	0	0	2	5	60	0	0	500
94-SJ-97	36	15	0	0	135	0	243	0	0	0	11	0	0	0	0	0	0	0	56	0	4	500
94-SJ-122	59	8	6	0	171	0	159	0	0	0	6	0	0	0	0	0	4	6	79	0	2	500
94-SJ-158	128	19	26	3	43	5	5	16	0	11	0	1	1	0	0	1	0	62	165	0	14	500
94-SJ-170	141	10	3	33	116	0	128	0	2	5	7	0	0	0	0	0	1	8	46	0	0	500
94-SJ-172	139	22	12	36	100	0	100	8	0	7	14	0	2	0	0	0	1	29	28	0	2	500
94-GS-4	36	1	2	0	248	1	124	0	0	1	1	0	0	0	0	0	1	22	63	0	0	500
94-GS-14	0	2	0	0	44	0	307	2	0	0	0	0	0	0	0	3	6	25	93	0	18	500
94-GS-72	25	10	3	0	159	0	176	1	0	48	20	0	0	0	0	0	0	8	50	0	0	500
94-GS-80	15	3	2	9	14	0	68	4	0	1	7	0	0	0	0	0	1	8	12	0	5	149
94-GS-86	115	3	3	28	120	3	18	7	8	7	5	2	2	0	0	0	1	171	0	0	10	503
94-GS-94	17	18	8	15	152	0	213	3	0	0	5	2	1	0	0	0	0	34	31	0	1	500
94-ED-1	1	19	0	0	64	50	287	0	0	0	13	0	0	0	0	0	12	0	37	0	18	501
94-ED-3	1	0	0	0	54	11	389	0	0	0	0	0	0	0	0	0	5	0	24	0	16	500
94-ED-4	1	0	0	1	94	1	257	0	0	0	0	0	0	0	0	9	1	0	134	0	3	501
94-ED-8	0	0	0	0	106	14	314	0	0	0	0	0	0	0	0	0	6	0	31	0	29	500
94-ED-26	5	5	0	1	15	42	304	10	0	0	0	0	0	21	0	0	0	4	36	0	58	501
95-NT-2	3	9	1	8	188	0	197	13	0	0	1	0	0	0	0	4	0	8	65	0	3	500
95-NT-3	1	0	6	0	0	28	180	35	0	18	0	0	0	0	0	0	0	0	24	0	6	298
95-NU-8	18	2	0	0	6	4	279	2	0	0	0	0	0	0	0	1	1	173	9	0	5	500
95-NU-4	13	1	0	2	48	0	289	10	0	1	0	0	6	0	0	12	0	85	32	0	0	499
95-NU-3	3	0	0	0	0	62	327	1	0	0	0	0	0	0	0	0	0	90	14	0	3	500

TABLE 4. RAW SANDSTONE POINT-COUNT DATA (continued)

Sample	Qm	Qp	Chert	K	P	Fu	Lv	Ls	CO_3	Lm	Lu	bt	ms	chl	mu	opq	hv	cmt	mtx	por	U	N
Devonian and Carboniferous (continued)																						
95-NU-2	0	0	0	0	5	2	147	0	0	0	0	0	0	0	0	2	0	27	1	0	0	184
95-NU-15	8	1	0	0	0	88	235	1	0	0	0	0	0	0	0	5	0	1	168	0	0	507
95-NU-24	1	0	0	0	45	2	325	0	0	0	0	0	0	0	0	3	5	90	28	0	1	500
95-DN-207	7	3	2	0	0	129	212	9	0	0	0	0	0	0	0	0	0	18	99	0	21	500
95-DN-208	2	0	0	0	0	143	185	9	0	0	0	9	0	0	0	0	8	91	27	0	26	500
95-DN-212	17	3	0	6	101	0	137	5	13	0	0	0	0	0	0	0	19	129	0	0	70	500
95-DN-1	8	5	0	0	0	118	220	4	0	0	0	0	0	0	0	0	0	84	56	0	5	500
95-DN-2	19	3	2	17	208	0	63	1	0	1	0	7	0	0	0	4	17	34	113	0	11	500
95-DN-5	4	5	0	3	123	0	186	13	0	0	15	0	0	0	0	0	0	13	136	0	2	500
95-DN-8	103	0	1	2	143	0	143	1	0	0	3	8	0	0	0	1	0	5	90	0	0	500
95-DN-9	13	2	2	5	85	0	276	5	0	3	0	1	0	0	0	0	0	2	106	0	0	500
95-DN-12	19	3	1	18	49	7	331	0	0	0	0	0	0	0	0	0	0	0	71	0	1	500
95-BN-1	6	2	0	0	89	5	239	0	0	0	0	0	0	0	0	2	0	0	134	0	23	500
95-BN-7	0	0	0	0	0	354	81	0	0	0	0	0	0	0	0	0	0	26	38	0	1	500
95-NM-214	8	3	1	0	0	255	119	1	0	0	0	0	0	0	0	4	0	17	88	0	4	500
95-NM-201	28	17	1	7	139	1	113	20	0	0	0	4	0	0	0	7	19	10	119	0	15	500
95-NM-202	57	4	2	17	215	2	9	2	0	0	0	#	0	0	0	0	14	29	85	0	35	500
95-NM-4	25	5	3	47	122	3	97	24	0	1	0	0	0	1	0	0	31	3	117	0	21	500
95-NM-204	9	2	1	5	96	0	271	4	1	0	2	0	0	0	0	1	0	9	99	0	1	501
95-NM-207	16	2	1	3	8	136	211	3	0	0	0	0	0	0	0	1	4	13	94	0	8	500
95-NM-208	9	6	1	0	0	156	230	0	0	0	0	0	0	0	0	2	0	0	68	0	28	500
95-NM-210	19	4	1	0	0	94	230	1	0	0	0	0	0	0	0	1	6	4	137	0	3	500
95-NM-212	11	1	0	1	93	6	243	6	0	0	1	0	0	1	0	0	13	1	105	0	18	500
95-NM-6	24	5	2	9	94	0	251	5	6	0	2	0	0	0	0	1	3	2	88	0	8	500
95-NM-8	19	7	2	9	137	0	227	3	0	5	0	0	0	1	0	1	1	2	80	0	7	501
95-NM-10	8	6	0	12	111	6	230	15	0	2	0	0	0	0	0	0	11	0	91	0	8	500
95-IU-8	61	20	4	4	43	32	121	35	0	12	0	0	1	0	0	1	0	33	126	0	7	500
95-IU-3	28	13	1	1	102	0	206	24	0	2	0	0	0	0	0	1	4	0	89	0	29	500
95-IU-4	77	9	3	2	78	30	54	7	0	2	2	0	0	0	0	3	1	203	26	0	3	500
95-IU-9	29	4	0	18	96	19	190	12	0	0	14	1	0	0	0	5	1	2	102	0	7	500
95-IU-15	18	3	2	7	88	2	282	19	0	0	4	2	0	0	0	1	1	1	63	0	7	500
95-IU-16	2	6	10	21	89	0	224	7	21	0	0	0	0	1	0	2	1	0	95	0	21	500
95-IU-20	1	1	2	0	0	170	225	2	0	2	0	0	0	0	0	4	0	14	77	0	2	500
Permian																						
94-SJ-181	80	6	10	56	78	0	187	4	0	8	24	0	0	0			1	0	36	0	10	500
94-SJ-182	72	8	4	63	127	0	157	0	0	0	9	0	0	0	0	0	6	1	40	0	13	500
94-SJ-230	148	17	14	28	126	0	5	20	0	2	0	8	2	0	0	0	0	0	130	0	0	500
94-SJ-231	126	30	14	23	179	0	71	5	0	4	15	4	4	0	0	0	3	0	20	0	2	500
94-SJ-240	72	24	23	0	90	12	40	29	0	1	6	0	0	0	0	6	2	0	184	0	14	503
94-SJ-241	111	27	14	1	50	0	166	10	0	14	36	0	0	0	0	0	6	0	60	0	5	500
94-SJ-243	140	15	19	11	120	0	133	2	0	8	8	0	0	0	0	0	0	33	11	0	0	500
94-SJ-245	150	13	21	0	99	0	141	4	0	14	16	0	0	0	0	0	1	21	19	0	1	500
94-SJ-246	130	33	35	0	70	9	58	26	0	14	0	0	1	0	0	1	0	7	110	0	6	500
94-SJ-247	151	19	16	0	85	0	167	12	0	14	8	0	1	0	0	0	1	13	13	0	0	500
94-SJ-251	112	12	14	18	126	0	179	10	0	1	5	0	0	0	0	0	0	2	21	0	0	500

Note: Qm—monocrystalline quartz, Qp—polycrystalline quartz, Chert—chert, K—potassium feldspar, P—plagioclase, Fu—feldspar undifferentiated, Lv—volcanic lithic, Ls—siltstone lithic, CO_3—carbonate grain, Lm—metatmorphic lithic, Lu—unidentified lithic, bt—biotite, ms—muscovite, chl—chlorite, mu—mica undifferentiated, opq—opaque, hv—heavy mineral, cmt—cement, mtx—matrix, por—porosity, U—unidentified, and N—total.

TABLE 5. RECALCULATED DETRITAL MODES FOR ORDOVICIAN–PERMIAN SANDSTONE

Sample	Locality	Age	Formation	Latitude (N)	Longitude (E)	QM	F	LT	QT	F	L	Lm	Lv	Ls
Ordovician and Silurian														
94-SJ-19	E	S	Ulan Shan	44°21'31.6"	99°27'6.3"	68	10	22	87	10	3	25	0	75
94-SJ-111a	E	S	Ulan Shan	44°21'8.9"	99°35'55.5"	53	25	21	71	25	4	0	88	13
94-SJ-111b	E	S	Ulan Shan	44°21'8.9"	99°35'55.5"	57	28	15	71	28	1	0	0	##
94-SJ-112a	E	S	Ulan Shan	44°21'8.9"	99°35'55.5"	61	7	32	87	7	6	23	68	9
94-SJ-112b	E	S	Ulan Shan	44°21'54.8"	99°28'54.2"	59	11	31	84	11	5	11	17	72
94-UH-6	C	O	Haryn Shand	44°53'10.5"	99°49'31.6"	67	21	12	76	21	3	8	83	8
94-UH-8	C	O	Haryn Shand	44°53'10.5"	99°49'31.6"	67	17	16	77	17	6	22	39	39
94-UH-10	C	O	Haryn Shand	44°53'10.5"	99°49'31.6"	43	37	21	56	37	8	0	##	0
94-UH-31	C	O	Haryn Shand	44°53'10.5"	99°49'31.6"	17	28	55	32	28	39	3	97	0
94-UH-32	C	O	Haryn Shand	44°53'10.5"	99°49'31.6"	31	37	31	59	37	4	31	69	0
94-UH-36	C	O	Haryn Shand	44°53'15.3"	99°49'34.9"	41	19	40	79	19	1	33	67	0
94-UH-36	C	O	Haryn Shand	44°53'15.3"	99°49'34.9"	40	16	44	77	16	7	10	90	0
94-UH-37	C	O	Haryn Shand	44°53'15.3"	99°49'34.9"	53	13	34	85	13	2	43	43	14
94-UH-38	C	O	Haryn Shand	44°53'15.3"	99°49'34.9"	58	0	42	89	0	11	33	57	11
94-UH-46	C	S	unnamed	44°53'15.3"	99°49'34.9"	71	4	24	93	4	3	70	30	0
94-UH-51	C	S	unnamed	44°53'18.1"	99°49'47.3"	44	3	53	89	3	8	52	48	0
94-UH-59	C	O-S	unnamed	44°52'373"	99°52'10.9"	27	32	41	60	32	9	8	81	11
94-UH-58	C	O-S	unnamed	44°52'373"	99°52'10.9"	24	17	60	57	17	26	3	96	1
94-UH-17	C	O	Haryn Shand	44°55'29.7"	99°43'44.9"	30	40	29	51	40	9	3	85	13
95-HG-6	J	S	unnamed	44°40.989'	106°47.199'	83	10	7	90	10	0	##	0	0
95-HG-8	J	S	unnamed	44°40.989'	106°47.199'	83	7	10	92	7	1	##	0	0
Average						51	18	30	74	18	7	28	55	17
Standard Deviation						19	12	15	16	12	9	31	36	29
Devonian and Carboniferous														
94-KK-1	F	C1	Tol Bulag	44°12'57.7"	98°47'12.1"	7	27	65	16	27	56	4	84	12
94-KK-3	F	D1	unnamed	44°12'57.7"	98°47'12.1"	9	42	49	14	42	43	7	88	5
94-KK-8	F	D1	unnamed	44°12'57.7"	98°47'12.1"	6	82	11	9	82	8	53	40	7
94-SJ-97	E	D3	Indertey	44°21'31.9"	99°30'53.4"	8	31	61	12	31	58	0	##	0
94-SJ-122	E	D3	Indertey	44°21'54.8"	99°28'54.2"	14	42	44	18	42	40	0	##	0
94-SJ-158	E	C1	Bayan Sair	44°24'7.1"	99°26'53.6"	50	20	30	68	20	13	34	16	50
94-SJ-170	E	C1	Bayan Sair	44°24'7.1"	99°26'53.6"	32	33	35	35	33	32	4	96	0
94-SJ-172	E	C1	Bayan Sair	44°24'7.1"	99°26'53.6"	32	31	37	39	31	29	6	87	7
94-GS-4	K	S-D	Berkhe Uul	43°47'1.4"	103°7'20.1"	9	60	31	9	60	30	1	99	0
94-GS-14	K	D	Berkhe Uul	43°47'21.6"	103°7'16.3"	0	12	88	1	12	87	0	99	1
94-GS-72	K	D	Berkhe Uul	43°47'20.4"	103°2'10.3"	6	36	58	9	36	55	21	78	0
94-GS-80	K	C	In Shanhai	43°47'43.6"	102°49'5"	12	19	69	16	19	65	1	93	5
94-GS-86	K	C	In Shanhai	43°46'26.8"	102°52'29.2"	36	48	16	38	48	14	22	56	22
94-GS-94	K	C	In Shanhai	43°46'38.2"	102°49'15.4"	4	39	57	10	39	51	0	99	1
94-ED-1	G	C (?)	unnamed	44°1'54.7"	98°10'6.6"	0	26	74	5	26	69	0	##	0
94-ED-3	G	C (?)	unnamed	44°1'54.7"	98°10'6.6"	0	14	85	0	14	85	0	##	0
94-ED-4	G	C (?)	unnamed	44°1'54.7"	98°10'6.6"	0	27	73	0	27	73	0	##	0
94-ED-8	G	C (?)	unnamed	44°1'54.7"	98°10'6.6"	0	28	72	0	28	72	0	##	0
94-ED-26	G	C (?)	unnamed	43°59'20.3"	97°59'5.6"	1	15	84	3	15	82	0	97	3
95-NT-2	R	C	In Shanhay	43°40.482'	100°42.591'	1	47	53	3	47	50	5	90	6
95-NT-3	R	C	In Shanhay	43°40.482'	100°42.591'	0	10	89	3	10	87	3	75	22
95-NU-8	M	C3/P1?	Tost Uul	43°9.670'	102°3.999'	6	3	91	6	3	90	0	99	1
95-NU-4	M	C3/P1?	Tost Uul	43°9.670'	102°3.999'	4	14	83	4	14	82	0	96	3
95-NU-3	M	C3/P1?	Tost Uul	43°9.670'	102°3.999'	1	16	83	1	16	83	0	##	0
95-NU-2	M	C3/P1?	Tost Uul	43°9.670'	102°3.999'	0	5	95	0	5	95	0	##	0
95-NU-15	M	C2-3	Tost Uul	43°7.819'	101°53.821'	2	26	71	3	26	71	0	##	0

TABLE 5. RECALCULATED DETRITAL MODES FOR ORDOVICIAN–PERMIAN SANDSTONE (continued)

Sample	Locality	Age	Formation	Latitude (N)	Longitude (E)	QM	F	LT	QT	F	L	Lm	Lv	Ls
Devonian and Carboniferous (continued)														
95-NU-24	M	C2-3	Tost Uul	43°10.5′	101°54.2′	0	13	87	0	13	87	0	##	0
95-DN-207	N	C1	unnamed	43°5.187′	102°43.720′	2	36	62	3	36	61	0	96	4
95-DN-208	N	C1	unnamed	43°5.187′	102°43.720′	1	42	57	1	42	57	0	95	5
95-DN-212	N	C1	unnamed	43°5.187′	102°43.720′	6	38	56	7	38	55	0	96	4
95-DN-1	N	C1	unnamed	43°4.935′	102°43.253′	2	33	65	4	33	63	0	98	2
95-DN-2	N	C1	unnamed	43°4.935′	102°43.253′	6	72	22	8	72	21	2	97	2
95-DN-5	N	C1	unnamed	43°4.935′	102°43.253′	1	36	63	3	36	61	0	93	7
95-DN-8	N	C1	unnamed	43°4.935′	102°43.253′	26	37	37	26	37	37	0	99	1
95-DN-9	N	C1	unnamed	43°4.935′	102°43.253′	3	23	74	4	23	73	1	97	2
95-DN-12	N	C1	unnamed	43°4.935′	102°43.253′	4	17	78	5	17	77	0	##	0
95-BN-1	O	C1	unnamed	42°53.1034′	105°20.438′	2	28	71	2	28	70	0	##	0
95-BN-7	O	C1	unnamed	42°53.1034′	105°20.438′	0	81	19	0	81	19	0	##	0
95-NM-214	P	C1	unnamed	42°54.929′	105°13.812′	2	66	32	3	66	31	0	99	1
95-NM-201	P	C1	unnamed	42°55.654′	105°12.88′	9	45	46	14	45	41	0	85	15
95-NM-202	P	C1	unnamed	42°55.654′	105°12.88′	19	76	6	20	76	4	0	82	18
95-NM-4	P	C1	unnamed	42°54.929′	105°13.812′	8	53	40	10	53	37	1	80	20
95-NM-204	P	C1	unnamed	42°54.929′	105°13.812′	2	26	72	3	26	71	0	99	1
95-NM-207	P	C1	unnamed	42°54.929′	105°13.812′	4	39	57	5	39	56	0	99	1
95-NM-208	P	C1	unnamed	42°54.929′	105°13.812′	2	39	59	4	39	57	0	##	0
95-NM-210	P	C1	unnamed	42°54.929′	105°13.812′	5	27	68	7	27	66	0	##	0
95-NM-212	P	C1	unnamed	42°54.929′	105°13.812′	3	28	69	3	28	69	0	98	2
95-NM-6	P	C1	unnamed	42°54.159′	105°12.703′	6	26	68	8	26	66	0	98	2
95-NM-8	P	C1	unnamed	42°54.159′	105°12.703′	5	36	60	7	36	57	2	97	1
95-NM-10	P	C1	unnamed	42°54.159′	105°12.703′	2	33	65	4	33	63	1	93	6
95-IU-8	Q	C1	unnamed	42°54.900′	105°31.883′	18	24	58	26	24	51	7	72	21
95-IU-3	Q	C1	unnamed	42°54.900′	105°31.883′	7	27	65	11	27	62	1	89	10
95-IU-4	Q	C1	unnamed	42°54.900′	105°31.883′	29	42	29	34	42	25	3	86	11
95-IU-9	Q	C1	unnamed	42°54.900′	105°31.883′	8	35	58	9	35	57	0	94	6
95-IU-15	Q	C1	unnamed	42°54.443′	105°32.726′	4	23	73	5	23	72	0	94	6
95-IU-16	Q	C1	unnamed	42°54.443′	105°32.726′	1	29	71	5	29	66	0	97	3
95-IU-20	Q	C1	unnamed	42°53.787′	105°32.291′	0	42	58	1	42	57	1	98	1
Average						8	34	59	10	34	56	3	92	5
Standard Deviation						10	18	21	12	18	22	9	15	9
Permian														
94-SJ-181	E	P1	Hargan Hudag	44°24′	99°28′	18	30	53	21	30	49	4	94	2
94-SJ-182	E	P1	Hargan Hudag	44°24′	99°28′	16	43	40	19	43	38	0	##	0
94-SJ-230	E	P1	Hargan Hudag	44°30′45.2″	99°18′33.2″	41	43	16	50	43	8	7	19	74
94-SJ-231	E	P1-P2	Hargan Hudag	44°30′45.2″	99°18′33.2″	27	43	30	36	43	20	5	89	6
94-SJ-240	E	P1-P2	Hargan Hudag	44°31′28.9″	99°20′45.2″	24	34	41	40	34	26	1	57	41
94-SJ-241	E	P2	Jinst	44°31′28.9″	99°20′45.2″	26	12	62	35	12	53	7	87	5
94-SJ-243	E	P2	Jinst	44°31′38″	99°20′59.1″	31	29	41	38	29	33	6	93	1
94-SJ-245	E	P2	Jinst	44°31′38″	99°20′59.1″	33	22	46	40	22	38	9	89	3
94-SJ-246	E	P2	Jinst	44°31′38″	99°20′59.1″	35	21	44	53	21	26	14	59	27
94-SJ-247	E	P2	Jinst	44°31′38″	99°20′59.1″	32	18	50	39	18	43	7	87	6
94-SJ-251	E	P2	Jinst	44°30′6.6″	99°21′27.2″	23	30	46	29	30	41	1	94	5
Average						28	30	43	36	30	34	6	79	16
Standard Deviation						7	11	12	10	11	13	4	24	23

Note: S—Silurian, O—Ordovician, D—Devonian, D3—Upper Devonian, D1—Lower Devonian, C1—Lower Carboniferous, C—Carboniferous, P1—Lower Permian, P2—Upper Permian; F—P+K+Fu, Lt—Lv+Ls+Lm+Qp+chert+Lu, L—Lv+Ls+Lm+Lu, and Qt—Qm+Qp+chert.

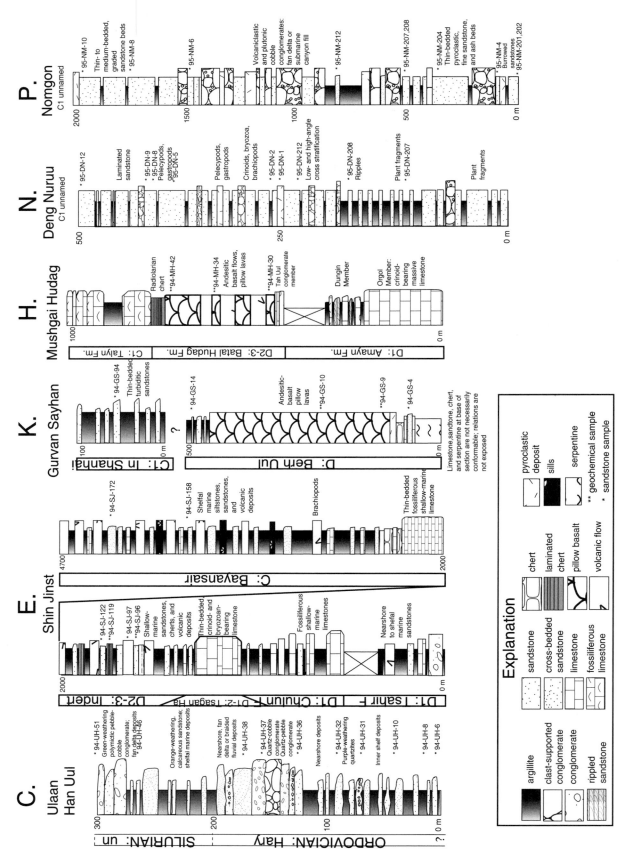

Figure 3. Representative stratigraphic columns from six of localities (see Fig. 2) examined. See Lamb and Badarch (1997) for full description of facies and complete set of stratigraphic columns. Asterisk denotes sandstone sample. Double asterisk denotes geochemical sample. Note that not all sandstone or geochemical samples are shown; localities typically have more than one measured column, but only one is shown here.

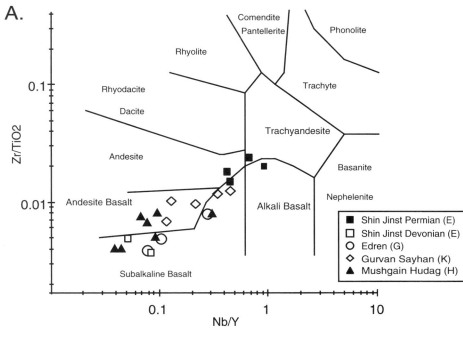

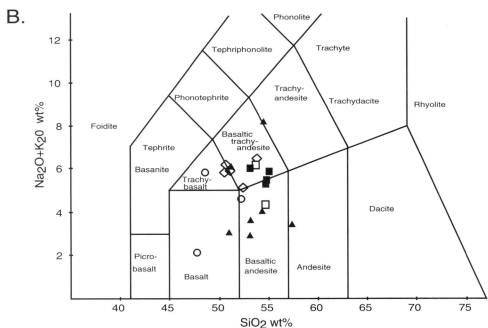

Figure 4. Rock classification diagrams for igneous samples. A: Samples plotted on Zr/TiO$_2$ vs. Nb/Y diagram of Winchester and Floyd (1977) B: Samples plotted on total alkali vs. silica diagram of Le Maitre et al. (1989).

Permian. Lower Permian basalt flows, samples 94-SJ-183 and 94-SJ-184, and what we interpret to be their associated feeder dikes, 94-SJ-203 and 94-SJ-275, were also analyzed (Tables 2 and 3; Fig. 5). In thin section, these rocks have minor amounts of chlorite.

On discrimination diagrams (Fig. 5), the Permian Shin Jinst samples consistently plot in WPB and alkalic fields. On the diagram by Pearce and Gale (1977; Fig. 5A), these samples plot as WPB. On the V-Ti diagram of Shervais (1982; Fig. 5B), the

Permian samples have a Ti/V ratio between 50 and 100, the range for alkalic basalts. On the Cr-Y diagram of Pearce (1982; Fig. 5C), the Permian samples plot within the overlapping fields of MORB, volcanic arc basalt, and WPB but are centered within the WPB field. On the Nb-Zr-Y diagram of Meschede (1986; Fig. 5D), the Permian samples all plot within the single field of within plate alkalic basalts.

The Permian Shin Jinst samples are more alkalic than the Devonian ones and have a strong within-plate signature. These

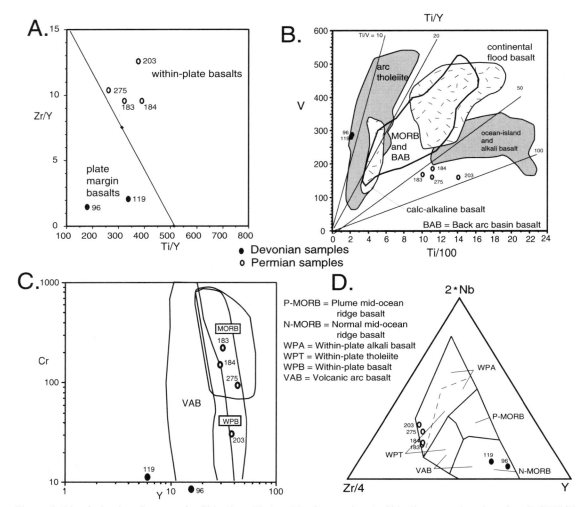

Figure 5. Discrimination diagrams for Shin Jinst (E) (see Fig. 2) samples. A: Shin Jinst samples plotted on Zr/Y-Ti/Y diagram of Pearce and Gale (1977). B: Shin Jinst samples plotted on V-Ti diagram of Shervais (1982). C: Shin Jinst samples plotted on Cr-Y diagram of Pearce (1982). D: Shin Jinst samples plotted on Nb-Zr-Y diagram of Meschede (1986).

flows suggest that arc volcanism ended prior to the Permian. They may be related to late Paleozoic non-arc-related volcanism in western China and thus, part of the final stage of consolidation of central Asia (cf. Coleman, 1989).

Edren: Locality G

Three Devonian and Permian samples from basalt to basaltic andesite flows at the Edren locality (Fig. 2, locality G; Tables 2 and 3) were analyzed. In thin section, these samples have minor chlorite but do not show any signs of alteration or metamorphism.

Trace element data from these samples plot as volcanic arc basalts on a number of diagrams (Fig. 6). On the diagram by Pearce and Gale (1977; Fig. 6A), the Edren samples plot as plate margin basalts. On the V-Ti diagram of Shervais

(1982; Fig. 6B), all three samples plot as arc tholeiites, although the Permian sample is also within the calc-alkaline basalt and MORB fields. On the Cr-Y diagram of Pearce (1982; Fig. 6C), the Edren samples plot solely as volcanic arc basalts. On the Mullen diagram (1983; Fig. 6D), all three points are within the calc-alkaline basalt arc field and have values similar to those of the Cliefden outcrop pillow basalts, New South Wales, Australia, which have been interpreted as mature island arc or continental margin volcanic arc basalts (Hellman et al., 1977).

Thus, the Edren samples plot primarily as volcanic arc basalts with some overlap into MORB fields and have values similar to those of basalts from mature island arc and continental arc settings. Overall, the diagrams support an arc setting for this area, but the data are too few to determine at what times the arc was calc-alkaline, tholeiitic, or transitional.

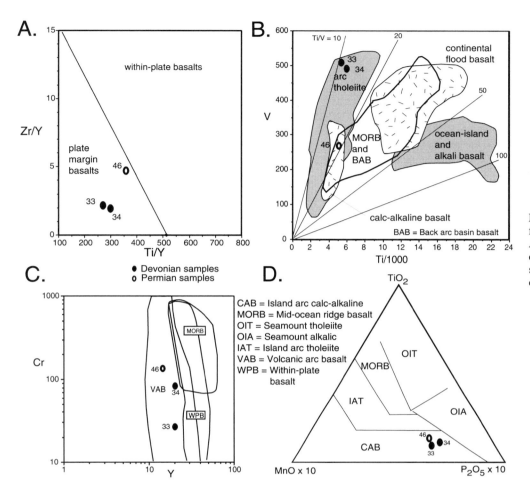

Figure 6. Discrimination diagrams for Edren (G) (see Fig. 2) samples. A–C: Edren samples plotted on same diagrams as in Figure 5. D: Edren samples plotted on TiO_2-MnO-P_2O_5 diagram of Mullen (1983).

Mushgai Hudag: Locality H

Analyzed samples from the Mushgai Hudag locality include Upper Devonian basalt flows (Figs. 3 and 7A) and what we interpret as their feeder dikes (Tables 2 and 3). Samples 94-MH-30, 94-MH-34, and 94-MH-42 are pillow basalts that have secondary chlorite and calcite-filled vesicles in thin section. They are part of a section that also contains marine siltstone, sandstone, limestone, and chert (Lamb and Badarch, 1997). These basalts may now be spillites. The remaining four samples are dikes that contain some secondary chlorite in thin section.

On discrimination diagrams, the samples are within normal MORB and volcanic arc fields, or within the overlap area between these two fields (Fig. 8, A–D). On diagrams that discriminate between alkalic and tholeiitic (see Lamb, 1998, for additional diagrams), these samples consistently plot as tholeiitic. On the diagram by Pearce and Gale (1977; Fig. 8A), the Mushgai Hudag samples plot as plate margin basalts with the exception of one sample that is near the boundary but within the WPB field. On the V-Ti diagram (Fig. 8B), the samples plot in the field for MORB and backarc basalts (BAB), with values

similar to those of the Shikoku and Lau backarc basins and MORB of the Pacific and Atlantic Oceans (Shervais, 1982). The Cr-Y and Ti/Y-Nb/Y diagrams (Fig. 8, C–D) demonstrate that the Mushgai Hudag samples may share features of volcanic arc basalts and MORB. A backarc basin setting for these samples is one explanation that accounts for the arc, MORB, and backarc basin geochemical signatures.

Gurvan Sayhan: Locality K

Samples from the Gurvan Sayhan Range were collected from Devonian pillow basalts (Figs. 3 and 7B) and Lower Carboniferous basalt flows (Tables 2 and 3). Samples 94-GS-9, 94-GS-10, and 94-GS-74 are from a 500-m-thick section of Devonian pillow basalts interbedded with and capped by volcaniclastic sandstone and siltstone. Samples 94-GS-76 and 94-GS-87 are from less well-exposed, basaltic-andesite Carboniferous flows. In thin section, the pillow basalts contain chlorite and calcite-filled vesicles, suggesting they may be spillites. The Carboniferous samples show no signs of major alteration or metamorphism.

Figure 7. Outcrop photos of lavas sampled for geochemistry. Note rock hammer for scale; hammer is 38 cm long. A: Pillow basalt from Mushgai Hudag (H) (see Fig. 2) locality. B: Pillow basalt from Gurvan Sayhan Range (K).

Trace and minor element abundances of these basalts most commonly plot as volcanic arc rocks that evolve into or are influenced by a within-plate component. When all five samples do not plot in the same field, they display an evolution from arc or plate margin fields to within-plate fields through time, e.g., the Zr/Y-Ti/Y and V-Ti diagrams (Fig. 9, A and B). On the Cr-Y diagram (Fig. 9C), the samples are within the overlap between the volcanic arc and WPB field but do not show an evolutionary trend. When several trace element values from these rocks are compared to several rocks suites from known tectonic settings, the Gurvan Sayhan basalt patterns are most similar to transitional volcanic arc basalt patterns (Pearce, 1982, 1983; Fig. 9D). The shape of the pattern is the same as typical volcanic arcs, but the entire pattern is shifted upward, indicating an enrichment in

all elements. Such an across the board enrichment is typical of WPBs (Pearce, 1982, 1983).

The rest of the Gurvan Sayhan major and trace element data are also similar to those of Grenada (Arculus, 1976; Thirlwall and Graham, 1984). DeLong et al. (1975) examined the Grenada basalts, as well as other alkaline mafic rocks within oceanic island arc settings, and found they all had one of two things in common: (1) they form where a linear feature, such as a fracture zone, is being subducted, or (2) they form at the lateral edge of a subduction zone, near a hinge fault, such as in Grenada or Samoa. In either case, the magma may be from areas of the asthenosphere not normally tapped for island arc magmas. These factors may also produce alkaline basalts in continental arcs (DeLong et al., 1975). This process is one possible explanation for the Gurvan Sayhan basalts and their geochemical signature. It is also possible that the Devonian samples represent arc volcanism and the Carboniferous samples represent within plate volcanism.

SANDSTONE PETROGRAPHIC RESULTS

Sandstone point-count data from southern Mongolia are presented by time period from the Ordovician and Silurian to the Devonian, Carboniferous, and Permian (Tables 4 and 5). A full sedimentologic and stratigraphic description, as well as discussion of depositional environments of the Ordovician through Carboniferous strata from each locality, was presented in Lamb and Badarch (1997). Representative stratigraphic columns are presented herein to provide a context for the geochemical and sandstone data (Fig. 3).

Ordovician and Silurian

Ordovician and Silurian sandstone was examined and collected in localities A–C, E, and H–J (Fig. 2). Point-count data from three of these localities, Shin Jinst (E), Ulan Han Uul (C), and Havtgai (J) are plotted and compared to Marsaglia and Ingersoll's (1992) modified QFL provenance fields of Dickinson (1985; Fig 10A) and Marsaglia and Ingersoll's (1992; Fig. 10B) LmLvLs diagram. The Shin Jinst area (E) contains a thick sequence of marine argillite and sandstone (Lamb and Badarch, 1997). Sandstone from this sequence is quartz rich and plots primarily within the continental block field of the QFL diagram, $Q_{80}F_{16}L_4$ (Fig. 10).

The Ulan Han Uul (C) samples are also quartz-rich sandstones from a marine sequence (Figs. 3 and 11A) and plot primarily in the continental block and recycled orogen fields on the QFL diagram, $Q_{70}F_{20}L_{10}$ (Fig. 10A). Samples from both localities (E and C) suggest a mix of sources when plotted on the LmLvLs diagram (Fig. 10B). Half of the samples from Ulan Han Uul (C) are within arc fields on the LmLvLs diagram, reflecting a high number of volcanic grains within the lithic component. Conglomerates from Ulan Han Uul (C; Fig. 3) also indicate mixed sources: Ordovician conglomerate of the

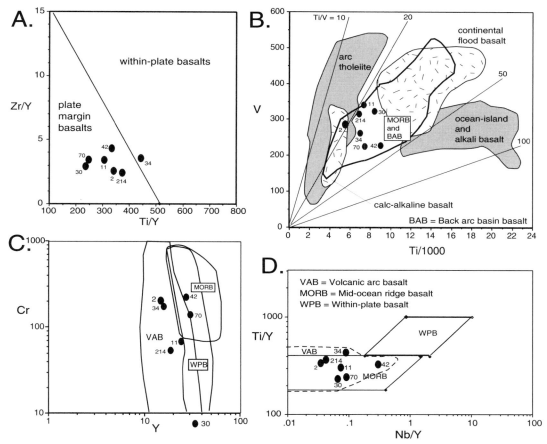

Figure 8. Discrimination diagrams for Mushgai Hudag (H) (see Fig. 2) samples. A–C: Mushgai Hudag (H) samples plotted on same diagrams as in Figure 5. D: Mushgai Hudag (H) samples plotted on Ti/Y-Nb/Y diagram of Pearce (1982).

Haryn Shand Formation consists of quartz cobbles and pebbles whereas the Silurian conglomerate contains volcanic, plutonic, and quartz clasts. Thus, the source area for the Ulan Han Uul (C) sandstone and conglomerate may have included an inactive, remnant arc as well as older sedimentary and metasedimentary strata.

The Silurian section at Havtgai (J) includes trough cross-stratified quartzite and quartz-pebble conglomerate interpreted to be nearshore to inner shelf deposits (Fig. 11B; Lamb and Badarch, 1997). Only two Silurian samples from Havtgai (J) were suitable for counting and these mature sandstones plot within the continental block field on the QFL diagram and on the Lm apex of the LmLvLs diagram (Fig. 10). Several samples from Mushgai Hudag (H) were examined and, although not suitable for point counting, are also mature and quartz rich (Lamb and Badarch, 1997).

Provenance data from the Ordovician and Silurian time periods point to a mature source, either an uplifted continental block or the eroded, exposed roots of a volcanic arc. There is no evidence of arc activity during this time, although the rapid shift in composition of the Ulan Han Uul (C) conglomerate and the

range of lithic grains from this locality suggest active uplift of an older arc system. Overall, these compositions suggest a continental crustal setting rather than an oceanic setting. Remnants of this continental crust may be represented by Vendian to Cambrian rocks now found to the north of the study area (Fig. 2).

Devonian and Carboniferous

The Devonian and Carboniferous samples show a dramatic shift in provenance from the older Silurian and Ordovician sandstone, which has an average of $Q_{74}F_{18}L_7$, to a much more lithic-rich composition of $Q_{10}F_{34}L_{56}$. All Devonian and Carboniferous sample means plot within the volcanic arc field of Marsaglia and Ingersoll's (1992) modified Dickinson (1985) QFL diagram (Fig. 12). Most samples also plot within the arc fields of Marsaglia and Ingersoll's (1992) LmLvLs diagram (Fig. 12).

Three localities presented in Figure 12A, Huvin Har (F), Shin Jinst (E), and Gurvan Sayhan (K), contain both Devonian and Carboniferous sandstone data. The Huvin Har (F) region contains Devonian and Carboniferous marine sandstone and

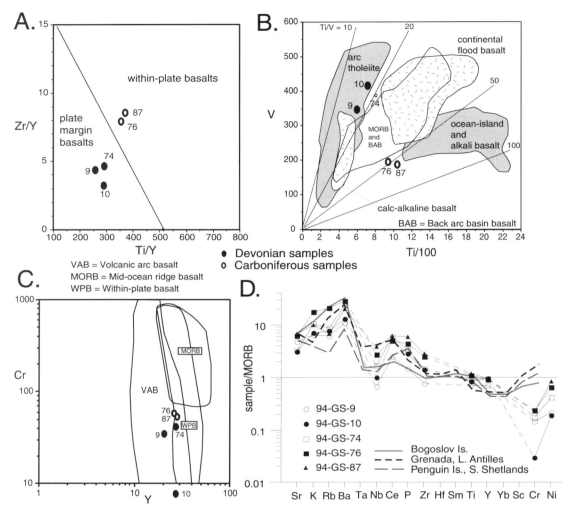

Figure 9. Discrimination diagrams for Gurvan Sayhan (K) (see Fig. 2) samples. A–C: Gurvan Sayhan (K) samples plotted on same diagrams as in Figure 5. D: Gurvan Sayhan (K) samples plotted on mid-ocean ridge basalt (MORB) normalized spider diagram with three anomalous volcanic arc patterns; spider diagram and anomalous volcanic arc patterns are all from Pearce (1983). Y-axis is dimensionless number obtained by dividing parts per million of element from sample in this study by typical value of same element from MORB; these MORB values are from Pearce (1983). Patterns from samples within this study do not have data from Ta, Hf, Sm, Yb, and Sc. Additional elements may be missing if amount of that element in sample was below detection limits. In both cases, data points on either side of missing element are connected with dashed line instead of solid line.

limestone that have not been mapped in any detail and are generally not well preserved. A small sampling of sandstone from this region plots within the volcanic arc and basement uplift fields of the QFL diagram, and within continental arc and mixed sources fields of the LmLvLs diagram (Fig. 12A).

The Shin Jinst area (E) contains a well-preserved series of Devonian and Carboniferous shallow-marine, fossiliferous sandstone and limestone interbedded with volcanic lava and pyroclastic flows (Fig. 3; Lamb and Badarch, 1997). Devonian sandstone from this sequence is slightly richer in volcanic lithics than Carboniferous samples and, thus, on the QFL diagram, plots as transitional arc whereas the Carboniferous

samples plot within dissected arc and recycled orogen fields (Figs. 12A and 13A). On the LmLvLs diagram (Fig. 12B), Devonian samples plot on the Lv apex, within the intraoceanic field, whereas the Carboniferous samples range from intraoceanic arc to mixed sources.

The Gurvan Sayhan Range (K) contains a Devonian sequence of basalt flows and pillow lavas interbedded with volcaniclastic sandstone (Fig. 3). The sandstone plots primarily within the undissected to transitional arc fields on the QFL diagram and intraoceanic arc fields on the LmLvLs diagram (Fig. 12A). The Carboniferous sandstone (Fig. 3) plots within the transitional to dissected arc field of the QFL diagram and in

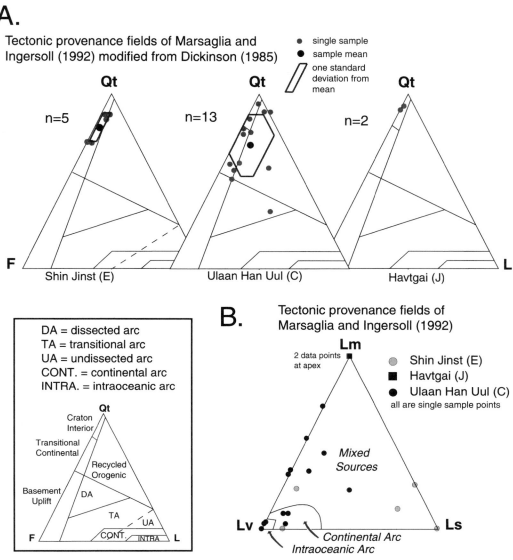

Figure 10. A: Sandstone compositional data for Ordovician and Silurian samples plotted on provenance fields of Marsaglia and Ingersoll (1992) modified from Dickinson (1985). B: Sandstone compositional data for Ordovician and Silurian samples plotted on provenance fields of Marsaglia and Ingersoll (1992). Sample locations are in Figure 2.

both the mixed sources and intraoceanic fields of the LmLvLs diagram (Fig. 12A).

Samples from all three of these localities, Huvin Har (F), Shin Jinst (E), and Gurvan Sayhan (K), suggest the initiation of arc activity in southern Mongolia during the Devonian. Many of the samples plot within transitional arc and mixed source fields. This is best explained by the mixing of two sources, active arc volcanism and continued input from an older, more mature source, most likely the same source that produced quartz-rich sandstones during the Ordovician and Silurian. Such a bimodal result could be produced in a backarc basin receiving sediment from the arc and the continental margin (e.g., the Japan Sea;

Marsaglia et al., 1992). The apparent shift in the Shin Jinst samples toward the quartz apex from the Devonian to the Carboniferous may also suggest that the locus of arc activity shifted southward during this time frame.

The remainder of the localities are in the southern half of the study area and contain Carboniferous, immature, volcaniclastic sandstones (Figs. 12B and 13B). Localities L through Q all contain thick sequences of volcaniclastic sandstone, lava flows, and pyroclastic deposits (Lamb and Badarch, 1997; Figs. 3 and 13, C and D). Volcanic units are primarily intermediate to felsic in composition. One additional locality, Nemegt (R), also contains Carboniferous marine sandstone. Sandstone

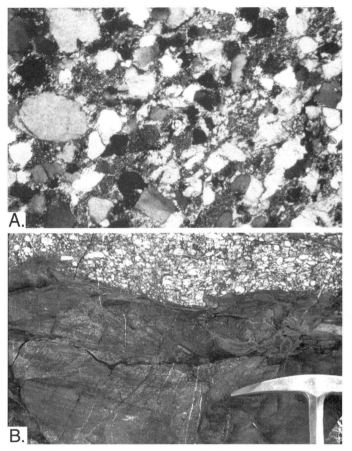

Figure 11. A: Photomicrograph of Silurian quartzose sandstone, sample 94-UH-46, from Ulan Han Uul (C) (see Fig. 2) locality, under crossed nichols. Field of view is ~4 mm across. B: Outcrop photo of Silurian, trough cross-stratified quartzite and quartz-pebble conglomerate from Havtgai (J) locality. Note rock hammer for scale; field of view is 45 cm across.

at Nemegt (R) is variably preserved and not mapped in detail, and therefore we only performed a reconnaissance study of this area. Provenance results from this area, however, are similar to the rest of the Carboniferous results. On QFL diagrams, all Carboniferous sample means plot within the undissected and transitional arc fields (Fig. 12B). On the LmLvLs diagram (Fig. 12B), all samples plot within the arc field.

Within the arc fields on these diagrams of the southern localities, sample means plot within both continental and intraoceanic arc fields (Fig. 12B). We interpret that this is in part a function of proximity to active volcanic centers; the most immature sandstones that plot within intraoceanic arc fields are interbedded with volcanic flow units and are thus likely very proximal to the volcanic source. Similarly, the samples that plot in the continental arc fields are from sequences with fewer volcanic flow units and may have been farther from the active volcanism. If so, they may have been deposited from sedimentary flows that incorporated material from a larger source area and/or underwent more weathering during transport, which may have preferentially eliminated volcanic lithic grains. It is also possible that the arc varied along strike between a continental arc and a very mature island arc.

Permian

Permian sedimentary strata in southern Mongolia record primarily nonmarine depositional environments, including fluvial and swamp facies, at a number of localities: Shin Jinst (E), Ih Uvgon (Q), Tsogt Tsisi (S), Naftgar Uul (T), and Noyon Uul (M) (Fig. 2; Amory, 1996; Hendrix et al., 1996; Zinniker, 1997). Two additional localities in far southeastern Mongolia, Lugin Gol (U) and Hobsgol (V), contain marine turbidites and shallow-marine deposits (Amory, 1996). Shin Jinst (E) sandstone samples contain a mix of quartz, feldspar, and lithics, $Q_{36}F_{30}L_{34}$, and plot within the dissected arc field on a QFL diagram (Fig. 14). Amory (1996) presented point-count data from localities M, T, and V. The sample means for these localities are within the undissected and transitional arc fields on a QFL diagram (Fig. 14).

One interpretation of these provenance results is that volcanism continued into the Permian within southern Mongolia, thereby providing a source for these sandstones. In the Shin Jinst area, certain volcanic units are not well dated, but may be Permian and may have been part of the source area. Another interpretation, however, is that the source material for Permian sandstone was the uplifted Devonian and Carboniferous arc deposits. It has been documented in nearby Xinjiang Province, China, that modern sands from the Junggar, Turpan, and Tarim basins plot within arc fields on Qm-F-Lt, Qt-F-L, and Qp-Lvt-Lsm diagrams, although they are not currently near an active arc; their source area contains uplifted, exposed Paleozoic arc deposits (Graham et al., 1993). Recent studies have also documented that global sandstone provenance fields are not always applicable to basins that receive sediment from a more localized source or a source composed of accreted terranes: sandstones from such basins plot within fields that reflect the relict tectonic setting of the source area rocks and not necessarily the tectonic setting of the source area at the time of deposition (e.g., Trop and Ridgway, 1997; Rumelhart and Ingersoll, 1997; Hendrix, 2000). Additional work on the Permian sandstones, possible local volcanic source rocks, and the basin geometries is needed to better interpret these provenance data. It is likely, however, that the relict Devonian and Carboniferous arc sequences were part of the source area.

DISCUSSION

Using the geochemical and petrographic data presented herein, combined with data presented in Lamb and Badarch (1997), we present new petrotectonic interpretations for the

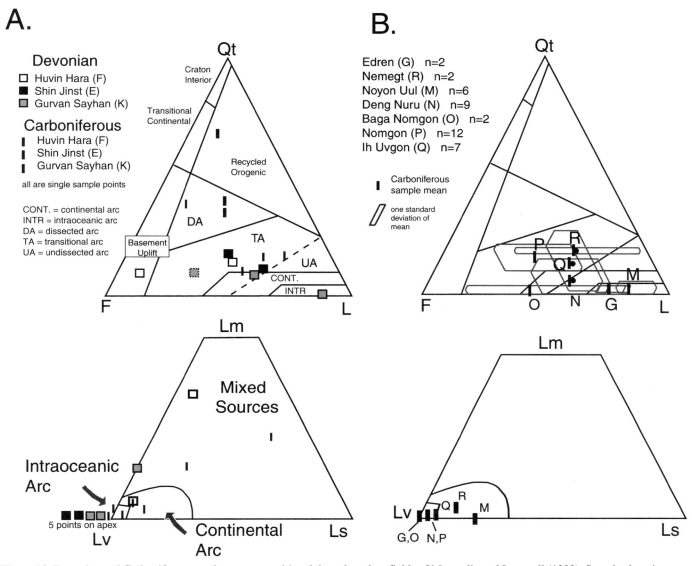

Figure 12. Devonian and Carboniferous sandstone compositional data plotted on fields of Marsaglia and Ingersoll (1992). Samples locations are in Figure 2. A: Devonian and Carboniferous data from northern portion of field area, localities F, E, and K. B: Carboniferous data from southern portion of field area, localities G, M, N, O, P, Q, and R.

study area, a modern-day analog for this region during the Devonian and Carboniferous, and one possible set of interpretations for the paleogeography of Central Asia (Figs. 15–21). The petrotectonic interpretations (Figs. 15, 18, and 20) are presented on a partially retrodeformed map of the study area from Lamb et al. (1999). Mesozoic and Cenozoic tectonic events caused deformation in southern Mongolia (Graham et al., 1996; Hendrix et al., 1996; Zheng et al., 1996; Johnson et al., 1997; Webb et al., 1997). These include early to middle Mesozoic contractile deformation (Hendrix et al., 1996; Graham et al., 1996; Zheng et al., 1996), late Mesozoic extension (Johnson et al., 1997; Webb et al., 1997), and Cenozoic transpressional defor-

mation (Baljinnyam et al., 1993; Cunningham et al., 1996, 1997). The palinspastic reconstruction of the study area (Lamb et al., 1999) retrodeforms one part of the Mesozoic–Cenozoic deformation, i.e., left-lateral strike-slip fault activity in southern Mongolia. Although it does not address all deformational events and is only a partial reconstruction, it provides a better base map on which to present our petrotectonic interpretations. We use these interpretations as well as previous work by others to present paleogeographic models for a larger area that includes portions of western and central China (Figs. 16, 19, and 21). These models include areas beyond our study area, and thus are more interpretative and somewhat speculative. Our interpretations are

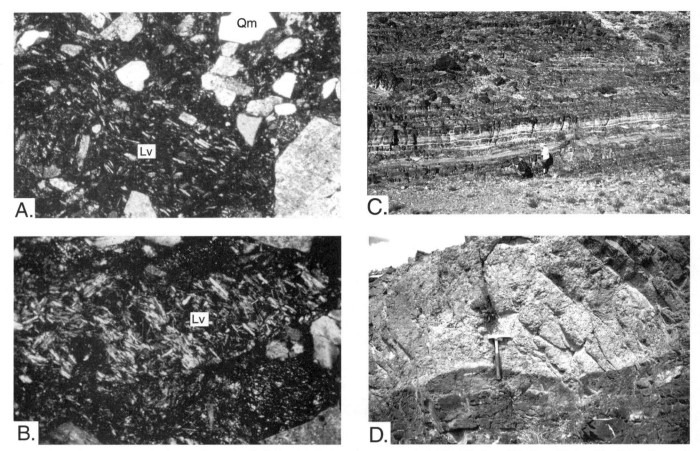

Figure 13. A: Photomicrograph of Devonian volcanic lithic-rich (Lv) sandstone, sample 94-SJ-122, from Shin Jinst (E) (see Fig. 2) locality, under crossed nichols. Qm = monocrystalline quartz. Field of view is ~4 mm across. B: Photomicrograph of Carboniferous, volcanic lithic-rich sandstone, sample 94-NM-207, from Nomgon (P) locality, under crossed nichols. Field of view is ~1.6 mm across. C: Interbedded volcaniclastic sandstone and volcanic pyroclastic deposits at Noyon Uul (M) locality. Note people for scale. D: Ignimbrite deposit overlying volcanic flow deposit at Noyon Uul (M). Note hammer for scale; hammer is 38 cm long.

necessarily simplistic given the sparsity of data and complexities of the geology, and should be viewed as working models.

Ordovician and Silurian

Provenance data presented here further support our previous interpretation (Lamb and Badarch, 1997) that the northern part of the study area was part of a marine basin that received compositionally mature sediment from a continental source or the eroded roots of an arc. The most likely candidates for the source area are represented by outcrops of Precambrian and Cambrian rocks found to the north of the study area (Fig. 1). These data also suggest that this area did not undergo active volcanism during the Ordovician and Silurian.

Devonian

Our petrotectonic interpretation of Devonian arc-related deposits within the study area is presented in Figure 15. We also infer that Devonian deposits in southernmost Mongolia and in the southwestern corner of Mongolia, outside of this study but observed by one of us (Badarch), record volcanic arc activity (Fig. 15). The data from this study and from Lamb and Badarch (1997) all suggest that in the region from Edren (G) to Gurvan Sayhan (K), the Devonian arc was an island arc near a continent, perhaps built on the outer edge of the sedimentary prism of a continental shelf, or may have been a continental arc that changed along strike to an island arc. Sedimentary facies and pillow lavas point to a marine environment, but Devonian provenance data, while mainly pointing to an arc source, also provide evidence for proximity to a continental source. Volcanic arc activity on the outermost edge of a continent could explain these data. To the southeast, in the vicinity of Tsagan Suvarga (Fig. 15, locality W), deposits have been interpreted to represent a continental to very mature island arc, based on the existence of porphyry copper deposits (Lamb and Cox, 1998).

In a larger geographical context, these Devonian deposits may represent a large arc system that trended from northwestern China and the Altai Mountains in western Mongolia (Windley et al., 1991, 1994; Bibikova et al., 1992; Ruzhentsev et al., 1992), across southern Mongolia through the Edren (G), Huvin Hara (F), and Gurvan Sayhan (K) field localities, to the Tsagan Suvarga (W) locality (Figs. 15 and 16). Figure 16B presents this arc as a long, continuous arc with a relatively simple geometry. In all likelihood, this arc system was probably much more complex. It may have been two arcs within one system: one in the Altai Mountains region, labeled the western Mongolia–Altai arc in Figure 16B, and one in the east. In this scenario, the eastern end of the western Mongolia–Altai arc is between the Edren (G) and Gurvan Sayhan (K) localities, the Gurvan Sayhan (K) basalts forming at the lateral edge of a subduction zone (Fig. 15). It is also possible that the arc looked more like the modern-day western Pacific, where subduction zones reverse polarity such as within the Philippine arc (Fig. 17A), or exhibit sharp bends, such as within the New Britain–Solomon–New Hebrides arc (Fig. 17B).

Uplifted Miocene volcanic arc facies within one portion of the New Hebrides arc are similar to Devonian deposits within the study area. The New Hebrides arc facies (Mitchell, 1970) consist primarily of marine volcaniclastic rocks, reworked limestone, and pelagic sediment with local lava flows, which Mitchell (1970) suggested may commonly be the only preserved remains of ancient island arcs. Typical arc facies, such as thick sequences of lava flows and associated ultrabasic, metamorphic, and accretionary material, are often eroded due to deformation and uplift that is common within arc settings (Mitchell, 1970). Thus, it is likely that currently exposed strata only represent one small portion of the Devonian arc system. In

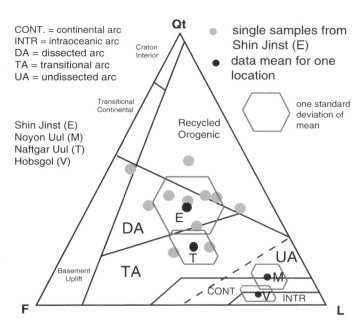

Figure 14. Permian sandstone compositional data plotted on fields of Marsaglia and Ingersoll (1992) modified from Dickinson (1985). Qt-F-L plot contains single sample and sample mean data from Shin Jinst (E) (see Fig. 2). Plots of data from other localities are from Amory (1996).

addition, the original geometry of the Devonian arc system has been obscured by late Paleozoic accretion, Mesozoic thrust faulting, and left-lateral strike-slip faulting in southern Mongolia and the Beishan in China (Hendrix et al., 1996; Zheng et al., 1996; Lamb et al., 1999).

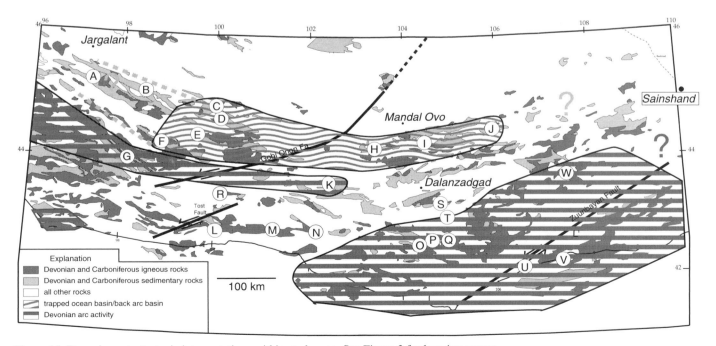

Figure 15. Devonian petrotectonic interpretations within study area. See Figure 2 for location names.

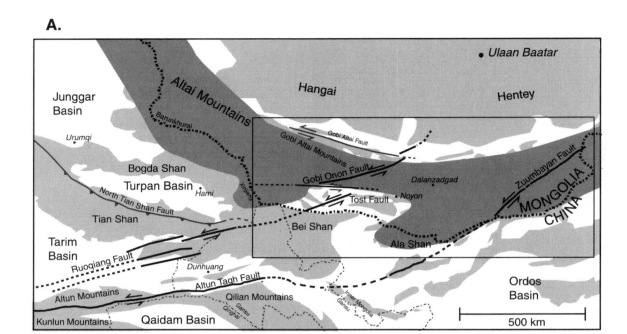

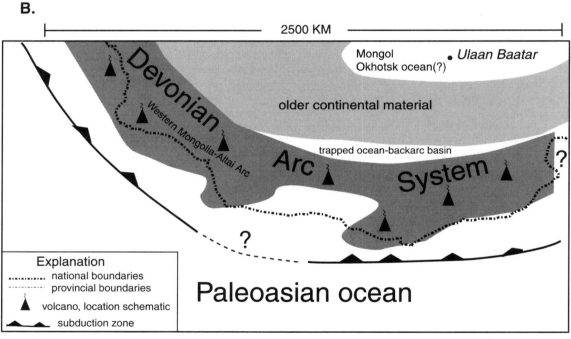

Figure 16. A: Map of Central Asia with generalized extent of Devonian deposits: white areas denote modern basins; light shaded areas denote current mountainous regions; and dark shaded areas denote approximate location of inferred Devonian arc system, including volcanic deposits, exposed plutons, accretionary material, and related metamorphic facies, all shown in present coordinates. B: Paleogeographic interpretation of same area shown in Figure 19A. Term Paleoasian ocean is from Dobretsov et al. (1994); this same ocean was also referred to as Paleotethys in Zorin et al. (1993, 1994). Light gray area denotes older material; dark gray represents Devonian interpretation.

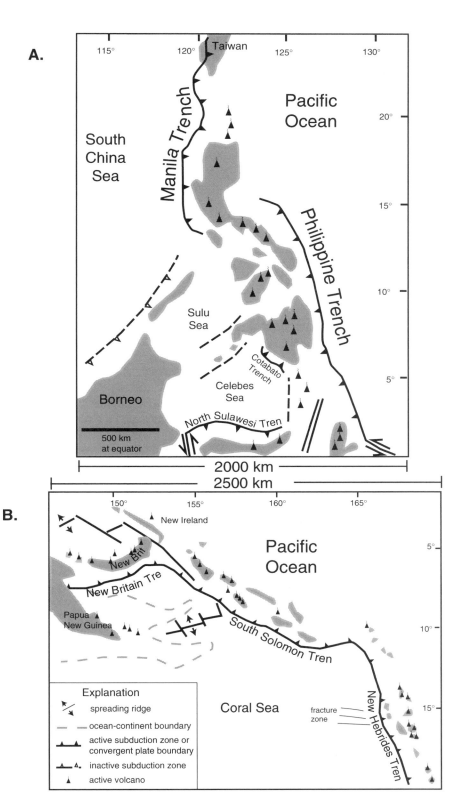

Figure 17. Two possible modern analogs for Devonian arc system within southern and western Mongolia. A: Philippine arc after World Mapping Project (1985), Katili (1991), and Packham (1996). B: New Britain–Solomon–New Hebrides arc, after World Mapping Project (1985) and Mitchell (1970), and explanation for both maps.

Coeval intrusive rocks are another common component of exposed ancient arc systems. Although there are a number of plutons exposed in southern Mongolia, neither their affinities nor ages have been well defined. Felsic intrusions near the Shin Jinst (E) and Mushgai Hudag (H) localities are mapped as late Paleozoic (Yanshin, 1989). Attempts to data these rocks more precisely using ^{40}Ar-^{39}Ar methods produced ages that are too young based on the relative age data, and probably have been affected by thermal overprinting. Further study is needed to determine if these rocks are associated with Devonian–Carboniferous arc activity or younger events.

In the northern part of the study area (Figs. 15 and 16), marine facies and geochemical data suggest that a trapped ocean basin received sediment and may have undergone backarc spreading during Late Devonian time. Evidence includes the pillow basalts at Mushgai Hudag (H), which plot consistently with a combined arc and MORB signature or as backarc basin basalts. Sedimentary deposits at Mushgai Hudag (H) and Narin Sahir Mountain (D) record a transition from shallow water to deeper water facies during the Middle to Late Devonian: the beach deposits of the Tah Uul conglomerate member at Mushgai Hudag (H; Fig. 3) grade upward into a thick sequence of pillow lavas and chert beds, whereas fossiliferous limestone at Narin Sahir (D) grades upward into fine-grained siliciclastic beds inferred to represent deeper water facies (Lamb and Badarch, 1997). Extension and subsidence within a backarc basin would account for the change to deeper water facies as well as the accommodation space needed to preserve several hundred meters of pillow basalts and lava flows. At Shin Jinst (E), Upper Devonian mafic volcanic rocks plot primarily as arc

basalts on discrimination diagrams, but the sequence also contains many rhyolitic units. A backarc basin setting is one explanation for a bimodal volcanic suite (Busby and Ingersoll, 1995; Fryer et al., 1990; Hochstaedter et al., 1990), although an arc setting for the Shin Jinst (E) locality is also a possibility.

Carboniferous

During Carboniferous time, one region of arc activity appears to have been focused in the south and, in part, built upon the Devonian arc system (Figs. 18 and 19). We refer to this as the Bogda–northern Tian Shan–southern Mongolian arc (Fig. 19B). Volcanic deposits and sandstone provenance results strongly suggest that localities L, M, N, O, P, and Q (Fig. 18) all record arc volcanism. Localities L and M contain thick sequences of welded tuff that most likely formed in a terrestrial setting (Lamb and Badarch, 1997). At localities P and Q, cobble and boulder conglomerate contains potassium feldspar–rich granite clasts. This suggests that a granitic source, which may have been part of the Devonian arc, was uplifted and exposed so that the arc was emergent in this area as well.

We infer that this arc extended westward into eastern Xinjiang Province, China, and included Carboniferous arc deposits within the Bogda Shan and northern Tian Shan (Fig. 19A). The Carboniferous Bogda arc has been interpreted as an oceanic island arc in the west, becoming more terrestrial toward the east, near the southwestern border of Mongolia (Carroll et al., 1990; Xinjiang Uygur Autonomous Region Bureau of Geology and Mineral Resources [XBGMR], 1993). Parts of this arc may also be preserved in the northernmost Beishan and Alashan regions

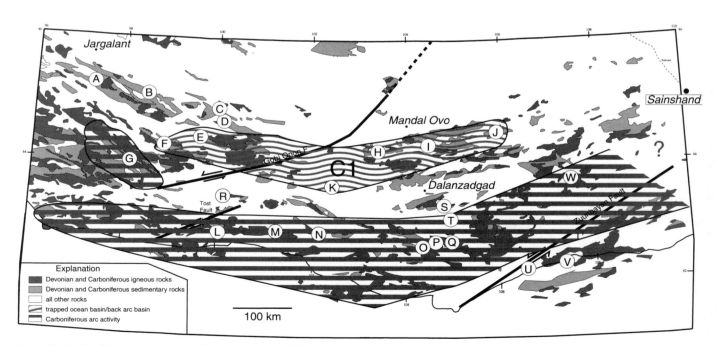

Figure 18. Carboniferous petrotectonic interpretations within study area. See Figure 2 for location names. C1 = Lower Carboniferous.

A.

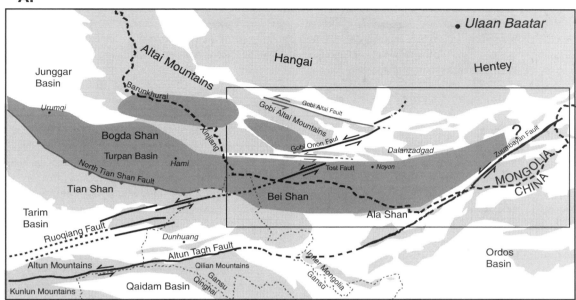

B.

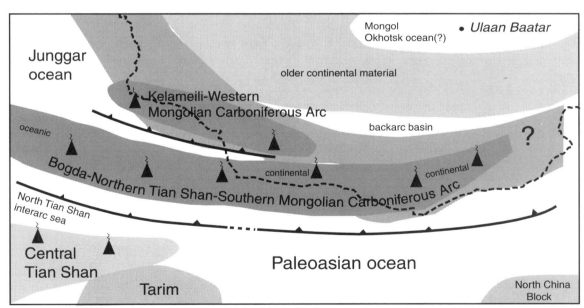

Figure 19. Map of Carboniferous deposits and paleogeographic interpretation. A: Dark shaded areas denote approximate location of inferred Carboniferous arc system; see Figure 16 caption for further explanation. B: Shapes of central Tian Shan, Tarim, and north China blocks, as well as Kelameili–Western Mongolian arc, are extremely schematic. Areas shown in light and medium grays denote older material; dark gray represents Carboniferous interpretation.

of Gansu and Inner Mongolia Provinces (Gansu Province Bureau of Geology and Mineral Resources [GBGMR], 1989; Nei Mongol Autonomous Region Bureau of Geology and Mineral Resources [NMBGMR], 1991). In our model, the arc extends eastward to Tsagan Suvarga (Fig. 18, locality W), but may actually continue eastward for several hundred kilometers (Yanshin, 1989; NMBGMR, 1991).

Volcanic deposits at Edren (G) and Shin Jinst (E; Fig. 18) also suggest continued arc activity into the Carboniferous, although better age and geochemical data are needed to determine the exact nature and timing of this volcanism. These deposits may be correlative with Carboniferous arc deposits in Xinjiang Province that Carroll et al. (1990) labeled the Kelameili arc. Our model incorporates the work of Carroll et al. (1990, 1995) and includes a second Carboniferous arc that formed through the closure of the eastern part of the Junggar ocean (Fig. 19). We infer that the Kelameili deposits were built upon part of the Devonian western Mongolia Altai arc to the north (Fig. 19), including Devonian volcanic arc and accretionary material in the Barunkhurai depression area (Fig. 19A) identified by Ruzhentsev et al. (1992), and within eastern Xinjiang (XBGMR, 1993).

The backarc basin we interpret for the northern part of the study area probably closed sometime during the Late Carboniferous or Early Permian, but this is poorly constrained. Early Carboniferous marine deposits are preserved at Huvin Har (F), Shin Jinst (E), Gurvan Sayhan (K), Mushgai Hudag (H), and Havtgai (J; Fig. 18; Lamb and Badarch, 1997). Shin Jinst (E) is the only locality of these five that has Permian deposits, and these are nonmarine.

The northern part of the Gurvan Sayhan Range (K) may contain further evidence for closure of this basin. The northern part of the range consists of poorly studied accretionary material, including chert, serpentine, and olistostromes. This melange may have developed during closure of the backarc basin as oceanic crust was subducted southward. One possible modern analog for this interpretation is the Cotabato trench on the western side of the Philippine arc (Fig. 17A; Rangin et al., 1996; Pubellier et al., 1996). The current tectonic setting of the Cotabato–eastern Celebes Sea region consists of a laterally restricted trench zone with accretionary material developed on the opposite side of a larger, more extensive arc. It is also possible, however, that the material in the Gurvan Sayhan Range (K) formed as part of the northward subduction regime, similar to the rest of the arc, and is correlative with other mafic material to the west in the Nemegt (R) region. The Gurvan Sayhan Range (K) is structurally complicated and many Paleozoic rocks are not well mapped or dated. Better age and structural data are needed throughout the range to discriminate between these two models and those of previous workers (e.g., Eenzhin, 1983; Ruzhentsev et al., 1985; Voznesenskaya et al., 1989).

This backarc basin is not related to Carboniferous marine clastic rocks found in the region of the proposed Mongol-Okhotsk ocean (Fig. 19; Lamb and Badarch, 1997). This latter, much larger basin was probably separated from the Carboniferous backarc basin by older, continental crustal material.

Permian

Most Permian deposits of southern Mongolia record nonmarine facies of local, intermontane, or foreland basins (Figs. 20 and 21; Traynor and Sladen, 1995; Amory, 1996; Hendrix et al., 1996). Continental Permian deposits within China suggest that the Junggar ocean retreated westward and that the Bogda arc and Kelameili–western Mongolian arc collided during the

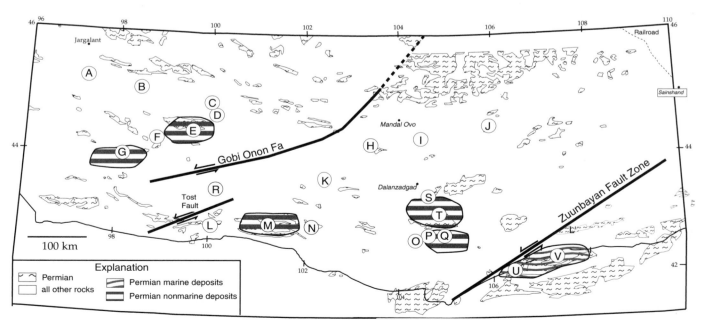

Figure 20. Permian petrotectonic interpretations within study area. See Figure 2 for location names.

A.

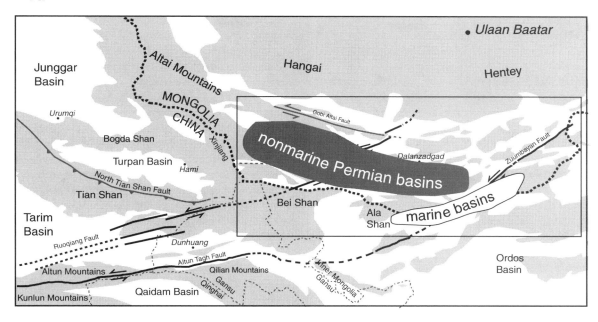

B.

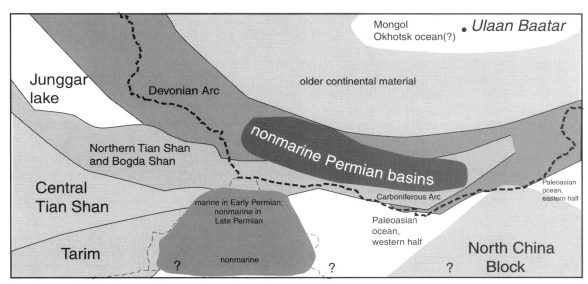

Figure 21. Map of Permian deposits and paleogeographic interpretation. A: Dark shaded area denotes limit of area within which we observed Permian nonmarine deposits. Additional area in white shows area that contains Permian marine deposits not observed as part of this study; see Figure 16 caption for further explanation. B: Areas shown in light and medium grays denote older material; dark grays represent Permian interpretation.

Early Permian (Fig. 21; Carroll et al., 1995). The northern Gansu Province also contains Early Permian marine deposits and Late Permian nonmarine deposits (GBGMR, 1989). Southeastern Mongolia and the Alashan region of Inner Mongolia, however, contain marine deposits from throughout the Permian (Fig. 20).

Most studies suggest that the Permian, and possibly Early Triassic, was the time of collision and amalgamation of southern Mongolia, the north China block, and the Tarim–Tian Shan region of China and the subsequent closure of the Paleoasian ocean (Ruzhentsev et al., 1989; Nie et al., 1990; Zhao et al., 1990; Mueller et al., 1991; Pruner, 1992; Dorjnamjaa et al., 1993; Zorin et al., 1993, 1994; Tomurtogoo, 1997). The timing of these events and the detailed paleogeographic history, however, is uncertain. Previous work indicates that, subsequent to the closure of the Junggar ocean basin in eastern Xinjiang, the north China block collided with southern Mongolia (Nie et al., 1990; Zhao et al., 1990; Zorin et al., 1994). It has been suggested that this represents a simple west to east diachronous, but geographically continuous, closure along the entire northern China–southern Mongolia political boundary (e.g., Zorin et al., 1994). Another possibility that we present here, however, is that the north China block collided initially with the southeastern corner of the Mongolian arc system, thereby dividing the Paleoasian ocean into two halves (Fig. 21B). As the collision ensued, the eastern portion of the ocean closed from west to east. The western half, most likely a small trapped ocean basin, closed east to west in latest Permian to earliest Triassic time. Upper Permian marine turbidites within the Lugin Gol Formation in southern Mongolia (observed by Badarch) and similar strata across the border in the Alashan region of Inner Mongolia may provide the youngest record of marine conditions; Triassic deposits in this region contain nonmarine fossil assemblages (NMBGMR, 1991). The closing of the western half of the ocean may be contemporaneous with the compression that resulted in the formation of an early Mesozoic foreland basin across southern Mongolia (Hendrix et al., 1996, and this volume).Until more detailed mapping, dating, and facies characterization are collected from southernmost Mongolia and northern Inner Mongolia, it will be difficult to distinguish between any of these models for the closure of the Paleoasian ocean.

CONCLUSIONS

In this chapter, we presented new geochemical and sandstone provenance data and combined these with sedimentary and stratigraphic data presented in Lamb and Badarch (1997) to present a number of paleogeographic models and outline a tectonic framework for southern Mongolia throughout most of the Paleozoic Era. Our models are necessarily simplistic but reflect these new geologic constraints and data available elsewhere in the literature.

During Ordovician and Silurian time, southern Mongolia contained a marine basin or basins that received composition-

ally mature siliciclastic sediment from a continental source or the eroded roots of an arc, most likely from a source to the north. The Ordovician and Silurian strata suggest that there was only minor tectonic activity during this time. There is no evidence of volcanism, only uplift of older strata that may be the result of incipient compression related to the development of a convergent margin during Devonian time.

The Devonian strata record a shift from relative tectonic quiescence to a period of major volcanism and tectonic activity. The range of sedimentary facies and geochemical data suggests that within southern Mongolia a complex arc system developed that varied temporally and spatially: there is evidence for both island arc and continental arc components as well as backarc and volcanic arc basin settings. The arc was built on or adjacent to Ordovician and Silurian deposits and may have connected to the northwest with the western Mongolian Altai arc.

Carboniferous rocks suggest continued volcanism and the development of a more continental arc built upon Devonian deposits. This arc system included a number a subaerial volcanic centers across southern Mongolia and may have been the eastern extension of the Bogda Shan and northern Tian Shan Carboniferous arc in China. Carboniferous arc deposits in the western portion of the study area may connect with the Kelameili arc in China.

Permian data indicate that arc volcanism stopped by this time, but that tectonic activity continued. Devonian and Carboniferous deposits were uplifted and formed the source of sediment deposited in localized intermontane basins. Volcanism continued into this time, but the geochemistry of these units suggests continental volcanism more than arc volcanism, which may be related to the consolidation of Central Asia. The marine to nonmarine transition for sedimentary rocks at this time reflects the closure of the Paleoasian ocean as north China, Tarim, the Tian Shan, and southern Mongolia collided and amalgamated.

Additional work in southern Mongolia and north-central China is crucial to test and refine this tectonic framework. Mesozoic and Cenozoic deformation needs to be better constrained in order to palinspastically reconstruct Paleozoic features, although a better understanding of Paleozoic features will help determine the history of younger deformation. Further work should include documenting sedimentologic facies and recording depositional environments; analyzing, characterizing, and dating volcanic deposits and other igneous rocks; and mapping and dating structural features of all ages.

ACKNOWLEDGMENTS

Research reported in this chapter was supported by National Science Foundation grant EAR-9315941, a Geological Society of America student grant, an American Association of Petroleum Geologists student grant, grants from the Shell Fund and McGee Fund (Stanford University), and the Stanford-Mongolia Industrial Affiliates Program, a consortium of companies that has included BHP, Exxon, Elf-Aquitaine,

Mobil, Nescor Energy, Occidental, Pecten, and Phillips. We gratefully acknowledge the technical support of the Mongolian Academy of Sciences Paleontological Center and its director, R. Barsbold. We thank Thomas Hickson, Juliet Crider, and David Zinniker for outstanding field assistance; Jerome Amory for permission to include his Permian sandstone data; and A. Chimedtseren and D. Tomorhuu for logistical support. We also thank S.A. Graham, R.G. Coleman, M.S. Hendrix, C.J. Busby, and S.M. DeBari for critical reviews that greatly improved earlier versions of this manuscript.

REFERENCES CITED

Amory, J.Y., 1996, Permian sedimentation and tectonics of southern Mongolia [M.S. thesis]: Stanford, California, Stanford University, 183 p.

Arculus, R.J., 1976, Geology and geochemistry of the alkali basalt-andesite association of Grenada, Lesser Antilles island arc: Geological Society of America Bulletin, v. 87, p. 612–624.

Baljinnyam, I., Bayasgalan, A., Borisov, B.A., Cisternas, A., Dem'yanovich, M.G., Ganbaatar, L., Kochetkov, V.M., Kurushin, R.A., Molnar, P., Philip, H., and Vashchilov, Y.Y., 1993, Ruptures of major earthquakes and active deformation in Mongolia and its surroundings: Geological Society of America Memoir 181, 62 p.

Bibikova, Y.V., Kirnozova, T.I., Kozakov, I.K., Kotov, A.B., Neymark, L.A., Gorokhovskiy, B.M., and Shuleshko, I.K., 1992, U-Pb ages for polymetamorphic complexes on the southern flank of the Mongolian and Gobi Altai: Geotectonics, v. 26, p. 166–172.

Busby, C.J., and Ingersoll, R.V., 1995, Tectonics of sedimentary basins: Cambridge, Massachusetts, Blackwell Science, 579 p.

Carroll, A.R., Liang Yunhai, Graham, S.A., Xiao Xuchang, Hendrix, M.S., Chu Jinchi, and McKnight, C. L., 1990, Junggar Basin, northwest China; trapped late Paleozoic ocean: Tectonophysics, v. 181, p. 1–14.

Carroll, A.R., Graham, S.A., Hendrix, M.S., Ying Xudong, and Zhou Da, 1995, Late Paleozoic tectonic amalgamation of northwestern China; sedimentary record of the northern Tarim, northwestern Turpan, and southern Junggar basins: Geological Society of America Bulletin, v. 107, p. 571–594.

Chang, E.Z., Coleman, R.G., and Ying Xudong, 1996, 1:2,000,000 geodynamic map of Asia (western part): Tulsa, Oklahoma, American Association of Petroleum Geologists, scale 1:2 000 000.

Coleman, R.G., 1989, Continental growth of northwest China: Tectonics, v. 8, p. 621–635.

Cunningham, W.D., Windley, B.F., Dorjnamjaa, D., Badamgarov, J., and Saandar, M., 1996, Late Cenozoic transpression in southwestern Mongolia and the Gobi Altai–Tien Shan connection: Earth and Planetary Science Letters, v. 140, p. 67–81.

Cunningham, W.D., Windley, B.F., Owen, L.A., Barry, T., Dorjnamjaa, D., and Badamgarav, J., 1997, Geometry and style of partitioned deformation within a late Cenozoic transpressional zone in the eastern Gobi Altai Mountains, Mongolia: Tectonophysics, v. 277, p. 285–306.

DeLong, S.E., Hodges, F.N., and Arculus, R.J., 1975, Ultramafic and mafic inclusions, Kanaga Island, Alaska, and the occurrence of alkaline rocks in island arcs: Journal of Geology, v. 83, p. 721–736.

Dickinson, W.R., 1970, Interpreting detrital modes of graywacke and arkose: Journal of Sedimentary Petrology, v. 40, p. 695–707.

Dickinson, W.R., 1985, Interpreting provenance relations from detrital modes of sandstones, in Zuffa, G.G., ed., Provenance of arenites: NATO Advanced Study Institutes Series C; Mathematical and Physical Sciences, v. 148, p. 333–361.

Dobretsov, N.L., Coleman, R.G., and Berzin, N.A., 1994, Geodynamic evolution of the Paleoasian Ocean: Russian Geology and Geophysics, v. 35, p. 233.

Dobretsov, N.L., Sobolev, N.V., Shatsky, V.S., Coleman, R.G., and Ernst, W.G., 1995, Geotectonic evolution of diamondiferous paragneisses, Kokchetav Complex, northern Kazakhstan; the geologic enigma of ultrahigh-pressure crustal rocks within a Paleozoic foldbelt: The Island Arc, v. 4, p. 267–279.

Dorjnamjaa, D., Badarch, G., and Orolmaa, D., 1993, The geodynamic evolution of the mobile fold belts of the territory of Mongolia, in Coleman, R.G., ed., Reconstruction of the paleo-Asian ocean: The Netherlands, VSP International Science Publishers, p. 63–76.

Eenzhin, G., 1983, The southern Mongolian Hercynian eugeosynclinal zone (Dzolen Range—Mandal-Ovoo) during Early Devonian time: Geotectonics, v. 17, p. 326–334.

Enkin, R.J., Yang Zhenyu, Chen Yan, and Courtillot, V., 1992, Paleomagnetic constraints on the geodynamic history of the major blocks of China from the Permian to the present: Journal of Geophysical Research, v. 97, p. 13953–13989.

Fournelle, J.H., Marsh, B.D., and Myers, J.D., 1994, Age, character, and significance of Aleutian arc volcanism, in Plafker, G., and Berg, H.C., eds., The geology of Alaska: Boulder, Colorado, Geological Society of America, Geology of North America, v. G-1, p. 723–757.

Fryer, P., Taylor, B., Langmuir, C.H., Hochstaedter, A.G., 1990, Petrology and geochemistry of lavas from the Sumisu and Torishima backarc rifts: Mariana Trough: Earth and Planetary Science Letters, v. 100, p. 161–178.

Gansu Province Bureau of Geology and Mineral Resources, 1989, Regional geology of Gansu province: Beijing, Geological Publishing House, Geological Memoirs, Series 1, no. 19 scale 1: 1 000 000, 690 p.

Golonka, J., Ross, M.I., and Scotese, C.R., 1994, Phanerozoic paleogeographic and paleoclimatic modeling maps, in Embry, A.F., et al., eds., Pangea; global environments and resources: Canadian Society of Petroleum Geologists Memoir 17, p. 1–47.

Graham, S.A., Hendrix, M.S., Wang, L.B., and Carroll, A.R., 1993, Collisional successor basins of western China; impact of tectonic inheritance on sand composition: Geological Society of America Bulletin, v. 105, p. 323–344.

Graham, S.A., Hendrix, M.S., Badarch, G., and Badamgarav, D., 1996, Sedimentary record of transition from contractile to extensional tectonism, Mesozoic, southern Mongolia: Geological Society of America Abstracts with Programs, v. 28, no. 7, p. A-68.

Hall, A., 1987, Igneous petrology: New York, Longman Scientific and Technical, 573 p.

Hellman, P.L., Smith, R.E., and Henderson, P., 1977, Rare earth element investigation of the Cliefden Outcrop, N.S.W., Australia: Contributions to Mineralogy and Petrology, v. 65, p. 155–164.

Hendrix, M.S., 2000, Evolution of Mesozoic sandstone compositions, southern Junggar, northern Tarim and western Turpan basins, northwest China: A detrital record of the ancestral Tian Shan: Journal of Sedimentary Research (in press).

Hendrix, M.S., Graham, S.A., Amory, J.Y., and Badarch, G., 1996, Noyon Uul syncline, southern Mongolia; lower Mesozoic sedimentary record of the tectonic amalgamation of Central Asia: Geological Society of America Bulletin, v. 108, p. 1256–1274.

Hochstaedter, A.G., Gill, J.B., Kusakabe, M., Newman, S., Pringle, M., Taylor, B., and Fryer, P., 1990, Volcanism in the Sumisu Rift; I, Major element, volatile, and stable isotope geochemistry: Mariana Trough: Earth and Planetary Science Letters, v. 100, p. 179–194.

Ingersoll, R.V., Bullard, T.F., Ford, R.L., Grimm, J.P., Pickle, J.D., and Sares, S.W., 1984, The effect of grain size on detrital modes; a test of the Gazzi-Dickinson point-counting method: Journal of Sedimentary Petrology, v. 54, p. 103–116.

Johnson, C.L., Graham, S.A., Webb, L.E., Badarch, G., Beck, M., Hendrix, M.S., Lenegen, R., and Sjostrom, D., 1997, Sedimentary response to late Mesozoic extension, southern Mongolia: Eos (Transactions, American Geophysical Union), v. 78, p. F175.

Katili, J.A., 1991, Tectonic evolution of eastern Indonesia and its bearing on the occurrence of hydrocarbons: Marine and Petroleum Geology, v. 8, p. 70–83.

Lamb, M.A., 1998, Paleozoic sedimentation, volcanism and tectonics in southern Mongolia [Ph.D. thesis]: Stanford, California, Stanford University, 207 p.

Lamb, M.A., and Badarch, G., 1997, Paleozoic sedimentary basins and volcanic-arc systems of southern Mongolia: New stratigraphic and sedimentologic constraints: International Geology Review, v. 39, p. 542–576.

Lamb, M.A., and Cox, D., 1998, New $^{40}Ar/^{39}Ar$ age data and implications for porphyry copper deposits of Mongolia: Economic Geology, v. 93, p. 524–529.

Lamb, M.A., Hanson, A.D., Graham, S.A., Webb, L., and Badarch, G., 1999, Left-lateral sense offset of Upper Proterozoic to Paleozoic features across the Gobi Onon, Tost, and Zuunbayan faults in southern Mongolia and implications for other central Asian faults: Earth and Planetary Science Letters, v. 173, p. 183–194.

Le Maitre, R.W., 1976, The chemical variability of some common igneous rocks: Journal of Petrology, v. 17, p. 589–637.

Le Maitre, R.W., Bateman, P., Dudek, A., Keller, J., Lemeyre, J., Le Bas, M. J., Sabine, P.A., Schmid, R., Sorensen, H., Streckeisen, A., Wooley, A.R., and Zanettin, B., 1989, A classification of igneous rocks and glossary of terms: Oxford, Blackwell Scientific Publications, 193 p.

Lowe, T.K., 1995, Petrogenesis of the Minarets and Merced Peak volcanic-plutonic complexes, Sierra Nevada, California [Ph.D. thesis]: Stanford, California, Stanford University, 157 p.

Marsaglia, K.M., and Ingersoll, R.V., 1992, Compositional trends in arc-related, deep-marine sand and sandstone; a reassessment of magmatic-arc provenance: Geological Society of America Bulletin, v. 104, p. 1637–1649.

Marsaglia, K.M., Ingersoll, R.V., and Packer, B.M., 1992, Tectonic evolution of the Japanese Islands as reflected in modal compositions of Cenozoic forearc and backarc sand and sandstone: Tectonics, v. 11, p. 1028–1044.

McBirney, A.R., 1993, Igneous petrology: Boston, Jones and Bartlett Publishers, Inc., 508 p.

Meschede, M., 1986, A method of discrimination between different types of mid-ocean ridge basalts and continental tholeiites with the Nb-Zr-Y diagram: Chemical Geology, v. 56, p. 207–218.

Mitchell, A.H.G., 1970, Facies of an early Miocene volcanic arc, Malekula island, New Hebrides: Sedimentology, v. 14, p. 210–243.

Mossakovsky, A.A., Ruzhentsev, S.V., Samygin, S.G., and Kheraskova, T.N., 1993, Central Asian fold belt: Geodynamic evolution and formation history: Geotectonics, v. 27, p. 445–474.

Mueller, J.F., Rogers, J.J.W., Jin Yugan, Wang Huayu, Li Wenguo, Chronic, J., and Mueller, J.F., 1991, Late Carboniferous to Permian sedimentation in Inner Mongolia, China, and tectonic relationships between north China and Siberia: Journal of Geology, v. 99, p. 251–263.

Mullen, E.D., 1983, $MnO/TiO_2/P_2O_5$; a minor element discriminant for basaltic rocks of oceanic environments and its implications for petrogenesis: Earth and Planetary Science Letters, v. 62, p. 53–62.

Natal'in, B.A., Sengör, A.M.C., and Itü, M.F., 1996, Arc-slicing and arc-shaving strike-slip faults in the Paleozoic collage of the Altaids, Central Asia: Geological Society of America Abstracts with Programs, v. 28, no. 7, p. A67-68.

Nie Shangyou, Rowley, D.B., and Ziegler, A.M., 1990, Constraints on the location of the Asian microcontinents in Palaeo-Tethys during the late Palaeozoic, in McKerrow, W.S., and Scotese, C.R., eds., Palaeozoic palaeogeography and biogeography: Geological Society [London] Memoir 12, p. 397–409.

Nei Mongol Autonomous Region Bureau of Geology and Mineral Resources, 1991, Regional geology of Nei Mongol (Inner Mongolia) Autonomous Region: Beijing, Geological Publishing House, Geological Memoirs, Series 1, no. 25 scale 1:2 000 000, 725 p.

Norrish, K., and Chappell, B.W., 1977, X-ray fluorescence spectrometry, in Zussman, J., ed., Physical methods in determinative mineralogy: London, Academic Press, p. 210–272.

Norrish, K., and Hutton, J.T., 1969, An accurate X-ray spectographic method for the analysis of a wide range of geological samples: Geochimica et Cosmochimica Acta, v. 33, p. 431–453.

Packham, G., 1996, Cenozoic SE Asia; reconstructing its aggregation and reorganization, in Hall, R., and Blundell, D.J., eds., Tectonic evolution of Southeast Asia: Geological Society [London] Special Publication 106, p. 123–152.

Pearce, J.A., 1982, Trace element characteristics of lavas from destructive plate boundaries, in Thorpe, R.S., ed., Andesites; orogenic andesites and related rocks: Chichester, U.K., John Wiley & Sons, p. 525–548.

Pearce, J.A., 1983, Role of the sub-continental lithosphere in magma genesis at active continental margins in Hawkesworth, C.J., and Norry, M.J., ed., Continental basalts and mantle xenoliths: Nantwich, U.K., Shiva Publications, p. 230–249.

Pearce, J.A., and Gale, G.H., 1977, Identification of ore-deposition environment from trace-element geochemistry of associated igneous host rocks, in Volcanic processes in ore genesis: London, Institution of Mining and Metallurgy, p. 14–24.

Pruner, P., 1992, Palaeomagnetism and palaeogeography of Mongolia from the Carboniferous to the Cretaceous; final report: Physics of the Earth and Planetary Interiors, v. 70, p. 169–177.

Pubellier, M., Quebral, R., Aurelio, M., and Rangin, C., 1996, Docking and post-docking escape tectonics in the southern Philippines in Hall, R., and Blundell, D.J., eds., Tectonic evolution of Southeast Asia: Geological Society [London] Special Publication 106, p. 511–523.

Rangin, C., Dahrin, D., Quebral, R., and Party, M.S., 1996, Collision and strike-slip faulting in the northern Molucca Sea (Philippines and Indonesia); preliminary results of a morphotectonic study, in Hall, R., and Blundell, D.J., eds., Tectonic evolution of Southeast Asia: Geological Society [London] Special Publication 106, p. 29–46.

Reynolds, R.C., Jr., 1967, Estimation of mass absorption coefficients by Compton scattering: Improvements and extensions of the method: American Mineralogist, v. 52, p. 1493–1502.

Rollinson, H.R., 1993, Using geochemical data; evaluation, presentation, interpretation: Harlow, U.K., Longman Scientific & Technical, 352 p.

Rumelhart, P.E., and Ingersoll, R.V., 1997, Provenance of the Upper Miocene Modelo Formation and subsidence analysis of the Los Angeles Basin, southern California: Implications for paleotectonic and paleogeogrpahic reconstructions: Geological Society of America Bulletin, v. 109, p. 885–899.

Ruzhentsev, S.V., and Pospelov, I.I., 1992, The south Mongolian Variscan fold system: Geotectonics, v. 26, p. 383–395.

Ruzhentsev, S.V., Badarch, H., and Voznesenskaya, T.A., 1985, Tectonics of the Trans-Altai Zone of Mongolia (the Gurvansaykhan and Dzolen ranges): Geotectonics, v. 19, p. 276–284.

Ruzhentsev, S.V., Pospelov, I.I., and Badarch, G., 1989, Tectonics of the Mongolian Indosinides: Geotectonics, v. 23, p. 476–487.

Ruzhentsev, S.V., Pospelov, I.I., and Badarch, G., 1992, Tectonics of the Barunkhurai Depression of Mongolia: Geotectonics, v. 26, p. 67–77.

Scotese, C.R., and McKerrow, W.S., 1990, Revised world maps and introduction, in McKerrow, W.S., and Scotese, C.R., eds., Palaeozoic palaeogeography and biogeography: Geological Society [London] Memoir 12, p. 1–21.

Sengör, A.M.C., and Natal'in, B.A., 1996, Paleotectonics of Asia: Fragments of a synthesis, in Yin, A., and Harrison, M., eds., Tectonic development of Asia: Cambridge, Cambridge University Press, p. 486–641.

Sengör, A.M.C., Burke, K., and Natal'in, B.A., 1993a, Asia: A continent built and assembled over the past 500 million years: Geological Society of America Short Course Notes, 262 p.

Sengör, A.M.C., Natal'in, B.A., and Burtman, V.S., 1993b, Evolution of the Altaid tectonic collage and Palaeozoic crustal growth in Eurasia: Nature (London), v. 364, p. 299–307.

Shervais, J.W., 1982, Ti-V plots and the petrogenesis of modern and ophiolitic lavas: Earth and Planetary Science Letters, v. 59, p. 101–118.

Thirlwall, M.F., and Graham, A.M., 1984, Evolution of high-Ca, high-Sr C-series basalts from Grenada, Lesser Antilles; the effects of intra-crustal contamination: Geological Society of London of Journal, v. 141, p. 427–445.

Tomurtogoo, O., 1997, A new tectonic scheme of the Paleozoides in Mongolia, *in* Xu Zhiqin, Ren Yufeng, and Qiu Xiaoping, eds., 30th International Geological Congress Proceedings: The Netherlands, VSP, p. 75–82.

Traynor, J.J., and Sladen, C., 1995, Tectonic and stratigraphic evolution of the Mongolian People's Republic and its influence on hydrocarbon geology and potential: Marine and Petroleum Geology, v. 12, p. 35–52.

Trop, J.M., and Ridgway, K.D., 1997, Petrofacies and provenance of a Late Cretaceous suture zone thrust-top basin, Cantwell basin, central Alaska Range: Journal of Sedimentary Research, v. 67, p. 469–485.

Vallier, T.L., Scholl, D.W., Fisher, M.A., Bruns, T.R., Wilson, F.H., von Huene, R., and Stevenson, A.J., 1994, Geologic framework of the Aleutian Arc, Alaska, *in* Plafker, G., and Berg, H.C., eds., The geology of Alaska: Boulder, Colorado, Geological Society of America, Geology of North America, v. G-1, p. 367–388.

VanderPlas, L., and Tobi, A.C., 1965, A chart for judging the reliability of point counting results: American Journal of Science, v. 263, p. 87–90.

Voznesenskaya, T.A., and Badarch, G., 1991, Sedimentation and volcanism in the Dzhinsetsk subzone of the southern Mongolian Variscides: Lithology and Mineral Resources (USSR), v. 26, p. 135–148.

Voznesenskaya, T.A., Rateyev, M.A., Kheirov, M.B., Kalashnikova, N.L., and Shabrova, V.P., 1989, Fine-grained tephrogenic and sedimentary rocks from the axial zone of the southern Mongolian Variscides: Lithology and Mineral Resources (USSR), v. 24, p. 114–124.

Webb, L.E., Graham, S.A., Johnson, C.L., Badarch, G., Beck, M., Hendrix, M.S., Lenegen, R., and Sjostrom, D., 1997, Characteristics and implications of the Onch Hayrhan metamorphic core complex of southern Mongolia: Eos (Transactions, American Geophysical Union), v. 78, p. F174–F175.

Willis, J.P., 1989, Compton scatter and matrix correction for trace element analysis of geological materials, *in* Ahmedali, S.T., ed., X-ray fluorescence analysis in the geological sciences; advances in methodology: Geological Society of Canada Short Course 7, p. 91–140.

Wilson, M., 1989, Igneous petrogenesis; a global tectonic approach: London, Unwin Hyman, 466 p.

Winchester, J.A., and Floyd, P.A., 1977, Geochemical discrimination of different magma series and their differentiation products using immobile elements: Chemical Geology, v. 20, p. 325–343.

Windley, B.F., Li Jingyi, Guo Jinhui, and Zhang Chi, 1991, The Devonian Altai orogenic belt, Chinese Central Asia: Terra Abstracts, v. 3, p. 376–377.

Windley, B.F., Guo Jinhui, Li Jingyi, and Zhang Chi, 1994, Subdivisions and tectonic evolution of the Chinese Altai: Russian Geology and Geophysics, v. 35, p. 98–99.

World Mapping Project 1985, Tectonic map of the World: Houston, Texas, Exxon Production Research Company, scale 1:10 000 000.

Xinjiang Uygur Autonomous Region Bureau of Geology and Mineral Resources, 1993, Regional geology of Xinjiang Uygur Autonomous Region: Beijing, Geological Publishing House, Geological Memoirs, Series 1, no. 32, scale 1:1 500 000, 841 p.

Yanshin, A.L., ed., 1989, Geologic map of the Mongolian People's Republic: Moscow, Akademia Nauk, scale 1:1 500 000.

Zhang, Z. M., Liou, J.G., and Coleman, R.G., 1984, An outline of the plate tectonics of China: Geological Society of America Bulletin, v. 95, p. 295–312.

Zhao Xixi, Coe, R.S., Zhou Yaoxiu, Wu Haoruo, and Wang Jie, 1990, New paleomagnetic results from northern China: Collision and suturing with Siberia and Kazakhstan: Tectonophysics, v. 181, p. 43–81.

Zheng Yadong, Zhang, Q., Wang, Y., Liu, R., Wang, S.G., Zuo Guochao, Wang Shizheng, Lkaasuren, B., Badarch, G., and Badamgarav, Z., 1996, Great Jurassic thrust sheets in Beishan (North Mountains); Gobi areas of China and southern Mongolia: Journal of Structural Geology, v. 18, p. 1111–1126.

Zinniker, D.A., 1997, Reconnaissance palynology and evidence of climate change across the Permian/Triassic boundary, Ih Uvgon, southern Mongolia: Geological Society of America Abstracts with Programs, v. 29, no. 6, p. A-97.

Zonenshayn, L.P., Kuzmin, M.I., and Natapov, L.M., 1990, Geology of the USSR: A plate-tectonic synthesis: American Geophysical Union, Geodynamics Series, v. 21, 242 p.

Zorin, Y.A., Belichenko, V.G., Turutanov, E.K., Kozhenvnikov, V.M., Ruzhentsev, S.V., Dergunov, A.B., Filippova, I.B., Tomurtogoo, O., Arvisbaatar, N., Bayasgalan, T., Biambaa, C., and Khosbayar, P., 1993, The South Siberia–Central Mongolia transect: Tectonophysics, v. 225, p. 361–378.

Zorin, Y.A., Belichenko, V.G., Turutanov, E.K., Turutanov, Y.K., Mordinova, V.V., Kozhenikov, V.M., Khozbayar, P., Tomurtogoo, O., Arvisbaatar, N., Gao, S., and Davis, P., 1994, Baikal-Mongolia transect: Russian Geology and Geophysics, v. 35, p. 78–92.

MANUSCRIPT ACCEPTED BY THE SOCIETY JUNE 5, 2000

Geological Society of America
Memoir 194
2001

Tectonic significance of early Paleozoic high-pressure rocks in Altun-Qaidam-Qilian Mountains, northwest China

Jingsui Yang
Zhiqin Xu
Jianxin Zhang
Institute of Geology, Chinese Academy of Geological Sciences, 26 Baiwanzhuang Road, Beijing, 100037, China
Ching-Yen Chu
Ruyuan Zhang
Juhn-Guang Liou
Department of Geological and Environmental Sciences, Stanford University, Stanford, California 94205-2115, USA

ABSTRACT

Several early Paleozoic eclogite-bearing very high-pressure (P) belts have been recognized in Alpine-type orogens in the north Qiadam Mountains, the Altun Mountains, and the Beishan Mountains of northwest China. These eclogitic rocks together with minor garnet peridotites occur as blocks or lenses in amphibolite facies gneissic terranes of continental affinity, and have been subjected to extensive retrogression. On the basis of mineral chemistries and phase assemblages, peak P-temperature (T) conditions were estimated as follows: for the north Qaidam eclogites, $T = 720 \pm 120$ °C and $P > 22$ kbar and for garnet peridotites, 720–850 °C and 25 kbar; for the Altun eclogites, $T = 730$–810 °C and $P > 15$ kbar; and for the Beishan eclogites, $T = 700$–800 °C and $P = 16$–18 kbar. U-Pb and Ar-Ar isotopic determinations for the north Qaidam eclogite gave a peak metamorphic age of 495 ± 6 Ma and a retrograde age of 470–480 Ma. Sm-Nd and U-Pb isotopic data indicate that the peak metamorphic age of the Altun eclogite is 500–504 Ma. Zircon separates from the Beishan eclogites provide a U-Pb upper intercept at 861 ± 50 Ma and a lower intercept at 440 ± 50 Ma. Similarities in the modes of occurrence, mineral parageneses, P-T estimates, and radiometric dates of eclogites suggest that the Altun very high P belt can be correlated with that of north Qaidam, whereas the Beishan belt may have been displaced to the north of the Alxa massif. These Alpine-type eclogite-bearing orogens were resulted from continental subduction and collision; they are different from the blueschist- and ophiolite-bearing Pacific-type orogen of the northern Qilian, where an early Paleozoic calc-alkaline volcanic arc was well developed due to northward subduction of the Qilian oceanic lithosphere. The disposition of several recognized Pacific- and Alpine-type orogens supports the hypothesis of 400 km of left-lateral displacement on the Altyn Tagh fault. A preliminary model for early Paleozoic tectonic evolution involving subduction, accretion, and collision for the formation and emplacement of both Alpine- and Pacific-type orogenic belts in the vicinity of the Altun-Qaidam-Qilian Mountains is proposed.

Yang, J.S., et al., 2001, Tectonic significance of early Paleozoic high-pressure rocks in Altun-Qaidam-Qilian Mountains, northwest China, *in* Hendrix, M.S., and Davis G.A., eds., Paleozoic and Mesozoic tectonic evolution of central Asia: From continental assembly to intracontinental deformation: Boulder, Colorado, Geological Society of America Memoir 194, p. 151–170.

INTRODUCTION

Eclogite and garnet peridotite constitute a volumetrically minor component in orogenic belts, but provide invaluable information about orogenic processes; two distinct eclogite types have been recognized in the high pressure-temperature (*P-T*) belts of China (e.g., Liou et al., 1989). Coesite-bearing Alpine-type eclogite in the Dabie-Sulu ultrahigh-*P* belt is associated with garnet peridotite, occurs as blocks, lenses, or layers in paragneiss, is retrograded to amphibolite, and has distinctly high *P-T* peak metamorphic conditions; $P > 25$ kbar and $T > 700$ °C (Q. Wang et al., 1992; Liou et al., 1994, 1996; Coleman and Wang, 1995; X. Wang et al., 1995). In contrast, Pacific-type high-*P* eclogite is associated with ophiolite, blueschist, and prehnite-pumpellyite facies rocks in the Qilian Mountains (Wu et al., 1993; Wu and Tian, 1994), is overprinted by blueschist-greenschist facies assemblages, and has peak metamorphic *P-T* conditions < 14 kbar and < 500 °C. As described in detail by Maruyama et al. (1996), such contrasting high *P-T* belts reflect different tectonic settings in terms of protoliths, subduction, and exhumation (e.g., Ernst and Liou, 1995).

Several newly discovered very high *P* eclogites in the vicinity of the Altun-Qaidam-Qilian Mountains are intercalated with continental sequences of sedimentary and igneous origin, and are indicative of continental subduction and/or collision. These Alpine-type eclogites are enclosed in similar country-rock gneisses, share analogous decompression *P-T* paths, and may have been formed in different parts of a similar or the same tectonic unit. Their occurrences can be used to constrain suture geometry and to determine the magnitude of slip along the Altyn Tagh fault, one of the largest strike-slip fault systems in the world.

The Altun-Qaidam-Qilian region shown in Figure 1 at the northern border of the Qinghai-Tibet Plateau is bounded by the Qaidam basin to the south, the Tarim basin to the west, and the Sino-Korean craton to the east. This region includes several orogenic belts. For example, the Qilian orogen, >300 km wide, extends from the Altyn Tagh fault southeastward for ~1000 km to the Qinling-Dabie orogen, and forms a major geographic-tectonic boundary between north and south China. The north Qilian early Paleozoic orogen has been regarded as one of the best exposed orogenic belts in the world (e.g., Xu et al., 1994; Feng and He, 1996) and contains Pacific-type margin petrotectonic assemblages, including lawsonite-bearing high-*P* belts and ophiolitic sequences (e.g., Wu et al., 1993; Wu and Tian, 1994). However, the petrotectonic assemblages of the southern border between the central Qilian block and the Qaidam basin are not well known, and their correlation across the Altyn Tagh fault to the west has not been investigated. Recent findings of eclogites and garnet peridotites in an Alpine-type collision belt along the northern border of the Qaidam basin (e.g., Yang et al., 1994, 1998) and the findings of similar eclogite in the Altun Mountains (e.g., Che et al., 1995a, 1995b; Hanson et al., 1995) and the Beishan Mountains (Mei et al., 1998) provide new data for the petrotectonic evolution of northwestern China.

The purposes of this chapter are to characterize the petrochemical and geochronological features of these very high *P* eclogites, garnet peridotites, and their enclosing gneissic rocks from both the north Qaidam and the Altun Mountains, and to document their significance regarding the regional tectonics. These data are compared and contrasted with analogous data from Pacific-type high-*P* rocks of the north Qilian terrane. A Paleozoic tectonic evolution scenario involving subduction, accretion, and collision involving the formation and emplacement of both Alpine- and Pacific-type orogenic belts in the Altun-Qaidam-Qilian Mountains is proposed.

REGIONAL GEOLOGIC SETTING

A number of intracontinental orogens separate several major Precambrian cratons and record magmatic and metamorphic events associated with subduction, collision, and subsequent tectonic movements of the cratonic blocks as eastern Eurasia was assembled. The broad Qilian orogen on the southern borders of the Sino-Korean craton (Fig. 1) extends eastward for ~1000 km and is connected with the Qinling-Tongbai-Dabie Mountains; its west end merges with the Altun Mountains. Bounded on the south by the Qaidam basin and on the northwest by the Altyn Tagh fault system, these remote, arid mountains comprise a series of west-northwest–trending subparallel tectonic units. Between cratons, several microcontinental blocks, including the Qaidam block and the central Qilian block, and early Paleozoic metamorphic belts are present in the regions between the Tarim and Qaidam basins and the Sino-Korean craton. The Qilian and adjacent orogens contain distinctive lithotectonic assemblages, suggesting early Paleozoic convergence, including subduction of oceanic lithosphere and continent-continent collision. High *P-T* rocks in these orogens therefore are useful in probing the various stages of plate interactions around the margins of the paleo-Eurasian continent.

On the basis of recent geologic mapping and detailed study of lithotectonic characteristics in the Qilian-Qaidam region (e.g., Bureau of Geology and Mineral Resources of Qinghai Province [BGMQ], 1991; Xu et al., 1994; Chen et al., 1995), several fault-bounded major tectonic units have been recognized (Fig. 2). From north to south, they are (1) the Alxa massif, (2) the north Qilian arc, (3) the north Qilian subduction complex, (4) the central Qilian block, (5) the north Qaidam subduction complex, and (6) the Qaidam block (approximately coincident with the Qaidam basin in Fig. 2). These units can be correlated with those in the Altun Mountains (Table 1), assuming 400 km of left-lateral displacement for the Altyn Tagh fault (Yue and Liou, 1999).

The Altun Mountains extend northeast for ~800 km, are between the Tarim and Qaidam basins and the Kunlun orogen, and are bounded on the northwest by strands of the north Altyn Tagh fault system and on the south by the left-lateral strike-slip Altyn Tagh fault (Fig. 2). The extent and displacement of the Altyn Tagh fault are still poorly known, partly because the region is

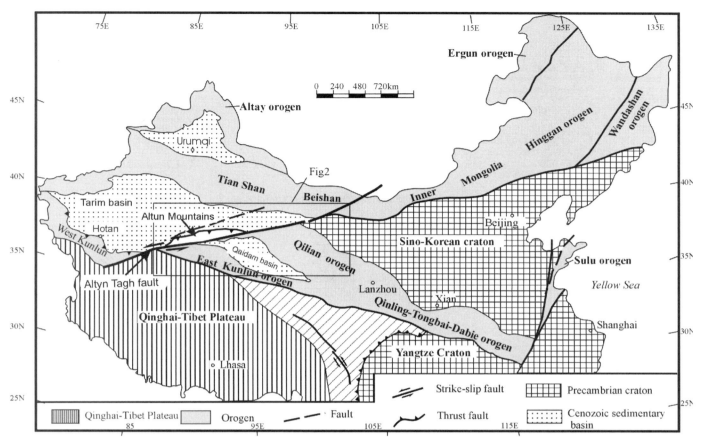

Figure 1. Regional tectonics of China showing distribution of Sino-Korean, Tarim, and Yangtze cratons, several major intracratonal mountain belts, and location of Figure 2. This map is modified after Ma et al. (1996).

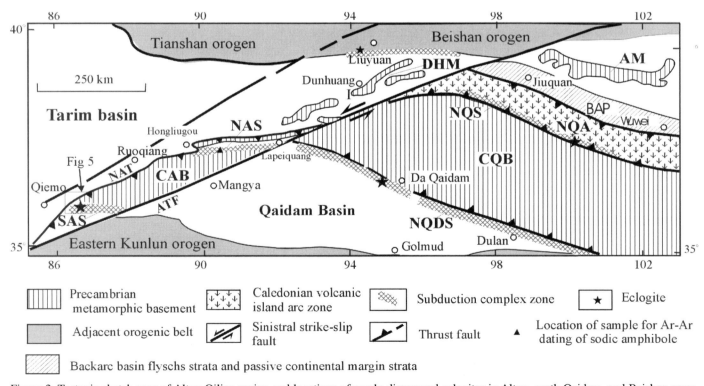

Figure 2. Tectonic sketch map of Altun-Qilian region and locations of newly discovered eclogites in Altun, north Qaidam, and Beishan areas, modified after Bureau of Geology and Mineral Resources of Gansu Province (BGM) (1989), BGM of Qinghai Province (1991), BGM of Xinjiang Uygur Autonomous Region (1993), Che et al. (1995a), Yang et al. (1998), and Yue and Liou (1999). Abbreviations: NAT—north Altun thrust; ATF—Altyn Tagh sinistral strike-slip fault; DHM—Dunhuang massif, NAS—north Altun subduction complex; CAB—central Altun block; SAS—south Altun subduction complex; AM—Alxa massif; BAP—backarc basin flysch strata and passive continental margin strata; NQA—north Qilian arc; NQS—north Qilian subduction complex; CQB—central Qilian block; NQDS—north Qaidam subduction complex.

TABLE 1. CORRELATION BETWEEN TECTONIC UNITS ON BOTH SIDES OF THE ALTYN TAGH FAULT

Qilian-North Qaidam Mountains		Altun Mountains	
Tectonic units	Petrotectonic assemblages	Tectonic units	Petrotectonic assemblages
Alxa Massif	Ar_3(?)-Pt_1 gneiss, graphite-bearing marble, amphibolite, migmatite	Dunhuang massif	Ar-Pt_1 amphibolite, granulite, marble, migmatite
Northern Qilian arc	O_{1-2} calc-alkalic volcanics	?	?
Northern Qilian subduction complex	Ophiolite and ophiolitic melange, blueschist (440-462 Ma), eclogite	Northern Altun subduction complex	Ophiolite, ophiolitic melange, sodic-amphibole-bearing metabasalt (457 Ma)
Central Qilian block	Amphibolite, marble, migmatite, pelitic schist, felsic gneiss	Central Altun block	Marble, quartz-mica schist, amphibole schist, migmatite, felsic gneiss
North Qaidam subduction complex	Early Paleozoic garnet-peridotite, eclogite (495 Ma), felsic gneiss	South Altun subduction complex	Early Paleozoic eclogite (500–504 Ma), amphibolite, felsic gneiss

Abbreviations: Ar_3, Late Archean; Pt_1, Early Proterozoic; O_{1-2}, Early-Middle Ordovician.
References: Bureau of Geological and Mineral Resources of Gansu Province, 1989; Bureau of Geological and Mineral Resources of Qinghai Province, 1991; Liu and Wang, 1993; Wu et al., 1993; Xia et al., 1995; Che and Sun, 1996.

extremely remote and only limited geologic data are available (for a summary of recent studies, see Yue and Liou, 1999). Major lithologic units within the Altun Mountains include the Dunhuang massif and the Early Proterozoic Altun Group, separated by two subduction complexes (Fig. 2). The regional geology of the Altun Mountains has been significantly modified by large Cenozoic displacement of the Altyn Tagh fault system due to tectonic extrusion during the Eurasia-India continent collision (Tapponnier and Molnar, 1977). Major lithotectonic units across and bordering the Altyn Tagh fault are described in the next section. Analytical details and uncertainties for age determination using Ar-Ar, U-Pb, and Nd-Sm isotopic systematics are described, respectively, in Chen et al. (1996), Lu and Li (1991), and Z. Zhang et al. (1999).

DESCRIPTION OF MAIN TECTONIC UNITS

Alxa-Dunhuang massif

The western margin of the Sino-Korean craton, named the Alxa massif (Fig. 2), consists mainly of an Early Proterozoic metamorphic complex including a variety of gneiss, schist, graphite-bearing marble, amphibolite, and migmatite. These basement rocks were intruded by granite at 1719 ± 50 Ma and gabbro at 1365 ± 50 Ma (Liu and Wang, 1993). The Longshou fault separates this basement complex from a 60–100-km-wide belt of Cambrian to Silurian passive margin strata in the Qilian Mountains to the south (Wu et al., 1993; Wu and Tian, 1994). The sedimentary strata include Cambrian marginal clastics and minor platform carbonate rocks, overlain by Ordovician sandstone, shale, quartzite, and limestone with thin flows of andesite and rhyolite and pyroclastics, which are in turn overlain by Upper Silurian molasse.

In the Altun Mountains, the Archean basement of the Dunhung massif (Fig. 2) includes high-grade amphibolite, granulite, and migmatite; Sm-Nd dating of these rocks yielded an age of 2789 ± 110 Ma (Che and Sun, 1996). The Paleozoic passive margin sequence is not well preserved.

North Qilian arc

The northern Qilian arc (Fig. 2) terrane extends for more than 300 km and is about 50–100 km wide; it consists of rhyolite, dacite, and andesitic flows and volcaniclastic strata. These volcanics have geochemical characteristics of the calc-alkalic series and yield U-Pb ages of 466–495 Ma (Xu et al., 1997; Zhang et al., 1997). Backarc basin (Fig. 2) flysch strata overlying a 454–469 Ma ophiolitic sequence (Xia et al., 1995) were developed to the north of the island arc complex. A thin molasse deposit of Middle to Late Silurian age represents the closure of the backarc basin resulting from the collision of the Qilian arc and the Alxa massif, whereas an overlying thick Devonian nonmarine molasse sequence was developed due to final collision of the Alxa massif with the central Qilian block (Zuo and Liu, 1987) (described in the following).

North Qilian–North Altun subduction complex

The north Qilian subduction (Fig. 2) complex is composed largely of ophiolite and ophiolitic melange, and has been divided into southern high-grade and northern low-grade blueschist belts (Zhang and Liou, 1987; Wu et al., 1993; Wu and Tian, 1994). The ophiolitic sequence consists of harzburgitic ultramafics, cumulate gabbro, pillow basalt, and radiolarian chert of Late Cambrian to Early Ordovician age (Xiao et al., 1978; Zhang and Xu, 1995; Feng and He, 1996); these in turn are overlain by pelagic to semipelagic flysch deposits and limestone. Blueschist and eclogite are widespread; they occur as tectonic blocks in serpentinite matrix and are respectively overprinted by greenschist facies and epidote amphibolite facies assemblages. Radiometric age determinations by U-Pb and Ar-Ar systematics yield consistent ages of 440–462 Ma for blueschist-eclogite facies metamorphism and 410–420 Ma for the greenschist facies overprint (Wu et al., 1993; Xu et al., 1994; Zhang et al., 1997). The disposition of an early Paleozoic high P-T belt and nearly coeval island arc and backarc basin sequences successively northward suggests a northward subduction of the Qilian oceanic lithosphere.

The northern Altun subduction (Fig. 2) complex crops out discontinuously along the northeastern edge of the Altun Mountains between Hongliugou and Ruoqiang. Ultramafics, cumulate gabbro, pillow lava, and radiolarian chert occur as blocks in ophiolitic melange. Transitional blueschist-greenschist with sodic amphiboles has been described in the Lapeiquang area of Ruoqiang (BGMQ, 1991). We obtained a ^{39}Ar–^{40}Ar plateau age of 457 ± 0.7 Ma for sodic amphibole from metabasalt in the Hongliugou area of Ruoqiang (Fig. 3) (also see Sobel and Arnaud, 1999). This preliminary age for blueschist facies metamorphism and the nature of ophiolitic assemblages suggest that the northern Altun complex can be correlated with the ophiolitic and high *P-T* rocks of the northern Qilian complex. However, the distribution, nature, and age of the Altun subduction complex remain to be investigated.

Qilian-Altun block

The central Qilian block (Fig. 2) is equivalent to the Early Proterozoic Huangyuan Group (BGMQ, 1991) and comprises mainly Precambrian metamorphic basement consisting of gneiss, low- to high-grade amphibolite, migmatite, marble, and pelitic schist (Wu et al., 1993; Wu and Tian, 1994). The basement rocks are unconformably overlain by upper Paleozoic shallow-marine sedimentary strata. Zircons from migmatitic gneiss give a U-Pb age of 2469 ± 110 Ma (Wang and Chen, 1987), whereas K-Ar amphibole and biotite ages and Rb-Sr whole-rock ages cluster around about 1700–1600 Ma. These basement rocks are intruded by 1416 ± 50 Ma granite and 874 ± 50 Ma diorite. This terrane has been considered to be a fragment of a much larger continent that rifted to form the Tarim, Qaidam, and Alxa massifs (Wu et al., 1993).

The central Altun block (Fig. 2) is equivalent to the Early Proterozoic Altun Group (Bureau of Geology and Mineral Resources of Xinjiang Uygur Autonomous Region [BGMX], 1993) and consists of low-grade amphibolite facies marble, quartz-mica schist, amphibole schist, migmatite, and felsic gneiss. Radiometric dating of these basement rocks has not been done.

North Qaidam–south Altun subduction complex

The north Qaidam Mountains along the northeastern rim of the Qaidam basin have long been considered to be exposed Qaidam basement (BGMQ, 1991). Recent discovery of high-*P* garnet peridotite (Yang et al., 1994) and eclogite (Yang et al., 1998) within deformed gneiss of the Early Proterozoic Dakandaban Group led to establishment of a new subduction complex unit in the north Qaidam Mountains (Figs. 2 and 4). The Early Proterozoic Dakendaban Group occurs throughout the range, and consists of various amphibolite facies lithologies including felsic gneiss, quartz schist, garnet-bearing gneiss, amphibolite, and minor eclogite and garnet peridotite. Details of the high *P-T* rocks are described in a later section.

A similar eclogite-bearing high *P-T* belt about 200 km long was identified by Che et al. (1995a, 1995b) on the southern margin of the Altun Mountains. Eclogitic cobbles of various sizes were discovered as detritus in Pleistocene conglomerates and as boulders

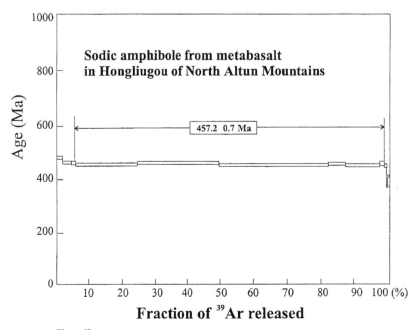

Figure 3. ^{39}Ar–^{40}Ar plateau spectra for sodic amphibole from metabasalt in Hongliugou area of Ruoqiang, Altun Mountains.

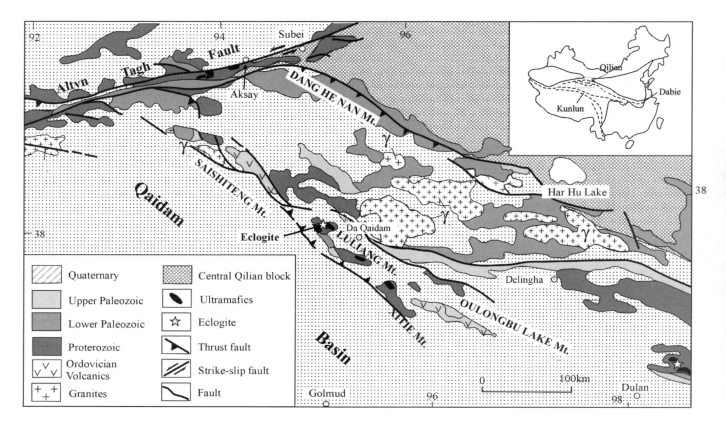

Figure 4. Schematic geologic map of northern Qaidam region, showing distribution of eclogite, ultramafics (including garnet peridotite), and postcollisional granite in northern Qaidam Mountains. γ—Paleozoic and Mesozoic granite.

in modern stream channels within the foothills near Qiemo (Hanson et al., 1995). Che et al. (1995a, 1995b) described eclogite lenses and blocks as garnet-quartz-clinopyroxenite within an Early Proterozoic mylonitized amphibolite, amphibole schist, garnet-bearing granitic gneiss, and muscovite-quartz schist sequence. Our reconnaissance study in the vicinity of the approximately north-south–trending stream channel (Jianggalesayi Creek) yielded the occurrence of eclogitic lenses in strongly deformed granitic gneisses (see Fig. 5). Consistent lithologies, P-T estimates, and timing of eclogite facies metamorphism described in a later section led us to correlate the eclogite-bearing South Altun complex (Fig. 2) with the north Qaidam subduction complex, as shown in Figure 2.

Qaidam block

Similar to the Tarim block, the basement of the Qaidam basin has been considered as part of a rigid, stable, Precambrian craton (BGMQ, 1991). However, extensive geological and geophysical studies indicate that the Qaidam basement is a complex of accreted tectonic fold belts (Wang and Coward, 1990). Basement units consist of a Paleozoic fold belt in the southwest, and a Proterozoic to Paleozoic complex in the northeast, based on the published borehole, geophysical, and field data from the surrounding mountains (BGMQ, 1991).

EARLY PALEOZOIC HIGH-PRESSURE METAMORPHIC ROCKS

Several high P-T rocks have been identified in the Altun–north Qaidam–Qilian regions. Petrotectonic characteristics, including index assemblages, P-T estimates, and metamorphic ages of these high P-T rocks, are described in the following. Representative compositions of the analyzed garnet and omphacite from these eclogitic rocks are listed in Table 2 and plotted in Figure 6. The eclogite and garnet peridotite from these regions consitute <10 vol% of the lithic section but preserve the best very high P record. Quartzofeldspathic gneisses are volumetrically predominant rock types and occur as country rocks for many very high-P eclogites and garnet peridotites. They contain abundant quartz, albitic plagioclase, phengitic mica, and K-feldspar together with biotite, hornblende, clinozoisite, minor garnet, and other accessory phases such as zircon, rutile, topaz, and tourmaline. Such mineral assemblages are stable over a broad P-T range, and appear to have been thoroughly reequilibrated under retrograde conditions. Hence, gneissic rocks lack a clear record of the peak very high-P event: this lack led to an early hypothesis that eclogites and country rocks were metamorphosed separately, under different P-T regimes, and were subsequently juxtaposed by faulting. According to this

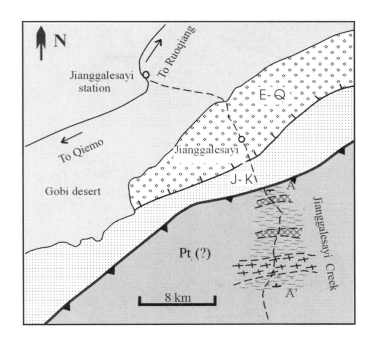

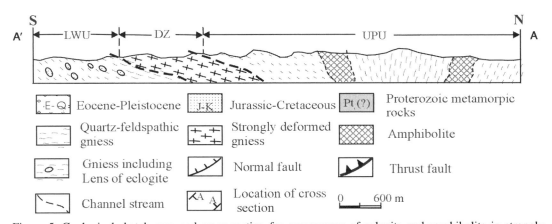

Figure 5. Geological sketch map and cross section for occurrences of eclogite and amphibolite in strongly deformed felsic gneisses in Altun Mountains. LWU—lower lithotectonic unit; DZ—decollement zone; UPU—upper lithotectonic unit.

hypothesis, the extent of very high P metamorphism was no larger than eclogite boudins and lenses now exposed at the surface, which usually measure no more than a few meters in length and thickness.

However, many new observations from the study of country rocks in very high P terranes worldwide indicate that country rocks and eclogites are not all fault bounded, and that many contacts retained structural coherence throughout subduction, metamorphism, and exhumation. Mineralogical indicators of very high P metamorphism have now been found in a number of country rocks, including gneiss, quartzite, and marble. For example, tiny coesite inclusions have been reported from other collisional belts in zircons from felsic gneisses, in dolomite and garnet from calc-silicate rocks and dolomite-bearing eclogite,

and in garnet and jadeite from jadeite-bearing quartzite. Hence, the in-situ model has been accepted (for details, see Liou et al., 1996; Chopin and Schertl, 1999; Carswell et al., 1999). Several newly discovered very high P terranes in northwestern China described in this chapter require detailed study of the country rocks in order to address this issue.

Qilian Pacific-type eclogite and blueschist

Reported occurrences of prehnite + pumpellyite– and lawsonite-bearing blueschist and jadeite-bearing blueschist + eclogite record the likely existence of parallel subduction zone complexes in the north Qilian, i.e., lower grade blueschist on the north, and higher grade blueschist + glaucophane-bearing

TABLE 2. REPRESENTATIVE COMPOSITIONS OF GARNET AND OMPHACITE FROM THE ALTUN, BEISHAN, AND DA QAIDAM ECLOGITES

| Rocks | Altun eclogite | | | | | | Beishan eclogite* | | | Da Qaidam eclogite | | | | | |
Minerals	Grt	Grt	Grt	Grt	Omp	Omp	Grt	Grt	Omp	Grt-R	Grt-M	Grt-C	Omp	Omp	Omp
SiO_2	39.32	40.65	39.29	40.65	54.36	55.85	39.57	39.47	56.02	39.32	37.52	38.71	55.72	54.61	54.79
TiO_2	0.03	0.00	0.08	0.04	0.12	0.17	0.00	0.26	0.26	0.01	0.13	0.08	0.08	0.07	0.04
Al_2O_3	21.78	21.92	22.63	22.74	8.67	8.81	21.38	20.83	7.57	22.83	22.36	21.46	10.48	9.38	8.43
Cr_2O_3	0.04	0.06	0.04	0.03	0.00	0.05	0.00	0.00	0.00	0.01	0.07	0.00	0.00	0.00	0.01
FeO	20.74	20.13	21.63	19.30	6.04	4.89	24.58	23.64	6.69	22.09	21.39	23.03	2.65	3.03	3.71
MnO	0.38	0.35	0.43	0.37	0.02	0.00	0.50	0.55	0.10	0.42	0.40	0.54	0.00	0.04	0.02
MgO	7.24	6.88	7.81	8.10	9.20	9.40	5.55	4.84	9.35	7.72	6.01	4.35	10.16	10.45	10.00
CaO	10.46	9.81	9.36	9.79	15.80	15.39	7.67	9.44	14.88	8.46	11.28	10.80	15.37	16.09	16.29
Na_2O	0.00	0.02	0.00	0.05	5.26	5.12	0.02	0.00	4.65	0.03	0.03	0.04	6.25	5.74	5.18
K_2O	0.00	0.00	0.00	0.00	0.00	0.00	0.10	0.02	0.00	0.00	0.00	0.00	0.00	0.00	0.00
Total	99.98	99.76	100.98	101.04	99.47	99.69	100.41	99.61	100.36	100.88	99.12	99.01	100.71	99.41	98.47
Si	3.01	3.11	2.97	3.05	1.97	2.01	3.04	3.06	2.01	2.98	2.91	3.04	0.95	1.95	1.99
Ti	0.00	0.00	0.01	0.00	0.00	0.01	0.00	0.02	0.01	0.00	0.01	0.01	0.00	0.00	0.00
Al	1.96	1.98	2.01	2.01	0.37	0.37	1.94	1.90	0.32	2.04	2.04	1.98	0.43	0.39	0.36
Cr	0.00	0.00	0.00	0.00	0.00	0.00	0.00	0.00	0.00	0.00	0.00	0.00	0.00	0.00	0.00
Fe	1.33	1.29	1.24	1.29	0.18	0.15	1.58	1.54	0.20	1.40	1.39	1.52	0.08	0.09	1.01
Mn	0.03	0.02	0.03	0.02	0.00	0.00	0.03	0.04	0.00	0.03	0.03	0.04	0.00	0.00	0.00
Mg	0.83	0.79	0.88	0.91	0.50	0.51	0.64	0.56	0.51	0.87	0.69	0.51	0.53	0.56	0.54
Ca	0.86	0.81	0.76	0.79	0.61	0.60	0.63	0.79	0.57	0.69	0.94	0.91	0.58	0.61	0.63
Na	0.00	0.00	0.00	0.01	0.37	0.36	0.01	0.00	0.32	0.00	0.01	0.01	0.43	0.40	0.36
K	0.00	0.00	0.00	0.00	0.00	0.00	0.00	0.00	0.00	0.00	0.00	0.00	0.00	0.00	0.00

Note: C—core; M—middle; R—rim; Grt—garnet; Omp—omphacite.
* Data from Mei et al., 1998.

eclogite on the south (Wu et al., 1993; Wu and Tian, 1994; Tian and Wu, 1994). These high P-T rocks have been extensively investigated since the first description of sodic amphibole by Xiao et al. (1974). The northern low-grade blueschist belt occurs in an Early to Middle Ordovician ophiolite complex, whereas the southern eclogite-bearing high-grade blueschist belt occurs in a Middle Cambrian ophiolitic complex. Pacific-type eclogites of the southern high P-T belt occur as tectonic blocks and have distinctly different compositions of garnet ($Alm_{53-77}Gro_{11-37}Pyr_{4-13}Spe_{2-3}$) and omphacite ($Jd_{27-41}$) compared to those from Alpine-type eclogites described here (Fig. 6). The P-T estimates for the northern low-grade blueschist are 150–250 °C and 4–7 kbar, whereas those for the high-grade blueschist are $T > 380$ °C and $P = 6$–7 kbar; both P-T estimates are similar to those for classic Franciscan Pacific-type blueschists (Ernst, 1971). Both Qilian high P-T belts are associated with graywacke, ophiolitic rocks, and deep-sea chert with radiolarian fossils of Early Ordovician age (Xiao et al., 1978); the closure of intervening oceanic-crust–capped basins prior to arc-generating collisional events seems plausible. Preliminary radiometric data suggest that the more northerly suture zone may have closed during the Silurian, whereas the more

southerly belt was the site of lithospheric underflow and concomitant high-P recrystallization and arc volcanism in Cambrian–Ordovician time (ca. 443–462 Ma)(Wu et al., 1993; Wu and Tian, 1994; Zhang et al., 1997).

North Qaidam eclogite

Mode of occurrence. More than 20 eclogite blocks have been described in the Dakendaban Group from an area ~40 km northwest of the town of Da Qaidam (Yang et al., 1998; Fig. 4). The eclogite occurs as pods, to 40–50 m long and 20–30 m wide, but most less than 20 × 10 m (Fig. 7A), in garnet-bearing felsic gneiss consisting mainly of plagioclase and quartz, with minor muscovite (5%–10% by volume) and garnet (1%–2%). Eclogite pods are boudinaged, and are parallel to the regional foliation. Some are massive et al. are extensively retrograded to garnet-amphibolite, which also occurs as blocks or layers concordant with the foliation of country-rock gneiss. Eclogite is best exposed in the Luliang Mountains near Da Qaidam (Fig. 4).

Mineral parageneses and pressure-temperature estimates. Eclogites with various degrees of retrogression are medium to fine grained, and consist mainly of garnet and om-

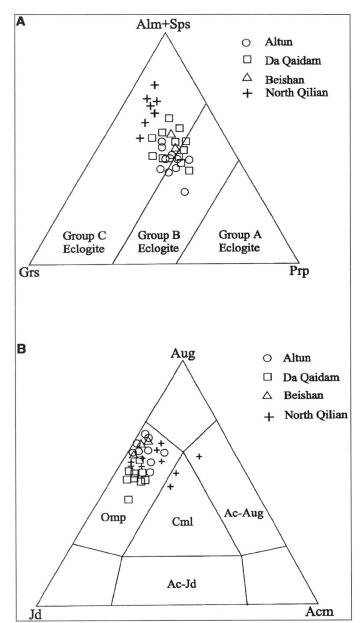

Figure 6. Compositional plots for (A) garnet, and (B) clinopyroxenes in eclogites from Altun, Da Qaidam, Beishan, and north Qilian areas. Analytical error is within size of symbol.

phacite (each 40–50 vol%), with minor amphibole, phengite, quartz, titanite, and rutile (Fig. 8, A–C). Garnets are euhedral, even grained, and mostly 1–2 mm in size; some contain minor inclusions of quartz, omphacite, phengite, and rutile, and are replaced by symplectic amphibole + plagioclase (Fig. 8C). Analyzed garnets ($Alm_{44–62}Gr_{15–33}Py_{12–30}$) are plotted within the compositional fields of group C and B garnets of Coleman et al. (1965) (Fig. 6A); they have compositions similar to those of eclogite garnets in the gneisses from the Altun Mountains described here, but differ considerably from group C eclogitic gar-

nets in the Qilian Mountains (Wu et al., 1993). Microprobe analyses of omphacite yield 37%–46% jadeite (Jd) and 50%–57% augite (Aug), and plot within the omphacite compositional region (Fig. 6B).

Fine-grained phengite (<5 vol%) occurs as oriented laths of 0.2–1 mm size in the matrix (Fig. 8C) or as fine-grained inclusions in garnet, and is in textural equilibrium with omphacite and garnet. Some late-formed tabular muscovite flakes are coarse grained (to 2–5 mm size) and randomly oriented in eclogite. Analyzed phengite varies in Si content from 6.6 to 6.9, Mg from 0.6 to 0.8, and Fe from 0.21 to 0.25 atoms/formula unit based on a stoichiometry of 22 oxygen atoms, belonging to a typical high-pressure phase (Liu et al., 1997).

As a result of varying degrees of retrograde metamorphism, hornblende and plagioclase occur as various grain sizes of symplectite around omphacite grain margins (Fig. 8C). Fine- to medium-grained euhedral amphiboles are in textural equilibrium with omphacite and garnet, and were likely formed during eclogite-stage recrystallization. However, both coarse grained amphiboles with fine-grained garnet and omphacite inclusions, and amphiboles along garnet fractures and symplectic amphiboles are retrograde. Eclogitic amphiboles contain a considerable amount of the glaucophane component, whereas retrograde amphiboles are either pargasite or hornblende.

On the basis of mineral chemistries and phase assemblages, peak *P-T* conditions of these rocks are estimated to be about 720 ± 120 °C and >22 kbar (Yang et al., 1998) (Fig. 9). Using the phengite-garnet-omphacite geobarometer described by Carswell et al. (1997), some of these eclogitic rocks have *P* estimates to 33 kbar. On the basis of these preliminary data, Yang et al. (1998, 2000) concluded that these eclogites and garnet peridotites in the north Qaidam terrane are similar to ultrahigh-*P* rocks in the Dabie-Sulu terrane in terms of occurrence, rock type, country-rock gneiss, and *P-T* conditions. The north Qaidam Mountains may be a coherent, very high *P* metamorphic terrane, constituting one of the major tectonic boundaries of the north Tibetan Plateau. These very high *P* (>22 kbar) rocks are the first ever recognized in this region.

Age of metamorphism. U-Pb concordia plot of zircons from an eclogite sample yield metamorphic ages of 494.6 ± 6.5 Ma, as shown in Figure 10A. Phengite separates from eclogites gave an $^{40}Ar/^{39}Ar$ plateau age of 467 ± 1 Ma and an isochron age of 466 ± 5 Ma (Fig. 11); these data represent the cooling age of the eclogite during exhumation. In addition, muscovite from the country-rock gneiss yields a disturbed, saddle-shaped age spectrum, indicating the presence of excess argon. No meaningful plateau can be defined (Fig. 11). The central portion of six contiguous steps, corresponding to more than 90% of the total ^{39}Ar released, define a linear array on the isochron diagram ($^{40}Ar/^{36}Ar$ vs. $^{39}Ar/^{36}Ar$). Although the linear relationship is not apparent, the data imply an isochron age of 477.7 ± 17.7 Ma with a trapped $^{40}Ar/^{36}Ar$ ratio of 1012 ± 425 (Fig. 11), which is significantly higher than the atmospheric value of 295.5, suggesting the existence of excess argon in the mineral.

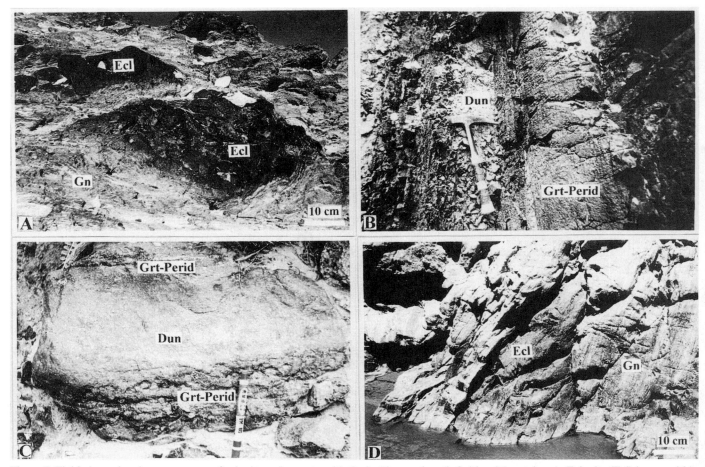

Figure 7. Field views showing occurrence of eclogite and garnet peridotite in Altun and north Qaidam Mountains. A: Eclogite (Ecl) lenses within felsic gneiss (Gn) in north Qaidam Mountains. B, C: Garnet peridotite (Grt-Perid) and dunite (Dun) occurring as blocks in gneiss of the Dakendaban Group in north Qaidam Mountains. D: Eclogite lens within felsic gneiss in Altun Mountains.

Nevertheless, the isochron age of the muscovite (477.7 ± 17.7 Ma) is in good agreement with that of phengite in the eclogite sample within the uncertainty. The consistent age data for both eclogites and their country rocks suggest that basaltic protoliths together with supracrustal rocks of the continental lithosphere were subducted and recrystallized ca. 495 Ma and subsequently exhumed to mid-crustal level by ca. 466 Ma.

North Qaidam garnet peridotite and associated ultramafic rocks

Various ultramafic rocks including garnet-bearing and garnet-free peridotite, pyroxenite, and dunite occur as lenses and blocks ranging from a few meters to more than 100 meters long in gneisses of the Dakendaban Group (Yang et al., 1994, 1998) (Fig. 7, B and C). These ultramafic rocks are extensively serpentinized, particularly along fractures or near contacts with country rocks. Some outcrops display compositional bands parallel to the regional foliation. Thus far, eclogitic rocks have not been found within these ultramafic rocks.

Garnet-free peridotite and dunite are dominant, and consist mainly of olivine (80–95 vol%) with subordinate orthopyroxene (5–15%), clinopyroxene (<5%), chromite (<1%), and minor sulfide. Garnet peridotite is characterized by porphyroblastic texture; porphyroblastic (2–3 cm) garnets (5–10 vol%; $Pyr_{73-68}Alm_{20-15}Gro_{8-6}Uva_{2-5}Sps_{1-0.5}$) are set in a fine-grained (< 0.2 cm) matrix of olivine (Fo_{89-90}), orthopyroxene (En_{85-91}; 1.37 wt% Al_2O_3), and clinopyroxene. Some garnets contain inclusions of chromite and have a wide kelyphitic rim (Fig. 8F); many have been entirely replaced by chlorite and amphibole.

Garnet-pyroxenite is significantly retrograded and contains more clinopyroxene than orthopyroxene; exsolution lamellae of orthopyroxene from clinopyroxene are common. Garnets have thick kelyphitic rims; most clinopyroxenes are replaced by amphiboles. In some samples, retrograded amphibole (to 50 vol%) and phlogopite (to 10 vol%) are dominant.

Equilibrium *P-T* conditions for recrystallization of the garnet-peridotite were calculated, using the garnet-orthopyroxene-olivine geobarometry (Stroh, 1976) and the garnet-olivine thermometry (Neill and Wood, 1979), as 837 °C and to 25 kbar

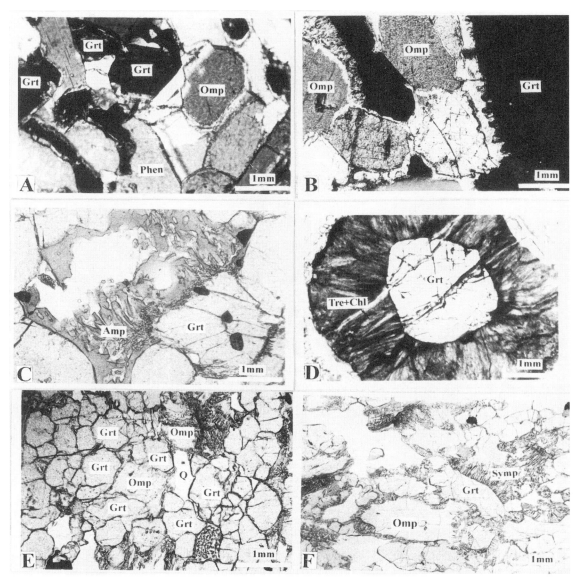

Figure 8. Photomicrographs of eclogite and garnet peridotite in Altun and north Qaidam Mountains. A, B: Eclogite from north Qaidam Mountains showing granoblastic garnet (Grt) and omphacite (Omp) with phengite (Phen) (cross polarizers). C: Eclogite from north Qaidam Mountains showing amphibole (Amp) riming garnet (plain polarized light). D: Tremolite + chlorite (Tre + Chl) kelyphite rim around garnet of garnet peridotite from north Qaidam Mountains (plain polarized light). E: Eclogite from Altun Mountains showing granoblastic garnet and omphacite with quartz (Q) (plain polarized light); F: Eclogite from Altun Mountains showing fine-grained symplectite (Symp) of clinopyroxene (+ amphibole) and plagioclase after omphacite (plain polarized light).

(Yang et al., 2000). This estimate is consistent with other estimates of 720–850 °C and 25 kbar by Yang et al. (1994) for garnet peridotite from this area. A K-Ar age of 490 Ma was obtained for phlogopite from garnet peridotite. Because phlogopite occurs as a retrograde phase, Yang et al. (1994) concluded that the peak metamorphic age of garnet peridotite should be >490 Ma. This suggestion is consistent with the age of 495 Ma for eclogites from this high *P-T* belt.

Altun eclogitic and gneissic rocks

Mode of occurrence. Eclogite occurs as lenticular blocks or layers, tens of centimeters to a few meters long, and a few centimeters to a few meters wide (Fig. 7D); they are concordant with the structural trends of the enclosing country-rock gneisses, which include mainly granitic gneiss and schist, and minor marble, calc-silicates, and amphibolite (Fig. 5). Granitic

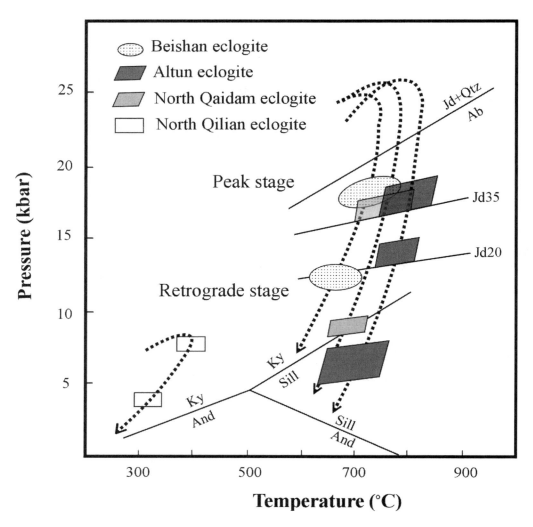

Figure 9. Pressure-temperature (*P-T*) diagram showing *P-T* estimates for peak and retrograde metamorphism of eclogites from Da Qaidam, Altun, Beishan, and north Qilian. Size of boxes and circles plotted in this figure correspond to uncertainty of *P-T* estimates.

gneisses are foliated and mylonitized, and contain quartz, K-feldspar, plagioclase, biotite, green amphibole, opaque minerals, and chlorite. In the mylonitic rocks, fractured and deformed K-feldspar porphyroblasts (3–10 mm) are surrounded by a matrix of quartz, plagioclase, and biotite with well-developed pressure shadows. Both green amphibole and biotite are partially replaced by chlorite. Stretched quartz, biotite, and chlorite define the regional foliation. Some gneisses contain kyanite, indicating high-pressure recrystallization. Margins of eclogite lenses are strongly retrograded to garnet-amphibolite or amphibole schist. Less-deformed amphibolite blocks (2–5 m long) also occur; these amphibolite blocks and layers appear to be retrograde products of eclogite.

Mineral parageneses and pressure-temperature estimates. Altun eclogites are partially to strongly retrograded, and contain medium- to coarse-grained omphacite, garnet, phengite, and quartz and fine-grained rutile (Fig. 8, E and F). Omphacites ($Jd_{34-45}Aug_{56-63}Acm_{0.1-3.4}$) (Fig. 6) are fractured, coarse grained (1.1–2.6 mm), and are surrounded by symplectite rims of clinopyroxene with relatively low Jd components (Jd_{21-25}) + plagioclase (An_{10-20}) (Fig. 8B). Garnets (0.45–0.88 mm) contain

inclusions of quartz, omphacite, and rutile in the core. Most garnets are fractured and have rather uniform compositions ($Al_{43-49}Py_{24-28}Gr_{27-33}Sps_1$) (Fig. 6). However, garnets of strongly retrograded eclogite and amphibolite have lower almandine and pyrope and higher grossular components than those from less retrograded eclogite. Rutile is rimmed by ilmenite.

Strongly retrograded eclogite is characterized by the total absence of omphacite and a well-preserved granoblastic texture. Widespread retrogression includes the replacement of omphacite by symplectitic intergrowth of clinopyroxene (Jd_{10}) + plagioclase (An_{10}) + minor pargasitic amphibole, rutile by ilmenite, garnet by amphibole-plagioclase kelyphite, and by the growth of matrix biotite and Ca-amphibole. With advanced retrogression, garnet-amphibolite occurs and contains porphyroblastic garnet (~1 mm) replaced by green amphibole, quartz, plagioclase, and chlorite. Matrix green amphiboles are elongate and wrap around relict garnets; this and quartz and ilmenite define the rock foliation.

Three distinct stages of metamorphic recrystallization were identified in the eclogites. Mineral assemblages and Fe-Mg distributions between garnet and omphacite yield $T = 730$–$810\,°C$

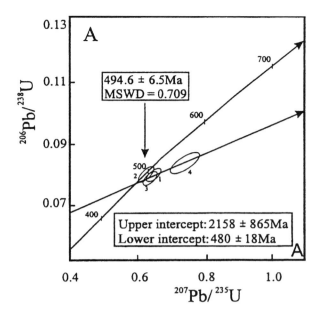

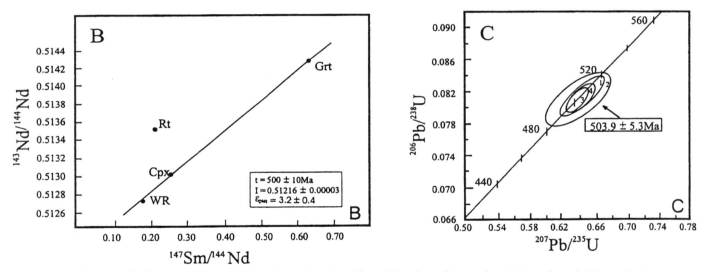

Figure 10. U-Pb concordia diagram of zircons and whole-rock–mineral Sm-Nd isochron diagram for eclogites from Qaidam and Altun Mountains. A: Eclogite sample from Da Qaidam of north Qaidam Mountains. B, C: Eclogite samples from Altun Mountains. MSWD—mean square of weighted deviates.

and $P > 15$ kbar for eclogite facies metamorphism (J. Zhang, et al., 1999). *P-T* estimates of 660–830 °C and 14–18.5 kbar were obtained by Che et al. (1995a, 1995b). Decompression is indicated by the breakdown of omphacite to symplectites of less sodic clinopyroxene (Jd_{21-25}) and plagioclase (An_{11-16}), followed by symplectites consisting of worm-like plagioclase + green amphibole intergrowths. The second-stage assemblage is represented by pargasitic amphibole, plagioclase (An_{21-38}), quartz, ilmenite, and clinopyroxene ($Jd < 5$). Such amphibolite facies parageneses are most common in country-rock gneiss and garnet amphibolite. The amphibole-plagioclase geothermome-

ter of Blundy and Holland (1990) was employed for the coarse- to medium-grained amphibole and fine- to medium-grained plagioclase, and yields $T = 700 \pm 75$ °C at $P = 5$ kbar; the pressure estimate was obtained by Al-in-hornblende (Schmidt, 1992) for felsic gneiss that contains two feldspars + quartz + amphibole + ilmenite. A late-stage greenschist facies overprint is indicated by the occurrence of biotite, chlorite, albite, and titanite in both eclogite and country rocks.

Age of metamorphism. Eclogite of the Altun Mountains has been recently dated by two independent methods as shown in Figure 10 (J. Zhang et al., 1999). A Sm-Nd mineral isochron

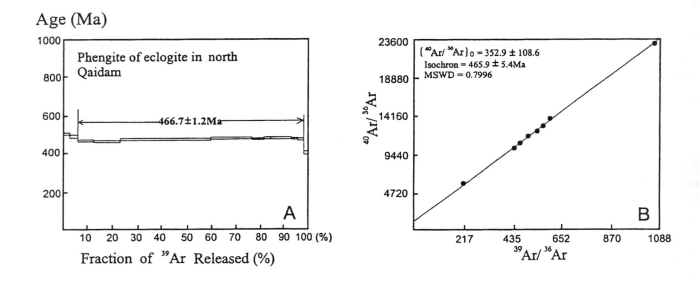

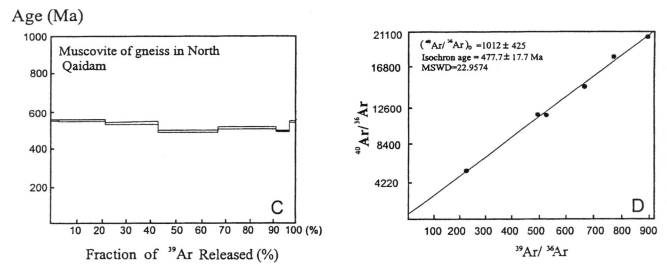

Figure 11. Ar-Ar age spectra and isochron of white micas from eclogite and granitic gneiss from north Qaidam Mountains. MSWD—mean square of weighted deviates.

for eclogitic garnet, omphacite, and the whole rock yields an age of 500 ± 10 Ma, whereas an U-Pb isochron for zircon separates from eclogite gives an overlapping age of 504 ± 5 Ma. A similar Sm-Nd mineral isochron age of 519 Ma was obtained for mylonitized amphibolite from this belt (Liu et al., 1996). Muscovite and biotite from Altun gneisses and schists yield $^{40}Ar/^{39}Ar$ ages of 430 ± 20 Ma, whereas $^{40}Ar/^{39}Ar$ age of 575 ± 20 Ma for phengite from eclogite-associated metapelite was interpreted to overestimate the age of high-P metamorphism (Sobel and Arnaud, 1999). Thus, the south Altun high-P

terrane involves eclogite-bearing granitic gneiss with consistent 500–504 Ma U-Pb and Sm-Nd ages, and represents an early Paleozoic subduction complex similar to that in the northern Qaidam basin.

Beishan Alpine-type eclogite

Numerous eclogitic blocks and lenses enclosed within granitic gneisses were discovered in a southern belt of the Beishan orogen (see Fig. 1 for location) (Mei et al., 1998). The largest extends for 500 m and is 100 m wide, larger than most eclogitic bodies described herein. These eclogitic lenses are concordant with foliation of country rocks, and are in a belt consisting mainly of granitic gneisses together with quartz schist, marble, amphibolite, and garnet peridotite. Details of the garnet peridotite have not been described. The eclogite contains equal amount of garnet ($Alm_{50-55}Grs_{22-29}Pyr_{19-22}Sps_{1-2}$) and omphacite ($Jd_{31-35}$), and minor rutile and retrograded phases including amphibole, ilmenite, and quartz. Garnet is well preserved and has only thin taramitic amphibole rims; some garnet grains contain inclusions of quartz aggregates and exhibit radial fractures around the inclusions. Mei et al. (1998) suggested that these textures were related to retrograded coesite developed during exhumation of the eclogitic rocks, and coesite was a stable phase of peak eclogite facies metamorphism. Omphacite shows various degrees of symplectite replacement of clinopyroxene (Jd_{17}) + amphibole + plagioclase (An_{13}); rutile is rimmed by ilmenite. Three distinct stages of metamorphic recrystallization were identified: peak eclogite, retrograded amphibolite, and later greenschist facies overprints. The *P-T* estimates yield peak stage recrystallization at 700–800 °C and 16–18 kbar, and amphibolite stage at 650 °C and 10–12 kbar. A U-Pb concordia plot of several zircon separates from granitic gneiss yields protolith ages of 1756–2056 Ma and metamorphic ages of 467 ± 50 Ma. Zircon separates from eclogite provides a U-Pb upper intercept at 861 ± 50 Ma and a lower intercept at 440 ± 50 Ma (H. Mei, 1999, personal commun.). Mineral parageneses, *P-T* estimates, and the possible occurrence of coesite together with petrotectonic assemblages of granitic gneiss, quartz schist, and marble as country rocks for eclogites and garnet peridotite blocks suggest A-type subduction of Early Proterozoic supracrustal rocks and an early Paleozoic continental collision.

TECTONIC CORRELATION AND DISCUSSION

Significance of Pacific- and Alpine-type subduction complexes

High-pressure eclogites, represented by the Franciscan Complex of western California, are products of Pacific-type subduction and are associated with blueschist and epidote amphibolite of oceanic affinity (Ernst et al., 1994; Ernst and Liou, 1995). Ultrahigh-*P* coesite- and/or microdiamond-bearing eclogites in the Dabie-Sulu terrane of east-central China are products of Alpine-type intracontinental collision, and are associated with gneiss, paraschist, and marble of continental affinity. Petrotectonic studies of eclogites and associated rocks establish the pressure-temperature-time (*P-T-t*) path, which can be used to interpret the crustal evolution, including subduction, collision, and exhumation. Moreover, these high-*P* and ultrahigh-*P* eclogites and associated garnet peridotites rarely crop out in subduction complexes, which generally occur in linear zones. These high- and ultrahigh-*P* metamorphic belts provide useful constraints in regional tectonic studies.

The north Qilian subduction complex, and its lawsonite-bearing blueschist and low-T eclogite associated with ophiolitic melange and graywacke, is similar to many Pacific-type subduction orogens around the circum-Pacific (see Maruyama et al., 1996). This complex has been intensively investigated (e.g., Wu et al., 1993; Wu and Tian, 1994; Tian and Wu, 1994) and is characterized by the occurrence of coeval calc-alkalic volcanics due to the inferred northward early Paleozoic subduction of the Qilian oceanic lithosphere.

Several newly documented eclogite terranes in the north Qaidam, the Altun Mountains, and the Beishan are within major continental collision belts in northwest China, extending several hundred kilometers or more, and are confined to Alpine-type orogens. They share common lithologic and geochronologic characteristics. (1) These terranes have been considered to be Precambrian basement because they occur mainly within amphibolite facies gneiss, schist, migmatite, marble, amphibolite, and serpentinized ultramafics at the margins of continental blocks (Bureau of Geology and Mineral Resources of Gansu Province [BGMG], 1989; Bureau of Geology and Mineral Resources of Qinghai Province [BGMQ], 1991). (2) Very high *P* records occur in minor but significant rocks—eclogite and garnet peridotite—included as pods and slabs within quartzofeldspathic gneissic units. The occurrence of inclusions of coesite pseudomorphs has been suggested, but they have not been positively identified. However, independent estimates using various thermobarometers indicate eclogite facies recrystallization took place at *P* > 20 kbar, and some estimates are >30 kbar (J. Yang, 1999, personal comun.). (3) Eclogites and peridotites have been subjected to various extents of retrograde metamorphism, first under amphibolite facies, and later under greenschist facies conditions. (4) Protolith lithologies have continental and subcontinental geochemical and petrological characteristics, including granitic gneiss, aluminous pelite, minor quartzite, marble, and mafic-ultramafics, and have Proterozoic isotopic ages. (5) Coeval calc-alkalic volcanic and plutonic rocks do not occur, whereas postcollisional or late-stage granitic plutons are common in some occurrences. These features are similar to many ultrahigh-*P* terranes, including the Dabie-Sulu belt (Liou et al., 1996; Ernst and Liou, 1999; Liou, 1999). Trace ultrahigh-*P* minerals would be best preserved in strong containers such as zircon or garnet in eclogite and their enclosing gneiss, and will be the subject of future research (Liou et al., 1998). Such coesite-bearing (ultrahigh *P*) or coesite-free (very high *P*) Alpine-type terranes have resulted from deep subduction of continental lithosphere prior to continent collision.

U-Pb data for zircon separates and Sm-Nd isochrons of minerals from very high *P* eclogitic rocks (Fig. 10) indicate that the peak eclogite facies metamorphism occurred ca. 495–504 Ma in the south Altun and north Qaidam eclogites. The $^{40}Ar/^{39}Ar$ plateau ages of amphibole and phengitic mica from these eclogitic rocks are 460–470 Ma (Fig. 11), which probably represents the timing of exhumation and overprinting of retrograde recrystallization. However, the Qilian high *P-T* blueschists and eclogites have a peak metamorphic age of 443–462 Ma and a retrograde age of 410–420 Ma (Wu et al., 1993; Wu and Tian, 1994; Zhang et al., 1997). These data suggest that the subduction and collision of the Qaidam block with the central Qilian block occurred prior to the convergence of the Qilian ocean beneath the margins of the Alxa massif.

Nature of the Altyn Tagh fault

The Altyn Tagh fault truncates and bounds many orogenic belts, including the Beishan, Altun, Kunlun, and Qilian Mountains (Fig. 2). This fault has been investigated extensively; understanding its evolution is essential for an accurate reconstruction of northwestern China (see Yue and Liou, 1999, for a summary). The proposed total left-lateral displacements by re-aligning critical petrotectonic assemblages on both sides of the fault range from 200 to 1200 km (Peltzer and Tapponnier, 1988; Ge et al., 1991; Yue and Liou, 1999).

The Altyn Tagh fault was identified as a key element of the escape tectonics model for the Eurasia-India continent-continent collision (Tapponnier and Molnar, 1977; Peltzer and Tapponnier, 1988). Zhou and Graham (1996) proposed a wedge extrusion model and suggested that: (1) the Altun Mountains at the tip and southern edge of the wedge are part of the west Kunlun Mountains, and (2) the Beishan and Qilian Mountains belong to the same orogenic belt. Yue and Liou (1999) proposed a two-stage evolution for the Altyn Tagh fault, which continues northeast to the inactive Alxa–east Mongolia fault: according to them, this continuous fault slipped about 400 km, and separated the Beishan orogen from the Inner Mongolia orogen, beginning in the Oligocene and continuing until the middle Miocene. The Alxa–east Mongolia fault then became inactive, and the displacement was compensated for by the shortening of the Qilian Mountains and the Qaidam basin from the middle Miocene to the present. This model indicates that (1) the Altun and Qilian Mountains are part of the same belt, and (2) the Beishan and Inner Mongolia orogens belong to the same late Paleozoic subduction and/or collision complex.

The similarities in the occurrences, associated country rocks, *P-T* estimates, and age data for the very high *P* eclogitic rocks in both the Altun and north Qaidam Mountains described herein attest to the plausibility of the correlation proposed by Yue and Liou (1999). Moreover, identification of the north Altun subduction complex, which has an ophiolitic melange and a transitional blueschist-greenschist assemblage of ca. 457 Ma, as being equivalent to the north Qilian Pacific-type subduction complex provides another piecing point for the Altyn Tagh fault. More geochronologic data are required for the high-*P* lithologies and their host rocks to test this hypothesis.

However, as shown in Figure 2, several tectonic units of the Qilian Mountains have not been identified in the Altun Mountains, including passive margin deposits at the southern margin of the Dunhuang massif, and the early Paleozoic island arc and its associated backarc basin deposit. Moreover, the central Altun terrane is much narrower than its equivalent in the central Qilian terrane. Such differences could be due in part to the different width for identical units along their strike and in part to the significant displacement along the Altyn Tagh fault system. Moreover, despite of the tectonic significance of the Altyn Tagh fault, the magnitude and direction of displacement along this fault are still poorly constrained, partly because there have only been limited field mapping and petrotectonic studies done in the Altun Mountains. The different models for the Altyn Tagh fault result from the lack of definitive geological data.

Occurrence of the Beishan very high pressure belt

According to Yue and Liou (1999), the Beishan orogen could be correlated with the Inner Mongolia orogen to the east of the Altyn Tagh fault; details of such correlation are documented in Yue et al. (this volume). The occurrence of the early Paleozoic, eclogite-bearing, A-type subduction complex as a suture between the Beishan terrane and the Dunhung massif leads to the question, what is the equivalent suture to the east, across the Altyn Tagh fault? The eclogite-bearing gneissic terrane is in a southern belt of the Beishan terrane, and on the northern margin of the Dunhuang massif. Its eastern extension may exist in the northern margin of the Alxa massif, but this is unsubstantiated. The Beishan very high *P* belt contains numerous eclogitic and ultramafic lenses enclosed within granitic gneiss, dolomitic marble, and garnet-bearing quartzite. However, regional geology, the distribution and the contact relation of this very high *P* belt with the adjacent tectonic units, the age of eclogitic metamorphism, and the nature of ultramafics are not known. Therefore, speculation regarding its correlation eastward to Inner Mongolia or westward to the Tian Shan is premature.

Tectonic model

The Qilian-Qaidam-Altun region of northwest China is a remote mountainous and high plateau area where geological studies are limited compared to other orogens in central and east China. We initiated a petrotectonic study of high-*P* metamorphic belts in this area, and our preliminary model for the tectonic evolution of the Qilan-Qaidam region is described in the following.

Figure 12 illustrates a tectonic evolution model for the northern Qilian Mountains based on our field and geochronological data (Xu et al., 1994, 1997; Zhang et al., 1997) and results from Tian and Wu (1994). Our discussion does not include the Tian Shan eclogite belt because its study is still in an early stage. Major chronological events are summarized in the following.

1. Three Precambrian continental fragments for the Alxa massif, central Qilian, and Qaidam basin could have rifted from a single continent (e.g., Wu et al., 1993) or might represent independent microcontinents (e.g., Z.M. Zhang et al., 1984). For example, the central Qilian terrane rifted from the Alxa massif

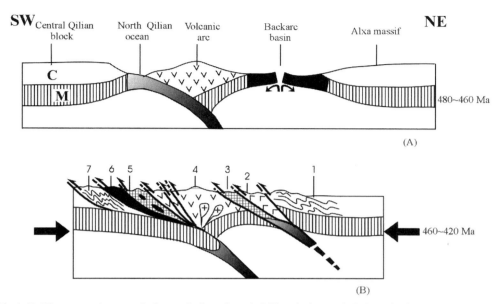

Figure 12. A, B: Plate tectonic scenario for evolution of north Qilian during early Paleozoic time. A: Northward subduction of north Qilian ocean resulting in formation of high-pressure (*P*) metamorphic rocks and volcanic arc. This was followed by rifting of backarc basin 460–480 Ma. B: Closures of north Qilian ocean and backarc basin resulted in two different continent-arc collision zones. 1—Early Paleozoic fold belt of continental margin of Alxa massif. 2—Backarc volcanic rocks. 3—Northern high-*P* metamorphic zone. 4—North Qilian volcanic arc. 5—South high-*P* metamorphic zone. 6—Ophiolite zone. 7—Early Paleozoic fold belt of central Qilian block. C—crust, M—mantle.

in Late Proterozoic time, and created a north Qilian ocean; the oceanic crust subsequently was obducted to form the Zhulongguan late Sinian ophiolitic suite (e.g., Tian and Wu, 1994). Although similar lithologies occur in the basement complex and overlying strata of these massifs, available data, particularly those from the Qaidam basin, do not support such a suggestion.

2. An Alpine-type convergent margin, including possible northward subduction of the Qaidam continental lithosphere beneath the Qilian terrane, may have occurred in Cambrian-Ordovician time (ca. 500 Ma). Available microstructural analyses of rocks from the Qaidam Mountains exclude southward subduction (Xu et al., 1999). During this period of time, a north Qilian ocean may have widely separated the Qilian terrane to the south and the Alxa massif and the Sino-Korean craton to the north.

3. The disposition of major lithologic units in the Qilian Mountains suggests two distinct stages of tectonic evolution, as shown in Figure 12: (a) subduction stage at 500–440 Ma, and a collision stage at 440–400 Ma. The nearly coeval occurrence of the north Qilian arc complex (466 to ca. 495 Ma, Xu et al., 1997; Zhang et al., 1997) to the north and the Pacific-type high *P-T* complex (440–462 Ma, Wu et al., 1993; Xu et al., 1994) to the south and kinematic data of Xu et al. (1994) suggest northward subduction of the north Qilian oceanic lithosphere ca. 500–440 Ma. Calc-alkaline volcanic activity extended to the transition region of the

Alxa massif. During this period of time, the Qilian ophiolite (454–469 Ma, Xia et al., 1995) was emplaced and a backarc basin was formed to the north of the north Qilian arc complex.

The collision stage (440–400 Ma) included two parts; the closure of the north Qilian ocean to the south, and that of the backarc basin to the north of the north Qilian arc complex. The former resulted in the collision of the central Qilian block and the north Qilian volcanic arc ca. 440 Ma, whereas the latter resulted in the collision between the north Qilian volcanic arc with the Alxa massif after 410 Ma. The collision time was determined on the basis of greenschist facies retrograded ages of 410–420 Ma for blueschists from the southern high *P-T* belt (Xu et al., 1994; Zhang et al., 1997) and Ar-Ar data (443 ± 3 Ma) for phengite from a Qilian eclogite in an ophiolite melange (Wu et al., 1993). A possible southward subduction of the backarc oceanic lithosphere has been suggested to have occurred in the Silurian (409 Ma) to form the low-grade lawsonite-bearing blueschist zone (e.g., Tian and Wu, 1994), and flysh strata were deposited in the backarc basin.

A thin Devonian molasse, which contains blueschist pebbles derived from the south, and unconformably overlies the folded early Paleozoic sequence, represents the collision of the north Qilian arc and the Alxa massif after 410 Ma. This suggestion is consistent with the deposition of passive-margin strata within the transition region along the southern margin of the Alxa massif, and the younger age of the low-grade blueschist compared with the southern eclogite-bearing epidote-blueschist zone.

4. Amalgamation of the united Qilian-Qaidam terrane with the north Qilian arc and Alxa massif may have occurred in Devonian time, when a widespread, thick, nonmarine molasse was deposited in the Qilian-Qaidam regions.

5. The Altyn Tagh fault–east Mongolia fault system may have begun around Oligocene time as a result of the Cenozoic collision between the Indian and Asian continents. This megafault system displaced all early Paleozoic tectonic units of the Qilian-Qaidam regions from the Altun Mountains northeastward for ~400 ± 50 km (Yue and Liou, 1999).

The proposed chronological sequence for tectonic evolution for this area is consistent with the petrological and geochronological data of major lithologic units, including both Pacific- and Alpine-type subduction complexes described herein. Further petrotectonic and geochronologic studies of basement, ophiolite, arc volcanics, and subduction-zone complexes in these remote regions are necessary to refine this new tectonic model.

CONCLUSIONS

Several Pacific-type and Alpine-type high P-T metamorphic belts of early Paleozoic age occur in the Qilian–north Qaidam–Altun blocks. The Pacific type was documented in the north Qilian block and is regarded as the result of subduction of oceanic crust. The peak high-P metamorphism occurred at 440–462 Ma, at $T > 380\ °C$ and $P = 6$–7 kbar, and the exhumation occurred at 410–420 Ma. The Alpine type occurs in the north Qaidam and south Altun Mountains, and is regarded as the result of continental subduction having a peak metamorphic age of 495–504 Ma, and higher pressure ($P > 22$ kbar) and temperature ($720 ± 120\ °C$), suggesting deeper subduction than that of the north Qilian blueschist belt.

According to the similarities in many aspects between the blocks on opposite sides of the Altyn Tagh fault, including disposition of lithological units, occurrence of both Pacific- and Alpine-type high P-T belts, P-T conditions, and ages of formation and exhumation, it is suggested that 400 km of left-lateral displacement has occurred along the Altyn Tagh fault (Xu et al., 1999).

ACKNOWLEDGMENTS

This paper was prepared by a joint effort of a Stanford-Chinese Academic Geologic Science cooperative project supported by National Science Foundation grant EAR-9814468, and the Stanford China Industrial Affiliates Program, and was supported by the National Natural Science Foundation (49732070) and the National Key Project for Basic Research on the Tibetan Plateau (G1998040800). We thank Liu Liang and Mei Hualing, respectively, for information regarding Altun and Beishan eclogites, and Gary Ernst, Edmund Chang, and Yongjun Yue for useful discussions. The manuscript was critically reviewed and materially improved by Gary Ernst, X.C. Xiao, Shige Maruyama, Edward Sobel, Hakon Austrheim, and Marc Hendrix. We thank these individuals and institutions for their support and assistance.

REFERENCES CITED

Blundy, J.D., and Holland, T.J.B., 1990, Calcic amphibole equilibria and a new amphibole-plagioclase geothermometer: Contributions to Mineralogy and Petrology, v. 104, p. 208–224.

Bureau of Geology and Mineral Resources of Gansu Province (BGMG), 1989, Regional geology of Gansu Province: Beijing, Geological Publishing House, p. 1–35 (in Chinese with English abstract).

Bureau of Geology and Mineral Resources of Qinghai Province (BGMQ), 1991, Regional geology of Qinghai Province: Beijing, Geological Publishing House, 661 p. (in Chinese with English abstract).

Bureau of Geology and Mineral Resources of Xinjiang Uygur Autonomous Region, 1993, Regional geology of Xinjiang Uygur Autonomous Region: Beijing, Geological Publishing House, 841 p. (in Chinese with English abstract).

Carswell, D.A., O'Brien, P.J., Wilson, R.N., and Zhai, M., 1997, Thermobarometry of phengite-bearing eclogites in the Dabie Mountains of central China: Journal of Metamorphic Geology, v. 15, p. 239–252.

Carswell, D.A., Cuthbert, S.J., and Krough Ravna, E.J., 1999, Ultrahigh-pressure metamorphism in the Western Gneiss Region of the Norwwegian Caledonides: International Geology Review, v. 41, p. 955–966.

Che, Z., and Sun, Y., 1996, The age of the Altun granulite facies complex and the basement of the Tarim basin: Regional Geology of China, v. 56, p. 51–57 (in Chinese with English abstract).

Che, Z., Liu, L., Liu, H., and Luo, J., 1995a, The discovery and occurrence of high-pressure metapelitic rocks from Altun Mountain area, Xinjiang Autonomous Region: Chinese Science Bulletin, v. 40, p. 1298–1300 (in Chinese with English abstract).

Che, Z., Liu, L., Wang, Y., and Luo, J., 1995b, Metamorphic evolution and tectonics: Exhumation geodynamics of ultra-high-pressure and high-pressure metamorphic rocks in Altun: Chinese Science Bulletin, v. 40, p. 131–133 (in Chinese with English abstract).

Chen, B.W., Wang, Y.B., and Zuo, G.C., 1995, Terrain subdivision of the Northern Qinghai–Xizang Plateau and tectonic evolution: Acta Geophysica Sinica, v. 38, p. 113–128 (in Chinese with English abstract).

Chen, W., Li, H.B., Xu, Z.Q., Arnaud, N., Han, D., and Shen, J., 1996, Using the laser microprobe $^{40}Ar/^{39}Ar$ dating technique and $^{40}Ar/^{39}Ar$ stepwise incremental heating technique to date the ages of Xidatan shear zone: Acta Geoscientia Sinica, special issue, p. 228–232.

Chopin, C., and Schertl, H.P., 1999, The UHP unit in the Dora-Maira Massif, western Alps: International Geology Review, v. 41, p. 765–780.

Coleman, R.G., and Wang, X., 1995, Overview of the geology and tectonics of UHPM: in Coleman, R.G., and Wang, X., eds., Ultrahigh pressure metamorphism: Cambridge, Cambridge University Press, p. 1–32.

Coleman, R.G., Lee, D.E., Beatty, L.B., and Brannock, W.W., 1965, Eclogites and eclogites: Their differences and similarities: Geological Society of America Bulletin, v. 76, p. 483–508.

Ernst, W.G., 1971, Do mineral parageneses reflect unusually high-pressure conditions of Franciscan metamorphism?: American Journal of Science, v. 270, p. 81–108.

Ernst, W.G., and Liou, J.G., 1995, Contrasting plate tectonic styles of the Qinling-Dabie-Sulu and Franciscan metamorphic belts: Geology, v. 23, p. 353–356.

Ernst, W.G., and Liou, J.G., 1999, Overview of UHP metamorphism and tectonics in well-studied collisional orogens: International Geology Review, v. 41, p. 477–493.

Ernst, W.G., Liou, J.G., and Hacker, B.R., 1994, Petrotectonic significance of high- and ultrahigh-pressure metamorphic belts: Inferences for subduction-zone histories: International Geology Review, v. 36, p. 213–237.

Feng, Y.M., and He, S.P., 1996, Geotectonics and orogeny of the Qilian Mountains, China: Beijing, Geological Publishing House, 135 p. (in Chinese with English abstract).

Ge, X.H., Duan, J., Li, C., Yang, H., and Tian, Y., 1991, A new recognition of the Altun fault zone and geotectonic pattern of Northwest China: International Geological Correlation Programme Project 321; Proceedings of first international symposium on Gondwana dispersion and Asian accretion; geological evolution of eastern Tethys: Beijing, Geological Publishing House, p. 125–128.

Hanson, A., Chang, E., Zhou, D., Ritts, B., Sobel, E., Graham, S., Chu, C., Liu, J., Zhang, R., and Liou, J.G., 1995, Discovery of eclogite blocks in the Altun Mountains, SE Tarim basin, NW China [abs.]: Eos (Transactions, American Geophysical Union), v. 76, p. S283.

Liou, J.G., 1999, Petrotectonic summary of less-intensively studied UHP regions: International Geology Review, v. 41, p. 571–586.

Liou, J.G., Wang, X., Coleman, R.G., Zhang, Z.M., and Maruyama, S., 1989, Blueschists in major suture zones of China: Tectonics, v. 8, p. 609–619.

Liou, J.G., Zhang, R., and Ernst, W.G., 1994, An introduction to ultrahigh-pressure metamorphism: The Island Arc, v. 3, p. 1–24.

Liou, J.G., Zhang, R.Y., Wang, X.M., Eide, E.A., Ernst, W.G., and Maruyama, S., 1996, Metamorphism and tectonics of high-P and ultrahigh-P belts in the Dabie-Sulu region, eastern central China, in Yin, A., and Harrison, M.T., eds., Tectonic development of Asia: Rubey Volume 13: Cambridge, Cambridge University Press, p. 300–344.

Liou, J.G., Zhang, R., Ernst, W.G., Rumble, D., and Maruyama, S., 1998, High pressure minerals from deeply subducted metamorphic rocks, in Mao, H.K., and Hemley, R.J., eds., Ultrahigh pressure mineralogy: American Mineralogical Society Reviews in Mineralogy, v. 37, p. 33–96.

Liu, L., Ce, Z.C., Luo, J.H., Wang, Y., and Gao, Z.J., 1996, Identification of the eclogite at the west part of Altyn Tagh Mountains and its geological implications: Chinese Science Bulletin, v. 41, p. 1485–1488 (in Chinese with English abstract).

Liu, L., Che, Z.C., Luo, J.H., Wang, Y., and Gao, Z.J., 1997, Recognition and implication of eclogite in the western Altun Mountains, Xinjiang: Chinese Science Bulletin, v. 42, p. 931–934 (in Chinese with English abstract).

Liu, X.Y., and Wang, Q., 1993, Tectonics of the Longshoushan ancient rift and Hexi Corridor: Chinese Academy of Geological Sciences Bulletin, v. 27, p. 1–14 (in Chinese with English abstract).

Lu, S.N., and Li, H.M., 1991, A precise U-Pb single zircon age determination for volcanics of Dahongyu formation, Changcheng System in Jixian: Chinese Academy of Geological Sciences Bulletin, v. 22, p. 137–146 (in Chinese with English abstract).

Ma, L., Ding, X., and Fan, B., 1996, Geological map of China: Beijing, Geological Publishing House, scale 1:12 000 000.

Maruyama, S., Liou, J.G., and Terabayashi, M., 1996, Blueschists and eclogites of the world and their exhumation: International Geological Review, v. 38, p. 485–594.

Mei, H.L., Yu, H.F., Li, Q., Lu, S.N., Li, H.M., Zuo, Y.C., Zuo, G.C., Ye, D.J., and Liu, J.C., 1998, First discovery of eclogite and Paleo-Proterozoic granitic rock in the Beishan region, Gansu province, China: Chinese Science Bulletin, v. 43, p. 2105–2111 (in Chinese with English abstract).

Neill, H.S.C., and Wood, B.J., 1979, An experimental study of Fe-Mg partitioning between garnet and olivine and its calibration as a geothermometer: Contributions to Mineralogy and Petrology, v. 70, p. 59–70.

Peltzer, G., and Tapponnier, P.G., 1988, Formation and evolution of strike-slip faults, rifts, and basins during the India-Asia collision. An experimental approach: Journal of Geophysical Research, v. 93, p. 15085–15117.

Schmidt, M.W., 1992, Amphibole composition in tonalite as a function of pressure: An experimental calibration of the Al-in-hornblende barometer: Contributions to Mineralogy and Petrology, v. 110, p. 304–310.

Sobel, E.R., and Arnaud, N., 1999, A possible middle Paleozoic suture in the Altyn Tagh, NW China: Tectonics, v. 18, p. 64–74.

Stroh, J.M., 1976, Solubility of alumina in orthopyroxene plus spinel as a geobarometer in complex system: Application to spinel-bearing Alpine-type peridotite: Contributions to Mineralogy and Petrology, v. 54, p. 173–188.

Tapponnier, P.G., and Molnar, P., 1977, Active faulting and tectonics in China: Journal of Geophysical Research, v. 82, p. 2905–2930.

Tian, B., and Wu, H., 1994, Geochemistry of the two blueschist belts and implications for the early Paleozoic tectonic evolution of the north Qilian Mountains, China, in Wu, H., et al., eds., Very low grade metamorphism: Mechanisms and geological applications: Beijing, Seismological Press, p. 92–116.

Wang, Q., and Coward, M.P., 1990, The Chaidam Basin, NW China: Formation and hydrocarbon potential: Journal of Petroleum Geology, v. 13, p. 93–112.

Wang, Q., Nishidai, T., and Coward, M.P., 1992, The Tarim basin, NW China: Formation and aspects of petroleum geology: Journal of Petroleum Geology, v. 15, p. 5–34.

Wang, X., Zhang, R., and Liou, J.G., 1995, Ultrahigh pressure metamorphic terrane in eastern central China, in Coleman, R.G., and Wang, X., eds., Ultrahigh pressure metamorphism: Cambridge, Cambridge University Press, p. 356–390.

Wang, Y.S., and Chen, J.N., 1987, Metamorphic zones and metamorphism in Qinghai Province and its adjacent areas: Beijing, Geological Publishing House, p. 42–56 (in Chinese with English abstract).

Wu, H., and Tian, B., 1994, New achievements on the study of the low-temperature metamorphism in north Qilian Mountains: Discovery of lawsonite in southern high-grade blueschist belt and metamorphism of ophiolite in Yushigou, Qilian County, China, in Wu, H., et al., eds., Very low grade metamorphism: Mechanisms and geological applications: Beijing, Seismological Press, p. 156–167.

Wu, H., Feng, Y., and Song, S., 1993, Metamorphism and deformation of blueschist belts and their tectonic implications, north Qilian Mountains, China: Journal of Metamorphic Geology, v. 11, p. 523–536.

Xia, L.Q., Xia, Z.C., and Xu, X.Y., 1995, Dynamics of tectono-volcano-magmatic evolution from north Qilian Mountains, China: Northwest Geosciences, v. 16, p. 1–28 (in Chinese with English abstract).

Xiao, X.C., Chen, G.M., and Zhu, Z.Z., 1974, Some knowledge about the paleo-plate tectonics of the Qilian Mountains: Geological Science and Technology, v. 3, p. 73–78 (in Chinese with English abstract).

Xiao, X.C., Chen, G.M., and Zhu, Z.Z., 1978, A preliminary study on the tectonics of ancient ophiolites in the Qilian Mountains, Northwest China: Acta Geologica Sinica, v. 52, p. 287–295 (in Chinese with English abstract).

Xu, Z.Q., Xu, H.F., Zhang, J.X., and Li, H.B., 1994, The Zoulang Nanshan Caledonian subduction complex in the Northern Qilian Mountains and its dynamics: Acta Geologica Sinica, v. 68, p. 225–241 (in Chinese with English abstract).

Xu, Z.Q., Zhang, J.X., and Xu, H.F., 1997, Ductile shear zones in the main continental mountain chains of China and their dynamics: Beijing, Geological Publishing House, p. 71–108 (in Chinese).

Xu, Z.Q., Yang, J.S., Zhang, J.X., Jiang, M., Li, H.B., and Cui, J.W., 1999, A comparison between the tectonic units on the sides of the Altun sinistral strike-slip fault and the mechanism of lithospheric shearing: Acta Geologica Sinica, v. 73, p. 193–205 (in Chinese with English abstract).

Yang, J.J., Zhu, H., Deng, J.F., Zhou, T.Z., and Lai, S.C., 1994, Discovery of garnet-peridotite at the northern margin of the Qaidam basin and its significance: Acta Petrologica et Mineralogica, v. 13, p. 97–105 (in Chinese with English abstract)).

Yang, J.S., Xu, Z.Q., Li, H.B., Wu, C.L., Cui, J.W., Zhang, J.X., and Chen, W., 1998, Discovery of eclogite at northern margin of Qaidam basin, NW China: Chinese Science Bulletin, v. 43, p. 1755–1760 (in Chinese with English abstract).

Yang, J.S., Xu, Z.Q., Li, H.B., Wu, C.L., Zhang, J.X., and Shi, R.D., 2000, A Caledonian convergent border along the southern margin of the Qilian terrane, NW China: Evidence from eclogite, garnet-peridotite, ophiolite, and S-type granite: Geological Society of China Journal, v. 43, p. 142–164.

Yue, Y., and Liou, J.G., 1999, A two-stage evolution model for the Altyn Tagh fault, China: Geology, v. 27, p. 227–230.

Zhang, J.X., and Xu, Z.Q., 1995, Caledonian subduction-accretionary complex/volcanic arc zone and its deformation features in the middle sector of north Qilian Mountains: Acta Geoscientia Sinica, v. 69, p. 154–163 (in Chinese with English abstract).

Zhang, J.X., Xu, Z.Q., Chen, W., and Xu, H.F., 1997, A tentative discussion on the ages of the subduction-accretionary complex/volcanic arcs in the middle sector of north Qilian mountains: Acta Petrologica Mineralogica, v. 16, p. 112–119 (in Chinese with English abstract).

Zhang, J.X., Zhang, Z.M., Xu, Z.Q., Yang, J.S., and Cui, J.W., 1999, The ages of U-Pb and Sm-Nd for eclogite from the western segment of Altyn Tagh tectonic belt—The evidence for existence of Caledonian orogenic root: Chinese Science Bulletin, v. 44, p. 1109–1112 (in Chinese with English abstract).

Zhang, Z.M., and Liou, J.G., 1987, The high *P-T* metamorphic rocks of China, *in* Leith, E.C., and Scheibner, E., eds., Terrane accretion and orogenic belts: American Geophysical Union Geodynamics Series, v. 19, p. 235–247.

Zhang, Z.M., Liou, J.G., and Coleman, R.G., 1984, An outline of the plate tectonics of China: Geological Society of America Bulletin, v. 95, p. 295–312.

Zhang, Z., Lu, J., and Tang, S., 1999, Sm-Nd ages of the Panxi layered basic-ultrabasic intrusion in Sichuan: Acta Geologica Sinica, v. 73, p. 263–271 (in Chinese with English abstract).

Zhou, D., and Graham, S.A., 1996, Extrusion of the Altyn Tagh wedge: A kinematic model for the Altyn Tagh fault and palinspastic reconstruction of northern China: Geology, v. 24, p. 427–430.

Zuo, G.C., and Liu, J.G., 1987, The evolution of tectonics of early Paleozoic in north Qilian Range, China: Scientica Geologica Sinica, v. 1, p. 14–24 (in Chinese with English abstract).

MANUSCRIPT ACCEPTED BY THE SOCIETY JUNE 5, 2000

Geological Society of America
Memoir 194
2001

Mesozoic tectonic evolution of the Yanshan fold and thrust belt, with emphasis on Hebei and Liaoning provinces, northern China

Gregory A. Davis
Department of Earth Sciences, University of Southern California, Los Angeles, California 90089-0740, USA
Zheng Yadong
Department of Geology, Peking University, Beijing 100871, China
Wang Cong and Brian J. Darby
Department of Earth Sciences, University of Southern California, Los Angeles, California 90089-0740, USA
Zhang Changhou
China University of Geosciences, Beijing 100083, China
George Gehrels
Department of Geosciences, University of Arizona, Tucson, Arizona 85721, USA

ABSTRACT

The Yanshan (Yan Mountains) of northern China extend westward at about lat 40°N from Bohai Bay and Liaoning province to the border between Hebei province and Inner Mongolia. It is likely, but not unequivocally demonstrated, that the Archean-floored Yanshan continue farther westward under a cover of Neogene strata to emerge as the Yinshan belt of Inner Mongolia. Mesozoic terrestrial sedimentation, magmatism and deformation—including multiple phases of folding and contractional, extensional, and strike-slip faulting—characterize the Yanshan fold and thrust belt. Field studies and radiometric dating (U-Pb, ^{40}Ar-^{39}Ar) of plutonic and volcanic rock units in Beijing Municipality and northern Hebei and western Liaoning provinces have revealed that the complexity of Mesozoic deformation in these areas is in large part (1) a consequence of profound earlier deformation of Permian(?) to early Mesozoic age, and (2) an unusual younger Mesozoic history of alternating northward and southward tectonic vergence of major structures.

Major south-directed low-angle thrust faulting of pre-Middle Jurassic age (>180 Ma) in the Yanshan involved Archean basement rocks and their Proterozoic and Phanerozoic cover, and developed south of a Permian–Triassic magmatic arc. Thrusting could have been (1) a consequence of the collisional suturing of Paleozoic Mongolian arcs against an Andean-style continental arc along the northern margin of the North China plate, or (2) an expression of a backarc, foreland fold and thrust belt of U.S. Cordilleran type formed during southward subduction beneath the North China Archean "craton." This episode of thrusting and folding was followed by widespread erosion of upper plate rocks and subsequent terrestrial deposition of Middle Jurassic volcanic and sedimentary strata across both plates.

E-mails: Davis, gdavis@usc.edu; Zheng, ydzheng@geoms.geo.pku.edu.cn; Wang, wangcong@usc.edu; Darby, briand@usc.edu; Zhang, changhou@ CUGB.EDU.CN; Gehrels, ggehrels@geo.Arizona.edu.

Davis, G.A., et al., 2001, Mesozoic tectonic evolution of the Yanshan fold and thrust belt, with emphasis on Hebei and Liaoning provinces, northern China, *in* Hendrix, M.S., and Davis, G.A., eds., Paleozoic and Mesozoic tectonic evolution of central Asia: From continental assembly to intracontinental deformation: Boulder, Colorado, Geological Society of America Memoir 194, p. 171–197.

Our studies indicate that Late Jurassic and Early Cretaceous contractional deformation in the Yanshan was also much more intense than generally believed, and that it followed a heretofore unrecognized phase of Middle Jurassic or early Late Jurassic east-west extension in northern Hebei province. Prior to our studies, popular views of east- to east-northeast–trending Yanshan contractional deformation proposed that (1) involvement of Archean basement rocks in faulting indicates a thick-skinned tectonic style analogous to the U.S. Laramide Rocky Mountains; (2) thrust faults steepen downward into the basement; (3) vertical movements predominated in the development of the belt; and (4) Mesozoic contraction across the belt was only a few tens of percent; some workers have believed that the Yanshan developed independently of plate interactions. We question all of these assumptions. We have identified a major, synformally folded, thin-skinned thrust plate just south of Chengde, Hebei province, that is of Late Jurassic age and had a minimum northward displacement of ~40 km. In earlier studies in the Yunmeng Shan area of the Yanshan north of Beijing we defined major south-vergent Late Jurassic–Early Cretaceous ductile structures involving Archean basement rocks and their cover. These structures include a recumbent, basement-cored anticlinal nappe and a lower limb ductile thrust fault within a 6-km-thick gneissic shear zone.

The Jurassic–Cretaceous Yanshan belt, and its probable western continuation, the Yinshan belt of Inner Mongolia, appear to be reflect regional north-south intraplate shortening. However, some Yanshan patterns of deformation, e.g., ductile nappe formation in the Yunmeng Shan and northeast structural trends of the belt in Liaoning province, were influenced by thermal regimes related to magmatism ($\leq$ 180–190 Ma) accompanying westward or northwestward Pacific plate subduction beneath eastern Asia. We thus believe it likely that two contrasting modes of plate interaction occurred synchronously in the Yanshan segment of the Yinshan-Yanshan belt during Middle Jurassic through Early Cretaceous time.

Jurassic–Cretaceous collision of an amalgamated North China–Mongolian plate with the Siberian plate is widely believed to have accompanied closure of a Mongolo-Okhotsk sea more than 800–1100 km north of the Yanshan belt. This collision might have been responsible for Yanshan (and Yinshan) intraplate contractional deformation farther south, but such a hypothesis is severely complicated by reports of widespread, basin-forming Late Jurassic and Early Cretaceous extension in the terranes between the Mongolo-Okhotsk suture and the Yinshan-Yanshan belt. Following Early Cretaceous contraction in both the Yinshan and Yanshan belts, the northern margin of the Archean-floored plate was also the site of major northwest-southeast regional extension beginning soon after 120 Ma. Subducting plate rollback or postorogenic collapse are only two of several possible explanations for the development of extensional metamorphic core complexes in northern Inner Mongolia and the Yanshan belt farther east.

INTRODUCTION

The Great Wall of northern China snakes across a strategic mountain barrier between the fertile North China plain and the vast and once hostile lands of Inner Mongolia, Mongolia, Manchuria, and Siberia. North of Beijing, this topographic barrier, the Yanshan (shan = mountains), is not a lofty mountain range by Chinese standards. Its highest elevation is <2500 m, but because of its proximity to China's capital, the Yanshan of Beijing Municipality (including the Western Hills) and Hebei and Liaoning provinces has often been referred to as the cradle of Chinese tectonics. It was the first of China's major mountain systems to be studied in the late 1800s and early 1900s from the

standpoint of both structure and stratigraphy (e.g., Wong, 1929). Those studies gave rise to the term Yanshanian deformation, subsequently applied to all Jurassic–Cretaceous deformation throughout China.

There is apt to be confusion regarding terms such as Yanshanian deformation, Yanshan (or Yenshan) movement, and Yanshan phase in China with respect to the geographically defined Yinshan-Yanshan belt. Chinese workers have divided the tectonic history of China into time periods (phases) defined by major unconformities in the stratigraphic record resulting from deformational episodes (movements). Deformational phases and movements such as Indosinian, Yanshanian, or Himalayan are temporally defined, not geographically as is the usual case for American orogenic terminology (e.g., Laramide, Sevier, and Nevadan, using western U.S. examples). Discussions in the Chinese literature of Yanshan or Yanshanian movement or deformation characteristically refer to Jurassic–Cretaceous deformation throughout China and thus blur the nature and causes of widely separated regions of such deformation in this time interval (cf. Wong, 1929; Zhang et al., 1984; Xu, 1990).

The Yanshan is widely considered, perhaps inappropriately (see following), to be the eastern segment of the east-west–trending Yinshan orogenic system of Jurassic–Cretaceous age (Fig. 1). As generally defined, this several-hundred-kilometer-wide system extends eastward >1100 km from the Yinshan (sensu stricto) belt of Inner Mongolia and through the Yanshan in the western part of Liaoning province north of Bohai Bay (BB, Fig. 1). Portions of the Yinshan belt west of Baotou in Inner Mongolia (~110°E) are largely concealed beneath sand and gravel of the Gobi Desert along the Mongolia-China border (Figs. 1 and 2). Structures in the Yinshan belt of Inner Mongo-

lia and in western parts of the Yanshan north and northeast of Beijing Municipality have predominantly east trends; east-northeast to northeast trends characterize more easterly parts of the Yanshan in eastern Hebei and western Liaoning provinces (Fig. 2). Although guardedly adopted in this chapter, this geologic and geographic depiction of the Yinshan (sensu lato) system may not be correct. Tectonic continuity between the Yinshan belt (sensu stricto) of Inner Mongolia and the more easterly Yanshan belt cannot be unequivocally demonstrated because of widespread Neogene cover in an intervening border region between Inner Mongolia (Nei Mongol) and Hebei province (area of question mark in Fig. 2). The tectonic implications of this uncertainty are discussed later in this chapter.

The southern margin of a Yinshan (sensu lato) orogenic system is defined by the stable Mesozoic Ordos basin in the west (O, Fig. 1) and concealed by the late Mesozoic–Cenozoic Huabei basin in the east (beneath the letters BB, Fig. 1). Northeast-trending Mesozoic folds and faults in the Taihang Shan between the two basins (TS, Figs. 1 and 2) merge northeastward and interact with Yinshan belt structures in the Western Hills west of Beijing. The northern boundary of the Jurassic–Cretaceous Yanshan belt in Hebei province is not well defined. It is considered by some Chinese workers to lie about 40 km north of Chengde City (Figs. 2 and 3) on the northern edge of the Archean-defined craton (Fig. 2). "Upper Jurassic and Cretaceous" strata in this area appear to be significantly less folded and faulted than in areas to the south (Hebei Bureau of Geology, 1989, p. 723, 726), although these poorly dated strata may be younger than believed and were simply not present to record the more intense deformations seen to the south (described subsequently as tectonic phases III and IV).

The Yinshan-Yanshan system lies along the northern margin of the Archean-floored North China craton (a misleading term, given its widespread Phanerozoic tectonic reactivation),

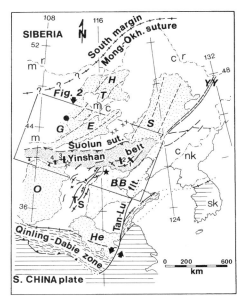

Figure 1. Tectonic setting of Yinshan belt (sensu lato, see text), northern China, and its relationships to Permian–Triassic(?) Suolun(-Linxi) suture and major convergent plate boundaries of Mesozoic age—Qinling-Dabie suture zone to south and Mongolo–Okhotsk (Mong.–Okh.) suture to north; west- to northwest-dipping paleo-Pacific subduction zone to east is not shown. Heavy numbered lines (1–4) in Yinshan belt show locations of specific study areas and/or cross sections described in text. Dotted areas are Mesozoic basins discussed in text: G-E—Gobi-Erlian; T-H—Tamsag-Hailar; S—Songliao; YY—Yilong-Yitang graben; BB—Bohai Bay (Mesozoic?–Cenozoic); He—Hehuai (Hefei); and O—Ordos. Heavy dashed lines in northern basins are trends of Jurassic–Cretaceous extensional faults. Black circle near Mongolia-China border is Unegt subbasin mentioned in text. Row of xs north of Suolun suture denotes average trend of Jurassic–Cretaceous plutons between basins. Bold lines in Yinshan belt and Taihang Shan (TS) are Mesozoic fold trends, except for fault in TS with double-arrow slip uncertainty. Closed star—Beijing; open star—Hohhot, Inner Mongolia. Dash-dot lines are national boundaries: c, China; m, Mongolia; r, Russia; nk and sk, North and South Korea.

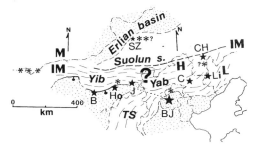

Figure 2. Location map of Yinshan belt sensu stricto (Yib) and Yanshan belt (Yab) and their relationships to Taihang Shan (TS); see Figure 1 for location. Dashed lines in Yib, Yab, and TS represent trends of Mesozoic contractional structures. Dash-dot lines are political boundaries within China; large bold letters: M—Mongolia; IM—Inner Mongolia; H and L—Hebei and Liaoning provinces. Stars are cities (from west to east): B—Baotou; Ho—Hohhot; J—Jining; SZ—Sonid Zouqi; BJ—Beijing; C—Chengde; CH—Chifeng; Li—Lingyuan. Mesozoic–Cenozoic basins are stippled. Asterisks indicate known and suspected extensional metamorphic core complexes of Cretaceous age; Yagan-Onch Hayrhan complex (Zheng et al., 1991; Webb et al., 1999; Johnson et al., this volume) is in far west.

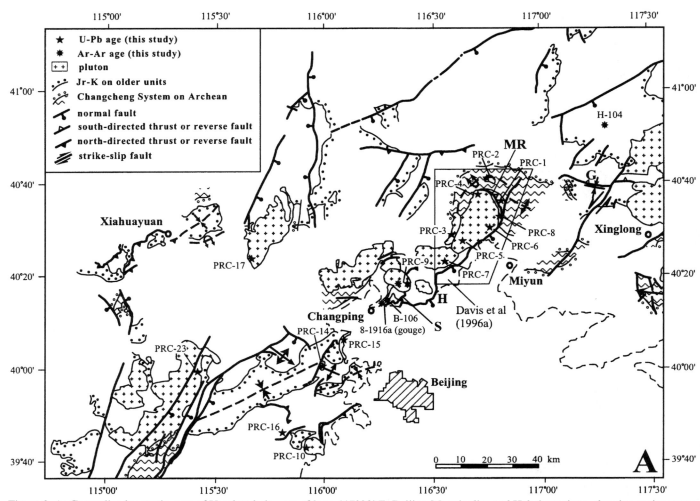

Figure 3. A: Generalized tectonic map of Yanshan belt west of long 117°30' E, Beijing Muncipality and Hebei province, showing major structures and localities for U-Pb, and ^{40}Ar-^{39}Ar age determinations reported here (cf. Tables 1 and 2; Fig. 4A) and in Davis et al. (1996a, 1998a, 1998b). Structures: H—Hefangkou low-angle normal fault; MR—Miyun Reservoir thrust; G—Gubeikou reverse fault; S—Shisanling thrust. B (facing page): Generalized tectonic map of Yanshan belt east of ~long 117° E, Hebei and Liaoning provinces, showing major structures and localities for U-Pb and ^{40}Ar-^{39}Ar age determinations reported here (cf. Tables 1 and 2; Fig. 4B) and in Davis et al. (1996a, 1998a, 1998b). Structures: C—Chengde thrust; CC—Chengde County thrust; G—Gubeikou reverse fault; L—unnamed Liaoning thrust faults; MA—Malanyu anticline; X—Xinglong thrust. Modified from Davis et al. (1998b, their Fig. 3).

but during Jurassic and Cretaceous time it lay within an amalgamated Mesozoic North China–Mongolian plate (Fig. 1). That plate, as defined here, consisted of two major elements: (1) the Archean North China craton; and (2) to the north, the Mongolian (or Tumen) accretionary fold belt. The Mongolian belt is a broad and poorly understood assemblage of multiple Ordovician to Early Permian oceanic arcs, Paleozoic blueschist-bearing melanges, Paleozoic ophiolites, and possible microcontinental blocks (Zhang et al., 1984; Zonenshain et al., 1990; Şengör and Natal'in, 1996; Lamb and Badarch, 1997; Xu and Chen, 1997).

The timing and style of amalgamation of the two elements of the Mesozoic North China–Mongolian plate is controversial, in part because of limited exposures between them. Many workers favor Late Permian to Early Triassic(?) amalgamation of the

Mongolian belt with the Archean craton along the Suolun-Linxi suture (Fig. 1; Wang and Liu, 1986; Wang and Mo, 1995; Yin and Nie, 1996; Zheng et al., 1996). We have adopted the terminology of Wang and Mo (1995), i.e., Suolun-Linxi, for this suture zone. It has been called by other names, among them the Solon Obo–Linxi suture (Wang and Liu, 1986), the Tian Shan–Ying Shan suture (Yin and Nie, 1996), and the Solonker suture (Şengör et al., 1993; Şengör and Natal'in, 1996). The age of final suturing and even the direction or directions of accompanying subduction are controversial (cf. Lamb and Badarch, 1997). Zhang et al. (1984) and Xu and Chen (1997) believed that it was a middle Paleozoic subduction zone because of the occurrence of early to middle Paleozoic ophiolites, blueschists, and magmatic rocks along it. Şengör and Natil'in (1996) de-

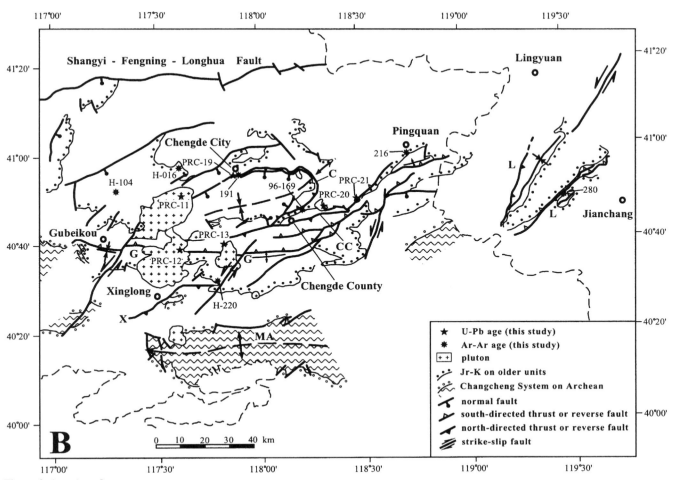

Figure 3. (*continued*)

noted it as a Late Carboniferous–Permian suture with, perhaps, final collision in the Triassic. Wang and Liu (1986) and Wang and Mo (1995) favored Late Permian amalgamation of North China and Mongolian terranes along it. Wang and Liu (1986) concluded that final suturing in the Late Permian was accompanied by subduction of oceanic crust to both north and south. Southward Permian–Triassic subduction beneath the Archean craton is supported by the widespread occurrence south of the suture of numerous plutons ranging in age from 285 to 217 Ma (e.g., Cui and Wu, 1997, their Fig. 6; methods of dating were not given).

If the intracontinental Yinshan-Yanshan belt were within North America, it would be a major and intensely studied zone of deformation, but because it is within the suture-dominated tectonic collage of eastern Asia, it has escaped the geologic attention that it deserves. Until very recently, geologic interest in the Yinshan-Yanshan belt had fallen behind the interest shown other Chinese orogenic areas and systems, e.g., Tibet, the Tian Shan, and the Kunlun-Qinling-Dabie suture zone between the early Mesozoic North and South China plates. For example, of

several hundred papers discussing structure and tectonics of China at the 1996 International Geological Congress in Beijing, only a dozen or so treated aspects of the tectonics of the Yinshan-Yanshan belt. Nevertheless, the belt is a key component of the eastern Asian tectonic collage. As such, any proper understanding of the late Paleozoic and Mesozoic assembly of Asian tectonic elements requires an improved understanding of the tectonic history of the Yinshan belt, surrounded as it is by many of those components.

GEOLOGIC OVERVIEW OF THE YANSHAN BELT

Stratigraphy

The Jurassic–Cretaceous Yanshan belt overlies the northern margin of the North China (or Sino-Korean) craton, an extensive region in eastern Asia underlain by Archean crystalline basement rocks ranging in age from ca. 3.9 to 2.5 Ga. Proterozoic shallow-water marine strata (ca. 1850–800 Ma) are widespread, but of variable total thickness (0 to >10 km). Upper Proterozoic

(Sinian) strata, ca. 800–615 Ma, are missing. Phanerozoic strata on the craton are represented by (1) Cambrian through Middle Ordovician units, mainly shallow-marine carbonates; (2) Upper Carboniferous through Lower Permian alternating marine and terrestrial sequences characterized, respectively, by carbonates and coal-bearing clastics; (3) Upper Permian and Triassic redbeds and conglomerates; and (4) Jurassic and Cretaceous terrestrial volcanic and clastic strata.

Jurassic coal-bearing clastics and continental volcano-sedimentary units unconformably overlie older units; Late Triassic volcanic activity has been identified, but is restricted geographically. Early to Late Jurassic mafic to intermediate volcanism was widespread across northern parts of the North China–Mongolian plate, including the Yanshan area, but reached its maximum intensity in the Late Jurassic and Early Cretaceous (Xu, 1990). A rhyolite-dacite-andesite association characterized much of the Late Jurassic in northern parts of the Yinshan-Yanshan belt (Xu, 1990). Strata intercalated with Jurassic and, locally, Lower Cretaceous volcanic rocks include coal, conglomerate, sandstone, tuff, and other volcaniclastic rocks. Unfortunately, age controls for the Mesozoic strata are typically lacking due to the terrestrial nature of the sedimentary record (abrupt facies changes, and unfossiliferous or poorly fossiliferous units) and the paucity of radiometric ages for plutons and volcanic units in the Yinshan belt.

Our geochronologic studies have already indicated that some age assignments for Mesozoic strata in the Yanshan segment of the belt are grossly in error (Davis et al., 1998b). With regard to Jurassic–Cretaceous stratigraphy, generalized regional correlations of stratigraphic units should be questioned because of (1) insufficient age controls for volcanic units, and (2) rapid facies changes in the Jurassic–Cretaceous terrestrial environment. Consider, for example, the regionally widespread Jurassic Jiulongshan and Tiaojishan Formations. The volcanic-rich Tiaojishan Formation is assigned a Middle Jurassic age by the Hebei Bureau of Geology (1989) and Dong (1996), but an early Late Jurassic age by the Beijing Bureau of Geology and Mineral Resources (1991), Chen et al. (1996), and Cui et al. (1996). However, an andesite we dated from low in a sequence of volcanic strata previously mapped as the Jiulongshan and Tiaojishan Formations east of Xinglong in Hebei province (Hebei Bureau of Geology, 1989; Figs. 3B and 4B) yields a latest Late Jurassic hornblende ^{40}Ar-^{39}Ar plateau age of 147.6 ± 2.6 Ma.

Facies changes in Jiulongshan-Tiaojishan sections, like those of any terrestrial volcanic arc, can occur over short distances along strike and can complicate regional correlations that are based on lithology alone. For example, two sections of Jiulongshan and Tiaojishan strata north of Chengde County and only 6 km apart, one along the Luan River and the other to the east along the paved highway leading north to Chengde (city), are dramatically different. The western river section consists primarily of massive volcanic-clast sedimentary breccias, whereas the eastern highway section is mainly mafic lavas, lithic tuffs (some welded), and water-lain tuffaceous sediments. The upper part of this eastern section (just below the Chengde thrust plate; see following) contains a biotite-bearing lithic tuff with an integrated ^{40}Ar-^{39}Ar early Late Jurassic age of 160.7 ± 0.8 Ma (Fig. 4B). More detailed mapping and dating of Mesozoic units in the Yanshan belt are needed to adequately define the chronology of belt events.

Traditional views of the structure of the Yinshan (sensu lato) fold and thrust belt

Long-recognized major structural elements in the belt include multiple generations of folds, thrust and reverse faults, extensional faults, strike-slip faults, and Mesozoic plutons (the latter not confined to the belt). Synclines (and, as we have discovered, a synformal klippe) filled with Jurassic strata and separated by anticlines cored by Archean basement rocks and Middle Proterozoic strata are the most obvious fold structures and economically the most important because of coal beds contained within them. Thrust and reverse faults bordering the synclines and dipping away from their hinges are common structures in the Yinshan belt, and are typically viewed as having limited displacement (e.g., Zhang, 1996; Chen, 1998). Our studies in the Daqing Shan near Baotou, Inner Mongolia (area 4, Fig. 1), indicate that some of the reverse faults represent inverted normal faults of late Paleozoic and/or Mesozoic age (cf. Darby et al., this volume); a similar conclusion has been reached elsewhere in the Yinshan-Yanshan belt by others (e.g., Wang, G., et al., 1995; Dong, 1996; Wang, Y., 1996). In some Yanshan belt areas, e.g., the Western Hills region west of Beijing, multiple generations of overprinted Mesozoic structures with nonparallel trends are very much in evidence (Beijing Bureau of Geology and Mineral Resources, 1991). There is abundant evidence for pre-Yanshanian (Indosinian) Mesozoic deformation in the region (see following).

Common interpretations of Yanshan belt deformation (e.g., Hebei Bureau of Geology, 1989, p. 738; Chen, 1996, 1998; Zhang, 1996; Zhang et al., 1996; Wang, 1996) propose that: (1) involvement of Archean basement rocks in contractional faulting indicates a thick-skinned tectonic style like that of the Laramide Rocky Mountains; (2) most thrust faults steepen downward into the basement, thus limiting horizontal components of slip; (3) block faulting and vertical movements have predominated in the development of the orogen; (4) Mesozoic contraction across the belt was only moderate (20%–30%, according to Wang, 1996, p. 140); and (5) this intracontinental, intraplate belt formed independently of plate interactions in eastern Asia (e.g., Cui and Wu, 1997). We believe, as discussed in the following, that all these interpretations are mistaken.

Our methodology in studying Yanshan belt tectonics

Many papers have been published over the years on the Yanshan segment of the Yinshan belt, and a wealth of geologic and geophysical information can be found in the reports of the

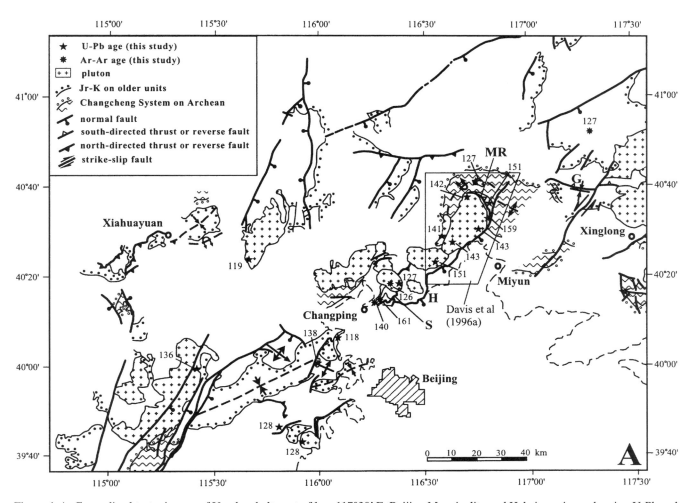

Figure 4. A: Generalized tectonic map of Yanshan belt west of long117°30' E, Beijing Muncipality and Hebei province, showing U-Pb and ^{40}Ar-^{39}Ar age determinations reported here (Tables 1 and 2) and in Davis et al. (1996a, 1998a, 1998b). Structures: H—Hefangkou low-angle normal fault; MR—Miyun Reservoir thrust; G—Gubeikou reverse fault; S—Shishanling thrust. B (page 178): Generalized tectonic map of Yanshan belt east of ~long117° E, Hebei and Liaoning provinces, showing U-Pb and ^{40}Ar-^{39}Ar age determinations reported here (Tables 1 and 2) and in Davis et al. (1996a, 1998a, 1998b), and locations of Figures 6 (A-A', B-B') and 7. Structures: C—Chengde thrust; CC—Chengde Country thrust; G—Gubeikou fault; L—unnamed Liaoning thrust; MA—Malanyu anticline; X—Xinglong thrust. Modified from Davis et al. (1998b, their Fig. 4).

Bureaus of Geology and Mineral Resources of Hebei province (1989), Liaoning province (1989), and the Beijing Muncipality (1991). Mapping of this belt by government-organized teams of Chinese geologists since the founding of the People's Republic has defined rather well the distribution of its major rock units, but in some instances reported here has failed to identify major structural features of the belt. Crustal deformation along the length of the Yinshan belt has been, as outlined in the following, far more intense and complicated than perceived by previous workers. Deciphering the origin of the belt requires an accurate determination of its structural style, kinematics, and timing of deformational events.

Our approach in the Yanshan belt (and in the Yinshan belt of Inner Mongolia as well) has been to focus on structurally complicated regions defined by published geologic mapping, usually at scales of 1:200 000. We have concentrated on selected areas where folds and faults involve Mesozoic strata and cut, or are cut by, Mesozoic plutons. In such areas, radiometric dating of Mesozoic plutonic rocks (U-Pb, Table 1; Figs. 4 and 5) and volcanic rocks (^{40}Ar-^{39}Ar, Table 2; Figs. 4 and 5) has enabled us to establish a preliminary chronology of Yanshanian events (see following). Most of our study areas contain grabens, half-grabens, synclines, or synforms, because Mesozoic strata are preserved in such structures. In intervening horst and anticlinal areas, Mesozoic strata have commonly either been removed by erosion, or were never deposited. We have had greater success in dating plutons than volcanic rocks, in part because many of the Jurassic and Cretaceous volcanic units have been extensively altered.

Some lessons learned from our Yinshan belt field studies may prove helpful to future workers. Future field studies should pay special attention to shallowly dipping contacts. We have

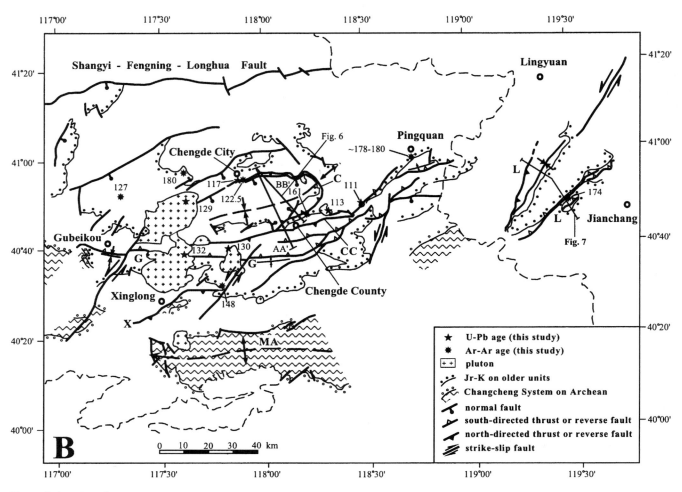

Figure 4. (*continued*)

TABLE 1. U-Pb AGES FROM YANSHAN BELT PLUTONS, NORTH CHINA

PRC #	Pluton name	Age (Ma)	PRC #	Pluton name	Age (Ma)
1	Shatuozi	151+/-2	13	Panjiadian	130.0+/-1.5
2	Dadonggou	127+/-2	14	Shangweidian	138.2+/-1.5
3	Wudaohe	141+/-2	15	Yangfang	118.0+/-1.5
4	Yunmengshan	142+/-2	16	Nanjiao	128.4+/-1.5
5	Shimenshan	143+/-3	17	Dahaituo	119+/-2
6	Yunmengshan	143+/-3	18	Naobaoshang*	119+/-2
7	Changyuan	151+/-2	19	Daguikou	117+/-3
8	Beishicheng	159+/-2	20	Jiashan	113+/-2
9	Xuejiashiliang	127.0+/-1.5	21	Guozhangzi	111+/-4
10	Fangshan	128.5+/-1.5	22	Luozidian†	253+/-3
11	Wulingshan	128.8+/-1.5	23	Wang'anzhen	136+/-2
12	Wulingshan	131.7+/-1.5			

Note: See Davis et al. (1996a) location maps for PRC 1-8. See Figure 3 for all pluton locations by number except for PRC-18 and PRC-22 (see locations below). All analyses by George Gehrels, University of Arizona. U-Pb isochrons and analytical data for all samples are available upon request from G.A. Davis.
*North of Hohhot, Nei Mongolia; N41° 08', E111° 46'.
†Southeast of Chifeng, Nei Mongolia; N42° 00', E119° 05'.

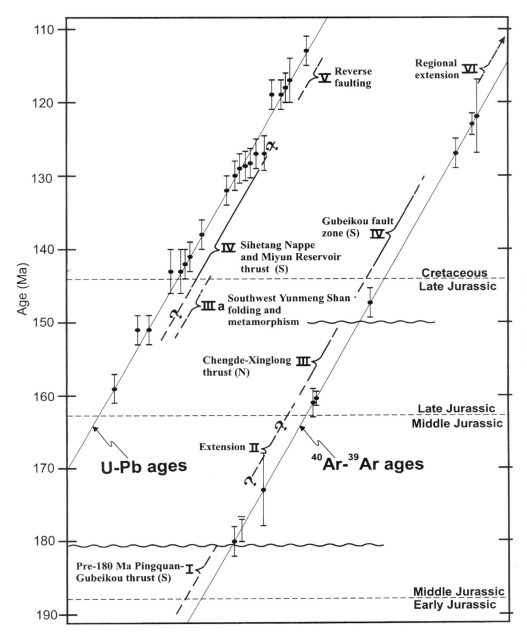

Figure 5. Graphical summary of U-Pb (left) and ^{40}Ar-^{39}Ar (right) age data illustrated in Figure 4. Plot shows age controls on tectonic phases I–VI in Yanshan belt of northern Hebei and western Liaoning provinces.

found numerous examples of unconformities mapped as faults, and faults mapped as unconformities. Folded thrust faults are much more common than previously thought. Because of this, it is an error to assume that the direction of thrusting is always updip. In addition, some shallow-dipping faults are not thrust faults; e.g., the Hefangkou normal fault north of Miyun Reservoir (H, Fig. 3A) dips as shallowly as 20° and is locally characterized by older (Archean) over younger (Proterozoic) rock units. This juxtaposition, interpreted in the past as due to thrust faulting, results from the normal faulting of previously recumbently folded rock units (Davis et al., 1996a, 1996b).

PRELIMINARY TECTONIC CHRONOLOGY OF THE YANSHAN IN NORTHERN HEBEI AND WESTERN LIAONING PROVINCES

Latitudinal deformation in the Yanshan began, at least locally, in the late Paleozoic and Triassic and has been assigned to the Indosinian deformational phase (e.g., Hebei Bureau of Geology, 1989; Wan, 1996; Chen et al., 1996; Cui and Wu, 1997). Widespread Jurassic and Cretaceous (Yanshanian) deformation overprinted and obscured the earlier Indosinian events. A widely accepted view of Jurassic–Cretaceous terrestrial sedimentation

TABLE 2. ^{40}Ar/^{39}Ar AGES FROM MESOZOIC IGNEOUS ROCKS, YANSHAN BELT, NORTH CHINA

Sample #	General locality	Age (Ma)	Rock type (mineral)
H-104*	Daijiagou	127.2 ± 1.5	Perlitic glassy tuff (sanidine)
H-106*	Xiaodonggou	180.2 ± 1.8	Silicic tuff (biotite)
H-220*	Huangtuliang	147.6 ± 1.6	Andesite (hornblende)
96-169*	Zhangjiadian	160.7 ± 0.8	Lithic tuff (biotite)
B-106†	Shisanling	161.1 ± 1.9	Andesite (hornblende)
PRC-9†	Xuejiashiliang	125.8 ± 1.3	Diorite (hornblende)
PRC-14†	Shangweidian	135.2 ± 3.8	Granodiorite (biotite)
37	Inner Mongolia (Yinshan)	123.3 ± 1.2	Rhyolite (whole rock)
191*	Chengde	122.5 ± 5.5	Trachyte (K-spar)
216*	Pingquan	~178-180	Silicic tuff (biotite)
280*	Jianchang	173 ± 6	Andesite (whole rock)
8-19-16a†	Shisanling	140 ± 6	Yellow gouge along Shisanling thrust fault

Note: ^{40}Ar/^{39}Ar age spectra and analytical data for all samples are available upon request from G.A. Davis.
*See Figure 3B for sample locations.
†See Figure 3A.

and volcanism in Yanshan areas of Figures 2, 3, and 4 is that deposition was punctuated or interrupted by major phases of deformation at the ends of the Early, Middle, and Late Jurassic (e.g., Hebei Bureau of Geology, 1989). Unfortunately, well-defined age controls on these phases of deformation have been lacking. K-Ar ages for igneous rocks are numerous (e.g., Tables 2 and 3 for the Beijing Municipality, *in* Chen et al., 1996), but many of the Jurassic–Cretaceous volcanic rocks in the Yinshan belt are too altered for reliable K-Ar dating. In addition, despite the fact that K-Ar pluton ages are cooling, not crystallization, ages, we have found a number of instances in which K-Ar ages on biotite (and one hornblende) from plutons in the area of Figure 3 are older than our U-Pb (zircon) crystallization ages of the same plutons (Table 3). Clearly, there is considerable uncertainty regarding the age and correlation of igneous rocks in the Yinshan belt.

We have begun to clarify the history of Yanshan deformation by dating Mesozoic plutons and volcanic strata using U-Pb

TABLE 3. DISCREPANCIES BETWEEN EARLIER YANSHAN BELT PLUTON AGE DETERMINATIONS AND RESULTS FROM THIS STUDY

Pluton name	Earlier work*	Method	This study†
Shangweidian (PRC-14)	159.4	U-Pb, Zircon	138.2 ± 1.5 135.2 ± 3.8
Yangfang (PRC-15)	124.3	K-Ar, biotite	118.0 ± 1.5
Xuejiashiliang (PRC-9)	152.8	K-Ar, biotite	127.0 ± 1.5 125.8 ± 1.3
Fangshan (PRC-10)	133.9–142.9 133.6 ± 0.7	K-Ar, biotite Ar-Ar, biotite	128.5 ± 1.5
Dahaituo (PRC-17)	132.7	K-Ar, biotite	119 ± 2
Nanjiao (PRC-16)	207	K-Ar, biotite	128.4 ± 1.5

Note:

*See Beijing Bureau of Geology and Mineral Resources, 1991.
†All the ages listed here are U-Pb (zircon) determinations except Shangweidian 135.2 Ma Ar-Ar (biotite), and Xuejiashiliang 125.8 Ma Ar-Ar (hornblende).

and ^{40}Ar-^{39}Ar age determinations, respectively. As of January 1999, 20 U-Pb age determinations were made by one of us (Gehrels) on plutons within the Yanshan (including the Western Hills); all analyses were conducted using conventional isotope dilution and thermal ionization mass spectrometry, as described by Gehrels and Boghossian (2000). In addition, 12 ^{40}Ar-^{39}Ar age determinations were completed in K. Hodge's CLAIR facility (Cambridge Laboratory for Argon Isotopic Research) at the Massachusetts Institute of Technology (Table 2), 9 on volcanic rocks, 2 on plutons, and 1 on fault gouge; the analyses were performed by incremental heating of mineral separates and 1 whole-rock sample in a resistance furnace as described in Hodges et al. (1994).

These age determinations, and field work through August 1999, have documented a fascinating history of extensional and contractional deformation in the Yanshan, the latter characterized by alternating vergence. We contend that the following chronology of events, although preliminary, represents a greatly improved understanding of the history of Yanshan deformation within the northern Hebei and western Liaoning provinces (Figs. 3B, 4B, and 5). It is important to emphasize that diachronism is to be expected in orogenic belts, and that the chronology reported here may not be applicable to more westerly Yanshan and Yinshan (sensu stricto) areas.

I. Pre-Middle Jurassic (Indosinian or early Yanshanian) south-vergent thrust faulting (pre-180 Ma)

A large region of northern Hebei province and western Liaoning province experienced south-directed thrusting and folding in what was perhaps the most intense and, to date, least understood phase of Mesozoic deformation in the Yanshan belt. Field work in the area of Chengde County (a municipality; Fig. 3B) led to our recognition of a major pre-Middle Jurassic south-directed thrust fault, which we called the "Unnamed thrust" (Davis et al., 1998a, their Fig. 2). The amount of southward displacement along this thrust fault (Fig. 1, area 2; Fig. 6, Unnamed thrust) is not known, but it must have been very large. The existence of the thrust is necessitated by a profound north-south difference in the regional geology of the Yanshan belt, as well as by the southward vergence of overturned upper plate and lower plate strata. To the south, in the lower plate of the thrust, all Proterozoic, Paleozoic, and early Mesozoic stratigraphic units of the Yanshan belt are widely preserved, but to the north, in the upper plate, all Proterozoic and Paleozoic strata are missing.

Unbeknownst to Davis et al. (1998a), Zhao (1990) recognized the importance of "Indosinian" south-directed thrusting in this region along a structure he called the Pingquan-Kubeiko (Gubeikou) fault, named for localities along its inferred eastern and western trace, respectively (Fig. 3B). In Zhao's words (1990, p. 12), "the large-scale overthrusting of the hanging wall of the Pingquan-Kubeiko fault resulted in the complete denudation of the covers [sic] and extensive exposure of the crystalline basement." Across a broad region in northern Hebei province, lower Middle Jurassic and younger strata directly overlie allochthonous Archean basement, a relationship that can be explained by early Mesozoic (pre-180 Ma) erosion of all

supracrustal units above Archean basement rocks in the thrust sheet (Figs. 7 and 8). For example, west of Chengde (Figs. 2 and 3B) a silicic volcanic unit only a few tens of meters above Archean basement (see Ren et al., 1996, near their stop 4) yields an early Middle Jurassic age of 180.2 ± 1.8 Ma (biotite ^{40}Ar-^{39}Ar). Basal Middle Jurassic volcanic strata of similar age (ca. 178–180 Ma) overlie deformed lower plate rocks as well, thus establishing an upper age limit for the Unnamed thrust (Fig. 5). Near Pingquan (Fig. 3B), isoclinally folded and locally south-overturned Triassic sandstone and conglomerate in the footwall of the thrust are overlain unconformably by steeply dipping early Middle Jurassic strata. Biotite from a silicic tuff that is just a few meters above the conglomeratic base of the Jurassic section has a "very complex" Ar-Ar spectrum "with a very crude mean age of about 178–180 Ma" (K. Hodges, 1999, written commun., Fig. 3B). An andesite of similar age in the lowermost Jurassic unit (Nandaling Formation) of the Western Hills, west of Beijing, appears to substantiate the regionality of onset of early Middle Jurassic volcanism in the Yanshan belt (177.5 Ma, whole-rock K-Ar, Beijing Bureau of Geology and Mineral Resources, 1991; Chen et al., 1996).

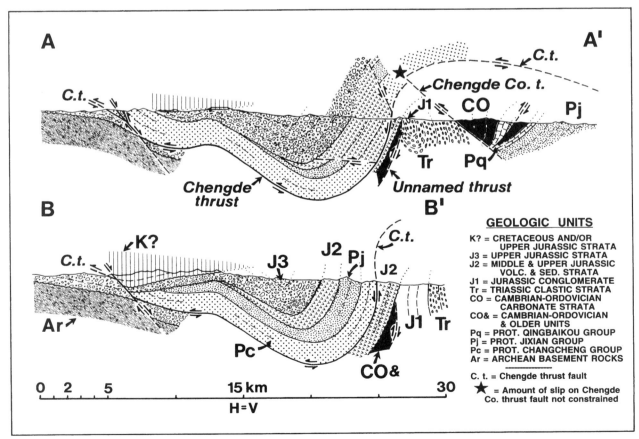

Figure 6. True-scale geologic cross sections across synformally folded Chengde thrust fault. For locations of sections see Figure 4B. Sections are based on prior mapping (Hebei Bureau of Geology, 1989, scales 1:500 000; 1:200 000) and our field work along lines of section. Highest elevation along section B–B′ =1000 m. From Davis et al. (1998a).

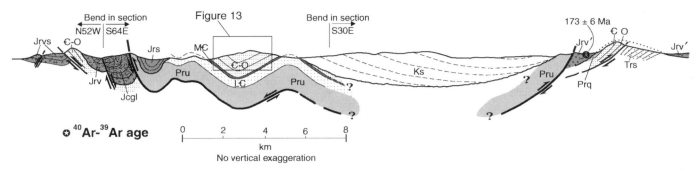

Figure 7. Geologic cross section across unnamed north- and south-directed thrust faults (L, Fig. 4B; section 1, Fig. 1) in western Liaoning province between Lingyuan and Jianchang. Geologic units from northwest to southeast include: Jrvs—Jurassic sedimentary and volcanic rocks; Є-O—Cambrian and Ordovician carbonate strata; Jrv—Jurassic volcanic rocks; Jcgl—Jurassic carbonate clast sedimentary breccia and conglomerate; Pru—undifferentiated Proterozoic strata; Jrs—Jurassic sedimentary rocks; Prq—Proterozoic Qinbaikou strata; LЄ and MЄ—Lower Cambrian and Middle Cambrian strata; Ks—Cretaceous sedimentary rocks; Jrv′—Middle Jurassic volcanic rocks (may or may not be equivalent to northwestern Jrv); Trs—Triassic(?) conglomerate and redbed sandstone. Modified from Davis et al. (1998b).

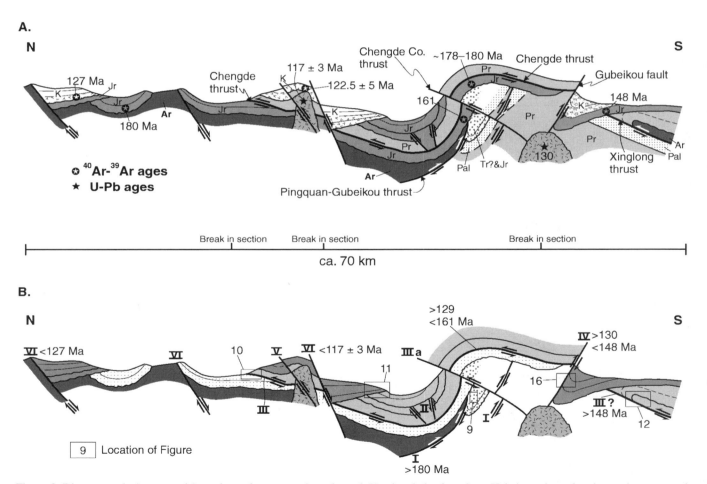

Figure 8. Diagrammatic (not to scale) north-south cross sections through Yanshan belt of northern Hebei province showing major structural and stratigraphic units. Ar—Archean crystalline rocks; Pr—Proterozoic strata; Pal—Cambrian and Ordovician carbonate strata; Tr ? and Jr—Triassic(?) and Lower Jurassic conglomerate and sandstone; Jr—Middle and Upper Jurassic strata; K—Lower Cretaceous strata. Central third of sections show relationships seen in Figure 6. Northern third represents geology west of Chengde (see Fig. 4B for location of U-Pb and ^{40}Ar-^{39}Ar localities). Southern third crosses Gubeikou fault in vicinity of Xinglong (Fig. 4B). A: Simplified composite geologic section showing major fault and fold structures and U-Pb and Ar-Ar age controls on these structures (Davis et al., 1996a, 1998a, this paper). B: Bracketing ages for major Yanshan belt structures formed during deformational phases I through VI (this paper).

Zhao (1990, his Fig. 1) mistakingly located the Pingquan-Kubeiko fault just south of Chengde, along or near the structural boundary we designate as the leading edge of the Late Jurassic, north-directed Chengde thrust (see following; Davis et al., 1998a). The Proterozoic strata Zhao assigned to the footwall of the Pingquan-Gubeikou fault in this area are allochthonous, and are in the upper plate of the younger Chengde thrust (Figs. 4B and 6). In recognition of Zhao's regional insights in early Mesozoic deformation, we adopt the term Pingquan-Gubeikou fault for our "unnamed thrust" (Davis et al., 1998a) of Figure 6. The inferred trace of this fault is hidden beneath Middle and lower Upper Jurassic strata several kilometers north and northwest of Chengde County municipality and the still younger, north-directed Chengde thrust plate (see following). Its general location in this area is, however, defined by the close proximity of synclinally folded, southward-overturned footwall strata (Early Jurassic and older) and anticlinally folded, southward-overturned hanging-wall units (Cambrian-Ordovician and Proterozoic strata, Archean basement). To the west near Gubeikou and to the east between Chengde County and Pingquan (Fig. 3B), the original geometry of the Pingquan-Gubeikou thrust has been further obscured by probable Early Cretaceous, south-directed reverse faulting (Gubeiko zone, G, Fig. 3B).

Thick (>3 km) pre-Middle Jurassic (>178–180 Ma) foredeep deposits of conglomerate, which we and Zhou (1990) regard as syntectonic, are strongly overturned beneath the south-directed Pingquan-Gubeikou thrust (Figs. 6, A-A′ and 9). These footwall conglomerates (mapped as Triassic and Lower Jurassic on the Hebei Bureau of Geology map, 1989, scale 1:500 000) are widely exposed in a footwall syncline between Chengde County and Pingquan to the east. Well-rounded cobbles and boulders of Archean basement rocks, Proterozoic quartzite, and Paleozoic carbonate in the conglomerate were almost certainly derived from the Pingquan-Gubeikou thrust plate to the north, because Archean basement in the footwall of the thrust to the south still has its Proterozoic (and higher) sedimentary cover. The conglomerate contains an inverted clast stratigraphy that reflects erosional unroofing of the thrust fault's upper plate (e.g., DeCelles et al., 1987). Jurassic(?) volcanic and Paleozoic and Proterozoic carbonate clasts are more abundant in lower parts of the conglomeratic sequence, whereas higher strata contain, almost exclusively, cobbles and boulders of lower units; i.e., Proterozoic quartzites and Archean gneisses (Fig. 9).

Another south-directed thrust, possibly correlative with the Pingquan-Gubeikou fault, also unconformably underlies Middle Jurassic volcanic strata in western Liaoning province west and northwest of Jianchang (Figs. 1, area 1, and Fig. 7). As seen in an erosional inlier along the southeastern end of the section, an overturned section of Neoproterozoic, Cambrian, and Ordovician carbonate has been thrust southward over Triassic conglomerate and sandstone; the plate is overlain by Middle Jurassic andesite (Lanqi Formation, 173 ± 6 Ma; plagioclase ^{40}Ar-^{39}Ar isochron; K. Hodges, 1999, written commun.) and basalt (Figs. 4B and 7).

Figure 9. Triassic or Lower Jurassic syntectonic conglomerates in lower plate of Pingquan-Gubeikou thrust along road 2–3 km east of Chengde County. View is perpendicular to vertical bedding in plane of photograph. Cobbles and boulders are derived from Proterozoic quartzite and Archean gneiss. See Figure 8B for representative location.

These Jurassic strata have in turn been overriden by a younger southeast-directed thrust plate (see following).

The Zhujiagou thrust fault south of Xinglong was described by Chen (1998) as a north-directed thrust within folded Proterozoic strata. This thrust or reverse fault of limited displacement lies in the complexly folded and faulted northern limb of the Archean basement-cored Malanyu anticline (MA, Fig. 3B). Its significance to this discussion is that it and the deformed northern limb of the anticline are intruded by the Wangpingshi granitic pluton (K-Ar, 180 Ma, Hebei Bureau of Geology, 1989, magmatic rocks map; Chen, 1998). Although the crystallization age of the Wangpingshi pluton is in question, Zhao et al. (1999) reported various Rb-Sr and model Nd early Middle Jurassic ages from other plutons that intrude the anticline. This regional anticline and its northern limb contractional structures may be candidates for phase I deformation.

Early Middle Jurassic to Cretaceous transitions in the magmatic character of the Yanshan belt

Most synopses of Yanshan belt magmatism report it as beginning in the Early Jurassic, ca. 195–185 Ma (e.g., Xu, 1990; Beijing Bureau of Geology and Mineral Resources, 1991), climaxing in the Late Jurassic (ca. 149–144 Ma, Hebei Bureau of Geology, 1989) or Late Jurassic to Early Cretaceous (147–132 Ma, Xu, 1990), and waning in the middle to Late

Cretaceous. Li (1996) reported that volcanic and plutonic activity in western Liaoning province began at 211 and 214 Ma, respectively. These various chronologies are almost exclusively based on K-Ar and Rb-Sr dating of both plutons and volcanic rocks. K-Ar ages for Yanshan plutons are particularly suspect because they are not crystallization ages. For example, several plutons we dated in the Yunmeng Shan that have U-Pb ages of 141–143 Ma yield younger K-Ar biotite ages of 124–114 Ma (Beijing Bureau of Geology and Mineral Resources, 1991). However, a number of other published K-Ar pluton ages are too old; e.g., the Xuejiashiliang pluton north-northeast of Changping and the Nanjiao pluton in the Western Hills of Beijing give K-Ar ages that are considerably older than our U-Pb zircon ages (Figs. 3, 4, and 5; cf. Table 3).

As illustrated in Figures 4 and 5, most of the granitic plutons we have dated in the Yanshan are Cretaceous in age (ca. 138–113 Ma), although we have found a plexus of Late Jurassic and Jurassic–Cretaceous plutons ranging in age from 159 to 141 Ma in the Yunmeng Shan. One explanation for the apparent localization of Jurassic plutons in the Yunmeng Shan is simply that Jurassic and Jurassic–Cretaceous intrusions in the area of Figure 4 were not widespread. Another explanation, which we prefer, is that there may be depth controls on the exposure of Jurassic and Cretaceous plutons in the Yanshan belt. We see the deepest structural levels of the belt in the Yunmeng Shan because that range is in the footwall of a Cretaceous metamorphic core complex (Davis et al., 1996a, 1996b). Major crustal extension along the Hefangkou detachment fault (see following) has tectonically exhumed mid-crustal rocks, including ductilely deformed Archean basement gneiss and its Proterozoic cover in the Sihetang nappe (described in the following; deformational phase IV). The Late Jurassic and Jurassic–Cretaceous plutons of the Yunmeng Shan (ca. 159–141 Ma) are also ductilely deformed below the Sihetang nappe, as is an Early Cretaceous stock in the overturned limb of the nappe (127 ± 2 Ma; PRC-2, Fig. 3A; Table 1); foliation and lineation in this Cretaceous pluton, however, are nonpenetratively developed. In contrast, Cretaceous plutons (132 Ma and younger) elsewhere in the area of Figure 4 cut across higher level structures of deformational phases I and III and lack internal ductile deformation.

Two Yunmeng Shan plutons (159–151 Ma, U-Pb; Davis et al., 1996a; Fig. 3B), including the Changyuan pluton described herein (deformational phase IIIa), are mafic (dioritic to gabbroic), primitive, and presumably mantle derived, but five granodioritic plutons in the range (151–127 Ma) exhibit considerable contamination by the crust. This major change in magma chemistry ca. 151 Ma—previously recognized by the Beijing Bureau of Geology and Mineral Resources (1991), but placed by it at 145 Ma on the basis of K-Ar dating—may be a response to tectonically induced crustal thickening in the Yanshan related to contractional deformational phases III and/or IV (see following). Geologists of the Beijing Municipality previously concluded that such changes represent a chemical evolution of Yanshan granitoids from melts of subcrustal or lower crustal

derivation to later anatectic melts from the upper crust (Beijing Bureau of Geology and Mineral Resources, 1991, p. 579–580). The youngest Cretaceous plutons in the Chengde–Chengde County–Pingquan area (ca. 120–110 Ma; Fig. 4B) are generally more alkalic (syenite-syenogranite) than the Yunmeng Shan granodiorites—possibly a reflection of a temporal relationship to crustal extension (deformational phase VI).

It is surprising that we have not yet identified Yanshan plutons that are age equivalent to the early Middle Jurassic volcanic rocks (ca. 180–173 Ma) we have dated in Hebei and Liaoning provinces, but their existence has been reported by others relying on K-Ar and Rb-Sr dating (e.g., ca. 185–170 Ma, Hebei Bureau of Geology, 1989; Zhao, 1990; Chen, 1998; Zhao et al., 1999). Although a minority of geologists proposed that eastern China was unaffected by continental margin subduction in the Mesozoic (cf. Zhang et al., 1989), most accept that the overall patterns of Mesozoic magmatism in eastern China indicate such interaction (e.g., Xu, 1990; Zhao et al., 1994; Deng et al., 1999). Isozaki (1997) reported that Pacific oceanic subduction beneath proto-Japan and northeastern China had begun by Middle Jurassic time, ca. 180 Ma, a conclusion broadly compatible with the literature-based chronology of Yanshan belt magmatism and our own geochronological findings.

II. Middle Jurassic and/or early Late Jurassic extensional faulting

Our studies have defined a previously unrecognized episode of Middle Jurassic and/or early Late Jurassic extension in northern Hebei province south of Chengde and north of Chengde County. Here, the synformally folded Late Jurassic Chengde thrust plate (event III) carries an allochthonous sequence of Proterozoic and Middle and Upper Jurassic strata (C, Figs. 3B and 6). However, on the south limb of the synform (Fig. 6, A-A') the upper plate Proterozoic section is locally missing, and Middle Jurassic to lower Upper Jurassic volcanic strata of the plate are in fault contact with broadly age-equivalent strata of the lower plate; the fault is exposed on the west bank of the Luan river north of Chengde County. The upper plate strata are within a previously unrecognized major north-trending graben that is now confined to the Chengde plate. Its bounding normal faults have stratigraphic throws >3 km (Fig. 6, A-A'). The westernmost, northwest- to north-northwest–striking fault is well exposed on a trail above the village of Jiangjiazhuang, about 10 km westsouthwest of Chengde County, and is bordered by hanging-wall gouge, locally at least 15 m thick. It is our interpretation that the graben contained Middle and lower Upper Jurassic and Proterozoic strata prior to formation of the Chengde thrust. During subsequent thrust faulting the graben was truncated from below, thus cutting out the downdropped Proterozoic strata. Upper plate Upper Jurassic rocks of the Houcheng Formation appear to postdate graben formation. Following thrusting, the Chengde plate and the graben within it were rotated ~90° by postthrusting folding (Chengde synform, Fig. 6).

III. Late Jurassic–earliest Cretaceous(?) northward displacement of the Chengde thrust, northern Hebei province (post-161 Ma, pre-148 Ma?, pre-131 Ma)

Major contraction occurred within the Yinshan-Yanshan fold and thrust belt in Late Jurassic and possibly Early Cretaceous time. We have recently discovered two major north-northwest–directed, low-angle thrust faults with displacements in excess of 25–30 km: the Chengde thrust in northern Hebei province (area 2, Fig. 1; C of Fig. 4B; Fig. 6); and the Daqing Shan thrust in Inner Mongolia (area 3, Fig. 1; Davis et al., 1998c; Zheng et al., 1998). Both of these thrust faults have been folded and both are now separated from southern root zones by the effects of later deformation. The folded Chengde thrust was not recognized as such by earlier workers, and portions of the Daqing Shan thrust had been erroneously mapped as Jurassic over Proterozoic unconformities (Nei Mongol Bureau, 1991).

Chengde thrust fault, Hebei province. The Chengde thrust fault (C, Fig. 3B) carries Proterozoic and Jurassic strata (J2, J3, Fig. 6), overrides Middle and early Late Jurassic J2 strata (ca. 180–161 Ma; ^{40}Ar-^{39}Ar volcanic ages, Figs. 5 and 8), and is intruded in western areas by plutons as old as 132 Ma (Fig. 4B). Displacement of this allochthon was to the north-northwest, according to abundant kinematic indicators along the northern trace of the fault (Fig. 10). Along much of its exposed trace, the brittle Chengde thrust was stratigraphically controlled by initial detachment of Proterozoic and younger strata from lowermost units of the lower Middle Proterozoic Changcheng System (Figs. 5, 8, and 9). The most favored and widespread detachment horizon was in Proterozoic Chuanlinggou pelites that

are above the thick (to 860 m) Changzhougou basal quartzite. Overall, the Chengde thrust resembles in its stratigraphically controlled geometry the sled-runner, thin-skinned thrust faults typical of foreland fold and thrust belts. This geometry is complicated, however, by prethrusting deformation of now allochthonous units, e.g., the contractional deformation of phase I and the extensional deformation of phase II. It should not be surprising, therefore, that the thrust plate has a complicated internal geometry. Locally, for example, in southwesternmost exposures of the synformal klippe, the allochthon contains the entire Changcheng section and the Archean basement on which it was deposited (Figs. 3B and 4B).

Following its emplacement, the Chengde thrust plate was folded in a major east- to east-northeast–trending asymmetric synform with a very steep to steeply overturned southern limb. The synform is a major Yanshan structure with a north-south width of 40 km or more. Its southern limb is offset by the younger north-directed Chengde County thrust (CC, Fig. 3B; Figs. 6, A-A′, and 8). The age of the Chengde County thrust and the amount of displacement along it are not known, but the synform is probably a Late Jurassic or earliest Cretaceous structure. It is intruded to the west by granitic plutons with U/Pb ages ranging from 128.8 ± 1.5 Ma (PRC-11) to 131.7 ± 1.5 Ma (PRC 12). Its gently south dipping northern limb is overlain with angular unconformity by gently north dipping Cretaceous volcanic and clastic units (Figs. 6, 8, and 11). K-feldspar from a trachyte flow in this section gives a ^{40}Ar-^{39}Ar plateau age of 122.5 ± 5.5 Ma (Fig. 4B; Table 2).

The folded Chengde thrust must root south of its footwall units in the southern limb of the synform (Figs. 6, A-A′, and 8).

Figure 10. View east of leading edge of Chengde thrust plate, southwest of Chengde and ~1 km north of Daguikou pluton (117 ± 3 Ma; Fig. 3B). See Figure 8B for representative location. Proterozoic carbonate rocks of Changcheng System (Prc; Tuanshanzi and Dahongyu Formations) overlie horizontal lower Upper Jurassic sandstone and pebbly conglomerate strata (J; Tuchengzi Formation). Carbonate strata of Dahongyu Formation are subhorizontal beneath highest peak, but dip northward at ~55° into shallow south-dipping(~15°) thrust along left skyline. Height of hill is ~100 m.

Figure 11. View to east of angular unconformity across northern limb of Chengde synform. In foreground are south-dipping (~20°) sandstone and pebbly conglomerate strata of Upper Tuchengzi Formation (UJ). In background are north-dipping Cretaceous volcanic and volcaniclastic units (K; ca. 120 Ma). This Cretaceous section has been previously mapped as Zhangjiakou Formation of Upper Jurassic age (Ren et al., 1996). See Figure 6, B–B′ for this relationship and Figure 8B for its representative location.

Cross sections through the klippe and its southern footwall indicate that the minimum northward displacement of the folded Chengde thrust plate is 40–45 km (the line length of the folded thrust, Fig. 6, A-A'). The eventual rooting of such a major thrust into Archean basement rocks is a necessary consequence of the large displacement of its upper plate and does not constitute a basis for considering that the Yinshan structural style is thick-skinned (Davis, 1999). Davis et al. (1998a, 1998b) inferred that the well-known Xinglong thrust (X, Figs. 3B and 8) is the root structure of the Chengde thrust. The Xinglong thrust is unconformably overlain by Jurassic strata. A hornblende andesite several hundred meters above the unconformity has a latest Late Jurassic ^{40}Ar-^{39}Ar age of 147.6 ± 1.6 Ma (Figs. 4B and 8; Davis et al., 1998a, 1998b). If our Chengde-Xinglong thrust correlation is correct, the fault is Late Jurassic in age (<161 Ma, >148 Ma); age controls on the Chengde thrust alone are not as well defined (<161 Ma, >132 Ma; Fig. 8).

Our correlation of the Chengde and Xinglong thrust faults is controversial. Zhang (1996; Davis et al., 1998a, 1998b) and Chen (1998) interpreted the Xinglong thrust to be a thick-skinned reverse fault similar to those of the Laramide Rocky Mountains because it carries Archean basement rocks that, with their Proterozoic and Paleozoic sedimentary cover, compose a large anticline overturned to the north (Fig. 12). The original (and present) angle of southward dip of this thrust beneath its upper plate Archean basement rocks is not known with certainty, although borehole data through cover units in the northern overturned limb of the anticline indicate that here the underlying thrust fault dips <15° to the south (Jiang et al., 1997; Chen, 1998).

Western Liaoning province thrust faults. Late Jurassic and/or Early Cretacous thrust faulting in the area between Luanping and Jianchang in western Liaoning province is represented by several northwest- and southeast-directed thrust plates carrying Proterozoic and younger strata over Jurassic units (Figs. 3B, 7, and 13). Davis et al. (1998b, their Fig. 5) and Davis (1999, his Fig. 9) erroneously considered that two divergent thrust faults between Lingyuan and Jianchang defined a synformally folded thrust plate with displacement to the northwest—a structure analogous to the Chengde synformal klippe. Additional studies by Zhang Changhou, Zheng Yadong, and Davis in 1999 refute this interpretation (Fig. 7).

The thrust plate southeast of the synclinal hinge (Figs. 4B and 13) exhibits spectacular southeast-vergence of asymmetric, overturned, and recumbent folds in upper plate Proterozoic carbonate strata both north and south of the line of section of Figure 7. Along the line of section (Fig. 7), the plate overrides Middle Jurassic strata that unconformably overlie an older south-directed thrust plate of deformational phase I. Dating of plagioclase in a pillowed andesite porphyry (Lanqi Formation) that is ~100–150 m above the Middle Jurassic–Proterozoic unconformity yields a "very reliable" Ar-Ar isochron age of 173 ± 6 Ma (Figs. 4B and 7; K. Hodges, 1999, written commun.). The younger southeast-directed thrust and both of its plates are overlain unconformably by widespread Cretaceous clastic strata (Liaoning Bureau, 1989).

Two southeast-dipping thrusts lie northwest of the synformal hinge (Figs. 7 and 13), one carrying Cambrian and Ordovician carbonate strata, and a higher, very steep fault carrying Proterozoic strata of the Changcheng Group. Both thrust plates were displaced northwestward across probable Jurassic units, and both appear to have been involved in the synformal folding. A spectacularly thick section of carbonate clast sedimentary breccia and conglomerate is in the footwall of the thrust that carries Proterozoic strata (Fig. 7). This section, which we interpret as syntectonic foredeep deposits derived from the northwest-advancing thrust plate, is tightly synclinally folded and locally

Figure 12. View east of overturned contact between basal quartzite cobble conglomerates of Proterozoic Changcheng System (Prc; Changzhougou Formation) and Archean gneiss (Ar). Contact is essentially a sheared nonconformity as seen by discordant dips of quartzite strata into it along highest ridge. This overturned fold is in upper plate of Xinglong thrust fault, which is at shallow depth in this area; see Figure 8B for representative location. Height of hill in foreground is ~125 m.

Figure 13. View to northwest of synclinal hinge in Ordovician carbonate strata, western Liaoning province between Lingyuan and Jianchang (see Figs. 1 [area 1], 3B, and 7 for location). Width of field of view ~2 km.

overturned beneath the thrust. Geometric and temporal relationships between the divergent thrusts just described are not known. Unfortunately, the area to the southwest where they converge is a restricted area closed to geologic investigation.

Shisanling thrust. The Shisanling thrust north of Changping in Beijing Municipality (S, Fig. 3A) is probably another fault formed during deformational phase III. Upper plate Proterozoic strata (Wumishan Formation) override Middle Jurassic andesitic rocks (161.1 ± 1.9 Ma, ^{40}Ar-^{39}Ar, hornblende; Fig. 3B). This north- to northeast-directed thrust fault is cut by a pluton we dated as 127.0 ± 1.5 Ma (U-Pb, zircon; PRC-9, Fig. 3B; Table 1), and possibly by the 151 ± 2 Ma Changyuan pluton (PRC-7, Fig. 3B). Gouge from one locality along the thrust (Fengshan quarry; Cui et al., 1996, their stop 2) has yielded a "very reliable" ^{40}Ar-^{39}Ar whole rock plateau age of 140 ± 6 Ma (K. Hodges, 1999, written commun.), but does not necessarily define the time of thrusting.

IIIa. Late Jurassic folding and metamorphism, southwestern Yunmeng Shan (<151 Ma, >143 Ma)

Davis et al. (1996a) presented evidence for a Yanshan-atypical folding and metamorphic event in the southwestern Yunmeng Shan that appears to be assignable to deformational phase III. In this area, ~7–12 km west-southwest of Hefangkou, is the Late Jurassic Changyuan pluton, now a gneissic biotite-hornblende metadiorite (Davis et al., 1996a, their Fig. 13.11). The Changyuan pluton (PRC-7, Fig. 3A) has a U-Pb crystallization age of 151 ± 2 Ma (Table 1) and shares tectonite fabric elements with its isoclinally folded country rocks. A southwest-plunging isoclinal to subisoclinal anticline borders and lies parallel to the southeastern contact of the pluton. Proterozoic Jixian(?) System marble and calc-silicate rocks compose the flanks of the anticline, which is cored by Archean(?) amphibolitic gneiss. A steep gneissic foliation and a southwest-plunging hornblende lineation in the metadiorite are parallel, respectively, to the axial surface foliation of the fold and a fold hinge–parallel tremolite lineation in its marble.

This deformation is well determined to be Late Jurassic because the foliated metadiorite is intruded sharply by the Wudaohe (or Mutianyu) and Yunmeng Shan granodioritic plutons (PRC 3-6, 141–143 Ma; Fig. 3A); its contact with the Yunmeng Shan pluton is strongly discordant (cf. Davis et al., 1996a, their Fig. 13.11). Although these younger plutons have localized mylonitic and weak tectonite fabrics, they are not penetratively deformed. The knife-sharp contact between undeformed granitic rocks of the Wudaohe granodiorite and the Changyuan dioritic gneiss is well exposed along the old highway to the Mutianyu Great Wall just north of the village of Weidian.

The regional extent and significance of this deformation has not yet been determined. However, this terrane of amphibolite-grade metamorphism and tight folding is in the exhumed footwall of the low-angle Hefangkou detachment fault (see following) and is overprinted by mylonitic fabrics in that footwall (Davis et al., 1996a, their Fig. 13.11). This Late Jurassic terrane predates the

metamorphism, recumbent folding, and thrusting of the Sihetang nappe (deformational phase IV) that involves the 141–143 Ma Yunmeng Shan plutons. The terrane appears to be in the upright limb of that nappe. It and the younger Sihetang event provide important clues to the middle Mesozoic thermal state of levels of the Yanshan crust deeper than are normally seen.

Evolution of the contractional structures of deformational phases I and III

Figure 14 is a preliminary interpretation of the events leading to the complicated geometry of older south-directed and younger north-directed thrust faulting seen in the cross sections of Figures 6 and 8. The major southward thrusting of pre-Middle Jurassic age (pre-180 Ma) in the Yanshan belt of northeastern Hebei and western Liaoning provinces may have developed as (1) a backarc, foreland fold and thrust belt of U.S. Cordilleran type formed during southward subduction along the northern edge of the Archean craton, or as (2) a consequence of the collision of a Mongolian arc (or arcs) against an Andean-style arc along the Suolun-Linxi zone. Numerous workers have concluded that subduction preceding Late Permian to Triassic amalgamation of the Mongolian arc terranes with the Archean

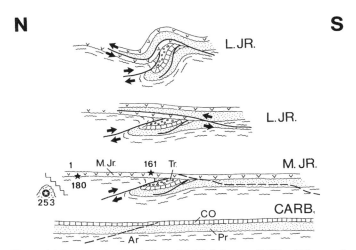

Figure 14. Interpretive evolution of folded thrusts (I, III) of Figures 6 and 8. In lower panel for Carboniferous (CARB.) time, dashed line is future trace of Pingquan-Gubeikou south-directed thrust involving Archean basement and supracrustal cover units (Pr =Proterozoic strata; CO=Cambrian and Ordovician strata). In Middle Jurassic time (M.JR.), pre-180 Ma displacement along thrust fault is accompanied by foredeep deposition of thick conglomerates (Tr =Triassic and/or Lower Jurassic strata; Fig. 9). The 253 Ma Luozidian pluton, located 100 km north of section line near Chifeng, Inner Mongolia (Table 1), is part of Permian–Triassic magmatic belt south of Suolun-Linxi suture. Middle Jurassic volcanic rocks (M. Jr.) were deposited across both deeply eroded upper plate and southern, lower plate. The dashed line in M. JR. panel is future trace of north-directed Chengde thrust. Formation of this thrust in Late Jurassic time (L.JR.)was followed by north-vergent synformal folding of both major thrusts. Northward offset of steep southern limb of synform by Chengde County thrust (Figs. 6,A-A′ and 8) is not shown.

Figure 15. View north of recumbent anticlinal fold on attenuated lower limb of Sihetang nappe, Yunmeng Shan, north of Beijing and east of village of Daguanqiao (cf. Fig. 13.5 in Davis et al., 1996a, for geologic map of this area). Upright, ductilely deformed Changzhougou Formation basal conglomerate (Prc) caps top of hill and unconformably overlies slope-forming Archean gneiss (Ar). Recumbent fold closes to left (west) and Great Wall watchtower (below arrow) was built on ledge of overturned basal Changzhougou conglomerate. Archean gneiss in core of fold was intruded by 151 Ma granodiorite pluton (gr; PRC-1, Table 1) that supports rocky cliffs in middle and right ground. Ductile Miyun Reservoir thrust underlies ledge of overturned, metamorphosed Changzhougou strata and separates it from highly foliated lower plate metadiorites (159 Ma; PRC-8, Fig. 3A). All rocks in this view are strongly foliated parallel to axial surface of fold. They and Miyun Reservoir fault are within 6-km-thick Sihetang gneissic shear zone. Great Wall watchtower is ~10–12 m high.

North China craton along the Suolun-Linxi suture was southward (Fig. 1; Zhang et al., 1984; Wang and Liu, 1986; Zhao et al., 1994; Yin and Nie, 1996; Zheng et al., 1996). This conclusion is based on the widespread occurrence of plutons ranging in age from 285 to 217 Ma (Cui and Wu, 1997) south of the suture in Hebei and Liaoning provinces and Inner Mongolia. One of us (Gehrels) dated one of these plutons by U-Pb (zircon) and obtained an age of 253 ± 2 Ma. Zhao et al. (1994, p. 119) described these plutons as composing a Triassic alkaline syenite belt.

The extent of this Indosinian (or early Yanshanian) deformation along the length of the Yinshan belt is not yet known. However, Darby et al. (this volume; see area 4, Fig. 1) have demonstrated in the southwestern Daqing Shan of the Yinshan belt (sensu stricto) that prior to deposition of Jurassic strata, Cambrian and unconformably overlying Permian strata were isoclinally folded and Archean basement was at least locally thrust over Permian strata.

IV. Late Jurassic to Early Cretaceous formation of the south-directed Sihetang nappe, its inverted limb ductile shear zone, and the ductile Miyun Reservoir thrust (pre-143 Ma to ≤127 Ma)

Jurassic–Cretaceous deformation in the Yanshan belt of Hebei province is predominantly brittle at present levels of ex-

posure. An important regional exception is present in the Yunmeng Shan, ~160 km southwest of Chengde and ~70 km north-northeast of Beijing (Fig. 3A). Here, major contractile structures include the recumbent, southeast-directed Sihetang anticlinal nappe and a 6-km-thick gneissic shear zone that developed across the nappe's inverted and attenuated lower limb and underlying plutonic rocks (Davis et al., 1996a). The inverted limb, estimated by Davis et al. (1996a) to be about 15 km across, contains recrystallized Archean basement rocks and their now up-side-down cover of metamorphosed Proterozoic (Changcheng and Jixian) strata (Fig. 15). A ductile, southeast-directed thrust fault, the Miyun Reservoir thrust (MR, Fig. 3A), lies in the middle of the gneissic shear zone and separates the folded Archean-Proterozoic sequence in its upper plate from a lower plate Jurassic–Cretaceous plutonic complex; its amount of displacement is unknown.

Nappe development, ductile shearing, and thrust faulting were contemporaneous with the intrusion and recrystallization of a Jurassic–Cretaceous granodioritic suite as young as 127 ± 2 Ma (Davis et al., 1996a; Fig. 4A). This south-vergent ductile deformation appears to have begun prior to 143 ± 3 Ma because a Yunmeng Shan pluton of this age displays lit par lit injection into foliated metadiorite rocks (U/Pb = 159 ± 2 Ma; Davis et al., 1996a) of the Sihetang ductile shear zone. This deformation is younger than the Late Jurassic brittle displacement of the north-directed Chengde thrust. Chengde thrusting ended prior to 148 Ma if it is correlative, as we believe, with the Xinglong thrust; the Chengde thrust is intruded by undeformed Cretaceous plutons ranging in age from ca. 131 to 129 Ma. A possible clue pointing to nappe development beginning ca. 151 Ma is the change in Yunmeng Shan magma composition from primitive diorites and gabbros (158–151 Ma) to granodiorites (151–127 Ma). As mentioned here, crustal thickening accompanying regional contraction and nappe formation might have led to lower crustal anatexis and the contamination of mantle-derived magmas. Given the close spatial and temporal relationships between plutons and nappe, it is tempting to relate nappe formation and accompanying inverted limb ductile shear and thrust faulting to magma-elevated geotherms. That similar deformation is not observed in other major areas of Yanshan plutonism may be a function of their higher level of exposure. Deeper crustal levels are exposed in the Yunmeng Shan nappe area than elsewhere in the Yanshan because the nappe is in the antiformally arched footwall of a major Cretaceous extensional detachment fault (the Hefangkou fault, H, Fig. 3A; Davis et al., 1996a, 1996b).

IVa. Gubeikou fault zone (<148 Ma, >132 Ma)

Basin-bounding, south-vergent reverse faults of the Gubeikou zone (G, Figs. 3, 4, and 8) at Gubeikou, Simatai, and north and northeast of the Xinglong thrust have thick footwall sequences of syntectonic, Early Cretaceous carbonate clast sedimentary breccia and conglomerate. (Note that the Gubeikou reverse fault zone [IV, Fig. 8B] is south of the Pingquan-Gubeikou thrust fault [I, Fig. 8B] of pre-Middle Jurassic age; however, to the west near Gubeikou and to the east near Pingquan [cf. Figs. 3B and 4B] the younger zone appears to overprint and obscure the older.) The sedimentary breccia and conglomerate cannot be of Late Jurassic age (Houcheng Formation) as designated on Hebei province maps. These strata must be Early Cretaceous because they are younger than 148 Ma (the age of an underlying andesite) and they and the Gubeikou fault are cut by granitic plutons as old as 132 Ma (PRC-12, PRC-13; Fig. 4B). Reverse faulting along the Gubeikou zone is therefore clearly younger than northward displacement of the Xinglong-Chengde(?) thrust (pre-148 Ma), and may be a high-level tectonic companion of the south-directed Sihetang nappe and Miyun Reservoir thrust fault.

Figure 16 illustrates the Gubeikou fault northeast of Miyun (Fig. 3) at the Simatai Great Wall. Steeply dipping Proterozoic carbonate strata lie above the 60° north-dipping fault. North-dipping carbonate-clast conglomerates and sedimentary breccias several hundred meters thick are in the footwall of the fault above a sandstone-dominated section. The dip of the section shallows from about 50° to 30° toward the fault and then is abruptly and strongly overturned beneath it in a tight footwall syncline. The carbonate-clast content of the footwall section increases upward, and these sediments are therefore interpreted by us as syntectonic deposits derived from Proterozoic strata in the Gubeikou hanging wall. However, the characteristic small clast size of the conglomerates and sedimentary breccias, generally <10–15 cm even in close proximity to the steep Gubeikou fault, is somewhat puzzling—unless, the displacement on the overlying fault is much larger than it might first appear (i.e., with displacement of the hanging wall across coarser grained, once more proximal footwall facies). Given this clue as to displacement and the geometric relationships between footwall strata and the Gubeikou fault described, we propose that the fault was initially a low-angle thrust, that its syntectonic footwall strata were initially subhorizontal, and that upper and lower plates were both subsequently rotated northward into their present steeper dips.

V. Localized late contraction (ca. 115–120 Ma)

The youngest contractional structure that we have recognized in the Yanshan belt of Hebei province is a south-dipping reverse fault southwest of Chengde that offsets the 117 ± 2 Ma Daguikou syenite stock (Figs. 3B, 4B, and V in Fig. 8). This shallow-level pluton intrudes and domes Cretaceous volcanic rocks that appear petrochemically compatible with the stock; K-feldspar phenocrysts from a trachytic flow yield an $^{40}Ar-^{39}Ar$ plateau age of 122.5 ± 5.5 Ma (K. Hodges, 1999, written commun.) that overlaps the U-Pb (zircon) age of the pluton. The domed volcanic strata are displaced along a south-dipping normal fault that we assign to deformation phase VI (although we

have not established by field work the relative ages of the reverse fault and the normal fault).

VI. Widespread mid-Cretaceous extensional deformation (post 118–115 Ma)

Widespread regional extension occurred within the Yanshan belt in Cretaceous time (and to the west in the Yinshan belt of Inner Mongolia). The majority of the normal faults of this deformational phase have east-northeast to north-northeast strikes and dip to the southeast. Most faults border half-grabens containing Cretaceous strata; at least one reactivates a preexisting fault (the Shangyi-Fengning-Longhua zone, Fig. 3). The most important extensional structure is the Hefangkou normal detachment fault (H, Figs. 3A and 17), which borders the southern margin of the Yanshan from near Xiwengzhuang (north of Miyun) to just east of Changping; the portion of this southeast-dipping fault west of Baiyachang (south of Hefangkou) has previously been mismapped as an unconformity between Jurassic and Proterozoic strata (Hebei Bureau of Geology, 1989; Beijing Bureau of Geology and Mineral Resources, 1991). The Hefangkou fault and mylonitic granitic gneiss in its ductilely deformed footwall—the Shuiyu (Dashuiyu, Huairou) shear zone—constitute a U.S. Cordilleran-type metamorphic core

Figure 16. View to west of Gubeiko reverse fault at Simatai Great Wall, west of Gubeiko (Fig. 3A). Shattered, steeply dipping Proterozoic carbonate rocks of Jixian System (Prw; Wumishan Formation) are displaced across Early Cretaceous carbonate clast conglomerate and sedimentary breccia (K) that form recessive slopes and dip into fault (see text). See Figure 8B for representative location, but note that view for Figure 8 is to east. Ridge in foreground is about 400 m high.

Figure 17. Exhumed surface of Hefangkou low-angle normal fault, 3 km east of Daguanqiao village, eastern Yunmeng Shan. Fault dips 20° to east-southeast. See Figure 13.5 in Davis et al. (1996a) for geologic map of this area.

complex (Davis et al., 1996a, 1996b). Dip-slip displacement on this fault was at least 10 km, although the timing of displacement is uncertain. Rapid cooling of the footwall Yunmeng Shan granodiorite between 118 and 116 Ma (K-feldspar ^{40}Ar-^{39}Ar ages; Davis et al., 1996a, 1996b) may have resulted from its rapid exhumation during displacement on the Hefangkou fault. This age is compatible with the age of the normal fault that offsets Cretaceous strata deformed by the 117 ± 2 Ma Daguikou stock (PRC-19) near Chengde city. Illite-rich gouge samples from along the Hefangkou fault yield a range of K-Ar ages, the youngest and most reliable of which is 72 Ma (Davis et al., 1996a, 1996b). If meaningful, this age could either represent the age of the Hefangkou fault, or its reactivation by Late Cretaceous or younger displacement along it. Cui and Wu (1997) apparently regarded deformational phase (VI) as Cenozoic in age, beginning in the Eocene, but a mid-Cretaceous age appears more likely to us.

VII. Cretaceous? sinistral strike-slip faulting in eastern Hebei and western Liaoning provinces

This late phase of deformation is characterized by numerous sinistral strike-slip faults in eastern Hebei and western Liaoning provinces that range in strike from north-northeast to northeast (~20°–45°), offset plutons as young as 128–132 Ma, and become more prevalent to the east (Fig. 3B). We have identified one major fault between Lingyuan and Jianchang in Liaoning province that has a possible left slip of 40–50 km based on offset of Archean to Middle Jurassic units.

The origin of the faults of phase VII remains unresolved. It is tempting to relate the faults to displacement across three northeast-striking northern branches of the Tancheng-Lujiang (Tan-Lu) fault of eastern China that are in Liaoning province

east of Figure 3. From east to west the branches are Dunhua-Mishan (also Dunmi), Yilan-Yitong, and Siping-Dehui; the first two have N40°–60°E strikes. According to Wan (1996, p. 29, his Fig. 7), all three branches underwent sinistral transtension between 135 and 52 Ma, although the strike-slip components of displacement on the Dunhua-Mishan (Dunmi) and Yilan-Yitong faults were "slight." In contrast, Jin and Wang (1996, p. 133) reported "Intensive sinistral strike slip . . . in Dumni fault in Late Mesozoic era, which dissected the northern margin fault of the Sino-Korean plate." Wang et al. (1997) described the Dumni fault as forming mainly in the Early Cretaceous and stated (p. 246–247) that "left-lateral strike-slip displacement up to 100–150 km took place in the Dumni fault zone."

PLATE TECTONIC SETTING OF THE JURASSIC–CRETACEOUS YANSHAN FOLD AND THRUST BELT

The plate tectonic setting of the Yanshan fold and thrust belt in Hebei province during contractional events III, IV, and V is uncertain. This uncertainty for the time period between the late Middle Jurassic (ca. 160 Ma) and the late Early Cretaceous (ca. 120–115 Ma) stems from two possible configurations for the Yanshan belt at that time: (1) the Yanshan belt was the eastern continuation of the east-trending Yinshan belt of Inner Mongolia; or (2) the Yinshan and Yanshan belts were separate tectonic entities and the western Yanshan belt turned southwestward in Beijing Municipality into the north-south–trending Taihang Shan (Figs. 1 and 2). The two alternatives require different plate tectonic scenarios. Alternative 1, which we favor at this time, identifies a Yinshan-Yanshan (Yinshan sensu lato) orogenic belt as a genuine intraplate construct, with considerable, but not exclusive independence from Pacific–eastern Asian plate interactions. Alternative 2 dissociates the Yanshan from the Yinshan and thus defines the Yanshan belt after ca. 180 Ma as a segment of a continental margin orogen related to Pacific plate (Izanagi)–North China plate interaction.

The choice between the two tectonic scenarios is obscured by geologic relationships in the border area between northern Hebei province and easternmost Inner Mongolia (cf. the area of the question mark in Fig. 2). The border area is a +100-km-wide topographic lowland with an extensive cover of Neogene mafic lavas and sedimentary strata that obscure bedrock relationships. We favor (but cannot yet prove) continuity of the Yinshan and Yanshan belts beneath this Neogene lowland for several reasons. (1) There are several east-striking Yanshan thrust faults just east of the Hebei–Inner Mongolia border area in the vicinity of Shangyi County; at least three major south-directed thrusts in this area and a north-northwest–directed thrust farther south juxtapose Archean basement gneiss atop Jurassic clastic strata. (2) The major Late Jurassic to Early Cretaceous Daqing Shan thrust fault north of Hohhot is recognized in Inner Mongolia (Fig. 2) (Zhu, 1997; Davis et al., 1998c; Zheng et al., 1998); this east-striking, north-northwest–directed thrust also involves Archean and Proterozoic(?) crystalline rocks and lower plate Middle and/or Upper Jurassic strata (Zhu, 1997). (3) The Daqing Shan thrust is intruded by the Cretaceous Naobaoshang pluton (119 ± 2 Ma; Table 1), which is within the age range of Yanshan belt plutons farther east (cf. Fig. 5). (4) Both the Yinshan and Yanshan areas contain extensional metamorphic core complexes of probable middle Cretaceous age; complexes north of Beijing (Yunmeng Shan; Davis et al., 1996a, 1996b) and Hohhot (newly recognized) were both characterized by extension along low-angle, southeast- to south-southeast–dipping normal faults.

The tectonic dilemma of a continuous Jurassic–Cretaceous Yinshan-Yanshan fold and thrust belt (Yinshan belt sensu stricto) is that the belt would have been an intraplate orogen, lying at considerable distance from Mesozoic northern and southern boundaries of the amalgamated North China–Mongolian plate and extending westward at a high angle from the west- or northwest-dipping subduction zone along the plate's eastern margin. North-south shortening across the belt might be attributable to far-field effects of collision of the North China–Mongolian plate with either the continental South China plate to the south and/or the continental Siberian plate to the north (Fig. 1). These two collisional boundaries are now represented, respectively, by the Permian–Triassic–Jurassic(?) Kunlun-Qinling-Dabie and Jurassic–Cretaceous Mongolo-Okhotsk suture zones.

Kunlun–Qinling–Dabie Shan collisional zone (Late Permian–Middle Jurassic?)

The Kunlun–Qinling–Dabie Shan zone separates the South and North China–Mongolian plates and has been a focus of geologic attention because of ultrahigh pressure metamorphic assemblages found in its eastern areas (e.g., Dabie Shan). There has previously been widespread agreement among both Chinese and western geologists that collision of the two continental plates was completed by Late Triassic time, and that the exhumation of ultrahigh pressure metamorphic rocks formed at depths in excess of 100 km began in the Early Triassic (ca. 240 Ma) and was largely completed by the end of the Triassic, ca. 210 Ma (Hacker et al., 1996, 1998, 2000; Liou et al., 1996; Yin and Nie, 1996).

However, a paleomagnetic study of Middle and Upper Jurassic strata in the southern part of the Hehuai (Hefei) basin on the southern edge of the North China craton (He, Fig. 1) concludes that apparent polar wander paths (APWPs) for this basin and the South China plate do not become coincident until near the Middle and Late Jurassic boundary, ca. 159 Ma (Gilder and Cortillot, 1997). Hence, Gilder and Cortillot (1997, p. 17725) concluded that the North and South China plates were not "fully sutured" until about 50 m.y. later than reported by others. Geologically, the case for such late suturing seems suspect. Gilder and Cordillot (1997) reported no radiometric age controls for

the Jurassic terrestrial section from Chinese sources, and evidence for strong deformation in the basin appropriate to continent-continent collision at the close of the Middle Jurassic seems scant. Quoting Gilder and Cordillot (1997, p. 17716), various stratigraphic sections in the basin "*suggest*" [our emphasis] that an angular unconformity separates Middle from Upper Jurassic strata."

Yin and Nie (1996) also postulated that contraction related to and following south China–north China collision may have continued in the North China–Mongolian plate in Jurassic time. Others (Liou et al., 1996; Han et al., 1989), however, have proposed an extensional rift origin for central and northern parts of the Hehuai basin just north of the Qinling-Dabie Shan suture (He, Fig. 1) from Early Jurassic through Early Cretaceous time—broadly synchronous with Yanshan events III and IV. Han et al. (1989) reported that this basin did not undergo contractile deformation until the Late Cretaceous, and Hacker et al. (1998) interpreted the northern half of the Dabie Shan to be a Cretaceous magmatic complex intruded between 137 and 126 Ma that was accompanied by ~100% crustal stretching. During this same time period, the south-vergent ductile Sihetang nappe and Miyun Reservoir thrust were being formed in the Yunmeng Shan north of Beijing (Yanshan deformational phases IV and IVa). In addition, Mesozoic contractional deformation in the Taihang Shan (TS, Fig. 1) between the Qinling–Dabie Shan zone and the Yanshan belt has north to northeast trends, more or less parallel to the subduction zone along the eastern edge of the continent and at a high angle to Yanshan belt structures. It thus seems unlikely to us for several reasons that far-field stresses from the Qinling-Dabie collisional zone could be responsible for major contractional deformation (phases III and IV) in the Yanshan belt during Late Jurassic and Early Cretaceous time.

Mongolo-Okhotsk collisional zone (Jurassic and Cretaceous)

The Mongolo-Okhotsk collisional zone (Fig. 1) lies within a geographically isolated region crossing southeastern Siberia and Mongolia. It has been the focus of recent geologic and paleomagnetic investigations concerning the closure of an oceanic realm once separating the Siberian plate and an amalgamated North China–Mongolian plate (e.g., Nie et al., 1990; Zhao et al., 1994; Zonenshain et al., 1990; Enkin et al., 1992; Nie and Rowley, 1994; Courtillot et al., 1994; Gilder and Courtillot, 1997; Halim et al., 1998; Zorin, 1999). The eastern end of the suture is geologically well defined at the Sea of Okhotsk (Halim et al., 1998), but the suture's location west of about lat 118°E near the Mongolian-Siberian border is less well constrained. Zorin (1999) extends the suture through Mongolia to latitudes west of 100°E by identifying precollisional Siberian and Mongolian terranes now adjacent to each other. Zonenshain et al. (1990) presented geologic arguments that subduction of the Mongolo-Okhotsk ocean preceding plate collision was northward, and Van der Voo et al. (1999) reported that deeply subducted slab remnants from this closure can be recognized beneath southern Siberia through the use of seismic tomography.

East Asian collision of the Siberian and North China–Mongolian plates across a Mongolo-Okhotsk ocean is generally believed to have proceeded from west to east in Jurassic and Early Cretaceous time (cf. Ziegler et al., 1996; Yin and Nie, 1996; Halim et al., 1998; Zonenshain et al., 1990). According to Zonenshain et al. (1990, their Fig. 96), ocean closure north of the Yanshan belt had occurred by the Late Jurassic (ca. 150 Ma), but more eastern parts of the seaway remained open until the Early Cretaceous (140–120 Ma). There are other interpretations. Zorin (1999), in the most complete discussion of the suture zone to date, contends that the main collision in western zones of the suture occurred at the Early Jurassic-Middle Jurassic boundary and ended in the Late Jurassic. He considers Early Cretaceous extensional deformation west of 120°E latitude on both sides of the suture to represent a separate, postcollisional episode. Somewhat contrarily, Gilder and Cortillot (1997) concluded from paleomagnetic studies that the Eurasian and North China–Mongolian plates were still ~2000 km apart in the Late Jurassic, but by the Early Cretaceous (ca. 130 Ma) paleomagnetic poles for the two plates were coincident.

The timing of plate collision, based largely on comparisons of the APWP of the North China–Mongolian plate with respect to the Eurasian reference path (see preceding references), seems broadly compatible with the timing of deformational events III and IV in the Yanshan belt (Fig. 5). Collision of the Siberian and the amalgamated North China–Mongolian plates might then account for north-south Jurassic and Early Cretaceous contractional deformation in the Yanshan belt, despite the fact that the distance between the suture and the belt is in excess of 1000 km. Present-day contractional deformation in the intraplate Tian Shan region is occurring more than 1600 km north of the Himalayan collision zone (at long 85° E) and is widely considered related to it.

Preexisting crustal anisotropies and the Yinshan-Yanshan belt. It is commonly believed that preexisting crustal anisotropies might focus far-field stresses and control or influence intraplate deformation distant from active plate boundaries. Such an explanation could apply to the Yinshan-Yanshan belt. There is no scarcity of candidate structures and lithic belts that had east-west strikes or trends in the basement beneath the Jurassic–Cretaceous Yinshan-Yanshan belt. Archean basement rocks along the northern edge of the North China plate from Baotou on the west to the Bohai Bay on the east are separated into northern amphibolite and southern granulite facies belts (Qian, 1996). Yu et al. (1996), Cui et al. (1996), and Chen et al. (1996), among others, described an east-west–trending Paleoproterozoic intracontinental rift system beneath the Yinshan-Yanshan belt that was ~500 km long, 50–100 km wide, and bordered by extensional faults. Major faults in and bordering the rift were said by Cui et al. (1996) and Chen et al. (1996) to have controlled the thickness and sedimentary facies of the late Paleo-

proterozoic Changcheng System (1.85–1.60 Ga) and to have localized coeval magmatism, both plutonic and volcanic (1.74–1.62 Ga; Yu et al., 1996). These Proterozoic-initiated fault zones are reported to include the Fengning-Longhua zone, the Chicheng-Gubeiko zone, and the Miyun-Xinglong zone. The Gubeikou zone is possibly the most obvious example of pre-Mesozoic structures controlling the east-west strike and trend of Jurassic–Cretaceous structural elements in the Yanshan belt. Elongate Proterozoic anorthosite-gabbro and granitic plutons (1.702–1.696 Ga, U-Pb zircon ages; Yu et al., 1996) are localized along the zone, as are the western ends of the pre-Middle Jurassic Pingquan-Gubeiko thrust and the Cretaceous Gubeikou reverse or thrust fault zone. The Fengning-Longhua zone controlled Yanshanian contractional deformation of Proterozoic strata and was reactivated, at least locally, as a Cretaceous normal fault. The Miyun-Xinlong fault was described by Yu et al. (1996) as the major fault within the Proterozoic rift, and it may have controlled or influenced southward rooting of the Chengde-Xinglong thrust in the Late Jurassic.

Permian-Triassic(?) subduction along the northern edge of the North China craton prior to and during amalgamation of the Mongolian arcs along the Suolun suture produced an east-trending magmatic arc and orogen, mostly north of Chengde, including folds, thrusts, and ductile shear zones (Zhao, 1990; Cui and Wu, 1997), but we include the Pingquan-Gubeikou thrust (I, Fig. 8) and possibly the Melanyu anticline (Fig. 3B) in this tectonic assemblage. Thus, the case for reactivation of older stuctures during Jurassic–Cretaceous intraplate deformation is strong.

Jurassic–Cretaceous extension between the Yinshan belt and the Mongolo-Okhotsk suture. At present, the principal problem in relating Yinshan-Yanshan intraplate deformation to collision far to the north between the Siberian plate and the amalgamated North China–Mongolian plate is the widespread occurrence of extensional basins within the Mongolian arcs terrane that are reported by some workers to be of Jurassic–Cretaceous age. These basins lie north of the Yinshan-Yanshan belt, south of the Mongolo-Okhotsk suture and within a northeast-trending magmatic arc of uncertain parentage (Fig. 1). The north-northeast–to northeast-trending basins include, from east to west, the Yilong-Yitang graben (YY, Fig. 1) and the Songliao, Hailar (China)-Tamsag (Mongolia), and Erlian (China)-Gobi (Mongolia) basins.

The prevailing consensus is that extensional (or transtensional) rifting in all of the northeast- to north-northeast–trending Mongolian basins occurred within a broad Andean type magmatic arc developed above an active northwest-dipping Pacific subduction zone (e.g., Watson et al., 1987; Hsü, 1989). Traynor and Sladen (1995), noting the early presence in the basins of synrift volcanic rocks, favored subduction rollback of a Pacific plate as the best explanation for rifting. They stated that extensional rifting in northeastern China appears to have migrated eastward toward the Pacific. Deng et al. (1999) also related the Early Jurassic to Early Cretaceous volcanic rocks of northeastern China and adjacent Inner Mongolia to Pacific plate subduction, but noted that there is no obvious compositional polarity in K_2O/SiO_2 ratios across this volcanic province. Unfortunately, these authors did not discriminate between volcanic rocks of differing age and thus may have blurred possible K_2O/SiO_2 polarity for narrower time intervals.

Traynor and Sladen (1995) reported that extensional rifting and synrift deposition of volcanic and clastic strata were widespread across southern and eastern Mongolia and northeastern China from the Middle Jurassic through the Early Cretaceous. Seismic sections across these basins show clear graben and half-graben geometries with typical patterns of synrift sedimentation, according to Traynor and Sladen (1995, their Fig. 10, A-A'). Hsü (1989, p. 221) and Traynor and Sladen (1995, p. 45) described similar patterns of deposition for the Hailar-Tamsag and Erlian-Gobi basins (H-T, E-G, Fig. 1). According to Traynor and Sladen (1995, p. 45), "On a basin scale there are typically early synrift alkaline volcanic rocks (basalts and tuffs) and coarse alluvial clastic sediments which interdigitate with, and are overlain by, synrift lacustrine mudstones in basin centres and coal-rich alluvium towards the basin margins." Hsü (1989, p. 221) reported that the Upper Jurassic rocks in the Erlian basin are "some 5 km thick, consist mainly of basic volcanics and pyroclastics." Extensional faulting and synchronous sedimentation and mafic volcanism are reported to have begun in Late Jurassic time in the Songliao basin; clastic sedimentation peaked there in the Early Cretaceous (Ma et al., 1989), but continued into the Tertiary.

There is widespread agreement in the literature that basin formation spanned Late Jurassic and Early Cretaceous time, although there are few well-documented age dates for basin-fill units. Volcanic rocks, felsic ash and basalt, in a half-graben within the Unegt subbasin of the East Gobi–Erlian basin (Fig. 1) have yielded ^{40}Ar-^{39}Ar ages of 156 ± 1.5 and 133 ± 2 Ma, respectively, and are interpreted as synrift by Graham et al., 1996; and Webb et al., 1999 (ages also reported in Johnson et al., this volume, their Fig. 16). These ages, if genuinely synrift, overlap the 160 to <127 Ma period of contractional deformations III and IV in the Yanshan belt to the southeast, and clearly complicate the interpretation that Yanshan contraction was related to plate collision north of the synchronously extending terrane.

There are possible solutions to this dilemma. Zorin (1999) states unequivocally that metamorphic core complexes and rift basins in Mongolia south of the Mongol-Okhotsk suture are Early Cretaceous. The presence of Jurassic strata in the normal fault-bounded basins and subbasins north of the Yinshan-Yanshan belt does not in and of itself imply extensional faulting in the Jurassic. Preextensional Jurassic and Early Cretaceous (?) strata might have been downdropped into grabens and half-grabens of younger Cretaceous age and then covered by syntectonic Cretaceous sediments. For example, Deng et al. (1997, p. 95) report that Early Cretaceous clastic and volcanic strata in the Hailar basin (the latter with ages of 84.7–130 Ma) unconformably overlie a "lightly" metamorphosed section of Early to

Middle Jurassic pyroclastic and clastic rocks; an age range of 171.4–213.0 Ma is given for these volcanic rocks, but the dating method is not mentioned. Tian and Du (1987) reported that extension in the easternmost of the extensional basins, the Yilan-Yitong graben (YY, Fig. 1), was accompanied by both plutonic and volcanic activity from the Late Jurassic into the Late Cretaceous. However, Jin and Wang (1996) reported that rifting of the graben did not begin until the Paleocene and that at that time volcanism changed from regional ("orogenic") to rift-type alkali-olivine and tholeiitic basalt.

This possible solution may not apply to other basins. Recently acquired, still unpublished seismic reflection sections across the Unegt subbasin of the East Gobi (Erlian) basin in southern Mongolia (Fig. 1; Graham et al., 1996; Johnson et al., this volume) appear to support Early Cretaceous, if not Late Jurassic, normal faulting and synchronous basin sedimentation according to Stephan Graham (2000, personal commun.). Zheng et al. (1991) and Webb et al. (1999) described a major extensional metamorphic core complex (Yagan–Onch Hayrhan) farther west along the China-Mongolia border (Fig. 2). Webb et al. (1999) and Johnson et al. (this volume) report that extension in the complex was of Early Cretaceous age (ca. 129–126 Ma) and was thus in part synchronous with extension and sedimentation in the Unegt subbasin described herein.

Widespread major crustal extension began somewhat later in the Yanshan belt following contractional deformation phases III–V. Most of this extension occurred after ca. 120 Ma along northeast-striking, southeast-dipping faults in the Yanshan belt (with the same strike as most normal faults in the Inner Mongolian basins); this phase of deformation included formation of the Hefangkou fault and its Yunmeng Shan metamorphic core complex north of Beijing (Davis et al., 1996a, 1996b). The origin of Late Jurassic–Early Cretaceous contractional deformation in the Yinshan-Yanshan belt would be much less enigmatic if formation of the Mongolian basins to the north was essentially the same Cretaceous age as Yanshan phase VI, and did not overlap in time earlier contractional phases III, IV, and V. It appears to us that thorough reassessments of the timing of initial extension within the Hailar, Erlian, and Songliao basins are warranted, as well as the mechanism(s) of basin formation.

Pacific plate subduction beneath eastern China and Yanshan belt

Although the east-west trend of Yinshan-Yanshan belt structures into Inner Mongolia at a high angle to the Pacific rim suggests considerable tectonic independence of the belt from Pacific-Eurasian plate interactions (Fig. 1), the regional patterns of Mesozoic magmatism in eastern parts of China cannot be ignored. Northeasterly tectonic trends of Mesozoic age in the Taihang Shan and in the Yanshan of western Liaoning province are reflected by regional topography and coincide spatially with a Pacific margin magmatic belt (Xu, 1990; see Figs. 1 and 2; Deng et al., 1999). Chinese geologists commonly relate these northeast-trending tectonic elements to northwest subduction of a Pacific basin plate (Izanagi) beneath eastern Asia (e.g., Deng et al., 1999), but disagree as to the time of commencement of such interactions following intraplate north-south shortening independent of Pacific interactions.

Zhao et al. (1994), for example, proposed that Indosinian and early Yanshanian Yinshan belt (sensu lato) deformation through Early Jurassic time was related to an Asian collision between the Siberian and North China–Mongolian plates, but that subsequent deformation beginning in the Middle Jurassic had Pacific margin origins. In contrast, we argue that collisional and subduction-related interactions were probably synchronous in Middle Jurassic, Late Jurassic, and Early Cretaceous time; this view is shared by Wang (1996). We believe it likely that Jurassic–Cretaceous contraction in the Yanshan belt was influenced by both north-south Eurasian intraplate deformation and synchronous thermal regimes resulting from northwestward Pacific basin subduction and attendant arc magmatism. For example, in the Yunmeng Shan north of Beijing, close spatial and temporal ties existed in the middle crust between south- to southeast-directed Late Jurassic–Early Cretaceous basement nappe formation (event IV) and voluminous plutonic intrusion. Our previous studies indicate the likelihood that during nappe formation, magma compositions changed ca. 151 ± 2 Ma from primitive, mantle-derived dioritic melts to crustal contaminated granodioritic magmas (Davis et al., 1996a). This change in composition suggests an interactive relationship between north-south intraplate contraction and Pacific subduction-related magmatism, an interaction that led to thermal softening and tectonic thickening of the Archean-floored "craton" through which the primitive magmas passed and became contaminated. Furthermore, our recent discovery of the Late Jurassic and/or Early Cretaceous Daqing Shan thrust north of Hohhot, Inner Mongolia (Zheng et al., 1998; Fig. 1, area 3), is an important indicator that major north-northwest–south-southeast intraplate contraction in western parts of the Yinshan-Yanshan belt was synchronous with Pacific basin subduction beneath more eastern parts (Yanshan). Archean and Proterozoic crystalline rocks in the upper plate of the Daqing Shan thrust have a minimum north-northwest displacement of 25 km (Zheng et al., 1998).

Whereas there seems little doubt of a Pacific margin influence in the Jurassic–Cretaceous development of Yanshan belt structures (phases III–VI), it is the source of the north-south contraction across the Yinshan-Yanshan belt and its relationships to surrounding regions that continue to constitute the principal enigma of this contractional orogen. The problem is less vexing if the Jurassic–Cretaceous Yanshan belt was tectonically independent of the Yinshan (sensu stricto) of Inner Mongolia. In that case, the Yanshan would simply be a central, east-trending component of a Pacific margin magmatic orogen lying between northeast-trending fold and thrust structures in western Liaoning province and to the southwest in the Taihang Shan (Figs. 1 and 2). In this case north-south Jurassic–Cretaceous contraction in a tectonically independent Yinshan belt farther west would be

isolated and truly intraplate—perhaps related to crustal shortening along the northern edge of a rigid Ordos basin crustal block (cf. Fig 1).

It is clear that some western Yanshan and northern Tiahang Shan structures merge in the area west of Beijing (Fig. 2). However, it is not at all clear why Yanshan contractional structural elements of deformational phases I, III, and IV (pre-180 Ma to ca. 125 Ma) have east to north-northeast trends across much of northern Hebei province unless they either (1) formed independently of North China–Pacific (Izanagi) plate interactions in response to north-south intraplate shortening, or (2) formed in response to North China–Pacific plate interactions and were subsequently rotated into an east-west orientation (with a fortuitous alignment to the Yinshan belt farther west). We know of no evidence that regional rotation of the Yanshan and its pre-Mesozoic east-west–trending structures (see preceding discussion) has occurred. The east-west linearity of the late Paleozoic–Early Triassic(?) Suolun-Linxi suture and the northern edge of Archean basement rocks of the Yinshan-Yanshan belt (Figs. 1 and 2) argues against major Cretaceous or younger regional rotation of the Yanshan segment.

CONCLUSIONS

The Yanshan fold and thrust belt of Beijing Municipality and Hebei and Liaoning provinces has long been recognized as a major Mesozoic tectonic element of northern China, although previous workers have generally underestimated both its geometric and kinematic complexity. Our studies have begun to document a complicated history of major superposed contractional deformations, the most important of which were: (1) pre-early Middle Jurassic (>180 Ma) southward thrusting of cover and Archean basement (Pingquan-Gubeiko thrust fault); (2) northward Late Jurassic thin-skinned thrust faulting of Mesozoic, Paleozoic, and Proterozoic strata and their Archean basement (Chengde thrust fault); and (3) latest Jurassic–Early Cretaceous formation of the Sihetang nappe, a south-vergent recumbent anticline cored by Archean gneissic units and underlain by a ductile thrust (Miyun Reservoir thrust) that is within a gneissic shear zone to 6 km thick.

The Pingquan-Gubeiko thrust developed south of a Permian–Triassic magmatic arc. Thrusting could either have been (1) a consequence of the collisional suturing of Paleozoic Mongolian arcs against an Andean-style continental arc along the northern margin of the North China plate, or (2) an expression of a backarc, foreland fold and thrust belt of U.S. Cordilleran type formed during southward subduction beneath the North China Archean "craton."

Late Jurassic and Jurassic–Cretaceous deformations (2 and 3 above) have opposing vergence and an uncertain relationship to plate interactions in eastern Asia. The Jurassic–Cretaceous Yanshan belt, and its probable western continuation, the Yinshan belt of Inner Mongolia, appear to be the consequence of regional north-south intraplate shortening—perhaps far-field effects of intra-Asian continental plate collisions to the north or south.

However, Yanshan patterns of deformation, e.g., ductile basement nappe formation in the Yunmeng Shan and northeast structural trends in Liaoning province indicate the influence of westward or northwestward Pacific plate subduction beneath the North China plate and the magmatism that accompanied it ($\leq$180–190 Ma). We thus believe it likely that Late Jurassic and Early Cretaceous contraction in the Yanshan belt was influenced by both north-south Eurasian intraplate deformation and northwestward Pacific basin subduction and attendant arc magmatism. A tectonic influence of Pacific-Asian plate interaction is not obvious farther west in Yinshan belt deformation of this age.

Jurassic–Cretaceous collision of an amalgamated North China–Mongolian plate with the Siberian plate is widely believed to have accompanied closure of a Mongolo-Okhotsk Sea more than 800–1100 km north of the Yanshan belt. Although sometimes considered the best candidate for causing north-south Yinshan-Yanshan intraplate contraction, this hypothesis is severely complicated by reports of widespread, basin-forming Late Jurassic and Early Cretaceous extension in the terranes between the Mongolo-Okhotsk suture and the Yinshan-Yanshan belt. We remain unconvinced that Late Jurassic rifting was as widespread as reported. The Jurassic strata preserved in some basins north of the Yinshan–Yanshan belt may be of pre-rifting age.

Portions of the Yanshan belt in Hebei province were also sites of major northwest-southeast regional extension beginning soon after 120 Ma. The Hefangkou extensional detachment fault, north of Beijing, and its footwall Yunmeng Shan metamorphic core complex were probably formed at this time. Rollback of a subducting Pacific oceanic plate or postorogenic collapse following Late Jurassic and Early Cretaceous contraction are the two most likely explanations for this extensional phase (V) of Yanshan belt development.

ACKNOWLEDGMENTS

We gratefully acknowledge U.S. National Science Foundation grants EAR-9903012, EAR-9627909, and EAR-8904985 to Davis. Early studies in the Yunmeng Shan (Davis et al., 1996a, 1996b) were also supported by the National Natural Science Foundation of China (NNSFC); we especially thank Zhang Zhi-Fei and Liu Caiquan of the Bureau of International Collaboration. Zheng Yadong and Zhang Jinjiang, both of Peking University, were financially supported in 1996 field studies by the NNSFC. Zhang Changhou, China University of Geosciences, Beijing, was funded by NNSFC grant 49702034; his summer 1999 field work (with Zheng and Davis) in Hebei and western Liaoning provinces was financially supported by the Bureau of International Collaboration (NNSFC).

A number of geologists have assisted our studies in the Yinshan-Yanshan belt. Most important among these is Qian Xianglin of Peking University, who first invited Davis to China in 1985, and who, with Zheng Yadong, encouraged the start of our collaborative Yinshan-Yanshan belt studies. Special thanks to Kip Hodges of the Massachusetts Institute of Technology for making the ^{40}Ar-^{39}Ar age determinations in his CLAIR facility, with the assistance of Michael Kroll. Others who have assisted this research include Tong Heng-Mao, Yu Hao, Muhammad Shaffiquallah, Joan E. Fryxell, Zhang Jinjiang, Hua Yonggang, and Liu Liqun. We thank Brad Hacker, Shangyou Nie, and Marc Hendrix for helpful and constructive reviews.

REFERENCES CITED

Beijing Bureau of Geology and Mineral Resources, 1991, Regional geology of Beijing Municipality: Ministry of Geology and Mineral Resources, Geological Memoirs, ser. 1, no. 27, 598 p. (in Chinese with English summary).

Chen, A., 1996, A preliminary study on deformational mechanism of basement-involved structure in the Yanshan Mountain: 30th International Geological Congress, Beijing, Abstracts, v. 2, p. 291.

Chen, A., 1998, Geometric and kinematic evolution of basement-cored structures: Intraplate orogenesis within the Yanshan orogen, northern China: Tectonophysics, v. 292, p. 17–42.

Chen, Z., Zhang, J., Li, Y., and Liu, Z., 1996, An outline of regional geology of the Beijing area, in International Geological Congress, 30th, Field trip guide: Beijing, Geological Publishing House, 12 p.

Courtillot, V., Enkin, R., Yang, Z., Chen, Y., Bazhenov, M., Besse, J., Cogne, J.-P., Coe, R., Zhao, X., and Gilder, S., 1994, Paleomagnetic constraints on the geodynamic history of the major blocks of China from the Permian to the Present (Enkin, R., et al.), Reply: Journal of Geophysical Research, v. 99, p. 18043–18048.

Cui, S., and Wu, Z., 1997, On the Mesozoic and Cenozoic intracontinental orogenesis of the Yanshan area, China, in Zheng, Y., et al., eds., Proceedings of the 30th International Geological Congress, v. 14: Utrecht, The Netherlands, VSP, p. 277–292.

Cui, S., Wu, G., Wu, Z., and Ma, Y., 1996, Structural features and stratigraphy of the Ming Tombs-Badaling area, Beijing, in International Geological Congress, 30th, Field trip guide: Beijing, Geological Publishing House, 17 p.

Davis, G.A., 1999, Challenging some widely held beliefs about thrust fault geometries—From field studies in the U.S. Cordillera and northern China: Earth Sciences Frontiers, v. 6, p. 49–66.

Davis, G.A., Qian, X., Zheng, Y., Yu, H., Wang, C., Tong, H.M., Gehrels, G.E., Shafiquallah, M., and Fryxell, J.E., 1996a, Mesozoic deformation and plutonism in the Yunmeng Shan: A Chinese metamorphic core complex north of Beijing, China, in Yin, A., and Harrison, T.M., eds., The tectonic evolution of Asia: Cambridge, Cambridge University Press, p. 253–280.

Davis, G.A., Qian, X., Zheng, Y., Tong, H.M., Yu, H., Wang, C., Gehrels, G.E., Shafiquallah, M., and Fryxell, J.E., 1996b, The Huairou (Shuiyu) ductile shear zone, Yunmengshan Mts., Beijing, in International Geological Congress, 30th, Field trip guide: Beijing, Geological Publishing House, 25 p.

Davis, G.A., Wang, C., Zheng, Y., Zhang, J., Zhang, C., and Gehrels, G.E., 1998a, The enigmatic Yinshan fold-and-thrust belt of northern China: New views on its intraplate contractional styles: Geology, v. 26, p. 43–46.

Davis, G.A., Zheng, Y., Wang, C., Darby, B.J., Zhang, C., and Gehrels, G.E., 1998b, Geometry and geochronology of Yanshan Belt tectonics, in Collected Works of International Symposium on Geological Science, 100th Anniversary Celebration of Peking University: Beijing, Peking University Department of Geology, p. 275–292.

Davis, G.A., Zheng, Y., Darby, B.J., Wang, C., and Hua, Y., 1998c, Geologic introduction and field guide to the Daqing Shan thrust, Nei Mongol: Nei Mongol Bureau of Geology and Mineral Resources, Yinshan—Yanshan

Major Thrust and Nappe Structures Field Conference, May 8–11, 1998, Hohhot, Nei Mongol, China, 23 p.

DeCelles, P.G., Tolson, R.B., Graham, S.A., Smith, G.A., Ingersoll, R.V., White, J., Schmidt, C.J., Rice, R., Moxon, I., Lemke, L., Handschy, J.W., Follo, M.F., Edwards, D.P., Cavazza, W., Caldwell, M., and Baragar, E., 1987, Laramide thrust-generated alluvial-fan sedimentation, Sphinx Conglomerate, southwestern Montana: American Association of Petroleum Geologists Bulletin, v. 71, p. 135–155.

Deng, J., Mo, X., Zhao, H., Luo, Z., and Zhao, G., 1999, Yanshanian magma-tectonic-metallogenic belt in east China of circum-Pacific domain (I): Igneous rocks and orogenic processes: China University of Geosciences Journal, v. 10, p. 21–24.

Deng, S., Ren, S., and Chen, F., 1997, Early Cretaceous flora of Hailar, Inner Mongolia, China: Beijing, Geological Publishing House, 116 p.

Dong, G., 1996, On Mesozoic Yanshanian movement in Yanshan Range, north China: 30th International Geological Congress, Beijing, Abstracts, v. 2, p. 303.

Enkin, R., Yang, Z., Chen, Y., and Courtillot, V., 1992, Paleomagnetic constraints on the geodynamic history of the major blocks of China from the Permian to the present: Journal of Geophysical Research, v. 97, p. 13953–13989.

Gehrels, G.E., and Boghossian, N.D., 2000, Reconnaissance geology and U-Pb geochronology of the west flank of the Coast Mountains between Bella Coola and Prince Rupert, coastal British Columbia, in Stowell, H.H., and McClelland, W.C., eds., Tectonics of the Coast Mountains, southeastern Alaska and coastal British Columbia: Geological Society of America. Special Paper 343, p. 61–75.

Gilder, S., and Courtillot, V., 1997, A high resolution middle to late Mesozoic apparent polar wander path from the north China block: Implications for the North-South China collision: Journal of Geophysical Research, v. 99, p. 17713–17727.

Graham, S.A., Hendrix, M.S., Badarch, G., and Badamgarav, D., 1996, Sedimentary record of transition from contractile to extensional tectonics, Mesozoic, southern Mongolia: Geological Society of America. Abstracts with Programs, v. 28, no. 7, p. A68.

Hacker, B.R., Wang, X., Eide, E.A., and Ratschbacher, L., 1996, The Qinling-Dabie ultrahigh-pressure collisional orogen, in Yin, A., and Harrison, T.M., eds., The tectonic evolution of Asia: Cambridge, Cambridge University Press, p. 345–370.

Hacker, B.R., Ratschbacher, L., Webb, L., Ireland, T., Walker, D., and Dong, S., 1998, U/Pb zircon ages constrain the architecture of the ultrahigh-pressure Qingling-Dabie orogen, China: Earth and Planetary Science Letters, v. 161, p. 15–230.

Hacker, B.R., Ratschbacher, L., Webb, L., McWilliams, M.O., Calvert, A., Shuwen, D., Wenk, H.-R., and Chateigner, D., 2000, Exhumation of ultrahigh-pressure continental crust in east-central China: Late Triassic–Early Jurassic tectonic unroofing: Journal of Geophysical Research, v. 105, p. 13339–13364.

Halim, N., Kravchinsky, V., Gilder, S., Cogne, J-P., Alexyutin, M., Sorokin, A., Courtillot, V., and Yan, C., 1998, A palaeomagnetic study from the Mongol-Okhotsk region: Rotated Early Cretaceous volcanics and remagnetized Mesozoic sediments: Earth and Planetary Science Letters, v. 159, p. 133–145.

Han, J., Zhu, S., and Xu, S., 1989, The generation and evolution of the Hehuai Basin, in Zhu, X., ed., Chinese sedimentary basins: Sedimentary basins of the world, Volume 1: Amsterdam, Elsevier, p. 125–135.

Hebei Bureau of Geology and Mineral Resources, 1989, Regional geology of Hebei Province: Ministry of Geology and Mineral Resources, Geological Memoirs, ser. 1, no. 15, 1741 p. (in Chinese with English summary).

Hodges, K.V., Hames, W.E., Olszewski, W.J., Burchfiel, B.C., Royden, L.H., and Chen, Z., 1994, Thermobarometric and $^{40}Ar/^{39}Ar$ geochronologic constraints on Eohimalayan metamorphism in the Dinggyl area, southern Tibet: Contributions to Mineralogy and Petrology, v. 117, p. 151–163.

Hsü, K.J., 1989, Origin of sedimentary basins of China, in Zhu X., ed., Chinese sedimentary basins: Sedimentary basins of the world, Volume 1: Amsterdam, Elsevier, p. 207–227.

Isozaki, Y., 1997, Jurassic accretion tectonics of Japan: The Island Arc, v. 6, p. 25–51.

Jiang, B., Wang, G., Liu, H., Wu, Q., and Li, Z., 1997, Structure of the Xinglong double imbricate fan: Scientia Geologica Sinica, v. 32, p. 165–172.

Jin, H., and Wang, G., 1996, The taphrogeny and inversion of Jiayi, Dunmi faults and their control on oil, gas, and coal, in Wu, Z., and Chai, Y., eds., Proceedings of the 1995 Annual Conference of Tectonics in China: Beijing, Geological Publishing House, p. 131–143.

Lamb, M.A., and Badarch, G., 1997, Paleozoic sedimentary basins and volcanic-arc systems of southern Mongolia: New stratigraphic and sedimentologic constraints: International Geology Review, v. 39, p. 542–576.

Li, Z., 1996, Mesozoic magmatic rock and its geodynamics in the west of Liaoning province, China: 30th International Geological Congress, Beijing, Abstracts, v. 2, p. 364.

Liaoning Bureau of Geology and Mineral Resources, 1989, Regional geology of Liaoning Province: Ministry of Geology and Mineral Resources, Geological Memoirs, ser. 1, no. 14, 858 p. (in Chinese with English summary).

Liou, J.G., Zhang, R.Y., Wang, X., Eide, E.A., Ernst, W.G., and Maruyama, S., 1996, Metamorphism and tectonics of high-pressure and ultrahigh-pressure belts in the Dabie-Sulu region, China, in Yin, A., and Harrison, T.M., eds., The tectonic evolution of Asia: Cambridge, Cambridge University Press, p. 300–344.

Ma, L., Yang, J., and Ding, Z., 1989, Songliao Basin—An intracratonic continental basin of combination type, in Zhu, X., ed., Chinese sedimentary basins: Sedimentary basins of the world, Volume 1: Amsterdam, Elsevier, p. 77–87.

Nei Mongol Bureau, 1991, Regional geology of Nei Mongol (Inner Mongolia) Autonomous Region: Ministry of Geology and Mineral Resources, Geological Memoirs, ser. 1, no. 25, 725 p. (in Chinese with English summary).

Nie, S., and Rowley, D.B., 1994, Paleomagnetic constraints on the geodynamic history of the major blocks of China from the Permian to the present: Comment: Journal of Geophysical Research, v. 99, p. 18035–18042.

Nie, S., Rowley, D.B., and Ziegler, A.M., 1990, Constraints on the locations of Asian microcontinents in Palaeo-Tethys during the late Paleozoic, in McKerrow, W.S., and Scotese, C.R., eds., Paleozoic palaeogeography and biogeography: Geological Society [London] Memoir 12, p. 397–409.

Qian, X., 1996, Major tectonic events in evolution in the Archean of the north China craton and its implications, in Wu, Z., and Chai, Y., eds., Tectonics of China: Proceedings of the 1995 Annual Conference of Tectonics in China: Beijing, Geological Publishing House, p. 31–38.

Ren, D., Jia, Z., and Lu, L., 1996, Mesozoic stratigraphy and faunae in the Luanping-Chengde region, Hebei province, in International Geological Congress, 30th, Field trip guide: Beijing, Geological Publishing House, 25 p.

Şengör, A.M.C., and Natal'in, B.A., 1996, Paleotectonics of Asia: Fragments of a synthesis, in Yin, A., and Harrison, T.M., eds., The tectonic evolution of Asia: Cambridge, Cambridge University Press, p. 486–640.

Şengör, A.M.C., Natal'in, B.A., and Burtman, V.S., 1993, Evolution of the Altaid tectonic collage and Palaeozoic crustal growth in Eurasia: Nature, v. 364, p. 299–307.

Tian, Z., and Du, Y., 1987, Formation and evolution of the Yilan-Yitong graben: Tectonophysics, v. 133, p. 165–173.

Traynor, J.J., and Sladen, C., 1995, Tectonic and stratigraphic evolution of the Mongolian People's Republic and its influence on hydrocarbon geology and potential: Marine and Petroleum Geology, v. 12, p. 35–52.

Van der Voo, R., Sparkman, W., and Bijwaard, H., 1999, Mesozoic subducted slabs under Siberia: Nature, v. 397, p. 246–249.

Wan, T., 1996, Evolution of the Tancheng-Lujiang fault zone and paleostress fields, *in* Wan, T., ed., Formation and evolution of the Tancheng-Lujiang fault zone: Beijing, China University of Geosciences Press, p. 23–37.

Wang, G., Shao, Z., Xu, F., Jing, H., Zheng, M., and Lik, H., 1996, Structural inversion of Mesozoic–Cenozoic coal basins in eastern China, *in* Wu, Z., and Chai, Y., eds., Tectonics of China: Beijing, Geological Publishing House, p. 88–95.

Wang, H., and Mo, X., 1995, An outline of the tectonic evolution of China: Episodes, v. 18, p. 6–16.

Wang, Q., and Liu, X., 1986, Paleoplate tectonics between Cathaysia and Angaraland in Inner Mongolia of China: Tectonics, v. 5, p. 1073–1088.

Wang, X., Li, Z.J., Chen, B.L., Xing, L.S., Chen, X.H., Zhang, Q., Chen, Z.L., Dong, S.W., Wu, H.M., and Huo, G.H., 1997, Evolution of Tan-Lu strike-slip fault system and its geologic implications, *in* Zheng, Y., et al., eds., Proceedings of the 30th International Geological Congress, Volume 14: Utrecht, The Netherlands, VSP, p. 227–250.

Wang, Y., 1996, Tectonic evolutional (sic) processes of Inner Mongolia-Yanshan orogenic belt in eastern China during the late Paleozoic–Mesozoic: Beijing, Geological Publishing House, 142 p. (in Chinese with English summary).

Watson, M.P., Hayward, A.B., Parkinson, D.N., and Zhang, M., 1987, Plate tectonic history, basin development and petroleum source deposition onshore China: Marine Petroleum Geology, v. 4, p. 205–225.

Webb, L.E., Graham, S.A., Johnson, C.L., Badarch, G., and Hendrix, M.S., 1999, Occurrence, age, and implications of the Yagan–Onch Hayrhan metamorphic core complex, southern Mongolia: Geology, v. 27, p. 143–146.

Wong, W.H., 1929, The Mesozoic orogenic movement in eastern China: Geological Society of China Bulletin, v. 8, p. 33–44.

Xu, B., and Chen, B., 1997, Framework and evolution of the middle Paleozoic orogenic belt between Siberian and north China plates in northern Inner Mongolia: Science in China, ser. D, v. 40, p. 463–479.

Xu, Z., 1990, Mesozoic volcanism and volcanogenic iron-ore deposits in eastern China: Geological Society of America. Special Paper 237, 46 p.

Yin, A., and Nie, S., 1996, A Phanerozoic palinspastic reconstruction of China and its neighboring regions, *in* Yin, A., and Harrison, T.M., eds., The tectonic evolution of Asia: Cambridge, Cambridge University Press, p. 442–485.

Yu, J., Xia, X., and Fu, H., 1996, Proterozoic anorogenic rapakivi granites and related potassic alkaline volcanics in eastern Beijing, *in* International Geological Congress, 30th, Field trip guide: Beijing, Geological Publishing House, 20 p.

Zhang, C., 1996, Laramide orogen in western United States and Yanshan orogen in north China: Comparison of some tectonic features and the significance to geodynamics of intraplate deformation, *in* Annual report, Laboratory of Lithosphere Tectonics and its Dynamics: Beijing, Geological Publishing House, p. 81–91.

Zhang, C., Song, H., Chen, A., and Wu, Z., 1996, Mesozoic thrust tectonics in Yanshan intraplate orogenic belt, *in* Wu, Z., and Chai, Y., eds., Tecton-ics of China: Proceedings of the 1995 Annual Conference of Tectonics in China: Beijing, Geological Publishing House, p. 77–82.

Zhang, Z.M., Liou, J.G., and Coleman, R.G., 1984, An outline of the plate tectonics of China: Geological Society of America. Bulletin, v. 95, p. 295–312.

Zhang, Z.M., Liou, J.G., and Coleman, R.G., 1989, The Mesozoic and Cenozoic tectonism in eastern China, *in* Ben-Avraham, Z., ed., The evolution of the Pacific Ocean: Oxford Monographs on Geology and Geophysics No. 8: New York, Oxford University Press, p. 124–139.

Zhao, H., Deng, J., Li, K., and Xu, L., 1999, Magmatism, deep processes and gold deposits in eastern Hebei, China: China University of Geosciences Journal, v. 10, p. 67–71.

Zhao, Y., 1990, The Mesozoic orogenies and tectonic evolution of the Yanshan area: Geological Review, v. 36, p. 1–13 (in Chinese with English summary).

Zhao, Y., Yang, Z., and Ma, X., 1994, Geotectonic transition from Paleoasian system and Paleotethyan system to Paleopacific active continental margin in eastern Asia: Scientia Geologica Sinica, v. 29, p. 105–119 (in Chinese with English summary).

Zheng, Y., Wang, S.Z., and Wang, Y.F., 1991, An enormous thrust nappe and extensional metamorphic core complex newly discovered in Sino-Mongolian boundary area: Science in China, ser. B, v. 34, p. 1145–1152.

Zheng, Y., Zhang, Q., Wang, Y., Liu, R., Wang, S.G., Zuo, G., Wang, S.Z., Lka-asuren, B., Badarch, G., and Badamgarav, Z., 1996, Great Jurassic thrust sheets in Beishan (North Mountains)-Gobi areas of China and southern Mongolia: Journal of Structural Geology, v. 18, p. 1111–1126.

Zheng, Y., Davis, G.A., Wang, C., Darby, B.J., and Hua, Y., 1998, Major thrust system in the Daqing Shan, Inner Mongolia, China: Science in China, ser. D, v. 41, p. 553–560 (also published in Chinese).

Zhu, S., 1997, Nappe tectonics in Sertenshan-Daqingshan, Inner Mongolia: Geology of Inner Mongolia, tot 84, p. 41–48 (in Chinese with English abstract).

Ziegler, A.M., Rees, P.M., Rowley, D.G., Bekker, A., Li, Q., and Hulver, M.L., 1996, Mesozoic assembly of Asia: Constraints from fossil floras, tectonics, and paleomagnetism, *in* Yin, A., and Harrison, T.M., eds., The tectonic evolution of Asia: Cambridge, Cambridge University Press, p. 371–400.

Zonenshain, L.P., Muzmin, M.I., and Natapov, L.M., 1990, Mongol-Okhotsk foldbelt, *in* Page, B.M., ed., Geology of the USSR: A plate tectonic synthesis: American Geophysical Union Geodynamic Series, v. 21, p. 97–120.

Zorin, V.A., 1999, Geodynamics of the western part of the Mongolia–Okhotsk collisional belt, Trans-Baikal region (Russia) and Mongolia: Tectono-physics, v. 306, p. 33–56.

MANUSCRIPT ACCEPTED BY THE SOCIETY JUNE 5, 2000

Geological Society of America
Memoir 194
2001

Structural evolution of the southwestern Daqing Shan, Yinshan belt, Inner Mongolia, China

Brian J. Darby*
Gregory A. Davis[†]
Department of Earth Sciences, University of Southern California, Los Angeles, California 90089-0740, USA
Zheng Yadong
Department of Geology, Peking University, Beijing, China 100871

ABSTRACT

The southwestern Daqing Shan lie ~500 km west of Beijing along the northern edge of the North China craton, within the amalgamated North China plate. The southwestern Daqing Shan are part of the Yinshan belt, an east-west–trending, ≥1100-km-long zone of folding and thrusting mainly of Jurassic–Cretaceous age. Unlike other areas of the Yinshan belt, the southwestern Daqing Shan lack the large-scale, low-angle thrust faults and plutons that obscure and overprint much of the belt's early history. Thus, this area is an ideal locality to understand how and when the belt began to form.

Detailed mapping, construction of numerous cross sections, and analysis of relations between structures and stratigraphic units allow for the distinction of multiple deformational events in the southwestern Daqing Shan. The structural chronology began in post-Cambrian–Ordovician time with the broad folding and downdropping of Cambrian–Ordovician strata. Following deposition of a Permian clastic sequence, Cambrian–Ordovician and Permian strata were folded into an isoclinal anticline. Also during this post-Permian–pre-Early Jurassic contractional deformation, north-directed, low-angle, basement-involved thrusts juxtaposed Archean gneiss over portions of the Permian section. In Early Jurassic time, strata were deposited syntectonically in an east-west–trending half-graben, with its master fault along the southern margin. The half-graben was inverted in the Late Jurassic, resulting in a second generation of basement involved folds (some subisoclinal in geometry), high-angle reverse faulting, and lesser low-angle thrusting. Dominant tectonic vergence appears to be to the north. Minimum Jurassic north-south crustal shortening in the southwestern Daqing Shan is estimated at ~30%. Following Late Jurassic contraction, the southwestern Daqing Shan underwent minor extension, which may be of Cretaceous age. Currently, the southern margin of the Daqing Shan is delineated by an active, south-dipping, normal fault of major proportions.

The tectonics of the southwestern Daqing Shan may be related to far-field effects of plate interactions. Post-Cambrian–Ordovician–pre-Permian broad folding and extension might be related to either Devonian collision along the southern margin of the North China craton, or Permian backarc spreading during south-directed subduction

*E-mail: briand@earth.usc.edu.
[†]E-mail: gdavis@usc.edu.

Darby, B.J., Davis, G.A., and Yadong Z., 2001, Structural evolution of the southwestern Daqing Shan, Yinshan belt, Inner Mongolia, China, *in* Hendrix, M.S., and Davis, G.A., eds., Paleozoic and Mesozoic tectonic evolution of central Asia: From continental assembly to intracontinental deformation: Boulder, Colorado, Geological Society of America Memoir 194, p. 199–214.

along the northern side of the North China craton. Accretion of Paleozoic Mongolian arcs to the North China craton along this subduction boundary (the Suolon suture) in Permian–Triassic time could be responsible for post-Permian–pre-Early Jurassic iso-clinal folding and low-angle thrusting. Late Jurassic contractional deformation in the southwestern Daqing Shan and elsewhere in the Yinshan belt is enigmatic, but is most likely an intraplate response to closure of the Mongolo-Okhotsk ocean 800–1000 km to the north, or accretion of crustal blocks to the south along the Bangong suture.

INTRODUCTION

The Asian continent is perhaps best described as a tectonic collage, composed of several continental blocks and island arc complexes that have been assembled over the Phanerozoic period (Yin and Nie, 1996; Şengör and Natal'in, 1996). Currently, the southwestern Daqing Shan, located ~500 km west of Beijing (Fig. 1), lie within an amalgamated North China plate (Davis et al., 1998a), which includes the North China craton and the Paleozoic Mongolian arcs terrane (Lamb and Badarch, 1997), or the Altaids arcs of Şengör et al. (1993) and Şengör and Natal'in

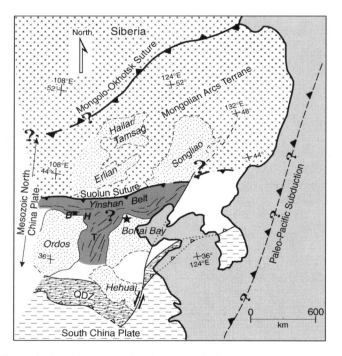

Figure 1. Location map showing tectonic setting of southwestern Daqing Shan (small black box) and Yinshan belt. Dark gray shading indicates areas of Jurassic–Cretaceous contractional deformation and structural trends (black lines). Stippled pattern represents late Mesozoic basins (Hailar-Tamsag, Erlian, Songliao, Bohai Bay, Hehuai, Ordos). Sutures from north to south include Jurassic–Cretaceous Mongolo-Okhotsk, Permian-Triassic Suolon, and Permian(?)–Triassic Qinling-Dabie zone (QDZ). Location of paleo-Pacific subduction zone is approximate. Star—Beijing, H—Hohhot, B—Baotou, T—Taihang Shan. Modified from Davis et al. (1996, 1998a).

(1996). Located just to the north of the southwestern Daqing Shan, the Suolon (or Suolon-Linxi) suture (Fig. 1) separates the North China craton from the Mongolian arcs terrane and is interpreted to record the amalgamation of two of the major crustal fragments that made up much of Asia in Late Permian to Early Triassic time (Zhang et al., 1984; Wang and Mo, 1995; Yin and Nie, 1996; Zheng et al., 1996). The amalgamated North China plate is bordered by the Qinling-Dabie suture zone to the south and the Mongolo-Okhotsk suture to the north (Fig. 1; Davis et al., 1998a). The Qinling-Dabie suture zone separates the North and South China plates with closure of the intervening ocean basin in Permian–Triassic time (Yin and Nie, 1993, 1996). Closure of the Mongolo-Okhotsk ocean and collision of Siberia with the Mongolian arcs terrane from west to east in Jurassic–Cretaceous time formed the Mongolo-Okhotsk suture (Nie et al., 1990; Zonenshain et al., 1990; Ziegler et al., 1996; Yin and Nie, 1996).

The Yinshan belt lies within the North China plate. It is an ~1100-km-long, east-west–striking fold and thrust belt of mainly Jurassic–Cretaceous age. Extending from north of Bohai Bay, the belt can be followed west to Inner Mongolia, where it is concealed by Cenozoic sediments. The Mesozoic Ordos basin defines the southern margin of the belt in the west, and northeast-southwest–trending structures of Mesozoic age in the Taihang Shan coalesce with structures in the Yinshan belt just west of Beijing (Fig. 1; Davis et al., 1998a). This chapter focuses on the structural evolution of the southwestern Daqing Shan in the western portion of the Yinshan belt.

Recent work in the central and eastern Yinshan belt has documented the existence of several large, low-angle thrust faults of Indosinian (Triassic–Early Jurassic?) and Yanshanian (Late Jurassic) age (Davis et al., 1996, 1998a, 1998b, this volume; Zheng et al., 1998). The southwestern Daqing Shan near Baotou lack the large low-angle thrust faults and large plutons found in more eastern parts of the range that obscure and overprint much of the early history of the belt. This makes the southwestern Daqing Shan an ideal locality to study the early history of the Yinshan belt, providing insights of how and when the belt began to form. Detailed mapping of a portion of the southwestern Daqing Shan for this study has revealed a complicated deformational history that began in the early Paleozoic and includes multiple extensional and contractional events.

The original mapping of the southwestern Daqing Shan documented the existence of several coal-bearing basins (Wang,

1928). Regional geologic surveys conducted in the late 1960s and early 1970s by mapping crews from the Regional Geologic Survey Brigade of the Nei Mongol Autonomous region produced 1:200 000 scale maps for the southwestern Daqing Shan and attempted to establish a regional stratigraphy while defining the general structural trends. Although these maps are useful for establishing general rock units, the ages assigned to units sometimes differ from map sheet to map sheet. In 1991, the Nei Mongol Bureau published 1:1 500 000 scale maps with updated unit ages. Wang and Yang (1986) concluded that coal-bearing basins in the southwestern Daqing Shan are bound by thrusts, which have the opposite sense of vergence and no more than 10 km of displacement. More detailed, 1:50 000 scale maps and/ or mapping in progress by the Nei Mongol Bureau of the southwestern Daqing Shan were not available to us.

Several workers have reported major Jurassic–Cretaceous thrust structures to the east and west of the southwestern Daqing Shan. In the Bei Shan and Gobi areas, several hundred kilometers west of the southwestern Daqing Shan, Zheng et al. (1991, 1996), reported the existence of Jurassic thrust sheets with more than 100 km of displacement. Closer to the southwestern Daqing Shan but ~80–100 km to the east of the field area, the Daqing Shan thrust has been documented by several workers (Zhu, 1997; Zheng et al., 1998). Minimum displacement on the Daqing Shan thrust, which involves Precambrian basement rocks, is ~22 km, with transport to the north-northwest (Zheng et al., 1998).

In a summary of Inner Mongolian structure and tectonics, Wang (1996) suggested that deformation in the Daqing Shan is long-lived, with several episodes of deformation. This deformation began in the Permian when basins formed as a result of backarc extension during south-directed subduction along the Suolon suture (Wang, 1996). The Permian basins were then deformed by collision of the North China craton with the Mongolian arcs terrane in Permian–Triassic time (Wang, 1996). Wang (1996) reported that downwarped, coal-bearing basins filled the troughs of synclines formed in Late Triassic to Early Jurassic time, and were subsequently shortened in late Early Jurassic to Middle Jurassic time.

FIELD OBSERVATIONS

Our work attempts to establish a stratigraphy for the field area and to map structures and relations between units in order to decipher the structural chronology and regional tectonic history of the southwestern Daqing Shan. The ages of rock units were established by discussion of the field area with researchers who work in the Yinshan belt at a May, 1998, conference in Hohhot and from recent 1:1,500,000 scale maps (Bureau of Geology and Mineral Resources of the Nei Mongol Autonomous Region [BGMRNMAR], 1991). Mapping was carried out with the use of a hand-held global positioning system (GPS) unit that recorded the position of each station. Stations and data collected were plotted on a latitude-longitude grid, which produced the geologic map in this chapter (Fig. 2).

Stratigraphy

Archean. The oldest of four mapped pre-Cenozoic units is undifferentiated Archean gneiss (Arg, Fig. 2) that boldly crops out over a large portion of the field area and generally forms the highest peaks. This unit is the basement of the southwestern Daqing Shan and includes amphibolitic, granitic, and garnet-bearing gneisses, all of which are strongly layered and foliated. Transects across the Archean gneiss along roads between Heimaban and Yangkeleng and Weijun and Adaohai reveal that foliation in the gneiss typically strikes north-south (±15°) and dips ~80°E (±15°; Fig. 2). This trend is nearly orthogonal to the post-Archean structural trends in the southwestern Daqing Shan, and makes it difficult to explain how the basement gneiss deformed when the overlying sedimentary cover was folded into east-west–trending synclines and anticlines of at least two generations.

Cambrian–Ordovician. Cambrian–Ordovician strata (BGMRNMAR, 1991) are preserved in the northern portion of the field area (Fig. 2). Cambrian–Ordovician strata unconformably overlie the Archean gneiss and the contact is exposed at several localities in the northwestern portion of the field area. In contrast to areas to the east near Hohhot, Proterozoic strata are absent in the southwestern Daqing Shan, suggesting that they may have been eroded prior to Cambrian time. Cambrian–Ordovician strata are mostly buff- to tan-weathering carbonate, although the base of the section has minor glauconitic sandstone, green micaceous shale, and local pebble conglomerate along the nonconformity. Cambrian–Ordovician strata within the study area are folded into an east-west–trending, isoclinal anticline (Figs. 2 and 3, C–C′, F–F′, G–G′, and I–I′).

Permian. A Permian section containing conglomerate, sandstone, shale, and coal rests variably on both Cambrian–Ordovician strata (Figs. 2, 3, and 4A) and Archean gneiss. Permian strata are most abundant in the northern portions of the field area and are highly deformed. The age of this unit is inferred from plant fossils found in coal beds (BGMRNMAR, 1991). The basal or lower portion of the Permian section consists of massive buff-colored, quartzite-cobble conglomerate interbedded with thick coal seams and sandstone. This basal section is usually ~20–40 m thick and overlain by a 10–20-m-thick coal seam. Above the coal seam, thick conglomerate beds, generally cropping out as ridges, are interbedded with much thinner coal seams and sandstone. Above the stratigraphically highest, major thick-bedded, quartzite-cobble conglomerate (a few hundred meters from the base of the section), the Permian section changes gradationally both in color and in lithology, becoming pink to red and containing much more fine-grained strata. Pebble- to cobble-conglomerate beds in the upper part of the section, usually several meters in thickness, are rhythmically interbedded with red mudstone and pink cross-bedded sandstone. The clast composition of the conglomerate changes up-section from quartzite to a mixture of gneiss, quartzite, minor felsic volcanics, and rare carbonate clasts. Carbonate nodules

Figure 2. Simplified geologic map of
southwestern Daqing Shan, Inner Mongolia, China.

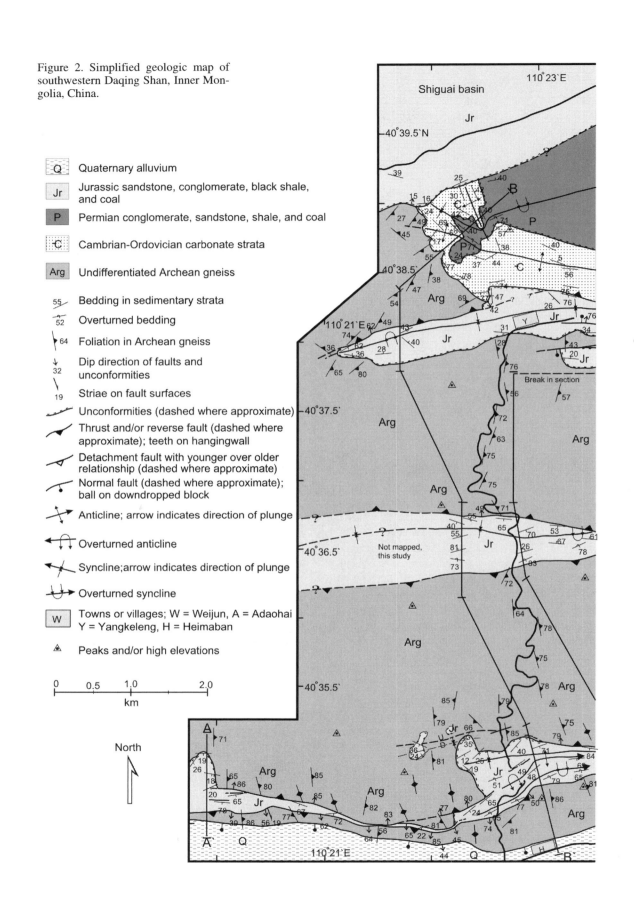

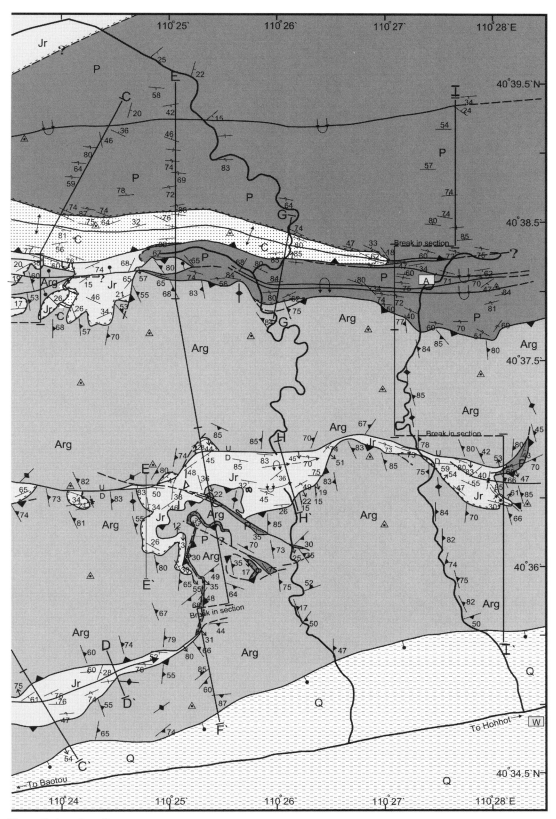

Figure 2. (*continued*)

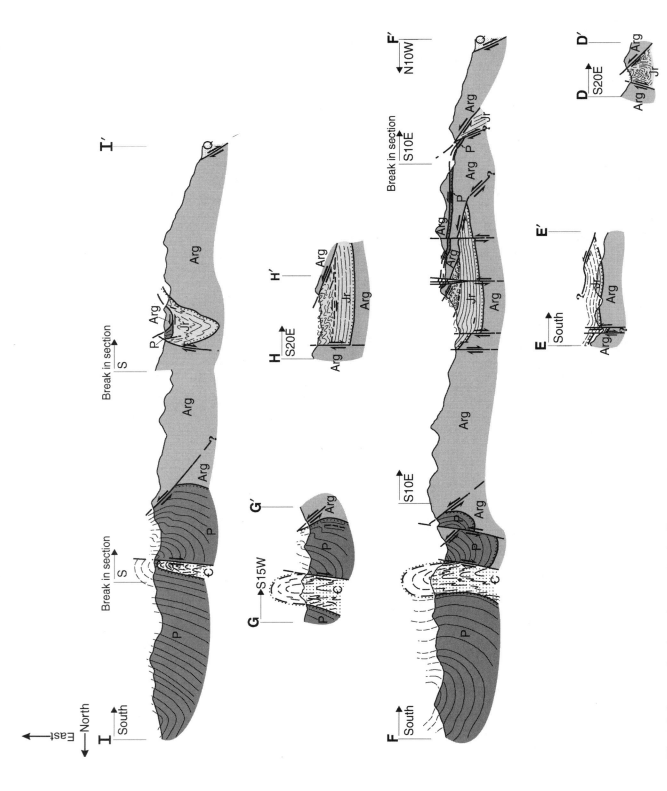

Figure 3. Cross sections in southwestern Daqing Shan, Inner Mongolia, China. (See Fig. 2 for locations.)

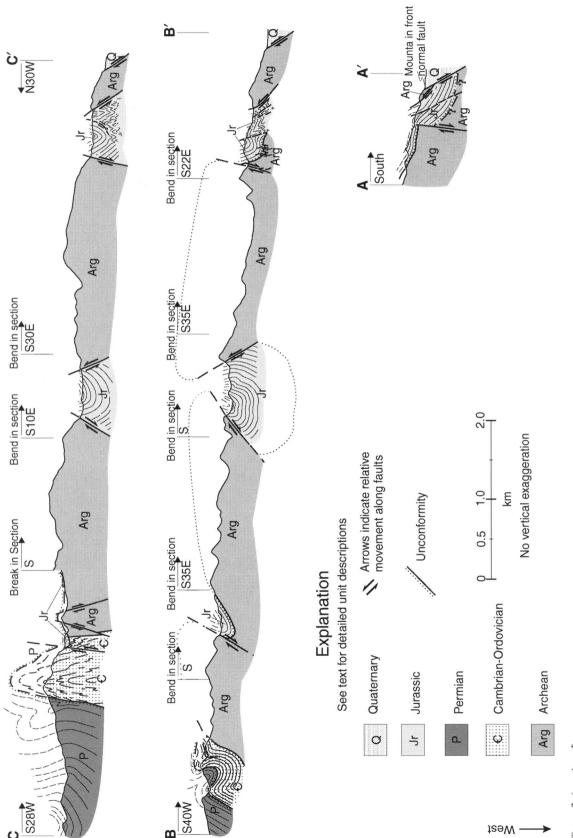

Figure 3. (continued)

Explanation

See text for detailed unit descriptions

Arrows indicate relative
movement along faults

Unconformity

	Quaternary	Q
	Jurassic	Jr
	Permian	P
	Cambrian-Ordovician	Ꞓ
	Archean	Arg

No vertical exaggeration

km
0   0.5   1.0   2.0

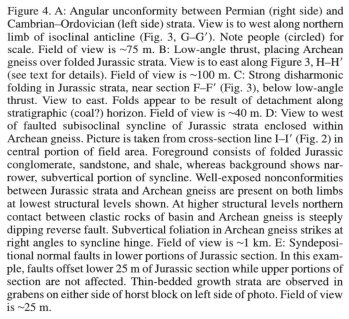

Figure 4. A: Angular unconformity between Permian (right side) and Cambrian–Ordovician (left side) strata. View is to west along northern limb of isoclinal anticline (Fig. 3, G–G'). Note people (circled) for scale. Field of view is ~75 m. B: Low-angle thrust, placing Archean gneiss over folded Jurassic strata. View is to east along Figure 3, H–H' (see text for details). Field of view is ~100 m. C: Strong disharmonic folding in Jurassic strata, near section F–F' (Fig. 3), below low-angle thrust. View to east. Folds appear to be result of detachment along stratigraphic (coal?) horizon. Field of view is ~40 m. D: View to west of faulted subisoclinal syncline of Jurassic strata enclosed within Archean gneiss. Picture is taken from cross-section line I–I' (Fig. 2) in central portion of field area. Foreground consists of folded Jurassic conglomerate, sandstone, and shale, whereas background shows narrower, subvertical portion of syncline. Well-exposed nonconformities between Jurassic strata and Archean gneiss are present on both limbs at lowest structural levels shown. At higher structural levels northern contact between clastic rocks of basin and Archean gneiss is steeply dipping reverse fault. Subvertical foliation in Archean gneiss strikes at right angles to syncline hinge. Field of view is ~1 km. E: Syndepositional normal faults in lower portions of Jurassic section. In this example, faults offset lower 25 m of Jurassic section while upper portions of section are not affected. Thin-bedded growth strata are observed in grabens on either side of horst block on left side of photo. Field of view is ~25 m.

occur near the top of some mudstone horizons and may represent calcisol development. The minimum thickness of the Permian section is 1.5 km, and its top was not seen in the study area.

Jurassic. The Jurassic section, consisting of buff, tan, and/or gray sandstone, conglomerate, black shale, and coal, is very well exposed and preserved in narrow synclinal basins bordered by steeply dipping reverse faults and unconformities (Figs. 2 and 3); the latter are locally overturned. Jurassic strata are in both fault contact with, and unconformably overlie, all older units. Precise age control on Jurassic strata is lacking, although an Early Jurassic age has been assigned by the BGMRNMAR (1991) on the basis of regional correlation and plant fossil ages derived from coal beds. In southern exposures of Jurassic strata, the section is coarser grained and lacks coal and abundant fine-grained strata. A boulder conglomerate facies, with clasts as large as 2 m, is present along the southernmost outcrops of the Jurassic section and extends only a few hundred meters to the north. Pebble conglomerate and sandstone dominate the rest of the southern exposure of the Jurassic. Central exposures of Jurassic strata in the study area contain less pebble conglomerate, more sandstone and shale, and several coal beds. Jurassic strata continue to fine northward; exposures contain mostly sandstone, shale, coal, and minor pebble conglomerate. The thickness of the Jurassic section is variable, but appears to be 500–1000 m, although the top of the section is not exposed.

The clast composition of Jurassic conglomerate and sandstone varies slightly from exposure to exposure, but is mostly dominated by Archean detritus. In south and southwestern exposures, cobbles and boulders of a distinctive red-brown Permian(?) syenitic pluton occur. The provenance for these plutonic clasts is not known, although a pluton of this composition lies about 10 km to the west. To the north, Jurassic strata locally contain clasts and fragments of Cambrian–Ordovician sandstone, green shale, and minor carbonate. This is not surprising, given that Jurassic strata rest along a strongly discordant angular unconformity with the Cambrian–Ordovician. Quartzite cobbles can also be found in the Jurassic conglomerate and may be reworked from the Permian section.

Structures

Reverse and thrust faults. Reverse and thrust faults in the southwestern Daqing Shan trend ~east-west. Most contractional faults in the field area dip steeply (Figs. 2 and 3) and include the reverse faults that bound Jurassic exposures, as well as a fault that places Archean gneiss over the Permian in the northern portion of the field area (Fig. 3, F–F', G–G', and I–I'). The basement-involved faults are exposed in many localities and generally have well-preserved striae and rare tension cracks. The dip of the reverse faults tends to vary along strike, as shown by the southernmost reverse fault that bounds Jurassic strata (e.g., 39°–85° dip variation; cf. Figs. 2 and 3). All of the steep reverse faults cut Jurassic strata, and at least some of these steep faults may represent inversion of normal faults or the reactivation of preexisting basement anisotropies such as faults.

There are two, possibly three, low-angle thrust faults in the field area, two of which are located in the central portion of the map (Figs. 2 and 3, F–F'). The higher of these two thrusts places Archean gneiss on top of a thin sequence of Permian sandstone and conglomerate that has a fairly consistent thickness of ~30–40 m. Permian strata in the lower plate of this thrust unconformably overlie Archean gneiss. The thrust is exposed at several places; one locality has preserved tension cracks and striae that suggest upper plate movement to the north. This upper thrust can be tracked to the south, where it is cut by a steep fault that borders the southern belt of Jurassic strata (Figs. 2 and 3, F–F'). The consistent thickness of the Permian lower plate section may be a function of the overlying thrust following the lowest major coal seam within the Permian section, located ~30 m from the base of the section. Minimum displacement on this thrust based upon structural overlap is ~1.6 km.

The lower of the two thrust faults carries the upper thrust in its upper plate and places Archean gneiss over a sequence of highly folded Jurassic strata (Figs. 2, 3 [F–F'], and 4B). The thrust dips gently to the south and is well exposed at several locations. Striae and tension cracks on the fault surface suggest upper plate transport to ~N10°W, a direction nearly perpendicular to the strike of fold hinges in the footwall of the thrust. The thrust can be followed to the south until Jurassic strata in the footwall are cut out, making it difficult to follow a thrust that juxtaposes upper and lower plate Archean gneiss. A detachment horizon in the Jurassic section is present in the footwall of this thrust, above which strata have undergone strong disharmonic folding (Figs. 2 and 3, F–F'). Below the detachment horizon, which is probably a coal bed, Jurassic strata are only broadly folded (Fig. 3, F–F'). Minimum displacement along this low-angle thrust is ~1.2 km.

A third low-angle thrust fault is illustrated in cross section I–I', in the east-central portion of the field area (Fig. 3). The thrust fault places Archean gneiss and Permian conglomerate and/or sandstone on top of footwall Jurassic strata, all of which are folded into a synform. This thrust is thought to be north directed, despite a small syncline, overturned to the south, located along the southern margin of the thrust (Fig. 3, I–I'). The syncline is interpreted to have formed during folding of the thrust. The complicated geometry in the hanging wall of this folded thrust could be the result of multiple generations of thrust faulting.

Folds: Cambrian–Ordovician–cored isoclinal anticline. The most dramatic example of folding is an isoclinal anticline in the northern part of the field area (Figs. 2 and 3, C–C', F–F', G–G', I–I') that involves Cambrian–Ordovician and Permian strata. Disharmonic folds can be found throughout the isoclinal structure. The Cambrian–Ordovician and Permian strata are separated by an angular unconformity (Fig. 2A) that can be followed from the north limb of the structure, over the hinge of the fold, to the southern limb, where it is only locally preserved and cut by a steep reverse fault. A syncline to the north is most likely related to the formation of the isoclinal anticline (Figs. 2 and 3, C–C', F–F', I–I'). It is likely that the folds to the south of the isoclinal anticline involving the Permian section are related to isoclinal fold formation, although development related to north-directed, basement-involved faulting is possible (Figs. 2 and 3, F–F', G–G', I–I'). Shortening across the isoclinal anticline is ~60%, calculated by removing displacement along faults (constrained by offset unconformities) and flattening the folded unconformity between Cambrian–Ordovician and Permian units.

The Permian clastic sequence directly overlies the Archean basement south of the isoclinal anticline, suggesting that Cambrian–Ordovician strata occupied a structurally low position prior to deposition of Permian sediments. Cambrian–Ordovician strata might have been preserved in the trough of a syncline or in a graben or half-graben. A graben or half-graben requires a normal fault to juxtapose the Cambrian–Ordovician sequence and more southerly Archean basement. This fault may be exposed in the study area along section line C–C' (Fig. 3), just south of the isoclinal structure (see following discussion). Borehole data located at N40°40' and E110°24' (~1 km north of the study area) provided by the Nei Mongol Geological Bureau indicate that Cambrian–Ordovician strata are not present north of the isoclinal anticline. Boreholes begin in Jurassic strata and, at ~150 m depth, penetrate the Archean basement without encountering Permian or Cambrian–Ordovician strata. We favor a graben or half-graben model to explain relations between Cambrian–Ordovician strata and Archean basement.

In the northwestern part of the field area (Fig. 2), the exposed unconformity between the Cambrian–Ordovician and Archean units has not been folded into an isoclinal anticline, as has the Cambrian–Ordovician section above it (Fig. 3, B–B'). This contrast in the amount of folding between the Archean basement and the overlying Cambrian–Ordovician and Permian cover implies that there must be an accommodating structure between Cambrian–Ordovician strata and the basement. This structure, most likely a low-angle thrust that allows the basement to deform differentially from the Cambrian–Ordovician cover, could dip to the north, and is now covered by Permian and Jurassic strata. Alternatively, the low-angle thrust could dip to the south, but was not seen because time constraints did not allow for the detailed mapping of the Archean basement in that area. The intense folding of the Cambrian–Ordovician carbonates might also be aided by slip along detachment horizons in the Cambrian–Ordovician section.

Folds: Jurassic disharmonic folding. Strongly disharmonic folding can be observed in east-central and southern exposures of Jurassic strata (Fig. 3, C–C', D–D', F–F', H–H', I–I'; Fig. 4C). Many of these structures are tight and overturned with the dominant sense of tectonic vergence to the north and gentle plunges to the east. Some of these folds, including some recumbent structures, may have developed as a result of detachment along coal beds or shale horizons. It is interesting to note that an axial planar cleavage is found in only one locality in Jurassic strata in the east-central field area, just west of cross section line I–I' (Fig. 3) along the road between Weijun and Adaohai (Fig. 2). Spacing of the weakly developed cleavage is ~10 cm. The lack of cleavage in the rest of the field area may be

a function of deformation taken up in abundant small faults and shear zones (not shown in Fig. 2) and slip along detachment horizons as discussed herein.

Folds: Basement-involved folding. The basement–sedimentary cover contact is clearly folded throughout the field area (Fig. 3, B–B', F–F', G–G', I–I'), although the strike and dip of the foliation in the basement (~north-south, ~75° E) is at a high angle to structural trends. This geometry complicates the deciphering of basement rock behavior during folding (i.e., folding is at a high angle to the Archean foliation and cannot, therefore, be facilitated by slip on favorably oriented surfaces of anisotropy in the basement gneiss). The tight folding of the Jurassic-Archean and Permian-Archean unconformities into narrow synclines (some subisoclinal in geometry; Fig. 3, B–B', G–G', I–I') requires profound shallow-level folding of the Archean gneiss as well. Figure 4D illustrates the extreme narrowness of a Jurassic sequence and the vertical orientation of its strata; the southern (left) contact is a nonconformity, and the northern contact is a nonconformity at the lowest topographic levels (only locally preserved) and a steep fault at higher levels. It is important to note that the foliation of the bordering Archean gneiss is nearly vertical and strikes at nearly right angles to the fold hinge (Fig. 2).

Although fabric in the basement has been shown to influence the development, orientation, and geometry of basement involved folds in the Laramide Rocky Mountains of western North America (e.g., Chase et al., 1993; Miller and Lageson, 1993; Schmidt et al., 1993), that does not seem to be the case with folds in the southwestern Daqing Shan. Miller and Lageson (1993) reported that when the discordance between the foliation in the basement and overlying sedimentary cover is <25°, the basement folds (due to flexural slip along foliation planes) similarly to the overlying cover. If the angle of discordance between the foliation in the basement and bedding in the overlying cover is great (similar to that in the study area), Miller and Lageson (1993) concluded the basement does not fold, but rather deforms as several rigid blocks separated by steep reverse faults. This seems contrary to mapped relations in the southwestern Daqing Shan (Fig. 2)

In the southwestern Daqing Shan, little to no shear is observed along the basement cover nonconformity where it is exposed. It seems that it is not possible to accommodate folding by allowing for slip to occur along foliation planes. Although steep faults can be seen in the Archean basement, the age of movement along such structures is difficult, if not impossible, to determine because there is no overlying sedimentary cover. Lineations in the Archean gneiss were not abundant and therefore cannot help to constrain basement behavior and determine if block faulting or bulk rotation plays a major role in basement deformation. Field observations (e.g., steep folding of nonconformities) suggest that the basement has been folded, despite the orientation of the fabric in the Archean gneiss.

Normal faults. Several normal faults can be found in the southwestern Daqing Shan, including an active structure at the mountain front. The mountain front fault surface is exposed at several localities, dips ~45°–55°, and truncates and offsets Quaternary alluvium. Fault surfaces typically contain striae that indicate pure dip-slip movement. Fault scarps cutting modern alluvial fans can be seen at many locations along the mountain front.

Normal faults are also found at several localities in lower portions of the Jurassic section (Figs. 3, B–B′, E–E′, and 4). The best-exposed example of these faults occurs in the southern exposures of Jurassic strata along section line B–B′ (Fig. 3). These small normal faults cut the lower 25 m of the Jurassic section, extend into the Archean basement, and have associated growth strata (Fig. 4E). Higher Jurassic strata overlie the faults and are not displaced. This suggests to us that the normal faults are synsedimentary in nature and were active during the early depositional history of the Jurassic strata.

Other potential normal faults include the fault in the north-central field area that separates Cambrian–Ordovician carbonates from Archean gneiss (Fig. 3, C–C′). This fault has limited exposure due to cover by Jurassic strata, but may have had normal displacement. Several small normal faults can be seen cutting folded Jurassic and Permian strata (Fig. 2). These structures tend to be small, with limited displacement (Fig. 3, C–C′, F–F′).

DISCUSSION

Timing relationships

Analyzing field relations between stratigraphic units and structures allows for the relative timing of folding and faulting events to be determined. Cambrian–Ordovician strata unconformably overlie the Archean basement. The lack of Proterozoic strata in the study area suggests that there was a late Precambrian event that allowed for the assumed erosional removal of Proterozoic strata still present ~80 km to the east. An angular unconformity (up to 25°) between the Permian and the Cambrian–Ordovician implies that Cambrian–Ordovician strata were broadly folded before the Permian. A small overturned fold (~1 m across) can be seen just below the angular unconformity near section line I–I′. Just to the south of Cambrian–Ordovician exposures, Permian conglomerate unconformably overlies the Archean basement (Figs. 2 and 3, F–F′, G–G′, I–I′). This relationship requires that a pre-Permian structure juxtaposes the Cambrian–Ordovician and Archean because significant topographic relief is not seen on the base-Permian unconformity. The structure—currently a high-angle fault—is exposed near section line C–C′ (Fig. 3) in the northern part of field area, just south of the isoclinal anticline. It is most likely a normal fault, downdropping Cambrian–Ordovician strata against the Archean basement. It is unclear if the broad folding of the Cambrian–Ordovician section, prior to deposition of the Permian, is related to displacement along this fault or another deformational event.

Portions of the pre-Permian normal fault that juxtaposes Cambrian–Ordovician carbonate strata and the Archean basement appear to be reactivated (see Fig. 3, F–F′). If displacement along the northernmost reverse fault, which dips steeply to the north in section F–F′ (Fig. 3), is backed-out until the unconformity between the Permian and Cambrian–Ordovician, and Permian and Archean is at the same structural level, the steep fault appears to be a normal fault, juxtaposing Cambrian–Ordovician and Archean units. The inversion of portions of this fault is thought to be post-Early Jurassic in age, because along strike to the west the steep reverse fault also deforms Lower Jurassic strata.

Pre-Jurassic–post-Permian (Indosinian) contractional deformation in the southwestern Daqing Shan can be seen in the spectacular isoclinal anticline and related syncline in the northern mapped area (Figs. 2 and 3, B–B′, C–C′, F–F′, G–G′, I–I′). Along the southern edge of the isoclinal anticline, gently dipping Early Jurassic sandstone and shale unconformably overlie nearly vertical Cambrian–Ordovician strata on the south limb of the isoclinal anticline (Fig. 3, C–C′). Because Permian strata (see Fig. 3, I–I′) are also folded in the anticline, the structure is post-Permian and pre-Early Jurassic.

A low-angle thrust, which places Archean gneiss on a thin section of Permian conglomerate and sandstone which, in turn, unconformably overlie the Archean basement (Fig. 3, F–F′), is interpreted to be *pre*-Jurassic (Indosinian) in age. This age is based on the unconformity between the Jurassic and the Archean units. In most places in the field area, the entire Permian section appears to have been stripped or eroded prior to deposition of Jurassic strata: if not, then we would expect to see Jurassic strata on Permian, and Permian strata unconformably overlying the Archean. The nonconformity between the Permian and Archean units, which is preserved in the lower plate of the upper, low-angle thrust (Figs. 2 and 3, F–F′), suggests that thrust faulting occurred prior to stripping of the Permian section from the central and southern portions of the field area and before deposition of Early Jurassic strata.

All high-angle reverse faults cut Lower Jurassic strata. The folded thrust located along section I–I′ (Fig. 3) and the structurally lowest, low-angle thrust (Fig. 3, F–F′) also cut Lower Jurassic strata.

Normal faulting in the southwestern Daqing Shan is thought to have occurred in Early Jurassic time as syndepositional faults and later, following Jurassic contractional deformation. Steeply dipping normal faults cut folded Jurassic and Permian strata in several locations (Figs. 2 and 3, C–C′, F–F′). The age of these structures is poorly constrained because they involve strata no younger than Jurassic. It is unclear if the aforementioned normal faults are related to the present episode of extensional deformation seen at the mountain front or represent collapse and/or relaxation of the crust after contractional deformation; the latter alternative is favored for reasons discussed in the following.

North of the southwestern Daqing Shan, in Inner Mongolia and southern Mongolia, several basins, such as the Erlian basin (Fig. 1), actively extended in late Mesozoic and early Cenozoic time (Traynor and Sladen, 1995; Graham et al., 1996). Zheng et al. (1991), Webb et al. (1999), and Johnson et al. (this volume) have described a major extensional structure, the Yagan–Onch Hayrhan metamorphic core complex, ~400 km west of the study area along the China-Mongolia border. Extension occurred in the Yagan–Onch Hayrhan core complex in Early Cretaceous time (ca. 129–126 Ma; Webb et al., 1999). North of Beijing in the Yunmeng Shan, 500 km to the east of the study area, Davis et al. (1996, this volume) reported formation of a probable mid-Cretaceous extensional metamorphic core complex. Near Hohhot, 80 km to the east of the study area, normal faults down-drop the Cretaceous Lisangou Formation in grabens and cut the older Daqing Shan thrust (Zhu, 1997; Zheng et al., 1998). This information, combined with data from the Erlian basin of Inner Mongolia and Mongolia, the Yagan–Onch Hayrhan area, and the Yunmeng Shan, suggest that late Mesozoic extension was a regional phenomenon. On the basis of regional relationships, normal faults that cut folded Jurassic strata within the south-western Daqing Shan are most likely late Mesozoic in age.

Jurassic reconstruction

Geologic relations in the field area, combined with removal of crustal shortening, can be used to interpret the prefolding and prefaulting geometry of the Jurassic exposures. Constraints on the preshortening geometry include localized syndepositional normal faulting, sediment facies relations, and the location of unconformities. Normal faults, with synextensional growth strata, cut lower portions of the Jurassic section (Fig. 4E). The presence of an unconformity along portions of all the Jurassic exposures suggests that now-isolated Jurassic sections in the study area were once continuous across the Archean basement. With this interpretation, it is possible that the currently observed synclines containing Jurassic strata are separated by anticlines, cored by Archean basement. Due to reverse faulting and anti-clinal folding, the Jurassic–Archean nonconformity between the now-separated Jurassic exposures was at higher structural levels and has since been removed by erosion.

In an attempt to calculate shortening of Jurassic strata across the study area, the unconformity at the base of the Juras-sic has been projected below and above current exposures of the Archean basement (e.g., Fig. 3, B–B′, dashed line). Shortening can then be calculated by flattening the projected unconformity and by estimating its original line length as compared to its present geometry (Fig. 3, B–B′). Using this method, a north-south shortening of ~30% was calculated. This amount is probably a minimum because there are inadequate controls for projecting the unconformity above the Archean.

A half-graben model has been constructed for the Jurassic, removing 30% of post-Jurassic crustal shortening and consider-ing the available structural and sedimentary data (Fig. 5). The northernmost exposure of Jurassic strata, just south of the Cambrian–Ordovician–cored anticline, was used as a pinning point, shortening removed to the south. This location was picked because Jurassic strata unconformably overlie nearly vertical Cambrian–Ordovician strata and because displace-ment along the steep reverse fault separating the Cambrian–Ordovician and Jurassic (Fig. 3, C–C′) is probably <0.5 km (see Fig. 3, F–F′). The isoclinal anticline in the Cambrian–Ordovician might have formed a sedimentologic barrier, or structural high, between Jurassic strata exposed to the south and to the north in the Shiguai basin.

We interpret the southernmost, Jurassic-bounding fault as the master fault controlling formation of a half-graben. This is supported by the presence of a boulder border conglomerate fa-cies (boulders up to 2 m in diameter) that only extends several hundred meters northward into the Jurassic exposure. Boulders in this facies most likely did not travel great distances (<10 km) and were probably derived from a nearby mountain front. Abrupt facies changes occur near some faults such as the north-ern fault in cross section A–A′ (Fig. 3). North of this fault, the Jurassic section consists mostly of sandstone and minor pebble conglomerate, whereas south of the structure, the section con-sists mostly of cobble and small boulder conglomerate, and lesser amounts of sandstone and pebble conglomerate (not shown in Figs. 2 or 3). Additional sedimentologic evidence, such as the general fining of grain-size northward and the re-striction of coal horizons to only central and northern exposures of Jurassic strata also support the reconstruction.

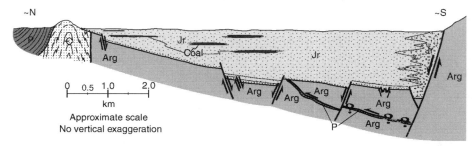

Figure 5. North-south Jurassic basin reconstruction with 30% shortening removed. Jurassic strata unconformably overlie Archean gneiss, nearly vertical Cambrian–Ordovician carbonates (north side), and pre-Jurassic low-angle thrust (see text for details). Note diagrammatized boul-der conglomerate facies along southern margin and coal horizons in central and northern por-tions of basin. Exact location and number of coal horizons are unknown. Jr, Jurassic; Arg, Archean gneiss; P, Permian; C, Cambrian–Ordovician. Approximately true scale.

We use the half-graben model to explain the present architecture of Jurassic exposures. Normal faults, active in the Early Jurassic, are presumed to have been structurally inverted and rotated and now form the steep reverse faults that bound the various Jurassic exposures. If this is correct, preexisting normal faults essentially controlled the geometry of later contractional structures. The reactivation of extensional faults so that they undergo reverse slip, i.e., the process of inversion (Cooper and Williams, 1989), has been documented in many orogenic belts throughout the world, including the Alps (Butler, 1989), the Canadian Cordillera (McClay et al., 1989), and the Yinshan belt (Wang et al., 1996).

Tectonic chronology

The southwestern Daqing Shan have undergone multiple deformational events, beginning in the Paleozoic, thus shedding light on the early history of the Yinshan belt (Fig. 6). The timing of each event is not specific due to lack of radiometric age control, but is bracketed by unit ages. The first Paleozoic deformational event involves Cambrian–Ordovician strata and is pre-Permian in age. During this event, Cambrian–Ordovician strata were broadly folded and juxtaposed with Archean gneiss during normal faulting. It is unclear if broad folding is related to normal faulting, although folding may represent a separate, earlier structural event. Following deposition of the Permian and prior to deposition of Jurassic strata, the southwestern Daqing Shan underwent major north-south shortening (present-day coordinates) that led to isoclinal folding of Cambrian–Ordovician and Permian strata (Fig. 6). To the south, at least one low-angle thrust developed in the basement and propagated into the Permian section (Fig. 3, F–F'). Minimum northward displacement

on this thrust is 1.6 km. Almost all of the Permian strata south of the isoclinal anticline were stripped away during and/or after this event, because Lower Jurassic strata unconformably overlie the basement throughout most of the study area.

Subsequently, and probably in Early Jurassic time, the southwestern Daqing Shan of the study area underwent north-south extension and formation of a half-graben (Figs. 5 and 6). Lower Jurassic strata, which contain local syndepositional normal faults, have a boulder conglomerate southern border facies. The grain size of these strata fine to the north, in support of our interpretation that the master fault of the half-graben was on the southern side of the basin.

Following deposition of the Jurassic sequence, the basin was inverted, reactivating and rotating preexisting normal faults as steep reverse faults and creating the synclinal folds seen in the Jurassic section. Low-angle, north-directed thrusting also occurred during this event as documented by two thrust faults, one of which has a minimum northward displacement of 1.2 km (Fig. 3, F–F'). The other has been folded into a synform (Fig. 3, I–I'). Both carry evidence of post-Permian–pre-Jurassic thrusting in their upper plates (Fig. 3, F–F', I–I'). Another example of low-angle thrust faulting can be found in the southern Jurassic exposure, ~5 km west of the field area and just north of Dongyuan (a town at the mountain front). Here, the low-angle fault offsets the southern Jurassic-bounding fault obliquely by ~0.75 km.

The basement-involved structures described here predate and now lie in the lower plate of the north-northwest–directed Daqing Shan thrust (see Zhu, 1997; Zheng et al., 1998; Davis et al., 1998b) 80 km to the east. The timing of the Daqing Shan thrust is constrained by folded Middle Jurassic strata in its lower plate and a Cretaceous pluton, which intrudes the thrust, dated as

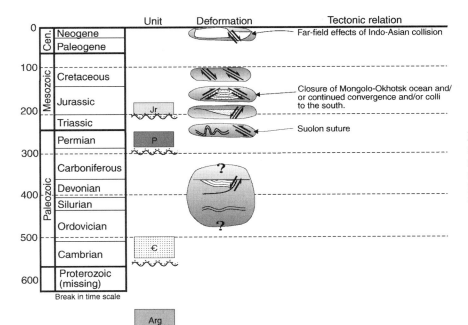

Figure 6. Time scale showing age of strata, deformational events, and possible plate tectonic relationships of structural elements in southwestern Daqing Shan study area. Wavy lines with dots represent unconformities.

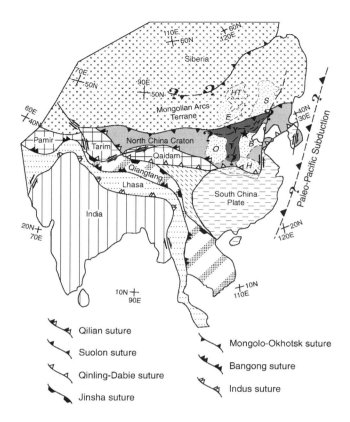

Figure 7. Tectonic map of Asia showing major crustal entities and sutures. Sutures discussed in text include (from north to south) Jurassic–Cretaceous Mongolo-Okhotsk, Permian–Triassic Suolon, Devonian Qilian, Permian(?)–Triassic Qinling-Dabie zone, and Jurassic–Cretaceous Bangong (Zhang et al., 1984; Şengör and Natal'in, 1996; Wang and Mo, 1995; Yin and Nie, 1996). Dark gray shading indicates location of Yinshan belt (Y) and Taihang Shan (T). Black lines in dark gray shading indicate Mesozoic structural trends in Yinshan belt (~east-west) and in Taihang Shan (~northeast). Stippled areas indicate late Mesozoic basins (*HT*, Hailar-Tamsag; *E*, Erlian; *S*, Songliao; *O*, Ordos; *B*, Bohai Bay (latest Mesozoic and Cenozoic], *H*, Hehuai). Location of paleo-Pacific subduction zone is estimated. Modified from Yin and Nie (1996) and Davis et al. (1996, 1998a).

119 ± 2 Ma (Ub-P zircon, Zheng et al., 1998). The folded Middle Jurassic strata in the lower plate have a structural style similar to that of folds in the southwestern Daqing Shan (i.e., basement involved). The lower plate of the thrust might also contain some Lower Cretaceous strata (Zhu Shenyu, 1998, personal commun.), which would further constrain movement along the thrust. On the basis of these timing relations, contractional deformation in the southwestern Daqing Shan is thought to be Late Jurassic or Early Cretaceous in age.

Following Late Jurassic time, the southwestern Daqing Shan have undergone at least two phases of extension (Fig. 6). There are several normal faults that cut Late Jurassic reverse faults and associated folds, although the timing of movement cannot be further constrained. Major normal faults cut the

Daqing Shan thrust and offset the Cretaceous Lisangou Formation 80 km to the east of the field area (Zhu, 1997; Zheng et al., 1998). Based upon these regional trends, we favor a Cretaceous age. The second phase of extension is the active normal faulting observed at the mountain front.

Tectonics

The complicated accretionary tectonics of Asia, which began in the Paleozoic (Şengör and Natal'in, 1996; Yin and Nie, 1996), make it difficult to ascribe each of the deformational events described above to an active plate margin(s) of the appropriate age. Plate tectonic events called upon to account for deformational events in the southwestern Daqing Shan become more questionable with increasing age for two reasons: (1) there is a lack of radiometric age control on deformational events, and (2) in general, less is known about Paleozoic plate interactions in Asia. Figure 7 is a tectonic map of Asia showing the major crustal blocks, accretionary complexes, and the sutures that separate them. Paleozoic and Mesozoic sutures shown in Figure 7 from oldest to youngest are the Devonian Qilian suture, the Permian–Triassic Suolon suture, the Permian–(?)–Triassic–Early Jurassic(?) Qinling-Dabie suture, the Late Triassic–Early Jurassic Jinsha suture, the Late Jurassic–Early Cretaceous Bangong suture, the Late Jurassic–Early Cretaceous Mongolo-Okhotsk suture, and the early Tertiary Indus suture (Zhang et al., 1984; Wang and Mo, 1995; Şengör and Natal'in, 1996; Yin and Nie, 1996).

The oldest event(s) in the southwestern Daqing Shan, post-Cambrian–Ordovician–pre-Permian broad folding, may be a result of the collision of the Qaidam block along the southern margin of the North China craton in Devonian time (Qilian suture) (Figs. 6 and 7). Alternatively, downdropping of Cambrian–Ordovician carbonates prior to deposition of the Permian could be a result of subduction along the northern margin of the North China craton in late Paleozoic time.

Pre-Jurassic (Indosinian) isoclinal folding of the Paleozoic strata in southwestern Daqing Shan and low-angle, basement-involved thrust faulting are probably the result of the collision and amalgamation of the North China craton and the Mongolian arcs terrane along the Permian–Triassic Suolon suture (Figs. 1, 6, and 7). This paleoplate boundary is only a few hundred kilometers north of the southwestern Daqing Shan, and the east-west trend of post-Permian–pre-Jurassic structures in the field area is compatible with that of the suture.

Late Jurassic contraction in the southwestern Daqing Shan and in the Yinshan belt is difficult to relate to any one plate margin. Traditional Chinese views of the Yinshan belt consider it to be an intraplate orogen with limited north-south shortening not related to plate interactions (Wang, 1996; Cui and Wu, 1997). The Jurassic–Cretaceous Yinshan belt may be the result of: (1) closure of the Mongolo-Okhotsk ocean 800–1000 km to the north (Davis et al., 1996, 1998a; Yin and Nie, 1996; Zheng et al., 1998), (2) accretion of crustal blocks to the south, including continued shortening between North and South China (Yin and Nie, 1996) or far-field effects of the collision between the Lhasa and Qiangtang blocks along the Bangong suture. Thermal effects related to westward subduction of a paleo-Pacific plate (Davis et al., 1996, 1998a; Yin and Nie, 1996) appear to have influenced tectonic development of eastern (Yanshan) portions of the Yinshan belt. The distances from paleoplate boundaries such as the Mongolo-Okhotsk and Bangong sutures are great, but observable far-field effects of the Indo-Asian collision (e.g., Molnar and Tapponnier, 1975) illustrate the possibility of far-field Jurassic intraplate responses as well (Hendrix et al., 1992; Ritts and Biffi, this volume).

North of the southwestern Daqing Shan and the Yinshan belt, but south of the Mongolo-Okhotsk suture, are several northeast-trending basins, such as the Erlian and Songliao (Figs. 1 and 7), which have been interpreted as Jurassic–Cretaceous extensional rift basins containing thick sequences of Jurassic–Cretaceous volcanic and clastic strata (Traynor and Sladen, 1995). Graham et al. (1996) and Webb et al. (1999) reported Ar-Ar and K-Ar ages of 156–125 Ma for synrift volcanic rocks in the Erlian basin, which is synchronous with contractional deformation in the Yinshan belt (Davis et al., this volume). Jurassic–Cretaceous extension north of the Yinshan belt makes it difficult to relate the Yinshan belt to the closure of the Mongolo-Okhotsk ocean. Alternatively, Jurassic volcanics found in southern and eastern Mongolian basins may represent a prerift sequence, with extension beginning in the Cretaceous (e.g., Zheng et al., 1991; Davis et al., 1996; Webb et al., 1999) and related to orogenic collapse after the cessation of contraction in the Yinshan belt (e.g., Yin and Nie, 1996).

CONCLUSIONS

Field studies in the southwestern Daqing Shan, in the western portion of the intraplate Yinshan belt, have revealed a complicated structural and tectonic history that began in the Paleozoic and included tectonic inversion as well as basement involvement. Deformation began with post-Cambrian–Ordovician–pre-Permian folding and juxtaposition of Cambrian–Ordovician carbonate against Archean gneiss. Post-Cambrian–Ordovician–pre-Permian deformation might be related to Devonian collision along the southern margin of the North China craton or pre-Permian, south-directed subduction along the northern margin. Contractional deformation of post-Permian and pre-Early Jurassic age is represented by isoclinal folding of Cambrian–Ordovician carbonates and Permian clastic strata. Low-angle, basement-involved thrusting also occurred in this time interval. This contractional deformation was most likely related to the collision and amalgamation of the North China craton and the Mongolian arcs terrane along the Suolon suture to the north.

Lower Jurassic sandstone, conglomerate, black shale, and coal were deposited across the study area in an east-west–trending half-graben with the master, graben-controlling fault along its southern margin. This basin was inverted in Late Jurassic time, resulting in basement-involved structures that include high-angle reverse faults, upright folds (some subisoclinal in geometry), and lesser, low-angle thrusting. Although some of the major folds are more or less symmetrical, smaller folds suggest that the dominant transport direction was to the north. Basement-involved folding at such shallow crustal levels is difficult to explain, given that the orientation of the subvertical foliation in the Archean gneiss is nearly perpendicular to the trend of fold hinges and could not have controlled flexural-slip deformation utilizing basement fabric anisotropy. The cause of Late Jurassic intraplate contraction is problematic and could be related to any of several convergent east Asian plate margins, some of which lie at great distances from the Yinshan belt.

ACKNOWLEDGMENTS

This research was supported by the National Science Foundation (grant EAR-9627909 to Davis), the Geological Society of America (grant to Darby), the American Association of Petroleum Geologists Grants-in-Aid fund (Darby), and the University of Southern California, Department of Earth Sciences graduate student research fund (Darby). We thank Ma Mingbo for his assistance in the field, Wang Jianmin of the Nei Mongol Bureau for logistical support, and Shao Heming for providing borehole data. Critical reviews were given by Marc Hendrix, Steve Graham, and An Yin.

REFERENCES CITED

Bureau of Geology and Mineral Resources of the Nei Mongol Autonomous Region, 1973, Geological map of the Nei Mongol Autonomous Region, People's Republic of China (Guyang sheet): Nei Mongol Geological Bureau, scale 1:200 000.

Bureau of Geology and Mineral Resources of the Nei Mongol Autonomous Region, 1983, Geological map of the Nei Mongol Autonomous Region, People's Republic of China (Tumute Youqi sheet): Nei Mongol Geological Bureau, scale 1:200 000.

Bureau of Geology and Mineral Resources of the Nei Mongol Autonomous Region, 1991, Geological map of the Nei Mongol Autonomous Region, People's Republic of China: Beijing, Geological Publishing House, Scale 1:1 500 000.

Butler, R.W.H., 1989, The influence of pre-existing basin structure on thrust system evolution in the Western Alps, *in* Cooper, M.A., and Williams, G.D., eds., Inversion tectonics: Geological Society [London] Special Publication 44, p. 105–122.

Chase, R.B., Schmidt, C.J., and Genovese, P.W., 1993, Influence of Precambrian-Ordovician rock compositions and fabrics on the development of Rocky Mountain foreland folds, *in* Schmidt, C.J., et al., eds., Laramide basement deformation in the Rocky Mountain foreland of the western United States: Geological Society of America Special Paper 280, p. 45–72.

Cooper, M.A., and Williams, G.D., eds., 1989, Inversion tectonics: Geological Society [London] Special Publication 44, 376 p.

Cui Shengqin, and Wu Zhenhan, 1997, On the Mesozoic and Cenozoic intra-continental orogenesis of the Yanshan area, China, *in* Zheng Yadong, et al., eds., Proceedings of the 30th International Geologic Congress, Volume 14: Utrecht, the Netherlands, VSP p. 277–292.

Davis, G.A., Qian Xianglin, Zheng Yadong, Tong Heng-Mao, Wang Cong, Gehrels, G.E., Shafiquallah, M., and Fryxell, J.E., 1996, Mesozoic deformation and plutonism in the Yunmeng Shan: A metamorphic core complex north of Beijing, China, *in* Yin An, and Harrison, T.M., eds., The tectonic evolution of Asia: Cambridge, Cambridge University Press, p. 253–280.

Davis, G.A., Wang Cong, Zheng Yadong, Zhang Jinjiang, Zhang Changhou, and Gehrels, G.E., 1998a, The enigmatic Yinshan fold-and-thrust belt of northern China: New views on its intraplate contractional styles: Geology, v. 26, p. 43–46.

Davis, G.A., Zheng, Y., Wang, C., Darby, B.J., Zhang, C., and Gehrels, G.E., 1998b, Geometry and geochronology of Yanshan Belt tectonics, *in* Collected works of International Symposium on Geological Science, 100th Anniversary Celebration of Peking University: Department of Geology, Peking University, Beijing, p. 275–292.

Graham, S.A., Hendrix, M.S., Badarch, G., and Badamgarav, D., 1996, Sedimentary record of transition from contractile to extensional tectonism, Mesozoic, southern Mongolia: Geological Society of America Abstracts with Programs, v. 28, no. 7, p. A-68.

Hendrix, M.S., Graham, S.A., Carroll, A.R., Sobel, E.R., McKnight, C.L., Schulein, B.J., and Wang, Z., 1992, Sedimentary record and climatic implications of recurrent deformation in the Tian Shan: Evidence from Mesozoic strata of north Tarim, south Juggar, and Turpan basins, northwest China: Geological Society of America Bulletin, v. 104, p. 53–79.

Lamb, M.A., and Badarch, G., 1997, Paleozoic sedimentary basins and volcanic arc systems of southern Mongolia: New stratigraphic and sedimentologic constraints: International Geology Reviews, v. 39, p. 542–546.

McClay, K.R., Insley, M.W., and Anderton, R., 1989, Inversion of the Ketchican Trough, northeastern British Columbia, Canada, *in* Cooper, M.A., and Williams, G.D., eds., Inversion tectonics: Geological Society [London] Special Publication 44, p. 235–257.

Miller, E.W., and Lageson, D.R., 1993, Influence of basement foliation attitude on geometry of Laramide basement deformation, southern Bridger Range and northern Gallatin Range, Montana, *in* Schmidt, C.J., et al., eds., Laramide basement deformation in the Rocky Mountain foreland of the western United States: Geological Society of America Special Paper 280, p. 73–88.

Molnar, P., and Tapponnier, P., 1975, Cenozoic tectonics of Asia: Effects of a continental collision: Science, v. 189, p. 419–426.

Nie, S., Rowley, D.B., and Ziegler, A.M., 1990, Constraints on the locations of Asian microcontinents in Palaeo-Tethys during the late Paleozoic, *in* McKerrow, W.S., and Scotese, C.R., eds., Paleozoic palaeogeography and biogeography: Geological Society [London] Memoir 12, p. 397–409.

Schmidt, C.J., Genovese, P.W., and Chase, R.B., 1993, Role of basement fabric and cover-rock lithology on the geometry and kinematics of twelve folds in the Rocky Mountain foreland, *in* Schmidt, C.J., et al., eds., Laramide basement deformation in the Rocky Mountain foreland of the western United States: Geological Society of America Special Paper 280, p. 1–44.

Şengör, A.M.C., and Natal'in, B.A., 1996, Paleotectonics of Asia: Fragments of a synthesis, *in* Yin An, and Harrison, T.M., eds., The tectonic evolution of Asia: Cambridge, Cambridge University Press, p. 486–640.

Şengör, A.M.C., Natal'in, B.A., and Burtman, V.S., 1993, Evolution of the Altaid tectonic collage and Palaeozoic crustal growth in Eurasia: Nature, v. 364, p. 299–307.

Traynor, J.J., and Sladen, C., 1995, Tectonic and stratigraphic evolution of the Mongolian People's Republic and its influence on hydrocarbon geology and potential: Marine and Petroleum Geology, v. 12, p. 35–52.

Wang, C.C., 1928, Geology of the Ta Ching Shan range and its coal fields: Geological Survey of China Bulletin, v. 10, p. 4–13.

Wang Guiliang, Shao Zhenjie, Xu Fengyin, Jing Huilin, Zheng Menglin, and Li Haiyu, 1996, Structural inversion of Mesozoic-Cenozoic coal basins in eastern China, *in* Wu Zhengwen, and Chai Yucheng, eds., Tectonics of China—Proceedings of the 1995 annual conference of tectonics in China: Beijing, Geological Publishing House, p. 88–95.

Wang Hongzhen, and Mo Xuanxue, 1995, An outline of the tectonic evolution of China: Episodes, v. 18, p. 6–16.

Wang, J.P., and Yang, Y.D., 1986, Discussion on the mechanism of the thrust-nappe tectonics in Daqingshan and its neighboring area, *in* Collection of structural geology: Beijing, Geological Publishing House, v. 6, p. 1–16 (in Chinese).

Wang, Y., 1996, Tectonic evolutional processes of Inner Mongolia–Yanshan orogenic belt in eastern China during the late Paleozoic–Mesozoic: Beijing, Geological Publishing House, 142 p. (in Chinese with English summary).

Webb, L.E., Graham, S.A., Johnson, C.L., Badarch, G., and Hendrix, M.S., 1999, Occurrence, age, and implications of the Yagan–Oncvh Hayrhan metamorphic core complex, southern Mongolia: Geology, v. 27, p. 143–146.

Yin An, and Nie Shangyou, 1993, An indentation model for North and South China collision and the development of the Tan Lu and Honam fault systems, eastern Asia: Tectonics, v. 12, p. 801–813.

Yin An, and Nie Shangyou, 1996, A Phanerozoic palinspastic reconstruction of China and its neighboring regions, *in* Yin An, and Harrison, T.M., eds., The tectonic evolution of Asia: Cambridge, Cambridge University Press, p. 442–485.

Zhang, Z.H., Liou, J.G., and Coleman, R.G., 1984, An outline of the plate tectonics of China: Geological Society of America Bulletin, v. 95, p. 295–312.

Zheng Yadong, Wang, S.Z., Wang, Y.F., 1991, An enormous thrust nappe and extensional metamorphic core complex newly discovered in Sino-Mongolian boundary area: Science in China, ser. B, v. 34, p. 1145–1154.

Zheng Yadong, Zhang, Q., Wang, Y., Liu R., Wang, S.G., Zuo, G., Wang, S.Z., Lkaasuren, B., Badarch, G., and Badamgarav, Z., 1996, Great Jurassic thrust sheets in Beishan (North Mountains)—Gobi areas of China and southern Mongolia: Journal of Structural Geology, v. 18, p. 1111–1126.

Zheng Yadong, Davis, G.A., Wang Cong, Darby, B.J., and Hua Yonggang, 1998, Major thrust system in the Daqing Shan, Inner Mongolia, China: Science in China, ser. D, v. 41, no. 5, p. 553–560.

Zhu Shenyu, 1997, Nappe tectonics in Sertenshan-Daqingshan, Inner Mongolia: Geology of Inner Mongolia, v. 84, p. 41–48 (in Chinese with English abstract).

Ziegler, A.M., Rees, P.M., Rowley, D.G., Bekker, A., Li Qing, and Hulver, M.L., 1996, Mesozoic assembly of Asia: Constraints from fossil floras, tectonics, and paleomagnetism, *in* Yin An, and Harrison, T.M., eds., The tectonic evolution of Asia: Cambridge, Cambridge University Press, p. 371–400.

Zonenshain, L.P., Muzmin, M.I., and Natapov, L.M., 1990, Mongol-Okhotsk foldbelt, *in* Page, B.M., ed., Geology of the USSR: A plate tectonic synthesis: American Geophysical Union Geodynamic Series, v. 21, p. 97–108.

MANUSCRIPT ACCEPTED BY THE SOCIETY JUNE 5, 2000

Geological Society of America
Memoir 194
2001

Fission-track constraints on Jurassic folding and thrusting in southern Mongolia and their relationship to the Beishan thrust belt of northern China

Trevor A. Dumitru
Department of Geological and Environmental Sciences, Stanford University, Stanford, California 94305, USA
Marc S. Hendrix
Department of Geology, University of Montana, Missoula, Montana 59812, USA

ABSTRACT

Across much of southern Mongolia, thick sequences of Permian through Middle Jurassic strata are generally strongly faulted, folded, and/or tilted, whereas Upper Jurassic and younger strata are much less deformed. This is evidence for as yet poorly documented Jurassic deformation in the region. To better constrain the magnitude and timing of this deformation, we have applied apatite fission-track thermochronology methods at seven localities. Our principal study area is the Noyon Uul syncline, a large (~30 × 10 km), upright, tightly folded, east-west–trending structure that exposes 5400 m of Permian through lowermost Jurassic(?) strata. Fission-track ages and track length distributions are nearly constant through the entire exposed section, with an average age of 138 Ma and mean track length of 13.6 μm. Simulation modeling of the fission-track data indicates major cooling in Late Jurassic time, with cooling below the 100 °C isotherm at 150 ± 10 Ma shallow in the section and 145 ± 10 Ma deep in the section. We infer that this cooling is related to a major episode of thrusting in the Beishan of northern China, ~100 km south of Noyon Uul. The thin-skinned Beishan thrust system accommodated inferred displacements of 140 km at some time between Middle Jurassic and Cretaceous time, and erosional klippen of the thrust system are preserved within 50 km of Noyon Uul. Associated basement-involved shortening of much smaller total magnitude apparently affected Noyon Uul, collapsing a preexisting east-west–trending depositional trough, tightly folding the syncline, and tectonically burying the currently exposed land surface at least 4–5 km. Subsequent rapid erosion of the resulting elevated area unroofed and cooled the fission-track samples in Late Jurassic time. Limited data from the Toroyt and White Mountain localities, ~200 km west and ~170 km southeast of Noyon Uul, respectively, yield very similar Late Jurassic cooling histories, attesting to the regional nature of this shortening episode in southern Mongolia.

In addition to the work at Noyon Uul, we collected fission-track samples near the Tost fault, a major east-northeast–trending structure that passes ~100 km west of Noyon Uul. Recent work has documented 95–175 km of left-lateral slip on the fault; offset and overlapping units indicate that motion occurred at some time between the Early Permian and Late Cretaceous. Fission-track data from four locations close to the fault record major Middle Jurassic cooling, with cooling through the ~100 °C

Dumitru, T.A., and Hendrix, M.S., 2001, Fission-track constraints on Jurassic folding and thrusting in southern Mongolia and their relationship to the Beishan thrust belt of northern China, *in* Hendrix, M.S., and Davis, G.A., eds., Paleozoic and Mesozoic tectonic evolution of central Asia: From continental assembly to intracontinental deformation: Boulder, Colorado, Geological Society of America Memoir 194, p. 215–229.

isotherm at 180 ± 20 Ma, significantly older than the cooling at Noyon Uul. We tentatively infer that these data record major rapid uplift and erosion during Middle Jurassic transpression along the Tost fault, supporting interpretations that motion along the Tost fault was related to an early phase of shortening in the Beishan. Given the east-northeast strike of the Tost fault, very strong transpression, uplift, and erosion would be expected along the fault if its left-lateral slip was related to north-directed shortening in the Beishan.

INTRODUCTION

Central Asia is a large tectonic mosaic consisting mainly of volcanic arcs, microcontinents, ancient subduction zones, and collapsed ocean basins that have assembled in piecemeal fashion throughout Phanerozoic time (e.g., Coleman, 1989; Şengör et al., 1993; Şengör and Natal'in, 1996; Fig. 1A). One consequence of this ongoing process of tectonic amalgamation is central Asia's long and complicated history of intracontinental deformation, caused by the propagation of continental margin tectonics far into the continental interior. Although the best documented intracontinental deformation in central Asia is associated with the continuing collision of India with the southern margin of Asia (Tapponnier and Molnar, 1977, 1979), many studies have described evidence of intracontinental deformation associated with earlier phases of continental growth (e.g., Hendrix et al., 1992; Davis et al., 1998a, 1998b; Darby et al., this volume; Webb et al., this volume; Vincent and Allen, this volume). Not surprisingly, intracontinental deformation in Asia has varied in form, ranging from strain partitioning in strike-slip systems (e.g., Sobel, 1999; Vincent and Allen, 1999; Johnson et al., this volume) to major compressional orogenesis (e.g., Xue et al., 1996; Zhu, 1997; Ritts and Biffi, this volume), to extensional deformation (e.g., Webb et al., 1999; Johnson et al., this volume). A major feature of intracontinental deformation appears to be the localization of orogenesis in perisutural regions between major elements of the Asian tectonic collage (e.g., Hendrix et al., 1992, 1996; Dumitru et al., this volume).

In southern Mongolia, deformation in an intracontinental setting likely began after Late Carboniferous–Permian time, when the north China block collided, causing arc volcanism to terminate and the active southern continental margin of Asia to jump southward to the south side of the north China block (Watson et al., 1987; Şengör et al., 1993; Lamb and Badarch, 1997, this volume; Fig. 1A). The earliest synorogenic sedimentation associated with intracontinental deformation in southern Mongolia appears to be Middle to Late Triassic in age, when >4 km of nonmarine conglomeratic strata were deposited at Noyon Uul (King Mountain; Fig. 1, B and C). These strata are now exposed in a spectacular doubly plunging syncline (Fig. 1D). Stratigraphic, paleocurrent, and provenance data indicate that synorogenic strata at Noyon Uul were deposited in a depositional trough that extended east-west for more than 80 km (Hendrix et al., this volume). Hendrix et al. (1996) related the subsequent strong folding and faulting of these Triassic deposits to a phase

of late Middle Jurassic compressional deformation, on the basis of the Late Jurassic or possibly Early Cretaceous age of strata overlapping deformed Triassic deposits and an inferred connection with a Middle to Late Jurassic phase of major north-vergent thrust faulting <100 km to the south in the Beishan (North Mountains) of northern China and adjacent parts of southernmost Mongolia (Zheng et al., 1996).

Possibly also related to shortening in the Beishan is the Tost fault, a major east-northeast–trending structure that passes ~100 km west of Noyon Uul (Fig. 1C). Lamb et al. (1999) documented 95–175 km of left-lateral strike-slip offset on the fault, with offset and overlapping units constraining motion to some time between the Early Permian and Late Cretaceous. They concluded that this left-lateral slip may have been related to north-vergent shortening in the Beishan to the south, and therefore preferred a Jurassic age for the slip. Given the east-northeast-trend of the Tost fault, north-south–oriented maximum compressive stresses implied by the Beishan thrusts would be expected to induce very strong transpression across the unfavorably oriented Tost strike-slip fault, producing localized folding, thrusting, and rock uplift.

Deformed Triassic through lowermost Jurassic strata crop out widely across southern Mongolia (Shuvalov, 1968, 1969; Yanshin, 1989; Hendrix et al., 1996, this volume; Fig. 1C) and are evidence that the Jurassic deformation inferred at Noyon Uul and along the Tost fault may have been quite widespread. To provide some basic constraints on the timing and magnitude of deformation in the area, we collected sandstone and conglomerate from seven localities for apatite fission-track thermochronology. The fission-track method may be used to date major rock exhumation and cooling that commonly accompany deformation. Our study was designed to address the following questions. (1) Is the inferred Middle to Late Jurassic timing for folding at Noyon Uul corroborated by apatite fission-track data, and do other areas yield fission-track records indicating widespread tectonism at this time? (2) What record of exhumation and cooling, if any, is associated with the Tost fault? Is the timing of any such exhumation consistent with the idea that strike-slip motion on the Tost fault was related to major thrusting in the Beishan to the south?

SAMPLE LOCALITIES

Samples for fission-track thermochronology were collected from seven localities of deformed upper Paleozoic and lower Mesozoic strata (Fig. 1C). Most samples were collected along

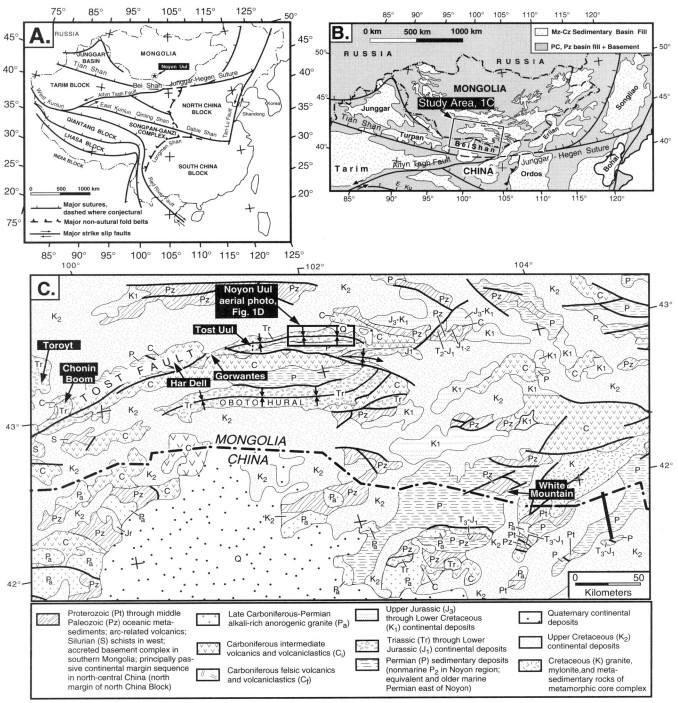

Figure 1. Location maps of study area. A: Simplified tectonic map of Asia showing major features discussed in text. B: Map of major Mesozoic–Cenozoic sedimentary basins and structural features in central and eastern Asia. C: Geologic map of southern Mongolia and adjacent parts of north-central China showing study locations discussed in this chapter. Simplified from Chen et al. (1985), Yanshin (1989), Zhou et al. (1989), Bureau of Geology and Mineral Resources (1991), and Zheng et al. (1991). D (following page): Aerial photograph mosaic of Noyon Uul syncline showing geologic transects. All samples from Noyon Uul discussed in this chapter were collected along Sain Sar Bulag transects and stratigraphically lowest part of Western transect on southern limb of syncline. White dots are original labels on each aerial photograph in mosaic and should be ignored.

. _ . ._ ._ .   Western Transect   ————   South Sain   _ _ _ _ _   South Goyot   ....   North Sain   . . . . .   North Goyot
                of Hendrix et al. (1996)   Sar Bulag section   Canyon section   Sar Bulag section   Canyon section

Figure 1. (*continued*)

the limbs of a very large, doubly plunging, tightly folded syncline at Noyon Uul. This locality exposes 5400 m of Permian through lowermost Jurassic(?) nonmarine siliciclastic strata described in detail by Hendrix et al. (1996, this volume). We sampled two transects on the south limb of the syncline at stratigraphic intervals of 500 m or less and combined them to obtain a composite sample suite through the entire section (Figs. 1D and 2). We also sampled the top and base of the exposed section on the adjacent north limb of the syncline. The stratigraphic positions of samples are shown on the measured sections of Hendrix et al. (1996, this volume).

Depositional ages at Noyon Uul are reasonably well known from floral and faunal remains, as summarized in Hendrix et al. (1996). Strata below the lowermost major conglomerate ridge have yielded Late Permian plant remains and are dated on this basis (Zaitsev et al., 1973; Vakrameyev et al., 1986), although Gubin and Sinitza (1993) placed the Permian-Triassic boundary ~300 m below this conglomerate. Triassic strata at Noyon Uul

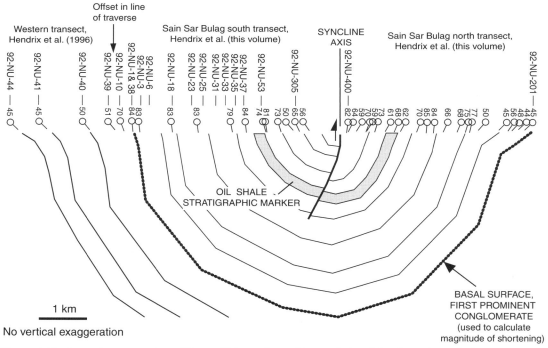

Figure 2. North-south cross section of Noyon Uul syncline, with sample localities. Strata located above and below basal Triassic conglomerate were projected from Sain Sar Bulag and western transects of Hendrix et al. (1996), respectively (Fig. 1D). Apparent dips (below sample numbers) are generally within 1°–2° of measured true dips. We estimate amount of shortening of basal Triassic conglomerate (highlighted) to be ~4.3 km or 34%.

are dated from floral elements, gastropods, insects, and a crustacean (Zaitsev et al., 1973; Badamgarav, 1985; Gubin and Sinitza, 1993). The occurrence of lowermost Jurassic strata at Noyon Uul is less certain. Zaitsev et al. (1973) suggested that the upper stratigraphic levels of the Noyon Uul section are lowermost Jurassic, but did not provide specific evidence for their interpretation.

In addition to Noyon Uul, we collected samples of Triassic and lowermost Jurassic(?) sandstone from six other areas to provide a more regional fission-track data suite and coverage of areas near the Tost fault (Fig. 1C). The depositional ages of these samples are based on lithostratigraphic correlations with the section exposed at Noyon Uul. These samples include the following. (1) Three samples are from the southern limb of Tost Uul, a doubly plunging syncline of Permian through lowermost Jurassic(?) strata located west of Noyon Uul along strike and in close proximity to the Tost fault. Approximately 4200 m of section are exposed in the southern limb of the Tost Uul syncline (see Fig. 3 in Hendrix et al., this volume). Sample 97-TU-1 was collected from Permian strata in the stratigraphically deepest part of the syncline, sample 97-TU-106/107 from Triassic strata ~1400 m higher in the section, and sample 97-TU-108 from the top of the exposed section, ~2250 m higher than TU-106/107. (2) One sample of coarse Triassic sandstone (97-GV-1) is from nonmarine conglomeratic strata at Gorwantes on the far western end of the Tost Uul syncline just east of the Tost fault. (3) One sample is from Har Dell (97-HD-1), a poorly exposed outcrop of Triassic(?) conglomerate near the Tost fault. (4) One sample is from Chonin Boom (97-CB-1A), a poorly exposed outcrop of deformed Triassic nonmarine conglomeratic strata just west of the Tost fault. (5) One sample is from inferred Triassic strata from Toroyt (97-TO-15), where a thick sequence of fluvial and lacustrine strata crop out ~40 km northwest of the Tost fault. (6) One sandstone sample is of inferred Triassic age from White Mountain, located ~170 km southeast of Noyon Uul in a region characterized by abundant erosional klippen of Proterozoic rocks thrust over younger rocks of various ages (Johnson et al., this volume). Zheng et al. (1996) inferred that thrusting at White Mountain occurred in the late Middle Jurassic, based on structural similarities with the Beishan to the south, where coaly Middle Jurassic strata are involved in the thrusting.

FISSION-TRACK METHODOLOGY

The use of fission-track methods for reconstructing time-temperature histories relies on the fact that new fission-tracks form at an essentially constant rate and with an essentially constant initial track length, while at the same time tracks are shortened in length and reduced in apparent number by annealing at elevated subsurface temperatures. The degree of annealing, track number reduction, and track length reduction are strong functions of temperature and are also weakly dependent on heating duration. In apatite, track annealing is slight at subsurface temperatures <60 °C, progressively more severe between about 60 °C and 125 °C, and total at temperatures >125 °C, assuming heating

durations on the order of a few million years (e.g., Green et al., 1989a, 1989b; Dumitru, 2000). Observed fission-track ages and fission-track length distributions may be compared to annealing calibrations to determine the range of time-temperature histories that are compatible with the observed data. This is done by using computer models that simulate the track creation and track annealing processes to test large numbers of possible thermal histories against the observed data (e.g., Green et al., 1989b; Gallagher, 1995). In this chapter we used the 1998 version of the simulation program of Gallagher (1995). The laboratory and modeling procedures used are listed in the footnote to Table 1 and the general approaches to data interpretation are discussed in more detail in Dumitru (2000) and Dumitru et al. (this volume).

FISSION-TRACK RESULTS

Timing and magnitude of shortening of the Noyon Uul syncline

The Noyon Uul syncline is a tightly folded, doubly plunging structure ~30 km long and 10 km wide (Fig. 1D). The core of the syncline contains a steeply dipping fault that is not well exposed but that we interpret on the basis of aerial photographs and field observations to be a north-vergent, out of core thrust resulting from the tight folding of the syncline (Beck, 1998; Fig. 2). The thrust fault does not appear to greatly offset the closure on the east end of the fold (Fig. 1D) and likely is not a major feature.

In order to estimate the general magnitude of shortening across the syncline, we constructed a cross section along the line of transect with no vertical exaggeration (Fig. 2). No subsurface data are available, so we simply used the measured stratigraphic thicknesses in the northern and southern limbs of the syncline (Hendrix et al., 1996, this volume) and assumed approximately concentric folding to project relations into the subsurface. Using this reconstruction, the original undeformed line length of the deepest Triassic conglomerate layer is ~12.5 km, whereas this layer's current separation in the exposed limbs on the surface is ~8.2 km, indicating a shortening of ~4.3 km, or 34%. This is a minimum estimate because of the assumption of concentric folding.

The fission-track samples from the southern limb of the Noyon Uul syncline yield ages that are almost constant, averaging 138 Ma, with only a slight decrease with increasing stratigraphic depth (Fig. 3A). Such a decrease with depth is expected because deeper samples needed to be exhumed farther to cool below a given temperature. Track length distributions are also uniform, with unimodal distributions and with mean lengths that increase very slightly with depth (Figs. 3B and 4).

To constrain the cooling histories of these samples, we used the 1998 version of Gallagher's (1995) program to model the 10 samples wherein large numbers of track lengths could be measured (Fig. 4). The results are very consistent (Fig. 5). Sample 92-NU-44, collected near the stratigraphic bottom of the section, indicates an episode of major cooling in Late Jurassic time, with cooling below the ~100 °C isotherm at 145 ± 10 Ma.

TABLE 1. FISSION TRACK SAMPLE LOCALITY, COUNTING, AND AGE DATA

Sample number	Irradiation number	Latitude (°N)	Longitude (°E)	Strat. (m)	No xls	Spontaneous		Induced		P(χ²) (%)	Dosimeter		Age ± 1σ (Ma)
						Rho-S	NS	Rho-I	NI		Rho-D	ND	
Noyon Uul Syncline, Southern Limb													
97-NU-305-2	SU047-06	43°15′	101°55′	SS-4100	40	0.7655	1076	1.5270	2147	5	1.5020	4390	144 ± 6
92-NU-53	SU016-33	43°14′	101°56′	SS-3480	23	1.0150	960	1.7630	1668	23	1.3820	4116	152 ± 7
92-NU-37	SU016-27	43°14′	101°56′	SS-3160	8	0.9164	253	1.7170	474	34	1.4030	4116	143 ± 11
92-NU-35	SU016-26	43°14′	101°56′	SS-2950	44	0.5243	846	1.0600	1710	60	1.4030	4116	133 ± 6
92-NU-33	SU016-25	43°14′	101°56′	SS-2710	43	0.5426	1680	1.1310	3503	22	1.4100	4116	129 ± 4
92-NU-31	SU016-24	43°14′	101°55′	SS-2500	14	0.2929	361	0.5768	711	19	1.4100	4116	137 ± 9
92-NU-25	SU016-23	43°14′	101°55′	SS-2190	33	0.4046	1078	0.7592	2023	22	1.4170	4116	144 ± 8
92-NU-23	SU016-22	43°13′	101°55′	SS-1920	21	0.5100	471	0.9269	856	33	1.4170	4116	149 ± 9
92-NU-18	SU016-20	43°13′	101°55′	SS-1450	46	0.4685	1064	0.9626	2186	12	1.4240	4116	132 ± 5
92-NU-6	SU016-14	43°13′	101°55′	SS-950	32	0.2802	938	0.5780	1935	61	1.4650	4116	136 ± 6
92-NU-3-2	SU016-13	43°13′	101°54′	SS-490	26	0.5887	1881	1.2980	4147	30	1.4650	4116	127 ± 4
92-NU-1-2	SU016-11	43°13′	101°54′	SS-350	17	0.6804	1131	1.5720	2613	<0.1	1.4710	4116	136 ± 16
92-NU-38	SU016-28	43°13′	101°44′	WT-1930	18	0.4698	1140	0.9277	2251	16	1.3960	4116	135 ± 5
92-NU-10*	SU016-16	43°13′	101°44′	SS- 0	18	0.2111	118	0.4258	238	79	1.4580	4116	138 ± 16
92-NU-39	SU016-29	43°13′	101°44′	WT-1540	7	0.3663	141	0.7040	271	40	1.3960	4116	139 ± 15
92-NU-40	SU016-30	43°13′	101°44′	WT-1090	28	0.4203	711	0.7999	1353	35	1.3890	4116	139 ± 7
92-NU-41†	SU016-31	43°13′	101°44′	WT-400	15	0.3233	88	0.5584	152	56	1.3890	4116	153 ± 21
92-NU-44	SU016-32	43°13′	101°44′	WT-10	46	0.3997	795	0.8090	1609	28	1.3820	4116	130 ± 6
Noyon Uul Syncline, Northern Limb													
97-NU-400	SU047-07	43°15′	101°56′		26	0.2595	557	0.3792	814	0.4	1.5290	4390	201 ± 16
97-NU-201	SU047-01	43°17′	101°56′		40	0.2999	674	0.5967	1341	75	1.4500	4390	139 ± 7
Tost Uul													
97-TU-106/107	SU047-18	43°16′	101°31′		12	0.2967	139	0.5485	257	64	1.6740	4390	172 ± 18
97-TU-108	SU047-19	43°17′	101°31′		9	0.3568	77	0.6302	136	13	1.6740	4390	180 ± 26
97-TU-1	SU047-17	43°14′	101°33′		4	0.2939	56	0.5616	107	95	1.6480	4390	164 ± 27
Other Areas													
97-To-15	SU047-16	43°13′	99°30′		20	0.4410	481	1.0160	1108	7	1.6480	4390	137 ± 8
97-CB-1A	SU047-20	43°06′	99°46′		30	0.3623	653	0.7196	1297	41	1.7010	4390	163 ± 8
97-HD-1	SU047-22	43°13′	100°26′		30	0.5822	755	1.1430	1482	34	1.7270	4390	168 ± 8
97-WM-3	SU047-14	42°00′	103°23′		18	0.8136	314	1.9930	769	20	1.6220	4390	127 ± 9
97-GV-1F	SU047-21	43°13′	101°04′		21	0.3167	266	0.6203	521	16	1.7010	4390	165 ± 13

Note: Abbreviations: Strat.—stratigraphic position (m) above faulted contact between Carboniferous and Permian for samples collected from Western transect (WT) (Hendrix et al., 1996) and above base of measured section for samples collected from south Sain Sar Bulag transect (SS) (Hendrix et al., this volume). No xls—number of individual crystals (grains) dated; Rho-S—spontaneous track density (x 10^6 tracks per cm²); NS—number of spontaneous tracks counted; Rho-I—induced track density in external detector (muscovite) (x 10^6 tracks per cm²); NI—number of induced tracks counted; P(χ²)—χ² probability (Galbraith, 1981; Green, 1981); Rho-D—induced track density in external detector adjacent to dosimetry glass (x 10^6 tracks per cm²); ND—number of tracks counted in determining Rho-D; Age—sample central fission track age (Galbraith and Laslett, 1993), calculated using zeta calibration method (Hurford and Green, 1983). Analyst: T.A. Dumitru.

The following is a summary of key laboratory procedures. Apatites were etched for 20 s in 5N nitric acid at room temperature. Grains were dated by external detector method with muscovite detectors. Samples were irradiated in well thermalized positions in Oregon State University reactor. CN5 dosimetry glasses with muscovite external detectors were used as neutron flux monitors. External detectors were etched in 48% HF. Tracks counted with Zeiss Axioskop microscope with 100x air objective, 1.25x tube factor, 10x eyepieces, transmitted light with supplementary reflected light as needed; external detector prints were located with Kinetek automated scanning stage (Dumitru, 1993). Only grains with c axes subparallel to slide plane were dated. Ages calculated using zeta calibration factor of 389.5. Confined tracks lengths were measured only in apatite grains with c axes subparallel to slide plane; only horizontal tracks measured (within ±~5°-10°), following protocols of Laslett et al. (1982). Track lengths were measured with computer digitizing tablet and drawing tube, calibrated against stage micrometer (Dumitru, 1993). Data reduction done with program by D. Coyle.

The following is a summary of fission-track modeling methods. Modeling done with February 11, 1998, version of "code_trax" program, an updated version of the Monte Trax program of Gallagher (1995). Modeling parameters: (1) used raw track length data (actual lengths of each track) and actual track counts (NS and NI of each grain), (2) used ± 10 Ma uncertainty on observed age, ± 0.35 µm uncertainty on observed mean track length, and ±0.5 µm uncertainty on observed standard deviation of track length distribution, (3) used least likelihood evaluation method, (4) used initial track length of 16.3 µm, (5) used Durango apatite annealing model of Laslett et al. (1987), (6) modeled with 100 simulated tracks and 500 to 2000 Monte Carlo runs (genetic algorithm not used), (9) output plots in Figures 5 and 7 show all runs that pass least likelihood test for both age and track length data. For samples from southern limb of Noyon Uul syncline only, age indicated by trend line in Figure 3 was used as modeling input rather than actual measured age, because trend line is a better estimate of the true age. This was implemented in the modeling by making a small adjustment to the input Rho-D value.

*No confined track lengths could be measured; omitted from Figures 3B and 4.
†Only one confined track length could be measured; omitted from Figure 3B.

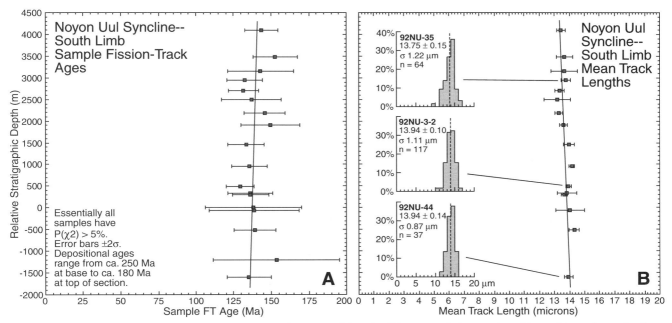

Figure 3. Plots of observed fission-track (FT) ages (A) and mean track lengths (B) versus relative stratigraphic positions of samples from southern limb of Noyon Uul syncline. Note nearly constant ages and mean lengths with depth. In B, three representative high-quality track length histograms are shown to illustrate highly consistent track length distributions.

Sample 92-NU-35, collected near the statigraphic top of the section, indicates similar cooling below the ~100 °C isotherm at 150 ± 10 Ma.

The shallowest fission-track sample on the south limb of the Noyon Uul syncline (97-NU-305; Figs. 2, 4, and 5) must have been subjected to burial temperatures of at least 110–125 °C in Late Jurassic time, because such temperatures are necessary to reset the fission-track clock to zero age. Assuming a nominal geothermal gradient of 25 °C/km, this indicates at least 4–5 km of burial of the shallowest sample and greater burial of the underlying section. Almost certainly this burial was tectonic, in which a thick body of rock was thrust or folded over the syncline during syncline formation, burying the areas that we sampled. This overburden probably consisted at least in part of the now eroded southern limb of the Noyon Uul syncline and perhaps its bounding basement rocks (Fig. 2). This would create a major mountain range, which would then be expected to erode rapidly, exhuming the fission-track samples and allowing them to cool. It is this cooling during postdeformational erosion that is recorded by the fission-track system, and thus the cooling is expected to somewhat postdate the actual folding event.

On the northern flank of the Noyon Uul syncline, the deepest sample (97-NU-201) yielded data indistinguishable from samples deep in the southern limb. However, the shallowest sample (97-NU-400) yielded a significantly older age of 200 Ma with a distinctly broad track length distribution (Fig. 6). This sample is stratigraphically higher than the highest sample on the southern limb of the syncline (Fig. 2). Unlike samples from the southern limb of the syncline, sample 97-NU-400 was not sub-

jected to sufficiently high burial temperatures to totally erase all fission tracks in Late Jurassic time and thus reset the fission-track clock completely to zero age (Fig. 5B). Assuming a nominal 25 °C/km geothermal gradient, these data suggest that this sample was buried only 3–4 km in Jurassic time. From this, we infer that the shallowest sample from the southern limb of the syncline was probably buried only ~4–5 km in Late Jurassic time, deep enough to completely erase all fission tracks, but not much deeper than the shallowest sample on the northern limb of the syncline in which track erasure was incomplete.

Results from other localities

Two broad groups of fission-track ages occur in our overall sample suite. The first group includes the samples from Noyon Uul, plus samples from Toroyt and White Mountain that yield very similar fission-track ages of 137 ± 8 and 127 ± 9 Ma and very similar track length distributions (Table 1; Fig. 6). Modeling of these two samples indicates major latest Jurassic–earliest Cretaceous cooling, with cooling below the ~100 °C isotherm at 145 ± 15 Ma at Toroyt and at 140 ± 15 Ma at White Mountain (Fig. 7), within statistical uncertainty of the modeled times of cooling below ~100 °C at Noyon Uul (Fig. 5).

The second clustering of fission-track data includes the samples from Chonin Boom, Gorwantes, Har Dell, and Tost Uul, all located in close proximity to the Tost fault (Fig. 1C). Of these localities, the data from Chonin Boom are of the highest quality and interpretation of this group of samples relies most heavily on the results from this sample. Sample 97-CB-1A

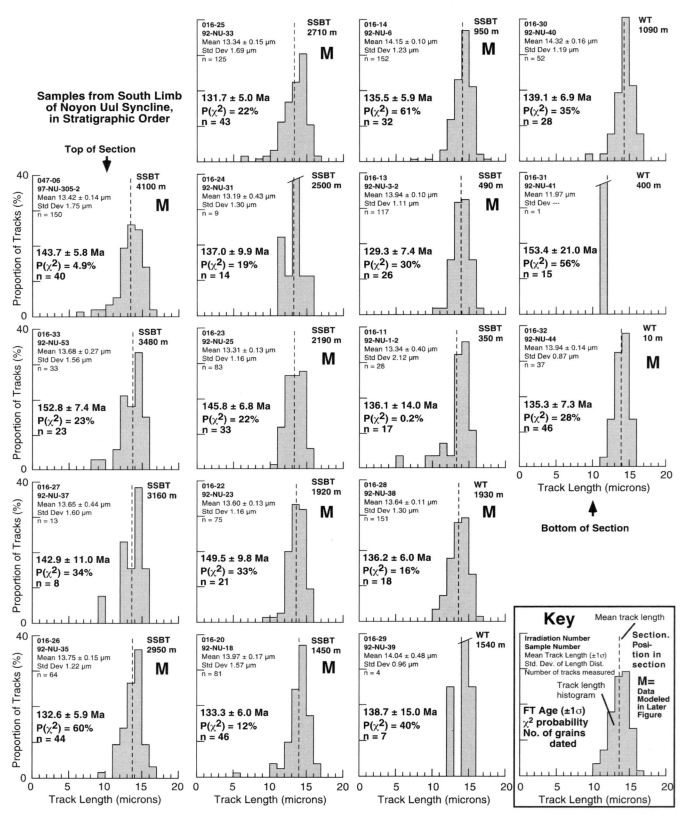

Figure 4. Fission-track length distributions for samples from southern limb of Noyon Uul syncline, in stratigraphic order. Samples with good quality track length data are labeled M and are modeled in Figure 5. These are samples for which more than 50 track lengths could be measured; sample 92-NU-44 (n = 37 lengths) was also modeled because of its critical position at bottom of section.

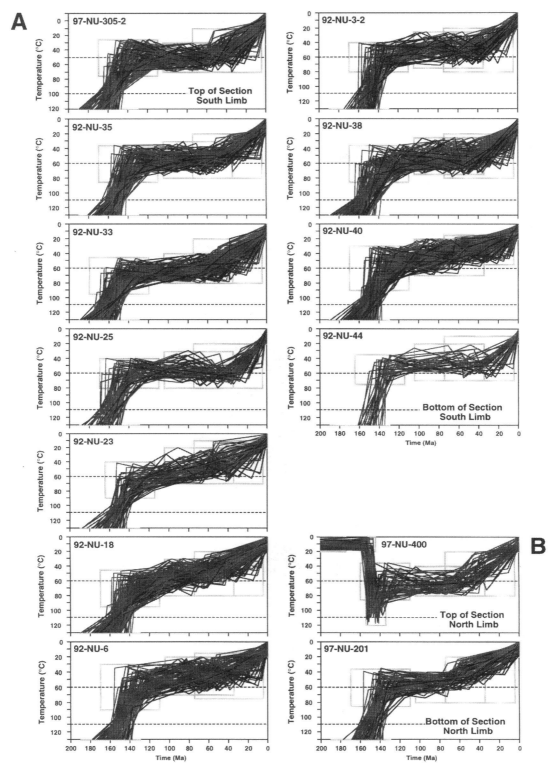

Figure 5. Modeling results for samples from southern (A) and northern (B) limbs of Noyon Uul syncline, displayed in stratigraphic order. Results were computed with 1998 version of computer program of Gallagher (1995), using parameters listed in footnote to Table 1. All cooling paths shown are consistent with observed fission-track age and track length data. Modeling runs for different samples yield highly consistent results, with major cooling in Late Jurassic time with cooling through 100 °C isotherm ca. 150 ± 10 Ma shallow in section and 145 ± 10 Ma deep in section. Samples then underwent additional very slow cooling from Cretaceous to present time. Unlike other samples, fission tracks in sample 97-NU-400 were not totally annealed before Late Jurassic cooling, so the 97-NU-400 model displays path segments for earlier times. Note that (1) fission-track data cannot determine histories at temperatures hotter than ~110°C, so portions of paths between 130 °C and 110 °C are not significant, and (2) pre-Late Jurassic portions of histories are not shown for any sample except 97-NU-400, because fission-track records from pre-Late Jurassic time have been totally overprinted.

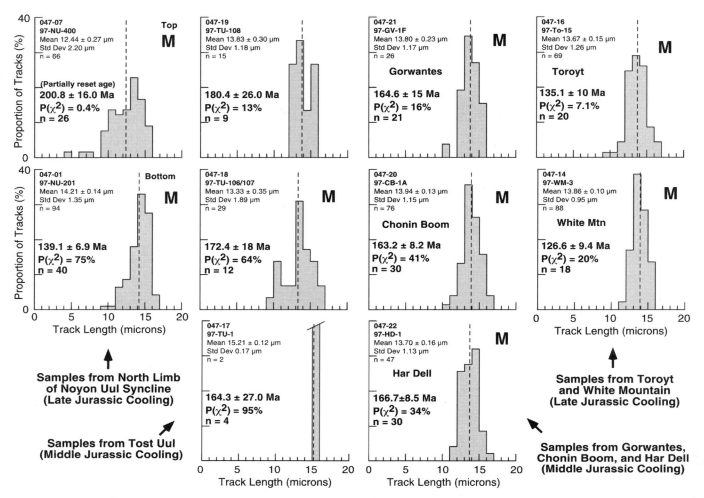

Figure 6. Fission-track length distributions for samples from northern limb of Noyon Uul syncline, Tost Uul, Toroyt, Chonin Boom, Har Dell, White Mountain, and Gorwantes. Samples with good quality track length data are labeled M and are modeled in Figure 5 or Figure 7. These are samples for which more than 50 track lengths could be measured plus three samples for which only 26–47 tracks could be measured, but which were included because no better samples were available in those areas.

yields an age of 163 ± 8 Ma, whereas the ages of the other samples range from 164 ± 15 to 180 ± 26 Ma (Fig. 6; Table 1). Modeling results from Chonin Boom, Gorwantes, and Har Dell suggest cooling below ~100 °C at 180 ± 20 Ma, distinctly older than the cooling at Noyon Uul (Fig. 7). Track length data for sample 97-TU-106/107 from Tost Uul were of marginal quality (only 29 track lengths could be measured) and modeling suggests cooling below 100 °C between 175 and 210 Ma. Given the poorer quality of data, this may be the same ca. 180 Ma cooling episode observed in the other samples.

Collectively, the separation of fission-track ages into two clusters suggests two separate cooling events. Samples from Chonin Boom, Tost Uul, Har Dell, and Gorwantes cooled below the 100 °C isotherm ca. 180 Ma (Late Triassic–Early Jurassic time). Samples from Noyon Uul, Toroyt, and White Mountain cooled below the 100 °C isotherm ca. 145 Ma. It should be noted that both cooling events could have affected each of the sampling areas, but would be recorded by the fission-track system only if they caused samples currently exposed at the surface to have cooled through the apatite fission-track temperature sensitivity window.

DISCUSSION AND CONCLUSIONS

Structural implications of fission-track data

The separation of our apatite fission-track data into two distinct groups is an unexpected and interesting result. We interpret the fission-track data from Noyon Uul, Toroyt, and White Mountain to reflect a period of Late Jurassic erosional unroofing and rapid cooling across the study area. The separation of the three localities by more than 350 km (Fig. 1C) attests to the regional nature of this deformational episode. Of particular interest is the structural setting and cooling history for sample 97-WM-3 from White Mountain. The White Mountain locality exposes klippen of Proterozoic sandstone thrust over Silurian carbonate. These

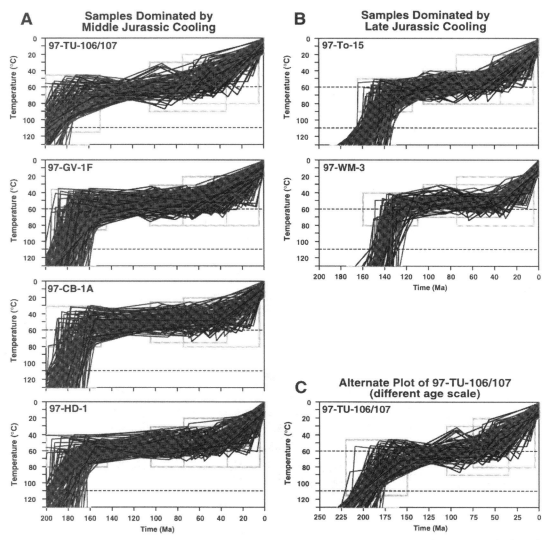

Figure 7. Modeling results for samples from all areas except Noyon Uul, computed using same methods as in Figure 5. Cooling histories define two groups of samples, one exhibiting major Middle Jurassic (A) and one major Late Jurassic (B) cooling. Samples exhibiting Late Jurassic cooling cooled at essentially same time as Noyon Uul samples in Figure 5. Among samples exhibiting Middle Jurassic cooling, cooling through 100 °C isotherm occurred ca. 180 ± 20 Ma. All samples then underwent additional very slow cooling in Late Jurassic to present time. Fission-track data cannot determine histories at temperatures hotter than ~110 °C, so portions of paths between 130 °C and 110 °C are not significant.

klippen were interpreted by Zheng et al. (1996) as the northern erosional remnants of a large-displacement, north-verging thrust belt that is well exposed in the Beishan of northern China and that was active during late Middle Jurassic time. Although we did not map the contact relations of Triassic strata in the White Mountain area, we interpret the period of rapid cooling of sample 97-WM-3 to reflect Late Jurassic to earliest Cretaceous erosional unroofing of Triassic strata involved in this regional late Middle Jurassic thrusting event.

The cooling history and structural geometry of the Noyon Uul syncline suggest that it formed by folding of Permian through lowermost Jurassic(?) strata during late Middle Jurassic time (Fig. 8). Depositional systems, paleocurrent, and prove-

nance analyses at Noyon Uul suggest that the Triassic and lowermost Jurassic(?) sediments were derived from source areas to the north and south of Noyon Uul and were funneled westward, toward a regional lake system (Hendrix et al., this volume). Combined with structural observations of the Noyon Uul syncline, our apatite fission-track results suggest that the depositional trough that governed sediment dispersal patterns during Triassic–earliest Jurassic(?) time collapsed during late Middle Jurassic time to form the Noyon Uul syncline (Fig. 8). During collapse, the currently exposed fission-track sampling areas were tectonically buried at least 4–5 km. Subsequent major erosion of the resulting elevated area produced the cooling recorded in fission-track data from the syncline.

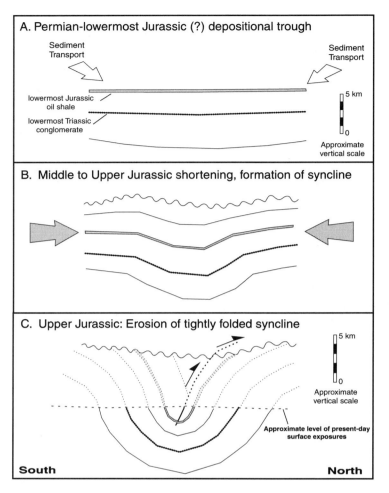

Figure 8. Schematic structural development of Noyon Uul syncline as inferred from this study. A: Permian through earliest Jurassic time. Noyon Uul syncline is broad, east-west–oriented depositional trough receiving sedimentary detritus from north and south and funneling it westward. B: Middle Jurassic time. Transition from sedimentation to structural collapse of Noyon Uul syncline and tectonic burial of fission track samples to 4–6 km. C: Late Jurassic time. Erosional unroofing of Noyon Uul syncline and cooling of fission-track samples through fission-track sensitivity window.

The regional east-west structural grain in southern Mongolia and the presence of at least two other east-west–trending synclines composed of folded Triassic strata (Fig. 1C) are consistent with our interpretation that Jurassic contractile deformation in southern Mongolia was regional in nature. According to Yanshin's (1989) geologic map of Mongolia, an east-plunging syncline involving Triassic and Lower Jurassic strata is present ~30 km east-southeast of the Noyon Uul structure (Fig. 1C). Likewise, ~50 km south of Noyon Uul, a poorly exposed sequence of Triassic coaly strata crop out on the flanks of the Oboto Hural syncline, a large fold that is exposed for ~150 km along strike. Although the poor quality of exposure at Oboto Hural precluded significant field documentation of this locality, we did observe several meter-scale north-vergent folds locally. Between the Oboto Hural syncline and the Noyon Uul–Tost Uul synclinal pair, Permian and Carboniferous strata are broadly folded into a regional antiform that is also likely related to Jurassic north-south contractile deformation.

We tentatively infer that the second group of older fission-track ages from Chonin Boom, Tost Uul, Har Dell, and Gor-

wantes may reflect Middle Jurassic unroofing associated with movement on the Tost fault, located in close proximity to each of these sampling localities (Fig. 1C). Our inference clearly would be buttressed if we had samples farther away from the fault with older ages and could thus demonstrate more definitely that the Middle Jurassic cooling is spatially associated with the fault. Lamb et al. (1999) identified three marker horizons that were offset 95–125 km in a left-lateral sense by the Tost fault. The youngest offset marker was a series of Carboniferous to Lower Permian arc-related volcanics, constraining the maximum age of the Tost fault as Early Permian. Unfortunately, Lamb et al. (1999) were only able to constrain the minimum age of the fault as Late Cretaceous, on the basis of Upper Cretaceous continental deposits that are the oldest strata to overlap the Tost fault (Fig. 1C). Lacking tighter constraints, they speculated that much of the movement on the Tost fault may have occurred during late Middle Jurassic time in association with large-magnitude shortening farther south in the Beishan (Zheng et al., 1996). The ca. 180 Ma cooling indicated by the fission-track data is slightly older than late Middle Jurassic, but is still compatible with Lamb et al.'s (1999) inferences, given the uncertainties in the actual time period over which thrusting occurred in the Beishan and the time of cooling indicated by the fission-track modeling.

If strike-slip motion on the Tost fault were related to generally northward thrusting in the Beishan, the east-northeast strike of the Tost fault would imply strong transpression across the structure. Major rapid uplift and unroofing along transpressional parts of strike-slip fault systems are common (e.g., Dumitru, 1991; Kamp and Tippett, 1993; Bürgmann et al., 1994; Dumitru et al., this volume). If our fission-track data from Chonin Boom, Gorwantes, and Tost Uul reflect erosional unroofing of Permian and Triassic strata uplifted due to slip on the Tost fault, that fault was active in Middle Jurassic time or perhaps somewhat earlier. This would not preclude other periods of motion, but the fission-track modeling indicates that total post-180 Ma uplift and unroofing in the sample areas was limited to no more than about 2 km (Fig. 7).

Implications for the early history of intracontinental deformation

Jurassic shortening and exhumation in the Noyon Uul region resulted directly from the propagation of continental margin tectonics into the interior of Asia. Following the late Paleozoic tectonic consolidation of southern Mongolia (Lamb and Badarch, 1997, this volume), ocean basin closure continued during Triassic and Jurassic time along the southern margin of the newly enlarged Asian continent (Şengör, 1984). The preserved record of this ocean basin and its closure now extends across China as the Dabie–Qinling–Songpan-Ganzi–Kunlun orogenic belt (e.g., Şengör et al., 1993; Yin and Nie, 1996; Zhou and Graham, 1996; Fig. 1A). The effects of Mesozoic collision propagated far northward into central Asia, in a manner analogous to the ongoing structural reactivation and uplift of the Tian Shan, located more than 1200 km north of the Indus suture (Tapponnier and Molnar, 1979; Avouac et al., 1993; Hendrix et al., 1994; Dumitru et al., this volume). The effects of Mesozoic tectonic amalgamation are perhaps best documented in western China, where the Junggar, Tarim, and Turpan basins (Fig. 1B) all contain subsurface thrust or reverse faults that involve strata as young as Triassic or Jurassic and are overlapped by strata as old as Upper Jurassic through Lower Cretaceous, depending on locality (Li and Jiang, 1987; Zhang et al., 1993; Hendrix et al., 1996; Greene et al., this volume). In addition, proximal conglomerates derived from the Tian Shan and its spur ranges such as the Bogda Shan were shed into each of these basins during the Late Triassic and Late Jurassic–Early Cretaceous (Hendrix et al., 1992; Greene et al., this volume; Vincent and Allen, this volume).

In southern Mongolia, evidence for Jurassic deformation is shown on the 1:1 500 000-scale geologic map by Yanshin (1989) as folded Lower and Middle Jurassic rocks that are overlapped by tilted molasse dated as Upper Jurassic and/or Lower Cretaceous, which in turn is covered by relatively undeformed

Upper Cretaceous strata. Earlier, Shuvalov (1968, 1969) had demonstrated on the basis of paleontologic and stratigraphic evidence that there is a major, locally highly angular unconformity between Middle Jurassic and Upper Jurassic strata across much of southern Mongolia. The fact that fission-track data from Noyon Uul, Toroyt, and White Mountain all record cooling below the 100 °C isotherm at ca. 145 Ma is regional evidence of major erosion and unroofing at about this time. Our fission-track data also suggest that this regional rapid cooling may have been preceded by unroofing and cooling associated with an early phase of movement on the Tost fault at ca. 180 Ma, although this hypothesis clearly needs to be further tested through systematic sampling of several transects across the fault.

We interpret our fission-track results from Noyon Uul to be related to north-vergent (and likely northward propagating) Middle-Late Jurassic thrusting and associated shortening in the Beishan, located about 170 km southeast of the field area (Figs. 1 and 9). Zheng et al. (1991) and Zou et al. (1992) documented a north-vergent thrust system south of Noyon Uul, where rocks as old as Proterozoic are locally thrust over Middle Jurassic nonmarine strata with a demonstrable displacement of 70 km and inferred displacement of 140 km. Approximately 500 km farther to the east along tectonic strike, Proterozoic metamorphosed carbonates of the Daqing Shan also have been thrust northward over Jurassic synorogenic conglomerate a minimum of 22 km (Zhu, 1997; Davis et al., 1998a, 1998b; Darby et al., this volume). The Proterozoic over Jurassic thrust relation is mapped as far north as 42°N in the Gansu Province of north-central China (Zou et al., 1992), within 100 km of Noyon Uul, and Proterozoic over Silurian thrust relations were mapped by Zheng et al. (1996) into the White Mountain locality of this study, providing a direct link between the thrust system(s) of northern China and the record of cooling in our study area. However, at present, there is little evidence to suggest that thrusting of Proterozoic rock occurred as far north as Noyon Uul. Rather, the rapid cooling at Noyon Uul probably resulted from rapid erosional unroofing of Jurassic and older strata that were folded by the rheologically controlled collapse of a gently warped Triassic depositional trough, itself formed during an early phase of intracontinental deformation (Hendrix, this volume).

ACKNOWLEDGMENTS

This research was supported by National Science Foundation grant EAR-9614555 and the Stanford-Mongolia Geosciences Industrial Affiliates program. We gratefully acknowledge the field support of the Mongolian Academy of Sciences Geological Institute and the Mongolian Centre for Paleontology and, in particular, thank R. Barsbold, D. Badamgarav, and G. Badarch for their support. We thank A. Chimitsuren, G. Badarch, S.A. Graham, C.J. Johnson, R. Lenegan, B. Ligden, D.J. Sjostrom, and L.E. Webb for outstanding field assistance. We are grateful to D.L. Miller for help with sample preparation and drafting, D.A. Coyle and

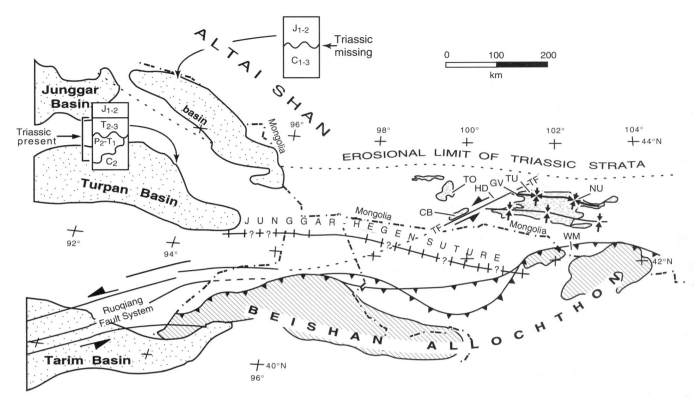

Figure 9. Regional tectonic relations in study area, highlighting structural features discussed in text. Note relative locations and orientations of Tost strike-slip fault, Noyon Uul syncline, and Beishan allochthon and fold-thrust belt. North-south–oriented maximum compressive stresses related to inferred shortening of 140 km in Beishan are inferred to have also caused several kilometers of shortening in Noyon Uul syncline and major transpression across Tost fault. TF, Tost fault; CB, Chonin Boom; TO, Toroyt; GV, Gorwantes; TU, Tost Uul; NU, Noyon Uul; HD, Har Dell; WM, White Mountain sample localities. Modified from Hendrix et al. (1996).

K. Gallagher for providing software, and the University of Oregon Radiation Center for sample irradiations. We thank S.A. Graham for helpful discussions during construction of this chapter and A.D. Hansen and P.J.J. Kamp for providing very useful reviews of an earlier version of the manuscript.

REFERENCES CITED

Avouac, J.P., Tapponnier, P., Bai, M., You, H., and Wang, G., 1993, Active thrusting and folding along the northern Tien Shan and late Cenozoic rotation of the Tarim relative to Dzungaria and Kazakhstan: Journal of Geophysical Research, v. 98, p. 6755–6804.

Badamgarav, D., 1985, Crustacean records: Kazacharthrids from the Triassic of Mongolia: Paleontological Journal, v. 19, p. 135–138.

Beck, M.A., 1998, Sedimentologic characterization of the Noyon Uul syncline, Mongolia: American Association of Petroleum Geologists, Annual Meeting Expanded Abstracts, 0094–0038, APGAB2, v. 1998, on CD.

Bureau of Geology and Mineral Resources, 1991, Regional geology of Nei Mongol (Inner Mongolia) Autonomous Region: People's Republic of China Ministry of Geology and Mineral Resources Geological Memoirs, v. 1, no. 25, 725 p.

Bürgmann, R., Arrowsmith, R., Dumitru, T., and McLaughlin, R., 1994, Rise and fall of the southern Santa Cruz Mountains, California, from fission-tracks, geomorphology, and geodesy: Journal of Geophysical Research, v. 99, p. 20181–20202.

Chen, Z., Wu, N., Zhang, D., Hu, J., Huang, H., Shen, G., Wu, G., Tang, H., and Hu, Y., 1985, Geological map of Xinjiang Uygur Autonomous Region, China: Beijing, Geological Publishing House, scale 1:2 000 000.

Coleman, R.G., 1989, Continental growth of northwest China: Tectonics, v. 8, p. 621–635.

Davis, G.A., Zheng, Y., Wang, C., Darby, B.J., and Yonggang, H., 1998a, Geologic introduction and field guide to the Daqing Shan thrust, Daqing Shan, Nei Mongol, China: Yinshan-Yanshan major thrust and nappe structures field conference, May 8–11: Peking University, Beijing, Department of Geology, Hohhot, Nei Mongol, China, 23 p.

Davis, G.A., Zheng, Y., Wang, C., Darby, B.J., Zhang, C., and Gehrels, G.E., 1998b, Geometry and geochronology of Yanshan belt tectonics, in Collected works of international symposium on geological science, 100th anniversary celebration of Peking University: Beijing, Department of Geology, Peking University, p. 275–292.

Dumitru, T.A., 1991, Major Quaternary uplift along the northernmost San Andreas fault, King Range, northwestern California: Geology, v. 19, p. 526–529.

Dumitru, T.A., 1993, A new computer-automated microscope stage system for fission-track analysis: Nuclear Tracks and Radiation Measurements, v. 21, p. 575–580.

Dumitru, T.A., 2000, Fission-track geochronology, in Noller, J.S., et al., eds., Quaternary geochronology: Methods and applications: American Geophysical Union Reference Shelf, v. 4, p. 131–156.

Galbraith, R.F., 1981, On statistical models for fission-track counts: Mathematical Geology, v. 13, p. 471–478.

Galbraith, R.F., and Laslett, G.M., 1993, Statistical models for mixed fission track ages: Nuclear Tracks and Radiation Measurements, v. 21, p. 459–470.

Gallagher, K., 1995, Evolving temperature histories from apatite fission-track data: Earth and Planetary Science Letters, v. 136, p. 421–435.

Green, P.F., 1981, A new look at statistics in fission-track dating: Nuclear Tracks and Radiation Measurements, v. 5, p. 77–86.

Green, P.F., Duddy, I.R., Gleadow, A.J.W., and Lovering, J.F., 1989a, Apatite fission-track analysis as a paleotemperature indicator for hydrocarbon exploration, *in* Naeser, N.D., and McCulloh, T.H., eds., Thermal history of sedimentary basins: Methods and case histories: New York, Springer-Verlag, p. 181–195.

Green, P.F., Duddy, I.R., Laslett, G.M., Hegarty, K.A., Gleadow, A.J.W., and Lovering, J.F., 1989b, Thermal annealing of fission-tracks in apatite, 4, Quantitative modeling techniques and extension to geological timescales: Chemical Geology, v. 79, p. 155–182.

Gubin, Y.M., and Sinitza, S.M., 1993, Triassic terrestrial tetrapods of Mongolia and the geological structure of the Sain-Sar-Bulak locality, *in* Lucas, S.G., and Morales, M., eds., The nonmarine Triassic: New Mexico Museum of Natural History and Science Bulletin, v. 3, p. 169–170.

Hendrix, M.S., Graham, S.A., Carroll, A.R., Sobel, E.R., McKnight, C.L., Schulein, B.J., and Wang, Z., 1992, Sedimentary record and climatic implications of recurrent deformation in the Tian Shan: Evidence from Mesozoic strata of north Tarim, south Junggar, and Turpan basins, northwest China: Geological Society of America Bulletin, v. 104, p. 53–79.

Hendrix, M.S., Dumitru, T.A., and Graham, S.A., 1994, Late Oligocene–Early Miocene unroofing in the Chinese Tian Shan: An early effect of the India-Asia collision: Geology, v. 22, p. 487–490.

Hendrix, M.S., Graham, S.A., Amory, J.Y., and Badarch, G., 1996, Noyon Uul syncline, southern Mongolia: Lower Mesozoic sedimentary record of the tectonic amalgamation of central Asia: Geological Society of America Bulletin, v. 108, p. 1256–1274.

Hurford, A.J., and Green, P.F., 1983, The zeta age calibration of fission-track dating: Chemical Geology, v. 41, p. 285–317.

Kamp, P.J.J., and Tippett, J.M., 1993, Dynamics of Pacific plate crust in the South Island (New Zealand) zone of oblique continent-continent convergence: Journal of Geophysical Research, v. 98, p. 16105–16118.

Lamb, M.A., and Badarch, G., 1997, Paleozoic sedimentary basins and volcanic-arc systems of southern Mongolia: New stratigraphic and sedimentologic constraints: International Geology Review, v. 39, p. 542–576.

Lamb, M.A., Hanson, A.D., Graham, S.A., Badarch, G., and Webb, L.E., 1999, Left-lateral sense offset of upper Proterozoic to Paleozoic features across the Gobi Onon, Tost, and Zuunbayan faults in southern Mongolia and implications for other central Asian faults: Earth and Planetary Science Letters, v. 173, p. 183–194.

Laslett, G.M., Kendall, W.S., Gleadow, A.J.W., and Duddy, I.R., 1982, Bias in the measurement of fission-track length distributions: Nuclear Tracks and Radiation Measurements, v. 6, p. 79–85.

Laslett, G.M., Green, P.F., Duddy, I.R., and Gleadow, A.J.W., 1987, Thermal annealing of fission-tracks in apatite, 2. A quantitative analysis: Chemical Geology, v. 65, p. 1–13.

Li, J., and Jiang, J., 1987, Survey of petroleum geology and the controlling factors for hydrocarbon distribution in the east part of the Junggar basin: Oil and Gas Geology, v. 8, p. 99–107.

Şengör, A.M.C., 1984, The Cimmeride orogenic system and the tectonics of Eurasia: Geological Society of America Special Paper 195, 82 p.

Şengör, A.M.C., and Natal'in, B.A., 1996, Paleotectonics of Asia: Fragments of a synthesis, *in* Yin, A., and Harrison, M. T., eds., Tectonic evolution of Asia: New York, Cambridge University Press, p. 486–640.

Şengör, A.M.C., Burke, K., and Natal'in, B.A., 1993, Asia: A continent built and assembled over the past 500 million years: Geological Society of America Short Course Notes, 262 p.

Shuvalov, V.F., 1968, More information about the Upper Jurassic and Lower Cretaceous in the southeastern Mongolian Altai: Academy of Sciences of USSR, Doklady Earth Sciences Section, v. 179, p. 31–33.

Shuvalov, V.F., 1969, Continental red beds of the upper Jurassic of Mongolia: Academy of Sciences of USSR, Doklady Earth Sciences Section, v. 189, p. 112–114.

Sobel, E.R., 1999, Basin analysis of the Jurassic–Lower Cretaceous southwest Tarim basin, northwest China: Geological Society of America Bulletin, v. 111, p. 709–724.

Tapponnier, P., and Molnar, P., 1977, Active faulting and tectonics in China: Journal of Geophysical Research, v. 82, p. 2905–2930.

Tapponnier, P., and Molnar, P., 1979, Active faulting and Cenozoic tectonics of the Tien Shan, Mongolia, and Baykal regions: Journal of Geophysical Research, v. 84, p. 3425–3459.

Vakrameyev, V.A., Lebedev, Y.L., and Sodov, Z., 1986, A cycad(?) *Guramsania* gen. nov. from the Upper Permian of south Mongolia: Paleontological Journal, v. 20, p. 95–101.

Vincent, S.J., and Allen, M.B., 1999, Evolution of the Minle and Chaoshui Basins, China: Implications for Mesozoic strike-slip basin formation in central Asia: Geological Society of America Bulletin, v. 111, p. 725–742.

Watson, M.P., Hayward, A.B., Parkinson, D.N., and Zhang, Z.M., 1987, Plate tectonic history, basin development and petroleum source rock deposition onshore China: Marine and Petroleum Geology, v. 4, p. 205–225.

Webb, L.E., Graham, S.A., Johnson, C.L., Badarch, G., and Hendrix, M.S., 1999, Occurrence, age, and implications of the Yagan-Onch Hayrhan metamorphic core complex, southern Mongolia: Geology, v. 27, p. 143–146.

Xue, F., Rowley, D.B., and Baker, J., 1996, Refolded syn-ultrahigh-pressure thrust sheets in the south Dabie complex, China: Field evidence and tectonic implications: Geology, v. 24, p. 455–458.

Yanshin, A.L., 1989, Map of geological formations of the Mongolian People's Republic: Moscow, Academia Nauk USSR, scale 1:1 500 000.

Yin, A., and Nie, S., 1996, A Phanerozoic palinspastic reconstruction of China and its neighboring regions, *in* Yin, A., and Harrison, M.T., eds., Tectonic evolution of Asia: New York, Cambridge University Press, p. 442–485.

Zaitsev, N.S., Mossakovsky, A.A., and Shishkin, M.A., 1973, Type section of upper Paleozoic and Triassic with first remains of Labyrintodonts, south Mongolia: Izvestia Akademii Nauk USSR, v. 7, p. 133–144.

Zhang, G.J., Wang, Z.H., Wu, M., Wu, Q.F., Yang, B., Yang, W.X., Yang, R.L., Fan, G.H., Zheng, D.S., Zhao, B., Peng, X.L., and Yung, T.H., 1993, Xinjiang petroleum province, Junggar basin, *in* Zhai, G., et al., eds., Petroleum geology of China, Volume 15: Beijing, Publishing House of Petroleum Industry, 390 p. (in Chinese).

Zheng, Y.D., Wang, S.Z., and Wang Y.F., 1991, An enormous thrust nappe and extensional metamorphic core complex newly discovered in Sino-Mongolian boundary area: Science in China, v. 34, p. 1145–1154.

Zheng, Y., Zhang, Q., Wang, Y., Liu, R., Wang, S.G., Zuo, G., Wang, S.Z., Lkaasuren, B., Badarch, G., and Badamgarav, Z., 1996, Great Jurassic thrust sheets in Beishan (North Mountains), Gobi areas of China and southern Mongolia: Journal of Structural Geology, v. 18, p. 1111–1126.

Zhou, D., and Graham, S.A., 1996, Songpan-Ganzi complex of the west Qinling Shan as a Triassic remnant-ocean basin, *in* Yin, A., and Harrison, M., eds., Tectonic evolution of Asia: New York, Cambridge University Press, p. 281–299.

Zhou, Z., Zhao, R., Mao, J., Lun, Z., and Zhu, Y., 1989, Geologic map of Gansu Province, China: Beijing, Geological Publishing House, scale 1:1 500 000.

Zhu, S., 1997, Nappe tectonics in Sertengshan—Daqingshan, Inner Mongolia: Geology of Inner Mongolia, v. 84, p. 41–48 (in Chinese with English abstract).

Zou, G., Feng, Y., Liu, C., and Zheng, Y., 1992, A new discovery of early Yanshanian strike-slip compressional nappe zones on middle-southern segment of Beishan Mts., Gansu: Scientia Geologica Sinica, v. 10, p. 309–316.

MANUSCRIPT ACCEPTED BY THE SOCIETY JUNE 5, 2000

Geological Society of America
Memoir 194
2001

Kinematics of exhumation of high- and ultrahigh-pressure rocks in the Hong'an and Tongbai Shan of the Qinling-Dabie collisional orogen, eastern China

Laura E. Webb*
Department of Geological and Environmental Sciences, Stanford University, Stanford, California, 94305-2115, USA
Lothar Ratschbacher
Institut für Geologie, Technische Universität, Bergakademie Freiberg, Bernhard von Cottastrasse 2, D-09596, Freiberg, Germany
Bradley R. Hacker
Department of Geological Sciences, University of California at Santa Barbara, Santa Barbara, California, 93106-9630, USA
Shuwen Dong
Institute of Geomechanics, Chinese Academy of Geological Sciences No. II, Ming zhu Xueyang South Road,
100081 Beijing, China

ABSTRACT

The Hong'an region offers an unique opportunity to investigate the tectonics of the continental collision event preserved in high-pressure (P) and ultrahigh-P metamorphic rocks in the Qinling-Dabie orogen of eastern China. Here, the extensive Cretaceous tectonic and thermal overprint observed in the Dabie Shan is weak. Normal-sense shear along the north-dipping Huwan detachment zone at the northern edge of the Hong'an block occurred ca. 235 Ma. This detachment facilitated the bulk of the exhumation of the high- and ultrahigh-P rocks as a penetratively deformed slab. The high- and ultrahigh-P rocks are exposed in a warped extensional footwall within which kinematic indicators in the high- and ultrahigh-P units show approximately top-to-north shear. Deformation was accompanied by retrograde metamorphism at amphibolite to greenschist facies conditions. Locally, younger northeast-southwest subhorizontal extension is recorded in ductile to brittle fabrics and the timing of deformation is defined by white mica recrystallization ca. 195 Ma. An Early Cretaceous dextral shear zone along the southwest boundary of the Tongbai Shan was synchronous with plutonism and normal to sinistral-oblique slip along the Xiaotian-Mozitang fault, which forms the northern boundary of the Dabie Shan. Coeval dextral and sinistral shear zones along the southwestern and northern margins of these blocks would have caused eastward lateral extrusion of the Tongbai, Hong'an, and Dabie Shan, perhaps driven by collision of the Lhasa block with Eurasia.

INTRODUCTION

Metamorphic diamond and coesite in ultrahigh-pressure (P) rocks of the Qinling-Dabie orogen attest to exhumation

of subducted continental crust from depths >100 km (e.g., Xu et al., 1992; Wang and Liou, 1991). Preservation of ultrahigh-P metamorphic assemblages requires cooling during decompression. This may be achieved by specific tectonic scenerios such as continued underthrusting beneath the ultrahigh-P rocks to suppress heating and/or exhumation of ultrahigh-P rocks in the lower plate of an extensional shear zone (Platt, 1986; Hacker

*Current address: Department of Earth Sciences, Heroy Geology Laboratory, Syracuse University, Syracuse, New York, 13244-1070, USA

Webb, L.E., et al., 2001, Kinematics of exhumation of high- and ultrahigh-pressure rocks in the Hong'an and Tongbai Shan of the Qinling-Dabie collisional orogen, eastern China, *in* Hendrix, M.S., and Davis, G.A., eds., Paleozoic and Mesozoic tectonic evolution of central Asia: From continental assembly to intracontinental deformation: Boulder, Colorado, Geological Society of America Memoir 194, p. 231–245.

and Peacock, 1994). Because of the wide variety and large volume of continental rocks metamorphosed at ultrahigh-*P* conditions, the Hong'an region of the orogen provides an excellent opportunity to address the kinematics of deformation within the deepest levels of the collision. In addition to the study of ultrahigh-*P* tectonics, determining the kinematics and timing of events in this complex region is important for understanding the Mesozoic tectonic evolution of Asia. In this chapter we present new structural data from the Hong'an and Tongbai Shan that are critical to determining which tectonic mechanisms were responsible for the exhumation of the ultrahigh-*P* rocks. The timing of deformation is determined by geochronological data discussed in detail in Webb et al. (1999), Hacker et al. (2000), and Ratschbacher et al. (2000). These data define the geometry, kinematics, and timing of tectonism.

REGIONAL GEOLOGY

The 2000 km northwest-trending Qinling orogen marks the boundary between the Precambrian Sino-Korean and Yangtze cratons (Fig. 1). This composite orogen is characterized by the juxtaposition of the ca. 400 Ma north Qinling–Tongbai metamorphic belt and the ca. 220 Ma south Qinling–Dabie metamorphic belt (Mattauer et al., 1985; Zhai et al., 1998). The south Qinling–Dabie complex formed during Triassic collision between the Yangtze and Sino-Korean cratons. Paleomagnetic data suggest that the Yangtze and Sino-Korean cratons did not fully converge until at least Late Triassic time (Lin et al., 1985; Opdyke et al., 1986) and that the Yangtze craton was oriented ~60° counterclockwise from its present orientation prior to the Middle Triassic. These data suggest that the collision progressed from east to west, closing the intervening oceanic basin as the Yangtze craton rotated with respect to the Sino-Korean craton (e.g., Enkin et al., 1992).

During collision, ultrahigh- and high-*p* metamorphic rocks formed in a north-dipping subduction zone. These rocks are exposed in the easternmost Tongbai, Hong'an, Dabie, and Su-Lu ranges, but transitional blueschist-greenschist facies rocks extend for ~600 km along the southern margin of the orogen through the Tongbai and Qinling areas (e.g., Dong, 1989; Ernst et al., 1991; Fig. 1). The high- and ultrahigh-*P* metamorphic rocks form the core of the Qinling-Dabie orogen in the Dabie Shan and Hong'an regions. North-dipping detachment fault zones define the northern topographic limit of both regions and separate high-grade rocks to the south from greenschist facies and amphibolite facies rocks to the north (cf. Hacker et al., 1996) (Fig. 1). Regionally, the high- and ultrahigh-*P* units trend subparallel to the northwest trend of the orogen. Inferred metamorphic pressures and temperatures increase northward and range from blueschist grade to coesite-eclogite facies rocks (Fig. 1; see Eide [1993] and Liou et al. [1996] for details of the petrologic data). Peak pressures are recorded in eclogite bodies that occur as outcrop-scale blocks or boudins within phengite and/or biotite quartzofeldspathic schists, gneisses, and marbles; locally, eclogites and paragneisses also contain the ultrahigh-*P*

indicator minerals coesite and diamond (e.g., Wang and Liou, 1991; Xu and Su, 1997). These features suggest that the eclogites and host paragneisses are part of a crustal sequence that was subducted, metamorphosed, and exhumed together. Each unit has undergone retrograde amphibolite and greenschist facies metamorphism, and all are intruded by Cretaceous plutons (Hacker et al., 1995). The Cretaceous plutonism is coeval with northwest-southeast extension, most pronounced in the Dabie Shan, where synkinematic plutons of intermediate compositions form the northern half of the range (Hacker et al., 1998).

The Tan-Lu fault marks the eastern boundary of the Dabie Shan (Fig. 1). Approximately 500 km of sinistral strike-slip motion during the Triassic–Jurassic has been postulated on the basis of the apparent offset of the Dabie and Su-Lu ultrahigh-*P* terranes (Okay et al., 1993; Yin and Nie, 1993) (Fig. 1). The tectonic significance and age of this fault are still a matter of debate. Although no Triassic–Jurassic structures associated with the Tan-Lu fault have been identified in the Dabie Shan, the Cenozoic Tan-Lu is morphologically well expressed by steep, eastward-dipping faults with normal to dextral-oblique displacement (Ratschbacher et al., 2000). Cenozoic west-northwest–striking, sinistral strike-slip faults conjugate to the Tan-Lu fault are related to the India-Asia collision (e.g., Zhang et al., 1995).

Radiometric data on the timing of collision

The timing of ultrahigh-*P* metamorphism resulting from collision between the Yangtze and Sino-Korean cratons is documented by a growing body of radiometric ages in the range of 245–170 Ma (Li et al., 1989, 1993; Chen et al., 1992; Okay et al., 1993; Eide et al., 1994; Hacker and Wang, 1995; Ames et al., 1996; Rowley et al., 1997; Xue et al., 1997; Hacker et al., 1998, 2000; Webb et al., 1999; Ratschbacher et al., 2000). Ames et al. (1996) and Rowley et al. (1997) reported U/Pb zircon ages of ca. 220 Ma from ultrahigh-*P* eclogites and host gneisses in the Dabie Shan and interpreted these to be the age of peak ultrahigh-*P* metamorphism. Hacker et al. (1998) found a second population of ca. 240 Ma metamorphic overgrowths on single zircon grains from the ultrahigh-*P* unit in the Hong'an and Dabie Shan using a sensitive high-resolution ion microprobe (SHRIMP). The Early Triassic zircon ages agree with Sm/Nd ages of ca. 245 Ma reported by Li et al. (1993) and Okay et al. (1993) and are postulated to represent the timing of ultrahigh-*P* metamorphism. Thus, the radiometric data suggest that the continental collision may have initiated by the end of the Permian.

DATA AND INTERPRETATIONS

Methods

Field work included several north-south transects across the structural grain of the Hong'an and Tongbai Shan to evaluate the development of structures with respect to metamorphism

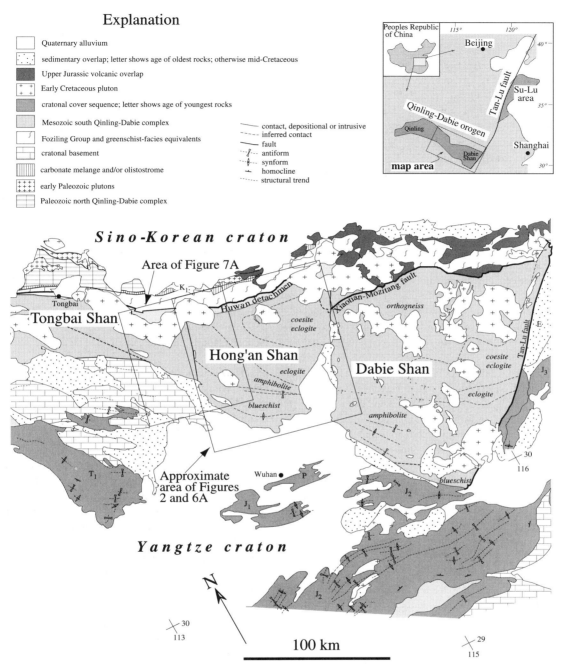

Figure 1. Regional geologic-tectonic map of Qinling-Dabie orogen comprising Tongbai, Hong'an, and Dabie Shan regions. Adapted from Hacker et al. (1995).

and plutonism. These transects included mapping, measurement, and documentation of foliations, lineations, faults and shear zones, and folds, and their relative ages and types. Shear sense was determined from features such as schistosité-cisaillement fabrics, shear bands, and offset dikes and veins. In domains of homogeneous deformation, shear sense was assessed from rotated and/or asymmetric features such as sigma and delta clasts. Mesoscopic fault-slip data were collected to understand fault arrays. The orientation and sense of slip on faults were used to calculate principal stress orientations and shape factors, qualitatively describing the stess geometry. (See Passchier and Trouw [1996] and Angelier [1994] for comprehensive summaries and critical discussions of these structural methods.) Oriented samples of igneous and metamorphic rocks were collected for further analyses, including petrography, microstructural analyses, and [40]Ar/[39]Ar geochronology.

Structural observations

Blueschist unit. The blueschist unit consists of blueschist and transitional blueschist-greenschist facies rocks that form the southern third of the Hong'an Shan (Fig. 1; Eide, 1993). The majority of rocks we observed in the blueschist unit were characterized by gneisses and schists of intermediate compositions including high-P assemblages of quartz + epidote + phengite + albite ± actinolite ± biotite ± garnet ± Na-amphibole ± titanite ± rutile.

Eide (1993) outlined the structure of the blueschist unit based on studies of exposures at Mulan Shan (Fig. 2A); that structural description is corroborated by our regional study. The blueschist unit is characterized by schist and gneiss that are both lineated and foliated (LS tectonites), in which S1 foliations are penetrative and defined by planar minerals and compositional banding. Mineral lineations (L1) are defined by glaucophane and muscovite. Both foliation and lineation are folded around shallowly southeast- or northwest-plunging B2 folds (Fig. 3A). No folds considered B1 generation were measured or observed in the field. The asymmetry of B2 folds indicates vergence that is generally to the north, with the exception of one locality where south vergence was observed (Figs. 3A and 4A). A second foliation (S2), or crenulation cleavage, is axial planar to B2 folds (Fig. 3A) and formed during progressive deformation of the blueschist unit (Eide, 1993). Locally, a secondary lineation (L2) is defined by quartz and chlorite and plunges gently to the southwest (Fig. 3). Late-stage veins filled with quartz and chlorite postdate B2 folds and indicate subhorizontal northeast-southwest extension.

In thin section, S1 foliations are typically openly folded and deformation is concentrated along layers of fine-grained quartz and mica. Quartz grains display dynamic recrystallization and in some rocks these fabrics are partially annealed (Fig. 4B). Syntectonic retrograde metamorphism resulted in the chlorite growth in the late stage of this progressive deformation.

Eide et al. (1994) reported phengite $^{40}Ar/^{39}Ar$ cooling ages of 225.1 ± 0.4 Ma and 222.0 ± 0.3 Ma (total fusion ages) from blueschists at Mulan Shan. A weighted mean $^{40}Ar/^{39}Ar$ cooling age of 171 ± 2 Ma was obtained from K-feldspar at locality D249 (Table 1; Webb et al., 1999).

Amphibolite unit. In Hong'an, greenschist to epidote-amphibolite facies rocks predominate in a northwest-trending, southwest-dipping zone between the blueschist unit and the eclogite-bearing schists and gneisses to the north (Fig. 1; Eide et al., 1994). Lithologically, this zone is dominated by epidote-chlorite-albite schists and quartzofeldspathic schists and gneisses. Mafic assemblages constitute fine layers and/or occur as blocks. Foliations dip predominately southwest and northeast, reflecting folding around approximately northwest-trending, tight, northeast-vergent B2 folds (Fig. 3B). Stretching lineations include early southeast-plunging lineations that are typically associated with the high-P mineral assemblage and subparallel to the B2 folds (Figs. 2 and 3B). They impose apparent dextral transpression on the southwest-dipping foliation. These L1 lineations display progressive overprinting by south- and southwest-plunging lineations associated with the retrogressive mineral assemblage; they impose apparent sinistral transpression (Figs. 2 and 3B). North-directed shear sense within the gneisses and mylonites from this zone and those along the southern margin of the high-P eclogite is indicated by shear bands, as well as by mesoscopic and microscopic σ and δ clasts (e.g., Fig. 4C). In felsic lithologies, albite is found as both synkinematic porphyroclasts, at times showing core-mantle structures and recrystallization (Fig. 4C), or as synkinematic to postkinematic porphyroblasts (Fig. 4D). This suggests that deformation occurred at temperatures of 400–500 °C (e.g., Passchier, 1982; Tullis and Yund, 1991), and that locally cooling may have postdated deformation. Mafic lithologies exhibit greenschist facies overprints, such as retrograde growth of chlorite on garnet.

White mica weighted mean $^{40}Ar/^{39}Ar$ cooling ages of 212 ± 1 Ma, 237 ± 2 Ma, and 247 ± 2 Ma were obtained from three sample localities within this zone (Table 1; Webb et al., 1999). In addition, Eide et al. (1994) reported a phengite $^{40}Ar/^{39}Ar$ total fusion age of 231 ± 2 Ma from the D317b locality.

High- and ultrahigh-P eclogite units. The high- and ultrahigh-P eclogite units are characterized by penetratively deformed, compositionally distinct lithologies that outline a series of northwest-trending synforms and antiforms that are overturned to the north (Figs. 2, A and B). The antiforms are topographically expressed as elliptical domes. The domes are characterized by cores of megacrystic, quartzofeldspathic gneiss mantled by metasedimentary lithologies (including quartz-albite-muscovite schists, garnet-muscovite-biotite gneisses, marble, and minor graphitic schists) that contain blocks and boudins of eclogite. This pseudostratigraphy may represent primary layering of subducted granitic basement of the Yangtze craton and its sedimentary and volcanic cover sequences.

Both high-P and ultrahigh-P eclogite units exhibit one dominant foliation (S1) that is common both to the host gneisses and schists and to the partially retrogressed eclogite and amphibolite blocks. In felsic lithologies, S1 foliations are characterized by parallel alignment of quartz + albite + phengite (± K-feldspar ± biotite ± garnet ± epidote ± titanite). Partially retrogressed and mylonitized eclogite bodies are typified by garnet + clinopyroxene + phengite + quartz + epidote (± hornblende ± biotite ± rutile ± kyanite). S1 predominately dips moderately to shallowly southwest and northeast, reflecting folding around around lineation-subparallel, isoclinal folds (B2) that indicate high-strain deformation; the mineral stretching lineations trend southwest or southeast (Fig. 5A). In the southern region of the high-P eclogite unit, shear bands dip southwest and kinematic indicators such as σ clasts, δ clasts, and S-C fabrics universally show top-to-north shear (Figs. 2, 4E, and 5A). Along the northern margin of the high- and ultrahigh-P eclogite units, the foliation rolls over and dips north

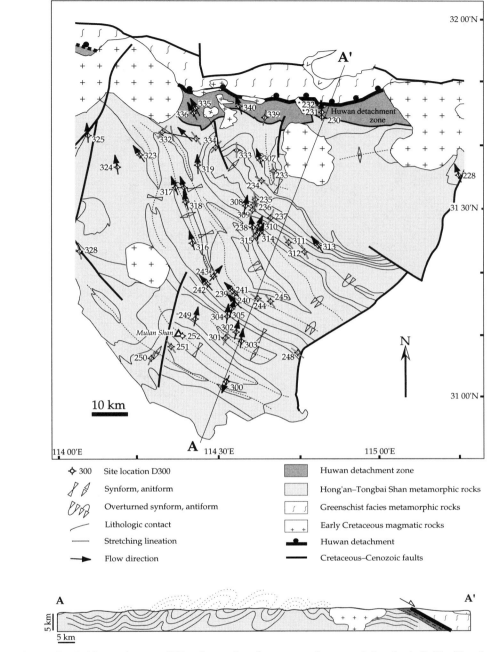

A

B

Figure 2. A: Triassic kinematic map of Hong'an region. Stereonets of structural data for individual localities are in Figures 3, 5, and 6. Black arrows portray flow direction during extensional ductile flow within high- and ultrahigh-pressure units. Lithological contacts outline large-scale folds. See Figure 1 for approximate locations of metamorphic boundaries. B: Schematic cross section across Hong'an Shan illustrating overturned folds and roll-over of foliation into north-dipping Huwan detachment. Note that section is at high angle to tectonic transport direction, which is subparallel to northwest trend of fold axes.

A

BLUESCHIST UNIT

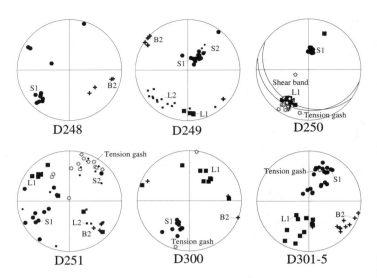

B

AMPHIBOLITE UNIT

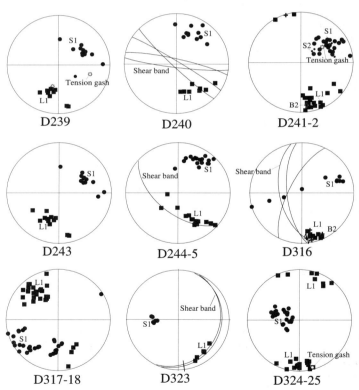

Figure 3. Structural data for (A) blueschist unit and (B) amphibolite unit represented in equal-area, lower hemisphere projections. Lettered annotations of data: S, pole to foliation (solid circles); L, stretching lineation (solid squares); B, fold axes. Number refers to generation of structure. Great circles represent shear bands. Whole, half, and headless arrows represent degree of certainty of shear-sense determination for that shear band where observed. See Figure 2 for location of structural data stations presented here.

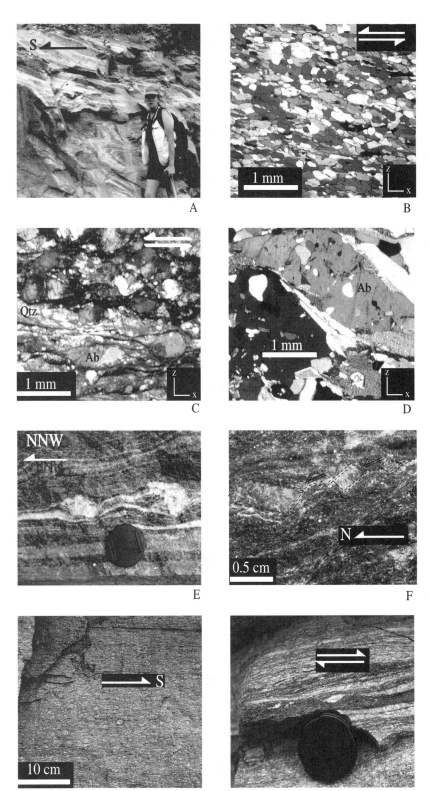

Figure 4. A: South-vergent, tight folds from locality D300 in blueschist unit. B: Partially annealed quartz grain-shape fabric parallel to incremental foliation (locality D301). Cut is parallel to finite lineation (X) and normal to finite foliation (Z). Inclination of quartz grains with respect to finite foliation indicates top-to-north shear sense (sinistral in reference frame of photo). C: Mantled albite porphyroclasts from mylonite in amphibolite unit indicating sinistral shear sense (sample D240b). D: Typical synkinematic to postkinematic albite porphyroblast is observed throughout amphibolite unit (sample D305b). E: Typical mesoscopic kinematic indicator in high- and ultrahigh-pressure eclogite units. Asymmetric boudinage of felsic layer in amphibolite facies gneiss gives top-to-north-northwest shear (locality D318). Diameter of lens cap is 50 mm. F: Outcrop photo of mylonitic eclogite from Huwan detachment zone (locality D335). Garnet boudins indicate top-to-north, normal-sense shear. G: Top-to-south sigma clasts formed by feldspars in gneissic core of dome (locality D253). View is parallel to lineation and normal to foliation. H: Dextral shear indicated by asymmetric boudinage of felsic layer in mylonite at locality D260. Diameter of lens cap is 50 mm.

TABLE 1. ^{40}Ar/^{39}Ar DATA SUMMARY

Sample number	Rock type	Mineral dated	TFA (Ma)	WMA (Ma)	IIA (Ma)	^{40}Ar/^{36}Ar Intercept	MSWD
Blueschist unit							
D249C	Schist	KFS	158 ± 2	171 ± 2	170 ± 3	319 ± 82	37
Amphibolite unit							
D241A	Schist	WM	236 ± 2	237 ± 2	235 ± 2	2419 ±1344	3.1
D317B	Eclogite	WM	243 ± 2	247 ± 2	243 ± 6	528 ± 460	29
D323A	Schist	WM	213 ± 2	212 ± 1	211 ± 2	506 ± 220	20
HP-UHP eclogite units							
D228A	Gneiss	WM	205 ± 2	206 ± 2	206 ± 2	294 ± 6	0.3
D238A	Gneiss	WM	201 ± 2	203 ± 2	203 ± 2	842 ± 129	0.5
D244B	Gneiss	KFS	163 ± 2	169 ± 2	163 ± 3	338 ± 19	5.7
D307A	Eclogite	WM	218 ± 2	222 ± 2	221 ± 2	365 ± 43	1
D307B	Gneiss	WM	195 ± 2	196 ± 2	195 ± 2	543 ± 262	26
D310C	Eclogite	WM	233 ± 1	234 ± 1	234 ± 1	299 ± 13	2.6
D312A	Gneiss	WM	208 ± 1	207 ± 1	206 ± 1	383 ± 17	0.8
D332A	Shear band	WM	222 ± 2	224 ± 2	225 ± 2	143 ± 61	4.4
D332C	Shear band	WM	232 ± 2	231 ± 2	229 ± 3	405 ± 110	11
Huwan detachment zone							
D232D	Mylonite	WM	235 ± 2	234 ± 2	234 ± 2	309 ± 35	17
D335B	Shear band	WM	232 ± 2	233 ± 2	211 ± 11	4471 ± 4691	10
Dawu dome							
D253A	Gneiss	WM	195 ± 2	194 ± 2	194 ± 2	499 ± 48	2.5
D254A	Schist	WM	196 ± 2	196 ± 2	197 ± 2	258 ± 23	0.5
D326A	Gneiss	WM	198 ± 2	198 ± 2	198 ± 2	344 ± 73	12
Tongbai shear zone							
D260B	Mylonite	WM	130 ± 1	131 ± 1	132 ± 1	224 ± 19	4.6
D260C	Mylonite	KFS	120 ± 1	110 ± 1	103 ± 1	649 ± 15	2.1
D345A	Gneiss	BIO	123 ± 1	124 ± 1	124 ± 1	367 ± 61	2
D345A	Gneiss	KFS	105 ± 2	97 ± 3	96 ± 4	318 ± 71	0.5
D347A	Shear band	BIO	120 ± 1	122 ± 1	121 ± 1	423 ± 74	1
D347B	Gneiss	BIO	119 ± 1	119 ± 1	119 ± 1	289 ± 8	0.2
Late Cretaceous sinistral faults							
D256A	Mylonitized granite	KFS	83.9 ± 0.8	83.4 ± 0.8	80.5 ± 2.5	430 ± 102	6.8
D256C	Pseudotachylite	WR	75.4 ± 0.7	74.9 ± 0.8	74.9 ± 0.8	590 ± 111	0.6
Cretaceous plutons							
D321C	Granite	BIO	121 ± 1	121 ± 1	121 ± 2	300 ± 110	9.6
D321C	Granite	KFS	98 ± 1	101 ± 1	87 ± 8	846 ± 641	1.6
D342.5	Granite	BIO	118 ± 1	130 ± 1	128 ± 2	552 ± 95	1.5

Note: TFA—total fusion age; WMA—weighted mean age; IIA—inverse isochron age; ^{40}Ar/^{36}Ar intercept—inherited argon component, 295.5 is atmospheric ^{40}Ar/^{36}Ar ratio; MSWD—mean square weighted deviation, expresses goodness of fit index for inverse isochron; HP—high pressure; UHP—ultrahigh pressure; KFS—K-feldspar; WM—white mica; BIO—biotite; WR—pseudotachylite. Data are from Webb et al., 1999.

or northeast, and stretching lineations plunge subhorizontally north or northwest (Fig. 5A). Kinematic indicators imply top-to-north shear (Fig. 2A). No tectonic contact between the high- and ultrahigh-*P* eclogite units was observed in the field; thus we interpret the change in metamorphic grade to represent an isograd rather than a tectonic boundary. In thin section, felsic lithologies display dynamic recrystallization of quartz and synkinematic to postkinematic albite porphyroblasts. In sam-

ples from eclogite bodies most garnet porphyroclasts contain prograde inclusion assemblages (Zhou et al., 1993) and show retrograde amphibole coronas or chlorite overgrowths.

White mica from eclogite, schist, and gneiss yielded weighted mean ^{40}Ar/^{39}Ar cooling ages ranging from 231 ± 2 Ma to 196 ± 2 Ma (Table 1; Webb et al., 1999).

Huwan detachment zone. The Huwan detachment forms the northern boundary of the high- and ultrahigh-*P* rocks in the

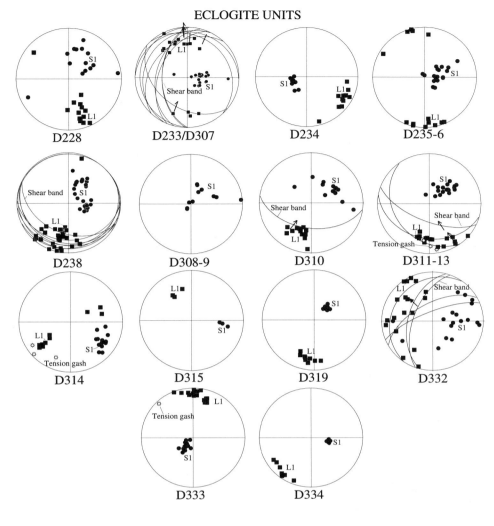

A ECLOGITE UNITS

D228 D233/D307 D234 D235-6

D238 D308-9 D310 D311-13

D314 D315 D319 D332

D333 D334

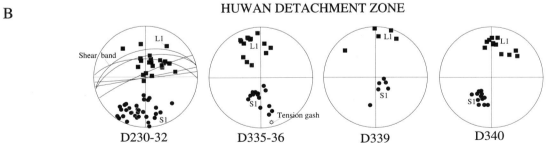

B HUWAN DETACHMENT ZONE

D230-32 D335-36 D339 D340

Figure 5. Structural data for (A) high- and ultrahigh-pressure eclogite units and (B) Huwan detachment zone represented in equal-area, lower hemisphere projections. Lettered annotations of data: S, pole to foliation; L, stretching lineation; B, fold axes. Number refers to generation of structure. Great circles represent shear bands. Whole, half, and headless arrows represent degree of certainty of shear-sense determination for that shear band where observed. See Figure 2 for location of structural data stations presented here.

Hong'an Shan, separating these rocks from greenschist facies metamorphic rocks to the north (Figs. 1 and 2A; Rowley and Xue, 1996; Webb et al., 1996). Rocks in the detachment zone, previously classified separately as the Sujiahe Group, were recently shown to comprise quartzofeldspathic schist and gneiss and high-*P* eclogite bodies correlating to the high- and ultrahigh-*P* units (Hacker et al., 1998). Deformation and retrogression is concentrated in an ~5-km-thick (structural thickness) deformation zone with generally north-dipping foliation and north-plunging stretching lineation (Figs. 2 and 5B). Shear bands, S-C fabrics, asymmetric boudinage, and σ and δ clasts all indicate top-to-north shear (e.g., Fig. 4F).

The detachment zone is intruded by undeformed Early Cretaceous granites. Synkinematic white mica ^{40}Ar/^{39}Ar cooling ages of 233 ± 2 Ma and 234 ± 2 Ma were obtained from a shear band and mylonite from the Huwan detachment zone (Table 1; Webb et al., 1999).

Dawu dome. A younger stage of deformation is observed throughout the Hong'an block and is best viewed in ductile to brittle structures at Dawu dome (Fig. 6A). The dome is characterized by a felsic gneissic core (Fig. 4G) mantled by greenschist facies metasedimentary rocks. These rocks are cut by north-dipping and south-dipping shear bands with normal-sense displacement (Fig. 6B). Ductile fabrics related to this coaxial deformation are cut by kinematically similar sets of ductile-brittle to brittle faults at and near Dawu, implying a continuum of exhumation through upper crustal depths during this tectonic episode (Fig. 6B). Late-stage quartz veins and fault arrays at localities in the blueschist and eclogite units in the Hong'an region indicate similar northeast-southwest subhorizontal extension during this stage of cooling and decompression (Fig. 6B).

Synkinematic white mica cooling ages of 196 ± 2 Ma, 195 ± 2 Ma, 198 ± 2 Ma were obtained from both the gneissic core and the metasedimentary cover of Dawu dome (Table 1; Webb et al., 1999).

Tongbai Shan. In the Tongbai Shan, a major northwest-trending shear zone involves both metasedimentary and igneous rocks, and appears to continue west toward the Qinling mountains (our observations; Regional Geological Survey [RGS] Henan, 1989). Orthogneiss at locality D347 along the southern margin of an Early Cretaceous pluton (Fig. 7A) was deformed at amphibolite facies conditions and exhibits relatively high grade, ductile fabrics. We did not observe a contact between the gneiss and the pluton in the field. A plausible explanation is that the orthogneiss is the deformed margin of a synkinematic pluton. Mylonitic, metasedimentary lithologies are found farther south and southwest, depicting lower temperature deformation farther from the pluton (Fig. 7A). Sense of shear in both the orthogneiss and metasedimentary rocks is dextral, as indicated by asymmetric boudinage, shear bands, and σ and δ clasts (Figs. 4H and 7B, localities D260, D345, and D347).

Synkinematic white mica and biotite yielded Early Cretaceous cooling ages ranging from 131 to 119 Ma (Table 1; Webb et al., 1999). K-feldspar from two localities gave weighted mean

cooling ages of 110 ± 1 Ma and 97 ± 3 Ma. Similar ages were obtained from Early Cretaceous plutons in the Hong'an and Tongbai Shan (Li and Wang, 1991; Eide et al., 1994; Webb et al., 1999).

A northwest-trending, sinistral strike-slip fault zone overprints the southern margin of the Tongbai dextral shear zone (Fig. 7, A and B, localities D256 and D259). The zone deforms blueschist facies rocks, Early Cretaceous plutons, and Cretaceous redbeds (RGS Hubei, 1990) under subgreenschist facies conditions. We correlate sinistral-transtensional faulting along the southern side of Dawu dome with this event (Fig. 7B). Late Cretaceous ages were obtained from a low-temperature deformed Early Cretaceous pluton and pseudotachylite cutting it (Table 1; Webb et al., 1999).

INFERRED MESOZOIC KINEMATIC EVOLUTION OF THE HONG'AN BLOCK

The collision between the Sino-Korean and Yangtze cratons is best demonstrated in the Hong'an region by the large volume of crustal rocks metamorphosed at blueschist facies to eclogite facies conditions and the preservation of coesite in ultrahigh-*P* metamorphic rocks in both the Hong'an and Dabie Shan. However, contractional structures of the appropriate age are cryptic within the interior of the orogenic belt (Xue et al., 1996). Here we interpret the overall structure of the high- and ultrahigh-*P* units to represent a warped extensional footwall, wherein top-to-northwest shear in south-dipping fabrics roll over into top-to-northwest shear in north-dipping fabrics. This implies a north-dipping continental subduction zone prior to exhumation and that all the described penetrative, Triassic–Jurassic deformation is related to exhumation along a crustal-scale normal-shear zone that encompasses the Hong'an and Dabie basement units. The extensional deformation within the interior of the orogen opposes contractional deformation in the foreland fold and thrust belt south and east of the Hong'an and Dabie Shan. The foreland structures record thrusting during the Triassic and the Jurassic, but a detailed understanding of the tectonic record is hampered by a combination of poor exposure and structural dismemberment during Cretaceous deformation (RGS Anhui, 1987; RGS Hubei, 1990; Ratschbacher et al., 2000; Schmid et al., 2000).

The oldest structures observed in the Hong'an region post-date the ultrahigh-*P* metamorphic event. The clearest link between ^{40}Ar/^{39}Ar ages and deformation comes from the Huwan detachment zone, where synkinematic white mica from the detachment zone yielded ages of ca. 235 Ma (Webb et al., 1999). Similarly old ^{40}Ar/^{39}Ar ages were reported from localities in the high- and ultrahigh-*P* rocks in both the Hong'an and Dabie Shan (Eide et al., 1994; Hacker and Wang, 1995; Webb et al., 1999). Tectonic denudation, defined as Middle to Late Triassic by ^{40}Ar/^{39}Ar white mica ages, was concomitant with thrusting in the foreland east and south of the Hong'an and Dabie Shan; there Middle Triassic sedimentary breccia and conglomerate,

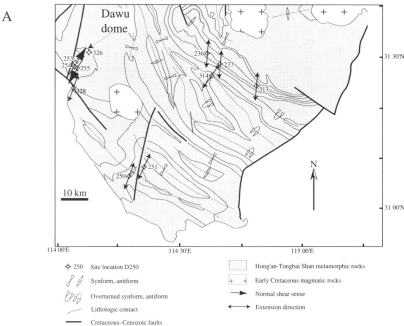

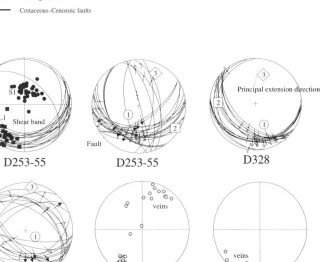

Figure 6. A: Kinematic map of Hong'an region for Early Jurassic deformation. B: Structural data plotted in equal-area, lower hemisphere projections. Lettered annotations of data: S, pole to foliation; L, stretching lineation. Number refers to generation of structure. Symbols for veins indicate poles to plane formed by tension gash. Whole, half, and headless arrows in fault data sets represent degree of certainty of slip sense determination on that fault. Principal stress directions (represented by 1, 2, and 3) were computed using fault-slip inversion techniques (see methods section in text).

deposited unconformably above Early Triassic limestone, are interpreted to date the main folding phase (Schmid et al., 2000).

Ages from synkinematic white mica in both the gneissic core and overlying metasedimentary rocks from Dawu dome indicate that the 198–194 Ma population of $^{40}Ar/^{39}Ar$ ages dates northeast-southwest extension. The subhorizontal extension direction in the ductile fabrics is identical to the extension direction during ductile-brittle to brittle faulting at Dawu (Fig. 6B) and to that determined from fault arrays and late-stage tension gashes elsewhere within the Hong'an block. The $^{40}Ar/^{39}Ar$ cooling ages and the continuum of faulting though the ductile and brittle regimes suggest that exhumation of the ultrahigh-*P* rocks through upper crustal levels occurred by the Early Jurassic in the

Hong'an block. It is likely that the suite of 200–180 Ma $^{40}Ar/^{39}Ar$ ages for white mica in the Dabie Shan reported by Hacker and Wang (1995) reflects cooling during this same tectonic episode.

Early Cretaceous tectonism is most evident in the Tongbai and Dabie Shan. Extension and sinistral-oblique slip occurred from ca. 140 to 120 Ma along the Xiaotian-Mozitang fault of the northern Dabie Shan and resulted in the exhumation of rocks from depths of ~15 km (Ratschbacher et al., 2000). We suggest that this fault is the eastward continuation of the Huwan detachment that was reactivated and overprinted during Cretaceous extension and plutonism. Consequently, high- and ultrahigh-*P* rocks that resided at deeper crustal levels for a longer period of time are now exposed in the Dabie Shan, as

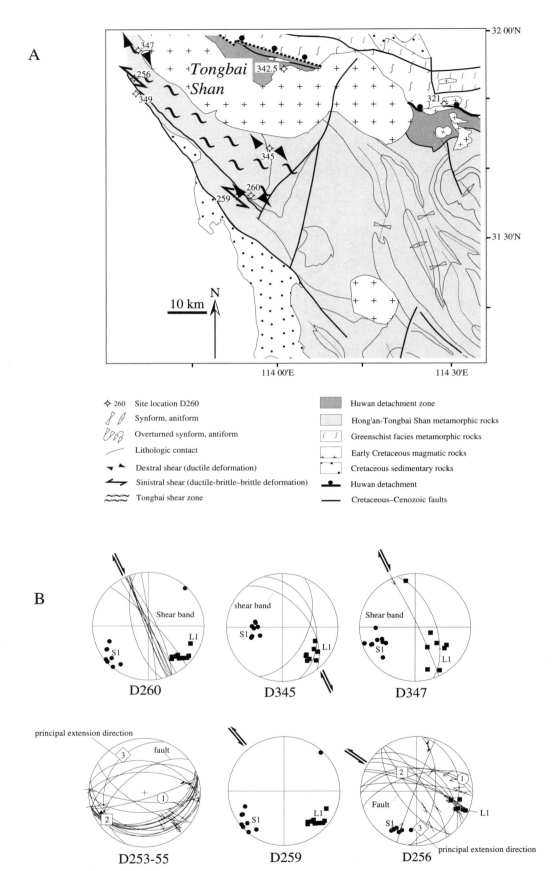

Figure 7. A: Cretaceous kinematic map of Tongbai and Hong'an Shan. B: Structural data are represented in equal-area, lower hemisphere projections. See Figures 3 and 6 for explanation of symbols and lettering used.

opposed to the Hong'an and Tongbai Shan, where Cretaceous deformation involved primarily strike-slip motion and little or no tectonic exhumation.

Dextral shear along the southwestern margin of the Tongbai Shan coeval with sinistral shear along Xiaotian-Mozitang fault of the northern margin of the Dabie Shan would have resulted in eastward lateral extrusion of the interior units of the Tongbai, Hong'an, and Dabie Shan. This tectonism may have been driven by the Jurassic–Cretaceous collision of the Lhasa block with Eurasia (Allegrè et al., 1984).

Late Cretaceous $^{40}Ar/^{39}Ar$ ages obtained for the low-temperature sinistral strike-slip faults along the southern margin of the Tongbai Shan are corroborated by Cretaceous–Eocene apatite fission-track data reported by Ames (1995). In addition, Hacker et al. (1995) reported that sinistral strike-slip faults form the boundary between blueschist facies rocks and the foreland fold and thrust belt in the southern Dabie Shan. Although not significant in the exhumation history of the ultrahigh-*P* rocks, this Late Cretaceous tectonism formed major block-bounding faults for the Dabie Shan, Hong'an, and Tongbai regions.

Implications for exhumation models

The structural data presented here suggest that high- and ultrahigh-*P* rocks of the Hong'an and Dabie Shan were exhumed as a single, penetratively deformed slab bounded at the top by a north-dipping detachment fault zone. Models incorporating this scenario were previously published for the Dabie Shan (e.g., Maruyama et al., 1994; Ernst and Peacock, 1996; Liou et al., 1996), but lacked substantiating structural evidence. In these models, reminiscent of the physical models by Chemenda et al. (1995, 1996), exhumation is driven by buoyancy forces due to density contrasts between the subducted crustal rocks and the overlying mantle lithosphere, and/or continued contraction during collision. This general model is attractive for explaining the preservation of ultrahigh-*P* rocks in that cooling of the exhumed slab is achieved across both upper and lower surfaces (Fig. 8;

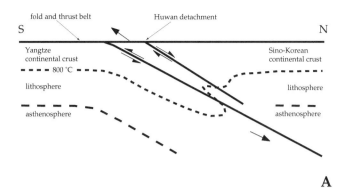

A

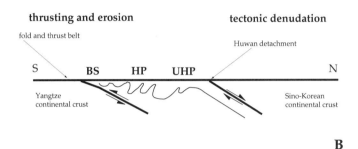

B

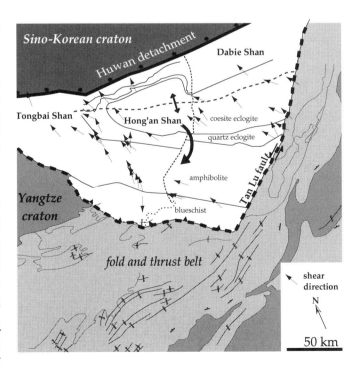

C

Figure 8. Schematic tectonic model for exhumation of blueschist (BS) unit, and high-pressure (HP) and ultrahigh-pressure (UHP) eclogite units. A: Figure is modified from Ernst and Peacock (1996). Bowed isotherm illustrates how cooling is achieved along upper and lower boundaries of exhuming slab as HP and UHP rocks come into contact with cooler rocks. B: Structures and relative locations of metamorphic units in Hong'an Shan resulting from exhumation. Unroofing is accompanied by thrusting and erosion within fold and thrust belt south and east of Hong'an and Dabie Shan and by normal shear within HP and UHP units and along Huwan detachment zone. Note that section is at an angle to tectonic transport direction, thus a component of exhumation occurs subnormal to section plane. C: Pre-Cretaceous restoration of Tongbai, Hong'an, and Dabie Shan modified after Hacker et al. (1998). Arrow represents clockwise rotation of slab during exhumation. In this model, Tan Lu fault originates in Middle to Late Triassic as sinistral transfer fault that accommodates extension and continued contraction in foreland as collision progresses westward.

Ernst and Peacock, 1996). Coeval extension along the Huwan detachment and shortening in the foreland require accommodation structures of the same age. One possibility is the initiation of the Tan Lu fault in the Middle to Late Triassic as a sinistral transfer fault (Fig. 8C). The paleomagnetic data suggesting clockwise rotation of the Tongbai, Hong'an, and Dabie Shan (Enkin et al., 1992) are compatible with this model.

The comprehensive structural studies and radiometric age data, summarized here and detailed in Hacker et al. (1998, 2000) and Webb et al. (1999), suggest that exhumation from mantle depths of 120–150 km to crustal depths occurred between ca. 245–240 Ma and ca. 235–230 Ma.

CONCLUSIONS

Evidence of multiple Mesozoic tectonic events affecting rocks in the Hong'an and Tongbai Shan is preserved in structural, petrologic, and radiometric age data. Early Triassic high- and ultrahigh-*P* metamorphism occurred during subduction of crustal rocks to mantle depths during collision. Middle to Late Triassic structures in the high- and ultrahigh-*P* units are associated with extension within the basement units of the Tongbai, Hong'an, and Dabie Shan and along the Huwan detachment zone that was synchronous with contraction in the foreland. Northwest-down, normal-sense shear along the Huwan detachment at the northern edge of the Hong'an block occurred ca. 235 Ma, on the basis of white mica cooling ages. The bulk of the mountain range is a warped extensional footwall, within which muscovites cooled by 205 Ma following Late Triassic–Early Jurassic retrograde metamorphism at mid-crustal levels. Younger, northeast-southwest subhorizontal extension is recorded by both ductile and brittle structures, best seen at Dawu. The associated white mica recrystallization at ca. 195 Ma and the continuum to low-temperature deformation suggest that exhumation of the high- and ultrahigh-*P* rocks through upper crustal levels occurred during Early Jurassic time. Early Cretaceous dextral shearing in the Tongbai Shan was synchronous with normal to sinistral-oblique shear along the Xiaotian-Mozitang fault of the northern Dabie Shan and syntectonic plutonism throughout the orogen. Late Cretaceous–Eocene deformation included strike-slip faulting and extension. Structures related to this deformation formed the major block-bounding faults for the Dabie Shan, Hong'an, and Tongbai regions.

ACKNOWLEDGMENTS

This work was supported by the U.S. National Science Foundation grant EAR-9417958, German National Science Foundation grants Ra442/4-1, Ra442/6-1/2, Ra442/9-1, Stanford-China Geosciences Industrial Affiliates, and a grant from the Stanford University McGee Fund. Special thanks to Mary Leech and Peng Lianhong for their assistance during the 1996 field season and to Marc Hendrix, Elizabeth Eide, and Steve Sheriff for thorough and constructive reviews of the manuscript.

REFERENCES CITED

Allegrè, C.J., and 35 others, 1984, Structure and evolution of the Himalaya-Tibet orogenic belt: Nature, v. 307, p. 17–22.

Ames, L., 1995, Geochronology and isotopic character of ultrahigh-pressure metamorphism with implications for collision of the Sino-Korean and Yangtze cratons, central China [Ph.D. thesis]: Santa Barbara, University of California at Santa Barbara, 124 p.

Ames, L., Zhou, G., and Xiong, B., 1996, Geochronology and isotopic character of ultrahigh-pressure metamorphism with implications for collision of the Sino-Korean and Yangtze cratons, central China: Tectonics, v. 15, p. 472–489.

Angelier, J., 1994, Fault-slip analysis and paleostress reconstruction, *in* Hancock, P.L., ed., Continental deformation: Tarrytown, New York, Pergamon, p. 53–100.

Chemenda, A.I., Mattauer, M., Malavieille, J., and Bokun, A.N., 1995, A mechanism for syn-collisional rock exhumation and associated normal faulting: Results from physical modeling: Earth and Planetary Science Letters, v. 132, p. 225–232.

Chemenda, A.I., Mattauer, M., and Bokun, A.N., 1996, Continental subduction and a mechanism for exhumation of high-pressure metamorphic rocks: New modeling and field data from Oman: Earth and Planetary Science Letters, v. 143, p. 173–182.

Chen, W., Harrison, T.M., Heizler, M.T., Liu, R., Ma, B., and Li, J., 1992, The cooling history of mélange zone in north Jiansu–south region: Evidence from multiple diffusion domain ^{40}Ar/^{39}Ar thermal geochronology: Acta Petrologica Sinica, v. 8, p. 1–17.

Dong, S., 1989, The general features and distributions of the glaucophane schist belts of China: Acta Geologica Sinica, v. 3, p. 273–284.

Eide, E.A., 1993, Petrology, geochronology, and structure of high-pressure metamorphic rocks in Hubei Province, east-central China, and their relationship to continental collision [Ph.D. thesis]: Stanford, California, Stanford University, 235 p.

Eide, E.A., McWilliams, M.O., and Liou, J.G., 1994, ^{40}Ar/^{39}Ar geochronologic constraints on the exhumation of high-pressure–ultrahigh-pressure metamorphic rocks in east-central China: Geology, v. 22, p. 601–604.

Enkin, R.J., Yang, Z., Chen, Y., and Courtillot, V., 1992, Paleomagnetic constraints on the geodynamic history of the major blocks of China from the Permian to the present: Journal of Geophysical Research, v. 97, p. 13953–13989.

Ernst, W.G., and Peacock, S.M., 1996, A thermotectonic model for preservation of ultrahigh-pressure phases in metamorphosed continental crust, *in* Bebout, G.E., Scholl, D.W., Kirby, S.H., and Platt, J.P., eds., Subduction top to bottom: American Geophysical Union Geophysical Monograph 96, p. 171–178.

Ernst, W.G., Zhou, G., Liou, J.G., Eide, E., and Wang, X., 1991, High-pressure and superhigh-pressure metamorphic terranes in the Qinling-Dabie mountain belt, central China; early to mid-Phanerozoic accretion of the western Paleo-Pacific Rim: Pacific Science Association Information Bulletin, v. 43, p. 6–15.

Hacker, B.R., and Peacock, S.M., 1994, Creation, preservation, and exhumation of coesite-bearing, ultrahigh-pressure metamorphic rocks, *in* Coleman, R.G., and Wang, X., eds., Ultrahigh pressure metamorphism: Cambridge, Cambridge University Press, p. 159–181.

Hacker, B.R., and Wang, Q., 1995, Ar/Ar geochronology of ultrahigh-pressure metamorphic rocks in central China: Tectonics, v. 14, p. 994–1006.

Hacker, B.R., Ratschbacher, L., Webb, L.E., and Dong, S., 1995, What brought them up? Exhumation of the Dabie Shan ultrahigh-pressure rocks: Geology, v. 23, p. 43–46.

Hacker, B.R., Wang, X., Eide, E.A., and Ratschbacher, L., 1996, Qinling-Dabie ultrahigh-pressure collisional orogen, *in* Yin, A., and Harrison, T.M., eds., Tectonic evolution of Asia: Englewood Cliffs, New Jersey, Prentice-Hall, p. 345–370.

Hacker, B.R., Ratschbacher, L., Webb, L.E., Ireland, T., Walker, D., and Dong, S., 1998, U/Pb zircon ages constrain the architecture of the ultrahigh-pressure Qinling-Dabie orogen, China: Earth and Planetary Science Letters, v. 161, p. 215–230.

Hacker, B.R., Ratschbacher, L., Webb, L.E., McWilliams, M., Ireland, T., Calvert, A., Dong, S., Wenk, H.-R., and Chateigner, D., 2000, Exhumation of the ultrahigh-pressure continental crust in east-central China: Late Triassic–Early Jurassic extension: Journal of Geophysical Research, v. 105, p. 13339–13364.

Li, S., and Wang, T., 1991, Geochemistry of granitoids in the Tongbaishan-Dabieshan, central China: Wuhan, China University of Geosciences Press, 208 p.

Li, S., Hart, S.R., Liu, D., Zhang, G.W., and Guo, A., 1989, Timing of collision between the north and south China blocks—The Sm-Nd isotopic age evidence: Scientia Sinica, ser. B, v. 32, p. 1393–1400.

Li, S., Xiao, Y., Liou, D., Chen, Y., Ge, N., Zhang, Z., Sun, S., Cong, B., Zhang, R., Hart, S.R., and Wang, S., 1993, Collision of the North China and Yangtze and formation of coesite-bearing eclogites: Timing and processes: Chemical Geology, v. 109, p. 89–111.

Lin, J.L., Fuller, M., and Zhang, W., 1985, Preliminary Phanerozoic polar wander paths for the North and South China blocks: Nature, v. 313, p. 444–449.

Liou, J.G., Zhang, R.Y., Wang, X., Eide, E.A., Ernst, W.G., and Maruyama, S., 1996, Metamorphism and tectonics of high-pressure and ultra-high-pressure belt in the Dabie-Sulu region, China, *in* Yin, A., and Harrison, T.M., eds., Tectonic evolution of Asia: Englewood Cliffs, New Jersey, Prentice-Hall, p. 300–344.

Maruyama, S., Liou, J.G., and Zhang, R., 1994, Tectonic evolution of the ultra-high-pressure (ultrahigh-*P*) and high-pressure (high-*P*) metamorphic belts from central China: Island Arc, v. 4, p. 112–121.

Mattauer, M.P., Matte, H., Maluski, Z., Xu, X., Lu, Y., and Tang, Y., 1985, Tectonics of the Qinling Belts, build-up and evolution of eastern Asia: Nature, v. 327, p. 496–500.

Okay, A.I., Şengör, A.M.C., and Satir, M., 1993, Tectonics of an ultrahigh-pressure metamorphic terrane: Dabie Shan, China: Tectonics, v. 12, p. 1320–1334.

Opdyke, N.D., Huang, K., Xu, G., Zhang, W.Y., and Kent, D.V., 1986, Paleomagnetic results from the Triassic of the Yangtze Platform: Journal of Geophysical Research, v. 91, p. 9553–9568.

Passchier, C.W., 1982, Mylonitic deformation in the Saint-Barithélemy Massif, French Pyrenees, with emphasis on the genetic relationship between ultramylonite and pseudotachylite: Amsterdam, GUA Papers of Geology, ser. 1, v. 16, p. 1–173.

Passchier, C.W., and Trouw, R.A.J., 1996, Microtectonics: New York, Springer, 289 p.

Platt, J.P., 1986, Dynamics of orogenic wedges and the uplift of high-pressure metamorphic rocks: Geological Society of America Bulletin, v. 97, p. 1037–1053.

Ratschbacher, L., Hacker, B.R., Webb, L.E., McWilliams, M., Ireland, T., Dong, S., Calvert, A., Chateigner, D., and Wenk, H.-R., 2000, Exhumation of the ultrahigh-pressure continental crust in east-central China: Cretaceous and Cenozoic unroofing and the Tan-Lu fault: Journal of Geophysical Research, v. 105, p. 13303–13338.

Regional Geological Survey Anhui, 1987, Regional geology of the Henan Province: Beijing, Geological Publishing House, 721 p.

Regional Geological Survey Henan, 1989, Regional geology of the Henan Province: Beijing, Geological Publishing House, 726 p.

Regional Geological Survey Hubei, 1990, Regional geology of the Hubei Province: Beijing, Geological Publishing House, 705 p.

Rowley, D.B., and Xue, F., 1996, Modeling the exhumation of ultra-high pressure metamorphic assemblages; observations from the Dabie/Tongbai region, China: Geological Society of America Abstracts with Programs, v. 28, no. 7, p. 249.

Rowley, D.B., Xue, F., Tucker, R.D., Peng, Z.X., Baker, J., and Davis, A., 1997, Ages of ultrahigh pressure metamorphism and protolith orthogneisses from eastern Dabie Shan: U/Pb zircon geochronology: Earth and Planetary Science Letters, v. 151, p. 191–203.

Schmid, J.C., Ratschbacher, L., Hacker, B.R., Gaitzsch, I., and Dong, S., 2000, How did the foreland react? Exhumation of the Dabie Shan ultrahigh-pressure continental crust and the Yangtze foreland fold-and-thrust belt (eastern China): Terra Nova, v. 11, p. 266–272.

Tullis, J.R., and Yund, A., 1991, Diffusion creep in feldspar aggregates: Experimental evidence: Journal of Structural Geology, v. 13, p. 987–1000.

Wang, X., and Liou, J.G., 1991, Regional ultrahigh-pressure coesite-bearing eclogitic terrane in central China: Evidence from country rocks, gneiss, marble and metapelite: Geology, v. 19, p. 933–936.

Webb, L.E., Hacker, B.R., Ratschbacher, L., and Dong, S., 1996, Structures and kinematics of exhumation; ultrahigh-pressure rocks in the Hong'an Block of Qinling-Dabie orogen, China: Geological Society of America Abstracts with Programs, v. 28, no. 7, p. 69.

Webb, L.E., Hacker, B.R., Ratschbacher, L., McWilliams, M.O., and Dong, S., 1999, Thermochronologic constraints on deformation and cooling history of high- and ultrahigh-pressure rocks in the Qinling-Dabie orogen, eastern China: Tectonics, v. 18, p. 621–638.

Xu, S., and Su, W., 1997, Raman determination on micro-diamond in eclogite from the Dabie Mountains, eastern China: Chinese Science Bulletin, v. 42, p. 87.

Xu, S., Okay, A.I., Ji, S., Şengör, A.M.C., Su, W., Liu, Y., and Jiang, L., 1992, Diamond from the Dabie Shan metamorphic rocks and its implication for tectonic setting: Science, v. 256, p. 80–82.

Xue, F., Rowley, D.B., and Baker, J., 1996, Refolded syn-ultrahigh-pressure thrust sheets in the south Dabie complex, China: Field evidence and tectonic implications: Geology, v. 24, p. 455–458.

Xue, F., Rowley, D.B., Tucker, R.D., and Peng, Z.X., 1997, U-Pb zircon ages of granitoid rocks in the North Dabie Complex, eastern Dabie Shan, China: Journal of Geology, v. 105, p. 744–753.

Yin, A., and Nie, S., 1993, An indentation model for the North and South China collision and the development of the Tanlu and Honam fault systems, eastern Asia: Tectonics, v. 12, p. 801–813.

Zhai, X., Day, H., Hacker, B.R., and You, Z., 1998, Paleozoic metamorphism in the Qinling orogen, Tongbai Mountains, central China: Geology, v. 26, p. 371–374.

Zhang, Y.Q., Vergely, P., Mercier, J., and Ben-Avraham, Z., 1995, Active faulting in and along the Qinling Range (China) inferred from SPOT imagery analysis and extrusion tectonics of South China: Tectonophysics, v. 243, p. 69–95.

Zhou, G., Liu, J., Eide, E.A., Liou, J.G., and Ernst, W.G., 1993, High-pressure/low-temperature metamorphism in northern Hube Province, central China: Journal of Metamorphic Petrology, v. 11, p. 561–574.

MANUSCRIPT ACCEPTED BY THE SOCIETY JUNE 5, 2000

Geological Society of America
Memoir 194
2001

Jurassic to Cenozoic exhumation history of the Altyn Tagh range, northwest China, constrained by $^{40}Ar/^{39}Ar$ and apatite fission track thermochronology

Edward R. Sobel
Institut für Geowissenschaften, Universität Potsdam, Postfach 60 15 53,14415 Potsdam, Germany
Nicolas Arnaud
CNRS UMR 652: Magmas et Volcans, 63000 Clermont-Ferrand, France
Marc Jolivet
UMR 5567, ISTEEM-Tectonique, Case courrier 58, Université Sciences Montpellier II,
1 place E. Bataillon 34095 Montpellier cedex 5, France
Bradley D. Ritts
Department of Geology, 4505 Old Main Hill, Utah State University, Logan, Utah 84322 USA
Maurice Brunel
UMR 5567, ISTEEM-Tectonique, Case courrier 58, Université Sciences Montpellier II,
1 place E. Bataillon 34095 Montpellier cedex 5, France

ABSTRACT

New $^{40}Ar/^{39}Ar$ and apatite fission-track thermochronologic results from the Altyn Tagh, northwest China, provide new constraints for the Mesozoic–Cenozoic cooling history of the range. Published muscovite and biotite argon ages from samples along a transect between Mangnai and Ruoqiang yield early-middle Paleozoic ages that are mirrored in this study by the highest temperature incremental step-heating results from K-feldspars from the same units. Multidomain diffusion modeling of the K-feldspar data from these samples shows Early Jurassic cooling ages, suggesting that the samples remained at temperatures of ~300–400 °C between ~400 and ~200 Ma followed by ~100 °C of rapid cooling. Samples then cooled slowly until final exhumation to the surface during the Cenozoic. A similar Jurassic to Cenozoic cooling history is obtained from a small pluton located south of the Altyn Tagh fault and ~350 km to the east near Lenghu. Cooling was likely linked to formation of Early to Middle Jurassic basins that developed above a regional unconformity in the Northeast and Northwest Qaidam basins. Similarities in Jurassic cooling ages from these two areas suggest a post-Middle Jurassic offset of about 350 km along the Altyn Tagh fault. This value agrees well with the 400 ± 60 km post-Bajocian offset along the fault established farther to the west by independent sedimentary basin studies. Only a few samples close to the Altyn Tagh fault and the North Altyn Tagh fault yield Eocene to Miocene apatite fission-track ages, suggesting that Cenozoic tectonism has had a relatively minor effect on cooling and hence exhumation of the region. The cooling histories from all of the samples, together with basin analysis of the adjacent Northeast Qaidam basin, provide a unifying picture. The timing of deformation and sedimentation within the basin, driven by southeast-directed contractile shortening within the Qilian Shan, broadly coincides with the uplift and cooling of samples farther to the northwest.

Sobel, E.R., et al., 2001, Jurassic to Cenozoic exhumation history of the Altyn Tagh range, northwest China, constrained by $^{40}Ar/^{39}Ar$ and apatite fission track thermochronology, *in* Hendrix, M.S., and Davis, G.A., eds., Paleozoic and Mesozoic tectonic evolution of central Asia: From continental assembly to intracontinental deformation: Boulder, Colorado, Geological Society of America Memoir 194, p. 247–267.

This contractional deformation is interpreted to cause uplift, and hence erosion and cooling of the area represented by the Mangnai-Ruoquiang transect. On a larger scale, the Jurassic cooling ages in the Altyn Tagh are tentatively linked to compressional or transpressional exhumation driven by accretionary tectonism to the south.

INTRODUCTION

Previous investigations of the northern margin of the Tibetan Plateau have concentrated on neotectonic studies related to the India-Asia collision (e.g., Peltzer and Tapponnier, 1988; Chinese State Bureau of Seismology, 1992; Wang, 1997; Meyer et al., 1996, 1998; Chen et al., 1999). This is notably the case for the Altyn Tagh, a mountain range between the Tarim basin to the north and the Qaidam basin, Kunlun Shan, and the Tibetan Plateau to the south (Fig. 1). The present topographic expression of the range is presumably linked to the sinistral Altyn Tagh fault, which roughly bounds the range on the south. While much attention has been focused on the late Cenozoic history of this fault, the older history of the range is poorly understood, hindering an understanding of the offset of the Altyn Tagh fault, and in turn, the dynamic history of the Cenozoic India-Asia collision. In particular, many models of the offset history of the Altyn Tagh fault assume that the fault has only been active during the Cenozoic (e.g., Bally et al., 1986; Wang, 1997; Meyer et al., 1998; Yue and Liou, 1999).

New geochronologic data presented herein and in a companion paper (Delville et al., this volume) document an important Jurassic cooling event in the central part of the range. While the dynamics associated with this episode of cooling are as yet poorly defined by structural geology, recent Jurassic basin analysis studies (Ritts, 1998; Ritts and Biffi, 2000, this volume) provide a framework in which to study this cooling. In addition, prediction of 400 ± 60 km of left-lateral post-Bajocian offset of the Altyn Tagh fault based on facies analysis constrained by new palynological results (Ritts, 1998; Ritts and Biffi, 2000) is supported by thermochronologic data presented in this study.

We collected a suite of samples roughly perpendicular to the Altyn Tagh fault between 89 °E and 90 °E for study by $^{40}Ar/^{39}Ar$ and apatite fission-track thermochronology. The higher temperature history of these samples preserves an early-middle Paleozoic tectonic event (Sobel and Arnaud, 1999), providing the starting point for the thermal evolution described in this study. The magnitude of Jurassic cooling observed primarily by K-feldspar multidomain diffusion modeling (Lovera et al., 1989, 1991) in this portion of the range is far greater than the magnitude of Cenozoic cooling observed in the same samples using apatite fission-track analysis. These Jurassic cooling ages can be related to the regional geology and thereby linked with exhumation. Cenozoic apatite fission-track ages are only obtained from samples collected close to major faults. A generally similar observation has been made for a suite of sample collected from a transect farther to the east within the range (Delville et al., this volume). Together, these two studies suggest

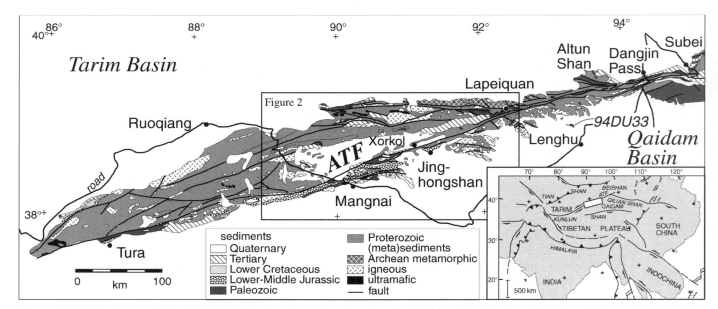

Figure 1. Geological map of Altyn Tagh, including locations of samples described in this study. Map is modified from Wang et al. (1993), Li et al. (1991), Liu (1988), and Che et al. (1995). There are several different spellings in literature for names of this range: herein we call entire range Altyn Tagh, and reserve name Altun Shan for specific peak in eastern part of range. Note that Tagh means mountain in Uygur while Shan means mountains in Chinese. ATF is Altyn Tagh fault. Inset is modified from Tapponnier et al. (1982) and shows location of main map. Location of sample 94DU33 is indicated.

that the tectonic history of this portion of the Tibetan Plateau, and by inference, a much larger area, included a complex Mesozoic history that must be better understood prior to a definitive understanding of the region's Cenozoic tectonism.

GEOLOGY OF THE ALTYN TAGH

The Altyn Tagh is within the Xinjiang Uygur Autonomous Region, in Gansu and Qinghai provinces. The most recent maps and descriptions are found within Wang et al. (1993), Liu (1988), Zhou et al. (1989), and Li et al. (1991). Older stratigraphic descriptions are available in Chinese Bureau of Geology (1981) and Chen (1985).

Rocks throughout the Altyn Tagh consist of oceanic material, including ophiolite as well as mafic and felsic high-pressure metamorphic rocks suggestive of a subduction and collision environment that collectively reflect a suture between the Tarim block and a block to the south. New geochronologic results combined with previously mapped lithologic relationships suggest that this suture formed during the middle Paleozoic (Sobel and Arnaud, 1999).

A Devonian through Early Jurassic unconformity separates Mesozoic and Cenozoic strata from older units that record Proterozoic and early to middle Paleozoic activity along the southern margin of the Tarim block. Scattered thin Lower Jurassic strata overlie this unconformity along the southern flank of the Altyn Tagh. However, Lower Jurassic strata are missing in Southeast Tarim, along the northern flank of the Altyn Tagh. In contrast, thick Middle Jurassic sequences are well exposed along portions of both sides of the range (Fig. 1). The eastern portion of the Altyn Tagh is much narrower and only small, isolated Mesozoic outcrops are mapped. Deformation within these Paleozoic and Precambrian units is high, and thermochronologic studies document a different suite of ages than are described herein (Delville et al., this volume). Relationships between the central and eastern portions of the Altyn Tagh are discussed in the following section.

Organic-rich lacustrine rocks Middle Jurassic strata in the Northwest Qaidam basin progressively grade upward into fluvial facies. There is also an eastward increase in fluvial and alluvial fan facies along the southern flank of the Altyn Tagh between Tura and Jinghongshan (Fig. 1). The Southeast Tarim basin has similar Middle Jurassic strata, including stratigraphic and areal distributions of lacustrine and alluvial facies. The Middle Jurassic sediments in the Southeast Tarim and Northwest Qaidam basins also share common sandstone compositions, concordant paleocurrent directions, and are biostratigraphically equivalent. On the basis of these lines of evidence, Ritts (1998) concluded that after the Early Jurassic the Southeast Tarim and Northwest Qaidam basins were contiguous and not segmented by the Altyn Tagh. Furthermore, Ritts and Biffi (2000) and Ritts (1998) correlated an Aalenian-Bajocian paleoshoreline facies tract across the fault, which indicated 400 ± 60 km of net, post-Bajocian left-lateral offset on the Altyn Tagh fault.

A poorly dated sequence of Upper Jurassic to Cretaceous red beds overlies the Middle Jurassic strata. Sandstone provenance and facies continued to be similar on both sides of the Altyn Tagh during the later part of the Mesozoic. Paleocurrent directions also remained concordant on each side of the range. However, between the Early to Middle Jurassic and the Late Jurassic to Cretaceous, paleocurrents in both the Southeast Tarim and Northwest Qaidam basins reversed, from generally northward to southward (Ritts, 1998; Ritts and Biffi, 2000). The tectonic significance of the Upper Jurassic to Cretaceous strata is not well understood; however, the paleocurrent reorganization and the presence of very thick (>1000 m) coarse conglomeratic sequences argues for continued tectonism in the region late into the Mesozoic. Ritts (1998) concluded that much of the tectonic activity during this interval occurred in the Kunlun Shan and Qilian Shan, and that the Altyn Tagh was not a physiographic feature late in the Mesozoic.

Cenozoic strata unconformably overlie older units along the range margins and within an intermontane basin east of Xorkol (Figs. 1 and 2). Facies analysis, paleocurrent indicators, and heavy mineral and sedimentary petrography show that the range had a positive relief commencing during Oligocene time (Hanson, 1998), in agreement with Oligocene growth strata along the southern margin of the Altyn Tagh (Bally et al., 1986) and isopach maps showing a thick Oligocene basin within the northwest Qaidam basin (Wang and Coward, 1990).

Offset estimates for the Altyn Tagh fault are controversial (e.g., Peltzer and Tapponnier, 1988; Chinese State Bureau of Seismology, 1992; Wang et al., 1993; Wang, 1997; Ritts and Biffi, 2000). For the portion of the fault east of 88 °E, one of the most rigorously determined estimates appears to be 200 km of left slip since 10 Ma (Meyer et al., 1998). Note that the post-Bajocian offset of 400 ± 60 km observed by Ritts and Biffi (2000) is based on a piercing point located to the west of data reported herein. Because the Altyn Tagh fault is believed to have propagated to the east during the late Cenozoic (e.g., Meyer et al., 1998), the net offset should decrease to the east along the south side of the fault. Ideally, this should be taken into account when making palinspastic reconstructions; however, a structurally justified retrodeformation of the northeastern portion of the Tibetan Plateau is beyond the scope of this study, and so a constant offset will be shown along the fault.

SAMPLING LOCALITIES

The majority of samples discussed herein were collected along a transect between the towns of Mangnai and Ruoqiang, hereafter called the Mangnai-Ruoqiang transect (Figs. 1 and 2). Here, the Altyn Tagh fault is near the southern side of the range. Along the transect north of the fault are Precambrian metasedimentary units intruded by Precambrian and early-middle Paleozoic igneous units (Wang et al., 1993). Ordovician and Jurassic sedimentary rocks and Mesozoic igneous units are present along the south side of the fault (Li et al., 1991; Wang

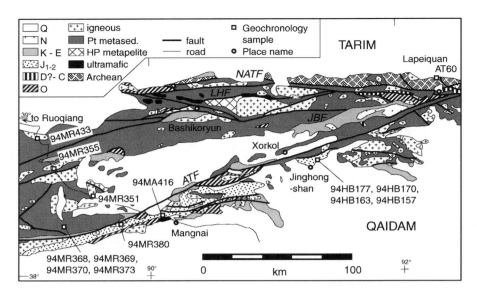

Figure 2. Enlargement of part of geologic map from Figure 1 showing sampling localities and faults. Abbreviations: ATF, Altyn Tagh fault; NATF, North Altyn Tagh fault; LHF, Lapeiquan Hongliugou fault; JBF, Jinyanshan-Bashikoryun fault; Q, Quaternary; N, Neogene; K-E, Cretaceous–Paleogene; J_{1-2}, Lower-Middle Jurassic; D-C, Devonian–Carboniferous; O, Ordovician; Pt, Precambrian; HP, high pressure. Units mapped as ultramafic were interpreted to be ophiolite fragments by Wang et al. (1993).

et al., 1993). The sampling localities are described from north to south; unit ages are from Wang et al. (1993), except as noted.

Sample 94MR433 was collected from a late Variscan large undeformed fine-grained leucogranite pluton intruding folded Middle Proterozoic marbles and schists on the northern margin of the range. Regional mapping suggests that the north-vergent North Altyn Tagh thrust is in the subsurface to the north (Meyer et al., 1998). Sample 94MR355 was collected from an undeformed Middle Proterozoic granodiorite 28 km to the southeast, and 32 km farther to the southwest, sample 94MR351 was collected from a Middle Proterozoic foliated leucogranite. Samples 94MR368, 94MR369, 94MR370, and 94MR373 were all collected just north of the Altyn Tagh fault, 30 km southwest of locality 94MR351. Gneissic schist sample 94MR373 was collected as float from the base of a small alluvial fan. The other three, 94MR368, granitic gneiss; 94MR369, amphibolitic gneiss; and 94MR370, mylonitic gneiss, were collected from float at the base of a slope located in a small drainage basin about 2 km farther south. The active strand of the Altyn Tagh fault is another 4 km to the south on the southern side of this ridge. The four samples from this locality were analyzed to test whether they were locally derived. This area is mapped as Lower Proterozoic lower amphibolite-grade metasedimentary rocks (Fig. 2). Indosinian granite sample 94MR380 was collected from the south side of the active trace of the Altyn Tagh fault. The granite is juxtaposed against Jurassic sedimentary rocks to the south by a vertical fault. Samples 94MA415 and 94MA416 were collected from an Aalenian-Bajocian sedimentary section (Ritts, 1998) just south of the Altyn Tagh fault; there, beds are subvertical and strike subparallel to the trend of the fault. Sample AT60 was collected from Archean granulite gneiss north of Lapeiquan, ~270 km east of the main transect. South of the Altyn Tagh fault, sample 94DU33 was collected 350 km farther east, near the town of Lenghu (Fig. 1). This fine-grained middle Variscan undeformed pluton intrudes metamorphosed Ordovician sedimentary rocks (Li et al., 1991). Samples 94HB157, 94HB163, 94HB170, and 94HB177 were collected from the basal 750 m of a 1400 m thick stratigraphic section of Triassic(?) through Middle Jurassic strata at Jinghongshan, about 6 km south of the Altyn Tagh fault. The apatite fission-track samples are stratigraphically below Hettangian shales dated by palynology (Ritts, 1998).

ANALYTICAL PROCEDURES AND INTERPRETATION METHODS

Argon geochronology

High-purity mineral separates were prepared using standard heavy liquid and gravity and/or magnetic separation methods. Final mineral separates were all hand-picked to remove impurities and then loaded into aluminum-foil packets. Samples were irradiated in the Siloé reactor, CEA (Commissariat à l'énergie atomique), Grenoble, for ~20 h with flux monitors distributed between every 5–10 samples. Both Caplongue hornblende, which has an assumed age of 344.5 Ma (Maluski and Schaeffer, 1982), and Fish Canyon Tuff sanidine, which has an assumed age of 27.55 ± 0.05 Ma (Lanphere et al., 1990), were used as monitors. Samples were irradiated along with CaF_2 and K_2SO_4 salts to account for interfering nuclear reactions. The CNRS UMR6524 (Centre National de la Recherche Scientifique, Unité Mixte de Recherche) $^{40}Ar/^{39}Ar$ laboratory has a Staudacher-type double vacuum resistance furnace, used to analyze K-feldspars, and a glass single-vacuum radio frequency furnace, used to analyze biotite 94DU33; both are coupled to a VG3600 mass spectrometer. Further details of the experimental system and the data reduction methodology were provided in

Arnaud et al. (1993) and Delville et al. (this volume). Quoted uncertainties are ±1σ; each step is given without error in the J-value; however, the latter error is included in the plateau age. K/Ca and Cl/K plots are given for comparison with the age spectra but are only qualitative, because $^{39}Ar/^{37}Ar$ and $^{38}Ar/^{39}Ar$ are used as proxies for the true elemental ratios. Full isotopic data for samples reported herein are presented in Table 1.

The biotite sample was analyzed by step heating and yielded an acceptable plateau age, reported as a mean weighted plateau age (WMPA). The following isotopic closure temperatures have been used in order to compare the thermal histories of different samples: 325 °C as an average for $^{40}Ar/^{39}Ar$ dating of biotite (Harrison et al., 1985), and 400 °C for $^{40}Ar/^{39}Ar$ dating of muscovite (Hames and Bowring, 1994).

Argon data from K-feldspars have long been considered difficult to interpret because they rarely exhibit plateaus in slowly cooled rocks (e.g., Harrison, 1990). However, recent work on argon diffusion in K-feldspars has revealed that they mostly behave as if composed of several subgrains (so-called domains), each characterized by a different closure temperature (Lovera et al., 1989, 1991). This explains why a flat plateau is rarely obtained: it is difficult to produce a homogenous isotopic distribution between domains losing argon at different temperatures. Therefore, the notion of a single closure temperature is rarely applicable to K-feldspars, and in most cases one observes a smoothly climbing age spectra with intermediate plateaus. However, this complexity can be overcome and used to derive a more complete cooling history than from other potassium-bearing minerals, which give only one temperature-time pair. Because the interplay between those multiple closure temperatures can be unraveled in laboratory experiments and then extrapolated to nature, one can model a cooling history that agrees with the observed complex age spectrum. This has led to the multidomain diffusion model (MDD) (Lovera et al., 1991). An important question about this modeling method concerns the precision and uniqueness of the resulting cooling curve. Monte Carlo modeling (Lovera et al., 1997) has shown that the solution is mathematically close to unique in each case, and that the errors on the model are essentially the result of the lack of bounding values both at the upper and lower bounds of the cooling history, and the errors on individual steps of the age spectra far from the extremities of the model. Notably, although the exact value of the cooling rate may be qualitative rather than quantitative, the timing of the inflection point of the cooling curve, marking significant changes in the cooling history, is only limited by the error on the age of individual steps around the time of this change. Although no precise treatment of error propagation has yet been established, most users of MDD modeling consider that errors in the model are ±25 °C in temperature and around 1.5% of the age at each point. In the following we have modeled the cooling history that would have produced the observed age spectra in the case of pure slow cooling following the methods and algorithms developed by Lovera et al. (1989, 1991).

Fission-track methodology

Laboratory procedures used in this study were essentially identical to those summarized in Dumitru et al. (1995). Samples analyzed at ETH, Zurich, were irradiated at Lucas Heights, Australia, and those analyzed at the University of Potsdam, Germany, were irradiated at Risø National Laboratory, Denmark. For apatite track length analysis, 100 horizontal confined tracks were measure in each sample (provided 100 were present). For age determinations, a minimum of 20 good-quality grains per sample were selected at random and dated (again assuming that sufficient grains were present). Following convention, all statistical uncertainties on ages and mean track lengths are quoted at the ±1σ level, but ±2σ uncertainties are taken into account for geologic interpretation. For certain apatite samples that yielded relatively tightly clustered single-grain age distributions, the genetic algorithm fission-track modeling program of Gallagher (1995) was used to determine the spectrum of time-temperature histories consistent with the observed data, using the Durango apatite annealing model of Laslett et al. (1987).

THERMOCHRONOLOGY RESULTS

Argon thermochronology

Early-middle Paleozoic mica ages from four of the units along the Mangnai-Ruoqiang transect (Fig. 2) were reported by Sobel and Arnaud (1999); K-feldspar ages from three of these units are discussed here. Because the mica data provide information on the high-temperature cooling history of the region, these data are briefly reviewed. Muscovite from the undeformed fine-grained leucogranite 94MR433 yielded a WMPA of 382.5 ± 7.4 Ma and the undeformed granodiorite 94MR355 yielded a WMPA of 413.8 ± 8.0 Ma on biotite. These two results are interpreted as cooling ages slightly postdating the intrusive age. On the southern side of the range, foliated leucogranite 94MR351 yielded a WMPA of 431.5 ± 7.8 Ma on muscovite, and mica schist 94MR368 yielded a WMPA of 453.4 ± 8.7 Ma on muscovite. These two results are interpreted as metamorphic ages.

The analytical results from step-heating multidiffusion domain analysis of eight K-feldspars are presented in Figure 3. Interpretations derived from thermal modeling of these results are discussed later, combined with the modeled apatite fission-track results.

Data from foliated leucogranite sample 94MR351 are good examples of the overall highly reproducible shape of most age spectra in this study (Fig. 3A). Duplicate analysis at each temperature yields a typical suite of alternating values, the first step at one temperature being always older than the second. This effect was attributed by Harrison et al. (1994) to the presence of excess argon, the second step being close to the real age. Taking those second steps into account, the age spectra slowly climb from minimum ages of roughly 150 Ma to a small plateau at

TABLE 1. ARGON ANALYTICAL DATA

Temperature (°C)	$^{40}Ar/^{39}Ar$	$^{38}Ar/^{39}Ar$	$^{37}Ar/^{39}Ar$	$^{36}Ar/^{39}Ar$ (10^{-3})	^{39}Ar $(10^{14}mol)$	$F^{39}Ar$ released*	$^{40}Ar*^{†}$ (%)	$^{40}Ar*/^{39}Ar$	Age (Ma)	± 1σ (Ma)
94MR433k		K-feldspar		$J^{§}$ = 0.0073150		weight = 23 mg			38°52′43.9″N, 89°06′29.8″E	
400	119.143	0.055	0.018	47.266	0.38	0.28	83.02	105.41	1031.19	15.82
400	17.403	0.029	0.011	15.142	0.32	0.52	41.40	12.98	163.68	4.81
450	22.258	0.024	0.010	7.562	0.63	1.00	72.72	20.04	246.74	4.05
450	13.874	0.021	0.012	4.145	0.55	1.41	54.97	12.64	159.58	2.54
500	20.736	0.021	0.020	4.374	1.09	2.24	80.63	19.44	239.87	3.64
500	15.921	0.019	0.025	1.758	0.89	2.91	69.59	15.38	192.40	2.84
550	19.575	0.019	0.025	2.150	1.85	4.31	87.38	18.92	233.91	3.53
550	18.464	0.018	0.018	1.369	1.78	5.66	83.13	18.04	223.61	3.33
600	20.084	0.019	0.013	2.398	2.97	7.90	90.32	19.36	238.94	4.28
600	19.289	0.018	0.007	0.901	2.71	9.95	88.31	19.00	234.77	3.83
650	19.578	0.018	0.007	0.975	3.16	12.34	92.18	19.27	237.86	4.41
650	19.183	0.018	0.006	0.439	2.33	14.10	86.98	19.03	235.09	3.33
700	19.488	0.018	0.009	0.850	2.63	16.09	90.87	19.21	237.24	3.57
700	18.951	0.018	0.009	0.405	1.63	17.32	81.96	18.80	232.51	3.47
750	18.651	0.018	0.011	0.906	1.95	18.80	87.58	18.36	227.34	2.21
750	18.787	0.018	0.010	0.241	1.44	19.89	79.59	18.69	231.17	4.34
800	19.001	0.019	0.012	0.887	1.98	21.38	87.58	18.72	231.48	3.67
800	19.092	0.018	0.010	0.434	1.53	22.54	79.98	18.94	234.06	3.35
800	19.176	0.019	0.008	0.151	1.37	23.57	73.47	19.10	235.98	3.80
800	19.527	0.018	0.007	0.147	2.62	25.56	64.36	19.45	240.04	3.24
700	18.324	0.015	0.000	0.000	0.05	25.59	12.47	18.29	226.58	3.32
750	18.305	0.017	0.000	0.000	0.14	25.70	27.78	18.27	226.35	2.36
800	19.574	0.019	0.006	0.000	0.35	25.97	45.48	19.55	241.08	4.21
850	20.097	0.019	0.010	0.823	1.26	26.92	76.68	19.83	244.37	3.35
900	20.443	0.019	0.011	1.285	2.39	28.72	84.95	20.04	246.81	3.44
950	20.627	0.019	0.010	0.837	3.94	31.70	90.09	20.36	250.42	4.20
1000	21.473	0.019	0.009	0.833	5.47	35.84	92.44	21.20	260.12	3.59
1050	21.366	0.019	0.009	1.284	6.67	40.88	92.75	20.97	257.41	5.19
1100	22.872	0.022	0.014	2.961	10.05	48.48	92.69	21.98	269.04	4.82
1150	24.702	0.024	0.020	5.471	12.73	58.10	90.93	23.09	281.53	5.61
1200	24.807	0.021	0.011	2.584	24.55	76.66	95.45	24.03	292.12	5.22
1400	25.063	0.019	0.005	1.013	27.50	97.44	97.30	24.74	300.09	4.73
1400	25.272	0.018	0.004	1.452	3.38	100.00	87.94	24.82	301.00	5.43
94MR351k		K-feldspar		J = 0.0073150		weight = 15 mg			38°31′27.5N, 89°31′20.8	
400	17.848	0.035	0.021	15.143	0.16	0.10	20.21	13.43	169.05	2.02
400	14.014	0.024	0.009	6.497	0.17	0.21	14.39	12.10	153.02	0.27
450	26.370	0.000	0.008	3.772	0.77	0.71	68.72	25.25	305.75	1.67
450	13.017	0.020	0.014	1.239	0.79	1.22	44.77	12.63	159.41	3.14
500	15.292	0.020	0.037	0.908	1.71	2.34	74.78	15.00	187.87	3.51
500	14.267	0.019	0.051	0.514	1.67	3.43	65.26	14.09	177.03	3.88
550	16.574	0.020	0.056	0.538	2.87	5.31	84.12	16.39	204.34	3.61
550	15.422	0.018	0.034	0.299	2.30	6.81	73.31	15.31	191.51	3.04
600	16.379	0.019	0.017	0.350	3.06	8.80	84.69	16.25	202.63	4.74
600	15.613	0.018	0.006	0.161	2.52	10.45	74.98	15.54	194.21	3.69
650	16.413	0.019	0.006	0.309	3.09	12.47	84.60	16.29	203.16	2.98
650	15.656	0.018	0.003	0.130	2.41	14.04	73.73	15.59	194.81	3.08
700	15.905	0.018	0.004	0.196	2.44	15.62	80.67	15.82	197.53	2.86
700	15.627	0.018	0.003	0.085	1.99	16.92	69.46	15.57	194.62	3.39
750	15.635	0.018	0.004	0.141	2.01	18.24	76.97	15.56	194.53	3.50
750	15.650	0.018	0.003	0.085	1.79	19.40	66.64	15.60	194.90	3.16
800	15.505	0.018	0.005	0.127	3.34	21.58	84.19	15.44	193.04	4.24
800	15.562	0.018	0.003	0.047	2.88	23.46	75.79	15.52	193.98	3.61
800	15.753	0.018	0.003	0.022	2.23	24.92	64.90	15.72	196.34	3.30
800	15.891	0.018	0.002	0.000	3.46	27.17	49.14	15.86	198.04	4.14
700	16.113	0.027	0.007	0.864	0.09	27.23	9.88	15.83	197.71	3.03
750	16.001	0.022	0.004	0.000	0.17	27.35	16.71	15.97	199.34	3.70
800	15.359	0.000	0.008	0.592	0.53	27.69	35.80	15.16	189.71	1.43
850	16.198	0.019	0.006	0.226	1.49	28.66	62.04	16.10	200.90	4.61
900	16.518	0.019	0.007	0.182	3.12	30.69	77.31	16.44	204.82	3.51
950	16.873	0.019	0.007	0.253	5.07	34.00	84.46	16.77	208.76	3.11

TABLE 1. ARGON ANALYTICAL DATA (continued)

Temperature (°C)	$^{40}Ar/^{39}Ar$	$^{38}Ar/^{39}Ar$	$^{37}Ar/^{39}Ar$	$^{36}Ar/^{39}Ar$ (10^{-3})	^{39}Ar $(10^{14}mol)$	$F^{39}Ar$ released*	$^{40}Ar*$ [†] (%)	$^{40}Ar*/^{39}Ar$	Age (Ma)	± 1σ (Ma)
94MR433k		K-feldspar	$J^{§}$ = 0.0073150			weight = 23 mg			38°52′43.9″N, 89°06′29.8″E	
1000	17.326	0.019	0.006	0.303	6.67	38.36	87.61	17.21	213.90	2.88
1050	17.106	0.019	0.004	0.346	8.07	43.62	89.06	16.98	211.17	4.67
1100	17.841	0.019	0.002	0.464	9.62	49.89	90.58	17.68	219.38	3.80
1150	18.696	0.020	0.001	0.623	11.71	57.53	91.98	18.49	228.80	6.26
1200	19.496	0.019	0.001	0.596	22.11	71.95	95.26	19.29	238.17	6.32
1400	20.675	0.020	0.001	0.438	40.72	98.50	97.16	20.52	252.29	5.08
1400	20.960	0.022	0.001	1.501	2.30	100.00	71.17	20.49	252.02	9.01
94MR373k		K-feldspar	J = 0.0073150			weight = 11 mg			38°19′N, 89°19′E	
400	25.247	0.094	1.057	47.814	0.01	1.14	3.24	11.49	145.54	5.99
400	20.702	0.019	0.460	9.292	0.01	1.96	2.56	18.04	223.60	5.44
450	17.414	0.021	0.523	12.557	0.01	3.44	4.83	13.81	173.62	4.83
450	14.558	0.018	0.500	1.418	0.01	4.95	3.47	14.18	178.07	4.15
500	16.661	0.024	0.901	11.926	0.02	6.87	5.68	13.29	167.37	3.61
500	17.610	0.049	1.438	17.371	0.02	9.67	5.28	12.73	160.60	3.94
550	19.079	0.025	2.053	12.982	0.03	13.53	11.73	15.55	194.41	4.26
550	19.119	0.019	1.936	10.485	0.03	17.53	8.90	16.30	203.27	3.47
600	20.973	0.020	1.612	14.137	0.05	23.19	16.70	17.06	212.14	3.80
600	20.206	0.019	1.005	9.135	0.04	27.80	10.41	17.66	219.20	4.66
650	19.292	0.021	0.480	7.137	0.04	33.02	15.53	17.26	214.46	2.76
650	18.614	0.022	0.295	4.049	0.03	36.50	7.74	17.45	216.72	4.21
700	19.535	0.019	0.514	6.341	0.03	40.12	11.23	17.73	220.05	3.52
700	18.891	0.014	0.516	3.301	0.02	42.27	4.88	17.97	222.84	4.18
750	21.226	0.023	0.918	11.302	0.02	44.77	7.88	18.04	223.64	5.89
750	20.658	0.017	0.859	9.942	0.02	46.67	4.13	17.86	221.52	4.57
800	25.993	0.023	0.852	18.835	0.02	49.47	9.38	20.62	253.42	4.83
800	21.595	0.019	0.624	3.023	0.02	51.36	4.60	20.77	255.22	4.61
800	20.742	0.013	0.480	0.000	0.01	52.90	2.86	20.78	255.26	4.64
850	22.771	0.020	0.771	7.987	0.02	55.25	5.39	20.53	252.42	4.50
900	22.692	0.019	0.676	5.715	0.03	58.67	7.55	21.10	258.91	3.81
950	24.740	0.012	0.469	7.125	0.03	62.66	9.02	22.71	277.25	4.78
1000	27.539	0.023	0.556	10.334	0.03	66.78	9.62	24.59	298.41	2.28
1050	31.162	0.027	0.559	6.102	0.04	71.42	12.27	29.44	351.90	3.94
1100	34.150	0.027	0.655	6.713	0.05	77.68	16.63	32.27	382.33	3.47
1150	33.832	0.026	0.879	15.683	0.07	86.28	19.10	29.38	351.23	4.73
1200	30.291	0.022	0.842	11.814	0.07	94.89	17.52	26.95	324.67	2.74
1400	37.875	0.026	0.892	22.156	0.04	100.00	11.66	31.55	374.64	4.66
94MR368k		K-feldspar	J = 0.0073150			weight = 15 mg			38°17.5′N, 89°19′E	
400	149.004	0.000	0.032	15.631	0.26	0.27	88.50	144.45	1300.88	14.75
400	20.049	0.023	0.016	1.619	0.15	0.44	35.17	19.55	241.16	3.90
450	30.780	0.029	0.014	1.953	0.57	1.05	78.00	30.19	359.93	5.98
450	15.807	0.020	0.021	0.564	0.54	1.62	54.97	15.62	195.14	4.35
500	14.755	0.000	0.082	1.194	1.08	2.77	64.65	14.39	180.55	1.38
550	18.161	0.020	0.088	0.558	1.52	4.38	80.61	17.98	222.93	5.73
550	15.610	0.018	0.064	0.168	1.37	5.83	69.10	15.54	194.23	4.27
600	17.727	0.020	0.040	0.349	1.82	7.77	81.27	17.60	218.50	3.45
600	15.895	0.018	0.017	0.089	1.40	9.25	67.21	15.84	197.81	3.12
650	16.813	0.019	0.010	0.216	1.44	10.78	74.71	16.72	208.19	3.47
650	6.119	0.000	5.009	483.709	0.00	10.79	-3.46	-134.28	0.00	0.00
700	16.228	0.018	0.006	0.207	1.49	12.37	72.66	16.14	201.32	4.05
700	15.822	0.018	0.005	0.181	1.19	13.63	58.21	15.74	196.61	3.79
750	16.125	0.018	0.006	0.184	1.51	15.24	71.01	16.04	200.18	2.99
750	15.952	0.018	0.005	0.093	1.32	16.64	58.72	15.90	198.45	3.58
800	15.914	0.018	0.007	0.074	1.76	18.51	72.25	15.86	198.06	3.19
800	15.955	0.018	0.005	0.000	1.48	20.08	59.47	15.93	198.80	3.11
800	16.031	0.018	0.003	0.000	1.27	21.43	48.83	16.00	199.69	3.06
800	16.250	0.018	0.002	0.000	2.40	23.98	37.85	16.22	202.28	3.23
700	16.290	0.004	0.000	0.000	0.04	24.02	4.27	16.26	202.75	3.08
750	15.199	0.017	0.000	0.000	0.12	24.14	10.88	15.17	189.83	1.59
800	15.411	0.018	0.001	0.000	0.37	24.54	26.36	15.38	192.36	2.00

TABLE 1. ARGON ANALYTICAL DATA (continued)

Temperature (°C)	$^{40}Ar/^{39}Ar$	$^{38}Ar/^{39}Ar$	$^{37}Ar/^{39}Ar$	$^{36}Ar/^{39}Ar$ (10^{-3})	^{39}Ar $(10^{14}mol)$	$F^{39}Ar$ released*	$^{40}Ar*$ † (%)	$^{40}Ar*/^{39}Ar$	Age (Ma)	± 1σ (Ma)
94MR368k		K-feldspar		J = 0.0073150		weight = 15 mg			38°17.5′N, 89°19′E	
850	16.682	0.018	0.006	0.000	1.01	25.61	49.33	16.65	207.38	3.10
900	16.982	0.018	0.009	0.106	2.06	27.80	65.05	16.92	210.55	2.85
950	17.215	0.018	0.012	0.153	3.27	31.26	73.52	17.14	213.13	2.88
1000	17.408	0.019	0.011	0.212	4.25	35.77	77.30	17.32	215.18	3.30
1050	17.910	0.019	0.007	0.264	5.07	41.15	79.65	17.80	220.88	3.54
1100	18.909	0.020	0.004	0.391	5.83	47.33	81.58	18.77	232.06	3.45
1150	18.818	0.020	0.003	0.482	8.40	56.23	85.49	18.65	230.71	4.44
1200	19.290	0.020	0.002	0.437	12.65	69.65	89.48	19.13	236.33	4.27
1400	20.441	0.019	0.003	0.341	27.38	98.69	94.04	20.31	249.92	4.04
1400	21.135	0.017	0.004	1.581	1.23	100.00	44.07	20.65	253.76	5.07
94MR369k		K-feldspar		J = 0.0073150		weight = 15 mg			38°17.5′N, 89°19′E	
400	18.893	0.022	0.018	11.035	0.30	0.27	48.86	15.67	195.73	3.65
400	11.242	0.020	0.005	5.758	0.31	0.55	31.68	9.54	121.73	1.41
450	12.542	0.018	0.004	2.415	0.69	1.17	64.34	11.81	149.52	1.64
450	10.943	0.018	0.012	2.494	0.61	1.73	49.22	10.19	129.72	2.26
500	13.274	0.018	0.028	1.890	0.80	2.45	69.13	12.70	160.26	1.81
500	12.338	0.018	0.043	1.787	0.70	3.09	56.13	11.80	149.30	2.42
550	15.184	0.018	0.070	1.474	1.10	4.09	77.85	14.74	184.68	2.55
550	13.058	0.017	0.054	1.098	0.82	4.83	61.72	12.72	160.47	2.23
600	13.878	0.018	0.042	0.911	1.25	5.97	78.79	13.59	170.98	2.41
600	12.712	0.017	0.019	0.640	1.03	6.90	66.43	12.50	157.84	2.50
650	13.371	0.018	0.015	0.581	1.50	8.27	81.25	13.17	165.99	2.30
650	12.799	0.017	0.010	0.365	1.14	9.30	68.69	12.66	159.83	2.14
700	13.028	0.017	0.009	0.332	1.56	10.72	81.60	12.90	162.72	2.39
700	12.892	0.017	0.005	0.226	1.24	11.84	70.51	12.80	161.44	2.23
750	13.311	0.018	0.004	0.290	1.64	13.33	82.42	13.20	166.26	2.51
750	13.445	0.018	0.005	0.274	1.27	14.48	71.48	13.34	167.93	2.35
800	13.834	0.018	0.006	0.397	1.73	16.06	83.33	13.69	172.18	3.03
800	13.802	0.018	0.004	0.251	1.69	17.59	76.96	13.70	172.29	4.51
800	14.042	0.018	0.002	0.271	1.41	18.86	68.24	13.93	175.10	2.78
800	14.279	0.018	0.000	0.211	2.74	21.35	58.85	14.19	178.14	2.72
700	13.546	0.014	0.000	0.064	0.07	21.42	12.36	13.50	169.87	2.18
750	13.624	0.016	0.000	0.018	0.17	21.57	26.33	13.59	170.97	1.37
800	14.020	0.018	0.000	0.218	0.47	22.00	49.11	13.93	175.02	1.99
850	15.476	0.018	0.004	0.823	1.52	23.38	76.08	15.21	190.30	3.23
900	15.675	0.018	0.005	0.793	2.98	26.08	85.59	15.42	192.76	3.45
950	15.619	0.018	0.004	0.875	4.07	29.78	88.37	15.34	191.82	2.96
1000	16.691	0.018	0.006	1.111	6.98	36.11	92.21	16.34	203.69	3.52
1050	18.211	0.019	0.007	1.340	8.13	43.48	93.12	17.79	220.74	3.37
1100	19.741	0.019	0.007	1.517	7.01	49.84	92.67	19.27	237.93	5.45
1150	22.402	0.019	0.007	1.982	9.85	58.77	94.10	21.80	266.92	5.71
1200	25.383	0.019	0.005	2.046	16.60	73.83	95.82	24.76	300.31	5.12
1400	30.403	0.019	0.004	1.672	28.85	100.00	97.41	29.89	356.73	4.30
94MR370k		K-feldspar		J = 0.0073150		weight = 17 mg			38°17.5′N, 89°19′E	
400	23.543	0.022	0.022	9.561	0.35	0.22	54.85	20.74	254.87	3.88
400	13.717	0.020	0.006	5.286	0.32	0.43	33.18	12.16	153.67	3.00
450	21.905	0.018	0.014	2.268	0.85	0.98	76.17	21.22	260.31	3.45
450	11.068	0.018	0.017	1.301	0.90	1.55	54.61	10.66	135.50	2.03
500	15.540	0.018	0.060	0.953	2.14	2.93	84.84	15.24	190.69	2.83
500	12.375	0.017	0.098	0.752	1.49	3.89	68.99	12.14	153.48	1.80
550	18.881	0.018	0.159	0.765	2.65	5.60	89.26	18.65	230.72	3.10
550	14.660	0.017	0.150	0.539	2.03	6.91	77.93	14.49	181.79	2.55
600	17.058	0.018	0.143	0.508	3.40	9.10	90.52	16.90	210.28	2.78
600	13.718	0.017	0.078	0.322	2.72	10.85	81.31	13.60	171.15	2.32
650	13.998	0.018	0.039	0.344	3.72	13.24	89.47	13.87	174.38	2.41
650	13.538	0.018	0.023	0.205	2.60	14.92	80.20	13.45	169.31	2.66
700	13.883	0.017	0.018	0.240	3.49	17.16	88.78	13.79	173.33	2.74
700	14.073	0.017	0.009	0.107	2.71	18.90	81.22	14.01	176.06	2.61
750	14.761	0.018	0.006	0.185	3.45	21.12	89.16	14.68	184.00	2.87

TABLE 1. ARGON ANALYTICAL DATA (continued)

Temperature (°C)	$^{40}Ar/^{39}Ar$	$^{38}Ar/^{39}Ar$	$^{37}Ar/^{39}Ar$	$^{36}Ar/^{39}Ar$ (10^{-3})	^{39}Ar $(10^{14}mol)$	$F^{39}Ar$ released*	$^{40}Ar^{*}$ [†] (%)	$^{40}Ar^{*}/^{39}Ar$	Age (Ma)	± 1σ (Ma)
94MR370k		K-feldspar		J = 0.0073150		weight = 17 mg			38°17.5′N, 89°19′E	
750	14.888	0.017	0.004	0.108	2.92	23.00	82.80	14.83	185.77	2.76
800	15.432	0.018	0.004	0.176	3.33	25.14	89.05	15.35	192.00	3.33
800	15.734	0.018	0.003	0.131	4.10	27.78	87.34	15.67	195.74	3.25
800	15.994	0.018	0.002	0.116	3.85	30.26	83.34	15.93	198.86	3.02
800	15.660	0.017	0.002	0.041	7.72	35.22	76.79	15.62	195.16	2.81
700	15.219	0.018	0.000	0.000	0.12	35.30	17.38	15.19	190.08	0.13
750	15.100	0.018	0.000	0.000	0.40	35.56	40.73	15.07	188.66	1.28
800	16.435	0.018	0.000	0.161	1.14	36.29	67.19	16.36	203.91	3.60
850	17.073	0.017	0.003	0.372	3.63	38.63	86.33	16.94	210.70	4.32
900	17.691	0.018	0.005	0.378	7.11	43.20	92.29	17.55	217.92	3.30
950	16.817	0.017	0.005	0.401	11.10	50.34	94.27	16.67	207.59	3.96
1000	18.026	0.017	0.005	0.468	13.66	59.12	95.28	17.86	221.53	3.31
1050	20.372	0.017	0.006	0.544	14.77	68.63	95.88	20.19	248.46	4.09
1100	24.038	0.018	0.009	0.743	10.14	75.15	94.99	23.79	289.49	4.56
1150	28.889	0.018	0.012	1.093	9.63	81.35	95.23	28.54	342.10	6.22
1200	33.457	0.019	0.007	1.233	13.22	89.85	96.52	33.07	390.86	8.21
1400	37.660	0.018	0.006	0.933	15.78	100.00	97.32	37.36	435.87	9.54
94MR380k		K-feldspar		J = 0.0073150		weight = 25 mg			38°21′N, 89°47′E	
400	13.137	0.031	0.039	33.628	0.10	0.10	10.07	3.36	43.80	2.86
400	9.133	0.022	0.000	23.415	0.10	0.20	6.23	2.31	30.26	1.18
450	6.523	0.020	0.008	6.393	0.43	0.65	41.53	4.64	60.22	0.94
450	3.924	0.020	0.007	3.273	0.42	1.07	26.28	2.95	38.46	0.42
500	4.677	0.017	0.018	2.631	0.62	1.71	48.11	3.89	50.57	1.32
500	4.107	0.019	0.052	1.400	0.74	2.47	43.87	3.68	47.89	0.42
550	5.316	0.019	0.079	1.845	0.80	3.29	58.90	4.76	61.75	0.04
550	4.815	0.017	0.065	0.911	0.74	4.04	48.63	4.53	58.79	0.21
600	6.178	0.018	0.056	1.828	1.13	5.20	68.50	5.62	72.74	0.10
600	5.675	0.017	0.030	0.679	0.99	6.22	59.42	5.45	70.54	0.44
650	7.099	0.018	0.022	1.739	1.21	7.45	67.80	6.57	84.65	3.26
650	6.582	0.017	0.012	0.398	0.96	8.44	56.22	6.44	83.03	2.66
700	7.223	0.019	0.019	0.903	0.95	9.42	60.93	6.93	89.26	3.09
700	7.290	0.018	0.016	0.291	0.84	10.28	50.03	7.18	92.32	2.63
750	7.773	0.019	0.024	1.307	0.98	11.28	58.23	7.37	94.69	5.24
750	7.766	0.019	0.011	0.676	0.65	11.95	40.19	7.54	96.87	3.49
800	8.290	0.018	0.019	1.020	0.64	12.61	46.72	7.97	102.18	2.70
800	8.188	0.018	0.010	0.150	0.65	13.27	38.35	8.12	104.05	0.82
800	8.490	0.017	0.001	0.202	0.58	13.87	30.32	8.40	107.59	0.65
800	9.134	0.017	0.000	0.065	1.03	14.93	21.73	9.09	116.08	2.95
700	8.074	0.007	0.000	0.000	0.03	14.96	3.85	8.04	103.15	1.09
750	8.227	0.014	0.000	0.000	0.06	15.02	6.18	8.20	105.06	1.37
800	9.015	0.017	0.000	0.000	0.17	15.19	15.13	8.99	114.85	0.46
850	10.156	0.018	0.010	1.771	0.42	15.62	28.95	9.61	122.61	2.64
900	10.818	0.020	0.018	2.942	1.09	16.73	47.72	9.94	126.59	2.50
950	10.744	0.019	0.012	2.107	1.86	18.64	58.59	10.10	128.65	2.50
1000	10.614	0.019	0.012	2.444	3.12	21.85	66.33	9.88	125.85	2.67
1050	10.606	0.020	0.014	2.601	4.49	26.45	70.89	9.82	125.19	2.44
1100	10.990	0.020	0.017	2.572	7.26	33.90	77.57	10.22	130.02	2.19
1150	10.939	0.020	0.015	2.088	12.11	46.32	83.45	10.31	131.12	3.66
1200	11.112	0.020	0.006	1.301	23.51	70.43	90.03	10.71	136.02	3.36
1400	11.814	0.019	0.002	0.805	28.83	100.00	91.52	11.55	146.33	3.67
94DU33k		K-feldspar		J = 0.0065780		weight = 8.65 mg			38°54′44″N, 93°14′37″E	
400	45.703	0.117	0.313	119.525	0.11	0.41	24.16	11.05	126.56	2.50
400	37.690	0.092	0.233	107.283	0.10	0.79	17.44	6.57	76.37	1.73
460	33.812	0.066	0.166	76.638	0.20	1.54	34.23	11.58	132.40	2.57
460	24.378	0.044	0.118	47.607	0.15	2.11	43.29	10.56	121.11	2.42
500	21.811	0.035	0.120	30.514	0.20	2.89	59.35	12.95	147.45	2.85
500	17.991	0.028	0.098	15.922	0.18	3.58	74.22	13.36	151.92	3.83
550	18.484	0.028	0.110	14.178	0.30	4.70	77.65	14.36	162.79	3.12
550	15.755	0.024	0.093	4.118	0.23	5.58	92.29	14.54	164.81	3.18

E.R. Sobel et al.

TABLE 1. ARGON ANALYTICAL DATA (continued)

Temperature (°C)	$^{40}Ar/^{39}Ar$	$^{38}Ar/^{39}Ar$	$^{37}Ar/^{39}Ar$	$^{36}Ar/^{39}Ar$ (10⁻³)	^{39}Ar (10¹⁴mol)	F^{39}Ar released*	$^{40}Ar^*$ [†] (%)	$^{40}Ar^*/^{39}Ar$	Age (Ma)	± 1σ (Ma)
94DU33k		K-feldspar		J = 0.0065780		weight = 8.65 mg			38°54′44″N, 93°14′37″E	
600	16.356	0.024	0.113	3.889	0.45	7.30	92.99	15.21	172.06	3.29
600	15.747	0.022	0.092	0.612	0.39	8.79	98.74	15.55	175.70	3.94
650	16.373	0.022	0.122	1.274	0.69	11.41	97.64	15.99	180.41	3.79
650	16.368	0.022	0.114	0.465	0.63	13.82	99.06	16.22	182.86	3.86
700	17.580	0.023	0.133	1.341	0.80	16.89	97.69	17.18	193.13	4.43
700	17.599	0.022	0.123	0.708	0.71	19.60	98.73	17.38	195.29	4.37
750	17.289	0.022	0.137	1.562	0.86	22.87	97.29	16.82	189.35	3.95
750	17.083	0.021	0.123	0.401	0.64	25.31	99.22	16.95	190.73	3.96
800	17.223	0.022	0.141	1.678	0.73	28.10	97.08	16.72	188.29	3.99
800	17.047	0.022	0.101	0.413	0.44	29.77	99.18	16.91	190.28	4.04
800	17.327	0.021	0.064	0.706	0.77	32.72	98.68	17.10	192.31	4.06
700	16.443	0.018	0.000	0.000	0.02	32.78	99.82	16.41	184.96	4.68
750	17.496	0.022	0.000	0.980	0.04	32.93	98.20	17.18	193.18	4.56
800	17.293	0.022	0.041	1.089	0.11	33.37	98.02	16.95	190.73	4.41
850	18.204	0.024	0.057	4.467	0.30	34.52	92.75	16.89	190.02	3.63
900	18.051	0.024	0.065	4.443	0.64	36.95	92.73	16.74	188.47	4.14
950	17.415	0.024	0.054	2.358	1.01	40.79	95.93	16.71	188.11	3.98
1000	17.698	0.025	0.052	2.884	1.40	46.14	95.13	16.84	189.51	4.02
1050	18.663	0.027	0.061	4.102	1.86	53.23	93.50	17.45	196.05	4.14
1100	19.951	0.028	0.082	6.166	2.29	61.97	90.93	18.14	203.40	4.31
1150	20.480	0.026	0.084	5.915	3.16	74.01	91.52	18.75	209.77	4.49
1200	19.929	0.024	0.073	3.165	4.20	90.04	95.28	18.99	212.36	4.75
1400	21.000	0.024	0.163	4.770	2.61	100.00	93.35	19.61	218.86	4.62
94DU33B		biotite		J = 0.0068130		weight = 6.55 mg			38°54′44″N, 93°14′37″E	
600	23.905	0.073	0.108	38.442	0.21	0.98	53.00	12.73	150.08	4.05
700	21.593	0.028	0.012	5.634	1.57	8.23	92.18	19.92	229.63	4.68
800	21.417	0.027	0.005	1.352	1.94	17.22	97.91	20.99	241.12	5.27
850	20.710	0.026	0.004	0.398	4.29	37.05	99.22	20.56	236.48	4.92
900	20.778	0.026	0.006	0.318	2.81	50.05	99.31	20.65	237.46	4.95
950	21.235	0.026	0.013	0.346	1.96	59.11	99.25	21.10	242.30	5.34
1000	20.756	0.025	0.015	0.278	2.84	72.23	99.36	20.64	237.36	4.88
1050	20.644	0.025	0.022	0.150	2.68	84.63	99.37	20.57	236.57	4.96
1100	20.789	0.025	0.069	0.105	2.64	96.85	99.30	20.73	238.34	5.12
1200	20.853	0.025	0.644	0.000	0.56	99.44	96.77	20.90	240.17	5.02
1400	31.310	0.033	1.148	35.720	0.12	100.00	55.93	21.07	241.96	10.17

Note: All K-feldspars were analyzed in a Staudacher-type double-vacuum resistance furnace; biotite was analyzed in a glass single vacuum radio frequency furnace.

* F^{39}Ar released = cumulative sum ^{39}Ar released.

[†] $^{40}Ar^*$ = percentage of radiogenic ^{40}Ar.

[§] J = Irradiation parameter.

180 Ma after 10%–30% of ^{39}Ar has been released. This small plateau is followed by a monotonic increase of ages over the last 70% of ^{39}Ar release to maximum ages of about 250 Ma. K/Ca and Cl/K plots show that these features in the age spectra are only weakly associated with chemical variations within the feldspar during degassing. The small age plateau ca. 180 Ma seems associated with a local maximum of the K/Ca ratio, although no recrystallization of the feldspars was observed in this sample. However, although the inverse isochron plot does not indicate large quantities of excess argon, the Cl/K plot from the duplicate analysis of low extraction temperature steps reveals a correlation between the high ages of each first step and the amount of Cl, suggesting that excess argon resides in fluid in-

clusions. Modeling the diffusion data reveals an excellent fit with the analytic data, with 10 domains of constant activation energy of 46.5 kcal/mol, a value in agreement with the statistical analysis of Lovera et al. (1997). The modeled age spectrum agrees well with the laboratory data and takes into account all steps below 50% of ^{39}Ar extracted; all other steps are above 1100 °C, the limit for incongruent melting of feldspars and consequent nonvolumetric diffusion for argon. The agreement is particularly good on the second steps of each duplicate at low extraction temperature steps.

Results from sample 94MR369 (Fig. 3B) are similar to those of 94MR351, except that excess argon has a greater effect in the low extraction temperature steps (leading to a less-well-

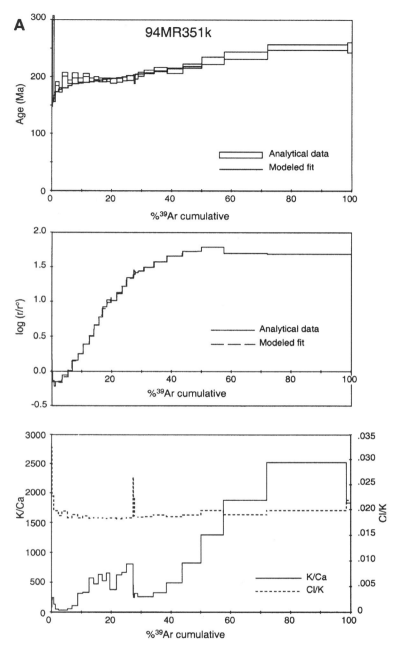

Figure 3. Argon thermochronologic data. A: Example of fit between analyzed and modeled K-feldspar data. Details of modeling are discussed in text. B (following page): Argon data for other samples analyzed in this study, arranged from north to south. Arhenius data and models are not shown. Analytical data are available in Tables 1 and 2. Letter at end of sample number indicates mineral analyzed; b is biotite, k and kspar are potassium feldspar.

developed plateau at 180 Ma) and also much higher ages in the last degassing steps. Because the older range of ages is known in the area (Delville et al., this volume; Sobel and Arnaud, 1999), the ages are probably geologically meaningful rather than the effect of excess argon. Both diffusion characteristics (with a constant activation energy of 44.6 kcal/mol and 8 domains) and age spectrum models agree extremely well with the original data.

Results from sample 94MR370 show the same pattern as those from 94MR369, with enhanced effects of excess argon at low extraction temperatures, and even higher ages at the end of the gas release, which are also interpreted to be significant.

Modeled diffusion characteristics and age spectra are both in close agreement with the analytical data, with the assumption that the lowest steps of each duplicate analysis represent a meaningful age.

The argon results from sample 94MR368 are almost identical to the results from the three previously described K-feldspars, with excess argon in the first steps. Modeling of both diffusion characteristics (10 domains, 45 kcal/mol) and the age spectrum yields a good agreement with analytic data, although the latter model is slightly discordant on the small plateau associated with a local maximum in K/Ca ratios.

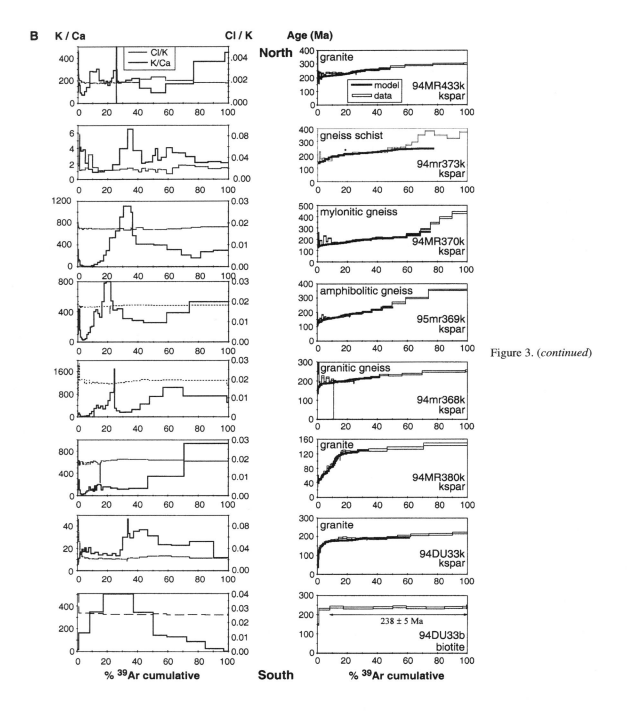

Figure 3. (continued)

The spectrum of sample 94MR373 is disturbed; K/Ca and Cl/K ratios indicate that this feldspar is probably affected by excess argon, at least in the first and last steps. The latter give a crude linear array on the inverse isochron plot, suggesting excess argon superimposed on original ages of ca. 250 Ma. Modeled diffusion characteristics agree well with laboratory data (8 domains with a constant activation energy of 45 kcal/mol), but the modeled cooling history and age spectrum are imprecise due to the spectrum's complexity.

Results from undeformed fine-grained leucogranite sample 94MR433 broadly show the same features as the previous samples, but although diffusion characteristics could be satisfactorily modeled, the first 30% of the age spectrum is dominated by excess argon, correlated with a Cl rich component. The cooling history can thus be only roughly modeled.

Undeformed granitic sample 94MR380 shows an age spectrum that contrasts with those previously described, with a strong increase in the first 20% of ^{39}Ar release, then a slow increase of ages culminating at 150 Ma. This sample appears quite retentive and more than 50% of the gas was released at temperatures above 1150 °C. Nevertheless, the diffusion characteristics can be modeled adequately (5 domains with a constant activation energy of 45 kcal/mol). The age spectra is also modeled, although it is not as precise as in other models.

Biotite from the undeformed, fine-grained granite sample 94DU33 yields a WMPA of 238 ± 5 Ma, interpreted to slightly postdate the intrusive age. K-feldspar argon data from the same sample show that cooling continued from the ~325 °C biotite age; the cooling rate increased ca. 200 Ma and decreased significantly ca. 175 Ma. The feldspar from 94DU33 shows a climbing age spectrum, starting from an imprecise Cenozoic age and increasing more gently from 200 to 230 Ma. There is no correlation with K/Ca and excess argon is not suspected. Diffusion parameters are well reproduced in the modeling (3 domains, 32 kcal/mol) and the age spectra are globally acceptable, although the finest variations of the age spectra are not reproduced. The activation energy implied by the modeling is lower than usual, and therefore the model is probably not very precise.

Fission-track data and modeling

Apatite fission-track thermochronology was used to determine the lower temperature cooling history. The results are described from north to south. All ages pass the chi squared test and are reported as pooled ages. Analytic data are shown in Figure 4 and summarized in Table 2.

Sample 94MR433 yields an age of 26 ± 2 Ma, suggesting cooling through the apatite partial annealing zone around this time. However, the confined track length distribution is short and broad, with a mean length of 11.26 ± 0.25 μm and a standard deviation of 2.69 μm, suggesting a prolonged residence within the partial annealing zone after 26 Ma followed by late-stage final cooling.

Samples 94MR355, 94MR351, and 94MR370 yield similar results; ages are 82 ± 6, 93 ± 5 and 83 ± 7 Ma, respectively. The first two samples have long track lengths (13.23 ± 0.14 μm and 13.34 ± 0.13 μm) that are tightly clustered (standard deviations of 1.39 μm and 1.29 μm). The third sample has track lengths that are slightly shortened and less tightly clustered than those of the other two samples; the mean length is 12.34 ± 0.21 μm and the standard deviation is 2.06 μm. All three results suggest Late Cretaceous cooling followed by final cooling starting in the early Cenozoic, as indicated by shortened track length distributions with the addition of a few longer tracks.

Samples 94MR380, 94MA415, and 94MA416 were collected close to the south side of the active strand of the Altyn Tagh fault. Although the former sample is from a Mesozoic granite and the latter two were collected from Jurassic sedimentary rock 25 km to the east, they all yield similar single grain ages of 20 ± 2 Ma, 19 ± 1 Ma, and 22 ± 2 Ma. Confined track lengths from the igneous rock are shorter and show greater spread than lengths from the sediment: mean lengths are

10.03 ± 0.69 μm versus 13.26 ± 0.52 μm and 13.50 ± 0.32 μm and standard deviations are 3.44 μm versus 2.31 μm and 1.52 μm, respectively. However, only a limited number of confined tracks could be measured in each sample. These results suggest that this portion of the Altyn Tagh fault zone underwent strong cooling ca. 21 Ma, although final cooling to surface temperature occurred more recently.

Sample AT60 was collected about 250 km to the east of these samples, on the north side of the range. Apatite and zircon were analyzed. The zircon results yield a tight cluster of single grain ages, with an age of 221 ± 26 Ma. corresponding to the time of cooling through $\sim 280 \pm 50$ °C (Tagami and Shimada, 1996). The apatite results yield an age of 167 ± 15 Ma. Confined track lengths are significantly shortened; the mean length is 11.73 ± 0.16 μm and the standard deviation is 1.64 μm. These data indicate significant Late Triassic to Early Jurassic cooling, and final cooling to surface temperature during the Cenozoic.

The single grain age of 109 ± 4 Ma from sample 94DU33 is similar to those from samples 94MR351, 94MR355, and 94MR370. Confined track lengths are also similar; the mean length is 12.99 ± 0.18 μm and the standard deviation is 1.81 μm. This sample also cooled during the Late Cretaceous, and final cooling to surface temperature occurred during the Cenozoic. In detail, the track length distribution has a broader, slightly bimodal peak, suggesting that the sample underwent a small amount of reheating during the early Cenozoic.

Samples 94HB157, 94HB163, 94HB170, and 94HB177 yield ages between 37 ± 3 and 47 ± 3 Ma; ages overlap at the 2σ level. Mean track lengths range between 12.52 ± 0.20 and 13.07 ± 0.24 μm; standard deviations are about 2 μm. Variations of track etch pit size suggest that the apatites represent a wide range of chemical compositions (e.g., Burtner et al., 1994); this likely explains the uneven track length distributions shown on histograms (Fig. 4). Such data are poorly suited to thermal modeling. The shortened mean track lengths suggest that the actual cooling ages are older than the measured ages. Vitrinite reflection values of 1.79 and 1.64 obtained from two Lower Jurassic samples overlying sample 94HB177 by ~100 m (Ritts, 1998) suggest a maximum temperature of 165–180 °C (Burnham and Sweeney, 1989). Because these vitrinite values are significantly higher than reported from Jurassic strata in other parts of the Qaidam basin (Ritts, 1998), it is unclear whether heating at this locality is due solely to sedimentary burial. The section likely cooled through ~100 °C during the Paleocene(?)–Eocene. The presence of confined tracks shorter than 10 μm in all four samples suggests either prolonged residence or reheating to temperatures between 90 and 60 °C prior to final cooling to surface temperatures.

IMPLICATIONS OF THERMAL MODELING RESULTS

Combining the argon data with modeled apatite cooling histories provides information on modeled cooling histories between ~400 and 60 °C (Fig. 5). Note that several parts of the

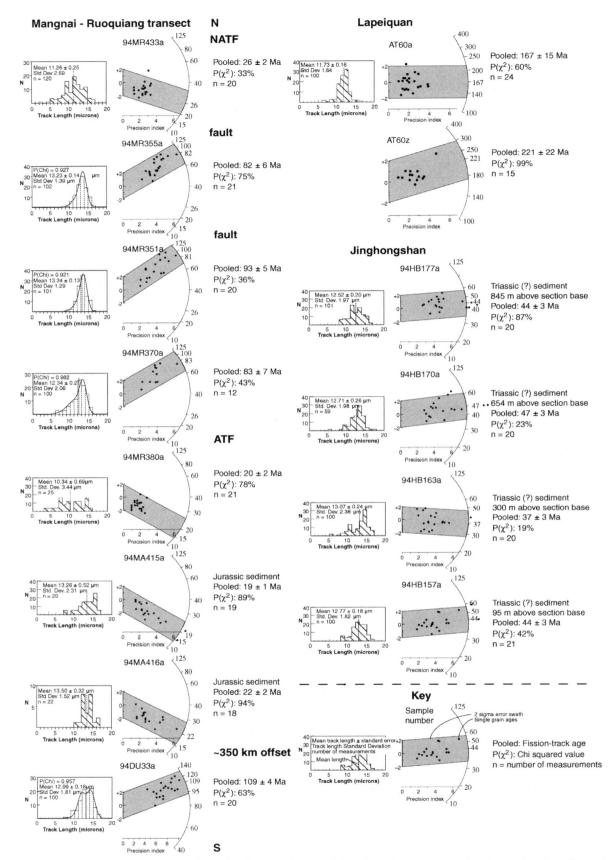

Figure 4. Fission-track data arranged in geographic order, from north to south. Single-grain ages are displayed on radial plots (for explanation of these graphs, see Dumitru et al., this volume). All uncertainties are quoted at ±1σ level. Curve superimposed on some track length histograms is distribution corresponding to best fit thermal model shown in Figure 5; P(χ^2) is chi squared probability that this modeled distribution matches observed track length histogram. Letter at end of sample number indicates mineral analyzed; a is apatite, z is zircon. NATF: North Altyn Tagh fault.

TABLE 2. APATITE FISSION-TRACK DATA

Sample number	Mineral	No xls*	Rho-S (x10⁶)[†]	NS[§]	Rho-I (x10⁶)[#]	NI**	χ² (%)[††]	Rho-D (x10⁶)[§§]	ND[##]	Age ± 1 σ (Ma)	Lab ***	Analyst[†††]
94MR433	ap	20	0.145	138	1.539	1466	33	1.551	3723	26.3 ± 2.4	ETH	ERS
94MR355	ap	21	1.195	493	3.097	1277	75	1.186	3988	82.1 ± 6.4	UP	ERS
94MR351	ap	20	0.466	656	0.904	1272	36	1.006	5322	92.9 ± 6.9	UP	ERS
94MR370	ap	12	0.965	251	3.159	822	42	1.531	3723	84.0 ± 7.8	ETH	ERS
94MR380	ap	21	0.115	77	1.670	1119	78	1.571	3723	19.5 ± 2.3	ETH	ERS
94MA415	ap	19	0.696	287	6.460	2664	89	1.003	5322	19.5 ± 1.6	UP	ERS
94MA416	ap	18	0.518	221	5.066	2163	94	1.178	3988	21.7 ± 2.0	UP	ERS
94DU33	ap	20	1.204	1831	2.342	3562	63	1.187	3988	109.2 ± 7.0	UP	ERS
AT60	ap	24	36.800	244	41.620	276	60	1.139	6891	167.0 ± 15.0	ETH	MJ
AT60	zr	15	0.197	1233	1.851	116	99	0.466	2821	220.8 ± 26.4	ETH	MJ
94HB177	ap	20	0.778	760	3.724	3638	87	1.181	3988	44.4 ± 3.1	UP	ERS
94HB170	ap	20	0.933	783	3.773	3167	23	1.046	4683	46.5 ± 3.2	UP	ERS
94HB163	ap	20	0.596	384	3.445	2220	19	1.182	3988	36.8 ± 2.9	UP	ERS
94HB157	ap	21	0.779	593	3.333	2538	42	1.046	4683	44.0 ± 3.2	UP	ERS

Note: The pooled age is reported for all samples as they pass the χ^2 test; error is 1 σ, calculated using zeta calibration method (Hurford and Green, 1983) with zeta of 361.5 ± 20 for apatite (E. Sobel, unpublished); 336 ± 5 for apatite (M. Jolivet, unpublished); and 90.7 ± 6.1 for zircon (M. Jolivet, unpublished). At least 12 standard samples were counted by each analyst in order to calculate personal zeta calibration factors. Durango, Fish Canyon Tuff, and Mount Dromedary apatites and Fish Canyon Tuff zircon were used for age standards.

*No xls is the number of individual crystals dated.

[†]Rho-S is the spontaneous track density measured (tracks/cm²).

[§]NS is the number of spontaneous tracks counted.

[#]Rho-I is the induced track density in the external detector (tracks/cm²).

**NI is the number of induced tracks counted.

[††] χ^2 (%) is the chi-square probability (Galbraith, 1981; Green, 1981). Values >5% are considered to pass this test and represent a single population of ages.

[§§]Rho-D is the induced track density in external detector adjacent to CN5 dosimetry glass (tracks/cm²).

[##]ND is the number of tracks counted in determining Rho-D.

***Lab is the laboratory where analysis was performed. ETH is Institute of Mineralogy and Petrography, Eidgenössische Technische Hochschule, Zürich, Switzerland; UP is Universitat Potsdam, Potsdam, Germany.

[†††]Analyst is the person performing analysis; ERS is E. Sobel, MJ is M. Jolivet.

K-feldspar history are distinguished on each model. Because modeling of the argon analytic data does not define diffusion characteristics above the start of incongruent melting of the feldspars, and diffusion and age data below 150 °C are likely to be the most affected by nonvolumetric diffusion and complex argon loss in nature (Arnaud and Kelley, 1997), the K-feldspar cooling histories are not constrained for ages obtained in the laboratory at temperatures above 1150 °C and below 150 °C. Therefore, dashes have been used in Figure 5B to show the poorly defined part of the cooling path, and solid lines indicate more robustly modeled data. The well-constrained K-feldspar analytical data correspond to geological temperatures between ~300 and 150 °C, although the actual temperature range varies between samples.

The most striking feature observed on the K-feldspar cooling models from the Mangnai-Ruoqiang transect is the presence of a transition from slower to much faster cooling between 207 and 181 Ma, with the exception of samples 94MR433 and 94MR73, in which faster cooling begins ca. 237. However, as discussed previously, the sample 94MR73 has a disturbed re-

lease spectrum that cannot be well modeled. It is noteworthy that $^{40}Ar/^{39}Ar$ ages of micas suggest that no major metamorphism affected these samples in the ~200 m.y. prior to the cooling recorded by K-feldspars (Sobel and Arnaud, 1999). Because the samples recording the earlier cooling also record higher temperatures, the difference in ages cannot simply reflect differences in structural depth between the samples at the onset of cooling. Rather, this implies heterogeneous unroofing across the range, even in restricted sampling areas. Sample AT60, located farther to the east, shows generally similar results, although cooling may initiate slightly earlier.

Sample 94MR380 varies notably from this regional pattern, with a much younger cooling history. However, it was sampled on the southern side of the active trace of the Altyn Tagh fault, it is leucocratic and undeformed, and its emplacement age, though not precisely known, is mapped as Yanshanian (late Mesozoic) (Wang et al., 1993). The sample also has an apatite fission-track age of 20 ± 2 Ma. The simplest explanation is that this granite was emplaced ca. 150 Ma, farther to the west relative to the rest of the transect, and that the modeled history

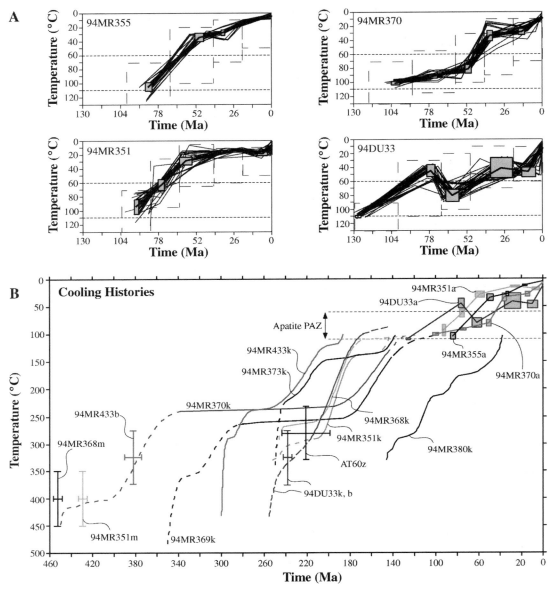

Figure 5. Thermal modeling of samples from Mangnai-Ruoquiang transect and Qaidam basin. A: Modeled time-temperature paths for four apatite samples, computed with genetic algorithm program of Gallagher (1995). Initial track length of 14.5 μm was used in constructing these models (Gleadow et al., 1986). Short dashed lines at 110 and 60 °C delineate apatite partial annealing zone, defined as temperature interval in which majority of track length shortening takes place (e.g., Fitzgerald et al., 1995). Dashed rectangles show time-temperature space searched by model. We modeled 1000 paths for each plot. Thin lines show best-fit solutions obtained for these model runs, and heavy lines and small boxes show average solutions obtained for same model run. Corrigan (1991) discussed robustness of data from this type of modeling. Modeling cannot determine cooling rate before ca. 110–130 Ma. B: Thermal histories combining all available argon and apatite fission-track data for selected samples, showing strong cooling during Early to Middle Jurassic. Late Cretaceous cooling seen in A is clearly minor compared to other cooling events. Large plus signs denote muscovite (m), biotite (b), and zircon (z) cooling data. Data from 94MR368m, 94MR351m, and 94MR433b are from Sobel and Arnaud (1999). K-feldspar model cooling curves are shown as solid where well defined and dashed where poorly defined. Error bars for these models are not shown; errors are considered to be ±25 °C and ~1.5% of age (ca. 2 Ma). Details of argon modeling are discussed in text. Apatite fission track (AFT) data shown are identical with those shown in A; for clarity, only average cooling paths have been plotted. PAZ is partial annealing zone. Note that samples with Neogene AFT ages are not shown.

essentially records conductive cooling within country rocks, and a probable faster cooling event during the early Cenozoic. One can speculate that this late cooling is tectonically linked to unroofing associated with motion along the Altyn Tagh fault.

Samples 94MR355, 94MR351, and 94MR370 from the Mangnai-Ruoquiang transect yield similar results from thermal modeling of the apatite cooling histories using the MonteTrax program (Gallagher, 1995) (Fig. 5A). At first glance, these thermal models all suggest Late Cretaceous cooling followed by final cooling beginning in the early Cenozoic, in agreement with the shortened track length distributions. Note that modeled cooling rates for all three samples are low, ~2.5 °C/m.y., and represent slower cooling rates than observed during the Jurassic (Fig. 5B). The track length distribution of sample 94MR370 has a shorter mean value than the other two samples, suggesting that this sample resided at a slightly higher temperature prior to cooling through the partial annealing zone slightly later than the other two samples. This interpretation is supported by modeling results. The modeled cooling histories show that samples 94MR351 and 94MR355 cooled at similar rates. The country rock hosting the northern granodiorite sample 94MR355 is composed of metamorphosed Middle Proterozoic carbonates, whereas sample 94MR351 was collected in a region characterized by Lower Proterozoic schists and gneisses (Wang et al., 1993), suggesting that the northern sample represents a much higher structural level. The two domains are separated by a subvertical fault trending parallel to the Altyn Tagh fault and mapped as extending for at least 160 km (Wang et al., 1993). Because the two samples cooled at the same rate, if Late Cretaceous exhumation had commenced synchronously, younger Altyn Tagh fault ages would reflect a greater amount of exhumation. However, the apparently structurally deeper sample cooled through the apatite partial annealing zone later than the structurally shallower sample. The two cooling histories can be explained by ~25 °C of Late Cretaceous cooling of sample 94MR351 commencing ~10 m.y. earlier than sample 94MR355. This suggests that the greatest vertical displacement along this fault, and hence the inferred difference in structural level, probably occurred prior to the Late Cretaceous.

Apatite fission-track results from sample 94DU33 are similar to those from samples 94MR355, 94MR351, and 94MR370. Although the apatite thermal history of the former sample is more complex than those from farther north, examination of the single grain ages and the track length histogram (Fig. 4) combined with MonteTrax modeling (Fig. 5A) show that this sample also cooled rapidly during the Late Cretaceous followed by final cooling beginning in the Paleocene–Eocene. The sample also underwent a small reheating event during the Paleocene–Eocene. This reheating event is not apparent in samples from the north side of the Altyn Tagh fault; however, it roughly coincides with the Altyn Tagh fault cooling ages from Jinghongshan.

Comparisons between the cooling histories from the Mangnai-Ruoquiang transect (particularly 94MR351, 94MR355, 94MR368, 94MR370) and sample 94DU33, located ~350 km to

the east along the trend of the Altyn Tagh fault, suggest that the two regions had common tectonic histories. The Altyn Tagh fault has left a clear late Cenozoic thermal overprint on some of the intervening units (e.g., 94MR380, 94MA415, 94MA416) that is absent from these two regions. Although the similar thermal histories may reflect the 400 ± 60 km post-Bajocian offset of the Altyn Tagh fault proposed by Ritts and Biffi (2000), the sampling density of this study clearly precludes a precise piercing point, as does the nature of the thermal data. Sample 94DU33 was collected from a small, fine-grained intrusion, and apparently similar intrusives are mapped along the south side of the Altyn Tagh fault. There is little evidence of ca. 240 Ma magmatism in the samples from the Mangnai-Ruoquiang transect as might be expected if 94DU33 was located nearby when it was emplaced, although similar units are mapped to the west of the transect (Wang et al., 1993). On the basis of similarities between middle Paleozoic ophiolites and high-pressure metamorphic units, Sobel and Arnaud (1999) suggested that the portion of the Altyn Tagh around Lapeiquan (Fig. 1) could be broadly correlated with the northern margin of the Qaidam basin. Restoration of ~400 km of offset along the fault does not provide an ideal match for the pre-Mesozoic geology. One can either postulate that post-240 Ma offset exceeded 350 km, that there was significant strike-slip offset between ca. 240 and 210 Ma, that similar-aged intrusives await identification within the region of the transect, or that our offset estimate is significantly in error. It is not possible to make definitive conclusions from such a small sample population. However, the similar thermal histories documented along the Mangnai-Ruoquiang transect and by sample 94DU33 are noteworthy and our suggestion that it represents a Mesozoic marker for offset along the Altyn Tagh fault provides a testable hypothesis for future research.

The sedimentary section at Jinghongshan cooled during the early Cenozoic, during roughly the same interval as reheating and subsequent cooling of sample 94DU33. Similar cooling ages are lacking from analyzed samples collected north of the Altyn Tagh fault, suggesting that there was an episode of Paleocene(?)–Eocene tectonism in the northern Qaidam basin that did not extend north into the portion of the Altyn Tagh discussed herein and that predates Oligocene tectonism documented in the northwestern Qaidam basin (Hanson, 1998). Future work could attempt to use these Paleogene cooling ages to determine post-Middle Jurassic, pre-Eocene (pre-Paleocene?) offset along the Altyn Tagh fault.

EXHUMATION HISTORY

Within the Altyn Tagh, the Jurassic–Cenozoic geothermal gradient is poorly defined. The current geothermal gradient is ~20–28 °C/km (Ulmishek, 1984). Because the Qaidam basin has been in an intracontinental foreland-style setting since the end of the Triassic (Ritts, 1998), it may be reasonable to assume that a similar geothermal gradient persisted from the Jurassic until the present. Maturation models for Jurassic coals from the

Qaidam basin are calibrated best to measured vitrinite reflectance using a constant 25 °C/km geothermal gradient (Ritts, 1998; Ritts and Biffi, this volume). Because the paleogeothermal gradient cannot be precisely known, it is difficult to make an exact conversion from temperature to depth. Nonetheless, reasonable assumptions allow a first-order interpretation. Almost all of the studied samples cooled by ~100–150 °C between ca. 210 and 140 Ma, the most rapid cooling being in the first half of this interval. Given the proximity of similar-age sedimentary basins (Li et al., 1991; Wang et al., 1993; Ritts, 1998), it is not unreasonable to assume that this cooling was balanced by erosion. For a geothermal gradient of 25 ± 5 °C and assuming purely conductive cooling, these values correspond to between 3.3 and 7 km of exhumation, in agreement with the prominent sub-Jurassic unconformity within the range. Note that the magnitude of Jurassic cooling, and hence presumably exhumation, far exceeds the 50–90 °C of early Cenozoic cooling recorded by the Altyn Tagh fault within the main portion of the Altyn Tagh. Significant Neogene cooling is only observed adjacent to the Altyn Tagh fault and the North Altyn Tagh thrust fault, suggesting that the <2000 m relief of the Altyn Tagh does not reflect significant Cenozoic exhumation.

It is notable that the cooling ages reported herein come from rocks that do not show obvious synchronous cooling ductile deformation and that are synchronous with sediment deposition in adjacent basins. In contrast, the slightly younger cooling ages from the Dangjin pass area are associated with ductile strike-slip deformation (Delville et al., this volume). These ages are not particularly well matched with significant deposition in nearby basins, suggesting that different deformation styles were responsible for cooling in the two study areas.

EXHUMATION PROCESSES AND LOCAL TECTONIC IMPLICATIONS

Given the observation of significant Jurassic exhumation within the Altyn Tagh, one must search for a causative mechanism. In the study area, only a slight spatial pattern is discernible for the onset of rapid cooling, with a tendency toward older ages on the northern side of the range. The only other published ther-

mochronologic data addressing Mesozoic cooling ages focuses on the region ~450 km to the east on the northern side of the Altyn Tagh fault, and documents rapid cooling between 160 and 140 Ma associated with horizontal deformation interpreted as strike-slip faulting subparallel to the present trace of the Altyn Tagh fault (Delville et al., this volume). Cooling ages for the interval between 210 and 160 Ma are notably absent in that study.

The occurrence of similar Jurassic cooling histories along the transect, including samples ranging from immediately adjacent to the Altyn Tagh fault to more than 70 km to the north, suggests that the cooling was not related to activity on an ancestral Altyn Tagh fault. This inference is in agreement with analyses of Jurassic sedimentary strata that were interpreted to have been deposited in a unified basin that was contiguous across the region of the present Altyn Tagh, including the present trace of the Altyn Tagh fault (Fig. 6; Ritts, 1998). Restoration of ~400 km of post-Bajocian sinistral slip on the Altyn Tagh fault places this integrated basin directly to the west of the transect.

Early to Middle Jurassic sedimentary rocks in the Qaidam and Tarim basins provide a good record of the local geodynamic setting that controlled exhumation of the basin-bounding ranges. A Middle Jurassic paleogeographic map that restores 400 km of post-Middle Jurassic left-lateral slip on the Altyn Tagh fault, is shown in Figure 6. Note that restoration of 400 km of left slip aligns the Mangnai-Ruoqiang transect with the foothills of the South Qilian Shan. This restoration suggests that the samples from this transect in the Altyn Tagh were spatially related to the Qilian Shan during the Early Jurassic when they were uplifted. Fortunately the sedimentary record of the Early to Middle Jurassic in the Northeast Qaidam basin is well preserved and well exposed, providing additional insight into the regional style of deformation.

Basin analysis of Lower to Middle Jurassic strata in the Northeast Qaidam basin supports a foreland-style basin setting for those strata, and a contractile structural regime for the Qilian Shan (Ritts and Biffi, this volume; Ritts, 1998). Like the Northwest Qaidam basin, Jurassic facies in the Northeast Qaidam basin are entirely nonmarine and nonvolcanogenic. Facies trends, sandstone provenance, and paleocurrent orientations in the proximal Northeast Qaidam basin all suggest that the

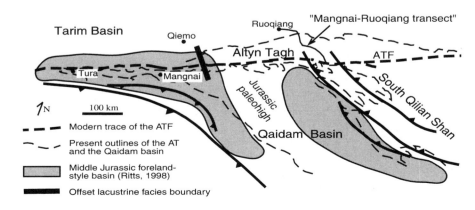

Figure 6. Jurassic paleogeographic map of northern Qaidam and Altyn Tagh region, including 400 km of restored sinistral slip on Altyn Tagh fault (ATF), based on offset Middle Jurassic lacustrine shoreline of Ritts and Biffi (2000). Positions of towns are shown, as is location of Mangnai-Ruoqiang transect. Thrust faults are inferred from this study and basin analysis of Jurassic strata (Ritts, 1998; Ritts and Biffi, this volume).

Qilian Shan was an uplifted sediment source during the Jurassic. Ritts and Biffi interpret an angular unconformity between Lower Jurassic and Middle Jurassic strata as evidence for basin-margin deformation during the Toarcian. This unconformity is immediately overlain by an ~1000-m-thick Middle Jurassic section, representing the most rapid subsidence episode for that basin during the Mesozoic. This evidence for structural disruption, followed by an acceleration in subsidence, is interpreted to represent contractional deformation in the Qilian Shan, resulting in flexural loading and subsidence in the Northeast Qaidam basin. The timing of this deformation and sedimentation broadly coincides with the cooling ages of samples farther to the north. Thus we propose that this contractional deformation accounts for exhumation and hence cooling of the area represented by the Mangnai-Ruoqiang transect.

The timing of rapid cooling recorded in the Mangnai-Ruoqiang portion of the Altyn Tagh agrees well with the depositional record of these basins during the Jurassic, suggesting a causative link. However, the inferred southwest-vergent thrusts (Fig. 6) are not readily apparent within the Altyn Tagh. The most likely candidate from a geometric point of view would be structures northeast of the transect and north of Xorkol (Fig. 1). Sinistral shear within the Altyn Tagh due to left-lateral offset along the Altyn Tagh fault can explain the more easterly trend of structures within the Altyn Tagh compared to the trend in the Qilian Shan. However, the required southwest transport on such faults within the Altyn Tagh would be expected to overthrust the study area, leading to heating rather than exhumation. A second west-northwest–trending fault is mapped farther west of the transect (Fig. 1). However, map relations show this structure cutting only different Proterozoic units (Wang et al., 1993), so it is difficult to assess whether this could have been a Mesozoic southwest-vergent thrust. It is clear that additional structural mapping combined with thermochronology is required to test this hypothesis.

REGIONAL TECTONIC IMPLICATIONS

South of the Qaidam basin, the Changtang block collided with northern Tibet during the Late Triassic–Early Jurassic, forming the Jinsha suture (Coward et al., 1988; Pearce and Mei, 1988). South of the western end of the Tarim basin, this collision is dated as Early to early-Middle Jurassic (Matte et al., 1996). The resulting deformation has been linked with basin formation in the south Junggar, north Tarim, and southwest Tarim basins (Hendrix et al., 1992; Sobel; 1999; Ritts and Biffi, this volume). The timing of this collision agrees well with the initiation of sedimentation in the northeast Qaidam basin and the cooling ages reported herein from the Altyn Tagh. The inferred Early Jurassic compressional structures in the Qilian Shan believed to be responsible for creation of this flexural basin are likely parallel to middle Paleozoic sutures on the north and south sides of this range (Song, 1996; Li et al., 1991). One can reasonably expect that the Qilian Shan would also have been af-

fected by this Mesozoic collision and that these favorably oriented older structures could have been reactivated. To date, there are little data to support Mesozoic shortening within the Qilian Shan; however, such data are required to understand both the Mesozoic and Cenozoic evolution of the region. Even if the suggested structural link between the Altyn Tagh and the Qilian Shan is disproven by future studies, the chronologic similarity between the closure of the Jinsha suture and the cooling ages in the central Altyn Tagh suggests a causative link that must be explained by subsequent tectonic models.

Similarities in Jurassic cooling ages from the Mangnai-Ruoquiang transect and the region near Lenghu suggest a post-Middle Jurassic offset of about 350 km along the Altyn Tagh fault. This value agrees well with the 400 ± 60 km post-Bajocian offset established farther to the west along the fault by Ritts and Biffi (2000). Meyer et al. (1998) suggested that Cenozoic shortening in the Qaidam basin and the Qilian Shan, compensated by sinistral offset along the eastward-propagating Altyn Tagh fault, commenced ca. 10 Ma and has accounted for ~200 km of offset along the Altyn Tagh fault. Therefore, an additional ~200 km of offset should have occurred between the Middle Jurassic and the middle Miocene. One possibility is that all of the offset occurred as a result of the India-Asia collision, and there is a significant Cenozoic, pre-10 Ma displacement history remaining to be documented. An equally plausible hypothesis is that the ductile strike-slip fabrics subparallel to the Altyn Tagh fault and dated as ca. 160 Ma in the eastern part of the Altyn Tagh (Delville et al., this volume) are associated with this missing displacement. The lack of Paleogene cooling ages on the north side of the Altyn Tagh fault supports this hypothesis. We hope that the rapidly growing regional database will resolve this question.

The hypothesis that significant Jurassic structures exist in the Qilian Shan and Altyn Tagh, which are influenced by Paleozoic structures, implies that the pattern of Cenozoic deformation in this region is at least partly controlled by these older fabrics. Therefore, the study of the pre-Cenozoic tectonic history is necessary to understand the offset history of the Altyn Tagh fault as well as the contribution of inherited structures to the development of the Tibetan Plateau. Given the likely importance of such Middle Jurassic structures in producing the thermochronologic signatures reported herein, it is equally likely that similar-age structures are likely to be important in the poorly studied region between the Qaidam basin and the Jinsha suture to the south. One of the first-order observations from studies along the northern margin of the Tibetan Plateau is how much more work remains before the polyphase geodynamic evolution of Central Asia is adequately understood.

CONCLUSIONS

1. On the basis of ^{40}Ar/^{39}Ar and apatite fission-track analysis, the region of the Mangnai-Ruoqiang transect north of the Altyn Tagh fault underwent ~100–150 °C of cooling during

the Early to Middle Jurassic, interpreted to have been caused by exhumation that also created a regional unconformity. The magnitude of Jurassic cooling greatly exceeds Cenozoic cooling recorded by the same samples, attesting to the importance of this Mesozoic tectonism.

2. A similar cooling history is obtained from a shallow pluton located ~350 km to the east and south of the Altyn Tagh fault. A plausible explanation for the similar cooling histories is that the two regions were formerly located close together, in agreement with published estimates of 400 ± 60 km of post-Middle Jurassic sinistral offset along the Altyn Tagh fault (Ritts and Biffi, 2000).

3. Cooling of the formerly unified region can be explained by exhumation linked to southeastward-vergent thrusting of the Qilian Shan over the Northeast Qaidam basin, in agreement with the depositional record of this basin (Ritts and Biffi, this volume). This contractile deformation is interpreted to be driven by the Late Triassic–Early Jurassic collision of the Changtang block with northern Tibet.

4. Only limited areas adjacent to the Altyn Tagh fault and the North Altyn Tagh thrust fault record Oligocene–Miocene apatite fission-track ages; some samples from south of the Altyn Tagh fault document Paleocene(?)–Eocene cooling, possibly preceded by reheating. Unraveling the displacement history of the Altyn Tagh fault, and hence the magnitude of the Cenozoic extrusion of Tibet, will require a better understanding of both the Cenozoic and pre-Cenozoic regional geology.

ACKNOWLEDGMENTS

Field work by Sobel and Ritts was supported by the Stanford-China Geoscience Industrial Affiliates Program; much of the analytical work was completed during a Chateaubriand postdoctoral fellowship to Sobel. Field work and analytical work by Arnaud, Jolivet, and Brunel was supported by the Institut National des Sciences de l'Univers. Diane Seward graciously made her fission-track laboratory available to Sobel and Jolivet. Da Zhou, Andrew Hanson, Steve Graham, and Edmund Chang provided valuable field assistance and/or advice, and Paul O'Sullivan, Marc Hendrix, and Ian Lange provided constructive reviews.

REFERENCES CITED

Arnaud, N.O., and Kelley, S.P., 1997, Argon behavior in gem-quality orthoclase from Madagascar: Experiments and some consequences for $^{40}Ar/^{39}Ar$ geochronology: Geochimica et Cosmochimica Acta, v. 61, p. 3227–3255.

Arnaud, N., Brunel, M., Cantagrel, J.M., and Tapponnier, P., 1993, High cooling and denudation rates at Kongur Shan (Xinjiang, China) revealed by $^{40}Ar/^{39}Ar$ K-feldspar thermochronology: Tectonics, v. 12, p. 1335–1346.

Bally, A.W., Ryder, R.T., Eugster, H.P., and Watts, A.B., 1986, Comments on the geology of the Qaidam Basin: Notes on sedimentary basins in China; report of the American Sedimentary Basins Delegation to the People's Republic of China: U.S. Geological Survey Open-File Report OF 86-0327, p. 42–62.

Burnham, A.K., and Sweeney, J.J., 1989, A chemical kinetic model of vitrinite maturation and reflectance: Geochimica et Cosmochimica Acta, v. 53, p. 2649–2657.

Burtner, R.L., Nigrini, A., and Donelick, R.A., 1994, Thermochronology of Lower Cretaceous source rocks in the Idaho-Wyoming thrust belt: American Association of Petroleum Geologists Bulletin, v. 78, p. 1613–1636.

Che, Z., Liu, L., Liu, H., and Luo, J., 1995, Discovery and occurrence of high-pressure metapelitic rocks from Altun Mountain areas, Xinjiang Autonomous Region: Chinese Science Bulletin, v. 40, p. 1988–1991.

Chen, W.-P., Chen, C.-Y., and Nábelek, J.L., 1999, Present-day deformation of the Qaidam basin with implications for intra-continental tectonics: Tectonophysics, v. 305, p. 165–181.

Chen Zhefu, ed., 1985, Geological map of Xinjiang Uygur Autonomous Region, China: Beijing, Geological Publishing House, scale 1:2 000 000.

Chinese Bureau of Geology, 1981, Stratigraphic tables for Xinjiang Autonomous Region: Beijing, Geological Publishing House, 496 p.

Chinese State Bureau of Seismology, 1992, The Altyn Tagh active fault system: Beijing, Seismology Publishing House, 319 p.

Corrigan, J., 1991, Inversion of apatite fission-track data for thermal history information: Journal of Geophysical Research, v. 96, p. 10347–10360.

Coward, M.P., Kidd, W.S.F., Pan, Y., Shackleton, R.M., and Zhang, H., 1988, The structure of the 1985 Tibet Geotraverse, Lhasa to Golmud: Royal Society of London Philosophical Transactions, ser. A, v. 327, p. 307–336.

Dumitru, T.A., Miller, E.L., O'Sullivan, P.B., Amato, J.M., Hannula, K.A., Calvert, A.T., and Gans, P.B., 1995, Cretaceous to Recent extension in the Bering Strait region, Alaska: Tectonics, v. 14, p. 549–563.

Fitzgerald, P.G., Sorkhabi, R.B., Redfield, T.F., and Stump, E., 1995, Uplift and denudation of the central Alaska Range: A case study in the use of apatite fission-track thermochronology to determine absolute uplift parameters: Journal of Geophysical Research, v. 100, p. 20175–20191.

Galbraith, R.F., 1981, On statistical models for fission-track counts: Mathematical Geology, v. 13, p. 471–478.

Gallagher, K., 1995, Evolving temperature histories from apatite fission-track data: Earth and Planetary Science Letters, v. 136, p. 421–435.

Gleadow, A.J.W., Duddy, I.R., Green, P.F., and Lovering, J.F., 1986, Confined fission-track lengths in apatite: A diagnostic tool for thermal history analysis: Contributions to Mineralogy and Petrology, v. 94, p. 405–415.

Green, P.F., 1981, A new look at statistics in fission-track dating: Nuclear Tracks, v. 5, p. 77–86.

Hames, W.E., and Bowring, S.A., 1994, An empirical evaluation of the argon diffusion geometry in muscovite: Earth and Planetary Science Letters, v. 124, p. 161–169.

Hanson, A.D., 1998, Organic geochemistry and petroleum geology, tectonics and basin analysis of southern Tarim and northern Qaidam basins, northwest China [Ph.D. thesis]: Stanford, California, Stanford University, 388 p.

Harrison, T.M., 1990, Some observations on the interpretation of feldspar $^{40}Ar/^{39}Ar$ results: Chemical Geology (Isotope Geoscience Section), v. 80, p. 219–229.

Harrison, T.M., Duncan, I., and McDougall, I., 1985, Diffusion of ^{40}Ar in biotite: Temperature, pressure and compositional effect: Geochimica et Cosmochimica Acta, v. 49, p. 2461–2468.

Harrison, T.M., Heizler, M.T., Lovera, O.M., Wenji, C., and Grove, M., 1994, A chlorine disinfectant for excess argon released from K-feldspar during step heating: Earth and Planetary Science Letters, v. 123, p. 95–104.

Hendrix, M.S., Graham, S.A., Carroll, A.R., Sobel, E.R., McKnight, C.L., Schulein, B.J., and Wang, Z., 1992, Sedimentary record and climatic implications of recurrent deformation in the Tian Shan: Evidence from Mesozoic strata of the north Tarim, south Junggar, and Turpan basins, northwest China: Geological Society of America Bulletin, v. 104, p. 53–79.

Hurford, A.J., and Green, P.F., 1983, The zeta age calibration of fission-track dating: Chemical Geology, v. 41, p. 285–317.

Lanphere, M.A., Sawyer, D.A., and Fleck, R.J., 1990, High resolution $^{40}Ar/^{39}Ar$ geochronology of Tertiary volcanic rocks, western USA, in Proceedings, Seventh International Conference on Geochronology, Cosmochronology and Isotope Geology: Canberra, Geological Society of Australia, p. 57.

Laslett, G.M., Green, P.F., Duddy, I.R., and Gleadow, A.J.W., 1987, Thermal annealing of fission-tracks in apatite, 2: A quantitative analysis: Chemical Geology (Isotope Geoscience Section), v. 65, p. 1–13.

Li, D.F., Lun, Z.Q., Guo, J.Y., and Zhu, Y.Y., eds., 1991, Regional geology of Qinghai Province: Bureau of Geology and Mineral Resources of Qinghai Province: Geological Memoirs Series 1, Volume 24: Beijing, Geological Publishing House, 662 p.

Liu, Z.Q., ed., 1988, Geological map of Qinghai-Xizang (Tibet) plateau and adjacent areas: Beijing, Geological Publishing House, 91 p., scale 1:1 500 000.

Lovera, O.M., Richter, F.M., and Harrison, T.M., 1989, The ^{40}Ar/^{39}Ar thermochronometry for slowly cooled samples having a distribution of diffusion domain sizes: Journal of Geophysical Research, v. 94, p. 17917–17935.

Lovera, O.M., Richter, F.M., and Harrison, T.M., 1991, Diffusion domains determined by ^{39}Ar released during step heating: Journal of Geophysical Research, v. 96, p. 2057–2069,

Lovera, O.M., Grove, M., Harrison, T.M., and Mahon, K.I., 1997, Systematic analysis of K-feldspar ^{40}Ar/^{39}Ar step heating results: I. Significance of activation energy determinations: Geochimica et Cosmochimica Acta, v. 61, p. 3171–3192.

Maluski, H., and Schaeffer, O.A., 1982, ^{39}Ar-^{40}Ar laser probe dating of terrestrial rocks: Earth and Planetary Science Letters, v. 59, p. 21–27.

Matte, P., Tapponnier, P., Arnaud, N., Bourjot, L., Avouac, J.P., Vidal, P., Liu, Q., Pan, Y., and Wang, Y., 1996, Tectonics of Western Tibet, between the Tarim and the Indus: Earth and Planetary Science Letters, v. 142, p. 311–330.

Meyer, B., Tapponnier, P., Gaudemer, Y., Peltzer, G., Guo, S., and Chen, Z., 1996, Rate of left-lateral movement along the easternmost segment of the Altyn-Tagh fault, east of 96 °E (China): Geophysical Journal International, v. 124, p. 29–44.

Meyer, B., Tapponnier, P., Bourjot, L., Métivier, F., Gaudemer, Y., Peltzer, G., Guo, S., and Chen, Z., 1998, Crustal thickening in Gansu-Qinghai, lithospheric mantle subduction, and oblique, strike–slip controlled growth of the Tibet Plateau: Geophysical Journal International, v. 135, p. 1–47.

Peltzer, G., and Tapponnier, P., 1988, Formation and evolution of strike-slip faults, rifts, and basins during the India-Asia collision: An experimental approach: Journal of Geophysical Research, v. 93, p. 15085–15117.

Pierce, J.A., and Mei, H., 1988, Volcanic rocks of the 1985 Tibet Geotraverse, Lhasa to Golmud: Royal Society of London Philosophical Transactions, ser. A, v. 327, p. 169–201.

Ritts, B.D., 1998, Mesozoic tectonics and sedimentation, and petroleum systems of the Qaidam and Tarim Basins, Northwest China [Ph.D. thesis]: Stanford, California, Stanford University, 691 p.

Ritts, B.D., and Biffi, U., 2000, Magnitude of post-Middle Jurassic (Bajocian) displacement on the Altyn Tagh fault, NW China: Geological Society of America Bulletin, v. 112, p. 61–74.

Sobel, E.R., 1999, Basin analysis of the Jurassic–Lower Cretaceous southwest Tarim basin, northwest China: Geological Society of America Bulletin, v. 111, p. 709–724.

Sobel, E.R., and Arnaud, N., 1999, A possible middle Paleozoic suture in the Altyn Tagh, Northwest China: Tectonics, v. 18, p. 64–74.

Song, S., 1996, Metamorphic geology of blueschists, eclogites and ophiolites in the North Qilian mountains, 30th IGC field trip guide T392: Beijing, Geological Publishing House, 40 p.

Tagami, T., and Shimada, C., 1996, Natural long-term annealing of the zircon fission-track system around a granitic pluton: Journal of Geophysical Research, v. 101, p. 8245–8255.

Tapponnier, P., Peltzer, G., Le Dain, A.Y., Armijo, R., and Cobbold, P., 1982, Propagating extrusion tectonics in Asia: New insights from simple experiments with plasticine: Geology, v. 10, p. 611–616.

Ulmishek, G., 1984, Geology and petroleum resources of basins in western China: Argonne National Laboratory Report ANL/ES-146, 131 p.

Wang, E., 1997, Displacement and timing along the northern strand of the Altyn Tagh fault zone, Northern Tibet: Earth and Planetary Science Letters, v. 150, p. 55–64.

Wang, G.P., Wu, G.T., Lun, Z.Q., and Zhu, Y.Y., eds., 1993, Regional geology of Xinjiang Uygur Autonomous Region: Bureau of Geology and Mineral Resources of Xinjiang Uygur Autonomous Region: Geological Memoirs Series 1, Volume 32: Beijing, Geological Publishing House, 841 p.

Wang, Q., and Coward, M.P., 1990, The Chaidam basin (NW China): Formation and hydrocarbon potential: Journal of Petroleum Geology, v. 13, p. 93–112.

Yue, Y., and Liou, J.G., 1999, Two-stage evolution model for the Altyn Tagh fault, China: Geology, v. 27, p. 227–230.

Zhou, D., and Graham, S., 1996, Extrusion of the Altyn Tagh wedge; a kinematic model for the Altyn Tagh fault and palinspastic reconstruction of northern China: Geology, v. 24, p. 427–430.

Zhou, Z.H., Zhao, R.S., Mao., J.H., Lun, Z.Q., and Zhu, Y.Y., eds., 1989, Regional geology of Gansu Province: Gansu Bureau of Geology and Mineral Resources: Geological Memoirs Series 1, Volume 19: Beijing, Geological Publishing House, 692 p.

Manuscript Accepted by the Society June 5, 2000

Geological Society of America
Memoir 194
2001

Paleozoic to Cenozoic deformation along the Altyn Tagh fault in the Altun Shan massif area, eastern Qilian Shan, northeastern Tibet, China

Nathalie Delville
Nicolas Arnaud*
Jean-Marc Montel
UMR 6524 CNRS, OPGC and Université Blaise Pascal, 63000 Clermont-Ferrand, France
Françoise Roger
UMR 7578 CNRS, U. P7 and IPGP, Paris, France
Maurice Brunel
UMR 5567, ISTEEM-Tectonique, Université Sciences Montpellier II, Montpellier, France
Paul Tapponnier
UMR 7578 CNRS, U. P7 and IPGP, Paris, France
Edward Sobel
Institut für Geowissenschaften, Universität Potsdam, Potsdam, Germany

ABSTRACT

The Altyn Tagh range shows evidence of a long history of deformation and metamorphism. Apart from the Cenozoic evolution largely associated with propagation of the Altyn Tagh fault, the area shows evidence of a Paleozoic orogeny. Mylonitic schist and granite from the northern Altyn Tagh range in the area between Subei, Lenghu, and the summit of Altun Shan show microscopic evidence of complex deformation that can be correlated with regional schistosity and folding. Ductile deformation started at ~350–400 ° C and was followed by an episode with a strong horizontal shear component at a maximum temperature of 300 °C. It is very difficult to assess whether this represents a continuum of deformation or two separate episodes. U/Th/Pb microprobe dating of monazite, Rb/Sr dating, and $^{40}Ar/^{39}Ar$ dating and modeling of cooling from micas and feldspar indicate a series of 280–230 Ma ages that weakly support a possible late Paleozoic–early Mesozoic event. This is tentatively linked either with suturing of the Qiantang and Kunlun blocks farther to the south, or accretionary tectonism farther to the north or east in the Qilian Shan and Bei Shan ranges. The most clearly identified deformation took place at 150 ± 10 Ma and could correspond to the whole series of deformation observed, which would then be interpreted as a continuum. Previously unrecognized, this significant event is probably associated with strike-slip movements along a fault that broadly coincides with the currently active trace of the Altyn Tagh fault, but interpretations concerning its origin remain speculative. Cenozoic brittle deformation overprints the entire area. The lack of a clear Cenozoic thermal and/or

*Corresponding author: Laboratoire Magmas et Volcans, UMR 6524 CNRS, 5 rue Kessler, 63000 Clermont-Ferrand, France.
E-mail: arnaud@opgc.univbpclermont.fr.

Delville, N., et al., 2001, Paleozoic to Cenozoic deformation along the Altyn Tagh fault in the Altun Shan massif area, eastern Qilian Shan, northeastern Tibet, China, *in* Hendrix, M.S., and Davis, G.A., eds., Paleozoic and Mesozoic tectonic evolution of central Asia: From continental assembly to intracontinental deformation: Boulder, Colorado, Geological Society of America Memoir 194, p. 269–292.

geochronological signature in the analyzed samples is likely due to the small amount of Cenozoic exhumation, which agrees well with the proposed distribution of shortening between pure strike-slip along the Altyn Tagh fault and thrusting and mountain building on linked thrusts in the Qilian Shan.

INTRODUCTION

The Altyn Tagh fault is the longest active (~1800 km km) strike-slip fault in Asia. Although estimates of Cenozoic left lateral strike-slip displacement differ, most authors agree that the fault accommodates a significant part of the continent-continent collision of India with Asia (Tapponnier and Molnar, 1977; Tapponnier et al., 1986; Copeland et al., 1987; Peltzer and Tapponnier, 1988; Peltzer et al., 1988; Avouac and Tapponnier, 1993; Meyer et al., 1996, 1998; Van der Woerd et al., 1998). Debate on the magnitude of Cenozoic movement largely proceeds from the lack of reliable data on pre-Cenozoic deformation along and near the fault. In many places the active fault follows a corridor of highly sheared rocks, deformation usually conforming to the present kinematics, but the ages of the shearing fabric are not always Cenozoic (Arnaud et al., 1998). Moreover, estimates of Cenozoic shortening and their geodynamical implications (e.g., Meyer et al., 1998) depend upon the inherited structure of the northern Tibetan crust and lithosphere prior to the Tertiary. It is thus of prime importance to assess in detail the history (age, regime) of deformation in various places along the geological trace of the fault. This should yield information essential for reconstruction of pre-Cenozoic activity along the fault, leading not only to a better knowledge of the Paleozoic and Mesozoic evolution of Asia but also, by comparison, to finite displacements of Tertiary age.

The fault is particularly active along its northern tip where it ends with a northeastward-vergent thrust system that transforms the strike-slip movement into crustal thickening and mountain building (Wittlinger et al., 1998). This area corresponds largely to the Altyn Tagh range, which is between the Tarim basin to the north, and the Qaidam basin, Kunlun Shan, and the Tibetan Plateau to the south (Fig. 1). During 1995 and 1996 Chinese-French and American traverses aimed at studying the area between Dunhuang to the north and Mangnai and Ruoqiang to the southwest, between ~90°E and 93°E. We present here petrographic and geochronologic data that define the style and age of mylonitic deformation events along the morphological trace of the active fault at the eastern end of the study area. Geochronologic results from farther west within the study area are presented in Sobel et al. (this volume).

GEOLOGICAL SETTING OF ANALYZED SAMPLES

The structure of the Altyn Tagh range as defined herein is strongly influenced by Cenozoic deformation associated with the Altyn Tagh fault. Most rock units within 20 km of the active trace (Fig. 1) are structurally juxtaposed screens of various sizes (as much as several tens of kilometers) juxtaposed in faulted contact. The most recent map (1:1 500 000) and regional geological descriptions of the Altyn Tagh range are those of within Wang et al. (1993) and Li et al. (1991). Older stratigraphic descriptions are available in Chinese Bureau of Geology (1981) and Chen (1985). The following descriptions are the result of our field observations and analysis of Chinese geological maps.

Although there are many petrographic rock types in the study area, the entire series has a common geodynamical underpinning resulting from one or more orogenic systems. Eclogite, marble, granulitic micaschist, amphibolite facies micaschist, and variably deformed granitoids constitute the basic lithologies. Most of these rocks are overprinted by a widespread greenschist facies metamorphism. Almost all units show a well-developed regional steep composite foliation. A first foliation, S1, is defined either by compositional banding of quartz and feldspars or by high-grade (at least amphibolitic facies) metamorphic paragenesis. S1 is typically refolded and a second foliation, S2, parallel to S1 is developed in the axial plane of folds with vertical axes. This composite S1-S2 foliation usually strikes N080 (locally varying from N070 to N110), and thus is regionally parallel to the Altyn Tagh fault. A retrograde metamorphic event is usually preserved by the recrystallization of muscovite flakes on foliation planes. The S1-S2 planes generally bear one strong horizontal lineation compatible both with S1-S2 minerals and the retrograde facies. Kinematic indicators on foliation planes indicate sinistral shear along lineation. However, in the Altun Shan massif area, one downdip lineation was identified (see following).

Contacts between the different lithologies are always faulted, steep, and typically show reverse motion. Geological maps of the Altyn Tagh range between 86° and 94°E (Fig. 1) indicate predominantly Precambrian basement and scattered Paleozoic outcrops, and igneous bodies of early Paleozoic age (Li et al., 1991; Wang et al., 1993). In the central part of the range, north of Mangnai (Fig. 1), a series of ophiolitic rocks, ultrahigh-pressure gneiss, and granite has been interpreted to reflect mid-Paleozoic closure of an ocean basin and subsequent formation of an accretionary belt at 400 ± 30 Ma (Sobel and Arnaud, 1999).

Traces of the active Altyn Tagh fault, indicating predominantly sinistral slip, are well documented on Quaternary river terraces (Van der Woerd et al., 1998). Where the fault involves basement rocks, kinematic indicators are scattered but dominantly show sinistral movement with a less important top-to-the-northeast thrust component. Therefore, it seems probable that some of the faulted contacts between different lithological units may be kinematically associated with the Cenozoic Altyn Tagh fault, although in places these may have reactivated older contacts. In order to test the hypothesis that multiple tectono-

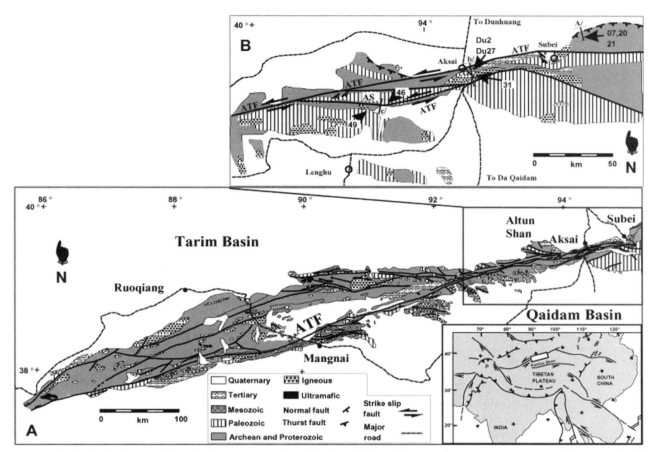

Figure 1. A: Geological map of Altyn Tagh range. Box shows location of B. B: Blow-up of area centered on Altun Shan summit with location of samples described in this study. Maps are modified from Jiao et al. (1988) and Li et al. (1991). Neotectonic structures are from Meyer et al. (1998). Abbreviations: ATF, Altyn Tagh fault; AS, Altun Shan summit. Inset is modified from Tapponnier et al. (1986) and shows location of main map. There are several different spellings in literature for name of this range. Herein, we call entire range Altyn Tagh, while we reserve name Altun Shan for specific peak in eastern portion of range. Locations of samples are shown and identified as XX for K96XX and DUXX for 95DUXX. Cross sections shown in Figure 2 are located as follows: a, Subei; b, Aksai; and c, Altun Shan.

metamorphic events are preserved, 12 samples were collected from various lithological units within a few kilometers of the active trace of the fault, along three north-south sections at Subei, Aksai, and in the Altun Shan massif (Fig. 1B). Samples from mylonitic micaschist, marble, and variably deformed granitoids were selected in the field because, on macroscopic examination, they were representative of one or several metamorphic and deformation events, in addition to containing mineral assemblages suitable for various geochronological methods, especially $^{40}Ar/^{39}Ar$.

A general overview of the present-day active structure at the range scale was proposed by Wittlinger et al. (1998). We modified this global sketch to focus on the structure of the range in the areas of Subei, Aksai, and the Altun Shan summit (Fig. 2A). Field cross sections are located in Figure 1 and also are projected along strike into the general section in Figure 2A. The detailed sections (Fig. 2, B, C, and D) are included to support the structural descriptions, but only depict field examina-

tions and measurements taken along the road or river terraces levels. They cannot be simply extrapolated at depth. We describe each section from north to south in the following.

The easternmost transect is located north of Subei (herein called the Subei section; Fig. 2B), on the margin of the Tarim basin (Fig. 1), just southeast of Dunhuang. It is a composite section made from two parallel fluvial valleys a few kilometers apart that show the same series of rocks; the quality of exposure and freshness of samples varies from one valley to the other. At the north end of the section, flat-lying Quaternary sedimentary rocks of the basin onlap a steep N80 mylonitic micaschist series, which is highly sheared, shows indications of thrust, and has a strong greenschist overprint. Sample K9607 comes from this unit. More than 200 m of less sheared micaschist and gneiss follow the strongly sheared mylonites at the northern end of the section. A major steep foliation (S1) occurs in all lithologies. It is locally refolded by structures of unclear orientation and further overprinted by a second foliation (S2) defined by white

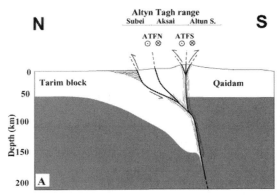

Figure 2. A: Composite sketch of Altyn Tagh range and fault system in study area (from Wittlinger et al., 1998, modified). Altyn Tagh fault is complex, with slip partitioning between sinistral strike slip and thusting. Altun Shan is push up between two active branches of Altyn Tagh fault, northern (ATFN) one and southern (ATFS) one. Dotted area shows Cenozoic and modern sediment engaged in active thusting; B, C, D (facing page): Field cross sections for areas of Subei (B), Aksai (C), and Altun-Shan summit (D). Each field cross section is fairly typical of one structural position within Altyn Tagh system as shown in A. Locations of samples discussed in this chapter are shown below each section. Question mark indicates that contacts are faulted but that senses of motion were unclear. Note that because these sections only reflect outcrops and cannot be extrapolated to depth, no vertical scale is given.

micas and associated with sinistral kinematic indicators. Contacts between schists and gneisses are faults with sinistral and/or reverse displacements. Most faults are vertical and strike N080 with slickensides with a 42° pitch to the west. Samples K9620 and K9621 were collected in a micaschist showing both foliations. The series continues to the south for several kilometers; amphibolite, gabbro, marbles, and acid metamorphic rocks such as granulite, pink granitoids, and leucogranitic dikes all show the major steep foliation. The series abuts the present trace of the Altyn Tagh fault, marked by offset river terraces and strong cataclasis of basement rocks.

The Aksai cross section (Fig. 2C) is located along the main road through the Dangjin pass and extends from Aksai city to the south toward Lenghu. Quaternary sedimentary rocks onlap or are cut by the trace of the northern branch of the active Altyn Tagh fault, which splits into two segments in the Altun Shan mountain area. The fault is characterized by a few tens of m of gouge rocks. Immediately south of the fault is a several-kilometer-thick series of metabasalt, acid granulitic micaschist, and leucocratic granite dikes, interlayered with quartzite, with a steep main foliation (S1) folded by vertical fold axis with S2 axial planar cleavage. This series passes with unclear contacts to a vertically sheared leucogranite sheet, which exhibits only a weak foliation in hand specimens (probably S2), from which sample K9631 was taken. The southern limit of the series is the southern branch of the active Altyn Tagh fault. South of this fault a thick series of weakly foliated pillow basalts and ultramafic rocks continues south to the Dangjin pass and the Qaidam basin.

The Altun-Shan cross section (Fig. 2D) is located north of Lenghu and essentially follows a north-south valley from the foot of the Altun Shan mountains south to the Qaidam desert. The southern flank of the Altun Shan is outlined by triangular facets along the southern branch of the active Altyn Tagh fault, suggesting that normal throw is an important component. North of the fault, marble and basalt are sheared with a locally well developed single foliation conformable to S1, but also containing a downdip lineation. Sample K9646 was taken from such a marble. South of the active fault, and after another layer of marbles, is a highly schistose granite in greenschist facies. Foliation is steep but the strike is highly variable, with an average of N40–N80 and a horizontal to southeast-dipping lineation. Because foliation is not associated with a strong metamorphic overprint and has a clear lineation, we tentatively relate it to the regional S2. Sample K9649 was taken from that schistose granitic body that apparently intrudes a thick series of acid granulites interlayered with marbles and flaser gabbros; the series extends over several kilometers with a strike roughly parallel to the Altyn Tagh fault.

MICROTECTONICS OF ANALYZED SAMPLES

Each sample was prepared and analyzed on thin sections perpendicular to the S1 or S2 (steep N80 planes) foliation and parallel to the stretching lineation identified in hand samples.

Subei cross section

Sample K9620 is a low-grade (greenschist facies) schistose mylonite with abundant white mica. At the microscopic scale, the main deformation (herein D1) produced a schistosity defined by muscovite and biotite corresponding to S1. Highly altered porphyroclasts of feldspar and kyanite constitute relics from a previous metamorphic assemblage, demonstrating that this rock was once equilibrated at relatively high pressure metamorphic conditions. Large muscovite flakes could also be relics of the high-pressure assemblage, although they appear to be much more in equilibrium with the major metamorphic texture than other relict minerals. We thus consider these muscovites to have grown early during the D1 tectonometamorphic event, implying temperatures of ~350–400 °C. These are sheared (Fig. 3A) and older cleavage planes acted as gliding planes where minor recrystallization can be distinguished. Limited muscovite recrystallization is also visible at grain boundaries (Fig. 3, A and B). These observations suggest that this rock has undergone retrograde metamorphism associated with shearing, referred to as D2, with a much lower intensity than D1. Many fractures, both perpendicular and parallel to the main foliation, are observed. They contain no secondary minerals that might suggest hydrothermal alteration. This characterizes the latest brittle deformation, referred to as D3.

About 10 monazites were observed in one thin section. Most of them are between 20 and 40 μm across. Larger grains

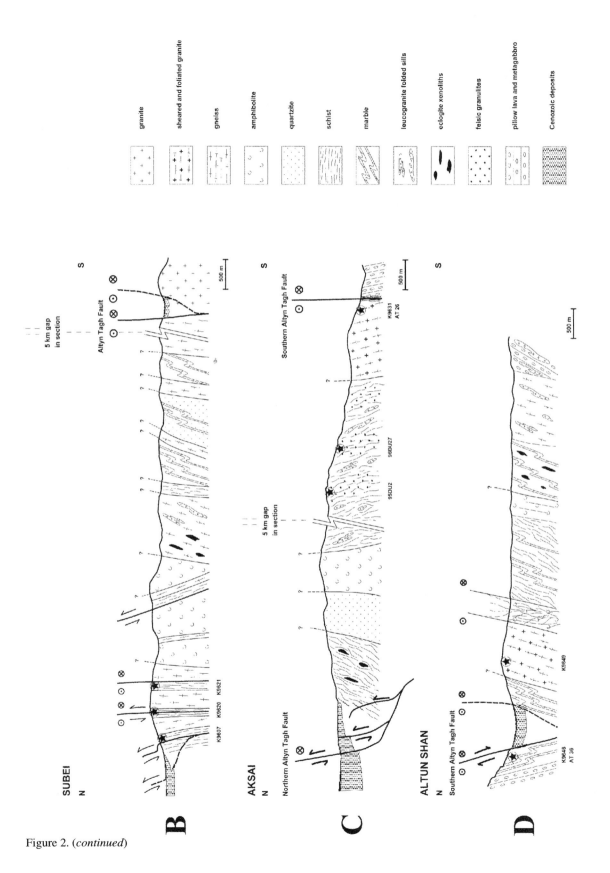

Figure 2. (*continued*)

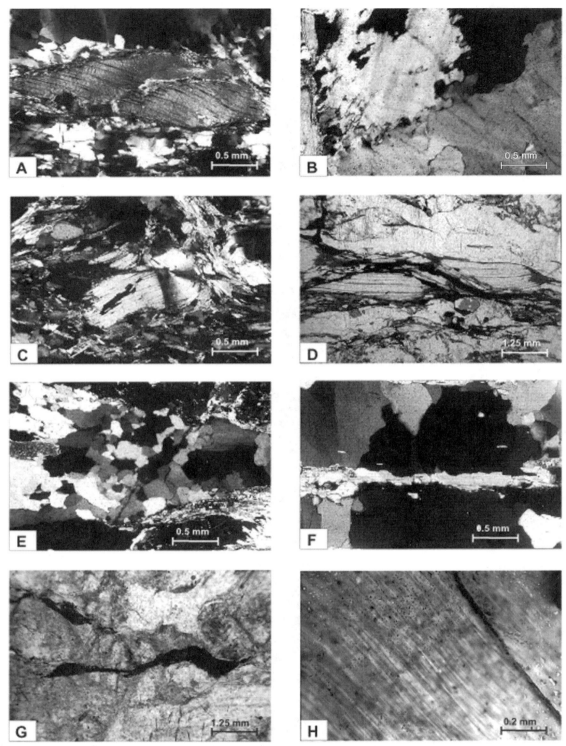

Figure 3: Crossed-polar microphotographs of texturally described and dated samples from study area. Scale is indicated in each photograph. A: Sample K9620: S1 foliation-forming muscovite crystal showing undulatory extinction. Although such muscovite crystals could be relicts of high-pressure metamorphic phase, they appear to be too well preserved. Together with their refolding, this suggests that they more likely grew early during D1 event. Later shearing of muscovite is very clear as well as minor recrystallization around margins, both suggesting second D2 deformation phase. B: Sample K9620: jagged grain boundaries of quartz resulting from relatively rapid grain-boundary migration during D1 phase. This suggests deformation temperatures in excess of 350 °C (Brunel and Geyssant, 1978). C: Sample K9621: folded muscovite exhibiting undulatory extinction. Folding of muscovites appears coeval with development of major S1 foliation during D1. D: Sample K9621: S1 forming muscovite sheared during D2 (shearing plane is denoted by opaque iron oxides). Note that shearing seems more developed than in sample K9620 (see A). E: Sample K9621: quartz texture showing newly formed polygonal grains superimposed on S1, and therefore probably formed during D2. This texture shows that quartz recrystallization in this sample is more advanced than in sample K9620 (A). Together with better developed shearing (D), this indicates that thermal effect of D2, especially on resetting of isotopic systems, should be more pronounced. F: Sample K9607: biotites lined up with foliation inside of quartz grains. Migrating grain boundaries have trapped biotites inside newly formed quartz during D1. G: Sample K9631: although deformation is weak at hand-specimen scale, it is more obvious in this thin section. Note that deformation during D2 in this granite has been concentrated in relatively weak biotites rather than in more competent feldspars. This results in plastic deformation of micas sheared and wrapped around feldspars. H: Sample K9631: microscopically visible subbasal deformation lamellae in quartz. Such textures are thought to develop during D2 phase at deformation temperatures of ~250–300 °C (Brunel and Geyssant, 1978).

rarely exceed 100 µm. All of them occur between or included within muscovite, plagioclase, and quartz. They often display an elongated shape and exhibit numerous fractures.

Sample K9621 from the Subei section is a mica-rich mylonite similar to the sample K9620. This rock has also undergone a major deformation and has similar characteristics, including a well-developed foliation defined by muscovite and biotite. Between foliation planes, a few large muscovites are folded and exhibit undulatory extinction (Fig. 3C). They are interpreted as having grown early in the deformation history, either in a pre-D1 event or in an early stage of D1. A few relatively small garnets are present along foliation planes and as inclusion inside feldspar crystals. These garnets and possibly the muscovites are interpreted as the remnants of the same high-pressure paragenesis as observed in sample K9620. Sample K9621 has also been affected by a retrograde event, D2, wherein biotites and muscovites forming the foliation were sheared (Fig. 3D), and the grain boundaries appear diffuse. This shearing deformation is more developed than in sample K9620, as indicated by quartz showing a more advanced reequilibration texture (Fig. 3E). Porphyroclasts are rotated and have asymmetric pressure shadows. A brittle, late-stage deformation is also observed in this sample, but with much greater evidence of fluid circulation: fissures are filled with quartz micrograins, massive epidote, or iron oxides presenting a widespread red color in natural light.

Sample K9607 shows a well-developed mylonitic texture with a distinct chemical segregation between quartz and micas (Fig. 3F). This greenschist facies rock is rich in chloritized biotite and epidote. The schistosity is dominated by mica and contains almost totally flattened porphyroclasts that define the main deformation event. Small isolated biotite crystals parallel to the foliation are incorporated inside quartz grains, indicating that quartz recrystallization occurred during or after the formation of foliation. No remnants of a relatively older, high-pressure assemblage were observed, nor is there evidence of a second deformation event: micas are not sheared, grain boundaries are sharp, and quartz is completely recrystallized to a homogeneous matrix of polygonal quartz grains. These observations suggest that quartz was not subsequently disturbed after D1. Late-stage brittle deformation was not observed in sample K9607.

Aksai section

Samples 95DU2 and 95DU27 were collected from acid schist. Petrographically, this series is characterized by a lack of muscovite and the presence of sparse sillimanite, suggesting possible granulite facies metamorphism. The regional foliation S1 is microscopically associated with quartz and feldspars ribbons or high-grade metamorphic paragenesis. It is generally folded about vertical fold axes and overprinted by a younger generation of rare muscovite and biotite, which define schistosity planes associated with regional foliation S2.

Sample K9631 shows a significantly lower strain than the previous samples: it is a sheared granite with little foliation visible in hand specimen. At the microscopic scale, one observes that deformation is mainly confined to micas, which show a plastic deformation, while more competent feldspars are unstrained (Fig. 3G). Large grains of quartz exhibit sweeping undulatory extinction, suggesting plastic behavior. These characteristics are taken as expressing D2 deformation in a lithology more competent than schist. Quartz recrystallization is estimated to begin at a temperature of ~300–350 °C (White, 1976), but the grains also show microscopically visible subbasal deformation lamellae (Fig. 3H) that are estimated to form at lower temperatures (250–300 °C, Brunel and Geyssant, 1978). Therefore, the deformation affecting this rock must have occurred at a temperature close to 250–300 °C. All these observations support the idea that this granite has undergone only the latest ductile deformation under greenschist facies conditions. Although no important fractures are observed, feldspars exhibit sericite micrograins within cleavage planes and grain joints. This feature could be the expression of the late-stage, brittle deformation, previously described in other samples.

Monazites exist but are rare in this sample. Only one small, unzoned euhedral monazite (~30 µm long) was observed in thin section. It occurs as an inclusion in a quartz grain.

Altun Shan section

Sheared granite sample K9649 shows a strong foliation defined by mica. As in sample K9631, mica grains accommodate deformation around K-feldspar porphyroclasts. The latter are highly altered, sericitized, and deformed by microfractures along which quartz recrystallized. Primary muscovite is sheared along foliation planes and secondary muscovite develops in or along feldspar rims. Rare primary biotite survived deformation and most grains are recrystallized along foliation planes. Strain characteristics seem intermediate between D1 and D2 but overall suggest an affinity with the D2 deformation. Many calcite-filled late-stage cracks developed, normal to the foliation. Monazite grains appear exclusively as inclusions in feldspar and quartz crystals, ranging in size from 100 to 300 µm. Monazite grains nearly always exhibit a distinct reaction corona structure, with a preserved monazite core in the center and a rim of apatite and allanite forming concentric growth rings.

Sample K9646 is a marble collected in the southern facets of the branch of the Altyn Tagh fault bordering the Altun Shan summit to the south (Fig. 2D). The texture is ultramylonitic, with plagioclase porphyroclasts, phlogopite, and biotite phenoblasts and sphene in a fine carbonate matrix. Plagioclase feldspar porphyroclasts are filled by mica and calcite, and are boudinaged and truncated (more rarely sigmoidal). Stretching is parallel to the mineral lineation as defined by phlogopite crystals and long axes of sphene crystals. Cracks in elongated feldspars are filled by oblique calcite fibers. It appears that this rock has been subjected to two successive deformation phases, one severe and ductile, probably D1, and the second under brittle conditions.

In conclusion, thin-section analysis indicates that the textures observed are characteristic of shearing deformation, and that the deformation was polyphase and polymetamorphic. Although it is often very difficult to discriminate between the various microstructures in the field or in thin sections, we have chosen to distinguish three episodes of tectonometamorphic deformation. (1) D1, the oldest and most severe deformation, is associated with S1 at the regional scale; temperatures of ~350–400 °C were deduced from the metamorphic paragenesis. Whether the high-pressure relicts are early facies of D1 or a completely distinct older facies is impossible to deduce solely on textural grounds. (2) D2 seems to be of lower intensity (250–300 °C from quartz fabrics) but seems responsible for most of the horizontal component of the shearing deformation; (3) D3 probably corresponds to the last deformation, which occurred under brittle conditions; there is little information available on the deformation temperature beyond an upper bound of 250 °C. The relations between D1 and D2 phases remain controversial. On the one hand they could constitute two successive phases of retrograde metamorphism associated with the same tectonic event: on the other hand, because the D2 phase is clearly associated with a strong sinistral shear while D1 is more pure flattening, they could be completely independent in time.

GEOCHRONOLOGY

Tables 1, 2, 3, and 4 and Appendix 1 describe the analytical techniques and detailed results. In order to analyze the results, the following isotopic closure temperatures have been used: 300–350 °C for Rb-Sr closure temperature of biotite (Purdy and Jäger, 1976; Harrison and Amstrong, 1978; Harrison et al., 1985), 325 °C as an average for $^{40}Ar/^{39}Ar$ dating of biotite (Harrison et al., 1985), and 400 °C for $^{40}Ar/^{39}Ar$ dating of muscovite (Hames and Bowring, 1994). Thermal modeling for K-feldspar is used as described in Tables 1–4 and Sobel et al. (this volume), and replicate modeling of the same sample (Lovera et al., 1991) indicates that with the assumption of volume diffusion, uncertainties on modeled temperatures at each point do not exceed 25 °C.

Subei section

We investigated 10 grains of monazite from sample K9620 mylonitic schist (Fig. 4, A and B), for a total of 17 analyses. U concentrations range between 2410 and 6300 ppm. Th concentrations vary between 29840 and 78530 ppm; there was one exceptionally low value of 8150 ppm. Pb abundances are low, and range between <128 ppm and 1060 ppm. Calculated ages range between 244 and 318 Ma and can be interpreted as a single population with a best estimate at 276 ± 18 Ma (mean square of weighted deviations, MSWD, = 1.12) (Fig. 4A). On the U-Th age diagram (Fig. 4B), all data points are arranged around the mean calculated age of 276 ± 18 Ma, which indicates that there is no relationship between individual ages and the U and Th abundance in monazites. In one grain there is no detectable Pb, suggesting that this grain was reset by, or formed during, an event younger than 65 Ma.

The $^{40}Ar/^{39}Ar$ age spectrum of muscovite from K9621 presents an imperfect plateau (due to a slightly higher intermediate step; Fig. 5A), yielding an approximate age of 227 ± 4 Ma calculated on the steps with highest K/Ca ratios and excluding the first lower two steps. The Cl/K curve shows that the gas released during these steps is enriched in chlorine, probably due to surficial alteration or a minor recrytallization. The last step corresponding to 1400 °C was not used in the plateau age calculation due to the minute quantity of gas released.

TABLE 1. U/Th/Pb MICROPROBE DATING OF MONAZITES

Sample	Th (ppm)	±	U (ppm)	±	Pb (ppm)	±	Age (Ma)	±
K9620								
	50630	602	4960	178	810	69	271	27
	78530	718	2410	169	1010	72	261	22
	61760	646	6300	182	1060	74	288	24
	29840	490	4100	172	470	68	244	40
	75010	709	3140	175	1000	75	262	23
	60470	649	2830	173	840	73	269	27
	49100	600	5470	182	870	72	291	29
	31960	510	4470	176	560	69	269	39
	33380	514	2860	171	480	69	251	42
	67950	679	3690	175	970	73	271	24
	30240	488	4430	173	< 128		<65	
	30740	487	4170	171	590	68	297	41
	31940	494	4420	171	600	69	289	39
	8150	340	5400	171	340	66	295	66
	39470	533	6130	179	790	71	297	31
	36490	519	4070	172	620	68	278	36
	35040	511	6050	178	780	70	318	34
K9631								
	82560	730	2840	170	1760	78	427	24
	69630	678	3630	173	1530	76	418	26
	64470	657	2970	171	1410	76	423	29
K9649								
	64100	656	730	163	430	68	144	25
	68640	673	2800	170	1080	73	310	25
	81120	712	500	150	220	61	58	17
	34950	518	880	161	470	68	277	47
	41530	552	1010	162	590	69	294	41
	41540	552	2620	168	770	70	343	38
	42540	557	4510	175	1050	72	408	35
	28380	481	530	160	390	67	289	59

TABLE 1. U/Th/Pb MICROPROBE DATING OF MONAZITES (continued)

Sample	Th (ppm)	±	U (ppm)	±	Pb (ppm)	±	Age (Ma)	±
K9649 (continued)								
	57140	627	2930	171	1150	72	384	30
	55930	621	3250	171	1170	73	392	31
	63100	652	2040	167	1230	73	393	29
	82110	729	2780	172	1640	77	400	24
	87420	746	2810	172	1750	78	403	23
	66640	667	2340	168	1330	74	398	28
	61840	648	3880	174	1420	75	424	29
	76770	710	6280	183	1910	79	437	23
	71400	688	3400	172	1490	76	402	26
	73320	697	5680	181	1790	79	433	24
	66970	669	5820	180	1690	78	437	26
	71690	689	3960	176	1450	76	381	25
	81400	727	3700	175	1610	77	383	23
	62250	649	5810	181	1440	75	395	26
	68680	677	5940	182	1630	77	411	25
	48170	588	600	161	640	69	285	36
	38240	538	360	160	450	68	255	45
	64100	659	1080	165	690	70	228	27
	34400	518	850	162	420	68	253	48
	66010	666	6860	185	1740	77	437	25
	58330	636	6300	183	1530	75	431	27
	68780	677	5800	181	1620	77	411	25
	52220	605	700	163	740	69	303	34
	81990	727	1880	168	690	70	175	20
	64280	657	990	162	840	70	277	28
	55260	621	1060	163	770	70	293	32
	75640	705	6220	183	1850	79	429	23
	64320	658	800	163	350	68	117	25

Note: Delville et al., 2000

TABLE 2. Rb/Sr RESULTS

		Rb (ppm)	Sr (ppm)	^{87}Rb/^{86}Sr *	^{87}Sr/^{86}Sr ± 2σ[†]
AT 26: Aksaï mylonite					
Feldspar		201	61.7	9.59	0.77412 ± 2
Whole Rock		164	29.9	16.21	0.77934 ± 3
Biotite 1	(600/400μm)	1034	8.88	357	1.41335 ± 10
Biotite 2	(300/400 μm)	1142	7.4	484	1.64368 ± 9
Biotite 3	(300/400 μm)	728	3.09	816	2.44788 ± 15
Biotite 4	(600/400 μm)	1174	4.16	936	2.54981 ± 25
AT 36: Altun Shan Marble					
Whole Rock		69	113	1.79	0.71963 ± 2
Biotite 1	(800/600 μm)	127	36.6	10.17	0.75213 ± 2
Biotite 2	(600/400 μm)	125	41.9	8.74	0.74641 ± 2
Biotite 3	(600/400 μm)	148	49.4	8.82	0.74826 ± 2

* The maximum error for (^{87}Rb/^{86}Sr) is ± 3%.
[†] Normalized to (^{86}Sr/^{88}Sr) = 0.1194

140 Ma or later at a temperature too low to affect the muscovite. Whether the decrease in age is associated with simple argon loss or recrystallization is difficult to assess because the chemical composition appear to be homogeneous.

The spectrum of muscovites from K9620 (Fig. 5C) is disturbed and probably reflects either a complex deformation history associated with partial recrystallization or argon loss. The first step is exceptionally large for typical degassing of a muscovite with 30% of the total released gas, and yields an age of 120 ± 2 Ma suggestive of slightly recrystallized muscovite. Between 30% and 60% of released gas, the spectrum shows a relatively flat segment with an apparent age similar to the plateau age of K9621 muscovite. Between 60% and 90% of the released gas, ages rise to about 400 Ma. The final two steps probably have no significance because an insufficient amount of gas was released. Regarding the highest ages recorded, it is significant that these steps correspond to the lowest ^{36}Ar/^{40}Ar ratio on the inverse isochron diagram, implying a possible artifact associated with excess argon.

Biotite from sample K9607 (Fig. 5D) exhibits a very disturbed spectrum that probably reflects widespread chloritization of biotites within the sample. This is well illustrated by the large quantity of chlorine measured in the released gas, and by a typical K/Ca correlation plot close to the shape described by Ruffet et al. (1991). The two maxima in the K/Ca curve and intermediate bulge in the age spectrum are effects of ^{39}Ar recoil during irradiation. Therefore, the maximum ages of ca. 350 Ma are probably higher than their true closure age. Nevertheless, two ages already obtained from previous spectra can be distinguished. On one hand, the first step corresponds to more than 20% of total released gas, yielding an age of ca. 190 ± 4 Ma. Other younger ages have also been obtained from both K9621 biotite (140 ± 2 Ma) and K9620 muscovite (120 ± 2 Ma). On the other hand, steps around the central bulge yield a flat segment around 283 ± 5 Ma, which we consider to be significant. This age is within the range of ages obtained from other nearby samples. The 1100, 1200, and 1400 °C steps have no significance because a low amount of argon was released.

The spectrum from biotite grains from the same sample is complex (Fig. 5B), but K/Ca and Cl/K correlation curves reflect no significant variation in composition, indicating very slight alteration. Nevertheless, the first 15% of released gas gives the lowest age, 140 ± 2 Ma. A regular increase of the age to about 230 ± 5 Ma can be observed on the rest of the spectrum, similar to the plateau age obtained from muscovite. Because the closure temperature of muscovite is higher than that of biotite, it is possible that sample K9621 has underwent deformation ca.

TABLE 3. ^{40}Ar/^{39}Ar RESULTS. ZERO VALUES INDICATE THAT MEASURED SIGNAL WERE EQUAL OR BELOW ANALYTIC NOISE

Temp (°C)	^{40}Ar/^{39}Ar	^{38}Ar/^{39}Ar	^{37}Ar/^{39}Ar	^{36}Ar/^{39}Ar (10^{-3})	^{39}Ar (10^{-14}mol)	F^{39}Ar released	%^{40}Ar*	^{40}Ar*/^{39}Ar	Age (Ma)	± 1s (Ma)
K9621		Muscovite	J = 0.0041600		wt = 4.6 mg					
600	32.207	0.020	0.016	9.092	0.33	3.42	91.72	29.54	209.10	3.97
700	32.727	0.018	0.018	6.601	0.50	8.59	94.06	30.78	217.39	4.10
800	32.807	0.013	0.004	2.037	1.87	27.91	98.11	32.19	226.69	4.26
850	32.638	0.012	0.002	0.869	2.90	57.80	99.13	32.36	227.82	4.29
900	33.779	0.014	0.004	1.420	1.10	69.14	98.69	33.34	234.30	4.47
950	32.422	0.013	0.006	1.305	0.61	75.44	98.74	32.01	225.55	4.25
1000	32.730	0.013	0.008	1.282	0.54	80.97	98.77	32.33	227.63	4.28
1100	32.508	0.013	0.009	0.892	1.07	92.01	99.11	32.22	226.91	4.30
1200	32.295	0.013	0.020	1.162	0.72	99.41	98.86	31.93	224.99	4.32
1400	34.687	0.017	1.074	12.600	0.06	100.00	89.56	31.09	219.44	4.41
K9621		Biotite	J = 0.0041600		wt = 3.5 mg					
600	22.265	0.024	0.015	10.518	0.55	14.75	86.16	19.18	138.52	2.67
700	29.043	0.020	0.013	7.461	1.47	54.30	92.45	26.85	191.03	3.66
750	31.642	0.019	0.014	5.444	0.34	63.38	94.92	30.03	212.39	4.02
800	33.708	0.021	0.023	7.278	0.17	68.00	93.65	31.57	222.61	4.22
850	33.198	0.020	0.018	5.891	0.11	71.02	94.76	31.46	221.89	4.28
900	33.284	0.020	0.019	6.548	0.19	76.18	94.20	31.36	221.19	4.21
1000	33.022	0.019	0.012	5.725	0.56	91.42	94.88	31.33	221.03	4.21
1100	33.755	0.018	0.012	3.955	0.28	99.08	96.51	32.58	229.29	4.37
1200	33.293	0.013	0.040	1.466	0.03	99.85	98.64	32.84	231.02	5.52
1400	29.543	0.000	0.000	0.000	0.01	100.00	99.90	29.51	208.91	8.96
k9620		Muscovite	J = 0.0041600		wt = 5.1 mg					
600	17.884	0.021	0.046	3.670	0.88	30.85	93.89	16.79	122.67	2.42
700	39.980	0.020	0.027	9.210	0.48	47.63	93.25	37.28	261.85	5.34
800	36.425	0.016	0.012	2.394	0.38	60.90	98.01	35.70	251.49	4.98
850	43.607	0.019	0.013	2.478	0.15	66.33	98.28	42.86	297.96	8.21
900	46.515	0.019	0.020	2.322	0.18	72.49	98.49	45.81	316.79	6.07
950	51.968	0.021	0.032	2.233	0.23	80.45	98.70	51.29	351.21	7.21
1000	55.809	0.022	0.036	3.042	0.15	85.80	98.37	54.90	373.51	7.09
1100	44.881	0.018	0.064	1.510	0.28	95.75	98.96	44.42	307.92	6.44
1200	71.342	0.029	0.413	5.770	0.04	97.01	97.65	69.69	462.16	23.79
1400	26.724	0.022	0.098	3.943	0.09	100.00	95.63	25.56	183.53	4.92
K9607		Biotite	J=0.0041600		wt= 3.5 mg					
600	30.102	0.064	0.164	12.743	0.56	21.02	87.65	26.39	189.20	3.83
700	43.884	0.050	0.063	13.131	0.60	43.46	91.26	40.05	279.87	5.35
750	54.800	0.058	0.095	13.733	0.16	49.38	92.69	50.80	348.11	7.72
800	56.189	0.069	0.092	17.939	0.12	53.98	90.70	50.97	349.17	10.96
850	50.823	0.063	0.089	14.074	0.16	59.98	91.92	46.72	322.53	10.42
900	44.430	0.059	0.070	12.733	0.33	72.34	91.63	40.71	284.15	6.39
1000	43.760	0.047	0.046	10.714	0.46	89.50	92.83	40.63	283.59	6.04
1100	54.474	0.042	0.275	10.770	0.07	92.16	94.24	51.35	351.55	8.57
1200	22.795	0.007	0.506	0.000	0.02	93.08	99.86	22.80	164.63	9.11
1400	8.017	0.011	0.010	0.000	0.18	100.00	99.61	7.99	59.39	1.35
K9631		Biotite	J = 0.0041600		wt = 3.5 mg					
600	6.358	0.094	0.018	4.238	0.71	11.06	80.20	5.10	37.87	0.76
709	19.960	0.082	0.022	3.183	1.02	26.91	95.23	19.01	137.29	2.65
754	18.767	0.076	0.012	0.797	0.47	34.25	98.61	18.51	133.80	2.59
798	19.388	0.071	0.010	0.517	0.38	40.10	99.07	19.21	138.69	2.68
843	20.416	0.076	0.010	0.673	0.48	47.52	98.89	20.19	145.50	2.80
902	20.928	0.084	0.010	0.547	1.09	64.48	99.10	20.74	149.30	2.87
1009	20.027	0.084	0.009	0.766	1.98	95.36	98.74	19.77	142.62	2.75
1107	24.387	0.077	0.051	17.035	0.27	99.63	79.63	19.42	140.16	2.79
1200	29.662	0.036	0.154	19.109	0.01	99.86	81.24	24.10	172.38	18.86
1405	11.163	0.078	0.194	0.000	0.01	100.00	99.72	11.15	81.76	23.30
95DU2		Muscovite	J = 0.0067960		wt = 4.7 mg					
600	8.948	0.065	0.104	7.372	0.11	0.76	73.31	6.78	81.32	1.59
700	11.328	0.030	0.156	2.237	0.20	2.13	92.47	10.66	126.18	2.45
800	11.645	0.026	0.047	0.848	0.18	3.35	95.53	11.37	134.25	2.60
950	13.874	0.022	0.004	0.518	3.26	25.56	98.53	13.69	160.45	5.49
1000	13.446	0.022	0.003	0.274	3.19	47.33	99.00	13.33	156.45	5.68
1050	13.447	0.022	0.004	0.270	2.49	64.30	98.80	13.33	156.46	4.63
1100	13.957	0.022	0.005	0.147	3.00	84.78	99.01	13.88	162.60	6.18
1200	14.571	0.022	0.010	0.073	2.02	98.57	98.60	14.51	169.71	4.78
1400	22.106	0.024	0.051	27.127	0.21	100.00	57.50	14.21	166.30	3.34

Temp (°C)	$^{40}Ar/^{39}Ar$	$^{38}Ar/^{39}Ar$	$^{37}Ar/^{39}Ar$	$^{36}Ar/^{39}Ar$ (10^{-3})	^{39}Ar (10^{-14}mol)	F^{39}Ar released	%$^{40}Ar*$	$^{40}Ar*/^{39}Ar$	Age (Ma)	± 1s (Ma)
95DU2		Biotite	J = 0.0067990		wt= 7.7 mg					
600	19.938	0.098	0.038	50.484	0.75	2.99	26.36	5.26	63.45	6.66
700	16.752	0.091	0.023	13.945	3.43	16.63	75.61	12.67	149.11	3.87
800	13.727	0.088	0.012	1.223	4.82	35.85	97.09	13.34	156.59	3.55
850	13.781	0.089	0.005	0.190	5.49	57.70	99.27	13.69	160.55	8.44
900	13.525	0.085	0.007	0.217	2.56	67.90	99.13	13.43	157.60	3.43
950	13.767	0.087	0.008	0.333	2.87	79.32	98.90	13.63	159.93	4.08
1000	13.675	0.088	0.012	0.305	3.04	91.43	98.95	13.55	159.00	3.49
1050	13.768	0.086	0.044	0.175	1.46	97.25	98.82	13.68	160.50	3.51
1100	15.117	0.093	0.120	0.000	0.41	98.89	97.05	15.09	176.25	4.18
1200	14.230	0.083	0.210	0.049	0.25	99.90	92.15	14.20	166.32	6.12
1400	60.557	0.114	0.295	167.873	0.03	100.00	14.62	11.87	140.03	31.15
94DU27		Muscovite	J = 0.0068450		wt = 5.0 mg					
700	7.052	0.037	0.079	2.762	0.86	4.86	87.82	6.22	75.24	1.54
800	10.091	0.026	0.042	0.836	0.61	8.29	96.75	9.82	117.32	2.36
900	12.125	0.023	0.010	0.653	2.28	21.18	98.01	11.90	141.27	4.14
950	13.199	0.022	0.004	0.406	3.36	40.20	98.74	13.04	154.29	6.26
1000	13.195	0.022	0.004	0.253	3.61	60.65	99.08	13.08	154.75	6.04
1050	13.433	0.022	0.009	0.195	2.72	76.04	99.04	13.34	157.65	5.10
1100	14.160	0.022	0.013	0.163	2.51	90.27	98.97	14.08	165.97	5.31
1200	13.819	0.022	0.032	0.419	1.51	98.83	97.59	13.66	161.31	3.82
1400	21.862	0.028	0.079	34.138	0.21	100.00	49.11	11.93	141.67	2.89
K9646		Muscovite	J = 0.0041600		wt = 4.3 mg					
600	11.696	0.029	1.389	8.156	0.20	5.41	80.22	9.39	69.64	1.94
700	51.909	0.036	0.409	17.899	0.18	10.30	89.99	46.73	322.57	8.86
800	69.319	0.039	0.106	13.140	0.08	12.36	94.47	65.49	437.44	20.31
850	56.070	0.030	0.049	3.647	0.23	18.45	98.06	54.99	374.04	9.38
900	54.491	0.029	0.046	2.860	0.25	25.06	98.43	53.63	365.72	9.19
950	48.489	0.026	0.025	1.540	0.84	47.39	99.02	48.01	330.69	6.33
1000	48.549	0.026	0.015	1.097	0.76	67.58	99.28	48.20	331.87	6.28
1100	49.816	0.026	0.020	1.143	0.79	88.38	99.27	49.46	339.74	6.64
1200	60.001	0.030	0.116	2.463	0.17	92.94	98.77	59.27	400.16	9.53
1400	26.648	0.021	0.047	3.948	0.27	100.00	95.60	25.48	182.98	3.78
K9649		K-feldspar	J = 0.0041600		wt = 8.5 mg					
450	81.666	0.092	0.246	44.531	0.03	1.61	84.17	68.75	456.67	8.27
450	19.357	0.029	0.248	16.659	0.02	2.45	74.96	14.51	106.50	4.37
500	35.584	0.040	0.263	15.879	0.03	4.25	87.02	30.97	220.12	4.91
500	16.154	0.019	0.339	8.068	0.03	5.57	85.45	13.81	101.47	2.87
550	28.674	0.028	0.898	11.311	0.05	8.24	88.64	25.43	182.69	3.67
550	19.460	0.017	1.895	4.839	0.03	10.05	93.21	18.17	132.34	2.97
600	25.679	0.017	1.436	7.550	0.06	13.02	91.69	23.57	169.91	3.33
600	22.154	0.014	0.782	2.461	0.05	15.43	96.85	21.47	155.40	3.28
650	23.241	0.021	0.442	12.212	0.08	19.75	84.74	19.70	143.10	3.06
650	23.758	0.014	0.259	5.720	0.06	22.77	92.95	22.09	159.68	3.21
700	25.225	0.012	0.172	1.943	0.07	26.60	97.68	24.64	177.28	3.47
700	45.357	0.028	0.212	1.002	0.05	29.46	99.32	45.06	311.98	11.47
750	40.427	0.029	0.184	3.847	0.13	36.31	97.19	39.30	274.99	10.77
800	44.802	0.033	0.292	8.226	0.12	42.42	94.64	42.41	295.09	8.14
800	44.572	0.027	0.352	4.023	0.06	45.45	97.36	43.41	301.47	10.32
800	51.576	0.029	0.454	4.967	0.05	47.90	97.20	50.15	344.08	19.26
800	98.520	0.085	0.099	203.341	0.02	48.98	40.12	39.53	276.49	5.64
700	23.592	0.028	0.173	0.000	0.02	49.98	99.87	23.57	169.94	4.85
800	22.458	0.018	0.215	2.015	0.07	53.73	97.32	21.86	158.10	3.43
900	22.836	0.016	0.276	3.789	0.14	61.08	95.12	21.73	157.18	3.12
1000	26.204	0.021	0.473	6.212	0.21	71.99	93.11	24.41	175.67	5.51
1050	28.957	0.025	0.548	7.053	0.12	78.10	92.94	26.92	192.84	4.36
1100	33.987	0.032	0.551	8.636	0.13	84.95	92.64	31.50	223.64	5.40
1200	50.896	0.048	0.713	11.480	0.18	94.39	93.48	47.60	328.11	10.48
1400	91.417	0.075	0.493	18.493	0.11	100.00	94.13	86.09	555.65	9.97
K9607		K-feldspar	J = 0.0041600		wt = 5.6 mg					
500	43.985	0.039	0.080	8.797	0.08	0.63	94.14	41.41	288.65	8.21
600	17.040	0.014	0.234	1.577	0.61	5.22	97.22	16.57	121.09	2.58
700	16.858	0.014	0.052	1.987	1.43	16.05	96.42	16.25	118.87	2.45
800	21.491	0.014	0.010	0.623	2.20	32.75	99.02	21.28	154.09	3.08
900	23.560	0.014	0.012	0.796	1.27	42.36	98.89	23.30	168.04	3.32
1000	29.055	0.015	0.012	1.007	1.42	53.16	98.89	28.73	205.08	4.06
1100	33.777	0.016	0.011	1.102	2.19	69.75	98.96	33.43	236.48	4.72
1200	40.344	0.017	0.006	1.319	3.33	95.01	98.98	39.93	279.10	5.39
1400	39.266	0.015	0.006	1.336	0.66	100.00	98.94	38.85	272.07	5.30

Temp (°C)	^{40}Ar/^{39}Ar	^{38}Ar/^{39}Ar	^{37}Ar/^{39}Ar	^{36}Ar/^{39}Ar (10^{-3})	^{39}Ar (10^{-14}mol)	F^{39}Ar released	%^{40}Ar*	^{40}Ar*/^{39}Ar	Age (Ma)	± 1s (Ma)
95DU2		K-feldspar		J = 0.0067930		wt = 5.47 mg				
400	56.207	0.042	0.000	14.409	0.04	0.14	90.09	51.99	545.65	9.61
400	4.261	0.027	0.000	2.744	0.08	0.39	62.02	3.43	41.52	2.29
450	7.319	0.023	0.000	0.849	0.25	1.20	92.71	7.04	84.23	1.70
450	2.246	0.022	0.000	0.536	0.26	2.04	77.88	2.05	24.99	0.82
500	3.353	0.021	0.015	0.461	0.46	3.53	90.89	3.18	38.60	0.80
500	2.518	0.022	0.025	0.360	0.39	4.82	85.64	2.38	28.92	0.71
550	3.073	0.021	0.051	0.339	0.74	7.23	92.87	2.94	35.71	0.72
550	2.756	0.021	0.039	0.226	0.75	9.70	91.76	2.66	32.28	0.68
600	3.131	0.021	0.034	0.209	1.22	13.69	95.19	3.04	36.83	0.73
600	3.088	0.021	0.008	0.146	0.96	16.85	94.03	3.01	36.51	0.74
650	3.563	0.021	0.010	0.204	0.96	19.99	95.26	3.47	42.00	0.84
650	3.851	0.021	0.003	0.048	0.73	22.39	94.95	3.80	45.98	0.94
700	4.391	0.021	0.010	0.147	0.70	24.67	95.83	4.31	52.08	1.04
700	4.756	0.021	0.005	0.213	0.46	26.17	93.06	4.66	56.19	1.17
750	5.422	0.021	0.007	0.261	0.49	27.77	95.15	5.31	63.92	1.28
750	5.902	0.022	0.008	0.251	0.40	29.08	93.53	5.79	69.63	1.43
800	6.590	0.021	0.013	0.227	0.58	30.99	96.46	6.49	77.81	1.54
800	6.969	0.022	0.001	0.028	0.44	32.43	95.63	6.92	82.91	1.67
800	7.437	0.022	0.000	0.118	0.45	33.91	94.57	7.37	88.08	1.79
800	7.994	0.022	0.000	0.605	0.85	36.68	90.93	7.78	92.93	1.98
700	9.548	0.024	0.000	6.602	0.02	36.75	52.28	7.60	90.76	7.45
750	8.565	0.025	0.007	2.885	0.04	36.90	68.63	7.69	91.88	4.17
800	8.434	0.023	0.000	1.389	0.12	37.28	84.68	7.99	95.39	2.33
850	8.359	0.021	0.000	0.609	0.39	38.56	93.81	8.14	97.15	1.95
900	8.624	0.021	0.000	0.486	0.83	41.28	96.14	8.45	100.65	1.97
950	9.015	0.021	0.004	0.540	1.39	45.84	96.77	8.82	105.00	2.27
1000	9.398	0.022	0.005	0.572	1.93	52.15	97.05	9.20	109.31	2.39
1050	9.994	0.022	0.007	0.651	2.54	60.47	97.12	9.77	115.91	2.78
1100	10.717	0.022	0.008	0.707	2.71	69.34	97.15	10.48	124.02	2.68
1150	11.430	0.022	0.006	0.862	3.01	79.19	96.94	11.14	131.64	3.07
1200	12.689	0.022	0.003	0.943	3.81	91.67	97.13	12.38	145.66	3.13
1400	15.870	0.022	0.003	1.824	2.54	100.00	95.80	15.30	178.43	4.02

Temp (°C)	Time (min)	f	D/r^2	1000/T (K^{-1})	-log(D/r^2)	log(r/r$_o$)
E = 41387 cal/mol +- 4674		log(Do/ro)= 5.35/s +- 1.38				
400	20	0.14	1.30E-09	1.486	8.886	0.400
400	30	0.39	5.83E-09	1.486	8.234	0.074
450	20	1.20	8.40E-08	1.383	7.076	-0.040
450	30	2.04	1.18E-07	1.383	6.928	-0.114
500	20	3.53	5.45E-07	1.294	6.263	-0.042
500	30	4.82	4.71E-07	1.294	6.327	-0.011
550	20	7.23	1.90E-06	1.215	5.721	0.042
550	30	9.70	1.82E-06	1.215	5.739	0.051
600	20	13.69	6.11E-06	1.145	5.214	0.103
600	30	16.85	4.20E-06	1.145	5.376	0.184
650	20	19.99	7.59E-06	1.083	5.120	0.337
650	30	22.39	4.43E-06	1.083	5.353	0.453
700	20	24.67	7.03E-06	1.028	5.153	0.605
E = 41387 cal/mol +- 4674		log(Do/ro)= 5.35/s +- 1.38				
700	30	26.17	3.32E-06	1.028	5.479	0.768
750	20	27.77	5.67E-06	0.978	5.247	0.879
750	30	29.08	3.24E-06	0.978	5.489	1.000
800	20	30.99	7.50E-06	0.932	5.125	1.024
800	30	32.43	3.98E-06	0.932	5.400	1.162
800	40	33.91	3.21E-06	0.932	5.493	1.208
800	120	36.68	2.13E-06	0.932	5.671	1.297
700	30	36.75	2.39E-07	1.028	6.621	1.339
750	30	36.90	4.63E-07	0.978	6.334	1.423
800	30	37.28	1.25E-06	0.932	5.901	1.412
850	30	38.56	4.24E-06	0.890	5.373	1.336
900	30	41.28	9.47E-06	0.853	5.024	1.333
950	30	45.84	1.73E-05	0.818	4.761	1.359
1000	30	52.15	2.70E-05	0.786	4.569	1.408
1050	30	60.47	4.30E-05	0.756	4.366	1.441
1100	30	69.34	5.73E-05	0.728	4.242	1.503
1150	30	79.19	8.73E-05	0.703	4.059	1.528
1200	30	91.67	2.06E-04	0.679	3.686	1.449

TABLE 3. ^{40}Ar/^{39}Ar RESULTS. ZERO VALUES INDICATE THAT MEASURED SIGNAL WERE EQUAL OR BELOW ANALYTIC NOISE (continued)

Temp (°C)	^{40}Ar/^{39}Ar	^{38}Ar/^{39}Ar	^{37}Ar/^{39}Ar	^{36}Ar/^{39}Ar (10^{-3})	^{39}Ar (10^{-14}mol)	F^{39}Ar released	%^{40}Ar*	^{40}Ar*/^{39}Ar	Age (Ma)	± 1s (Ma)
K9631		K-feldspar	J = 0.0041600							
450	28.822	0.034	0.020	17.500	0.12	0.56	82.29	23.72	170.92	7.93
450	7.302	0.021	0.014	6.620	0.09	0.99	73.29	5.35	40.01	3.10
500	4.235	0.014	0.016	2.640	0.22	2.00	81.21	3.44	25.81	0.54
500	3.876	0.015	0.062	1.970	0.22	3.01	84.56	3.28	24.61	0.67
550	3.731	0.013	0.082	1.016	0.63	5.93	91.40	3.41	25.60	0.54
550	3.987	0.013	0.024	0.535	0.74	9.33	95.37	3.80	28.52	0.64
600	4.569	0.013	0.010	0.321	1.31	15.35	97.29	4.45	33.29	0.67
600	4.929	0.013	0.005	0.206	1.15	20.66	98.17	4.84	36.21	0.75
650	5.588	0.013	0.006	0.888	0.99	25.23	94.84	5.30	39.62	0.81
650	6.216	0.013	0.005	0.432	0.70	28.43	97.49	6.06	45.23	0.93
700	7.554	0.012	0.004	0.296	0.70	31.66	98.46	7.44	55.36	1.14
700	13.096	0.013	0.004	0.855	0.68	34.81	97.87	12.82	94.37	1.90
750	10.932	0.013	0.006	0.402	0.60	37.57	98.65	10.79	79.74	1.65
800	12.013	0.013	0.005	0.282	0.53	40.01	99.06	11.90	87.78	2.27
800	13.899	0.013	0.004	0.453	0.41	41.89	98.83	13.74	100.96	1.99
800	14.926	0.013	0.003	0.386	0.32	43.36	99.04	14.78	108.42	2.18
700	33.897	0.039	0.007	0.000	0.03	43.50	99.91	33.87	239.39	22.20
800	14.975	0.013	0.005	0.691	0.44	45.53	98.46	14.74	108.14	2.27
900	15.457	0.013	0.006	0.993	0.83	49.36	97.94	15.14	110.95	2.23
1000	15.663	0.013	0.004	0.583	2.37	60.28	98.72	15.46	113.26	2.27
1050	16.541	0.013	0.003	0.630	1.47	67.05	98.71	16.33	119.38	2.42
1100	18.274	0.013	0.004	0.769	1.38	73.38	98.61	18.02	131.32	2.65
1200	20.012	0.013	0.002	0.967	3.09	87.60	98.44	19.70	143.09	2.87
1400	17.858	0.013	0.003	0.765	2.70	100.00	98.58	17.61	128.40	2.63

Temp (°C)	Time (min)	f	D/r^2	1000/T (K^{-1})	-log(D/r^2)	log(r/r$_o$)
		E=51846 cal/mol +- 618	log(Do/ro)= 7.98/s +- 0.17			
450	20	0.56	2.05E-08	1.383	7.689	-0.003
450	30	0.99	2.89E-08	1.383	7.539	-0.078
500	20	2.00	1.98E-07	1.294	6.703	0.011
500	30	3.01	2.21E-07	1.294	6.655	-0.013
550	20	5.93	1.71E-06	1.215	5.767	-0.012
550	30	9.33	2.26E-06	1.215	5.645	-0.073
600	20	15.35	9.73E-06	1.145	5.012	0.005
600	30	20.66	8.33E-06	1.145	5.079	0.038
650	20	25.23	1.37E-05	1.083	4.862	0.281
650	30	28.43	7.49E-06	1.083	5.126	0.413
700	20	31.66	1.27E-05	1.028	4.896	0.613
700	30	34.81	9.13E-06	1.028	5.039	0.685
750	20	37.57	1.31E-05	0.978	4.884	0.892
800	20	40.01	1.24E-05	0.932	4.906	1.161
800	30	41.89	6.73E-06	0.932	5.172	1.294
800	40	43.36	4.09E-06	0.932	5.388	1.402
700	120	43.50	1.26E-07	1.028	6.899	1.615
800	60	45.53	3.96E-06	0.932	5.403	1.409
900	20	49.36	2.38E-05	0.853	4.624	1.470
1000	20	60.28	8.15E-05	0.786	4.089	1.582
1050	20	67.05	6.31E-05	0.756	4.200	1.805
1100	20	73.38	7.21E-05	0.728	4.142	1.932
1200	20	87.60	2.58E-04	0.679	3.588	1.936

Note: % ^{40}Ar* denotes radiogenic argon

Alkali feldspars from the sample K9607 (Fig. 5E) supply a spectrum consistent with the biotite. The first 2% of the gas released is masked by excess argon and correlated with a Cl-rich phase (possibly fluid inclusions). The spectrum then exhibits a minimum age of 120 ± 2 Ma and regularly climbs to a maximum age of 280 ± 5 Ma. The entire range of ages is coherent with the biotite from the same sample, the highest ages being ca. 280 Ma, close to but lower than the biotite highest steps, as expected due to the lower closure temperature of feldspar. Moreover, if argon loss also affected the feldspar from this sample, then it is logical that the lowest ages are lower than in the biotite. The feldspar results also suggest that such a loss must be younger than ca. 125 Ma.

Aksai section

Three monazite age measurements were made for a single grain from the mylonitic granite K9631 (Fig. 4, C and D). U-Pb-Th concentrations are uniform and vary respectively

TABLE 4. INPUT DIFFUSION CHARACTERISTICS OF K-FELDSPARS USED FOR MODELING THEIR COOLING HISTORIES

Model for K9631

Number of domains	4					
Activation energy (Kcal/mol)	47.72					
Domain from least to most retentive	1	2	3	4		
D/r_o^2 (1/s)	8.19	6.97	4.78	3.35		
Size ratio to smallest domain	0.20	0.15	0.20	0.45		

Model for 95DU2

Number of domains	6					
Activation energy (Kcal/mol)	38.16					
Domain from least to most retentive	1	2	3	4	5	6
D/r_o^2 (1/s)	6.10	5.41	4.67	2.13	0.79	-0.39
Size ratio to smallest domain	0.08	0.10	0.06	0.56	0.09	0.10

between 2840 and 3630 ppm, 1410 and 1760 ppm, and 64470 and 82560 ppm. These three measurements yield calculated ages homogeneously distributed around the mean value of 423 ± 18 Ma (Fig. 4C). The data do not provide any evidence for a relationship between individual ages and U and Th content (Fig. 4D).

From the Aksai mylonitic granite (K9631 and AT26, both from the same outcrop), whole rock, feldspar, and four different size fractions of biotite were analyzed for Rb-Sr. Plotted on isochron diagrams (Fig. 6A), the six points yield an array with a date of 140 ± 2 Ma and an initial $^{87}Sr/^{86}Sr$ ratio of 0.7528 ± 6 ($\pm 2\sigma$). Argon dating shows a large flat portion of the spectra at 140 ± 2 Ma but no real plateau (Fig. 5F). A slight effect of secondary chlorite is apparent from the correlations between the K/Ca plot and the maximum age in an intermediate position on the age spectrum. The first step is significantly younger, ca. 40 Ma, and is also associated with a slightly different K/Ca ratio, suggestive of partial recrystallization or argon loss.

Muscovite from felsic micaschist south of Aksai (95DU2, 95DU27, Fig. 5, G and H) give almost identical concordant $^{40}Ar/^{39}Ar$ ages of 163 ± 4 and 160 ± 4, and biotite from 95DU2 yields an age of 159 ± 3 Ma (Fig. 5I). These results are remarkably similar for muscovite and biotite ages.

Feldspars from samples K9631 and 95DU2 are in agreement with their corresponding micas (Fig. 5, J and K). They show climbing age spectra ranging from ca. 25 Ma (the first steps are significantly masked by excess argon for 95DU2) to ages similar to the plateau from the micas (except the last step of 95DU2, which was most probably affected by an incorrect extraction blank correction or excess argon). Such a spectrum is typical of slowly cooled samples when subsequent recrystallization did not occur. Even if partial recrystallization cannot

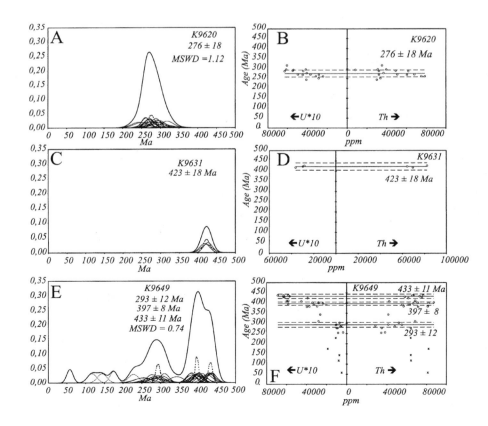

Figure 4. Weighted-histogram representation and U-Th-age relations for electron microprobe data for samples K9620 (A and B), K9631 (C and D), and K9649 (E and F). For weighted-histogram representation, results from each sample are represented by curve; thick curve is sum of all individual curves; thick dashed curves represent statistically calculated ages calculated by statistical procedure; thin-dashed curves indicate data that are not included in statistical analysis procedure. U-Th-age relations shown with x indicate data that are not included in statistical analysis procedure.

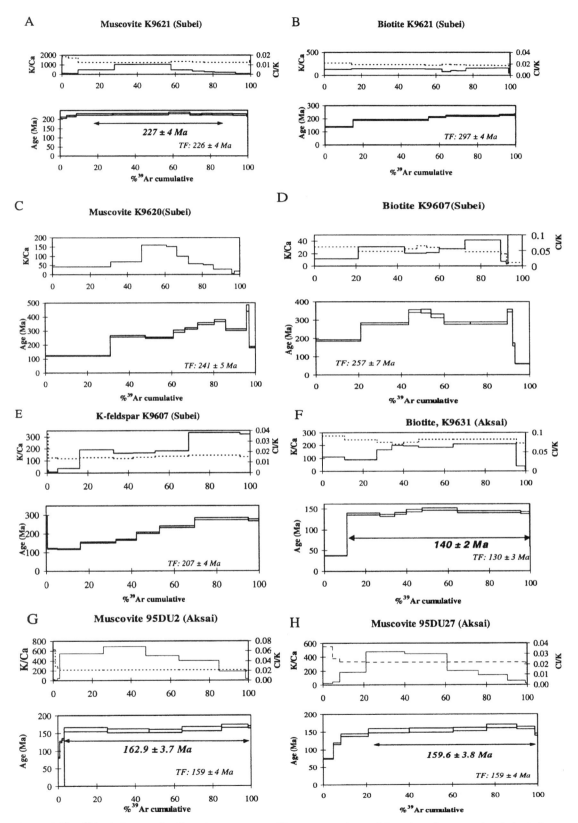

Figure 5. $^{40}Ar/^{39}Ar$ dating results. Samples are arranged by cross section and in order of presentation in text. Age spectra are shown with covariations of K/Ca (solid) and Cl/K (dashed) ratios when meaningful. Plateau age is indicated below steps used for calculation together with total fusion (TF) age. For K-feldspars, age models used to define cooling history of Figure 7 are shown in bold on top of original spectra. Detailed explanation of this type of modeling is described in Sobel et al. (this volume).

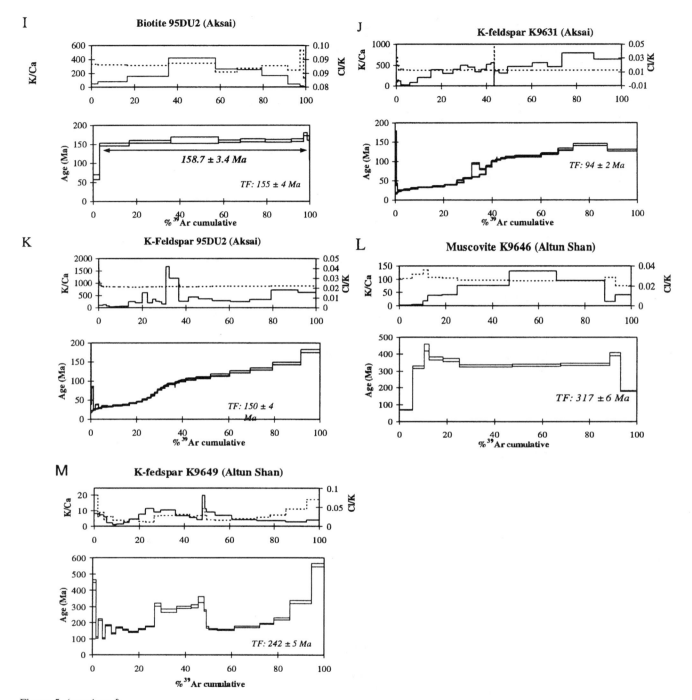

Figure 5. (*continued*)

be ruled out for the biotites from these samples (as discussed in the previous petrological section), the feldspars apparently were much stronger during deformation and are only brittlely deformed. We thus modeled the cooling history that would have produced such an age spectra in the case of pure slow cooling with the methods and algorithms developed by Lovera et al. (1989, 1991). Diffusion characteristics are easily modeled and imply for both samples a cooling history (Fig. 7) characterized by rapid cooling from the mica ages followed by a protracted period of slower cooling, then a significant increase

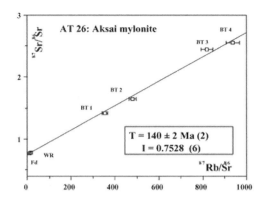

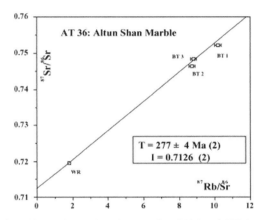

Figure 6. Rb/Sr isochron plots for samples AT26 and AT36.

of cooling rates ca. 20 Ma. The results from sample 95DU2 suggest a period of more rapid cooling at the beginning of the Tertiary, but this is not corroborated by the other sample, probably because the domain structure of the latter sample did not allow it to record temperatures in the range that would show that cooling.

Altun Shan section

The Altun Shan marble (K9646 and AT36 from the same outcrop) was dated by Rb/Sr and $^{40}Ar/^{39}Ar$. Whole rock and three fractions of biotite yield a best-fit regression line corresponding to a date of 277 ± 4 Ma with an initial $^{87}Sr/^{86}Sr$ ratio of 0.7126 ± 2 (± 2σ) (Fig. 6B). Argon dating was carried out on muscovite and yields a saddle-shape age spectra (Fig. 5L). The first step has a Late Cretaceous age of ca. 70 Ma followed by a concave-up spectra with a minimum age of 334 ± 6 Ma. It is clear that some excess argon is present in this sample, but the inverse isochron was not resolvable. The difference in age between the muscovite and biotite, however, may be significant because the closure temperature of Rb/Sr on biotite is lower than that of argon in muscovite (Cliff, 1985; Hames and Bowring, 1994).

We obtained 36 measurements for 12 monazite grains from the altered granite K9649 (Fig. 4, E and F) and yielded varying U (360–6860 ppm), Pb (220–1910 ppm), and Th (28380–87420 ppm) concentrations. The weighted-histogram from this sample is complex, and shows calculated ages ranging between 437 and 58 Ma (Fig. 4E). To obtain a statistically acceptable model (MSWD = 0.74), it is necessary to exclude the five youngest calculated ages (i.e., 58, 117, 144, 175, and 228 Ma). Statistical analysis of the remaining data yields a three-age model with populations at 293 ± 12 Ma, 396 ± 8 Ma, and 433 ± 11 Ma. The U-Th-age diagram (Fig. 4F) shows some interesting features: there is no relationship between Th and age, but a clear relationship between U content and age; i.e., the younger the age, the lower the U content.

The $^{40}Ar/^{39}Ar$ results from the K-feldspars of sample K9649 are very scattered (Fig. 5M). The coherent K/Ca and Cl/K plots together with the age spectrum are indicative of a strong fluid interaction, already noted in the petrological description of this sample. Excess argon is certainly present at various stages of degassing and only with difficulty can geological

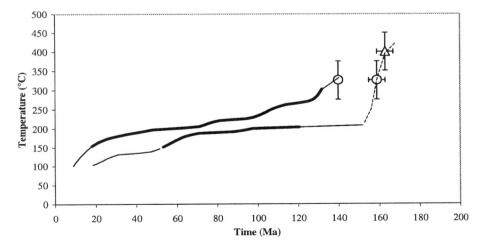

Figure 7. Thermal modeling results from feldspars from Aksai section. Note that because modeling does not define diffusion characteristics above start of incongruent melting of feldspars, and also because diffusion and age data below 150 °C are likely to be most affected by nonvolumetric diffusion and complex argon loss in nature (Arnaud and Kelley, 1997), cooling history is not determined for ages obtained in laboratory above 1150 °C and below 150 °C. Therefore, thick solid lines indicate most robustly modeled data and thin solid lines have been used to show poorly constrained part of cooling path. Extrapolation from mica ages to feldspar cooling curve are shown as long dashes.

information be recovered. However, one might tentatively suggest that the range of maximum ages are 100 Ma at the beginning of the spectrum, around 150 Ma in the middle portion, and probably to 200 Ma at the end.

ANALYSIS OF GEOCHRONOLOGICAL RESULTS

The variety of ages recorded by the various isotopic systems and minerals underlines the role of inheritance in these multiply deformed samples. The following critical analysis of the data is intended to identify the relevant groups of ages before geological interpretations can be drawn.

U/Th dating of monazite provides the oldest group of ages, ca. 423 and 433 Ma, both from deformed granites. Although it is impossible from the monazite data alone to decide whether these ages are indicative of metamorphism or emplacement, it is striking that the two are identical within errors while the Altun Shan granite (sample K9649) is far more deformed than the Aksai sample (sample K9631). Moreover, these ages are close to values from a growing database of geochronological data obtained from undeformed granite, suggesting a period of major granitic magmatism in northern Tibet (Harris et al., 1988; Arnaud, 1992; Xu et al., 1996; Yuquan et al., 1996) and especially in the Altyn Tagh range, where muscovite ages of 432 ± 8 are reported from pegmatites (Sobel and Arnaud, 1999). We conclude that those dates are significant and probably date granite emplacement. The interpretation of the population at 396 Ma in sample K9649 is more puzzling. In the weighted histogram representation, the group of ages >350 Ma actually seems to be bimodal. In the U-Th representation, the population at 396 Ma contains less U than the 433 Ma population. However, the statistical parameters indicate that the entire population of ages >350 Ma is very close to a single population, with an age of 417 Ma and MSWD = 1.73. If only one young or one old age is excluded from this population, it would give the same age with an acceptable MSWD of 1.61. Because we do not have any other age that would confirm the existence of a Devonian intrusive age, we think that the best interpretation is to consider that we have a single Silurian population (410 ± 7 Ma) that was slightly disturbed by later events. In this sample, a striking feature is the relationship that exists between age and U content. Because this relationship does not exist for Th, we must find a geological process that only affects U. The simplest explanation is a progressive change of the redox state of the environment. Under oxidizing condition, U become U^{6+}, making uranite ions that cannot be incorporated into the monazite structure. This progressive oxidation hypothesis is consistent with a progressively increasing exchange with the oxidizing atmosphere.

A second, younger, scattered group of ages ranges between 240 and 334 Ma. It is expressed in the monazite dating, especially in the schists with an average at 278 Ma (K9620), and in the deformed granite from Altun Shan at 293 Ma (K9649), indicating perturbation of the U/Th systematics (especially for the granites, because emplacement has been bracketed ca. 430 Ma), either by Pb loss or growth ages. Nevertheless, the age given by

U/Th dating is imprecise due to the limitations of the method. Rb/Sr dating of the Altun Shan marble (sample AT36, Fig. 6B) yields an isochron age of 277 Ma. The $^{40}Ar/^{39}Ar$ dating of muscovite from the Subei mylonitic micaschists (sample K9621) yielded plateau ages of 230 Ma (Fig. 5A), whereas biotite and feldspars have increasing age patterns from ca. 120 Ma to 230 Ma (Fig. 5, B, D, and E). This scattered population of ages is rather puzzling. On the one hand, one is tempted to consider that this cluster of ages from independent isotopic systems can hardly be random, with a group around 280 Ma. On the other hand, the range of ages spans more than 50 m.y. if the excess argon muscovite age is not considered; this does not strongly support one major resetting event. The former point suggests a weak or strongly polyphased event around or older than 280, the latter, a series of intermediate ages with little or no geological meaning, between a well-identified Paleozoic event and a later middle Mesozoic one.

An important group of Late Jurassic–Early Cretaceous ages are revealed by $^{40}Ar/^{39}Ar$, especially in the Aksai section (samples 95DU2, 95DU27, and K9631), with muscovite and biotite plateau ages at 160 Ma (Fig. 5, G, H, and I) and slightly younger plateau ages of 140 Ma from the granite, confirmed by a Rb/Sr isochron age of 140 Ma (Fig. 5F) on the same granite. Thus the two age groups at 160 Ma and 140 Ma both appear to be robust and significant. Micaschist from Subei also strongly supports a Jurassic resetting of the argon chronometer with many early degassing steps in the range 140–160 Ma. Monazite data suggest reopening of the system at an age that can only be determined to be younger than the Permian. K-feldspar modeling confirms the existence of a significant Jurassic event recorded by the $^{40}Ar/^{39}Ar$ chronometer and also suggests a Cenozoic cooling event only weakly identified by the other methods; the existence of this younger thermal event seems probable, although it is imprecisely dated.

In summary, four suites of ages are interpreted as geologically significant: a granitic emplacement event ca. 423–433 Ma, a resetting of the isotopic systems possibly ca. 230–280 Ma and more certainly ca. 140–160 Ma, and a probable Cenozoic event as identified both in $^{40}Ar/^{39}Ar$ data and by the reset monazites in K9630 and K9649.

DISCUSSION

Dating of deformation

When studying metamorphism, deformation, and geochronology, the question of whether one is dating the time of metamorphism, deformation, or subsequent cooling is paramount. Petrographic and textural examinations of deformational features are required to determine the relationship between temperature estimates for deformation and for isotopic closure of geochronological systems. In our case, observations of macroscopic and microscopic textures delineate three episodes of deformation, clearly superimposed in time, and characterized by different regimes, which have been characterized as D1, D2, and

D3. The oldest metamorphic event recorded is a high-pressure event seen as remnant phases in micaschist. It has proven impossible to date these relicts, but it is logical to associate them with an older orogenic event because it is overprinted by all other fabrics. Its relation with the granite ages is also poorly understood at present. It is entirely possible that it is related to the lower-middle Paleozoic subduction and collision event described farther to the southwest at Lapeiquan by Sobel and Arnaud (1999) that is associated with 430 Ma granitic magmatism, although there is no strong support for that interpretation.

The significant group of Jurassic ages (140–160 Ma) is the best defined. Because most of the mica from the granulite schist clearly grew during the formation of S2, and because the Aksai granite emplaced at 433 Ma only recorded D2 and D3 (brittle) deformation, this series of ages must characterize D2. Petrographic textures of quartz suggest temperatures of 250–350 °C, and thus indicate that muscovite argon ages should date deformation, whereas biotite (Rb/Sr or ^{40}Ar/^{39}Ar) and feldspar most likely record cooling. However, mica ages in the Aksai micaschist are undistinguishable within error, suggesting that deformation at 160 Ma was followed by rapid cooling. The origin of the younger group of ages at 140 Ma can be due either to a continuum of deformation with a slightly variable expression in space during 20 m.y., or to locally slower cooling or to a deeper position of the Aksai granite during deformation, relative to the neighboring schists. The latter would imply post-Jurassic vertical offset between those two series, the southern part of the section being uplifted relative to the northern one. Given the number of faulted contacts in those sections and the obvious effect of the nearby Altyn Tagh fault, this is a likely explanation. In summary, D2 deformation is dated as ca. 160 Ma, possibly extending to as young as 140 Ma. The feldspar model results confirm that cooling was very slow until the Late Cretaceous. Ages in the range of 160–140 Ma are also present in the Subei micaschist where the first ~10% of ^{39}Ar released always have younger ages than the bulk plateau. This is interpreted either as argon loss or as the result of degassing of recrystallized margins of the mica. Therefore, it seems that D2 as identified in the Subei micaschist is the same deformation event as the one identified more clearly at Aksai.

The interpretation of the Permian–Triassic (230–280 Ma) group of ages recorded in the micaschist muscovites from Subei and the marbles in the Altun Shan depends largely on whether these are interpreted as being significant. The age of D1 and the meaning of the Permian–Triassic ages are inferred to be closely linked. The mineral assemblage of the D1 deformation suggests equilibrium ca. 350–400 °C, temperatures that imply partial or total resetting of argon and Rb/Sr ages and possibly U/Th ages if Pb loss occurred via a fluid phase. Because D1 is clearly anterior to, and developed at higher temperature than, D2, the simplest solution would be to associate D1 to the 230–280 cluster of ages, and most simply to the 277 Ma Rb/Sr age. In that case D1 and D2 have no temporal relation whatsoever and the large range of ages is due to partial resetting of D1 isotopic equilibrium by D2. Alternatively, if that range of ages is geologically

meaningless, and because D1 must alter the isotopic systematics, then one has no other solution than to associate D1 with the Jurassic ages already clearly related to D2. This hypothesis implies that D1 and D2 are two phases of the same tectonic event on a retrograde path, and the Permian–Triassic ages are due to partial resetting of the Paleozoic high-grade metamorphic paragenesis by the Jurassic D1-D2 event. At present it is clearly impossible to choose between those two hypothesis.

The final D3 deformation highlights a significant event that occurred under brittle conditions. The age of this event is only poorly defined. It crosscuts all other fabrics and is thus younger than D2, but it could represent either the final phase in a continuum of deformation after D2, or a completely separate, much younger event. At the same time a significant but imprecise Cenozoic event is identified in geochronological data. D3 is clearly not associated with any significant recrystallization, but it is associated with pervasive fluid circulation taking place during or just after fracturing. Therefore, it is suggested that the low ages in many of the age spectra could be the result of argon loss enhanced at the borders of grains by that fluid flow. Although definitive proof of brittle Cenozoic deformation is lacking based on geochronologic data presented herein, this conclusion seems highly probable from field studies of macroscopic structures. In addition, preliminary apatite fission-track studies in this region have documented rapid late Cenozoic cooling below 100 °C (Arnaud and Sobel, 1998; M. Jolivet, 1999, personal commun.).

Geological implications for the pre-Cenozoic history of the Altun Shan range

Deformation within the Altyn Tagh range has obviously been long lasting and polyphase. The first geochronologically recorded event in the study area is a middle Paleozoic orogenic event, which probably started with the closure of an oceanic basin by south-dipping subduction and associated high-pressure metamorphism ca. 450 Ma (Sobel and Arnaud, 1999, and this work), and ended with formation of a tectonic collage with regional metamorphism and posttectonic granite emplacement between ca. 435 and 380 Ma. That suturing event has tentatively been related to other north Tibetan sutures of similar age in the east and west Kunlun ranges and in the southern Qilian Shan.

Another event is possibly identified near the Permian–Triassic boundary ca. 280 Ma. Dating of postorogenic undeformed granites north of Mangnai as 385 Ma (Sobel and Arnaud, 1999) implies that ages in the range of 250 Ma can hardly be associated with a late episode of the middle Paleozoic event. Late Permian ages are largely reported from all over northern Tibet (Harris et al., 1988; Arnaud, 1992; Xu et al., 1996; Yuquan et al., 1996; Sobel and Arnaud, 1999) and also farther east along the Qinling and Dabie Shan mountains (Eide et al., 1994; Ames et al., 1996; Chavagnac and Jahn, 1996; Hacker et al., 1998; Xue et al., 1997; Meng and Zhang, 1999), where they are interpreted to reflect the collision of the South and North China blocks. Retrodeforming the complex Tertiary and Mesozoic deformation, a difficult task, brings the Altun

Shan area back to the northeast roughly several hundred of kilometers relative to its present position, because most rocks described herein are north of the trace of the active fault (Meyer et al., 1998; Ritts and Biffi, 2000). This brings the study area closer to the Qilian Shan and Qinling and Dabie Shan. However, the westward continuation of the early Mesozoic Dabie and Qinling Shan orogenic belt is poorly defined, and the youngest ages for that event suggest that it could be as young as 245 Ma, and thus is unlikely to be associated with the deformation in the Altyn Tagh range. A key area to most of these hypotheses thus seems to be the Qilian Shan range. Geochronologically defined studies of that area, ideally coupled with new paleomagnetic data, are required to solve these problems.

This study does not provide evidence for cooling and deformation between ca. 250 and 160 Ma; however, ca. 210–180 Ma cooling linked to significant exhumation has been documented from the Altyn Tagh ~300 km to the southwest (Sobel et al., this volume).

The most clearly defined event is the 160–140 Ma deformation. That deformation developed a high horizontal strain in connection with probable sinistral strike-slip movements, and is associated with a restricted area along the trace of the present-day Altyn Tagh fault. Deformation of this age is documented in other places in Tibet, most notably along the Kunlun fault (Arnaud et al., 1997) and the eastern segment of the Altyn Tagh fault (Matte et al., 1996; Arnaud et al., 1997). Notably, the Kunlun fault is interpreted as having reused a corridor generated by strike-slip motion along a Late Cretaceous (ca. 120 Ma) fault, but even older movements along that fault are strongly supported by age data presented in this chapter. The origin of Early Cretaceous deformation and especially strike-slip faulting in the Altyn Tagh range is poorly defined, but the accretion of the Qiantang and Lhasa blocks and formation of the Jinsha suture (Coward et al., 1988) is a likely candidate. Xu et al. (1985) reported a U/Pb age of 121 Ma for leucogranitic bodies in southern Tibet as probably postdating the accretion of the Qiantang and Lhasa blocks. Recent studies (Yin et al., 1995; Matte et al., 1996; Murphy et al., 1997) confirm that some shortening occurred in southern Tibet before the India-Asia collision, and there is evidence for overthrusting at 150 Ma in the northeastern Pamir (Arnaud et al., 1993). Deformation also seems to have happened at that time farther north in the Tian Shan (Hendrix et al., 1992). However, unfolding Tertiary deformation places the Altun Shan to northeast, farther away from the then-active margin of southern Qiantang. Another possible driver for the Jurassic–Cretaceous deformation documented in this chapter is the poorly understood closure of the Mongol-Okhotsk ocean in Mongolia and adjacent Russia. New paleomagnetic evidence (Gilder and Courtillot, 1997; Halim et al., 1998) suggests that by 120 Ma this ocean in northern Mongolia was closed (Khramov, 1958; Enkin et al., 1992).

Recent work in the Bei Shan and the Mongolian Gobi Altai has documented a large nappe system that was active during late Middle Jurassic time (Zheng et al., 1996). Belts of Upper Proterozoic and Paleozoic units in Mongolia that show a consistent pattern of left-lateral offsets have been interpreted to have been caused by a pair of sinistral strike-slip faults of either Triassic to Jurassic or Early Cretaceous age; one possible interpretation is that these faults were active and linked to the subparallel Altyn Tagh and Ruoqiang faults and the Bei Shan thrusts during the Late Jurassic (Lamb et al., 1999). In addition, Late Jurassic to Early Cretaceous pull-apart basins related to strike-slip faulting have been documented to the north of the Qilian Shan (Vincent and Allen, 1999). Whatever the cause of this Mesozoic intracontinental deformation, the timing and sense of deformation appear to closely match the 160–140 Ma deformation reported herein.

Impact of this work on recovering Cenozoic deformation estimates

Our data define two episodes of pre-Cenozoic deformation in the Altyn Tagh. However, it is difficult to ascribe specific amounts of crustal shortening or strike-slip displacement to either of these events. Only weak evidence of Tertiary deformation is recorded in the samples from the Altyn Tagh. The morphological effects of the fault are obvious and local and large-scale tectonic studies (Avouac and Tapponnier, 1993; Van der Woerd et al., 1998; Meyer et al., 1998) imply significant offsets (between 120 and 150 km) during the Cenozoic along the segment of the Altyn Tagh fault near the Altun Shan. Nevertheless, it is extremely difficult to clearly identify related deformation fabrics in the field or in hand samples because an earlier Mesozoic deformation with apparently similar kinematic characteristics took place in the same area. It is interesting that currently available geochronological data only vaguely show the effect of Cenozoic tectonics, probably because of minor exhumation of the Tertiary mylonites. This lack of exhumation along the fault strongly supports a model of an almost pure strike-slip fault motion, and thus reinforces the idea of decoupling of movement between strike-slip and thrust faults in the Qilian Shan (Meyer et al., 1998). In conclusion, the existence of pre-Cenozoic sinistral strike-slip motion along the Altyn Tagh fault calls for caution when estimating the amount of Cenozoic finite displacement along the fault based on pre-Cenozoic piercing points. The limited amount of strong post-Paleozoic shortening prior to the Cenozoic implies that conservative estimates of Tertiary shortening are probably correct, but that more has to be learned about the Paleozoic and Mesozoic history in the Qilian Shan to definitively assess this question.

ACKNOWLEDGMENTS

The French-Chinese field work was made possible by funding through Institut National des Sciences de l'Univers, Centre National de la Recherche Scientifique program Intérieur de la Terre and the Chinese Ministry of Geology and Mineralogical

Resources. Thorough reviews by Marc Hendrix, Kip Hodges, and Laura Webb were greatly appreciated.

APPENDIX 1. ANALYTICAL TECHNIQUES

U/Th dating of monazite by electron microprobe

General features. This geochronological technique is based on chemically dating monazite using an electron microprobe. Because monazite is very rich in U and Th, radiogenic lead accumulates quickly and reaches a level at which a precise measurement can be made with an electron probe within ~50 m.y. The age is calculated from the U, Th, and Pb concentrations, assuming that (1) nonradiogenic lead is negligible and (2) no partial loss of lead occurred since the last resetting of the system (concordant behavior). The first hypothesis is entirely reasonable, because the amount of nonradiogenic lead in monazite is very low (<1 ppm; Parrish, 1990) compared to radiogenic lead. The second hypothesis is applied with less confidence. Although many monazites are nearly concordant on a concordia diagram, some case studies document significant Pb loss from this mineral (Parrish, 1990; Suzuki and Adachi, 1991). Moreover, because electron microprobe dating is an in situ method with a very high spatial resolution (5 μm or better), discordance caused by mixing of grains with different ages or discordance created by a grain that contains several domains with different ages is not a problem. The geochronological procedure is now well established and several laboratories have published reliable ages using the method (Suzuki and Adachi, 1991, 1994a, 1994b; Montel et al., 1994, 1996; Rhede et al. 1996; Cocherie et al. 1998; Finger et al., 1998; Finger and Hemly, 1998; Braun et al., 1998).

A three-step procedure is used: (1) measurement of all of the U-Th-Pb concentrations, and their associated 95% confidence interval; (2) calculation of the ages for each measurement (with associated error); and (3) calculation of a global age (with associated error) from the age population, using a statistical procedure. The statistical procedure was described by Montel et al. (1996), but the analytical procedure has been adapted to our new microprobe and is described hereafter.

Analytical description. The electron probe is a SX-100 Cameca microprobe, with four spectrometers. The complete chemical composition of the monazite is determined, so matrix effect corrections are properly calculated for each measurement. The analytical conditions are 100 nA probe current and 15 kV accelerating voltage. This voltage is lower than the value of 25 kV used by other laboratories. We prefer the lower value not only because the domain size analyzed is <1.5 μm diameter at 15 kV (at 25 kV it is about 3.5 μm, calculated according to Castaing, 1960), but also because the matrix corrections are much smaller. For example, at 25 kV, the correction factor is ~1.0 for Pb with our standard, but is ~1.25 with 25 kV. The only advantage of working at 25 kV would be to have a higher signal on rare earth element (REE) peaks. However, monazite is so rich in REE that even at 15 kV the L lines are easily measured. We simultaneously use 3 pentaery thritol (PET) crystal and 1 lithium fluoride (LIF) crystal. The Mα line for Pb is counted for 300 s on a high-sensitivity PET crystal. During this time, another PET crystal counts for 75 s on the Th Mα line and 225 s on the U Mβ line. Another PET and a LIF are used for La (Lα, 30 s), Ce (Lα, 30 s), Pr (Lβ, 45 s), Nd (Lβ, 45 s), Sm (Lβ, 60 s), Gd (Lβ, 60 s), Y (Lα, 75 s), Si (Kα, 75 s), Ca (Kα, 60 s), and P (Kα, 60 s). The standards for REE and Y are synthetic REE phosphates made in the experimental petrology laboratory in Clermont-Ferrand (Samyn, 1996); for Si it is a synthetic zircon provided by D. Ohnensetter (Cesbron et al., 1993); Durange apatite is used for Ca; synthetic NdPO$_4$ is used for P; UO$_2$ is used for U; ThO$_2$ is used for Th; and a synthetic CaO-Al$_2$O$_3$-SiO$_2$ doped in Pb is used for Pb. With this procedure, typical errors on Pb, calculated according to Ancey et al. (1978), are close

to 80 ppm, and the detection limit is ~120 ppm. At the beginning of each session, the complete procedure is carried out on a fragment of high-quality monocrystal from a pegmatite, dated by conventional U/Pb method as 474 ± 1 Ma (Seydoux et al., 1999).

Statistics and presentation of the data. Because electron-probe dating is rapid but relative imprecise, a statistical procedure is necessary in order to (1) obtain a more accurate age by averaging the data, and (2) check, on a rigorous mathematical basis, for the existence of a unique or multiple age populations. The method selected here is based on the least-square concept, because we want to compare a geological model with the data. For an age population obtained from a single sample (typically a thin section), the simplest hypothesis is that all of the analyzed domains were reset or formed at time τ during an event that can be considered to be instantaneous. The least-square method provides a best estimate of τ, and a confidence interval ±$2\sigma_\tau$. We can estimate to what extent the data fit the model by calculating a numerical parameter, here the mean square of weighted deviations (MSWD). We must also perform a statistical test, here the χ^2 test at the 5% confidence level, to ascertain whether this agreement is acceptable. We stress that, as with most statistical tests, the χ^2 test yields a nonambiguous answer only when it is negative (high MSWD), when we must reject the model. A positive test resul (low MSWD) only means that we have no statistical reasons to reject the model.

When the single population age model is not acceptable we must propose another model. The next simplest hypothesis is that there are two geological events, at times $\tau1$ and $\tau2$, each one having generated a population of domains. The procedure described in Montel et al. (1996) indicates how least-square fitting is carried out in this case. If the two population ages model is not acceptable, the procedure can be extended to 3 or more populations. In addition to a statistical test on the global model expressed by a global MSWD, we also check the quality of the individual populations identified, using a χ^2 test.

This statistical procedure provides a firm basis for discussing the results of electron microprobe dating of a sample, but it must be combined with other data for a robust geological interpretation. Other important elements to discuss are the age distribution, as shown by a weighted histogram, the geometrical relationships between age domains (core-rim relationships), the petrographical positions of the crystal (interstitial, included in another mineral, in a fracture), the composition-age relationships, and the relative importance of the populations.

In order to make the results easy to follow for the reader, the data are presented in the following way: analytical results are presented in Table 1, with all of the calculated errors. Age distributions are presented as weighted histograms (Fig. 4A). In these diagrams, each individual age obtained from one microprobe measurement is represented by a bell-shaped curve, which is its probability density function defined by the age and the uncertainty. Also presented on this diagram is a curve representing the sum of all the individual curves. The average ages, calculated as explained here, are also represented by a bell-shaped curve. A second useful source of information is the relationship between the age and the chemical composition of the crystal (Fig. 4B). Monazite composition is highly variable, even within a single crystal. Therefore, each point displays variable U and Th contents and therefore variable Pb contents. If the geochronological system is simple, all the points have the same age (within analytical errors), and plot as a horizontal line on the Th-age and U-age diagrams. In this case we can be sure that the geochronological system remained closed (concordant), because a partial opening of the system (discordance) cannot preserve such a regular distribution.

Rb/Sr dating

Mineral separates for Rb/Sr isotope analyses were obtained by processing 1–2 kg samples through a crushing mill, a disk-mill, a Frantz magnetic separator, and heavy liquids. Feldspar, whole rock, and micas were dissolved in two steps, using >50% HF followed by 6N HCl, both at 150 °C. Rb-Sr analyses were performed by the isotope dilution method using an ^{85}RB-^{84}Sr mixed isotope tracer; isotope ratios were measured on a cameca TSN 206 mass spectrometer, equipped with a single Faraday collector and a secondary electron multiplier for Rb and Sr concentrations, and a double Faraday collector for Sr. The Rb decay constants used are those recommended by the International Union of Geological Sciences (Steiger and Jäger, 1977). For Sr standard NBS 987, an average value of $^{87}Sr/^{86}Sr = 0.71024 \pm 0.00003$ (2σ, n = 25) was obtained. Sr ratios were normalized to $^{86}Sr/^{88}Sr = 0.1194$. Total blanks for Rb and Sr are negligible. Regression lines were derived after York (1969) modified for correlated errors by Minster et al. (1979). Results are given in Table 2 and shown in Figure 6 (A and B).

$^{40}Ar/^{39}Ar$ dating

Because deformation events did not lead to widespread recrystallization, it has proven impossible to separate minor recrystallized micas around phenocrysts, and these phases were also too small for laser ablation. Thus $^{40}Ar/^{39}Ar$ dating has been carried out on populations of grains, and a precise analysis of the age spectrum has been done to identify the various mineral generations. Whole rocks were crushed, sieved, and individual grains chosen under a binocular microscope. All separates were irradiated at the Nuclear Ford reactor of the University of Michigan or in the Siloée nuclear facility of Commissariat à l'Energie Atomique. The J factor was estiamted by replicate analysis of the Fish Canyon sanidine standard with an age of 27.55 ± 0.08 Ma (Lanphere and Baadsgaard, 1997) or the Caplongue amphibole standard with an age of 344.5 Ma (Maluski and Schaeffer, 1982) with values of 0.0041 ± 0.0002 and 0.0068 ± 0.0005, respectively, with 1% relative standard deviation. Interfering nuclear reactions on K and Ca were calculated by coirradiation of pure salts with values of $^{40}Ar/^{39}Ar_k = 0.031$, $^{37}Ar/^{39}Ar_{Ca} = 0.000205$, and $^{36}Ar/^{39}Ar_{Ca} = 0.000781$ for Michigan Ford and $^{40}Ar/^{39}Ar_k = 0.037$, $^{37}Ar/^{39}Ar_{Ca} = 0.000225$, and $^{36}Ar/^{39}Ar_{Ca} = 0.000349$ for Siloée. Samples were loaded in aluminium packets into a Staudacher type double-vacuum furnace and step heated in a classical fashion, usually from 600 °C to 1400 °C. Feldspars were heated using a more evolved cycled protocol, following that suggested by Lovera et al. (1989). The gas was purified by means of cold traps with liquid air and Al-Zr getters. Once cleaned, the gas was introduced into a VG3600 mass spectrometer, and 2 min were allowed for equilibration before static analysis was done. Signals were measured using a Faraday cup with a resistor of 10^{11} ohm for ^{40}Ar and ^{39}Ar while ^{39}Ar, ^{38}Ar, ^{37}Ar, and ^{36}Ar were analyzed with a photomultiplier after interaction with a Daly plate. Blanks at 500 °C, 1000 °C, and 1200 °C were systematically measured for each mass between samples and extrapolated, then subtracted directly from measured signals for each temperature. Gain between collectors was estiamted by duplicate analysis of ^{39}Ar on both collectors during each analysis and by statistical analysis over on a period of several years. This gain has an average value of 95 and is known at better than 1.5%. This error is included in the age calculation, along with analytical errors on each signal and errors on the blank values. Detailed analytical results are given in Table 3 and spectra are presented in Figure 5.

Plateau ages given are weighted mean plateaus that include the error on the J factor. To be considered for the calculation of a plateau age, steps must be contiguous, all individual ages must agree at the 1σ level, and the amount of total ^{39}Ar contained in all used steps must be no less than 70%. Individual errors on age and error on plateau ages are given at 1σ. Isochron ages are calculated using an inverse isochron diagram plotting $^{36}Ar/^{40}Ar$ versus $^{39}Ar/^{40}Ar$ (Roddick et al., 1980) which allows for identification and correction of homogeneous excess components. Errors on age and intercept are given at 1σ and include individual errors on each data point and linear regression following York's method

(1969). The goodness of fit relative to individual errors is measured with the mean square of weighted deviations (MSWD). K/Ca and Cl/K plots are given for comparison with the age spectra, but are only qualitative because they represent $^{39}Ar/^{37}$ and $^{38}Ar/^{39}Ar$ used as proxies for the true elemental ratios.

REFERENCES CITED

Ames, L., Zhou, G., and Xiong, B., 1996, Geochronology and isotopic character of ultrahigh-pressure metamorphism with implications for collision of the Sino-Korean and Yangtze cratons, central China: Tectonics, v. 15, p. 472–489.

Ancey, M., Bastenaire F., and Tixier, R., 1978, Application des méthodes statistiques en microanalyse, in Maurice, F., et al., eds., Microanalyse, microscopie electronique à balayage: Orsay, France, Les Editions du Physicien, p. 323–347.

Arnaud, N.O., 1992, Apports de la thermochronologie $^{40}Ar/^{39}Ar$ sur feldspath potássique à la connaissance de la tectonique cénozoïque d'Asie. Etude des mécanismes d'accommodation de la collision continentale [Thèse de Doctorat] Clermont-Ferrand II, Université Blaise Pascal, 263 p.

Arnaud, N.O., and Kelley, S.P., 1997, Argon behaviour in gem-quality orthoclase from Madagascar: Experiments and some consequences for $^{40}Ar/^{39}Ar$ geochronology: Geochimica et Cosmochimica Acta, v. 61, p. 3227–3255.

Arnaud, N.O., and Sobel, E.R., 1998, Mesozoic-Cenozoic cooling history of the Altyn Tagh Range, NW China: Eos (Transactions, American Geophysical Union), v. 79, p. F757.

Arnaud, N.O., Brunel, M., Cantagrel, J.M., and Tapponnier, P., 1993, High cooling and denudation rates at Kongur-Shan, Eastern Pamir (Xinkiang, China) revealed by $^{40}Ar/^{39}Ar$ alkali feldspar thermochronology: Tectonics, v. 12, p. 1335–1346.

Arnaud, N., Brunel, M., Roger, F., Schärer, Malavieille, J., and Tapponnier, P., 1997, Age, duration and thermal conditions associated with major strike-slip faults: Examples from northern Tibet: Terra Abstracts, v. 9, p. 486.

Avouac, J.P., and Tapponnier, P., 1993, Kinematic model of active deformation in central Asia: Geophysical Research Letters., v. 20, p. 895–898.

Braun, I., Montel, J.M., and Nicollet, C., 1998, Electron microprobe dating of monazites from high-grade gneisses and pegmatites from the Kerala Kondalite Belt, southern India: Chemical Geology, v. 146, p. 65–85.

Brunel, M., and Geyssant, J., 1978, Mise en évidence d'une déformation rotationnelle Est-Ouest par l'orientation optique du quartz dans la fenêtre des Tauern (Alpes Orientales); implications géodynamiques: Revue de Géographie Physique et de Géologie Dynamique, v. 2, p. 335–346.

Castaing, R., 1960, Advances in electronics and electron physics, New York, Academic Press, 240 p.

Cesbron, F., Ohnenstetter, D., Blanc, P., Rouer, O., and Sicher, M.C., 1993, Incorporation de terres rares dans les zircons de synthèse: étude par cathodoluminescence: Paris, Académie des Sciences Comptes Rendus, v. 316, p. 1231–1238.

Chavagnac, V., and Jahn, B.-M., 1996, Coesite-bearing eclogites from the Bixiling Complex, Dabie Mountains, China: Sm-Nd ages, geochemical characteristics and tectonic implications: Chemical Geology, v. 133, p. 29–51.

Chen, Z., ed., 1985, Geological map of Xinjiang Uygur Autonomous Region, China: Beijing, Chinese Bureau of Geology, scale 1:2 000 000.

Chinese Bureau of Geology, 1981, Stratigraphic tables for Xinjiang Autonomous Region: Beijing, Chinese Bureau of Geology, 496 p.

Cliff, R.A., 1985, Isotopic dating in metamorphic belts: Geological Society of London Journal, v. 142, p. 92–110.

Cocherie, A., Legendre, O., Peucat, J.J., and Kouamelan, A.N., 1998, Geochronology of polygenetic monazites constrained by in-situ electron microprobe Th-U-total lead determination: Implicatins for lead behaviour in monazite: Geochimica et Cosmochimica Acta, v. 62, p. 2475–2497.

Copeland, P.C., Harrison, T.M.,. Kidd, W.S.F., Ronghua, X., and Yuquan, Z., 1987, Rapid early Miocene acceleration of uplift in the Gangdese belt, Xizang (southern Tibet), and its bearing on accommodation mechanism of the India-Asia collision: Earth and Planetary Science Letters, v. 86, p. 240–252.

Coward, M.P., Kidd, W.S.F., Pan, Y., Shackleton, R.M., and Zhang, H., 1988, The structure of the 1985 Tibet Geotraverse, Lhasa to Golmud: Royal Society of London Philosophical Transactions, ser. A, v. 327, p. 307–336.

Eide, E.A., McWilliams, A.O., and Liou, J.G., 1994, Ar/Ar geochronologic constraints on the exhumation of HP-UHP metamorphic rocks in eastern China: Geology, v. 22, p. 601–604.

Enkin, R.J., Yang, Z.Y., Chen, Y., and Courtillot, V., 1992, Paleomagnetic constrains on the geodynamic history of China from Permian to present: Journal of Geophysical Research, v. 97, p. 13953–13989.

Finger, F., and Helmy, H.M., 1998, Composition and total-Pb model ages of monazite from high-grade paragneisses in the Abu Swayel area, southern Eastern Desert, Egypt: Contributions to Mineralogy and Petrology, v. 62, p. 269–289.

Finger, F., Broska, I., Roberts, M.P., and Schermaier, A., 1998, Replacement of primary monazite by apatite-allanite-epidote coronas in amphibolite facies granite gneiss from the eastern Alps: American Mineralogist, v. 83, p. 248–258.

Gilder, S., and Courtillot, V., 1997, Timing of the north-south China collision from new middle to late Mesozoic paleomagnetic data from the north China block: Journal of Geophysical Research, v. 102, 17713–17727.

Hacker, B.R., Ratschbacher, L., Webb, L., Ireland, T., Walker, D., and Shuwen, D., 1998, U/Pb zircon ages constrain the architecture of UHP Qinling-Dabie orogen, China: Earth and Planetary Science Letters, v. 161, p. 215–230.

Halim, N., Kravchinsky, V., Gilder, S., Cogné, J.P., Alexyutin, M., Sorokin, A., Courtillot, V., and Chen, Y., 1998, A palaeomagnetic study of the Mongol-Okhotsk region: Rotated Early Cretaceous volcanics and remagnetized Mesozoic sediments: Earth and Planetary Science Letters, v. 159, p. 133–145.

Hames, W.E., and Bowring, S.A., 1994, An empirical evaluation of the argon diffusion geometry in muscovite: Earth and Planetary Science Letters, v. 124, p. 161–169.

Harris, N.B.W., Ronhua, X., Lewis, C.L., Hawkesworth, C.J., and Yuquan, Z., 1988, Isotope geochemistry of the 1985 Tibet geotraverse, Lhassa to Golmud, in the geological evolution of Tibet: Royal Society of London Philosophical Transactions, ser. A, v. 327, p. 263–286.

Harrison, T.M., Armstrong, R.L., and Clarke, G.K.C., 1978, Thermal models and cooling histories from fission-track, K-Ar, Rb-Sr, and U-Pb mineral dates, northern Coast Plutonic Complex, British Columbia, in Zartman, R.E., ed., Short papers of the fourth international conference: Geochronology, cosmochronology, and isotope geology: Reston, Virginia, U.S. Geological Survey, Report OF78-0701, p. 167–170.

Harrison, T.M., Duncan, I., and Dougall, I., 1985, Diffusion of ^{40}Ar in biotite: Temperature, pressure and compositional effects: Geochimica et Cosmochimica Acta, v. 49, p. 2461–2468.

Hendrix, M.S., Graham, S.A., Carroll, A.R., Sobel, E.R., McKnight, C.L., Schulein, B.J., and Wang, Z., 1992, Sedimentary record and climatic implications of recurrent deformation in the Tian Shan: Evidence from Mesozoic strata of the north Tarim, south Junggar, and Turpan basins, northwest China: Geological Society of America Bulletin, v. 104, p. 53–79.

Jiao, S.P., Zhang, Y.F., Yi, S.X., Ai, C.X., Zhao, Y.N., Wang, H.D., Xu, J.E., Hu, J.Q., and Guo, T.Y., 1988, Geological map of Qinghai-Xizang (Tibet) plateau and adjacent areas: Beijing, Geological Publishing House, scale 1:1 500 000.

Khramov, A.N., 1958, Palaeomagnetic correlation of sediment formations: Leningrad, Godtechizdat, 218 p.

Lamb, M.A., Hanson, A.D., Graham, S.A., Badarch, G., and Webb, L.E., 1999, Left-lateral sense offset of Upper Proterozoic to Paleozoic features across the Gobi Onon, Tost, and Zuunbayan faults in southern Mongolia and implications for other central Asian faults: Earth and Planetary Science Letters, v. 173, p. 183–194.

Lanphere, M.A., and Baadsgaard, H., 1997, The Fish Canyon Tuff: A standard for geochronology: American Geophysical Union Abstract with Program, v. 78, 17, p. S326.

Li, D.F., Lun, Z.Q., Guo, J.Y., and Zhu, Y.Y., eds., 1991, Regional geology of Qinghai Province, Geological Memoires Series 1, Volume 24: Beijing, Geological Publishing House, 662 p.

Lovera, O.M., Richter, F.M., and Harrison, T.M., 1989, The ^{40}Ar/^{39}Ar thermochronometry for slowly cooled samples having a distribution of diffusion domain sizes: Journal of Geophysical Research, v. 94, p. 17917–17935.

Lovera, O.M., Richter, F.M., and Harrison, T.M., 1991, Diffusion domains determined by ^{39}Ar released during step heating: Journal of Geophysical Research, v. 96, p. 2057–2069.

Maluski, H., and Schaeffer, O.A., 1982, ^{39}Ar-^{40}Ar laser probe dating of terrestrial rocks: Earth and Planetary Science Letters, v. 59, p. 21–27.

Matte, P., Tapponnier, P., Arnaud, N., Bourjot, L., Avouac, J.P., Vidal, P., Qing, L., Yusheng, P., and Wang, Y., 1996, Tectonics of western Tibet, between the Tarim and the Indus: Earth and Planetary Science Letters, v. 142, p. 311–330.

Meng, Q.R., and Zhang, G.W., 1999, Timing of collision of the north and south China blocks: Controversy and reconciliation: Geology, v. 27, p. 123–126.

Meyer, B., Tapponnier, P., Gaudemer, Y., Peltzer, G., Guo, S., and Chen, Z., 1996, Rate of left-lateral movement along the easternmost segment of the Altyn Tagh fault, east of 96°E (China): Geophysical Journal International, v. 124, p. 29–44.

Meyer, B., Tapponnier, P., Bourjot, L., Métivier, F., Gaudemer, Y., Peltzer, G., Guo, S., and Chen, Z., 1998, Crustal thickening in Gansu-Qinghai, lithospheric mantle subduction, and oblique, strike-slip controlled growth of the Tibet Plateau: Geophysical Journal International, v. 135, p. 1–47.

Minster, J.F., Ricard, L.P., and Allegre, C.J., 1979, ^{87}Rb-^{87}Sr chronology of enstatite meteorites: Earth and Planetary Science Letters, v. 44, p. 420–440.

Montel, J.M., Veschambre, M., and Nicollet, C., 1994, Datation de la monazite à la microsonde électronique: Paris Académie des Sciences Comptes Rendus, v. 318, p. 1489–1495.

Montel, J.M., Foret, S., Veschambre, M., Nicollet, C., and Provost, A., 1996, Electron microprobe dating of monazite: Chemical Geology, v. 131, p. 37–53.

Murphy, M.A., Yin, A., Harrison, T.M., Duerr, S.B., Chen, Z., Ryerson, F.J., Kidd, W.S.F., Wang, X., and Zhou, X., 1997, Did the Indo-Asian collision alone create the Tibetan Plateau?: Geology, v. 25, p. 719–722.

Parrish, R.R., 1990, U-Pb dating of monazite and its application to geological problems: Canadian Journal of Earth Sciences, v. 27, p. 1431–1450.

Peltzer, G., and Tapponnier, P., 1988, Formation and evolution of strike-slip faults, rifts, and basins during India-Asia collision: An experimental approach: Journal of Geophysical Research, v. 93, p. 15085–15172.

Peltzer, G., Tapponnier, P., Gaudemer, Y., Meyer, B., Guo, S., Yin, K., Chen, C., and Dai, H., 1988, Offsets of late Quaternary morphology, rate of slip and recurrence of large earthquakes on the Chang Ma fault: Journal of Geophysical Research, v. 93, p. 7793–7812.

Purdy, J.W., and Jäger, E., 1976, K-Ar ages on rock forming minerals from the Central Alps: Padova Universita Istituto di Geologia e Mineralogia, Mémoire 30, 31 p.

Rhede, D., Wendt, I., and Forster, H-J., 1996, A three-dimensional method for calculating independent chemical U/Pb- and Th/Pb-ages of accessory minerals: Chemical Geology, v. 130, p. 247–253.

Ritts, B.D., and Biffi, U., 2000, Magnitude of post-Middle Jurassic (Bajocian) displacement on the Altyn Tagh fault, northwest China: Geological Society of America Bulletin.

Roddick, J.C., Cliff, R.A., and Rex, D.C., 1980, The evolution of excess argon in alpine biotites: Earth and Planetary Science Letters, v. 48, p. 185–208.

Ruffet, G., Féraud, G., and Amouric, M., 1991, Comparison of $^{40}Ar/^{39}Ar$ conventional and laser dating of biotites from the North Trégor batholith: Geochimica et Cosmochimica Acta, v. 55, p. 1675–1688.

Samyn, F., 1996, Elaboration de cristaux synthétiques de structures monazites et xénotime par la méthode des flux: Clermont-Ferrand, Université Blaise Pascal, Mémoire de Maîtrise, 45 p.

Seydoux, A.M., Montel, J.M., Paquette, J.L., and Marinho, M., 1999, Experimental resetting of the U-Pb geochronological system of monazite: Journal of Conference Abstracts, v. 4, issue 1, p. 800.

Sobel, E., and Arnaud, N., 1999, A possible lower Paleozoic suture in Eastern Kunlun, Altyn Tagh range, China: Tectonics, v. 18, p. 64–74.

Steiger, R.H., and Jäger, E., 1977, Subcommission on geochronology: Convention on the use of decay constants in geo- and cosmochronology: Earth and Planetary Science Letters, v. 36, p. 359–362.

Suzuki, K., and Adachi, M., 1991, Precambrian provenance and Silurian metamorphism of the Tsunosawa paragneiss in the South Kitakami terrane, northeast Japan, revealed by the chemical Th-U-total Pb isochron ages of monazite, zircon and xenotime: Geochemical Journal, v. 25, p. 357–376.

Suzuki, K., and Adachi, M., 1994a, Precambrian detrital monazites and zircon from Jurassic sandstones in the Japanese islands, revealed by "CHIME" geochronology: 8th International Conference on Geochronology, Cosmochronology and Isotope Geology: U.S. Geological Survey Circular 1107, 311 p.

Suzuki, K., and Adachi, M., 1994b, Middle Precambrian detrital monazite and zircon from the Hida gneiss on Oki-Dogo Island, Japan: Their origin and implications for the correlation of basement gneiss of southwest Japan and Korea: Tectonophysics, v. 235, p. 277–292.

Tapponnier, P., and Molnar, P., 1977, Active faulting and tectonics of China: Journal of Geophysical Research, v. 82, no. 20, p. 2905–2930.

Tapponnier, P., Peltzer, G., and Armijo, R., 1986, On the mechanics of the collision between India and Asia, in Coward, M.P., and Ries, A.C., eds., Collision tectonics: Geological Society [London] Special Publication 19, p. 115–157.

Van der Woerd, J., Ryerson, F.J., Tapponnier, P., Gaudemer, Y., Finkel, R., Mériaux, A.S., Caffee, M., Guoguang, Z., and Qunlu, H., 1998, Holocene left-slip rate determined by cosmogenic surface dating on the Xidatan segment of the Kunlun fault (Qinghai, China): Geology, v. 26, p. 695–698.

Vincent, S.J., and Allen, M.B., 1999, Evolution of the Minle and Chaoshui basins, China: Implications for Mesozoic strike-slip basin formation in Central Asia: Geological Society of America Bulletin, v. 111, p. 725–742.

Wang, G.P., Wu, G.T., Lun, Z.Q., and Zhu, Y.Y., editors, 1993, Regional geology of Xinjiang Uygur Autonomous Region: Bureau of Geology and Mineral Resources of Xinjiang Uygur Autonomous Region, Geological Memoirs Series 1, Volume 32: Beijing, Geological Publishing House, 841 p.

White, S.H., 1976, The effect of strain on the microstructures, fabrics, and deformation mechanisms in quartzites: Royal Society of London Philosophical Transactions, ser. A, v. 283, p. 69–86.

Wittlinger, G., Tapponnier, P., Poupinet, G., Mei, J., Danian, S., Herquel, G., and Masson, F., 1998, Tomographic evidence for localized lithospheric shear along the Altyn Tagh fault: Science, v. 282, p. 74–76.

Xu, R., Schärer, U., and Allègre, C.J., 1985, Magmatism and metamorphism in the Lhassa block (Tibet): A geochronological study: Journal of Geology, v. 93, p. 41–57.

Xu, R., Yuquan, Z., Yingwen, X., Vidal, P., Arnaud, N., Qiaoda, Z., and Dunmin, Z., 1996, Isotopic geochemistry of plutonic rocks, in Yusheng, P., ed., Geological evolution of the Karakoram and Kunlun mountains: Beijing, Seismologic Press, p. 137–186.

Xue, F., Rowley, D.B., Tucker, R.D., and Peng, Z.X., 1997, U-Pb zircon ages of granitoid rocks in the North Dabie Complex, Eastern Dabie Shan, China: Journal of Geology, v. 105, p. 744–753.

Yin, A., Murphy, M.A, Harrison, T.M., Durr, S.B., Chen Wang, Z., Zhou, X., Ryerson, F.J., and Kidd, W.S.F., 1995, Significant crustal shortening in the Lhasa Block (southern Tibet) predates the Indo-Asian collision: Geological Society of America Abstracts with Programs, v. 27, no. 6, p. A-335.

York, D., 1969, Least square fitting of a straight line with correlated errors: Earth and Planetary Science Letters, v. 5, p. 320–324.

Yuquan, Z., Yingwen, X., Ronghua, X., Vidal, P., and Arnaud, N., 1996, Geochemistry of granitoids rocks, in Yusheng, P., ed., Geological evolution of the Karakoram and Kunlun mountains: Beijing, Seismological Press, p. 94–123.

Zheng, Y., Zhan, Q., Wang, Y., Liu, R., Wang, S.G., Zuo, G., Wang, S.Z., Lkaasuren, B., Badarch, G., and Badamgarav, Z., 1996, Great Jurassic thrust sheets in Beishan (North Mountains) - Gobi areas of China and southern Mongolia: Journal of Structural Geology, v. 18, p. 1111–1126.

MANUSCRIPT ACCEPTED BY THE SOCIETY JUNE 5, 2000

Geological Society of America
Memoir 194
2001

Mesozoic northeast Qaidam basin: Response to contractional reactivation of the Qilian Shan, and implications for the extent of Mesozoic intracontinental deformation in central Asia

Bradley D. Ritts*
Department of Geology, Utah State University, 4505 Old Main Hill, Logan, Utah 84322-4505, USA
Ulderico Biffi*
Agip, LABO-STIG, Laboratori Eniricerche, Via Maritano 26, Bolgiano, 20097 San Donato Milanese, Milan, Italy

ABSTRACT

Cenozoic intracontinental tectonics of central Asia are dominated by reactivation of older structures and initiation of new structures in response to the far-field effects of collision between India and Asia. Similarly, the Mesozoic was a time of tectonic amalgamation along the southern margin of Asia, and geologic evidence suggests corresponding intracontinental deformation in localities as diverse as the Tarim basin and southern Mongolia. However, the extent to which Mesozoic intracontinental deformation was comparable in style and distribution to that of the Cenozoic is not well understood.

Newly characterized Mesozoic rocks in the northeast Qaidam basin provide evidence for rejuvenation of the Qilian Shan in response to accretionary tectonics on the southern margin of Asia. Jurassic and Cretaceous strata were deposited in an entirely nonmarine intracontinental foreland basin system that developed southwest of, and was subsequently involved in, the south Qilian Shan fold and thrust belt. Paleocurrent indicator measurements and provenance analysis indicate derivation of siliciclastic detritus from the Qilian Shan. Facies analysis, geohistory analysis, and structural relationships provide evidence that Jurassic and Cretaceous rocks currently exposed along the northeast margin of Qaidam were deposited in proximal foreland basin and thrust wedge-top (piggyback basin) positions.

This evidence extends an already documented series of linked Mesozoic foreland basins from the China-Kazakstan border into central China along the Tian Shan, Bei Shan, and Qilian Shan, and suggests that Mesozoic intracontinental deformation may have more closely resembled current structural patterns than previously recognized. Sedimentary evidence for a Cretaceous paleorain shadow south of the Qilian Shan, with humid lacustrine facies north of the range, and semiarid fluvial and overbank facies south of the range, provides evidence for sustained uplift of the mountain belt into the late Mesozoic.

INTRODUCTION

Western China has been a locus of intracontinental tectonism since at least Late Triassic time (Hendrix et al., 1992, 1996;

Sobel, 1999; Hendrix, 2000; Vincent and Allen, this volume). Today, intracontinental deformation in response to the Cenozoic collision between India and Asia is inferred to be responsible for seismicity from the Himalaya to Lake Baikal, and for producing lofty mountain ranges such as the 6000–7000 m Tian Shan and Kunlun Shan (Tapponnier and Molnar, 1979). Similarly, the

*E-mails: Ritts, ritts@cc.usu.edu; Biffi, ulderico.biffi@agip.it.

Ritts, B.D., and Biffi, U., 2001, Mesozoic northeast Qaidam basin: Response to contractional reactivation of the Qilian Shan, and implications for the extent of Mesozoic intracontinental deformation in central Asia, *in* Hendrix, M.S., and Davis, G.A., eds., Paleozoic and Mesozoic tectonic evolution of central Asia: From continental assembly to intracontinental deformation: Boulder, Colorado, Geological Society of America Memoir 194, p. 293–316.

Mesozoic tectonics of central Asia were dominated by terrane accretion and continental assembly along the southern margin of Asia. Specifically, this includes the Triassic amalgamation of the North China Block and South China Block, and the Triassic-Jurassic assembly of Tibet (Fig. 1) (Girardeau et al., 1984; Coward et al., 1986; Smith, 1988; Smith and Xu, 1988; Xia, 1990; Enkin et al., 1992; Yin and Nie, 1993). The far-field effects of the Mesozoic collisional tectonics on central Asia have been documented by previous and ongoing studies (Hendrix et al., 1992, 1996, this volume; Vincent and Allen, this volume). Notably, a regionally extensive foreland basin system has been recognized extending from the western border of China to southern Mongolia, associated with the contractional tectonics in the Tian Shan and Bei Shan of northwest China (Hendrix et al., 1992, 1996, this volume; Zheng et al., 1996). Even farther removed, Jurassic strike-slip faulting has been documented in Kazakstan along the Talas-Ferghana fault (Burtman, 1980; Tseysler et al., 1982), and has also been hypothesized to be the result of intracontinental deformation associated with tectonism on the southern edge of Asia (Sobel, 1999). However, the extent to which intracontinental deformation dominated central Asia during the Mesozoic, either by reactivation of older Paleozoic structures or far-field effects of Mesozoic collisional tectonics, is not fully understood.

The Qaidam basin, located on the northeastern margin of the Tibetan Plateau, is currently bounded on all sides by mountain belts: the Kunlun Shan on the south and west, the Qilian Shan on the northeast, and the Altun Shan on the north (Fig. 2). The geodynamic setting of Qaidam during the Mesozoic can be reasonably inferred from regional tectonic syntheses (Girardeau et al., 1984; Smith, 1988; Smith and Xu, 1988; Xia, 1990; Enkin et al., 1992). The Kunlun Shan is thought to have been an active, south-facing Andean-style orogen during the late Paleozoic and early Mesozoic because of its position on the southern margin of Asia, facing the allochthonous Qiantang and Lhasa blocks, which amalgamated to form Tibet during the Early-Late Jurassic, and because of the occurrence of late Paleozoic calc-alkaline plutonic and Triassic volcanic rocks along the length of the range (Bureau of Geology and Mineral Resources of Qinghai Province [BGMRQ], 1991; Bureau of Geology and Mineral Resources of Xinjiang Uygur Autonomous Region [BGMRX], 1993). This places Qaidam and southern Tarim in a retroarc foreland position (cf. Graham et al., 1993; Jordan, 1995; Ritts, 1995, 1998; Ritts and Biff; 2000) at least until the Middle Jurassic, and then in a more distal foreland position as the active margin stepped out to the Bangong suture between the Lhasa and Qiantang blocks (Fig. 1). The Qilian Shan is known to be a middle Paleozoic suture between North China and the basement of the Qilian Mountains and Qaidam (Zhang et al., 1984; Zhang, 1985); however, its Mesozoic expression, if any, has not been documented previously.

Owing to the remoteness of Qaidam and inaccessibility to non-Chinese citizens, previous work on the Qaidam basin is extremely limited in the Chinese and western literature. Western publications are limited to low-resolution syntheses of Chinese data (Ulmishek, 1984; Lee, 1984; Bally et al., 1986; Hsü, 1988; Gu and Di, 1989; Wang and Coward, 1990), and a few original studies of Qaidam petroleum geology (Huang et al., 1991, 1994; Ritts et al., 1999). Chinese work has focused primarily on regional geologic mapping (BGMRQ 1991), paleontology (He, 1984; Kang, 1984, Anonymous, 1990; BGMRQ, 1991), and, to a lesser extent, Cenozoic structural, sedimentary, and petroleum geology (Wang, 1981; Di, 1983; Wang and Chen, 1984, Anonymous, 1990; BGMRQ, 1991; Chinese State Bureau of Seismology, Aergin Active Fault Belt Research Group [CSBS], 1992). This lack of data has resulted in poor understanding of the geology and tectonics of Qaidam, even by frontier standards. Qaidam plays a central role in many models of the Mesozoic tectonics of China, so this lack of data results in an unfortunate limitation on the accuracy of these models (e.g., Yin and Nie, 1996).

This study represents a comprehensive basin analysis of Mesozoic strata in the Qaidam basin. As such, this is the first data set that includes detailed sedimentology and facies descriptions, paleocurrent and provenance data, and relatively well defined subsidence analysis. This data set allows the Mesozoic geology of Qaidam to be used with a reasonable level of confidence to address regional problems in geology and tectonics, and issues related to styles and processes of intracontinental deformation and tectonic inheritance.

STRATIGRAPHY

Jurassic and Cretaceous sedimentary rocks crop out discontinuously or are penetrated by wells along the Northeast Qaidam basin margin from Lenghu to Dameigou (Figs. 2 and 3). Minor occurrences of Triassic rocks also are mapped along the basin margin (BGMRQ, 1991; Fig. 3); however, the age of these strata is not well defined, and one outcrop that we checked in the field consisted of metasedimentary rocks.

Jurassic and Cretaceous strata are entirely nonmarine, nonvolcanogenic sedimentary rocks. Facies are laterally discontinuous, except for some coal and open-lacustrine facies shale, making lithostratigraphic correlations difficult. The division between Lower-Middle Jurassic and Upper Jurassic–Cretaceous rocks is easily determined by the presence and absence of coal and organic-rich shale, respectively (see following biostratigraphy discussion; Anonymous, 1990). The Northeast Qaidam basin fill, like previously studied areas of northwest China (such as the northern Tarim, Turpan, and Junggar basins; Graham et al, 1990; Hendrix et al., 1992), typically consists of green and gray fluvial and alluvial strata with coal, coaly mudstone, and well-developed organic-rich open lacustrine shale in the Lower and Middle Jurassic part of the section. Upper Jurassic and Cretaceous strata lack organic-rich rocks and are dominated by sandy and conglomeratic red beds. Fine-grained strata, where they occur in the Upper Jurassic and Cretaceous, are red fluvial overbank or marginal lacustrine strata, often with well-developed calcic paleosols.

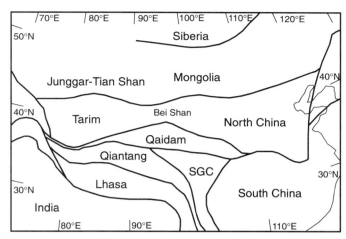

Figure 1. Terrane map of Asia showing key tectonic elements. Terranes and composite terranes in large font. SGC signifies Songpan-Ganzi remnant ocean basin. Mongol-Okhotsk suture (not labeled) separates Siberia from Mongolia, and Bangong suture separates Lhasa block from Qiantang block. Position of Bei Shan orogenic belt also is shown in smaller font. Lhasa and Qiantang blocks underpin most of what is today Tibetan Plateau (after Yin and Nie, 1996).

The ages of strata as mapped by Chinese workers (BGMRQ, 1991) are supported by existing paleontologic studies (Table 1; Anonymous, 1990; BGMRQ, 1991). The ages of all sections in this paper were confirmed and refined by our own palynology and lithostratigraphic correlation (discussed in the following).

The Lower Jurassic Xiaomeigou Formation (Table 1) is exposed at only one locality in northeast Qaidam (Dameigou).

TABLE 1. PREVIOUS CHINESE BIOSTRATIGRAPHIC CONSTRAINTS

Formation	Age	Evidence
Xiaomeigou	Early Jurassic	**plant fossils**
		Cladophlebis tsaidamensis
		C. ingens
		Equisetites colunmaus
		Clathropteris pekingensis
		Ciliatopteris pectinata
Dameigou	Middle Jurassic	**plant fossils**
		Coniopteris hymenophylloides
		C. tatungensis
		C. burejensis
		C. szeiana
		C. simplex
		C. spectabills
Hongshigou	Late Jurassic	**ostracodes**
		Darwinula sarytirmenensis
		Sjungarica
		Cetella
		Damenella
Quyagou	Cretaceous	**palynoflora**
		Lygodiumsporites
		Classopolis
		conchostrachans
		Estherites
		Orthestheriopsis loxoquadrata
		Yanjee estheria
		ostracodes
		Pinnocypridea
		Rhinocypris cirrata
		Cypridea unicostata
		C. vitimensis
		Clinocyris scolia
		Theriopsis

Reference: Anonymous (1990).

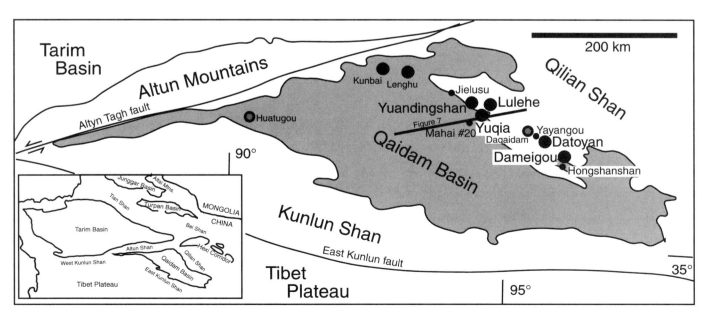

Figure 2. Location map of Qaidam region. Key localities studied are shown by black circles. Smaller circles indicate localities for which data were derived from other sources. Towns within Qaidam are indicated by gray circles. Approximate location of cross section in Figure 7 is indicated by black line.

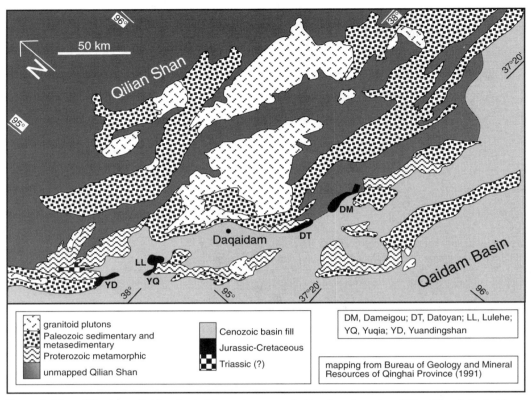

Figure 3. Map of Northeast Qaidam basin Mesozoic outcrop belt and southwestern Qilian Shan (adapted from Bureau of Geology and Mineral Resources of Qinghai Province, 1991).

Drilling has not been deep enough elsewhere on the margin of the basin to penetrate the base of the Middle Jurassic section to test continuity of the Xiaomeigou Formation, except at Lenghu, where Xiaomeigou strata are apparently penetrated (Anonymous, 1990). Where exposed, the Lower Jurassic is about 100 m thick and overlain with angular unconformity by Middle Jurassic rocks of the Dameigou Formation (Figs. 4 and 5).

Where exposure permits, lithostratigraphic correlations can be made within the Dameigou Formation on the basis of a thick, regionally extensive coal in the upper part of the Middle Jurassic, and possibly other organic-rich horizons (Fig. 6). These lithostratigraphic correlations are also supported by plant fossils (Wang, 1981; BGMRQ, 1991; Qinghai Petroleum Administration, proprietary data; this study; Table 1). The Qinghai Petroleum Administration (Ma, 1997, personal commun.; proprietary data, Qinghai Petroleum Administration, 1997; Anonymous, 1990) reported that Upper Jurassic rocks conformably overlie the Dameigou Formation. However, our data indicate that organic-rich deposition of the Dameigou Formation ceased in the early Bathonian. Therefore, a significant unconformity (representing at least middle-late Bathonian and Callovian) occurs between the top of Middle Jurassic organic-rich strata and overlying rocks in northeast Qaidam.

The Upper Jurassic Hongshigou and Cretaceous Quyagou Formations unconformably overlie the Middle JurassicDameigou

Formation. They are poorly exposed in the northeast Qaidam basin, but are penetrated by wells at Lenghu and Mahai (Fig. 6).

The distribution of Mesozoic rocks in the subsurface of Qaidam is unknown. Layered reflectors, inferred to be sedimentary rocks, are imaged between the base of the Cenozoic and top of acoustic basement on seismic reflection profiles (Fig. 7; Bally et al., 1986; Anonymous, 1990; Qinghai Petroleum Administration, 1997, personal commun.). However, thick Cenozoic cover (>5–7 km in northern Qaidam; Bally et al., 1986) has never been completely penetrated by drilling, so the age and character of underlying rocks are unknown. Three observations and inferences suggest that these sedimentary rocks of unknown, but pre-Cenozoic age, are Mesozoic rather than Paleozoic. First, Mesozoic sedimentary rocks crop out discontinuously and are penetrated by boreholes around the margin of northern Qaidam, whereas Paleozoic rocks are sparse (especially in the north and northwest) and are metamorphosed where exposed in the northern margins of the basin. Second, the western Qaidam basin is inferred to have been in a retroarc foreland basin position relative to the Kunlun Shan, and the northeast Qaidam basin was in an intracontinental foreland basin position relative to the Qilian Shan during the Mesozoic, suggesting that accommodation space was being created across much of Qaidam during the Mesozoic. The Paleozoic setting of Qaidam is undocumented, but thought to be dominated by accretion and amalgamation of the Qaidam region (Zhang et al., 1984; Zhang, 1985), which would be inconsistent with

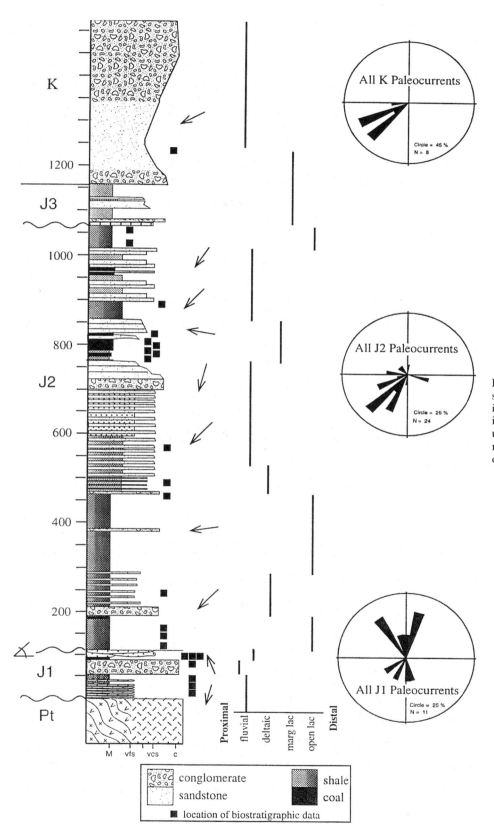

Figure 4. Measured section of Mesozoic strata from Dameigou locality. Arrows indicate mean paleocurrent vector within individual intervals. Bars to right of figure indicate depositional environment: marg lac, marginal lacustrine; open lac, open lacustrine.

All K Paleocurrents
Circle = 45 %
N = 8

All J2 Paleocurrents
Circle = 25 %
N = 24

All J1 Paleocurrents
Circle = 25 %
N = 11

conglomerate
sandstone
shale
coal
location of biostratigraphic data

Proximal Distal
fluvial deltaic marg lac open lac

Figure 5. Lower-Middle Jurassic angular unconformity at Dameigou. Unconformity (u/c) separates folded Pliensbachian-Toarcian sandstone and coal from gently dipping Aalenian shale. Labeled ridge has ~150 m of Middle Jurassic visible.

extensive basin formation. Third, Mesozoic rocks, where observed on the northeast Qaidam margin, do not appear to thin toward the basin center, suggesting that these strata are present in the subsurface of Qaidam.

Biostratigraphy

Palynological studies of the Dameigou and Yuandingshan localities (Figs. 3 and 4) yielded rich, age-diagnostic assemblages useful in refining the mapped ages (BGMRQ, 1991) for each section (Ritts, 1998; Ritts and Biffi, 2000). These age data

are used throughout this study for biostratigraphic correlations, geohistory modeling, and tectonic interpretations.

Identification of palynomorphs and age interpretation of individual samples were made by one of us (Biffi), on samples with geological context provided by the other (Ritts). These new data permit chronostratigraphic resolution beyond that previously available for the sections studied. Age-diagnostic assemblages were recovered from the Lower Jurassic Xiaomeigou section at the Dameigou locality, the Middle Jurassic sections at Dameigou and Yuandingshan, and the Lower Cretaceous section at Dameigou. These assemblages were interpreted in the context of previous biostrati-

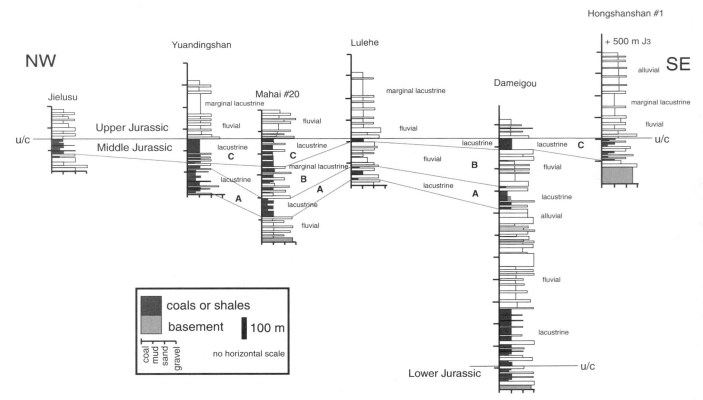

Figure 6. Correlation diagram of Jurassic strata along northeast Qaidam basin. Horizontal datum is Middle Jurassic–Upper Jurassic contact (sections for Jielusu, Lenghu, Mahai, Yayangou, and Hongshanshan localities adapted from Qinghai Petroleum Administration, unpublished data). Intervals A, B, and C are discussed in detail in text. u/c is unconformity.

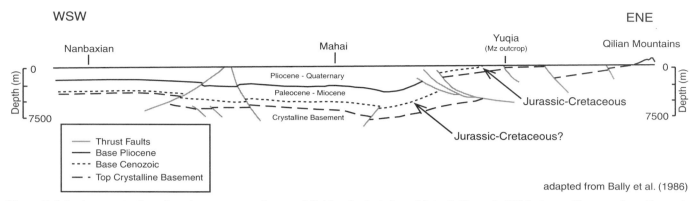

Figure 7. Seismic cross section of northeastern to north-central Qaidam basin (adapted from Bally et al., 1986). Interval between base Cenozoic and top basement has not been penetrated, but is inferred to be Mesozoic sedimentary rocks. Location of line in Figure 2 is approximate. Cross section is ~200 km long.

graphic studies in central Asia and northwest China (Vakhrameyev, 1982; Sun, 1989) and more complete Lower and Middle Jurassic sections in northeastern Qaidam and parts of Tarim (Hendrix, 1992; Sobel, 1995; Ritts, 1998), as well as proprietary data from the Tarim basin (Agip, proprietary database, 1998).

The Lower Jurassic Xiaomeigou section (Fig. 4) at the Dameigou locality is interpreted to be Sinemurian–early Pliensbachian to late Pliensbachian–early Toarcian based on the subsequent occurrence of *Chasmatosporites* spp., *Cerebropollenites thieghartii, Cerebropollenites mesozoicus* (=*macroverrucosus*), and *Callialasporites* spp. from the base of the section to the top (Fig. 4). Furthermore, the absence of *Classopollis* is consistent with an early Toarcian or older age (cf. Vakhrameyev, 1982).

The Xiaomeigou section is immediately overlain with angular unconformity by the Middle Jurassic–Cretaceous Dameigou section at the Dameigou locality (BGMRQ, 1991). Most of the section did not yield a rich or age-diagnostic palynologic assemblage. However, the lower 900 m, coal-bearing part of the section yielded enough information in the vertical succession of palynofloras to confidently estimate the age. The frequent occurrence of *Chasmatosporites* spp., *Cerebropollenites mesozoicus, Callialasporites* spp., and *Quadreculina* in the coal-containing part of the section suggests an Aalenian-Bajocian age for most of the Middle Jurassic part of section. The top of the inferred Middle Jurassic section is defined as Bajocian to early Bathonian on the basis of the abundance of *Quadreculina anellaeformis* and the absence of *Classopollis* (Vakhrameyev, 1982; Sun, 1989).

An interesting characteristic of the Lower Jurassic and, to a much lesser extent, the Middle Jurassic samples is the presence of reworked Triassic bisaccate palynomorphs (*Staurosaccites* sp., *Ovalispollis* sp., *Vesicaspora fuscus*). Although not the focus of this study, Triassic rocks are scarce and undocumented in the northern Qaidam basin. The closest mapped Triassic locality to the Dameigou section (near Yuandingshan) consists of metasedimentary rocks and hosts a gold ore body. The presence of reworked Triassic palynomorphs within Jurassic strata of

Qaidam may indicate more regionally extensive deposition of nonmarine or deltaic Triassic sedimentary rocks in the Qilian-Qaidam area that were subsequently cannibalized by uplift and reworked into younger sedimentary deposits.

The Middle Jurassic strata at the Yuandingshan section (BGMRQ, 1991) yielded a poorly age-diagnostic palynomorph assemblage. However, comparison with typical Middle Jurassic assemblages, both regionally and at the Dameigou section, suggests an Aalenian–Bajocian age for the section, roughly the same age span as Middle Jurassic strata at Dameigou.

Data from the Dameigou locality identify the youngest Lower Jurassic samples to be late Pliensbachian or earliest Toarcian and the oldest Middle Jurassic strata to be Aalenian. Therefore, the angular unconformity between Lower and Middle Jurassic strata at Dameigou records a 7–8 m.y. gap (time scale of Gradstein et al., 1995). Furthermore, the age bounds placed on the Lower and Middle Jurassic section indicate that each interval was deposited in about the same amount of time (Early Jurassic, ~9 m.y.; Middle Jurassic, ~10 m.y.). This requires a dramatic increase in the sedimentation and subsidence rates during the Middle Jurassic.

Palynomorph data from the Dameigou and Yuandingshan localities, separated by 60 km along structural strike and currently carried in different thrust sheets, indicate a Bajocian to early Bathonian age for the top of organic-rich Middle Jurassic strata. This consistent age from disparate localities suggests that deposition of Middle Jurassic organic-rich lacustrine muds and peat ceased at about the Bajocian-Bathonian boundary. Furthermore, it suggests that there was little or no subsequent uplift and erosion of Middle Jurassic strata prior to deposition of Upper Jurassic and Cretaceous sediments.

The Upper Jurassic to Cretaceous part of the Dameigou section (BGMRQ, 1991) is confined to mostly Early Cretaceous (Fig. 4) on the basis of the high abundance of "giant" *Alisporites* and *Callialasporites* sp. In addition, the absence of *Classopollis* may further define the age of the upper Dameigou section as late Early Cretaceous (Albian; Vakhrameyev, 1982; Sun, 1989).

UNCONFORMABLE RELATIONS

A prominent angular unconformity exists between Lower and Middle Jurassic strata at Dameigou, the only locality where the contact is exposed (Fig. 5). Gently dipping Middle Jurassic lacustrine shale of the Dameigou Formation overlie a truncated, southwest-vergent fold in Lower Jurassic Xiaomeigou Formation sandstone and coal. The southwest vergence of the fold is suggestive of thrust-related folding associated with the Qilian fold and thrust belt. If correct, this is direct evidence of Early-Middle Jurassic contractional deformation in the Qilian Shan. Furthermore, deformation of synorogenic Lower Jurassic strata and subsequent deposition of Middle Jurassic lacustrine and alluvial strata implies that the Dameigou locality was in at least a proximal foredeep position since the Early Jurassic and was probably actually in a wedge-top (piggyback) position (cf. DeCelles and Giles, 1996) during Middle Jurassic time.

Farther to the north, the Bei Shan (Fig. 1) displays some of the best stratigraphic evidence for Mesozoic thrust faulting and synorogenic sedimentation because appropriate age rocks are widely preserved. Field work by Zheng et al. (1996) and map analysis by Hendrix et al. indicate a Middle Jurassic age for thrust faulting in the Bei Shan (Hendrix et al., 1996, their Figs. 2 and 9). Although these data come from a separate mountain range well to the north of the Qilian Shan, they provide a more complete picture from which the significance of the Lower-Middle Jurassic unconformity at Dameigou may be inferred.

An angular unconformity also exists at the base of the Paleocene in the Northeast Qaidam basin, and is best documented between the Lulehe and Yuandingshan localities (Fig. 8). The Mesozoic–Cenozoic section is folded into a large east-southeast–plunging anticline. The southern limb displays Cretaceous rocks overlain concordantly by Paleocene strata. At the eastern end of the fold, the base Paleocene unconformity cuts downsection through the Cretaceous and Upper Jurassic. Over a distance of ~5 km, the base Paleocene unconformity goes from Paleocene on Cretaceous to Paleocene on the upper part of the Middle Jurassic Dameigou Formation. This relationship indicates significant nonuniform uplift and erosion of older rocks prior to deposition of Paleocene–Eocene conglomerate of the Lulehe Formation.

SEDIMENTOLOGY, DEPOSITIONAL ENVIRONMENTS, AND LITHOFACIES

Jurassic and Cretaceous depositional systems of the Northeast Qaidam basin were entirely nonmarine. Depositional environments ranged from alluvial fans, through coarse- and fine-grained fluvial systems, to deep, open lakes. Lithologically, these grade from boulder conglomerates to laminated shales (Figs. 9 and 10A).

Fluvial depositional systems

Fluvial depositional systems dominate Mesozoic strata of the study area and are of several different types, ranging from muddy, suspended load–dominated, meandering-anastomosed systems, to gravely, bedload-dominated braided systems (cf. Miall, 1996).

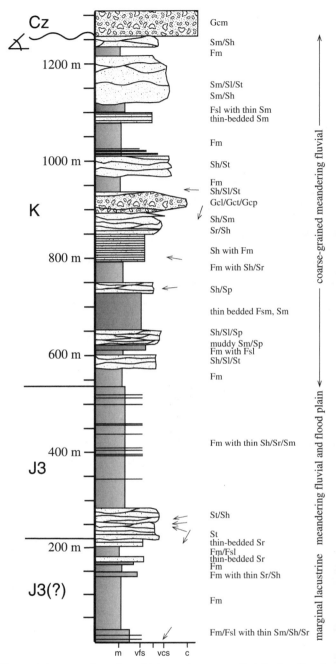

Figure 8. Measured section of Mesozoic strata at Lulehe locality. Arrows indicate mean paleocurrent vector in each interval. Patterns are same as in Figure 4. Facies codes are defined in Figure 9.

Braided fluvial. These deposits consist of coarse sandstone to pebble and cobble conglomerate. Sandstone is more common than conglomerate, and typically consists of meter-scale beds with plane lamination and low-angle and trough cross-stratification. Beds have erosional bases and commonly pebble lags. Little or no mud is preserved in the system, and the sandy character of the deposit is laterally consistent (Fig. 9A). Individual lithosomes are typically lenticular over hundreds of meters laterally.

Braided fluvial conglomerate is characterized by clast supported, 0.3–2-m-scale bedding, and is typically very well sorted and organized with clast imbrication or cross-stratification, indicating traction transport. Beds are typically erosive and lenticular; however, the conglomeratic character of the deposit is laterally and vertically consistent. This is the most common depositional style for conglomerate in Mesozoic northeast Qaidam.

Coarse-grained meandering fluvial. These fluvial deposits share many of the sedimentological characteristics just listed, including texture, organization, and erosive contacts. However, these deposits are considered to be meandering fluvial based on two criteria. First, some deposits can demonstrably be walked out laterally into overbank mudstones and thinly interbedded fine, rippled sandstone, siltstones, and calcic paleosols (cf. Machette, 1985) that are interpreted as levee and floodplain deposits. Furthermore, all of these deposits occur as lenticular coarse intervals within sections that dominantly consist of overbank mudstone. Second, very coarse, but well-organized horizons to 50 m thick can be found that fine upward from cobble conglomerate to pebble conglomerate and finally trough cross-stratified sandstone, before giving way to red mudstone with reduction spots, desiccation cracks, rare ripple marks, and calcite nodules interpreted as overbank fines and paleosols (cf. Machette, 1985). This upward-fining succession is interpreted to be produced by lateral migration and infilling of a coarse-grained channel. Coarse-grained meandering fluvial facies are common in the Cretaceous of northeast Qaidam (i.e., Lulehe; Figs. 9C and 10, B and C).

Fine-grained meandering fluvial. These deposits consist of relatively thin (<1–30 m) sections of trough cross-stratified, fine to pebbly sandstone, interbedded with massive mudstone (Figs. 9B and 10, D and E). Beds are typically ~0.5–1 m, have erosive bases (often with pebble or granule lags), and commonly fine upward. Large-scale low-angle cross-stratification and planar lamination is also commonly present and ripple cross-lamination is moderately common. Lateral accretion surfaces are documented within this facies. The sandstone bodies may be laterally discontinuous over tens of meters and grade into mudstone or may extend for hundreds of meters. Sandstone bodies are encased by overbank mudstone and organic-rich paleosols, similar to those described in the coarse-grained meandering fluvial deposits. These deposits are common in the Lower and Middle Jurassic strata of the study area, and commonly are interbedded with lacustrine deposits.

Lacustrine Depositional Systems

Open lacustrine. Freshwater open lacustrine systems are prevalent throughout the Lower and Middle Jurassic strata of the study area. These systems are dominated by dark gray or black, fissile to massive shale (Figs. 9D and 10F). Organic material in the shale varies from disseminated organic matter, with few macrofossils, to abundant terrestrial plant macrofossils. Open lacustrine facies do not occur in Upper Jurassic or Cretaceous rocks of northeast Qaidam.

The freshwater character of the open lacustrine deposits in the Lower and Middle Jurassic is indicated by palynoflora that include freshwater algae such as *Inaperturopollenites* spp., *Ovoidites* spp., *Botryococcus,* and *Tasmanites*-type, thick-walled cysts. The molecular organic geochemistry of the open lacustrine shales confirms the freshwater nature of Jurassic lacustrine systems. Saline lacustrine biological marker compounds (biomarkers), such as β-carotane, γ-carotane, or other indicators of saline or hypersaline environments such as gammacerane or C_{34} and C_{35} homohopanes (Peters and Moldowan, 1993), are absent (Ritts et al., 1999). The biomarker and nonbiomarker molecular organic geochemistry is dominated by higher plant organic matter indicators (Ritts et al., 1999), such as heavy *n*-alkanes (>C_{25}), an odd over even *n*-alkane preference, high pristane:phytane ratios, dominance of C_{29} steranes, and high concentrations of diterpane compounds (Peters and Moldowan, 1993). This indicates that the northeast Qaidam lacustrine systems were freshwater and surrounded by abundant terrestrial vegetation, consistent with the presence of extensive coals and well-preserved plant fossils.

Open lacustrine shale with these characteristics is typical of lakes that are in humid climates and overbalanced with respect to water (Carroll and Bohacs, 1999). Furthermore, these lacustrine systems are often intimately associated with progradational fluvial facies and coal (Bohacs and Suter, 1997; Carroll and Bohacs, 1999), as is the case in northeast Qaidam.

Marginal lacustrine. Marginal lacustrine deposits in the Lower and Middle Jurassic are typically marked by coal and mudstone with abundant plant fossils that are interbedded with laminated open lacustrine shale and isolated, progradational fluvial-deltaic sandstone bodies (Figs. 9E and 10G; cf. Remy, 1989). Most of the thickest coals in the study area were deposited in coal swamps in marginal lacustrine settings, although fluvial floodplain and oxbow lake coals are also common. Coals are distributed throughout the study area and reach maximum thicknesses in single intervals of 40 m in the Middle Jurassic and 8–10 m in the Lower Jurassic (Fig. 4). There is no coal or associated organic-rich mudstone in Upper Jurassic or Cretaceous marginal lacustrine sections.

Extensive marginal lacustrine systems tracts also occur in the Upper Jurassic, and to a lesser extent the Cretaceous, where they are associated with alluvial plain depositional systems (Fig. 10H). These deposits consist of centimeter- to decimeter-scale beds of mudstone and sandstone that are largely devoid of primary sedimentary structures; however, desiccation cracks, symmetric and climbing ripples, calcite nodules, and burrows (Fig. 9F) provide evidence for periodic emergence, oscillatory flow, and an oxic environment, respectively (cf. Picard and High, 1981).

Fan delta and alluvial fan depositional systems

Fan delta and alluvial fan depositional systems are volumetrically minor in northeast Qaidam, and are represented only by deposits in the Lower and lower Middle Jurassic section at the Dameigou locality. These depositional systems are characterized by rapid lateral facies changes (over hundreds of meters) from texturally and compositionally immature conglomerate to open lacustrine shale. Fan deposits consist of pebble to boulder

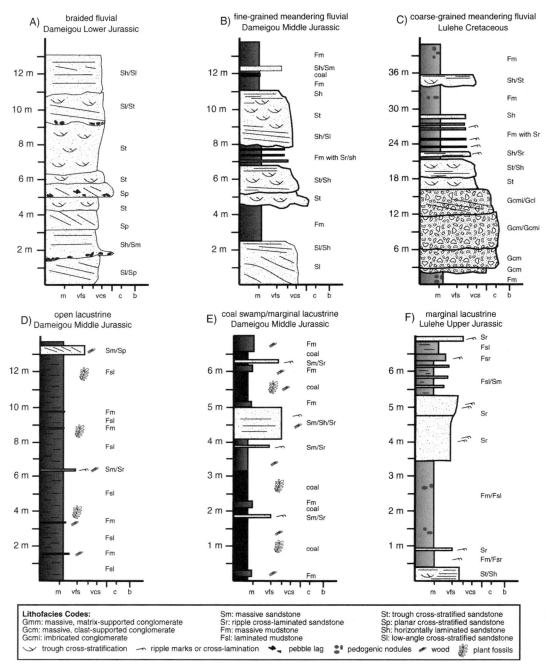

Figure 9. Detailed measured sections from Dameigou and Lulehe localities, representing key depositional environments. Patterns are same as for Figure 4.

conglomerate composed of clasts derived from local basement lithologies (granitoid intrusive rocks and metamorphic rocks), and minor interbedded sandstone. Beds are 50 cm to more than 1 m thick. They are clast supported and sometimes have crude horizontal or low-angle stratification and rare clast imbrication. These conglomerate beds are interpreted to be waterlain traction deposits within a channelized streamflow-dominated fan depositional system. Intervals with poorly organized, matrix-supported conglomerates are rare and are interpreted to represent relatively minor debris-flow deposition on alluvial fan or fan delta slopes.

Facies trends

The vertical succession of lithofacies and depositional systems in the Lower Jurassic through Cretaceous stratigraphy is

Figure 10. Key depositional facies in northeast Qaidam. A: Overview of Middle Jurassic section at Dameigou. B: Cretaceous coarse-grained meandering fluvial channel fill at Lulehe; top is to right, and section fines upward from boulder conglomerate at base to thin-bedded sandstone and overbank mudstone at top. C: Coarse-grained fluvial channel incision into overbank mudstone from Middle Jurassic of Dameigou. D: Lenticular meandering fluvial sandstone encased in overbank mudstone from Middle Jurassic of Dameigou. E: Overbank mudstone and thin-bedded sandstone from Upper Jurassic of Lulehe. F: Thick section of open lacustrine, laminated shale from Middle Jurassic of Dameigou. G: Thick coal from Middle Jurassic of Dameigou. H: Interbedded fine rippled, plane-laminated, and massive sandstone with massive mudstone in Upper Jurassic marginal lacustrine facies at Lulehe.

defined by one Lower Jurassic section (Xiaomeigou), several Middle Jurassic sections (Fig. 6), and three Upper Jurassic– Cretaceous sections (Dameigou, Lulehe, Yuqia). The Lower Jurassic Xiaomeigou Formation unconformably overlies older rocks. The presence of reworked Triassic palynomorphs within Lower Jurassic coal suggests that the Mesozoic history of Qaidam included deposition of some Triassic sedimentary rocks; however, these rocks are not preserved along the northeast margin of Qaidam, and their nature is not known.

The Lower Jurassic section is dominated by sandstone and conglomerate deposited in laterally discontinuous fluvial and alluvial fan environments. Lacustrine depositional systems are not well developed in the Lower Jurassic, and fine-grained rocks are limited to coal and coaly mudstone deposited in overbank and coal swamp environments (Fig. 4).

Above the 7–8 m.y. unconformity between Lower and Middle Jurassic rocks, the base of the Middle Jurassic Dameigou Formation consists of dark gray, organic-rich, laminated, open lacustrine shale (Ritts et al., 1999). Over more than 550 m, the lower part of the Middle Jurassic coarsens and shoals upward in facies from open lacustrine shale, through marginal lacustrine and deltaic sandstone and mudstone, into fluvial sandstone, conglomerate, and overbank mudstone. A prominent coal that was deposited in a long-lived coal swamp environment overlies the lower Middle Jurassic shoaling sequence (Fig. 4, 800 m). The upper part of the Middle Jurassic (above the coal) consists of a deepening- and fining-upward sequence that is more than 250 m thick. This part of the Middle Jurassic passes from rejuvenated conglomeratic and sandy channelized fluvial depositional systems into marginal lacustrine coaly mudstone and fine sandstone, and finally into more than 30 m of lacustrine shale (Fig. 4).

Middle Jurassic strata are disconformably overlain by Upper Jurassic rocks that are limited to a thin section of red mudstone and sandstone deposited in fluvial overbank or alluvial plain depositional systems (Fig. 4). The Upper Jurassic section passes into the Cretaceous section, which also consists of red mudstone and sandstone near the base of the formation, but contains thick channelized fluvial intervals that coarsen from sandstone to boulder conglomerate upward in the formation (Figs. 4 and 8). Cretaceous and older rocks are overlain with angular unconformity by Tertiary alluvial fan conglomerates (Fig. 8).

The lateral distribution of lithofacies can only be documented in Middle Jurassic rocks, for which data from several wells and outcrop localities exist. A lithostratigraphic correlation of strata along the northeast margin of Qaidam is shown in Figure 6. In general, facies become more distal to the west (toward the present-day basin) for at least the upper Middle Jurassic.

In the upper part of the Middle Jurassic section, three intervals are correlated on the basis of lithostratigraphic character (informally referred to as A, B, and C, and shown in Fig. 6) and supporting biostratigraphy (Fig. 6). The uppermost, C, unit is dominantly an open-lacustrine shale along the entire length of the section. The underlying unit, B, is distributed between Dameigou

and Yuandingshan, and has a dramatic proximal to distal change between those sections. At Dameigou, the B interval consists of an upward-coarsening section of sandstone and conglomerate with only minor, centimeter-scale coaly interbeds. This interval is interpreted as a proximal fluvial deposit tapping a relatively high relief source terrane. The section remains dominantly fluvial at Lulehe (to the northwest, obliquely across strike); however, most of the conglomeratic fraction is absent. The section is markedly more distal at the Mahai #20 well and at Yuandingshan (to the west, across strike). At Mahai #20, the B interval consists of mostly lacustrine shale and some interbedded sandstone and pebbly sandstone that were deposited by fluvial systems encroaching into a lacustrine system (cf. Remy, 1989). The same interval at Yuandingshan is a very thin lacustrine shale interval that probably pinches out somewhere in the vicinity. The lowest interval that can be correlated between several sections, A, consists of coal and interbedded lacustrine shale and sandstone that were deposited along the entire northeast Qaidam margin in a relatively uniform interval. This horizon contains the best-developed thick commercial coal accumulations in Qaidam. The A interval passes upsection from lacustrine shale deposition to marginal-lacustrine coal-swamp deposition.

The continuity of the A and C open lacustrine shales and marginal-lacustrine coal-swamp facies along the length of the northeast Qaidam margin suggests that deposition of these parts of the Middle Jurassic did not occur in isolated basins, but rather that the entire present northeast margin of the basin was drowned by a lacustrine system. In contrast, the B interval has a dramatic proximal to distal trend and probably pinches out near Yuandingshan. This geometry suggests that deposition of this interval was structurally confined.

Although there are only enough data to address facies continuity in the upper part of the Middle Jurassic, the data from that part of the section indicate that, although the thrust wedge bounding the northeast margin of the basin was actively deforming during the Jurassic (as suggested by the Early-Middle Jurassic unconformity at Dameigou), depositional systems periodically filled the Qaidam basin and onlapped the wedge top. On the basis of the inferred wedge-top setting and the lateral relationships in unit B, it is also likely that some periods of deposition were more segmented by active deformation into distinct, or at least interrupted, piggyback basins, as futher developed in the following.

PROVENANCE ANALYSIS

Paleocurrent indicators

Paleocurrent data were derived from fluvial and alluvial strata using trough cross-stratification (DeCelles et al., 1983), primary current lineation, planar cross-stratification, and clast imbrication.

Paleocurrent data from northeastern Qaidam indicate dominantly southwest-directed flow, transverse to and away from the

Qilian Shan, throughout the Jurassic and Cretaceous. Mean paleocurrent direction is toward 231° in the Lower-Middle Jurassic and 250° in the Upper Jurassic–Cretaceous (Fig. 11).

The Lower Jurassic locality (Dameigou) has a more complex paleocurrent signal that should be examined independently. Most of the Lower Jurassic section has southwest-directed paleocurrent indicators. However, the uppermost fluvial unit in the Lower Jurassic, which overlies a 9-m-thick coal, has northwest-directed paleocurrent indicators. Southwest paleocurrent directions are reestablished in the Middle Jurassic (Fig. 12). The Lower Jurassic paleocurrent deviation may be explained in several ways. First, it could reflect a better developed, northward axial drainage system in the Early Jurassic foreland. This interpretation is supported by the thick (at least 30 m), sandy nature of the braided fluvial deposystem. Alternatively, the paleocurrent deviation may reflect growth of the structure that ultimately folded Lower Jurassic strata at Dameigou and produced the angular unconformity. That sandstone composition (described in the following) does not vary systematically between the different paleodrainage regimes suggests that uplift of an entirely different drainage basin is not responsible for the reorganization of paleodispersal systems.

Sandstone petrography

Method. Provenance data were derived from petrographic examination of 45 sandstone samples in thin section. Half of each thin section was stained to facilitate identification of plagioclase and K-feldspar, then point-counted (at least 500 points per thin section) using the Gazzi-Dickinson point-counting method (Dickinson, 1970; Ingersoll et al., 1984); however, important rock fragments also were noted and documented separately. Point-counting and recalculated parameters are listed in Table 2 (Ingersoll et al., 1984).

Results. The point-count results for each thin section are shown in Table 2. Northeast Qaidam sandstones are quartzofeldspathic (mean = $Qm_{62}F_{24}Lt_{15}$). The mean and most individual samples plot within the mixed provenance field of Dickinson et al., (1983), between continental block and basement uplift provenance fields (Fig. 13). The majority of feldspar in the sandstone samples is K-feldspar, with a mean P/F ratio of 0.39 ± 0.2 (Fig. 13; Table 2). Quartzose grains are dominantly monocrystalline (Fig. 13), although a significant fraction of both Qm and Qp is strongly deformed metamorphic or mylonitized quartz. The lithic fraction is small, and consists of metamorphic, sedimentary, and volcanic lithics in roughly equal amounts. No secular or lateral trends in sandstone composition are observed in the data.

Interpretation. Mesozoic sandstone compositions in the northeast Qaidam basin are consistent with derivation from the Qilian Shan. The Qilian Shan, as previously discussed, consists of a middle Paleozoic orogenic belt that was rejuvenated in the Mesozoic. The Paleozoic–Triassic history of the source terrane allowed for much of the supracrustal sedimentary and volcanic carapace of the orogen to have been stripped by erosion, resulting in exposure of basement metamorphic and plutonic rocks. Rocks of these types, particularly granitoid plutonic rocks, are currently the dominant outcrops in adjacent parts of the Qilian Shan (Fig. 3). A source terrane of this type would result in quartzofeldspathic sandstone, with a strong K-feldspar signal (particularly microcline), and very minor amounts of supracrustal lithics (Graham et al., 1993), as is observed in northeast Qaidam sandstones. Furthermore, because most unroofing of the range would have occurred during the Paleozoic and Triassic, vertical compositional changes, such as unroofing signatures, would not be expected in Mesozoic sandstone, and are not observed in northeast Qaidam.

The presence of significant Ls fragments (the dominant lithic type) may indicate reworking of Mesozoic strata in response to uplift and cannibalization in the fold and thrust belt. This interpretation is consistent with the structural evidence for intra-Jurassic deformation at Dameigou, as well as recycled

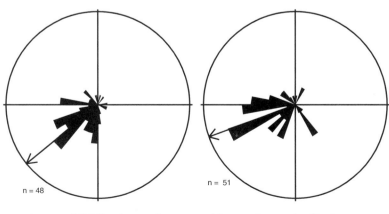

Figure 11. Paleocurrent rose diagrams for Lower Middle Jurassic (left) and Upper Jurassic–Cretaceous (right) strata of northeast Qaidam basin. Arrow indicates mean paleocurrent vector.

n = 48

n = 51

Lower-Middle Jurassic
Dameigou, Lulehe, Yuandingshan

Upper Jurassic-Cretaceous
Dameigou, Yuqia, Lulehe

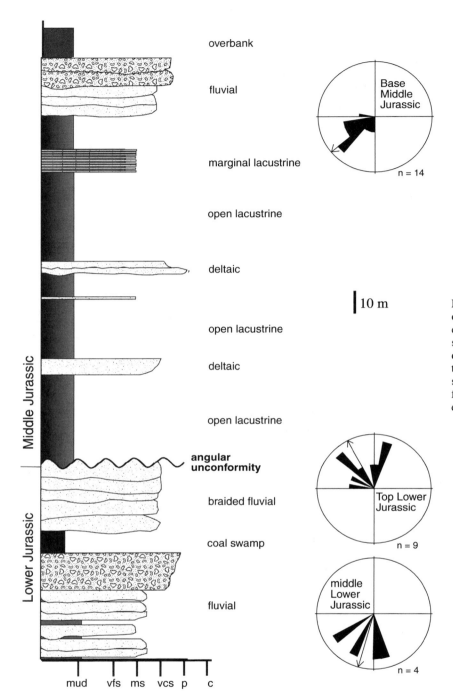

Figure 12. Detailed measured section over Lower-Middle Jurassic angular unconformity at Dameigou. Patterns are same as in Figure 4. Paleocurrent rose diagrams are for lower Middle Jurassic, top Lower Jurassic braided fluvial sandstone, and middle Lower Jurassic fluvial-alluvial sandstone and conglomerate (top to bottom).

Triassic palynomorphs that are observed in some Early and Middle Jurassic samples.

The mixed provenance field of Dickinson et al., (1983) and Dickinson and Suczek (1979) includes sandstone compositions that are mixed or transitional between quartz and lithic-rich recycled orogen (fold and thrust belt) provenance and quartz and feldspar-rich continental block (basement uplift) provenance. This setting accurately describes the inferred setting of the Mesozoic northeast Qaidam basin. Although we interpret the basin to be an intracontinental foreland basin system related to a southwest-vergent Qilian fold and thrust belt, we note that most of the orogen had been stripped of supracrustal cover prior to Mesozoic rejuvenation. This allowed most detritus to have a basement uplift provenance signature, with cannibalization of Mesozoic synorogenic rocks and Paleozoic sedimentary remnants providing a minor, lithic-rich, supracrustal provenance signature. The importance of tectonic inheritance and source terrane composition on sediment provenance was discussed by

TABLE 2. SANDSTONE PETROGRAPHY

Sample Information		Raw Framework Counts																Raw Non-Framework Counts			Recalculated	
AGE	SAMPLE #	Qm	Qp	Cht	P	K	Fu	Lm	Ls	Lv	Lu	other L	musc	Bt	chlor	Mu	H	matrix	cement	pore	P/F	%Mica
J1	95XM12	371	5	4	5	7	5	0	4	5	3	0	0	0	0	0	0	6	90	12	0.42	0.0
J1	95XM16	188	0	8	4	101	2	5	39	15	23	0	1	2	0	1	0	28	70	6	0.04	1.0
J1	95XM21	193	15	5	10	65	4	10	12	18	17	0	7	2	4	0	0	8	129	5	0.13	3.6
J1	95XM24	167	0	1	89	97	4	4	15	9	2	0	3	0	2	0	1	12	60	6	0.48	1.3
J1	95XM35	216	11	4	6	95	3	10	31	22	24	6	3	0	0	0	1	6	46	14	0.06	0.7
J1	95XM36	222	19	18	60	11	2	9	1	82	13	0	0	0	0	0	0	17	20	2	0.85	0.0
J2	95DM44	261	9	12	20	64	7	16	16	15	4	0	7	0	0	0	0	16	51	14	0.24	1.6
J2	95DM48	281	18	9	9	57	17	19	10	26	9	0	2	0	0	1	0	14	20	12	0.14	0.7
J2	95DM51	260	1	0	27	79	9	0	0	0	0	0	0	0	0	1	0	5	81	1	0.25	0.3
J2	95DM53	390	0	0	0	4	4	0	2	2	0	0	0	0	0	0	0	31	30	47	0.00	0.0
J2	95DM57	247	17	23	4	70	4	6	2	3	4	0	0	0	0	0	1	18	124	0	0.05	0.0
J2	95DM59	204	2	3	19	84	18	5	16	2	12	0	0	0	0	0	1	24	55	61	0.18	0.0
J2	95DM60	259	5	7	41	50	3	9	16	12	7	0	4	2	0	0	0	13	87	5	0.45	1.4
J2	95DM61	237	13	22	25	54	6	12	50	15	6	0	6	2	0	0	0	16	18	14	0.32	1.8
J2	95DM63	143	4	0	103	83	3	16	26	3	16	0	4	6	4	9	0	66	33	7	0.55	5.5
J2	95DM66	408	3	3	8	7	0	1	0	0	0	0	0	0	0	0	1	25	2	40	0.53	0.0
J2	95DM69	292	1	2	52	72	9	2	45	1	11	0	1	5	0	0	0	13	10	10	0.42	1.2
J2	95DM73	252	9	3	22	53	10	4	0	2	5	7	0	0	0	0	1	0	128	8	0.29	0.0
J2	95DM88	313	10	31	15	23	4	10	23	6	23	0	2	0	0	0	0	17	10	31	0.39	0.4
J2	95DM97	54	2	1	1	14	1	0	2	2	1	0	1	0	0	0	0	13	9	5	0.07	1.3
J2	95DM100	230	3	29	22	59	0	2	3	32	7	0	1	0	0	0	0	1	112	0	0.27	0.3
J2	95DM102	203	5	8	27	102	5	8	12	3	11	0	1	0	0	1	0	28	5	94	0.21	0.5
J3-K	95DM105	247	9	7	39	38	9	6	22	17	27	1	0	0	0	0	0	0	0	0	0.51	0.0
K	95DM107	312	8	4	25	35	18	21	23	16	15	0	1	0	0	0	1	0	0	0	0.42	0.2
J2	95DM110	224	4	7	6	12	6	0	6	2	10	76	0	0	0	0	0	34	122	3	0.33	0.0
J3	95LL141	197	3	8	99	36	9	14	15	3	35	0	1	2	0	0	1	43	44	8	0.73	0.7
J3	95LL142	223	5	11	58	22	10	14	34	4	20	3	6	1	7	2	0	27	84	8	0.73	3.8
J3	95LL145	207	7	0	92	21	14	10	18	1	5	0	3	1	0	0	1	20	124	4	0.81	1.1
J3	95LL146	293	5	12	40	59	15	3	17	18	29	0	0	0	0	0	1	0	0	0	0.40	0.0
K	95LL149	271	1	2	69	77	14	9	8	2	5	0	0	2	1	0	6	0	0	0	0.47	0.7
K	95LL150	291	1	1	56	55	11	18	12	3	10	1	0	4	0	1	2	0	0	0	0.50	1.1
K	95LL151	230	0	2	36	84	1	6	34	4	7	0	2	5	1	0	1	35	63	15	0.30	1.9
K	95LL152	257	5	1	59	20	6	14	9	21	7	2	1	10	4	0	1	7	92	12	0.75	3.6
K	95LL153	234	6	2	27	39	11	24	16	10	19	0	1	17	0	0	1	0	0	0	0.41	4.4
J2	95YD164	306	9	16	44	24	1	6	14	13	11	0	4	0	0	1	0	12	62	3	0.65	1.1
J2	95YD163	244	4	8	33	42	5	1	2	18	3	0	0	0	0	0	0	5	104	0	0.44	0.0
J2	95YD165	255	12	0	0	20	4	46	8	1	14	0	2	5	0	9	1	6	109	1	0.00	4.3
J2	95YD167	245	7	10	26	50	3	8	5	3	1	0	0	1	0	0	0	6	130	5	0.34	0.3
J2	95DT184	226	1	1	77	61	6	23	16	1	10	0	10	0	1	0	0	43	37	0	0.56	2.5
J3-K	95YQ135	166	3	1	20	70	11	20	10	5	8	0	0	0	0	0	2	0	187	3	0.22	0.0
J3-K	95YQ136	142	1	0	30	97	2	14	26	3	16	0	1	0	0	1	0	0	0	0	0.24	0.3
J2	95KB40	285	1	0	85	31	0	22	9	4	7	4	3	7	0	2	14	11	2	1	0.73	2.6
J2	96KB42	294	0	0	49	29	4	22	32	4	10	0	4	1	0	2	10	30	33	7	0.63	1.6
J2	95LT52	223	2	1	104	32	0	28	12	6	16	1	8	17	0	5	10	32	3	0	0.76	6.6

Note: Qm—monocrystalline quartz; Qp—polycrystalline quartz excluding chert; Cht—nonvolcanogenic chert; P—plagioclase; K—K-feldspar; Fu—indistinguished feldspar; Lm—metamorphic lithic; Ls—sedimentary lithic; Lv—volcanic lithic; Lu—indistinguished lithic; other L—other lithic (Ritts, 1998); musc—muscovite; Bt—biotite; chlor—detrital chlorite; Mu—other Mica; H—heavy minerals complete raw and recalculated petrographic parameters are available in tablular format in Ritts (1998).
Modal framework composition determined using Gazzi-Dickinson point-counting method (Dickinson, 1970; Ingersoll et al., 1984).

Graham et al., (1993), and has repeatedly proven to be the main factor controlling sandstone composition in the basins of northwest China (Sobel, 1995; Hendrix, 2000; this study).

GEOHISTORY ANALYSIS

Subsidence modeling was completed using Genex software (Ungerer et al., 1990; Institut Français du Petrole [IFP], 1991) on measured Jurassic sections from Dameigou and Lulehe. The Dameigou-Lulehe composite section is composed of Lower Jurassic through Cretaceous strata (Dameigou) and Upper Jurassic through Cretaceous strata (Lulehe). Data for lithology and age were derived entirely from the Dameigou section, as are Lower and Middle Jurassic thicknesses. The Lulehe section, located 50 km to the northeast, was used only to constrain total thickness for the Upper Jurassic and Cretaceous sections. Paleo-elevation was assumed to be zero (sea level) throughout the Mesozoic, based on the lack of data constraints and the known

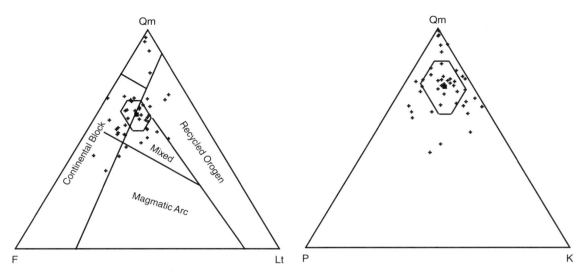

Figure 13. Modal framework composition of northeast Qaidam Jurassic and Cretaceous sandstone. Raw data are shown in Table 2. Raw data are recalculated according to Ingersoll et al. (1984).

proximity of the basin to the southern margin of the continent. The primary cause for uncertainty in the subsidence analysis is the poor understanding of the amount of missing section represented by unconformities within the Mesozoic section, particularly between the Lower and Middle Jurassic, and at the base of Cretaceous. The base-Cretaceous unconformity problem is compounded by a lack of our own biostratigraphic data on the time span of missing section.

Subsidence curve shape analysis is not emphasized in the interpretation for two reasons. Primarily, the lack of multiple age control points between unconformities results in straight curve segments, thereby masking details of curve shape. Furthermore, recent work on subsidence analysis in foreland basin systems (particularly wedge-top depozones, as in northeast Qaidam) casts doubt on the uniqueness of interpretations based on curve shape (DeCelles and Giles, 1996; DeCelles and Currie, 1996). However, the first derivative of the subsidence curve, the subsidence rate curve, is a quantitative, reliable measure of the rate and magnitude of subsidence and sedimentation. These data thus directly indicate periods of increased tectonic subsidence and relative quiescence, without addressing geodynamic mechanism. This analytical method was previously employed for the Mesozoic northern Tarim basin by Hendrix et al., (1992).

Northeast Qaidam tectonic subsidence rates more than doubled from the Early Jurassic to Middle Jurassic (Fig. 14). Following a hiatus, tectonic subsidence continued at modest rates in the Late Jurassic and, following another hiatus, increased slightly in the Cretaceous.

A flexural mechanism in response to a structural, topographic load is interpreted to be the cause of subsidence in northeast Qaidam because the preponderance of other evidence (angular unconformities, provenance, regional structure, and tectonic setting) supports a contractional setting for the Qilian

Shan. Therefore, peak Mesozoic subsidence rates in the Middle Jurassic are interpreted to indicate intense structural shortening and topographic loading in the Qilian Shan. This is supported by structural evidence related to the angular unconformity at Dameigou, as well as evidence on timing of peak deformation in other ranges of central Asia (Hendrix et al., 1996).

DISCUSSION

Contractile tectonic setting

The sedimentary and structural characteristics of Jurassic and Cretaceous strata in northeast Qaidam are all consistent with deposition in a disrupted foreland basin, or thrust-belt wedge-top setting, related to a rejuvenated, contractional Qilian Shan. These include (1) paleocurrents away from the range front; (2) sediment provenance reflecting uplifted and unroofed Qilian Shan basement lithologies, particularly sand and conglomerate clasts derived from Paleozoic plutons that are extensively exposed in the central and southern Qilian Shan (Fig. 3); (3) periods of increased subsidence rates corresponding to times of inferred structural shortening; (4) prominent angular unconformities between the Lower and Middle Jurassic and at the top of Cretaceous (unconformities followed by dramatic increases in subsidence rate); (5) facies trends that are more continuous along structural strike than across structural strike; and (6) interbedding of alternatively low-gradient lacustrine systems and high-gradient fluvial systems, suggestive of active tectonics, not just in the bounding range, but also in the basin substrate (also shown with direct evidence from the thrust-related folding of the Lower Jurassic rocks at Dameigou). When considered with the nonvolcanic character of the sedimentary fill and the regional context of Mesozoic Asia, a flexural foreland-style basin

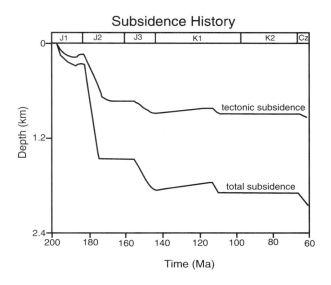

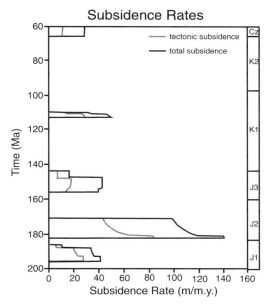

Figure 14. Subsidence history for Mesozoic Northeast Qaidam basin. Subsidence curve (left) and subsidence rate curve (right) indicate most pronounced tectonic subsidence during Middle Jurassic.

setting related to intracontinental deformation provides the best fit for the characteristics of the Northeast Qaidam basin.

The extent to which deposition in the wedge-top position (represented by current rock outcrops) occurred in distinct, structurally confined piggyback basins is not completely discernible. The lateral continuity of lacustrine facies in the A and C intervals of the upper Middle Jurassic (Fig. 6) suggests that, at times, much of the wedge top was sublacustrine and undergoing continuous deposition of lacustrine shale or marginal lacustrine coal. Fluvial intervals seem to have greater lateral discontinuity, and the B interval (Fig. 6) actually appears to pinch out, or lap out basinward into a condensed interval or unconformity. This may be due to heterogeneity of the depositional system, or may reflect periods of increased structural segmentation. Further downsection in the Middle Jurassic and Lower Jurassic, there is good evidence for active basin tectonics and wedge-top deposition; however, there are not sufficient exposures of these rocks to address issues of lateral continuity and stratal geometry.

The sedimentary basin characteristics of northeast Qaidam can be integrated to make a reasonable Lower Jurassic–Cretaceous paleogeographic and tectonic model of the region as an intracontinental flexural basin in the foreland of a reactivated orogen. This model is internally consistent, and is also consistent with the paleogeography and tectonic setting of other parts of northwest China, to the extent that they are known (Hendrix et al., 1992, 1996; Sobel, 1995, 1999; Hendrix, 2000; Greene et al., this volume).

Specifically, the late Sinemurian through earliest Toarcian (197–189 Ma) was characterized by modest amounts of subsidence. The accommodation space was filled by transverse allu-

vial fan and fan delta systems derived from the Qilian Shan that prograded into more distal lacustrine systems basinward (Fig. 15A). These transverse dispersal systems were replaced by a north-northwest–flowing axial braided fluvial system (Fig. 15B). Following Toarcian deformation of the proximal part of the basin, rapid Aalenian subsidence recreated a deep lacustrine basin in the foredeep and frontal wedge top of the Qilian Shan (Fig. 15C). Through the Aalenian-Bajocian this lacustrine system was progressively filled by prograding fluvial depositional systems derived from the Qilian Shan (Fig. 15D). Higher base level later in the Bajocian and early Bathonian resulted in retrogradation of the lacustrine shoreline. Following a hiatus, Upper Jurassic through Cretaceous deposits consist of an upward-coarsening sequence ranging from marginal lacustrine sandstone and mudstone, through dominantly overbank and alluvial plain mudstone, to coarse-grained meandering fluvial channel conglomerate and associated mudstone. Sediments were still derived from the uplifted Qilian Shan, and were delivered to the basin by transverse depositional systems (Fig. 15E). The lack of open lacustrine strata in the Upper Jurassic–Cretaceous progradational sequence is interpreted to be a consequence primarily of semiarid climates and a negative water budget (Hendrix et al., 1992; see following discussion), although reduced subsidence rates and maintenance of the basin in an overfilled state (cf. Carroll and Bohacs, 1999) were probably also factors in prograding fluvial systems across the basin.

Role of strike-slip deformation

Although contractional deformation has been previously documented in the Mesozoic of central Asia (Hendrix et al.,

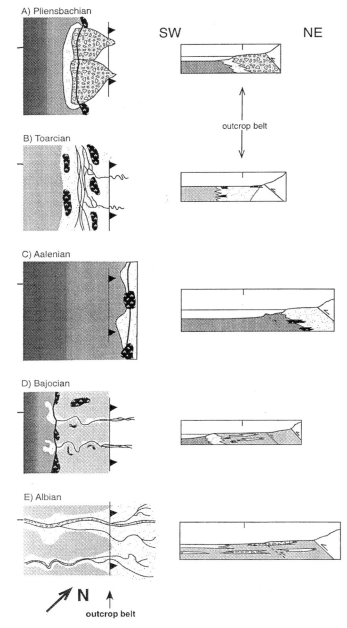

Figure 15. Paleogeographic sketches representing northeast margin of Qaidam during Early Jurassic through Cretaceous (top to bottom). Pliensbachian—sediment dispersal transverse to Qilian Shan with alluvial fans debouching into open lakes; early Toarcian—filled basin dominated by axial sandy braided fluvial system; Aalenian—rapid subsidence results in lacustrine transgression and frontal fold and thrust belt is drowned by lacustrine system; Bajocian—transverse fluvial and alluvial depositional systems prograde into basin; Albian—coarse-grained meandering fluvial systems dominate basin without major lacustrine deposition. Position of Dameigou-Yuandingshan outcrop belt is indicated by arrows. Patterns are same as Figure 4; increasing darkness of shading is intended to represent increasing inferred organic carbon content of mudrocks. Dark patterned areas indicate deposition of coals.

1992, 1996), the role of strike-slip faulting has not been thoroughly evaluated. The best evidence for strike-slip faulting during the Mesozoic in central Asia comes from the Talas-Ferghana fault in Kazakstan (Burtman, 1980; Tseysler et al., 1982). Largely on the basis of an along-strike projection of the Talas-Ferghana fault into northwest China, Sobel (1995, 1999) hypothesized that the western Tarim basin was a strike-slip–related basin between the Talas-Ferghana and Karakoram faults during the Jurassic. Transtension has also been suggested to account for small-scale normal faults imaged on seismic reflection profiles in the Hexi corridor, north of the Qilian Shan (Xu et al., 1989; Bally et al., 1986).

In spite of these hypothesized Mesozoic strike-slip systems and the importance of strike-slip faulting in central Asia during the Cenozoic (Tapponnier and Molnar, 1977), there are no characteristics of the sedimentary or structural geology of northeast Qaidam that either imply or require strike-slip faulting. To the contrary, the best evidence that there is not significant strike slip between Qaidam and the Qilian Shan is the provenance link between the basin and source terrane. Specifically, the dominance of granitoid-derived detritus in northeast Qaidam (quartzo-feldspathic sandstones, granitoid conglomerate clasts; Fig. 13) correlates well with the presence of a very large granitoid batholith source in the south Qilian Shan, directly northeast of the sections from which sandstones were sampled (Fig. 3). In addition, there are no extreme thicknesses of Mesozoic strata that might imply depositional offlap (shingling) in a pull-apart basin (e.g., Crowell and Link, 1982). Previous studies in Tarim, Junggar, Turpan, and southern Mongolia have found no explicit evidence requiring Mesozoic strike-slip deformation (Hendrix et al., 1992, 1996), but in no case can strike-slip deformation be ruled out.

Role of extensional deformation

Some earlier studies hypothesized an extensional origin for some Mesozoic basins of northwest China (Ulmishek, 1984; Wang and Chen, 1984; Jiao, 1984; Allen et al., 1991). However, most recent studies of the same basins favor a flexural origin related to contractional deformation on the basin margins (Hendrix, 1992; Hendrix et al., 1992, 1996; Gu, 1994; Sobel, 1999). Most of the data collected in this study also support a flexural origin for the Mesozoic Qaidam basin, related to contractile loading in the Qilian Shan. However, the nearly complete lack of subsurface data (particularly seismic reflection profiles) and poor understanding of Mesozoic structural geology in the Qilian Shan prevents the explicit demonstration of either contractional or extensional tectonics in Qaidam.

In the absence of convincing subsurface and structural data, the best interpretation of subsidence mechanism must be inferred from ambiguous data. The depositional systems and sediment provenance are not unique to either extensional or contractional settings. However, the lack of Mesozoic volcanic and hypabyssal intrusive rocks is problematic for an extensional interpretation. In addition, one would expect to see sediment dispersal from both sides of the basin in an extensional setting (Leeder and Jackson, 1993; Leeder, 1995), rather than the strong unimodal paleocurrents that are documented. Geohistory analysis (Fig. 14A) is also equivocal and could be interpreted either as Middle Jurassic rifting or flexural subsidence accelerating from Early Jurassic into Middle Jurassic, and again from Late Jurassic into Cretaceous. Although sediment dispersal patterns, geohistory modeling, and lack of volcanic rocks do not preclude a rift model, they favor a flexural model for the basin.

Perhaps the best direct evidence against an extensional origin for the Mesozoic Qaidam basin is the level of thermal maturity in Lower and Middle Jurassic rocks. These rocks have vitrinite reflectance values of only 0.8%–1.0% (Ritts et al., 1999) and were buried to a depth exceeding 6000 m by Mesozoic and Cenozoic strata. Fission-track data from the Mesozoic rocks at Lulehe indicate that uplift of the Mesozoic section was very recent, suggesting that rock outcrops underwent nearly the same Cenozoic burial history as Mesozoic rocks in the subsurface of Qaidam (E.R. Sobel, 1997, personal commun.). This amount of burial precludes high paleoheat flow in order to maintain the Mesozoic section at 0.8%–1.0% Ro (Ritts et al., 1999; Ritts, 1998). The relatively low heat-flow history model that best matches the measured maturity indicators in Jurassic rocks (Fig. 16) is typical of flexural foreland or intracontinental basins, and is low compared to most extensional basins worldwide (Sclater et al., 1980; Lysak, 1987).

In summary, the interpretation of Qaidam as a flexural basin related to contractile loading in the Qilian Shan is best supported by the Lower-Middle Jurassic angular unconformity

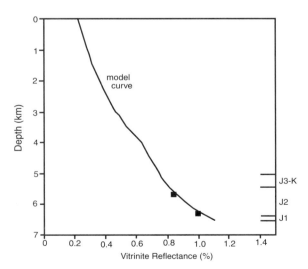

Figure 16. Calibration of modeled vitrinite reflectance with measured values on Lower and Middle Jurassic coals. Model uses same burial history as for Figure 14, which was primarily derived for Dameigou locality. Measured vitrinite reflectances also were derived from Dameigou locality (Ritts, 1998). Calibration shown was achieved by maintaining very low heat flow throughout Mesozoic and Cenozoic of 25 mW/m².

at Dameigou, low paleoheat flow, sediment dispersal patterns, absence of volcanic rocks, and the regional setting of Qaidam between the accretionary Tibetan margin and known contractional, intracontinental deformation to the north (Hendrix et al., 1992; 1996; Gu, 1994). All other characteristics of the basin are either nondiagnostic (depositional systems and provenance) or fit better a contractional tectonic setting (i.e., duration of rapid subsidence rates).

Regional tectonics

Several specific tectonic events have been identified for the Mesozoic of central and eastern Asia that may control intracontinental deformation, and thus basin subsidence. These include the accretion of the Qiantang, Lhasa, and Kohistan blocks to form Tibet in the Early Jurassic, Late Jurassic, and Cretaceous, respectively; the collision between the North China Block and South China Block in the Triassic; and the closure of the Mongol-Okhotsk ocean in the Middle-Late Jurassic (Figs. 1, 2, and 17). Various of these events have been associated with contractional deformation in east China (Davis et al., 1998, this volume), the Bei Shan (Zheng et al., 1996), southern Mongolia (Hendrix et al., 1996), the Tian Shan (Hendrix et al., 1992), and the Kunlun and Pamir (Sobel, 1999). In addition, extensional tectonics have been documented as gaining importance in the Late Jurassic and Cretaceous of southern and eastern Mongolia, and nearby regions of China (Johnson et al., this volume; Webb et al., 1999).

The temporal correlation between basin subsidence and tectonic events has been used to link individual periods of sedimentation to specific tectonic driving mechanisms. The most notable attempt at such a correlation was between basins along the Tian Shan, and the amalgamation of Tibet (Fig. 5; Hendrix et al., 1992). With the addition of data from western and southern Tarim, and Qaidam (Ritts, 1998; this study), the issue of tectonic driving mechanisms must be readdressed.

First, it is clear that when subsidence rate curves from each basin are superimposed for the Mesozoic, different basins have distinct histories, and the most significant spikes in subsidence rate were not synchronous throughout northwest China (Fig. 18). Furthermore, subsidence was maintained at significant rates throughout the Jurassic and Lower Cretaceous, suggesting that the central Asian continental interior was maintained in a tectonically active setting for the duration of most of the Mesozoic (Figs. 17 and 18). In more detail, it is apparent that peak subsidence rates in each basin were reached in the Jurassic, and, in general, subsidence rates and total magnitudes of subsidence were lower in the Cretaceous, particularly the Late Cretaceous (Fig. 18). This suggests a reduction in the intensity of tectonism, or at least less-sustained tectonism in the Cretaceous.

Although subsidence was maintained through the Mesozoic, dramatic increases in subsidence rate occurred, but not in all areas, and not strictly synchronously (Fig. 18). Subsidence rate spikes that occurred in more than one basin at the same time

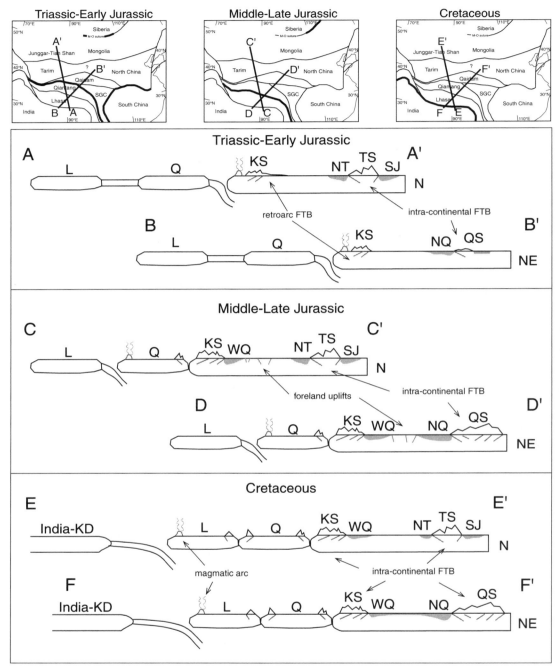

Figure 17. Mesozoic tectonic setting of western China. Sutures indicated in bold on tectonic maps indicate active plate margins (active subduction zones) for each time interval (note there is no attempt to palispastically restore terranes in maps; maps are merely to indicate which sutures were active in each period). Cross sections schematically indicate tectonic setting of subsiding basins (shaded) with respect to accreting terranes and active southern margin of Asia for each time interval. Cross sections indicate locations of intracontinental thrust belts (shown as mountains), foreland uplifts (shown as steep faults), and magmatic arcs and retroarc fold and thrust belts (shown as volcano). Abbreviations: SGC, Songpan-Ganzi remnant ocean basin; Q, Qiantang block; L, Lhasa block; KS, Kunlun Shan; WQ, northwest Qaidam and southeast Tarim basin (contiguous); NT, north Tarim basin; TS, Tian Shan; SJ, south Junggar basin; NQ, northeast Qaidam basin; QS, Qilian Shan; KD, Kohistan-Dras arc.

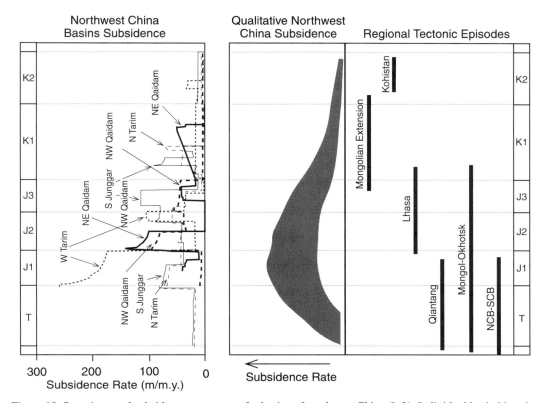

Figure 18. Superimposed subsidence rate curves for basins of northwest China (left). Individual basin histories are generalized to give qualitative subsidence rate curve (center) that represents general regional history of subsidence rate magnitudes. These curves are compared to timing of key tectonic episodes in Mesozoic assembly of central Asia (right). Bold solid line, Northeast Qaidam basin; bold dashed line, Northwest Qaidam basin; solid line, south Junggar basin; dashed line, north Tarim basin; dotted line, west Tarim basin. Timing of tectonic events comes from Girardeau et al. (1984), Coward et al. (1986), Smith (1988), Zonenshain et al. (1990), Searle (1991), Şengör and Okurogullari (1991), Yin and Nie (1993), Webb et al. (1997, 1999), and Johnson et al. (1997). Details of subsidence analysis, including data, sensitivity analysis, and assumptions, can be found in Ritts (1998). Models are based on data from Hendrix (1992), Zhou and Chen (1992), Sobel (1995), Ye et al. (1996), and Ritts (1998). Plots of individual curves from each basin are available in Ritts (1998). NCB, North China Block; SCB, South China Block.

are seen in the first half of the Early Jurassic, the first half of the Middle Jurassic, the late Middle Jurassic through middle Late Jurassic, and again in the middle of the Early Cretaceous. However, whether these increases can be attributed to regional accretionary tectonic events, or were merely the result of basins being influenced by the same orogen, is not clear.

Broadly speaking, the bulk of the amalgamation of central Asia occurred in the early Mesozoic, especially the Triassic–Middle Jurassic, including the accretion of the Qiantang and Lhasa blocks, and the closure of the Mongol-Okhotsk ocean (Figs. 2, 17, and 18). This corresponds to the periods of most rapid basin subsidence and, by inference, structural deformation, throughout northwest China (Figs. 17 and 18). The specific rate increases noted in the Early Jurassic, Middle Jurassic, Middle-Late Jurassic, and Early Cretaceous correspond less clearly to tectonic events. The only real increase in subsidence rate that roughly correlates with a collisional event in Tibet is the Early Jurassic rate increase and the collision of the Qiantang

block (Fig. 17; Smith, 1988; Şengör and Okurogullari, 1991). The Middle and Middle-Late Jurassic subsidence rate increases may overlap with the poorly determined closure of the Mongol-Okhotsk ocean. However, the variation in subsidence history for each basin, and the questionable correlation between basin geology and specific tectonic events, suggests that the basins of northwest China subsided more in response to a sustained contractional setting that persisted throughout the Jurassic and much of the Cretaceous as well as local deformation, and less in response to specific terrane collisions (Figs. 17 and 18; see also Vincent and Allen, this volume). This suggests that not only collision, but post-collisional convergence, and Andean-style plate margin tectonism are sufficient to drive large-scale intracontinental deformation, and associated sedimentary basin formation.

The Cretaceous reduction in relative rates of subsidence was more pronounced in the basins of northern Xinjiang and corresponded with extensional tectonics in Mongolia and eastern China. The reduction in contractional deformation and associated basin subsidence probably signals the relaxation of the long-term compressive stress field in central Asia (Fig. 18).

Paleoclimate

Paleoclimatic signatures of Mesozoic strata in northwest China were discussed by Hendrix et al., (1992). Triassic–Middle Jurassic rocks in northern Tarim were inferred to have been deposited in humid environments, based on the abundance of coal, plant fossils, and open lacustrine and meandering fluvial depositional systems (Hendrix et al., 1992). In contrast, Upper Jurassic and Cretaceous rocks of northern Tarim were interpreted to have been deposited in arid to semiarid environments based on thick sections of red beds, abundant calcic paleosols, presence of *Classopolis,* and the lack of coals, lacustrine strata, and plant fossils (Hendrix et al., 1992). This change from wetter to drier climates at the end of the Middle Jurassic was attributed to changes in oceanic and atmospheric circulation related to the restriction of the Tethys seaway due to plate tectonic reorganization. This reorganization apparently resulted in elimination of monsoonal weather patterns, and established a dominantly offshore (southward in present coordinates) wind direction that produced a rain shadow south of the Tian Shan (Hendrix et al., 1992).

Similar to northern Tarim, Lower and Middle Jurassic strata in northeast Qaidam were deposited under humid climatic conditions, whereas Upper Jurassic and Cretaceous rocks were deposited in semiarid to arid climatic conditions. The evidence for climatic differences is identical to that reported for northern Tarim (Hendrix et al., 1992). Wetter Early and Middle Jurassic climates are indicated by abundant coal, plant fossils, poorly drained organic-rich flood-plain deposits, abundant open lacustrine deposits, and meandering fluvial deposits. Evidence for drier Late Jurassic and Cretaceous climates is provided by monotonous red beds, abundant calcic paleosols, and the lack of organic-rich rocks, plant fossils, and open lacustrine depositional systems.

The arid climates of the Cretaceous in northeast Qaidam are interpreted as the result of a rain shadow cast by the relatively high Qilian Shan, similar to that proposed for the Tian Shan and northern Tarim by Hendrix et al., (1992). This interpretation follows from the analogous position that northeast Qaidam has relative to the Qilian Shan, as compared to north Tarim and the Tian Shan (Fig. 17). It is also supported by the presence of contrasting Cretaceous open lacustrine systems in the Hexi corridor north of the Qilian Shan (Xu et al., 1989).

CONCLUSIONS

Newly characterized strata of the Northeast Qaidam basin provide a clearer understanding of the long history of intracontinental deformation in central Asia. Mesozoic tectonism along the southern margin of Asia caused contractional reactivation of the Qilian Shan and subsidence in the adjacent Northeast Qaidam basin. Jurassic and Cretaceous rocks in the Northeast

Qaidam basin were deposited in proximal foreland basin and wedge-top positions in response to the prograding Qilian thrust belt. The sedimentary record of structural rejuvenation in the Qilian Shan preserved in the Northeast Qaidam basin indicates that the most intense tectonism occurred in the Middle and Late Jurassic. Evidence for sustained uplift of the Qilian Shan is provided by the severe climatic gradient observed across the Qilian Shan in the Cretaceous, as well as the continued delivery of coarse-clastic detritus to the foreland basin in the Cretaceous.

Evidence for recurrent uplift of the Qilian Shan during the Mesozoic, in response to the amalgamation of Asia, extends the region of known Mesozoic intracontinental deformation and orogenic rejuvenation into central China from other documented examples in the Tian Shan, Bei Shan, and Yinshan-Yahshan, and helps to determine the degree to which the pattern of Mesozoic intracontinental deformation resembled that of today. Integration of evidence from Northeast Qaidam and other basins of northwest China suggests that timing of structural and sedimentary episodes was nonuniform from basin to basin and did not have a direct temporal correspondence to tectonic collisions on the continental margin. However, the bulk of evidence suggests that peak tectonic amalgamation during the Jurassic correlated with the most intense periods of intracontinental deformation and basin formation.

ACKNOWLEDGMENTS

Ed Sobel provided valuable discussions and field assistance. Edmund Chang was instrumental in field-season organization, and provided translations of Chinese literature. Other discussions and insights were provided by Andrew Hanson, Steve Graham, Albert Bally, Calvin Cooper, Edmund Chang, Lisa Lamb, and Marc Hendrix. Funding came from the Stanford-China Geoscience Industrial Affiliates, a consortium consisting of Agip, Amoco, Anadarko, ARCO, British Petroleum, BHP, Chevron, Conoco, Enterprise, Exxon, Fletcher Challenge China, Japan National Oil Corporation, Mobil, Occidental, Shell, Phillips, Statoil, Texaco, Triton, Union Texas, and Unocal. Other funding was provided by the Geological Society of America (GSA), American Association of Petroleum Geologists, Sigma Xi, and the GSA coal division (Medlin award). The Stanford Molecular Organic Geochemistry Industrial Affiliates supported biomarker studies. This work was completed with the support of a National Science Foundation Graduate Research Fellowship. Agip is thanked for funding palynological analyses and for permission to publish. Reviews by Steve Graham, Brian Currie, Albert Bally, Marc Hendrix, Gary Ernst, and Mike McWilliams greatly improved this manuscript.

REFERENCES CITED

Allen, M.B., Windley, B.F., Zhang, C., Zhao, Z.Y., and Wang, G.R., 1991, Basin evolution within and adjacent to the Tien Shan Range, northwest China: Geological Society of London Journal, v. 148, p. 369–379.

Anonymous, 1990, The petroleum geology of Qinghai-Xizhang, *in* The petroleum geology of China, Volume 14: Beijing, Petroleum Industry Press, 423 p.

Bally, A.W., Chou, M., Clayton, R., Eugster, H.P., Kidwell, S., Meckel, L.D., Ryder, R.T., Watts, A.B., and Wilson, A.A., 1986, Notes on sedimentary basins in China—Report of the American sedimentary basins delegation to the People's Republic of China: U.S. Geological Survey Open-File Report 86–327, 107 p.

Bohacs, K., and Suter, J., 1997, Sequence stratigraphic distribution of coaly rocks; fundamental controls and paralic examples: American Association of Petroleum Geologists Bulletin, v. 81, p. 1612–1639.

Bureau of Geology and Mineral Resources of Qinghai Province, 1991, Regional Geology of Qinghai Province: Geological Memoirs, ser. 1, no. 24: Beijing, Geological Publishing House, 662 p., scale 1:1 000 000 (in Chinese).

Bureau of Geology and Mineral Resources of Xinjiang Uygur Autonomous Region, 1993, Regional geology of Xinjiang Uygur Autonomous Region: Geological Memoirs, ser. 1, no. 32: Beijing, Geological Publishing House, 841 p., scale 1:1 500 000.

Burtman, V.S., 1980, Faults of middle Asia: American Journal of Science, v. 280, p. 725–744.

Carroll, A.R., and Bohacs, K.M., 1999, Stratigraphic classification of ancient lakes; balanching tectonic and climatic controls: Geology, v. 27, p. 99–102.

Chinese State Bureau of Seismology, Aerjin Active Fault Belt Research Group, 1992, The Aerjin active fault belt: Beijing, Seismology Publishing House, 319 p.

Coward, M.P., Windley, B.F., Broughton, R.D., Luff, I.W., Petterson, M.G., Pudsey, C.J., Rex, D.C., and Asif, K.M., 1986, Collisional tectonics in the NW Himalayas, *in* Coward, M.P., and Ries, A.C., eds., Collisional tectonics: Geological Society [London] Special Publication 19, p. 203–219.

Crowell, J.C., and Link, M.H., 1982, Geologic history of Ridge basin, southern California: Pacific Section, Society of Economic Paleontologists and Mineralogists, 304 p.

Davis, G.A., Cong, W., Zheng, Y.D., Zhang, J.J., Zhang, C.H., and Gehrels, G.E., 1998, The enigmatic Yinshan fold-and-thrust belt of northern China: New views on its intraplate contractional styles: Geology, v. 26, p. 43–46.

DeCelles, P.G., and Currie, B.S., 1996, Long-term sediment accumulation in the Middle Jurassic–early Eocene Cordilleran retroarc foreland-basin system: Geology, v. 24, p. 591–594.

DeCelles, P.G., and Giles, K.A., 1996, Foreland basin systems: Basin Research, v. 8, p. 105–123.

DeCelles, P.G., Langford, R.P., and Schwartz, R.K., 1983, Two new methods of paleocurrent determination from trough cross-stratification: Journal of Sedimentary Petrology, v. 53, p. 629–642.

Di Hengshu, 1983, Overthrust belts on the northern margin of Qaidam basin and their oil prospecting: Oil and Gas Geology, v. 5, p. 79–88.

Dickinson, W.R., 1970, Interpreting detrital modes of greywacke and arkose: Journal of Sedimentary Petrology, v. 40, p. 695–707.

Dickinson, W.R., and Suczek, C.A., 1979, Plate tectonics and sandstone compositions: American Association of Petroleum Geologists Bulletin, v. 63, p. 2164–2182.

Dickinson, W.R., Beard, L.S., Brakenridge, G.R., Erjavec, J.L., Ferguson, R.C., Inman, K.F., Knapp, R.A., Lindberg, F.A., and Ryberg, P.T., 1983, Provenance of North American Phanerozoic sandstones in relation to tectonic setting: Geological Society of America Bulletin, v. 94, p. 222–235.

Enkin, R.J., Yang, Z., Chen, Y., and Courtillot, V., 1992, Paleomagnetic constraints on the geodynamic history of the major blocks of China from the Permian to the present: Journal Of Geophysical Research, v. 97, p. 13953–13989.

Girardeau, J., Marcoux, J., Allegre, C.J., Bassoullet, J.P., Tang, Y., Xiao, X., Zao, Y., and Wang, X., 1984, Tectonic environment and geodynamic significance of the Neo-Cimmerian Donqiao ophiolite, Bangong-Nujiang suture zone, Tibet: Nature, v. 307, p. 27–31.

Gradstein, F.M., Asterberg, F.P., Ogg, J.G., Hardenboe, J., van Veen, P., Thierry, J., and Huang, Z., 1995, A Triassic, Jurassic and Cretaceous time scale, *in* Berggren, W.A., ed., Geochronology, time-scales and global stratigraphic correlation: SEPM (Society for Sedimentary Geology) Special Publication 54, p. 95–126.

Graham, S.A., Brassel, A.R., Carroll, A.R., Xiao, X., Demaison, G., McKnight, C.L., Liang, Y., Chu, J., and Hendrix, M.S., 1990, Characteristics of selected petroleum source rocks, Xinjiang Uygur Autonomous Region, northwest China: American Association of Petroleum Geologists Bulletin, v. 74, p. 493–512.

Graham, S.A., Hendrix, M.S., Wang, L.B., and Carroll, A.R., 1993, Collisional successor basins of western China: Impact of tectonic inheritance on sand composition: Geological Society of America Bulletin, v. 105, p. 323–344.

Gu, J., 1994, Depositional facies and petroleum: Petroleum exploration of Tarim basin: Beijing, Petroleum Industry Press, 310 p.

Gu, S., and Di, H., 1989, Mechanism of formation of the Qaidam basin and its control on petroleum, *in* Zhu, X., ed., Chinese sedimentary basins: Amsterdam, Elsevier, p. 45–51.

Hendrix, M.S., 1992, Sedimentary basin analysis and petroleum potential, Mesozoic strata, northwest China [Ph.D. thesis]: Stanford, California, Stanford University, 562 p.

Hendrix, M.S., 2000, Evolution of Mesozoic sandstone compositions, southern Junggar, northern Tarim, and western Turpan basins, northwest China: A detrital record of the ancestral Tian Shan: Journal of Sedimentary Research.

Hendrix, M.S., Graham, S.A., Carroll, A.R., Sobel, E.R., McKnight, C.L., Schulein, B.J., and Wang, Z., 1992, Sedimentary record and climatic implications of recurrent deformation in the Tian Shan: Evidence from Mesozoic strata of north Tarim, south Junggar, and Turpan basins, northwest China: Geological Society of America Bulletin, v. 105, p. 53–79.

Hendrix, M.S., Graham, S.A., Amory, J.Y., and Badarch, G., 1996, Noyon Uul syncline, southern Mongolia: Lower Mesozoic sedimentary record of the tectonic amalgamation of central Asia: Geological Society of America Bulletin, v. 108, p. 1256–1274.

Hsü, K.J., 1988, Relict back-arc basins: Principles of recognition and possible new examples from China, *in* Kleinspehn, K.L., and Paola, C., eds., New perspectives in basin analysis: New York, Springer-Verlag, p. 245–264.

Huang, D., Li, J., Zhang, D., Huang, X., and Zhou, Z., 1991, Maturation sequence of Tertiary crude oils in the Qaidam basin and its significance in petroleum resource assessment: Journal of Southeast Asian Earth Sciences, v. 5, p. 359–366.

Huang, D., Zhang, D., and Li, J., 1994, The origin of 4-methyl steranes and pregnanes from Tertiary strata in the Qaidam basin, NW China: Organic Geochemistry, v. 22, p. 343–348.

Ingersoll, R.V., Bullard, T.F., Ford, R.L., Grimm, J.P., Pickle, J.D., and Sares, S.W., 1984, The effect of grain size on detrital modes: A test of the Gazzi-Dickinson point-counting method: Journal of Sedimentary Petrology, v. 54, p. 103–116.

Institut Français du Petrole, 1991, Genex Single Well for MS/Windows, version 2.3.1: Beicip-Franlab, Petroleum Software Division.

Jiao, S.P., 1984, The evolution of geodepression tectonics in Tarim and Chaidam regions, *in* Collected geological papers on Tibetan Plateau: Beijing, Geology Publishing House, p. 129–142.

Johnson, C.L., Graham, S.A., Hendrix, M.S., and Badarch, G., 1997, Sedimentary record of Jurassic–Cretaceous rifting, southeastern Mongolia: Implications for the Mesozoic tectonic evolution of central Asia: Geological Society of America Abstracts with Programs, v. 29, no. 6, p. 228.

Jordan, T.E., 1995, Retroarc foreland and related basins, *in* Busby, C.J., and Ingersoll, R.V., eds., Tectonics of sedimentary basins: Cambridge, Massachusetts, Blackwell Science, p. 331–362.

Lee, K.Y., 1984, Geology of the Chaidamu basin, Qinghai Province, northwest China: U.S. Geological Survey Open-File Report 84-413, 44 p.

Leeder, M.R., 1995, Continental rifts and proto-oceanic rift troughs, *in* Busby, C.J., and Ingersoll, R.V., eds., Tectonics of sedimentary basins: Cambridge, Massachusetts, Blackwell Science, p. 119–148.

Leeder, M.R., and Jackson, J.A., 1993, The interaction between normal faulting and drainage in active extensional basins, with examples from the western United States and central Greece: Basin Research, v. 5, p. 79–102.

Lysak, S.V., 1987, Terrestrial heat flow of continental rifts: Tectonophysics, v. 143, p. 31–41.

Machette, M.N., 1985, Calcic soils of the southwestern United States, *in* Weide, D.L., ed., Soils and Quaternary Geology of the Southwestern United States, Geological Society of America Special Paper 203, p. 1–21.

Miall, A.D., 1996, The geology of fluvial deposits: New York, Springer, 586 p.

Peters, K.E., and Moldowan, J.M., 1993, The biomarker guide: Englewood Cliffs, New Jersey, Prentice Hall, 363 p.

Picard, M.D., and High, L.R., 1981, Physical stratigraphy of ancient lacustrine deposits, *in* Etheridge, F.G. and Flores, L.R., eds., Recent and ancient nonmarine depositional environments: Society of Economic Paleontologists and Mineralogists Special Publication 31, p. 233–260.

Remy, R.R., 1989, Deltaic and lacustrine facies of the Green River Formation, southern Uinta basin, Utah, *in* Nummedal, D., and Remy, R.R., eds., Cretaceous shelf sandstones and shelf depositional sequences, Western Interior Basin, Utah, Colorado, and New Mexico: International Geological Congress, 28th, Field Trip T-119: Washington, D.C., American Geophysical Union, p. 1–11.

Ritts, B.D., 1995, Mesozoic tectonics of the Qaidam region, NW China, and the relationship between the Mesozoic Qaidam and Tarim basins: Geological Society of America Abstracts with Programs, v. 27, no. 6, p. A-456.

Ritts, B.D., 1998, Mesozoic tectonics and sedimentation, and petroleum systems of the Qaidam and Tarim basins, NW China [Ph.D. thesis]: Stanford, California, Stanford University, 691 p.

Ritts, B.D., and Biffi, U., 2000, Magnitude of post-Middle Jurassic (Bajocian) displacement on the Altyn Tagh fault, NW China: Geological Society of America Bulletin, v. 112, p. 61–74.

Ritts, B.D., Hanson, A.D., Zinniker, D., and Moldowan, J.M., 1999, Lower-Middle Jurassic nonmarine source rocks and petroleum systems of the northern Qaidam basin, NW China: American Association of Petroleum Geologists Bulletin, v. 83, p. 1980–2005.

Sclater, J.G., Jaupart, C., and Galson, D., 1980, The heat flow through oceanic and continental crust and the heat loss of the Earth: Reviews of Geophysics and Space Physics, v. 18, p. 269–311.

Searle, M.P., 1991, Geology and tectonics of the Karakorum Mountains: New York, Wiley, 358 p.

Şengör, A.M.C., and Okurogullari, A.H., 1991, The role of accretionary wedges in the growth of continents; Asiatic examples from Argand to plate tectonics: Eclogae Geologicae Helvetiae, v. 84, p. 535–597.

Smith, A.B., 1988, Late Paleozoic biogeography of East Asia and paleontological constraints on plate tectonic reconstructions: Royal Society of London Philosophical Transactions, v. 326, p. 189–227.

Smith, A.B., and Xu, J., 1988, Paleontology of the 1985 Tibet geotraverse, Lhasa to Golmud: Royal Society of London Philosophical Transactions, v. 327, p. 53–105.

Sobel, E.R., 1995, Basin analysis and appatite fission-track thermochronology of the Jurassic-Paleogene southwest Tarim basin, northwest China [Ph.D. thesis]: Stanford, California, Stanford University, 308 p.

Sobel, E.R., 1999, Basin analysis of the Jurassic–Lower Cretaceous southwest Tarim basin, northwest China: Geological Society of America Bulletin, v. 111, p. 709–724.

Sun, F., 1989, Early and Middle Jurassic sporopollenin assemblages of Quiquanhu coal-field of Turpan, Xinjiang: Acta Botanica Sinica, v. 31, p. 638–646.

Tapponnier, P., and Molnar, P., 1977, Active faulting and tectonics in China: Journal of Geophysical Research, v. 82, p. 2905–2930.

Tapponnier, P., and Molnar, P., 1979, Active faulting and Cenozoic tectonics of the Tien Shan, Mongolia, and Baykal regions: Journal of Geophysical Research, v. 84, p. 3425–3459.

Tseysler, V.M., Florenskiy, V.S., Vasyukov, V.S., and Turov, A.V., 1982, Tectonic structure of the northern Fergana Range: International Geology Review, v. 24, p. 881–890.

Ulmishek, G., 1984, Geology and petroleum resources of basins in western China: Argonne, Illinois, Argonne National Laboratory report ANL/ES-146, 131 p.

Ungerer, P., Burrus, J., Doligez, B., Chenet, P.Y., and Bessis, F., 1990, Basin evaluation by integrated two-dimensional modelling of heat transfer, fluid flow, hydrocarbon generation and migration: American Association of Petroleum Geologists Bulletin, v. 74, p. 309–335.

Vakhrameyev, V.A., 1982, Classopolis pollen as indicator of Jurassic and Cretaceous climate: International Geology Review, v. 24, p. 1190–1196.

Wang, Q., and Coward, M.P., 1990, The Chaidam basin (NW China): Formation and hydrocarbon potential: Journal of Petroleum Geology, v. 13, p. 93–112.

Wang, Y.S., and Chen, J.L., 1984, The formation and Evolution of Chaidam terrane, *in* Collected geological papers on Tibetan Plateau: Beijing, Geology Publishing House, p. 27–37.

Wang, Z.Z., 1981, The feature of Mesozoic sedimentary distribution in the northern margin of Caidam basin: Petroleum Exploration and Development, v. 2, p. 20–26.

Webb, L.E., Hacker, B.R., Ratschbacher, L., Leech, M., Dong, S., and Peng, L., 1997, Mesozoic tectonism in the Qinling-Dabie collisional orogen: New constraints on the multistage exhumation of ultrahigh-pressure rocks: Geological Society of America Abstracts with Programs, v. 29, no. 6, p. A-119.

Webb, L.E., Graham, S.A., Johnson, C.L., Badarch, G., and Hendrix, M.S., 1999, Occurrence, age, and implications of the Yagan-Onch Hayrhan metamorphic core complex, southern Mongolia: Geology, v. 27, p. 143–146.

Xia, B., 1990, Terranes of Xizang (Tibet), China, *in* Wiley, T.J., et al., eds., Terrane analysis of china and the Pacific rim: Houston, Texas, Circum-Pacific Council for Energy and Mineral Resources Earth Science Series, v. 13, p. 231–241.

Xu, W., He, Y., and Yan, Y., 1989, Tectonic characteristics and hydrocarbons of the Hexi Corridor, *in* Zhu, X., ed., Chinese sedimentary basins: Amsterdam, Elsevier, p. 53–62.

Ye, L., Wang, G., and Ahzi, G., 1996, Geology of the Kuqa River and Kalpin areas, Tarim Basin: International Geological Congress, 30th, Beijing, Geological Publishing House, p. 88.

Yin, A., and Nie, S., 1993, An indentation model for the North and South China collision and the development of the Tan-Lu and Honam fault systems, eastern Asia: Tectonics, v. 12, p. 801–813.

Yin, A., and Nie, S., 1996, A Phanerozoic palinspastic reconstruction of China and its neighboring regions, *in* Yin, A., and Harrison, T.M., eds., Tectonic evolution of Asia: Cambridge, Cambridge University Press, p. 442–485.

Zhang, Z.M., 1985, Plate tectonics and high P/T metamorphic rocks of China [Ph.D. thesis]: Stanford, California, Stanford University, 193 p.

Zhang, Z.M., Liou, J.G., and Coleman, R.G., 1984, An outline of the plate tectonics of China: Geological Society of America Bulletin, v. 95, p. 295–311.

Zheng, Y., Zhang, Q., Wang, Y., Liu, R., Wang, S.G., Zuo, G., Wang, S.Z., Lkaasuren, B., Badarch, G., and Badamgarav, Z., 1996, Great Jurassic thrust sheets in Beishan, Gobi areas of China and southern Mongolia: Journal of Structural Geology, v. 18, p. 1111–1126.

Zhou, Z., and Chen, P., 1992, Biostratigraphy and geological evolution of Tarim: Beijing, Science Press, 400 p.

Zonenshain, L., Kuzmin, M., Natapov, L., and Page, B., 1990, Geology of the USSR: A Plate Tectonic Synthesis, Page, B. (ed.): American Geophysical Union Geodynamics Series, v. 21, 242 p.

MANUSCRIPT ACCEPTED BY THE SOCIETY JUNE 5, 2000

Geological Society of America
Memoir 194
2001

Sedimentary record of Mesozoic deformation and inception of the Turpan-Hami basin, northwest China

Todd J. Greene*
Department of Geological and Environmental Sciences, Stanford University, Stanford, California, 94305-2115, USA
Alan R. Carroll*
Department of Geology and Geophysics, University of Wisconsin, 1215 West Dayton Street, Madison, Wisconsin 53706, USA
Marc S. Hendrix*
Department of Geology, University of Montana, Missoula, Montana 59812, USA
Stephan A. Graham*
Department of Geological and Environmental Sciences, Stanford University, Stanford, California 94305-2115, USA
Marwan A. Wartes*
Department of Geology and Geophysics, University of Wisconsin, 1215 West Dayton Street, Madison, Wisconsin 53706, USA
Oscar A. Abbink*
Laboratory of Palaeobotany and Palynology, Utrecht University, Utrecht, The Netherlands

ABSTRACT

The Turpan-Hami basin is a major physiographic and geologic feature of northwest China, yet considerable uncertainty exists as to the timing of its inception, its late Paleozoic and Mesozoic tectonic history, and the relationship of its petroleum systems to those of the nearby Junggar basin. To address these issues, we examined the late Paleozoic and Mesozoic sedimentary record in the Turpan-Hami basin through a series of outcrop and subsurface studies. Mesozoic sedimentary facies, regional unconformities, sediment dispersal patterns, and sediment compositions within the Turpan-Hami and southern Junggar basins suggest that these basins were initially separated between Early Triassic and Early Jurassic time.

Prior to separation, Upper Permian profundal lacustrine and fan-delta facies and Triassic coarse-grained braided-fluvial–alluvial facies were deposited across a contiguous Junggar-Turpan-Hami basin. Permian through Triassic facies were derived mainly from the Tian Shan to the south, as indicated by northward-directed paleocurrent directions. This is consistent with the sedimentary provenance of Triassic sandstone (mean $Qm_{29}F_{29}Lt_{42}$, $Qp_{23}Lvm_{49}Lsm_{28}$, and $Qm_{51}P_{25}K_{24}$) and conglomerate (~32% granitic clasts) in the northern Turpan-Hami basin. We interpret a relative increase in quartz and feldspar concentration and a relative decrease in volcanic lithic grains in the northern Turpan-Hami basin to reflect unroofing in the Tian Shan and exposure of late Paleozoic granitoid rocks. In addition, two basinwide unconformities of Late Permian–Early Triassic and Early Triassic–middle-Late Triassic age attest to deformation within the Turpan-Hami basin and associated continued uplift and erosion of the Tian Shan.

*E-mails: Greene, greene@pangea.stanford.edu; Carroll, carroll@geology.
wisc.edu; Hendrix, marc@selway.umt.edu; Graham, graham@pangea.stanford.
edu; Wartes, mwartes@geology.wisc.edu; Abbink, O.A.Abbink@bio.uu.nl.

Greene, T.J., et al., 2001, Sedimentary record of Mesozoic deformation and inception of the Turpan-Hami basin, northwest China, *in* Hendrix, M.S., and Davis, G.A., eds., Paleozoic and Mesozoic tectonic evolution of central Asia: From continental assembly to intracontinental deformation: Boulder, Colorado, Geological Society of America Memoir 194, p. 317–340.

By Early Jurassic time, the Turpan-Hami basin and the southern Junggar basin became partitioned by uplift of the Bogda Shan, a major spur of the Tian Shan. In contrast to the thoroughgoing northward-directed Permian-Triassic depositional systems, Lower through Middle Jurassic strata begin to reflect ponded coal-forming, lake-plain environments within the Turpan-Hami basin. These strata contain paleocurrent indicators reflecting flow off the intervening Bogda Shan. A basinwide Lower Jurassic–Middle Jurassic unconformity in the Turpan-Hami basin suggests continued uplift and erosion of the Bogda Shan, consistent with a return to more lithic-rich sandstone and volcanic-rich conglomerate compositions. These sedimentary facies, paleocurrent, and provenance data sets provide the best available constraints on the initial uplift of the Bogda Shan and the first documentary evidence of intra-Mesozoic shortening within the basin.

INTRODUCTION

The present-day Turpan-Hami basin is a major physiographic feature of central Asia. It contains the second lowest elevation on earth (−154 m), yet is bounded on its northern flank by the Bogda Shan, a major mountain range containing peaks as high as 5570 m (Fig. 1A). Few published data bear directly on the initiation of the Turpan-Hami basin or its premodern physiographic and depositional history. Post-Carboniferous sedimentary facies, paleocurrent dispersal patterns, and sandstone

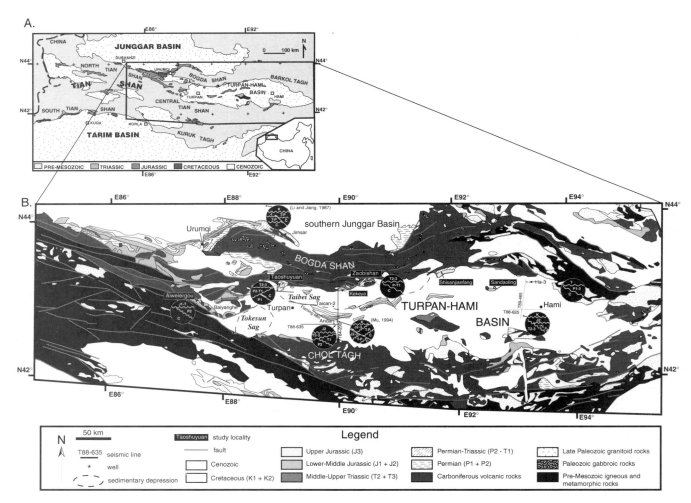

Figure 1. A: Location map of study area within Xinjiang Uygur Autonomous Region, northwestern China. Boxed area is shown in detail in B. B: Geologic map of Turpan-Hami basin, showing various study locations in basin, location of seismic lines and wells referred to in text, and two main western sedimentary depressions, Tokesun Sag and Tabei Sag (adapted from Chen et al., 1985). Solid black circles show general stratigraphic relationships and unconformities of Carboniferous through Cretaceous strata at various locations; data for each black circle are from this study unless otherwise noted. Fault locations are adapted from Chen et al. (1985) and Allen et al. (1993).

compositions from the southern Junggar and northern Tarim basins (Hendrix et al., 1992; Carroll et al., 1995) suggest that the Junggar and Tarim basins have existed as discrete physiographic features, separated by the intervening Tian Shan, since late Paleozoic time. However, previous interpretations of the time of inception and early structural style of the Turpan-Hami basin vary widely. Hendrix et al. (1992) suggested, on the basis of paleocurrents, sedimentary facies, and isopach data, that the Turpan-Hami basin was established as a discrete physiographic feature by Early Jurassic time, in response to compressional uplift of the Bogda Shan. Allen et al. (1995) proposed that the Turpan basin was one of several major depocenters created by transtensional rotation within a Late Permian–Triassic sinistral shear system. In determining the sedimentary provenance of northern Turpan-Hami deposits, Greene et al. (1997) and Greene and Graham (1999) suggested that granitic cobbles contained within Lower Triassic deposits were derived solely from the Central and South Tian Shan blocks, south of the Turpan-Hami basin, and inferred that the ancestral Bogda Shan had not been uplifted by that time.

Better timing constraints for the initiation of the Turpan-Hami basin are especially important because they bear directly on the lateral extent of thick, organic-rich Upper Permian lacustrine shale in the southern Junggar basin. Carroll et al. (1992) and Clayton et al. (1997) demonstrated that these and equivalent strata are the source of oils produced at the Karamay oil field and other fields along the northwestern Junggar basin margin, which collectively are thought to contain reserves of 2–3 billion barrel oil fields (Taner et al., 1988). If the initial separation of these two basins by the Bogda Shan occurred during post-Late Permian time, then it is possible that the rich lacustrine source rocks in the southern Junggar basin may extend into the Turpan-Hami basin (Fig. 1B).

At a more regional level, structural, geochronologic, and sedimentologic evidence suggests that, during Mesozoic time, the Turpan-Hami and Bogda Shan area existed within a broader zone of contractile deformation that underwent several episodes of shortening. Mesozoic contractile structures have been documented in the subsurface of the Junggar and Tarim basins (Liu et al., 1979; Liu, 1986; Wu, 1986; Li and Jiang, 1987; Peng et al., 1990; Zhao et al., 1991a, 1991b; Li, 1995; Hendrix et al., 1996; Wu et al., 1996), and Mesozoic fission-track ages from the core of the Tian Shan (Dumitru et al., this volume) suggest significant Mesozoic unroofing. Hendrix et al. (1992) documented several successions of coarse, alluvial conglomerate in each basin and inferred that they represent episodic renewed downcutting of the ancestral Tian Shan. Hendrix (2000) interpreted variations in the composition of Mesozoic sandstone from the southern Junggar and northern Tarim basins to reflect polyphase deformation of the ancestral range. Vincent and Allen (this volume) documented several Mesozoic angular unconformities and coarse conglomerate successions in the northeastern Junggar basin and interpreted them to reflect far-field deformation originating at the southern continental margin of Asia. Allen

et al. (1993) also identified two Mesozoic compressional episodes in the Turpan-Hami basin and suggested that Lower Permian marine facies in the northern Turpan basin were deposited as a foreland basin fill related to north-vergent thrust faults in the Tian Shan region.

Key to understanding the tectonic and sedimentary history of the Turpan-Hami basin is better documentation of the record of sedimentary fill within the basin, combined with information regarding the history of uplift and unroofing of the Bogda Shan. Unfortunately, very little information has been published about the structure of the interior of the Bogda Shan; access to key exposures is difficult, and there are no seismic lines that transect the range. Tectonic interpretations from the Turpan-Hami basin are largely limited either to specific reports on a focused petroleum-related topic (Huang et al., 1991; Wang et al., 1993, 1998; K. Cheng et al., 1996; L. Liu et al., 1998), or regional interpretive syntheses that lack detailed data sets and hence are impossible to evaluate critically (Fang, 1990, 1994; Li and Shen, 1990; Shen, 1990; Chen, 1993; Tao, 1994; Allen et al., 1995; Shao et al., 1999b). Shao et al. (1999b), for example, used basin subsidence curves to infer a period of Late Permian thermal subsidence that was followed by a Middle-Triassic to early Tertiary period of flexural subsidence for the Turpan-Hami basin. However, error bars for age control are absent, and paleo-elevation level is set at sea level for the past 250 m.y. in their analysis. Likewise, numerous Chinese authors have published reports on Early-Middle Jurassic depositional history and sequence stratigraphy, mainly due to its importance in petroleum trapping for Turpan-Hami oil production (Wu and Zhao, 1997; Li, 1997; Qiu et al., 1997; and many others). These reports, however, are difficult to assess because stratigraphic and sedimentologic data are not presented. Marine sequence stratigraphic nomenclature is often used to describe high-resolution lacustrine base-level changes during Early-Middle Jurassic time, without presenting accurately located seismic lines, well-log cross sections with appropriate well logs used for correlating flooding surfaces or sequence boundaries (e.g., gamma-ray log), or fossil assemblages to justify age control.

In order to better define the timing of initial formation of the Turpan-Hami basin and, by extension, the initial uplift of the Bogda Shan, we examined key outcrops of Mesozoic strata, oil well cores and electrical logs, and seismic lines during the summers of 1996, 1997, and 1998. We collected Permian through Jurassic stratigraphic and sedimentologic data at four different localities along the northern basin margin of Turpan-Hami (Fig. 1B): Aiweiergou (westernmost corner), Taoshuyuan (west), Zaobishan (north-central), north of the town of Shisan-jianfang (east), and Sandaoling (east). We measured and described ~20 stratigraphic sections from exposed Mesozoic strata, and we used those data as the basis for interpreting the environment of deposition for each section. We also examined nine previously unpublished reflection seismic profiles from the interior of the basin, along with data from oil well boreholes, and we used those to compliment our surface data set. In addition to providing data on the timing of initiation of the Turpan-Hami basin as a depositional entity, we sought to examine the structure of the basin and its record of fill for evidence of Mesozoic shortening, as has been suggested from other basins of western China. In this chapter, we summarize our data pertaining to Mesozoic sedimentary facies relations and sediment dispersal systems within the Turpan-Hami basin. We integrate those data with sandstone composition and conglomerate clast count data. Collectively, these data provide the best available constraints on Mesozoic shortening within the Turpan-Hami basin, and, indirectly, on the initial uplift of the Bogda Shan.

STRATIGRAPHY AND SEDIMENTOLOGY

Our sedimentary data set, described in detail in the following, indicates that several fundamental changes in depositional style occurred in the Turpan-Hami basin in Permian through Middle Jurassic time. This depositional history can be summarized as follows: Upper Permian strata are marked by fine-grained, mostly profundal lacustrine, fan-delta, and fluvial facies (Carroll et al., 1992; Wu and Zhao, 1997; Carroll, 1998; Wartes et al., 1998, 1999, 2000), whereas Triassic deposits contain coarse-grained braided-fluvial–alluvial facies. In sharp contrast are Lower through Middle Jurassic strata that reflect the ponding of water in coal-forming, lake-plain environments. Concomitant with this major change in Jurassic environments was a shift during Early and Middle Jurassic time in the main sedimentary depocenter, from the western Tokesun Sag to the north-central Tabei Sag (Fig. 1B: Huang et al., 1991; Wang et al., 1996; Qiu et al., 1997; Wu and Zhao, 1997).

Age control

Due to the lack of interbedded datable volcanic units and a paucity of paleomagnetic studies, age assignments in the Turpan-Hami basin are based solely on plant microfossil (spores, pollen), plant macrofossil, and vertebrate fossil assemblages. This study relies heavily on new palynological results (Abbink, 1999) derived from selected mudrocks within borehole core and outcrop samples (Fig. 2). Age interpretations are based on the first occurrence datum and/or last occurrence datum of spores and pollen (sporomorphs). Most of the stratigraphic ranges of the encountered sporomorphs are not calibrated against marine faunas from northwest China (e.g., Huang, 1993; Ouyang, 1996). We therefore conducted a data search to determine the range of the stratigraphic marker taxa for northwest China based on selected literature references from other regions. Furthermore, on the basis of the overall content of the sporomorph assemblages, the palynoflora of our samples are not considered to be endemic; rather, the floral associations closely resemble those of Europe and of the former USSR. On the basis of our palynological studies, along with independent age assignments published by Chinese paleontologists (Liao

et al., 1987; Z. Cheng et al., 1996; Wang et al., 1996), most formations are dated to the epoch level, sufficient for the tectonic interpretations in this paper (Fig. 2).

Triassic

Taodongou (east Taoshuyuan). At Taodongou (Figs. 1 and 3A), 144 m of Lower Triassic strata (T1s; see Fig. 2 for formation abbreviations herein) are truncated by an erosional unconformity and overlain by at least 70 m of Middle to Upper Triassic (T2-3k) conglomerate (Figs. 3B and 4A). The lower 80 m of T1s are mostly red to pink mudstone locally interbedded by 4–8 m sandy conglomerate interbeds that sharply fine to siltstone. The upper 64 m of the T1s section represent a series of stacked upward-fining sandy conglomeratic intervals ~8–16m thick that fine to siltstone; little to no mudstone is exposed. Erosional contacts are common at the base of each package. Toward the top of the T1s, the upward-fining sand packages pinch out laterally over 20–30 m. An erosional contact with ~25 m of relief separates the T1s deposits from the overlying higher energy T2-3k conglomerate (Fig. 3B). The latter consists of clast-supported conglomerate containing abundant tabular and trough cross stratification with >1 m relief. Cross-bedded sandy lenses, ~ 1 m thick, are common; little mudstone is present.

We interpret these deposits to represent two different depositional environments. The lower 144 m of T1s deposits record relatively low energy flow associated with a braided river flood plain (Nanson and Croke, 1992). The upper 64 m record increasing depositional energy and erosive power that produced minor channels over a braid plain. Following an ~25-m-deep incision, a gravel bed braided-fluvial system dominated, incising large channels and depositing a series of gravel bars (Miall and Gibling, 1978). Hendrix et al. (1992) described similar deposits from Upper Canfanggou Group (lowermost Triassic) strata at Taodongou (Fig. 3A). They described a detailed 50 m section of coarse conglomerate and lenticular sandstone and siltstone red beds deposited in a braided-fluvial–alluvial environment.

Zaobishan. Three measured sections were described on a meter scale at the Zaobishan locality (Fig. 4, B, C, and D). Sections 4B and 4C both record Lower Triassic and Middle-Upper Triassic deposits on the north and south flanks of a large east-west–trending Carboniferous-cored anticline (Figs. 5A and 6), and are chronostratigraphically similar to the measured section at Taodongou (Fig. 4A).

Both measured sections 4B and 4C can be divided into a lower fine-grained section and an upper coarse-grained section, presumably separated by an erosional unconformity at the 25 m mark (although we observed an erosional contact in section 4B, this contact is covered at section 4C). The lower finer grained portion consists of fine- to medium-grained sandstone beds with trough cross-beds and planar bed stratification. The lower half of section 4B is relatively lower energy than 4C, the former con-

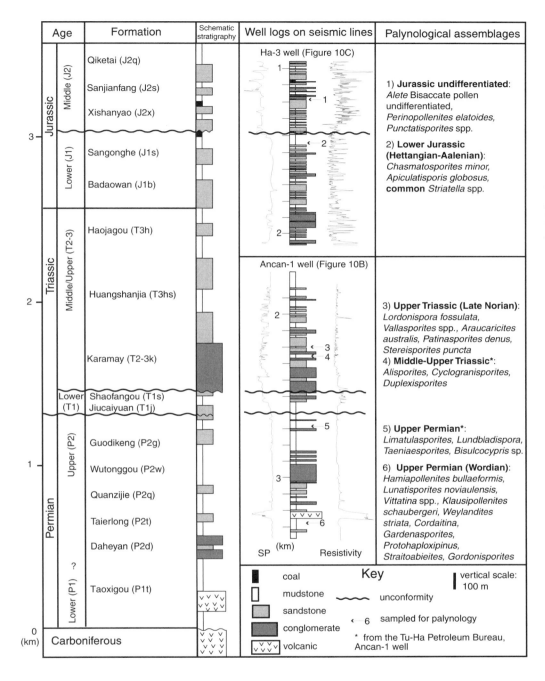

Figure 2. Generalized stratigraphic and lithologic chart of Turpan-Hami stratigraphy from Carboniferous through Middle Jurassic strata with reported fauna and flora assemblages and stratigraphic positions of major unconformities discussed in text. Palynological assemblages from Ha-3 and Ancan-1 wells serve as age control for seismic line drawings. Unless otherwise noted, all palynology-based age interpretations are from Abbink (1999). For Upper Permian age assignment of sample 6, note that association with *Limatulasporites* (=*Gordonisporites*) and *Taeniasporites* (=*Lunatisporites*) is not typical Late Permian. Although latest Permian samples from Salt Range (Pakistan) contain the taxa, these taxa are also typical for Early Triassic, in particular in north China (Ouyang and Norris, 1988).

taining paleosol horizons and the latter having more stacked trough cross-bedded sandstone beds. Paleocurrent indicators point north-northeast for section 4C. At the 25 m mark in both sections, a sharp increase in grain size occurs, above which are at least 30 m of polymictic conglomerate containing large (>1 m relief) trough cross-beds mantled by pebbles and small cobbles. The upper half of section 4B consists of sandy conglomerate with 1–2-m-thick medium- to coarse-grained trough cross-bedded sandstone lenses distributed throughout the section. The conglomerate consists of 2–10-m-thick shingled,

lenticular packages that grade from clast supported at their scoured bases to matrix supported.

We interpret sections 4B and 4C to be similar in depositional style to section 4A. The Lower Triassic (T1s) sections record lower energy deposition in a braided river flood plain or distal sheetflood environment (Nanson and Croke, 1992). Middle-Upper Triassic (T2–3k) strata recorded a major change in depositional energy represented by gravelly braided-fluvial systems (Miall and Gibling, 1978) with associated sandy overbank sheet flows. Hendrix et al.'s (1992) description of Triassic

A. Taodongou locality, east Taoshuyuan

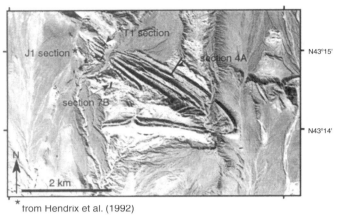

* from Hendrix et al. (1992)

B. Section 4A, Triassic deposits

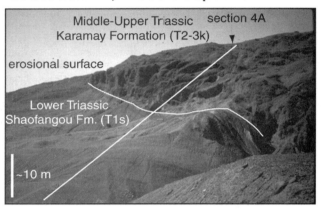

C. Section 7B, Lower Jurassic deposits

Figure 3. A: Corona Satellite image of Taodongou locality in east Taoshuyuan. Black lines show locations of sections 4A and 7B from this study, as well as Lower Triassic (T1) and Lower Jurassic (J1) measured sections from Hendrix et al. (1992). B: Intra-Triassic erosional surface at Taodongou, where braided-fluvial deposits of Karamay Formation (T2-3K) overlie Shaofangou Formation (T1s); stratigraphic "up" is to upper right of photo. White bar shows location of portion of measured section 4A. C: Entire 66 m measured section 7B (shown as white line) representing Lower Jurassic flood-plain deposits; stratigraphic "up" is to upper left of photo.

deposits at Qijiagou (southern Junggar basin) was similar to that for Triassic strata at Zaobishan and Taodongou. Namely, fine-grained red beds of lowermost Triassic deposits sharply grade to coarse braided fluvial conglomerate of the Middle-Upper Triassic Karamay Formation.

Section 4D was measured in a valley of Lower Triassic deposits (T1s) and is presented as a photomosaic in Figure 5B. Generally, section 4D consists of 2–10-m-thick lenses of trough cross-bedded conglomerate packages interbedded with 1–2-m-thick medium- to coarse-grained, trough cross-bedded sandstone. Cobbles commonly mantle the troughs within the conglomerate, and scoured basal contacts are ubiquitous. Measured paleocurrent indicators for trough cross-beds point north and northeast (Fig. 5C).

The excellent two-dimensional valley-wall exposure shown in Figure 5B contains all the major elements of a classic gravel-bed braided-fluvial system (Miall, 1996), i.e., abundant gravel bars, sandy bedforms, and stacked, channelized sandy conglomerate packages with numerous internal erosion surfaces, lack of fines, and few observed downstream accretion surfaces. Zhao et al. (1991) documented similar T1s deposits in the Dalongou area (southern Junggar basin) on the north side of the Bogda Shan, and also interpreted them to reflect a braided-fluvial environment. A modern analog for the coarse-grained Triassic deposits throughout Turpan-Hami occurs in the western Tian Shan in the Lake Issyk-kul area of Kyrgyzstan, where intermontane basins commonly contain gravel-dominated braided river environments with associated transverse bars (Sgibnev and Talipov, 1990; Rasmussen and Romanovsky, 1995).

Shisanjianfang. Section 4E depicts Middle-Upper Triassic conglomerate just north of the town of Shisanjianfang (Figs. 1 and 4E). Several 4–18-m-thick beds of clast- and matrix-supported conglomerate interfinger with medium- to coarse-grained trough cross-bedded sandstone. The mean paleocurrent indicator direction points north-northeast. We also interpret section 4E to represent deposition in a braided-fluvial system with occasional lower energy sand-rich, matrix-support deposition similar to those described herein.

Lower and Middle Jurassic

We studied exposures of Lower and Middle Jurassic deposits in the western part of the basin at Aiweiergou, in the north-central part of the basin at Kekeya, at Taodongou (east Taoshuyuan), and in the eastern Hami basin at Sandaoling (Figs. 1 and 7). We also studied Middle Jurassic strata in the central part of the basin at Flaming Mountain where thick sandstone packages of the Xishanyao and Sanjianfang Formations (J2x and J2s) and lacustrine deposits of the Qiketai Formation (J2q) are exposed (Huang et al., 1991; Schneider et al., 1992; C. Wang et al., 1996; Liu and Di, 1997; H. Wang et al., 1997; Wu and Zhao, 1997; L. Liu et al., 1998; Greene et al., 2000).

Aiweiergou. Near the Aiweiergou coal mine, we measured a 240 m section of Lower Jurassic Badaowan Formation (J1b)

Triassic outcrops of the Turpan-Hami basin

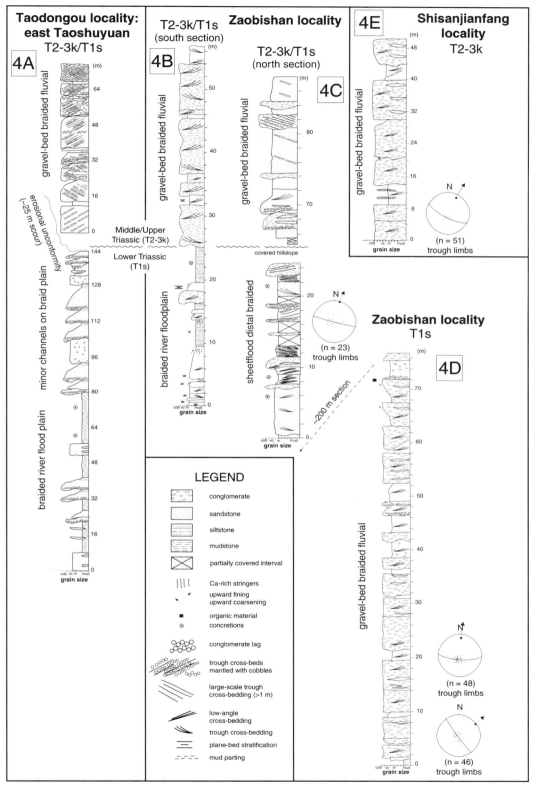

Figure 4. Summary of Triassic measured sections in Turpan-Hami basin. Note erosional unconformity between undifferentiated Middle-Upper Triassic conglomerate scouring into fine sandstone of Lower Triassic deposits in sections 4A, 4B, and 4C. This erosional surface (photo in Fig. 3B) is present at both Zaobishan and Taodongou (east Taoshuyuan) as well as in seismic line T88-635. Note that paleocurrent indicators are directed north to northeast in sections 4C, 4D, and 4E.

A. Zaobishan locality

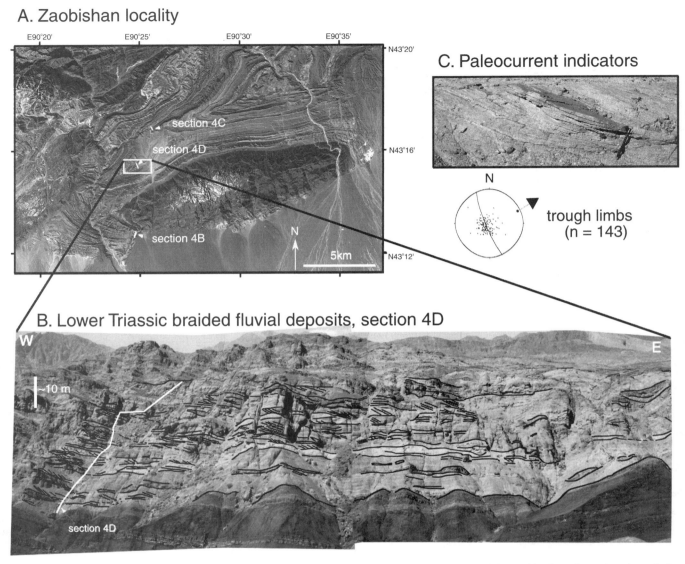

Figure 5. A: Corona Satellite image of Zaobishan locality showing location of sections 4B, 4C, and 4D. White box shows location of photomosaic pictured in part B. B: Photomosaic of Lower Triassic braided-fluvial deposits at Zaobishan. Black lines highlight interpreted depositional surfaces. White line represents 76-m-long measured section 4D. Stratigraphic "up" is to top of photo. Note numerous large-scale cross-beds indicating flow to right (northeast). C: Photo of ubiquitous trough cross-sets (pencil for scale) measured at locality pictured in B, along with corresponding stereoplot of measured paleocurrent indicators. Mean vector indicates northeast sediment-dispersal direction.

deposits (Fig. 7A). The lower 96 m consist mainly of 8–16-m-thick, clast-supported, polymictic cobble conglomerate with several associated 1–2-m-thick medium- to coarse-grained, trough cross-bedded and plane-bed laminated sandstone lenses. Conglomerate intervals commonly contain scoured basal contacts and are laterally continuous over several meters. Conglomerate clasts are 2–8 cm and are well rounded. Wood fragments are common, as are interbeds of 0.5–1-m-thick, rippled, fine-grained sandstone and siltstone.

We interpret this portion of the Badaowan Formation to have been deposited in a gravel-sand meandering fluvial environment (cf. Nijman and Puigdefábregas, 1978; Campbell and

Hendry, 1987). Conglomerate beds have crude horizontal stratification, and they commonly interfinger with large-amplitude cross-bedded sandy facies. The rippled siltstone beds are most likely interchannel overbank deposits or abandoned channels and meander scars on the flood plain that are preserved as silt plugs. Campbell and Hendry (1987) described modern gravel meander lobes on the Saskatchewan River that contain many of the same elements as this portion of the Badaowan Formation: horizontal gravel sheets, interfingering cross-bedded pebbly sands, and clay-silt plugs.

The remainder of the section (above the 96 m mark) is relatively finer grained and more organic rich than the underlying

Summary stratigraphic column for Zaobishan locality

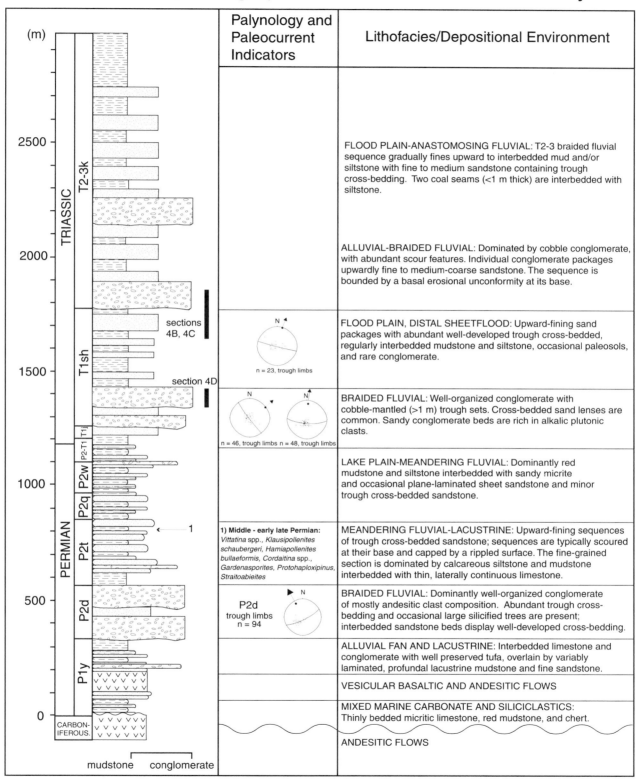

Figure 6. Measured stratigraphic section of Permian through Triassic deposits at Zaobishan with descriptions of lithostratigraphy and depositional environment (see Figs. 1 and 5 for locations). This generalized section at Zaobishan is typical along north rim of Turpan-Hami basin. Note that paleocurrent indicators for Upper Permian and Lower Triassic deposits are pointed northwest to northeast. Arrow (labeled 1) points to location of P2t mudstone from which we recovered mid-early Late Permian palynomorphs.

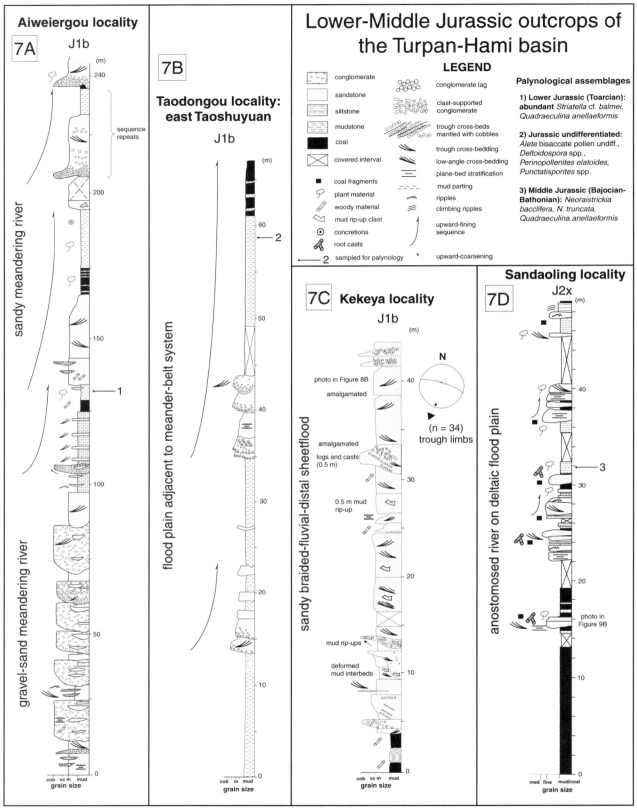

Figure 7. Summary of Lower and Middle Jurassic outcrops throughout Turpan-Hami basin (see Fig. 1B for localities). Measured section 7A, at Aiweiergou, represents only portion of thick Lower Jurassic strata in western depression of Turpan-Hami (Tokesun Sag). Section 7B (photo in Fig. 3C) represents contemporaneous deposits in north-central depression (Tabei Sag). Generally, Lower Jurassic deposits in Tabei Sag are not as thick and coarse as Lower Jurassic deposits in Tokesun Sag. Section 7C is from Kekeya locality. Paleocurrent indicators in measured section 7C are directed south to southeast, signifying reversal from previously north- to northeast-directed paleocurrent indicators in Permian and Triassic deposits. Middle Jurassic deposits from Sandaoling coal mine are described in section 7D. Numbered arrows signify positions of samples studied for palynology; interpreted ages and palynological assemblages for each sample are listed in legend (Abbink, 1999).

section (Fig. 7A). We interpret this section to represent a sandy meandering depositional environment with associated crevasse splays, flood-plain, and overbank deposits. This interval consists of three main upward-fining successions (30–60-m thick), each consisting of basal scouring cross-bedded conglomerate and sandstone capped by siltstone, carbonaceous mudstone, and coal with high mud content. In the lower portion (96–134 m), mudstone is interbedded with tabular fine-grained, trough cross-bedded sandstone beds (<1 m thick) and a couple of 3–5-m-thick high-mud humic coal seams, abundant wood fragments, and plant material. The tabular sandstone layers most likely record periodic flooding events or small chute channels deposited on a meander flood plain. The next overlying section (134–196 m) upwardly fines from medium-grained, trough cross-bedded sandstone with rare <1 m conglomerate lenses to coal and carbonaceous mudstone containing plant debris and numerous concretions. This section could either represent prograding sandy bars or a meandering fluvial channel. The final portion (205–236 m) consists of an upward-fining package with a scoured base and a basal conglomerate lag that abruptly fines to a medium- to coarse-grained, trough cross-bedded sandstone with dispersed cobbles. We interpret these repeating upward-fining cycles to reflect small meander channels.

Taodongou (east Taoshuyuan). Excellent lateral exposure of J1b deposits can be observed near the Taodongou coal mine in north-central Turpan-Hami (Fig. 3, A and C). Measured section 7B encompasses 65 m (represented by a white line in Fig. 3C) of mostly green-gray mudstone capped by a 4–5-m-thick coal seam. An 8-m-thick succession of amalgamated, fining-upward sandstone beds occurs midway through the section. These beds contain trough cross-beds mantled by pebbles, plane-bed stratification, and scoured basal contacts. Overall, the Taodongou J1b section records deposition on a flood plain, overbank environment with associated crevasse splays that are probably part of larger meander-belt system. Hendrix et al. (1992) also examined lowermost Jurassic deposits at Taodongou that they interpreted to be from a braided-fluvial flood plain; however, no attempt is made to correlate the two sections.

Kekeya. Near the Kekeya coal mine (Figs. 1B and 8) we measured a detailed 44-m-thick section of J1b deposits (section 7C). The lowermost 4 m consist of coal interbedded with dark gray, carbonaceous shale containing abundant coaly wood fragments. These units are overlain by 40 m of amalgamated 2–8-m-thick intervals of medium- to coarse-grained, trough cross-bedded sandstone. Silicified logs and casts to 0.5 m long, mud rip-up clasts, and basal conglomerate lags are all common throughout the section. Trough cross-bed orientations generally indicate paleoflow to the south-southwest (Fig. 8, B and C). On the basis of the lack of lateral accretion surfaces and abundant erosional features with ubiquitous mud rip-up clasts and woody debris, we interpret this succession to reflect sandy braided-fluvial or distal sheetflood sedimentation along a flood plain (Olsen, 1989; Miall, 1996).

Sandaoling. In the eastern Turpan-Hami basin, 50 km northwest of Hami, we measured Middle Jurassic coaly strata

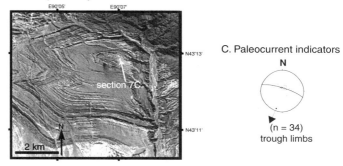

A. Kekeya locality

C. Paleocurrent indicators

(n = 34)
trough limbs

B. Trough cross-bedding in J1b deposits, section 7C

Figure 8. A: Corona Satellite photo of Kekeya locality showing location of section 7C. B: Trough cross-bedding associated with sand-rich braided-fluvial facies of Lower Jurassic Badaowan Formation (J1b) described in section 7C (photo is from 40 m mark of section 7C). Stratigraphic "up" is to upper right of photo. C: Stereoplot of measured paleocurrent indicators from section 7C. Mean vector indicates south-southwest sediment-dispersal direction.

in the Sandaoling coal mine (Figs. 1B and 9). The base of section 7D contains a 12-m-thick coal seam, overlain by organic-rich interbedded coal and laminated mudstone with small (1 m thick) channelized sand bodies containing cross-bedding, coal fragments, and root casts (Fig. 9B). The channels appear to be isolated, with minimal evidence for lateral migration, and are bounded by flood-plain deposits. We interpret the depositional environment as an anastomosed river deposited on a subaqueous, deltaic flood plain, as Smith (1986) described for the Magdalena River in northwestern Columbia.

REGIONAL UNCONFORMITIES

Pre-Tertiary uplift of the Bogda Shan undoubtedly would have affected the Mesozoic sedimentary fill of the Turpan-Hami basin on a regional scale, and therefore might be expressed as regional unconformities. To test this hypothesis, we examined surface exposures, as well as nine different regional seismic reflection lines throughout the Turpan-Hami basin (made available by the Chinese National Petroleum Corporation). The seismic lines were shot either parallel (east-west) or perpendicular (north-south) to the basin axis, and were recorded down to ~3.0 s (two-way traveltime). Two sets of two crossing seismic

A. Sandaoling locality

B. Root casts in J2x deposits, section 7D

Figure 9. A: Outcrop of anastomosed river deposits described in section 7D (black line) from Middle Jurassic Xishanyao Formation (J2x) at Sandaoling coal mine (Figs. 1B and 7D). Note person for scale at bottom of photo. B: Photo of root casts (indicated by arrows) from 16 meter mark of section 7D. Stratigraphic "up" is to top for both photos.

line drawings are presented (see Fig. 1B for locations): T84–200 and T88-635 in the west part of Turpan-Hami (Fig. 10, A and B), and T89-465 and T88-625 in the east (Fig. 10, C and D). Palynology from this study, as well as from the Turpan-Hami Petroleum Bureau, provide age control for Permian through Jurassic reflectors intersecting the Ancan-1 and Ha-3 wells (Figs. 1B, 2, and 10, B and C).

On the basis of seismic line interpretations and outcrop observations, at least four different regional unconformities can be identified (Fig. 1B): (1) Upper Permian–Lower Triassic (P2-T1); (2) Lower–Middle-Upper Triassic (T1-T2-3); (3) Lower-Middle Jurassic (J1-J2); and (4) Jurassic-Cretaceous (J-K). Because of the thick Jurassic through Tertiary section in the northern half of the basin (>5 km in the Tabei Sag; Wu and Zhao, 1997), no pre-Jurassic reflectors were imaged close to the Bogda Shan.

An Upper Permian–Lower Triassic (P2-T1) angular unconformity occurs at the Aiweiergou locality (Fig. 11) and on seismic line T88-635 (Fig. 10B). It is important to note that although an Upper Permian–Lower Triassic angular unconformity is present in the southern and western portions of the basin, localities to the north in the southern Junggar basin (e.g., Jimsar; Fig. 1B) show conformable P2-T1 stratigraphy (Yang et al., 1986; Liao et al., 1987; Hendrix et al., 1992; Carroll et al., 1995; Tang et al., 1997).

At both Taodongou (east Taoshuyuan) and Zaobishan, an erosional unconformity marks a major sedimentologic break from Lower Triassic sandstone and paleosols (T1) to deeply scoured Middle-Upper Triassic (T2-3) conglomeratic deposits (Fig. 4, A and B). The T1-T2-3 unconformity is also expressed in the eastern side of seismic line T88-635 (Fig. 10B). A slightly younger intra-Upper Triassic unconformity appears in seismic line T88-625 (Fig. 10D).

Although we observed no exposures of the Lower-Middle Jurassic (J1-J2) contact, the J1-J2 unconformity is a convincing erosional surface on all of the seismic lines (Fig. 10, A, B, C, and D). In map view, the discordance persists for as much as 50 km on both strike and dip seismic lines (Figs. 1B and 10). It is especially important to understand this feature because Turpan-Hami basin petroleum source-rock intervals and reservoirs are within Lower and Middle Jurassic strata (Huang et al., 1991; Wang et al., 1996; Wu and Zhao, 1997). Hence, this unconformity has direct ramifications for the lateral preservation of source rocks and the geometry of petroleum traps.

A prominent Jurassic-Cretaceous (J-K) angular unconformity appears in seismic line T89-465 in the eastern Hami area (Fig. 10C). Although we did not observe a J-K unconformity at Flaming Mountain, it is recognized by many Chinese publications in the western Turpan-Hami area (Li and Shen, 1990; Zhao et al., 1991a, 1991b; Mu, 1994; Wang et al., 1994).

PALEOCURRENT DATA

Paleocurrent indicator data from this study were collected from limbs of trough cross-beds and from imbricated clasts in fluvial deposits. After correcting for structural dip, the three-dimensional planes of trough cross-beds were plotted as poles on a stereographic plot (lower hemisphere projection). The pole to a statistical best-fit great circle through all of the trough pole data describes the average paleocurrent direction for the deposits (cf., DeCelles et al., 1983). For clast imbrication data, the pole of each restored imbrication plane was plotted on a rose diagram along with the mean vector (bin = 1°). Although the small number of paleocurrent indicators is not statistically significant, it is sufficient to define broad temporal and regional trends.

Figure 12 summarizes paleocurrent data from this study as well as five other reports for Permian through Jurassic deposits of the Turpan-Hami and southern Junggar basins (Carroll, 1991; Zhao et al., 1991; Hendrix et al., 1992; Schneider et al., 1992;

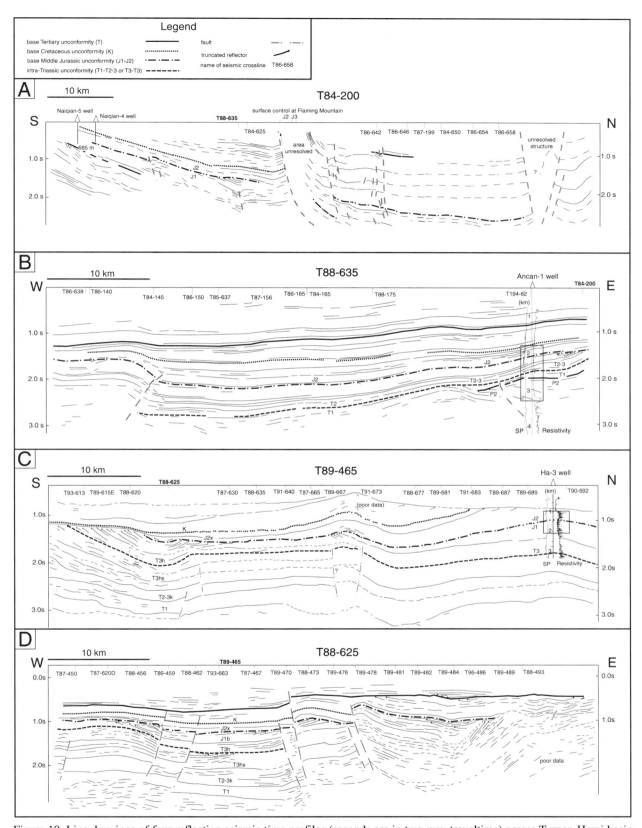

Figure 10. Line drawings of four reflection seismic time profiles (seconds are in two-way traveltime) across Turpan-Hami basin (see Fig. 1B for locations). Lines T84-200 (north-south trending) and T88-635 (east-west) are from western portion of basin, near Tabei Sag, and lines T89-465 (north-south) and T88-625 (east-west) are from eastern portion of basin, near Hami. Ancan-1 (total depth 4222 m) and Ha-3 (total depth 3001 m) wells provide stratigraphic control for lines T88-635 and T89-465, respectively. Formation abbreviations (see Fig. 2 for key) for specific reflectors are indicated where age control is sufficient. Boxed intervals within each well are expanded in Figure 2 with their corresponding palynological age interpretations. Tu-Ha Petroleum Bureau provided seismic ties to wells, based on extensive previous experience with seismic interpretations basinwide. Cross-lines are labeled at top of each seismic line drawing. Note Lower Jurassic–Middle Jurassic (J1-J2) unconformity that appears on all lines (parts A, B, C, D). Also note intra-Triassic deformation on lines T88-635 (part B) and T88-625 (part D), as well as Permian-Triassic (P2-T1) unconformity near Ancan-1 well (part B).

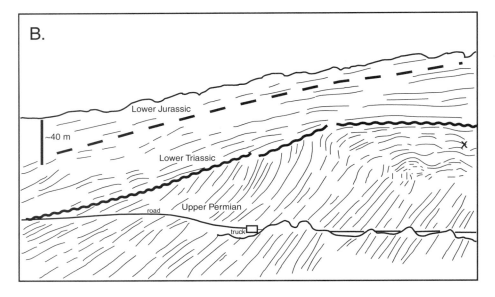

Figure 11. A: Angular unconformity at Aiweiergou (westernmost Turpan-Hami) between Upper Permian rocks (P2t) and Lower Triassic red beds (T1j). See Figure 1B for location. B: Line tracing interpretation of photo in part A. Squiggle lines denote approximate location of P2-T1 unconformity; truck (highlighted with box) hauling coal provides scale. X indicates sampled for palynology. Middle-Late Permian age is based on presence of *Klausipollenites schaubergeri* and of *Alisporites/Falcisporites* cpx. (Abbink, 1999). See Figure 7A (legend) for Lower Jurassic palynological assemblage (section 7A was measured along strike from this exposure).

Carroll et al., 1995; Yu et al., 1996). At the Taoshuyuan locality in northern Turpan-Hami, previous workers noted a reversal from northward-directed paleocurrent indicators in Permian strata (Carroll et al., 1995) to southward-directed paleocurrents in Lower Jurassic strata (Hendrix et al., 1992), suggesting an initial uplift of the Bogda Shan between Permian and Early Jurassic time. Yu et al. (1996) reported divergent paleocurrents in Lower Jurassic deposits on either side of the Bogda Shan, implying that the Turpan-Hami and Junggar basins were separated by Early Jurassic time or earlier. However, Shao et al. (1999b) presented paleocurrent data from Lin (1993) for both sides of the Bogda Shan that suggested convergent Lower Jurassic depositional systems that drained into a Bogda Shan depocenter.

This study presents paleocurrent indicator data from four localities in the Turpan-Hami basin. Upper Permian cobble clast imbrication at Aiweiergou suggests east-directed sediment transport (Fig. 12A). At Zaobishan, trough cross-bedding in Lower Triassic braided-fluvial deposits (Fig. 6) indicates a general northeastward paleocurrent direction, consistent with paleoflow off the ancestral Tian Shan, south of the Turpan-Hami basin. The Shisanjianfang locality also contains Lower to Middle Triassic braided-fluvial deposits with northeastward-directed paleocurrent indicators (Figs. 4E and 12A). These data differ markedly from measurements of trough cross-bedding from Lower Jurassic fluvial deposits at Kekeya that indicate transport to the south (Figs. 8C and 12B). Collectively, we interpret these paleocurrent data and data from

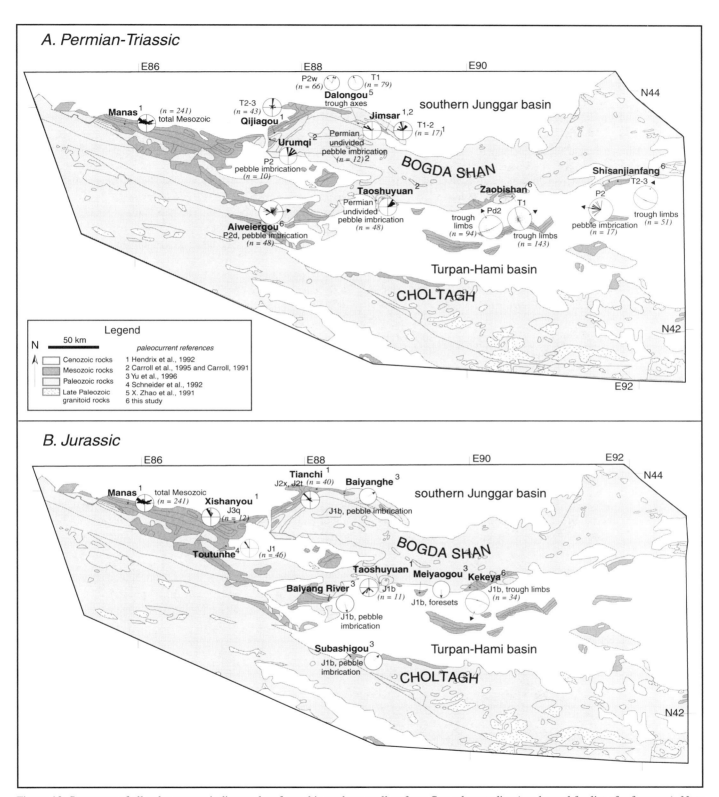

Figure 12. Summary of all paleocurrent indicator data from this study as well as from five other studies (see legend for list of references). Note reversal from north-directed paleocurrents in Permian-Triassic time (A) to south-directed paleocurrents during Jurassic time (B). See text for discussion of paleocurrent indicator data format.

previous reports to reflect northward pre-Jurassic transport in the Turpan-Hami basin, off the ancestral Tian Shan located to the south, followed by a reversal to south-directed paleoflow during Jurassic time as the ancestral Bogda Shan was uplifted.

SANDSTONE AND CONGLOMERATE PROVENANCE

Provenance studies of Mesozoic sedimentary strata within the Turpan-Hami basin are an especially powerful tool for paleogeographic reconstruction, because the Paleozoic rocks exposed in the ranges surrounding the Turpan-Hami basin are compositionally different from each other (Wen, 1991; Hopson et al., 1998). The Bogda Shan, located north of the Turpan-Hami basin, consists almost entirely of intermediate to mafic volcanics and related plutonic rocks (Chen et al., 1985). Felsic plutonic rocks are very minor in modern exposures of the Bogda Shan (Fig. 1B). In contrast, the central and south Tian Shan blocks to the south and west of Turpan-Hami (Fig. 1A) contain numerous exposures of late Paleozoic granitoids (Wen, 1991; Hopson et al., 1998), emplaced during the final phase of tectonic amalgamation of the Tian Shan (Tilton et al., 1986; Coleman, 1989; Feng et al., 1989; Kwon et al., 1989). In order to use the composition of Mesozoic sedimentary strata to constrain the initial timing of uplift of the Bogda Shan, we focused our efforts on the northern flanks of the Turpan-Hami basin, located closest to present-day exposures of the Bogda Shan and currently receiving erosional detritus from that mountain range (Graham et al., 1993). Our rationale was that these localities should have been the first to receive sediment unequivocally derived from the Bogda Shan. In our provenance studies, outlined in the following, we use sandstone petrography and conglomerate clast counts to infer that Lower Triassic deposits were derived solely from the central and south Tian Shan blocks, south of the Turpan-Hami basin, and that the ancestral Bogda Shan had not been uplifted by that time.

Sandstone framework grains

Sandstone point-count data from this study are compared with similar data from upper Paleozoic and Mesozoic sandstone from the Turpan-Hami basin (Carroll et al., 1995; Shao et al.,

1999a; Hendrix, 2000; Fig. 13). We point-counted a small set of Triassic sandstone samples (n = 7) from the Turpan-Hami basin, using a modified Gazzi-Dickinson method (see Graham et al., 1993, for detailed techniques; Dickinson and Suczek, 1979; Dickinson, 1985). The raw point-count data (Table 1) were normalized into detrital modes following the methods of Ingersoll et al. (1984) and plotted on standard ternary diagrams (Fig. 13) in order to provide direct comparison with previous provenance studies in the area (e.g., Carroll, 1991; Hendrix, 2000).

Despite the relatively small number of samples counted for this study (Table 1), when integrated with data from previous workers, several important trends appear to characterize Permian through Lower Jurassic sandstone samples from the Turpan-Hami basin. Carboniferous and Permian sandstone contains abundant volcanic lithic grains and plagioclase feldspar (Fig. 13; mean $Qm_4F_{41}Lt_{55}$, $Qp_1Lvm_{95}Lsm_4$, and $Qm_{10}P_{84}K_6$). We interpret the lithic-rich Carboniferous and Permian samples to reflect erosional unroofing of volcanic cover strata from the southern Tian Shan, a conclusion supported by north-trending paleocurrent indicators from these strata (Fig. 12A; Carroll et al., 1990; Shao et al., 1999b). Jurassic sandstone is also rich in lithic-volcanic grains (mean $Qm_{23}F_{21}Lt_{56}$, $Qp_{12}Lvm_{56}Lsm_{32}$, and $Qm_{53}P_{27}K_{20}$), but Hendrix (2000) interpreted these compositions to reflect erosional unroofing of volcanic strata in the Bogda Shan, consistent with south-directed paleocurrent measurements in the northern Turpan-Hami basin (Fig. 12B). In contrast to lithic volcanic-dominated Carboniferous, Permian, and Jurassic sandstone samples from the Turpan-Hami basin are Triassic sandstone compositions containing lower modal percentages of total lithic grains (Lt) and higher percentages of potassium feldspar (mean $Qm_{29}F_{29}Lt_{42}$, $Qp_{23}Lvm_{49}Lsm_{28}$, and $Qm_{51}P_{25}K_{24}$). Greene et al. (1997) interpreted these compositional characteristics to reflect Triassic erosional unroofing of late Paleozoic granites in the central and south Tian Shan, consistent with northeastward-directed Triassic paleocurrent indicators (Fig. 12A). Shao et al. (1999a) described a simple unroofing history based on point counts of Turpan-Hami sandstones that showed mean values in QFLt%Lt decreasing and QFLt%Q increasing from Permian through Tertiary time; notably, they showed no increase in quartz or feldspar content for Triassic samples.

TABLE 1. RAW POINT-COUNT DATA OF TRIASSIC SANDSTONE FROM THE TURPAN-HAMI BASIN

Sample	Formation	N. Lat-E.Long	N	Qm	Qp	cht	K	P	Lvm	Lslt	CO₃	Lm	unid L	bt	chl	heav	cmt	matr	por
96-ZB-501	T1s	43.15-90.25	500	132	13	46	48	19	91	35	10	8	10	0	0	0	87	1	0
96-ZB-502	T1s	43.15-90.25	500	100	17	31	49	40	56	14	9	12	4	0	0	6	160	2	0
96-ZB-402	T1j	43.15-90.25	500	148	21	21	14	29	90	24	14	14	7	1	1	0	108	8	0
96-ZB-201	T3hs	43.15-90.25	500	130	14	13	114	90	64	17	0	13	7	1	0	0	12	25	0
96-ZB-202	T3hs	43.15-90.25	500	112	20	9	79	65	91	44	0	15	19	1	0	0	16	12	17
96-ZB-203	T3hs	43.15-90.25	500	106	22	23	64	77	102	23	2	18	14	1	1	0	19	28	0
96-ZB-207	T3hs	43.15-90.25	500	105	10	23	62	130	63	22	4	24	8	8	1	0	36	23	0

Note: See Figure 2 for formation abbreviations. N.Lat-E. Long—North latitude-East longitude; Qm—monocrystalline quartz; Qp—polycrystalline quartz + chert; cht—chert; K—potassium feldspar; P—plagioclase; Lvm—volcanic and metavolcanic lithic fragments; Lslt—sedimentary lithic fragments; CO₃—carbonate grains; Lm—metamorphic lithic fragments; unid L—unidentified lithic fragments; bt—biotite; chl—chlorite; heav—heavy minerals; cmt—cement; matr—matrix; por—pore space.

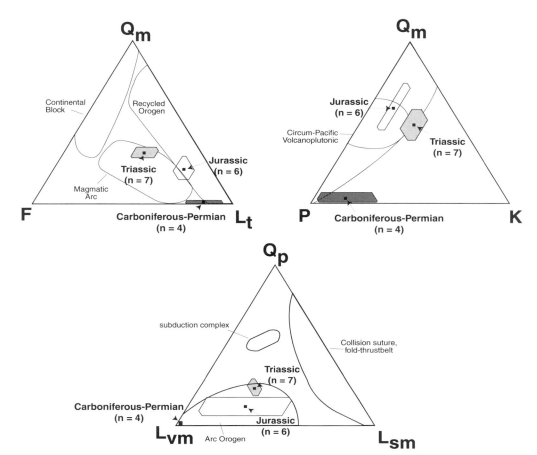

Figure 13. Summary of sandstone provenance point-count data of Turpan-Hami basin. Means (closed squares) and standard deviation fields (polygons; 1σ) are shown for each time slice. Global provenance suites proposed by Dickinson and Suczek (1979) are provided for comparison. Qm, monocrystalline quartz; F, total feldspars; Lt, total lithic grains; K, potassium feldspar; P, plagioclase; Qp, polycrystalline quartz + chert; Lvm, volcanic and metavolcanic lithics; Lsm, sedimentary + metamorphic lithics. All Mesozoic data, except Triassic data from Turpan-Hami basin (this study), are from Hendrix (2000); all Paleozoic data are from Carroll (1991). Although data are too few to be statistically significant, they are sufficient to define broad temporal and regional changes in sandstone composition. Note relative enrichment in QmFLt%Qm and QmPK%K for Triassic sandstone as well as overall enrichment in Qm for Mesozoic sandstone relative to Paleozoic sandstone.

Conglomerate data

We conducted conglomerate clast lithology counts at four localities along the northern basin rim: Taoxigou (west Taoshuyuan), Aiweiergou, Zaobishan, just north of the town of Shisanjianfang, and Kekeya (Figs. 1B and 14). Permian conglomerate at Taoxigou consists mainly of limestone clasts most likely derived from the underlying marine Upper Carboniferous deposits (Carroll et al., 1995). At Aiweiergou, Upper Permian conglomerate (P2d) consists mainly of intermediate and mafic volcanic clasts. At Zaobishan, conglomerate clasts in Lower Triassic fluvial deposits contain anomalously high percentages of pink granitic and aplitic dike rock cobbles (32%), whereas Middle-Upper Triassic conglomerate is dominated by vein quartz and mafic and intermediate volcanic compositions (Fig. 14). At Shisanjianfang, Middle-Upper Triassic clast lithology is dominantly intermediate and mafic and felsic volcanics as well as granitic clasts. Lower Jurassic conglomerate at Kekeya has a well-mixed population of clast lithologies, although granitic clasts are notably absent. Yu et al. (1996) also reported conglomerate clast percentages for Lower Jurassic strata at Baiyang River and Taoshuyuan: high proportions of volcanic clasts (80% and 95%, respectively) and low percentages of granitic clasts (0% and 15%, respectively).

High percentages of felsic plutonic clasts in Lower Triassic deposits at Zaobishan contrast sharply with Permian and Jurassic conglomerate containing few or no felsic plutonic clasts (Fig. 14). We interpret this to reflect deeper erosion and unroofing of exposed Paleozoic granites in the central and south Tian Shan, consistent with sandstone compositions and northward-directed paleocurrents. Granitic clasts are also very rare in Upper Permian strata at Aiweiergou and Zaobishan, where conglomerate is dominated by intermediate volcanic clasts most likely derived from erosion of Carboniferous andesitic cover strata (Carroll et al., 1995).

DISCUSSION

Tectonic setting of the North Tian Shan–Bogda Shan block

Details regarding the Paleozoic tectonic setting of the North Tian Shan–Bogda Shan (NTS/BS) block are uncertain, yet have obvious implications for the subsequent fill of the Turpan-Hami basin. For example, if the Bogda Shan existed as an active island arc until the Early Permian, as has been proposed by various authors (Hsü, 1988; Coleman, 1989; Wang et al., 1990; Allen et al., 1991, 1992; Fang, 1994; Mu, 1994), then it is unlikely that the Turpan and Junggar basins were depositionally linked during

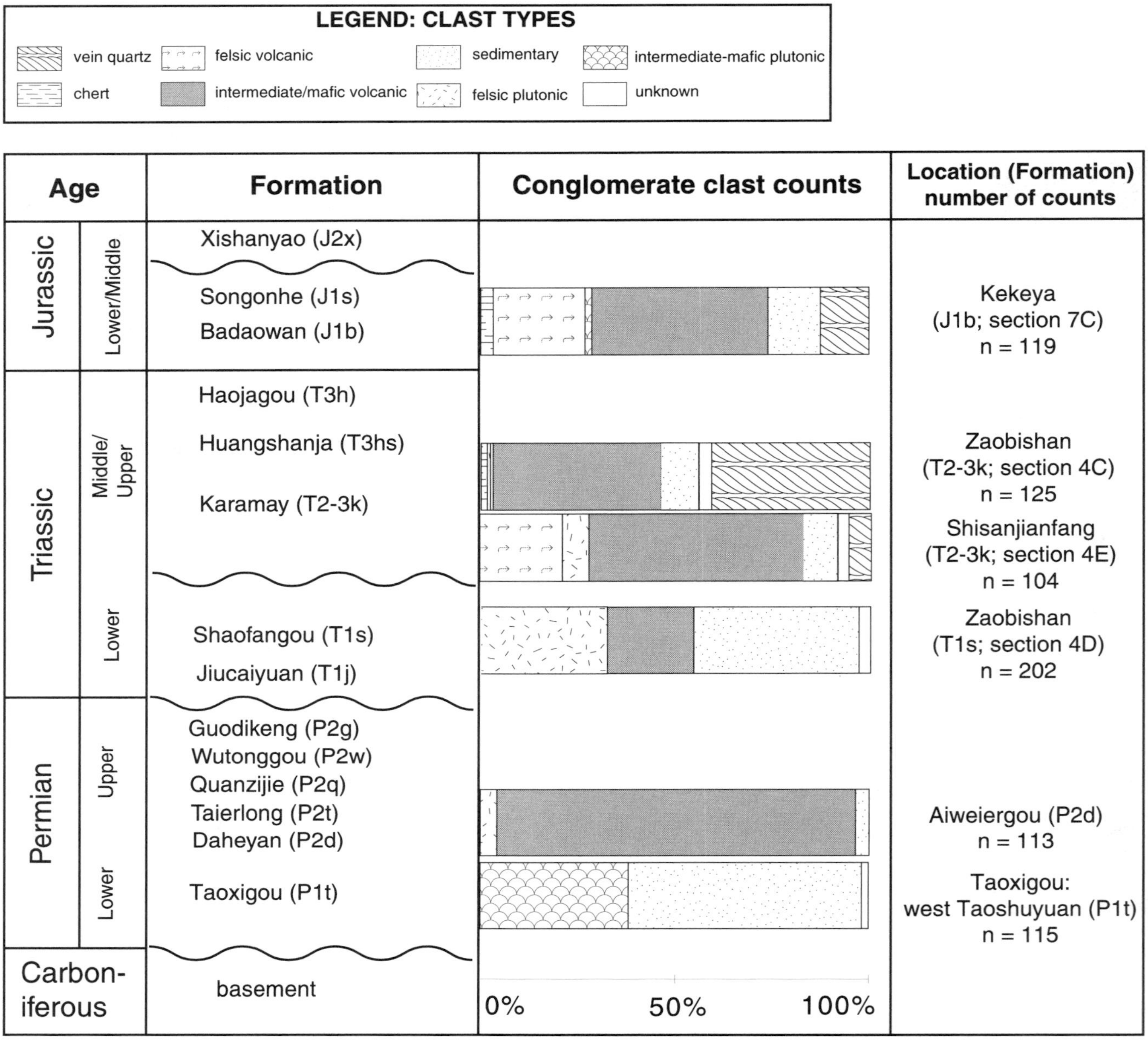

LEGEND: CLAST TYPES

vein quartz	felsic volcanic	sedimentary	intermediate-mafic plutonic
chert	intermediate/mafic volcanic	felsic plutonic	unknown

Age		Formation	Conglomerate clast counts	Location (Formation) number of counts
Jurassic	Lower/Middle	Xishanyao (J2x)		Kekeya (J1b; section 7C) n = 119
		Songonhe (J1s)		
		Badaowan (J1b)		
Triassic	Middle/ Upper	Haojagou (T3h)		Zaobishan (T2-3k; section 4C) n = 125
		Huangshanja (T3hs)		
		Karamay (T2-3k)		Shisanjianfang (T2-3k; section 4E) n = 104
	Lower	Shaofangou (T1s)		Zaobishan (T1s; section 4D) n = 202
		Jiucaiyuan (T1j)		
Permian	Upper	Guodikeng (P2g) Wutonggou (P2w) Quanzijie (P2q) Taierlong (P2t) Daheyan (P2d)		Aiweiergou (P2d) n = 113
	Lower	Taoxigou (P1t)		Taoxigou: west Taoshuyuan (P1t) n = 115
Carbon-iferous		basement	0% 50% 100%	

Figure 14. Composite stratigraphic chart with summary of conglomerate clast count data collected in Turpan-Hami basin. Although mostly mafic to intermediate clast compositions are dominant, felsic plutonic clasts in Lower Triassic conglomerate (T1s) at Zaobishan (measured section 4D) are higher than in underlying or overlying units.

Late Permian time. However, this scenario is in direct conflict with concordant, north-directed paleocurrent indicators in the Turpan-Hami and Junggar basins (Fig. 12A) and isotopic provenance data that support post-Permian uplift of the Bogda Shan (Greene et al., 1997; Greene and Graham, 1999). Moreover, on the basis of facies relations and organic geochemical attributes, many authors have reported contemporaneous Upper Permian organic-rich lacustrine deposits on both sides of the Bogda Shan in the southern Junggar and Turpan-Hami basins, implying a

unified Junggar-Turpan-Hami lake system (Liu et al., 1979; Taner et al., 1988; Nishaidai and Berry, 1991; Carroll et al., 1992; Ren et al., 1994; Brand et al., 1993; Wang et al., 1996; Greene, 1997; Wartes et al., 1998, 1999, 2000).

Many Chinese authors refer to the Bogda Shan as either a late Paleozoic intracontinental rift belt disrupting a Devonian–Middle Carboniferous passive margin (Huang et al., 1991; Fang, 1990; Chen, 1993; Lin, 1993; Wang et al., 1996; Wu et al., 1996; Wu and Xue, 1997; Wu and Zhao, 1997; Y. Liu et al.,

1998), or as a focal point of extension due to mantle diapirism beneath the Bogda Shan (Zhu and Zhao, 1992; Tao, 1994). These works, however, are largely conceptual in nature and present no outcrop, subsurface, or geochemical data that support a late Paleozoic Bogda rift.

Others agree that oceanic crust of unknown width and an associated island arc complex, termed the North Tian Shan–Bogda Shan block, was being subducted under the northern margin of the Central Tian Shan block (also referred to as the Yili microcontinent) during Carboniferous time; arc-related magmatism ceased during the Late Carboniferous (Z. Wang et al., 1986; C. Wang et al., 1990; Zhou, 1987; Coleman, 1989; Hopson et al., 1989; Carroll et al., 1990, 1995; Windley et al., 1990; Allen et al., 1991, 1992; Wen, 1991; Şengör et al., 1993; Z. Cheng et al., 1996; Wu et al., 1996; Wu and Zhao, 1997). The precise temporal and spatial distribution and the polarity of subduction represented by the North Tian Shan–Bogda Shan volcanic rocks, however, are uncertain (see Carroll et al., 1990, 1995; Windley et al., 1990; Allen et al., 1991, for discussion of possible paleogeographic scenarios). Carroll et al. (1995) interpreted 1 km of shallow-marine deposits and andesitic and dacitic volcanic flow rocks in the south and southwest Bogda Shan as representing an emergent Late Carboniferous Bogda Shan island arc. If arc-related magmatism shut off soon after, this allows a time span of 20–30 m.y. for any topographic highs to be eroded before Late Permian deposition. We infer that such an episode of erosion is manifested in the basinwide Carboniferous–Lower Permian angular unconformity observed throughout the Turpan-Hami basin (Fig. 1B; Liao et al., 1987; Carroll et al., 1999). This interpretation permits the Bogda Shan to be a part of a Late Carboniferous island arc chain without being a physiographic barrier between the Turpan and southern Junggar basin during Late Permian and Early Triassic time.

Allen et al. (1995) proposed that Permian–Triassic magmatism in the Turpan-Hami basin was the result of transtensional rotation within a Late Permian–Triassic sinistral shear system. In the Turpan-Hami basin, they based this on mapped (but undated) Lower Permian mafic volcanic rocks and tholeiitic dike rocks intruding late Paleozoic turbidites. Gabbroic bodies in the western Bogda Shan were interpreted as evidence of Permian magmatism by Windley et al. (1990) and Allen et al. (1991, 1995). However, an alternative interpretation is that the mapped gabbroic bodies of Chen et al. (1985) are hypabyssal intrusives from diorite-trondhjemite magmas that are typical of arc magmatism (Clifford Hopson, 1999, personal commun.). These rocks were analyzed by Hopson et al. (1989) and yielded a U-Pb radiometric age of 328 ± 10 Ma (Carboniferous); confirming the pre-Permian arc-related history of the western Bogda Shan (Carroll et al., 1995).

Carroll et al. (1999) reported several north-south–trending mafic dike rocks in the Lower Permian Taoxigou Group at the Taoshuyuan locality (Fig. 1B). Their inferred Early Permian age derives from crosscutting relationships with shallow-marine

Upper Carboniferous rocks, which are in turn overlapped by Upper Permian lacustrine rocks. In addition, on the basis of stratigraphic and sedimentologic relationships, Carroll et al. (1999) reported a large north-south–trending Lower Permian normal fault with as much as 3 km of displacement. Both features are interpreted as indicators of Early Permian extension in a direction normal to earlier Devonian-Carboniferous east-west–trending compressional features.

Paleogeographic models

On the basis of the results of our study, along with data and interpretations by other workers, we offer the following scenario for the paleogeographic and tectonic evolution of the Turpan-Hami basin (Fig. 15).

Late Permian. Shortening and folding of Upper Permian strata occurred in the western and southern parts of Turpan-Hami (Fig. 15A). Subsequent latest Permian–earliest Triassic uplift of the Central Tian Shan resulted in beveling of Upper Permian strata, and deposition of Lower Triassic strata, as demonstrated in outcrop at Aiweiergou (Fig. 11) and by seismic line T88-635 (Fig. 10B). Zhou (1997) and Dumitru et al. (this volume) inferred a period of rapid Permian-Triassic cooling in the Central Tian Shan, farther to the west, based on apatite fission-track data. Notably, the northern part of Turpan-Hami, as well as a southern Junggar locality (Jimsar), contain continuous Permian through Triassic sections (Fig. 1B; Yang et al., 1986; Liao et al., 1987; X. Zhao et al., 1991; Tang et al., 1997), suggesting that these more distal localities were not affected by the deformation.

Several lines of evidence point to continuous Permian deposition across the Turpan-Hami and Junggar basins. Wartes et al. (1998, 1999, 2000) reported the existence of a large Late Permian Junggar-Turpan-Hami lake system represented by a complex association of lake marginal and basinal facies on both sides of the Bogda Shan. There are other reports of widespread Upper Permian lacustrine deposition within at least four main depocenters in the Turpan-Hami basin; however, no sedimentologic or stratigraphic data have been presented (Allen et al., 1993, 1995; Mu, 1994; Wang et al., 1996; Wu and Zhao, 1997). Paleocurrent indicators suggest sediment dispersal from south to north, through the present-day Bogda Shan (Fig. 12A; Carroll, 1991; X. Zhao et al., 1991; Hendrix et al., 1992; Carroll et al., 1995). Coeval sandstone and conglomerate is extremely volcanic rich, indicating erosion of the extinct Carboniferous arc (Figs. 13 and 14).

Although we infer a proximal foreland setting for the Upper Permian deposits in Figure 15A, isopach trends for Permian strata are ambiguous, given the currently available data. Wu and Guo (1991) published north-south seismic lines between Urumqi and Jimsar indicating northward thinning of Upper Permian deposits; in contrast, Carroll et al. (1992, 1995) and Wartes et al. (2000) reported Upper Permian thicknesses in southern Junggar that greatly exceed those of Turpan-Hami.

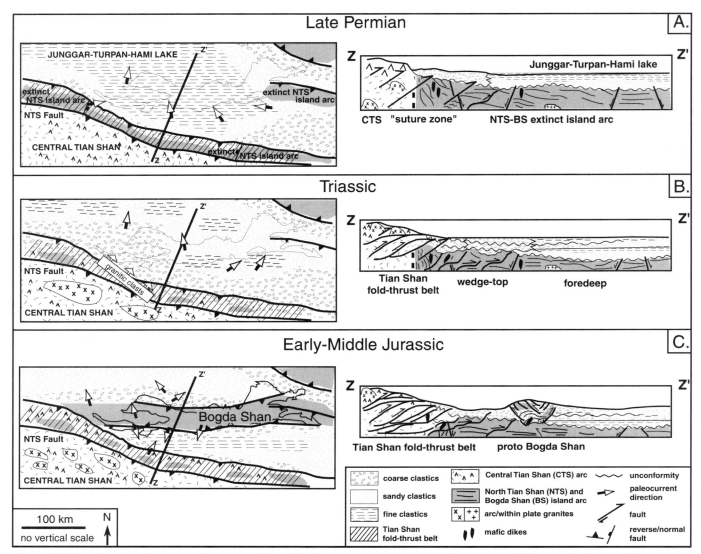

Figure 15. Schematic, nonpalinspastic paleogeographic model of study area from Late Permian to Early-Middle Jurassic time. A: Subsequent to Carboniferous collision between the Central Tian Shan arc and the North Tian Shan–Bogda Shan island arc, north-vergent Tian Shan fold-thrust belt began shedding debris into contiguous Turpan-Hami-Junggar basin. Large Late Permian lacustrine system spanned most of basin, with sediment dispersal generally directed north. During this time, Bogda Shan had not been uplifted (its present-day position is outlined). Greene and Graham (1999) also reported existence of Permian within-plate granites (cf. Pearce et al., 1984) in present-day Bogda Shan. See Wartes et al. (2000) and Carroll et al. (1995) for more detailed descriptions of Permian paleogeography. B: Triassic time was dominated by coarse-clastic deposition as compression in north-vergent Tian Shan fold-thrust belt continued. Several intra-Triassic unconformities in Turpan-Hami basin suggest repeated uplift and erosion of Tian Shan, although there is no break in sedimentation between Permian and Triassic time in southern Junggar basin. Dissection of Tian Shan fold-thrust belt unroofed Central Tian Shan granitoid rocks (e.g., T1s deposits at Zaobishan; Fig. 14) as sediment dispersal was still directed to north (Greene et al., 1997). C: By Early Jurassic time, shortening of Tian Shan fold-thrust belt continued, and Bogda Shan became significant physiographic feature partitioning southern Junggar basin from Turpan-Hami. As thrusting initiated in Bogda Shan area, amalgam of faulted Carboniferous island-arc basement, recycled Permian and Triassic deposits, mafic dike rock, and within-plate granites became structurally displaced to create proto-Bogda Shan. Paleocurrents diverged away from Bogda Shan as sediment was ponded within intermontane Turpan-Hami basin.

However, if we assume that the Turpan-Hami basin was in a wedge-top position relative to a northward-vergent Tian Shan fold-thrust belt, then basinward stratal thickening of Upper Permian deposits toward the southern Junggar basin would be entirely permissible (cf. DeCelles and Giles, 1996). This would explain the abundant regional and local unconformities near the orogenic wedge, the coarseness and extreme immaturity of Upper Permian deposits, localized lacustrine deposits that possibly formed in isolated piggy-back basins, and the northward-thickening of Upper Permian strata toward the Junggar foredeep (Fig. 15A).

Triassic. An abrupt change to coarser, more fluvial-alluvial–dominated environments characterizes most of the Triassic in the Turpan-Hami basin (Figs. 4 and 15B). As the central and south Tian Shan became more dissected, felsic plutons were unroofed and contributed erosional detritus to Lower Triassic fluvial deposits across the Turpan-Hami basin. The T1-T2-3 unconformity observed at Zaobishan (Fig. 4B), Taodongou (Figs. 3B and 4A), and in seismic line T88-635 (Fig. 10B) could reflect deformation due to renewed uplift of the Tian Shan to the south, although this idea remains to be further tested. At a more regional scale, Hendrix et al. (1992) inferred an episode of Late Triassic uplift of the Tian Shan that resulted in large coarse-clastic deposition in the southern Junggar and northern Tarim basin.

Conglomerate and sandstone provenance data also support Triassic unroofing of the Tian Shan via an increase of K-feldspar–rich felsic clasts and decrease in total lithic grains in Lower and Middle Triassic braided-fluvial deposits (Figs. 14 and 15). Based on isotopically enriched ε_{Nd} initial values of these felsic clasts, Greene and Graham (1999) inferred a Central Tian Shan provenance. In addition, paleocurrent indicators at Zaobishan and Shisanjianfang are directed north to northeast away from the ancestral Tian Shan (Fig. 12A). The paucity of granitic pebbles in modern drainages coming off the Bogda Shan at Zaobishan and Taoxigou (west Taoshuyuan) is further circumstantial evidence supporting a non-Bogda Shan Triassic provenance for the northern Turpan basin. All of these data together imply that the Bogda Shan was not a significant positive physiographic feature during Triassic time.

Early Jurassic through Middle Jurassic. Our analysis of Jurassic strata reveals a distinct change in depositional style in Lower Jurassic deposits (Figs. 7 and 15C) represented by an increase in swampy, lacustrine, and meander-belt systems. A major reorganization of basin physiography and possibly the onset of internal drainage in the basin is implied by the existence of a J1-J2 unconformity across much of the basin (Fig. 10). Figure 10 shows four examples of seismic lines where truncated Lower Jurassic reflectors are overlapped by Middle Jurassic reflectors. Our palynological studies provide age control both above and below the unconformity in the Ha-3 well (Figs. 2 and 10C); Jurassic age control for seismic lines T84-200 and T88-635 (Fig. 10, A and B) was provided by the Tu-Ha Petroleum Bureau for the Naiqian-4, Naiqian-5, and Ancan-1 wells. In addition,

based on total thickness of Lower and Middle Jurassic strata (provided by the Tu-Ha Petroleum Bureau), we infer a shift in Jurassic depocenters from the western Tokesun Sag to the north-central Taibei Sag (Fig. 1B). Notably, many other studies report a widespread Triassic-Jurassic unconformity throughout the Turpan-Hami basin, although no age control data are presented (Wang et al., 1994; D. Li, 1995; W. Li, 1997; Qiu et al., 1997; Wu and Xue, 1997; Wu and Zhao, 1997).

Southward-directed paleocurrent indicators in the northern Turpan-Hami basin contrast with northward-directed indicators in the southern Junggar basin and thus suggest divergent sediment-dispersal patterns on both sides of the Bogda Shan for Lower Jurassic strata (Fig. 12B; Hendrix et al., 1992; Schneider et al., 1992; Yu et al., 1996). Jurassic sandstone compositions contain higher percentages of lithic-volcanic detritus, reflecting Jurassic erosion of the Bogda Shan (Fig. 13). Hendrix (2000) noted increased percentages of radiolarian chert from Jurassic samples on the northern and southern flanks of the Bogda Shan and suggested that the chert may have been derived from the Jurassic unroofing of small, intraarc basins within the range. In addition, Chinese facies maps show Jurassic internal drainage patterns confined by the Bogda Shan to the north and the Tian Shan to the south (Huang et al., 1991; Qiu et al., 1997; Wu and Zhao, 1997), suggesting that by this time, the Turpan-Hami basin had become a single depositional basin, separated from the southern part of the Junggar basin.

CONCLUSIONS

1. Uplift of the Bogda Shan occurred later than Early Triassic time. The composition of sandstone and conglomerate in Lower Triassic strata on the north flanks of the Turpan-Hami basin suggests derivation from late Paleozoic granite in the Tian Shan to the south. This remains to be tested for southern Junggar Lower Triassic deposits. Because volumetrically significant granites are absent in the Bogda Shan to the north, we infer that the Bogda Shan was not being erosionally unroofed during Early Triassic time.

2. Pre-Early Jurassic initial uplift of the Bogda Shan is supported by paleocurrent indicators from Lower Jurassic strata that are north directed north of the Bogda Shan but south directed south of the Bogda Shan. Pre-Jurassic uplift of the Bogda Shan is also supported by lithic-volcanic–rich Jurassic sandstone exposed along the northern and southern flanks of the Bogda Shan, inferred to have been derived from unroofing of Carboniferous arc-related volcanics in the ancestral mountain range.

3. Regional seismic reflection lines and outcrop studies indicate several intra-Mesozoic shortening events affected the Turpan-Hami basin. We interpreted angular unconformities of Upper Permian–Lower Triassic (P2-T1), Lower-Middle-Upper Triassic (T1-T2-3), Lower-Middle Jurassic (J1-J2), and Jurassic-Cretaceous (J-K) age, indicating that the

intra-Mesozoic shortening recorded in other basins of western China also affected physiography in the Turpan-Hami basin.

4. Continued uplift of the Bogda Shan during Early Jurassic time caused a major reorganization of depositional systems in the Turpan-Hami basin and erosional beveling of Lower Jurassic and older strata. Jurassic depositional systems in the Turpan-Hami basin were inundated with the supply of sediment derived from erosion of the Bogda Shan, resulting in a basinwide J1-J2 angular unconformity. This feature could have been associated with a basinwide shift of Jurassic depocenters, from Early Jurassic deposition principally in the western depression (Tokesun Sag) to Middle Jurassic deposition principally in the north-central depression (Tabei Sag).

5. The post-Early Triassic constraints on initial uplift of the Bogda Shan described in this chapter suggest that organic-rich Upper Permian lacustrine deposits in the Turpan-Hami basin were likely contiguous with voluminous organic-rich lacustrine deposits in the southern and central Junggar basin. Because these deposits are the primary petroleum source rock in the Junggar basin, the suggestion that their updip equivalents are present in the Turpan-Hami basin increases the petroleum source-rock potential of that basin.

ACKNOWLEDGMENTS

The Chinese National Petroleum Corporation and the Turpan-Hami Petroleum Bureau arranged invaluable logistical and technical support as well as access to outcrop and subsurface data. K. Cheng, X. Zeng, W. Wang, T. Hu, A. Su, and X. Zhang assisted greatly in interpreting the stratigraphy of the Turpan-Hami basin as well as providing access to key regional seismic lines and wells. A. Hessler provided outstanding field assistance during the 1997 season. Discussions with G. Ernst, A. Hanson, C. Hopson, J. Hourigan, L. Hsiao, C. Johnson, M. McWilliams, B. Ritts, X. Ying, Y. Yue, and D. Zhou improved this manuscript. We are grateful to C. Cooper, M. Allen, S. Vincent, and T. Lawton for their constructive reviews. Acknowledgment is made to the donors of The Petroleum Research Fund, administered by the American Chemical Society, for partial support of this research (ACS-PRF32605-AC2). Additional funding came from the American Association of Petroleum Geologists and the Stanford University McGee Fund. Financial support during the term of this research project was also provided by the Graduate School of the University of Wisconsin and the Stanford-China Industrial Affiliates, an industrial consortium that has included Agip, Arco, Chevron, Exxon, Japan National Oil Corporation, Mobil, Phillips, Shell, Statoil, Texaco, Triton, and Unocal.

REFERENCES CITED

Abbink, O.A., 1999, Palynology of outcrop and core samples from the Turpan-Hami basin, northwest China: Utrecht, The Netherlands, Laboratory of Palaeobotany and Palynology Report 9916, 18 p.

Allen, M.B., Windley, B.F., Zhang, C., Zhao, Z.Y., and Wang, G.R., 1991, Basin evolution within and adjacent to the Tien Shan Range, northwest China: Geological Society of London Journal, v. 148, p. 369–379.

Allen, M.B., Windley, B.F., and Zhang, C., 1992, Paleozoic collisional tectonics and magmatism of the Chinese Tian Shan, central Asia: Tectonophysics, v. 220, p. 89–115.

Allen, M.B., Windley, B.F., Zhang, C., and Guo, J., 1993, Evolution of the Turpan basin, Chinese central Asia: Tectonics, v. 12, p. 889–896.

Allen, M.B., Şengör, A.M.C., and Natal'in, B.A., 1995, Junggar, Turfan, and Alakol basins as Late Permian to ?Early Triassic extensional structures in a sinistral shear zone in the Altaid orogenic collage, central Asia: Geological Society of London Journal, v. 152, p. 327–338.

Brand, U., Yochelson, E.L., and Eagar, R.M., 1993, Geochemistry of Late Permian non-marine bivalves: Implications for the continental paleohydrology and paleoclimatology of northwestern China: Carbonates and Evaporites, v. 8, p. 199–212.

Campbell, J.E., and Hendry, H.E., 1987, Anatomy of a gravelly meander lobe in the Saskatchewan River, near Nipawin, Canada, *in* Ethridge, F.G., et al., eds., Recent developments in fluvial sedimentology: Society of Economic Paleontologists and Mineralogists Publication 39, p. 179–189.

Carroll, A.R., 1991, Late Paleozoic tectonics, sedimentation, and petroleum potential of the Junggar and Tarim basins, northwest China [Ph.D. thesis]: Stanford, California, Stanford University, 405 p.

Carroll, A.R., 1998, Upper Permian lacustrine organic facies evolution, southern Junggar Basin, northwest China: Organic Geochemistry, v. 28, p. 649–667.

Carroll, A.R., Graham, S.A., Hendrix, M.S., Chu, J., McKnight, C.L., Xiao, X., and Liang, Y., 1990, Junggar basin, northwest China: Trapped late Paleozoic Ocean: Tectonophysics, v. 181, p. 1–14.

Carroll, A.R., Brassell, S.C., and Graham, S.A., 1992, Upper Permian lacustrine oil shales, southern Junggar basin, northwest China: American Association of Petroleum Geologists Bulletin, v. 76, p. 1874–1902.

Carroll, A.R., Graham, S.A., and Hendrix, M.S., 1995, Late Paleozoic amalgamation of northwest China: Sedimentary record of the northern Tarim, northwestern Turpan, and southern Junggar basins: Geological Society of America Bulletin, v. 107, p. 571–594.

Carroll, A.R., Wartes, M.A., and Greene, T.J., 1999, Sedimentary evidence for Early Permian normal faulting, southern Bogda Shan, northwest China: Geological Society of America Abstracts with Programs, v. 31, no. 7, p. A369–A370.

Chen, M., 1993, Structural styles of the Turpan-Hami basin: Petroleum Exploration and Development, v. 20, no. 5, p. 1–7.

Chen, Z., Wu, N., Zhang, D., Hu, J., Huang, H., Shen, G., Wu, G., Tang, H., and Hu, Y., 1985, Geologic map of Xinjiang Uygur Autonomous Region: Beijing, Geologic Publishing House, scale 1:2 000 000.

Cheng, K., Su, A., Zhao, C., and He, Z., 1996, Study of coal-generated oil in the Tuha basin: 30th International Geological Congress, Progress in geology of China: Beijing, Geological Publishing House, p. 796–799.

Cheng, Z., Wu, S., and Fang, X., 1996, The Permian-Triassic sequences in the southern margin of the Junggar basin, and the Turpan basin, Xinjiang, China, *in* International Geologic Congress, 30th, Field Trip Guidebook T394: Beijing, Geological Publishing House, 25 p.

Clayton, J.J., Yang, J., King, J.D., Lillis, P.G., and Warden, A., 1997, Geochemistry of oils from the Junggar basin, northwest China: American Association of Petroleum Geologists Bulletin, v. 81, p. 1926–1944.

Coleman, R.G., 1989, Continental growth of northwest China: Tectonics, v. 8, p. 621–636.

DeCelles, P.G., and Giles, K.A., 1996, Foreland basin systems: Basin Research, v. 8, p. 105–123.

DeCelles, P.G., Langford, R.P., and Schwartz, R.K., 1983, Two new methods of paleocurrent determination from trough cross-stratification: Journal of Sedimentary Petrology, v. 53, p. 629–642.

Dickinson, W.R., 1985, Interpreting provenance relations from detrital modes of sandstone, *in* Zuffa, G.G., eds., Provenance of arenites: Hingham, Massachusetts, D. Reidel Publishing Company, p. 333–361.

Dickinson, W.R., and Suczek, C.A., 1979, Plate tectonics and sandstone compositions: American Association of Petroleum Geologists Bulletin, v. 63, p. 2164–2182.

Fang, G., 1990, Initial studies about Bogda Mountains late Palaeozoic aulacogen: Xinjiang Geology, v. 8, p. 133–141.

Fang, G., 1994, Paleozoic plate tectonics of eastern Tianshan mountains Xinjiang, China: Acta Geologica Gansu, v. 3, p. 34–40.

Feng, Y., Coleman, R.G., Tilton, G., and Xiao, X., 1989, Tectonic evolution of the west Junggar region, Xinjiang, China: Tectonics, v. 8, p. 729–752.

Graham, S.A., Hendrix, M.S., Wang, L.B., and Carroll, A.R., 1993, Collisional successor basins of western China: Impact of tectonic inheritance on sand composition: Geological Society of America Bulletin, v. 105, p. 323–344.

Greene, T.J., 1997, Petroleum geochemistry of Upper Permian and Middle Jurassic source rocks and oils of the Turpan-Hami basin, northwest China: American Association of Petroleum Geologists Bulletin, v. 81, p. 1774.

Greene, T.J., and Graham, S.A., 1999, Isotopic provenance of Devonian-aged granitic cobbles and geochemistry of Bogda Shan granitoids marking the inception of the Turpan-Hami basin, northwest China: Geological Society of America Abstracts with Programs, v. 31, no. 7, p. A374.

Greene, T.J., Carroll, A.R., Hendrix, M.S., and Li, J., 1997, Permian-Triassic basin evolution and petroleum system of the Turpan-Hami Basin, Xinjiang Province, northwest China: American Association of Petroleum Geologists and Society of Economic Paleontologists and Mineralogists Annual Meeting Abstracts, v. 6, p. 42.

Greene, T.J., Carroll, A.R., Hendrix, M.S., Cheng, K., and Zeng, X.M., 2000, Sedimentology and paleogeography of the Middle Jurassic Qiketai Formation, Turpan-Hami basin, northwest China, *in* Gierlowski-Kordesch, E., and Kelts, K., eds., Lake basins through space and time: American Association of Petroleum Geologists Studies in Geology 46, p. 141–152.

Hendrix, M.S., 2000, Evolution of Mesozoic sandstone compositions, southern Junggar, northern Tarim, and western Turpan basins, northwest China: A detrital record of the ancestral Tian Shan: Journal of Sedimentary Research, v. 70, p. 520–532.

Hendrix, M.S., Graham, S.A., Carroll, A.R., Sobel, E.R., McKnight, C.L., Schulein, B.J., and Wang, Z., 1992, Sedimentary record and climatic implications of recurrent deformation in the Tian Shan: Evidence from Mesozoic strata of north Tarim, south Junggar, and Turpan basins, northwest China: Geological Society of America Bulletin, v. 105, p. 53–79.

Hendrix, M.S., Graham, S.A., Amory, J.A., and Badarch, G., 1996, Noyon Uul syncline, southern Mongolia: Lower Mesozoic sedimentary record of the tectonic amalgamation of central Asia: Geological Society of America Bulletin, v. 108, p. 1256–1274.

Hopson, C., Wen, J., Tilton, G., Tang, Y., Zhu, B., and Zhao, M., 1989, Paleozoic plutonism in east Junggar, Bogdashan, and eastern Tianshan, northwest China: Eos (Transactions, American Geophysical Union), v. 70, p. 1403–1404.

Hopson, C.A., Wen, J., and Tilton, G.R., 1998, Isotopic variation of Nd, Sr, and Pb in Paleozoic granitoid plutons along an east Junggar-Bogdashan-Tianshan transect, northwest China: Geological Society of America Abstracts with Programs, v. 30, no. 5, p. 20.

Hsü, K.J., 1988, Relict back-arc basins: Principles of recognition and possible new examples from China, *in* Kleinspehn, K.L., and Paola, C., eds., New perspectives in basin analysis: New York, Springer-Verlag, p. 245–263.

Huang, D., Zhang, D., Li, J., and Huang, X., 1991, Hydrocarbon genesis of Jurassic coal measures in the Turpan Basin, China: Organic Geochemistry, v. 17, p. 827–837.

Huang, P., 1993, An Early Jurassic sporopollen assemblage from the northwestern margin of the Junggar Basin, Xinjiang: Acta Micropalaeontologica Sinica, v. 10, p. 77–88.

Ingersoll, R.V., Fullard, T.F., Ford, R.L., Grimm, J.P., Pickle, J.D., and Sares, S.W., 1984, The effect of grain size on detrital modes; a test of the Gazzi-Dickinson point-counting method: Journal of Sedimentary Petrology, v. 54, p. 103–116.

Kwon, S.T., Tilton, G.R., Coleman, R.G., and Feng, Y., 1989, Isotopic studies bearing on the tectonics of the west Junggar region, Xinjiang, China: Tectonics, v. 8, p. 719–727.

Li, D., 1995, Hydrocarbon occurrences in the petroliferous basins of western China: Marine and Petroleum Geology, v. 12, p. 26–34.

Li, G., and Shen, S., 1990, On the formation and evolution of Turpan-Hami basin and its characteristic of bearing oil and gas, Xinjiang: Bulletin of Xi'an Institute Geological and Mineral Resources no. 28: Beijing, Chinese Academy of Geological Sciences, p. 25–36.

Li, J., and Jiang, J., 1987, Survey of petroleum geology and the controlling factors for hydrocarbon distribution in the east part of the Junggar basin: Oil and Gas Geology, v. 8, p. 99–107.

Li, W., 1997, Sequence stratigraphy of Jurassic in Taibei Sag, Turpan-Hami basin: Oil and Gas Geology, v. 18, p. 210–215.

Liao, Z., Lu, L., Jiang, N., Xia, F., Sung, F., Zhou, Y., Li, S., and Zhang, Z., 1987, Carboniferous and Permian in the western part of the east Tianshan Mountains: Beijing, Eleventh Congress of Carboniferous Stratigraphy and Geology, Guide Book Excursion 4, 50 p.

Lin, J.Y., 1993, Sedimentary sequence of the Bogda Rift: Discussion about the formation and evolution of the entire intercontinental sedimentary basin in the northern Xinjiang [Ph.D. thesis]: Xi'an, Northwest University, 104 p.

Liu, H., 1986, Geodynamic scenario and structural styles of Mesozoic and Cenozoic basins in China: American Association of Petroleum Geologists Bulletin, v. 70, p. 377–395.

Liu, H., Lian, H., Cai, L., Xia, Y., and Liu, L., 1979, Evolution and structural style of Tian Shan and adjacent basins, northwestern China: Earth Science, v. 19, p. 727–741.

Liu, L., and Di, S., 1997, Characteristics of Middle Jurassic sedimentation and reservoir pore evolution in Turpan Depression: Oil and Gas Geology, v. 18, p. 17–18.

Liu, L., Liu, Y., and Di, S., 1998, Sedimentation and diagenesis of Qiketai Formation, in north Turpan Depression: Oil and Gas Geology, v. 19, p. 238–243.

Liu, Y., Wu, T., Cui, H., and Feng, Q., 1998, Paleogeothermal gradient and geologic thermal history of the Turpan-Hami basin, Xinjiang: Science in China, ser. D, v. 41, p. 62–68.

Miall, A.D., 1996, The geology of fluvial deposits: Sedimentary facies, basin analysis, and petroleum geology: Berlin, Springer-Verlag, p. 208–211.

Miall, A.D., and Gibling, M.R., 1978, The Siluro-Devonian clastic wedge of Somerset Island, Arctic Canada, and some regional paleogeographic implications: Sedimentary Geology, v. 21, p. 85–127.

Mu, Z., 1994, Permian and Triassic formation distribution and palaeogeographical pattern of Turpan-Hami basin: Xinjiang Petroleum Geology, v. 14, p. 14–20.

Nanson, G.C., and Croke, J.C., 1992, A genetic classification of flood plains: Geomorphology, v. 4, p. 459–486.

Nijman, W., and Puigdefábregas, C., 1978, Coarse-grained point bar structure in a molasse-type fluvial system, Eocene Castisent sandstone formation, south Pyrenean Basin, *in* Miall, A.D., ed., Fluvial sedimentology: Canadian Society of Petroleum Geologists Memoir 5, p. 487–510.

Nishidai, T., and Berry, J.L., 1991, Geological interpretation and hydrocarbon potential of the Turpan basin (NW China) from satellite imagery: Journal of Petroleum Technology, v. 13, p. 35–58.

Olsen, H., 1989, Sandstone-body structures and ephemeral stream processes in the Dinosaur Canyon Member, Moenave Formation (Lower Jurassic), Utah, U.S.A.: Sedimentary Geology, v. 61, p. 207–221.

Ouyang, S., 1996, Spore-pollen assemblage from Bquinshan Group, Qinghai and its geological age: Acta Palaeontologica Sinica, v. 35, p. 1–25.

Ouyang, S., and Norris, G., 1988, Spores and pollen from the Lower Triassic Heshanggou Formation, Shaanxi Province, North China: Review of Palaeobotany and Palynology, v. 54, p. 187–231.

Pearce, J.A., Harris, N.B.W., and Tindle, A.G., 1984, Trace element discrimination diagrams for the tectonic interpretation of granitic rocks: Journal of Petrology, v. 25, p. 956–983.

Peng, X., Hu, B., and Liu, L., 1990, A restudy for pre-Bogda mountain folded zone: Xinjiang Petroleum Geology, v. 11, p. 276–295.

Qiu, Y., Xue, S., and Ying, F., 1997, Continental hydrocarbon reservoirs of China: Beijing, Petroleum Industry Press, p. 87–104.

Rasmussen, K.A., and Romanovsky, V.V., 1995, Late Holocene climate change and lake-level oscillation; Issyk-kul, Kyrgyzstan, Central Asia: Society of Economic Paleontologists and Mineralogists Program and Abstracts, v. 1, p. 103.

Ren, Z., Jiang, H., Liu, Y., and Li, W., 1994, Organic geochemical characterization of the Permian-Jurassic source rocks in Aiweiergou and Taoyuanzi sections on the bordering areas of the Tu-Ha basins: Experimental Petroleum Geology, v. 16, p. 1–9.

Schneider, W., Zhao, X., Long, N., Zhao, Y., and Falke, M., 1992, Sedimentary environment and tectonic implication of Jurassic in Toutunhe area, Junggar basin: Xinjiang Geology, v. 10, p. 192–203.

Şengör, A.M.C., Natal'in, B.A., and Burtman, V.S., 1993, Evolution of the Altaid tectonic collage and Paleozoic crustal growth in Eurasia: Nature, v. 364, p. 299–307.

Sgibnev, V.V., and Talipov, M.A., 1990, Evolution of the Issyk-kul sedimentation basin (Tien-Shan) during Quaternary [abs.]: 13th International Sedimentological Congress, v. 13, p. 202.

Shao, L., Li, W., and Yuan, M., 1999a, Characteristic of sandstone and its tectonic implications of the Turpan basin: Acta Sedimentologica Sinica, v. 17, p. 95–99.

Shao, L., Stattegger, K., Li, W., and Haupt, B.J., 1999b, Depositional style and subsidence history of the Turpan Basin (NW China): Sedimentary Geology, v. 128, p. 155–169.

Shen, J., 1990, The characteristics of petroleum geology in Chaiwopu basin: Xinjiang Petroleum Geology, v. 11, p. 297–310.

Smith, D.G., 1986, Anastomosing river deposits, sedimentation rates and basin subsidence, Magdalena River, northwestern Colombia, South America: Sedimentary Geology, v. 46, p. 177–196.

Taner, I., Kamen-Kaye, M., and Meyerhoff, A.A., 1988, Petroleum in the Junggar basin, northwestern China: Journal of Southeast Asian Earth Sciences, v. 2, p. 163–174.

Tang, Z., Parnell, J., and Longstaffe, F.J., 1997, Diagenesis and reservoir potential of Permian-Triassic fluvial/lacustrine sandstones in the southern Junggar basin, northwestern China: American Association of Petroleum Geologists Bulletin, v. 81, p. 1843–1865.

Tao, M.X., 1994, Tectonic environmental analysis of Turpan-Hami basin: On the genetic relationship between basin and orogenic belt of continental inner plate: Acta Sedimentologica Sinica, v. 12, p. 40–50.

Tilton, G.R., Kwon, S.T., Coleman, R.G., and Xiao, X., 1986, Isotopic studies from the West Junggar Mountains, northwest China: Geological Society of America Abstracts with Programs, v. 18, p. 773.

Wang, C., Ma, R., and Ye, S., 1990, Allochthonous terranes in eastern Tianshan, northwest China, *in* Wiley, T.J., et al., eds., Terrane analysis of China and Pacific rim: Circum-Pacific Council for Energy and Mineral Resources, Earth Science Series, v. 13, p. 257–260.

Wang, C., Luo, B., and Zheng, G., 1993, Organic geochemical characteristics and genesis of crude oils from the Terpan basin, China: Acta Sedimentologica Sinica, v. 11, p. 72–81.

Wang, C., Cheng, K., Xu, Y., and Zhao, C., 1996, Geochemistry of Jurassic coal-derived hydrocarbon of Turpan-Hami basin: Beijing, Petroleum Industry Press, 247 p.

Wang, C., Fu, J., Sheng, G., Zhang, Z., Xia, Y., and Cheng, X., 1998, Laboratory thermal simulation of liquid hydrocarbon generation and evolution of Jurassic coals from the Tu-Ha basin: Acta Geologica Sinica, v. 72, p. 276–284.

Wang, H., Liu, W., Chen, Y., Mou, Z., Li, B., and Zhu, H., 1997, Sedimentary microfacies and petroleum productivity of Sanjianfang Formation in Wenxi-I and Wen-V blocks: Oil and Gas Geology, v. 18, p. 252–256.

Wang, S., Zhang, W., Zhang, H., and Tan, S., 1994, Petroleum geology of China: Beijing, Petroleum Industry Press, p. 411–414.

Wang, Z., Wu, J., Lu, X., Zhang, J., and Liu, C., 1986, An outline on the tectonic evolution of the Tian Shan of China: Chinese Academy of Geological Sciences Institute of Geology Bulletin, v. 15, p. 81–92.

Wartes, M.A., Greene, T.J., and Carroll, A.R., 1998, Permian lacustrine paleogeography of the Junggar and Turpan-Hami basins, northwest China: American Association of Petroleum Geologists Annual Convention, Extended Abstracts, v. 2, p. A682.

Wartes, M.A., Carroll, A.R., and Greene, T.J., 1999, Permian sedimentary evolution of the Junggar and Turpan-Hami basins, northwest China: Geological Society of America Abstracts with Programs, v. 31, no. 7, p. A291.

Wartes, M.A., Carroll, A.R., Greene, T.J., Cheng, K., and Ting, H., 2000, Permian lacustrine deposits of northwest China, *in* Gierlowski-Kordesch, E., and Kelts, K., eds., Lake basins through space and time: American Association of Petroleum Geologists Studies in Geology 46, p. 123–132.

Wen, J., 1991, Geochronological and isotopic tracer studies of some granitoid rocks from Xinjiang, China: Constraints on Paleozoic crustal evolution and granitoid petrogenesis [Ph.D. thesis]: Santa Barbara, University of California, 163 p.

Windley, B.F., Allen, M.B., Zhang, C., Zhao, Z.Y., and Wang, G.R., 1990, Paleozoic accretion and Cenozoic redeformation of the Chinese Tien Shan range, central Asia: Geology, v. 18, p. 128–131.

Wu, C., and Xue, S., 1997, Sedimentology of petroliferous basins in China: Beijing, Petroleum Industry Press, p. 384–400.

Wu, T., and Zhao, W., 1997, Formation and distribution of coal measure oil-gas fields in Turpan-Hami Basin: Beijing, Petroleum Industry Press, 262 p.

Wu, T., Zhang, S., and Wang, W., 1996, The structural characteristics and hydrocarbon accumulation in Turpan-Hami coal-bearing basin: Acta Petrolei Sinica, v. 17, p. 12–18.

Wu, Z., 1986, Characteristics of evolution and division of tectonic structure in Junggar basin and the appraisal of gas and oil: Xinjiang Geology, v. 4, p. 20–34.

Wu, Z., and Guo, F., 1991, Second discussion of Bogda nappe tectonic and its oil-gas accumulation: Xinjiang Geology, v. 9, p. 40–49.

Yang, J., Qu, J., Zhou, H., Cheng, Z., Zhou, T., Hou, J., Li, P., Sun, S., Li, Y., Zhang, Y., Wu, X., Zhang, Z., and Wang, Z., 1986, Permian and Triassic strata and fossil assemblages in the Dalongkou area of Jimsar, Xinjiang: Ministry of Geology and Mineral Resources Geologic Memoir, ser. 2, no. 3: Beijing, Geological Publishing House, 262 p.

Yu, C., Jiang, Y., and Liu, S., 1996, Jurassic sedimentary boundary between the Junggar and Turpan-Hami basins in Xinjiang: Sedimentary Facies and Palaeogeography, v. 16, p. 48–54.

Zhao, W., Li, W., and Yan, L., 1991a, Types, characteristics of oil-gas pools and hydrocarbon distribution regularities in Turpan-Hami basin: Oil and Gas Geology, v. 12, p. 351–363.

Zhao, W., Yuan, F., and Zeng, X., 1991b, The structural characteristics of Turpan-Harmy basin: Acta Petrolei Sinica, v. 13, p. 9–18.

Zhao, X., Lou, Z., and Chen, Z., 1991, Response of alluvial-lacustrine deposits of Permian-Triassic Canfanggou Group to climate and tectonic regime in Dalonggou area, Junggar basin, Xinjiang: China Earth Sciences, v. 1, p. 343–354.

Zhou, D., 1997, Studies in the tectonics of China: Extensional tectonics of the northern margin of the south China sea; amalgamation and uplift of the Tian Shan; and wedge extrusion model for the Altyn Tagh fault [Ph.D. thesis]: Stanford, California, Stanford University, 354 p.

Zhou, R., 1987, The advance on isotope geochronology of Xinjiang: Xinjiang Geology, v. 5, p. 15–105.

Zhu, Y., and Zhao, J., 1992, Quaternary paleoglacier and neotectonic movement on the northern slope of Bogda Mountain: Xinjiang Geology, v. 10, p. 40–50.

MANUSCRIPT ACCEPTED BY THE SOCIETY JUNE 5, 2000

Geological Society of America
Memoir 194
2001

Sedimentary record of Mesozoic intracontinental deformation in the eastern Junggar Basin, northwest China: Response to orogeny at the Asian margin

Stephen J. Vincent*
Mark B. Allen
China Basins Project, CASP, Department of Earth Sciences, University of Cambridge, 181A Huntingdon Road, Cambridge CB3 0DH, UK

ABSTRACT

A nonmarine Mesozoic succession is exposed along the flanks of the Kelameili Shan, eastern Junggar Basin. Three angular unconformities occur within Upper Triassic to Lower Cretaceous strata: at the Triassic-Jurassic boundary, between the Lower Jurassic Badaowan and Sangonghe formations, and at the base of the Lower Cretaceous Tugulu Group. Localized landscape unconformities also mark the base of the Middle Jurassic Xishanyao Formation and the Middle to Upper Jurassic Shishugou Group. All of the unconformity surfaces are overlain by alluvial conglomerate. Unconformity generation resulted from the episodic reactivation of the Paleozoic west-northwest–orientated Kelameili Shan fault zone and other north-south–orientated intrabasinal faults during the Jurassic. Previous interpretations of the Mesozoic stratigraphy of northwest China have linked pulses of clastic sedimentation to continent-continent and arc-continent collisions at Asia's southern margin, namely the collision of the Qiangtang Block in the Late Triassic, the Lhasa Block in the latest Jurassic, and the Kohistan-Dras island arc in the Late Cretaceous. The Kelameili Shan stratigraphy contains too many unconformities, deformation events, and conglomeratic pulses for a one-to-one correlation to be made. Additional deformation events might have been caused by the collision of other smaller continental blocks at the southern side of Asia and phases of retroarc compressional deformation.

INTRODUCTION

The continental crust of northwest China was largely assembled during a series of late Paleozoic orogenies (Allen et al., 1993; Şengör and Natal'in, 1996), including the Altaid orogeny, which involved the generation of juvenile Paleozoic crust in giant subduction-accretion complexes. Upper Paleozoic flysch, melange, dismembered ophiolite, volcanics, and granitoids created by Altaid events crop out within the North Tian Shan and around the margins of the Junggar Basin (Figs. 1 and 2). Another major orogeny represented within the Tian Shan is the collision

_____
E-mail: stephen.vincent@casp.cam.ac.uk.

between Tarim and a continental fragment known variously as the Yili microcontinent, the Chinese Central Tian Shan and the Chu-Terskey fragment (Allen et al., 1993; Carroll et al., 1995). This fragment was a promontory at the eastern margin of a much larger, composite Paleozoic continent known variously as Kazakstania or the Kazak-Kyrgyz Massif (e.g., Zonenshain et al., 1990; Biske, 1995; Fig. 1). Tarim collided diachronously from east to west, between the latest Devonian to Late Carboniferous (Windley et al., 1990). The collision zone is preserved within the present southern Tian Shan (Fig. 1).

There are three major basins within and marginal to the Tian Shan; Tarim, Junggar, and Turfan. The Junggar Basin formed as a foreland basin in the Early Permian following closure of a branch

Vincent, S.J., and Allen, M.B., 2001, Sedimentary record of Mesozoic intracontinental deformation in the eastern Junggar Basin, northwest China: Response to orogeny at the Asian margin, *in* Hendrix, M.S., and Davis, G.A., eds., Paleozoic and Mesozoic tectonic evolution of central Asia: From continental assembly to intracontinental deformation: Boulder, Colorado, Geological Society of America Memoir 194, p. 341–360.

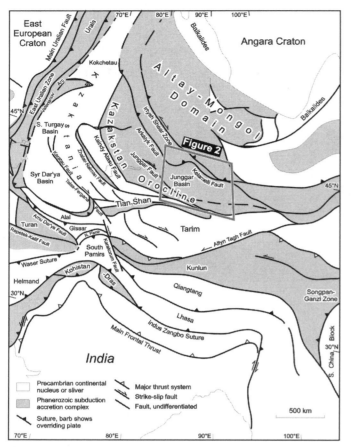

Figure 1. Major structures within Central Asia and position of Junggar Basin.

of the Turkestan Ocean (Şengör and Natal'in, 1996); there is possibly a cryptic suture within the basin interior. Lower Permian depositional facies shallow upward, from siliciclastic turbidites to alluvial strata near the boundary with the Upper Permian.

Late Permian tectonics are disputed. Carroll et al. (1995) proposed that compressional deformation operated throughout the Permian, and Allen et al. (1995) proposed a phase of Late Permian sinistral transtension. The Turfan Basin appears to have been the southern continuation of the Junggar Basin from the Permian to the Middle Triassic, but to have been separated from it by Late Triassic thrusting in the Bogda Shan–Barkol Tagh region (Greene et al., this volume; Fig. 2).

There is general agreement that the nonmarine, Mesozoic–Cenozoic clastic fill of these basins was generated as the result of thrusting and exhumation in the surrounding mountain ranges (Watson et al., 1987; Hendrix et al., 1992). These deformation events have been attributed to the intracontinental effects of collisional orogenies at the southern margin of Asia, specifically the accretion of continental fragments of Gondwana, resultant from the destruction of various branches of Palaeotethys and Neotethys (Şengör et al., 1988). However, the record of actual Mesozoic de-

formation in exposures in northwest China is surprisingly poor, as opposed to its inference from conglomeratic pulses in the sedimentary record (Hendrix et al., 1992) or its appearance in Chinese accounts of subsurface geology of the Junggar Basin, based on hydrocarbon exploration data (Zhang Jiyi, 1984; Li Xibin and Jiang Jianheng, 1987; Xie Hong et al., 1988; Yan Yugui, 1993; Zhang Xuejun et al., 1999). Overall, the effects of these Mesozoic orogenies on the interior of the Asian continent are currently poorly understood by comparison with the Cenozoic India-Asia collision. The intracontinental effects of this immense and active modern collision system can be studied through a combination of seismicity, geodetic, and remote sensing techniques not possible on older orogenic systems. In addition, major Cenozoic fault systems and sedimentary successions have overprinted and covered pre-Cenozoic structures and strata.

This chapter describes the Mesozoic geology of the Kelameili Shan, which is located at the northeastern margin of the Junggar Basin (Fig. 3). Our aims are to add to the volume of data on the Mesozoic rocks of this area, and to test, refine, and, where necessary, improve upon current models for the interaction of tectonics and sedimentation in this era. Reconnaissance-level field data are integrated with Chinese literature on the subsurface geology of the region. We focus on aspects of the stratigraphy that include evidence for active deformation in the Kelameili Shan and/or the eastern Junggar Basin during sedimentation.

PRE-MESOZOIC FRAMEWORK

Exposed basement rocks in the Kelameili Shan are Devonian slates unconformably overlain by subaerial Carboniferous volcanics (this survey). This assemblage appears to represent a southward-vergent subduction accretion complex, formed during the late stages of the elimination of oceanic crust in the Altaids (Şengör and Natal'in, 1996).

The main structural feature of the Kelameili Shan is the Kelameili fault zone (Figs. 2 and 3). This is a complex, west-northwest–trending braided shear zone, with numerous ultramafic fragments distributed along its exposed length. The main part of the fault system is linear, trends at 280°–290°, and is present along the length of the Kelameili Shan before disappearing beneath Quaternary cover to the west and east. Its exposed length is ~160 km. There is undeformed Quaternary cover in some parts, with no indication from field observations, satellite imagery or major historical seismicity that it is a major active structure. Subsurface data suggest that the fault system continues west of its exposed limit, where splays form the boundaries to several basement highs and lows in the Junggar Basin: the Dishuiquan Rise, Sannan Sag, and Dongdaohaizibei Sag (Fig. 2). The eastern portion of the fault may link with faults of similar orientation exposed in the eastern Tian Shan.

In the southwest of the Kelameili Shan, there is an imbricate zone that exposes interthrust Devonian flysch and Carboniferous volcanics (Fig. 3). The most continuous of these faults is ex-

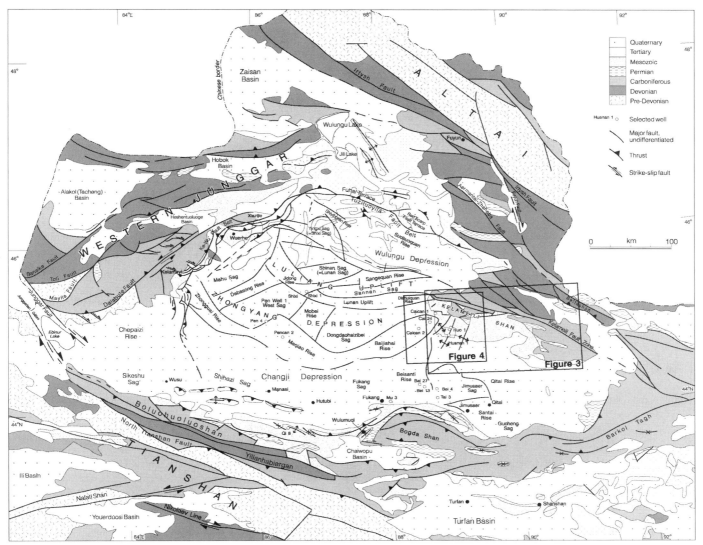

Figure 2. Structure of Junggar Basin and adjacent areas. Compiled principally from Xie Hong et al. (1988), Bureau of Geology and Mineral Resources of Xinjiang Uygur Autonomous Region (1993), Allen et al. (1995), and Wang Yilin (1996). Located in Figure 1.

posed for ~95 km, disappearing beneath Mesozoic cover at its western end, and swinging in strike to an east-west orientation at its eastern end before merging with the main Kelameili fault zone. The subsurface continuation of this fault to the west is termed the south Dishuiquan fault (Pu Renhai et al., 1994b), and we apply this name to the exposed portion of the fault.

Permian strata in the eastern Junggar Basin consist of mixed clastic alluvial to lacustrine facies, the deposition and distribution of which were controlled at least in part by north-south– to northeast-southwest–trending faults (Fig. 4A; Qin Subao, 1986). These faults do not appear to continue far into either the Kelameili Shan or the Bogda Shan regions, at the north and south of the eastern Junggar Basin. Some of the faults affecting Permian strata within the basin interior were extensional

in the Permian period, although the exact timing of this motion remains to be clarified (Allen et al., 1995).

MESOZOIC CHRONOSTRATIGRAPHIC FRAMEWORK

Mesozoic strata within the Junggar Basin are nonmarine, and the resolution of age dating is relatively poor. Age determinations presented in Figure 5 are from the Regional Survey Team of the Bureau of Geology of the Xinjiang Uygur Autonomous Region (RSTBGXUAR, 1980), the Compiling Team of the Xinjiang Uygur Autonomous Region Regional Stratigraphic Tables (CTXUARRST, 1981), the Bureau of Geology and Mineral Resources of Xinjiang Uygur Autonomous Region

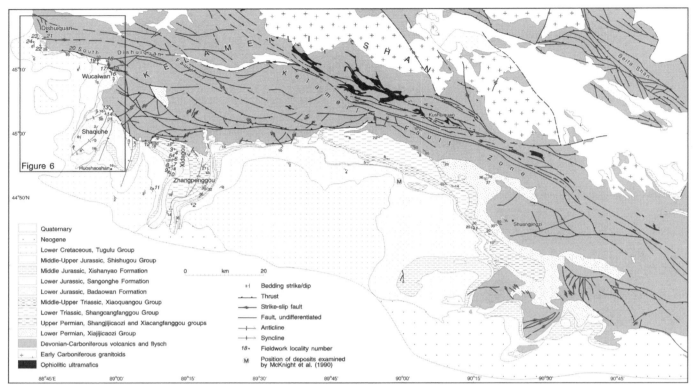

Figure 3. Geology of Kelameili Shan and adjacent areas. Compiled from data from Regional Survey Team of Bureau of Geology of Xinjiang Uygur Autonomous Region (1965, 1966, 1980), with additions from this study. Located in Figure 2.

(BGMRXUAR, 1993), and Zhang Guojun (1993); additional data from the Kelameili Shan region were taken from Shen Yanbin and Mateer (1992) and Liu Zhaosheng (1993). A recent combined seismic and borehole study in eastern Junggar by Pu Renhai et al. (1994a) placed the base of the Sangonghe Formation within the Pliensbachian stage of the late Early Jurassic.

Outcrop localities are shown in Figures 3 and 6 and a summary stratigraphic log from the Mesozoic succession of the Kelameili Shan region is presented as Figure 7. Relevant photo-documented sedimentary and structural geological information is presented in Figures 8 and 9. Sediment color is recorded using Munsell Soil Color Chart standard notation, comprising hue, value, chroma, and description (e.g., 2.5YR3/3; dusky red).

TRIASSIC

The Triassic of the Kelameili Shan region is divided into the Lower Triassic Shangcangfanggou (or Upper Cangfanggou) Group and the Middle and Upper Triassic Xiaoquangou Group. The Xiacangfanggou (or Lower Cangfanggou) Group is Late Permian age (Fig. 5).

Shangcangfanggou Group

Lower boundary, distribution, and thickness. Lower Triassic strata are exposed between 90°15′E and the core of the

Zhangpenggou anticline at 89°15′E (Fig. 3). They conformably or disconformably overlie Permian sedimentary rocks exposed at the southern side of the Kelameili Shan and, beyond the limit of Permian outcrops, unconformably onlap Devonian and Carboniferous rocks (RSTBGXUAR, 1966, 1980; BGMRXUAR, 1993). The CTXUARRST (1981) recorded 2 m of coarse conglomerate, composed of volcanic clasts, at the base of the succession.

The Shangcangfanggou Group is reported to be as thick as 620 m in the core of the Zhangpenggou anticline (CTXUARRST, 1981), and is 149 m thick in the Caican 2 well (Zhang Xuejun et al., 1999; located on Fig. 4A).

Sedimentology. The lower part of the Lower Triassic Shangcangfanggou Group in the Xidagou section (44°58.025′N, 89°12.191′E; Fig. 3, locality 3) comprises 2.5YR3/3 dusky red mudrock. The poorly exposed succession is uniformly reddened, and passes up into medium- to fine-grained pebbly sandstone 1.25 km to the south at 44°57.338′N, 89°11.962′E (Fig. 3, locality 4). The sandstone is poorly sorted and is either massive with a subhorizontal fabric or contains crude channelization and trough cross-bedding with foresets dipping toward 160°. Clasts comprise angular to subangular pebbles of red lithic sandstone, pale green tuff, volcanic rock, and rare quartzite. These strata probably represent disorganized, discontinuous, poorly channelized sheet flood units deposited in an arid distal alluvial fan or intermittent stream-playa environment.

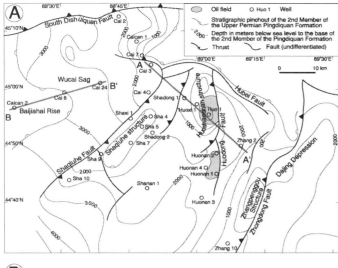

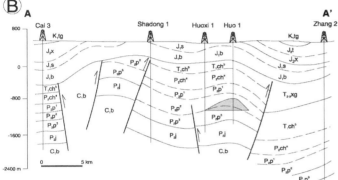

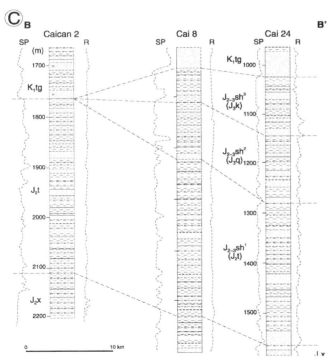

Figure 4. A: Structure and isopach map for base of Second Member of Upper Permian Pingdiquan Formation in region of Huoshaoshan and Huonan oil fields (located in Fig. 2; from Li Xibin and Jiang Jianheng, 1987). B: Structural cross section (from Li Xibin and Jiang Jianheng, 1987). C: Correlation of, and lateral facies variations in, Middle to Upper Jurassic Shishugou Group in subsurface of Cainan oil field (from Pu Renhai et al., 1994b). Stratigraphic abbreviations are listed in Figure 5.

		Bogda Shan	Kelameili Shan
QUATERNARY	Holocene	Holocene	Holocene
	Pleistocene — late	Xinjiang Gp	Xinjiang Gp
	Pleistocene — middle	Wusu Gp	Wusu Gp
	Pleistocene — early	Xiyu Fm	
NEO-GENE	Pliocene	Dushanzi Fm	
	Miocene	Taxihe Fm	Changjihe Gp
PALEO-GENE	Oligocene	Shawan Fm	
	Eocene	Anjihaihe Fm	Suosuoquan Fm
	Paleocene	Ziniquanzi Fm	
CRETACEOUS	Late	Donggou Fm — Subashi Fm	
		Donggou Fm — Kumutake Fm	
		Tugulu Gp — Lianmuqin Fm	
	Early	Tugulu Gp — Shengjinkou Fm	Tugulu Gp (K₁tg)
		Tugulu Gp — Hutubihe Fm	
		Tugulu Gp — Qingshuihe Fm	
JURASSIC	Late	Kalazha Fm (J₃k)	
		Qigu Fm (J₃q)	Shishugou Gp (J₂₋₃sh)
	Middle	Toutunhe Fm (J₂t)	
		Shuixigou Gp — Xishanyao Fm	Xishanyao Fm (J₂x)
	Early	Shuixigou Gp — Sangonghe Fm	Sangonghe Fm (J₁s)
		Shuixigou Gp — Badaowan Fm	Badaowan Fm (J₁b)
TRIASSIC	Late	Xiaoquangou Gp — Haojiagou Fm	
		Xiaoquangou Gp — Huangshanjie Fm	Xiaoquangou Gp (T₂₋₃xg)
	Middle	Xiaoquangou Gp — Kelamayi Fm	
	Early	Shangcangfanggou Gp — Shaofanggou Fm	Shangcangfanggou Gp (T₁ch)
		Shangcangfanggou Gp — Jiucaiyuan Fm	
PERMIAN	Late	Xiacangfanggou Gp — Wutonggou Fm	Xiacangfanggou Gp (P₂ch)
		Xiacangfanggou Gp — Quanzijie Fm	
		Shangjijicaozi Gp — Hongyanchi Fm	Shangjijicaozi Gp — Pingdiquan Fm (P₂p)
		Shangjijicaozi Gp — Lucaogou Fm	
		Shangjijicaozi Gp — Jingjingzigou Fm	Shangjijicaozi Gp — Jiangjunmiao Fm (P₂j)
		Shangjijicaozi Gp — Wulabo Fm	
	Early	Xiajijicaozi Gp / Aqikebulake Gp — Tashikula Fm	Xiajijicaozi Gp
		Aqikebulake Gp — Shirenzigou Fm	
		Aoertu Fm	Kongqueping Fm
CARBONIFEROUS	Late	Qijiagou Fm	Liukeshu Fm
		Liushugou Fm	Shiqiantan Fm
	Early		Batamayineishan Fm (C,b)
			Dishuiquan Fm
			Tamugang Fm

Figure 5. Carboniferous to Holocene stratigraphic chart for southern (Bogda Shan) and northern (Kelameili Shan) sides of eastern Junggar Basin. Data sources are cited in text. Gp, Group; Fm, Formation.

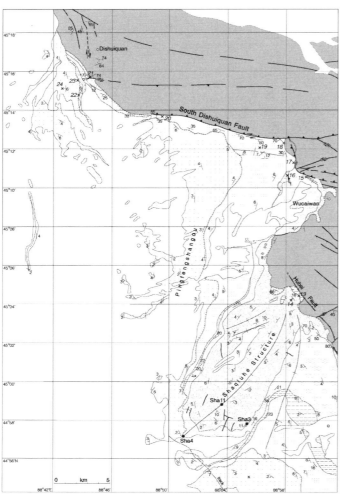

Figure 6. Detailed geological map of Shaqiuhe-Dishuiquan region (source anonymous). Note that displacement on south Dishuiquan fault, which thrusts Carboniferous basement obliquely over Middle and Upper Jurassic Shishugou Group, dies out to west; fault tip is close to 45°13.440'N, 88°49.603'E (Figs. 3 and 6, locality 20). To west of this point, original stratigraphic relationship between cover and basement is preserved; Lower Jurassic Badaowan Formation onlaps Carboniferous volcaniclastic sediments. Key to figure and its location are given in Figure 3.

The group is again dominated by poorly exposed reddened mudstones 7 km to the northwest. Just beneath the Jurassic angular unconformity, however, at 44°59.216'N, 89°07.575'E (Fig. 3, locality 12) a poorly sorted, matrix-rich subrounded conglomerate is exposed (Fig. 8A). Clasts are dominated by weathered light green-gray rhyolite and tuff up to 16 cm in diameter, along with reddened (possibly Permian) sandstone and Triassic red mudstone rip-up clasts. Thin fine-grained sandstone interbeds are weak red (10R5/2). Mild channelization and reworked glaebule horizons (pedogenic calcareous nodules) also occur in the uppermost part of the succession. These strata are best interpreted as debris-flow deposits that had undergone minor tractional stream flow reworking.

The Shangcangfanggou Group is reported to yield dinosaur remains including *Chasmatosaurus yuani* Young, *Lystrosaurus broami* Young, and *L. hedini* Young, and ostracods (RSTBGXUAR, 1965, 1966).

Xiaoquangou Group

Lower boundary, distribution, and thickness. The Middle and Upper Triassic Xiaoquangou Group crops out between 90°15'E and 89°20'E (the eastern flank of the Zhangpenggou anticline), and conformably overlies the Shangcangfanggou Group (RSTBGXUAR, 1966, 1980; CTXUARRST, 1981). It is up to 520 m thick (CTXUARRST, 1981), and is 71.5 m thick in the Caican 2 well (Zhang Xuejun et al., 1999).

Sedimentology. The Xiaoquangou Group at 44°58'N, 89°50'E is described as 88 m of interbedded yellow-brown conglomerate and carbonaceous silty mudstone and mudstone, with limonite nodules and cone-in-cone limestone horizons developed at its top; dinosaurs, fish, gastropods, and plant fossils are also reported (CTXUARRST, 1981). The uppermost part of the group at 44°45.760'N, 90°14.600'E, near Shuangjingzi (Fig. 3, locality 1), comprises pale gray siltstone interbedded with 20 cm-thick layers of cone-in-cone limestone.

Evidence for synsedimentary deformation. No direct evidence of synsedimentary deformation was observed during this study. The coarsening-upward nature of the Lower Triassic Shangcangfanggou Group and conglomeratic horizons within the succession recorded by the RSTBGXUAR (1965, 1966, 1980) and the CTXUARRST (1981) (Fig. 7) might indicate evidence for tectonic source-area rejuvenation through the Triassic.

Subsurface data from immediately south of the Kelameili Shan show Jurassic strata unconformably overlying Upper Carboniferous to Middle to Upper Triassic strata within structures that were active prior to the Jurassic (Fig. 4B). Triassic strata are absent on the Shaqiuhe structure and thicken from 205 m in the Huo 1 well on the Huoshaoshan structure to ~1230 m in the Zhang 2 well in the footwall of the Huodong fault. The exact timing of uplift on these structures and movement on the Huodong fault cannot be assessed, because Triassic intragroup markers were not shown on either side of the fault by Li Xibin and Jiang Jianheng (1987). It is highly unlikely that the fault remained inactive during the Triassic, because this would necessitate more than 1 km of uplift on the fault (and subsequent denudation) at the end of the Triassic. Instead, it is plausible that at least some Middle to Late Triassic movement occurred, resulting in the partial to complete removal of Lower Triassic strata on the Huoshaoshan and Shaqiuhe structures, respectively, and the deposition of Middle and Upper Triassic deposits in the Huodong thrust footwall (as represented in Fig. 10A). This is supported by the interpretation of Li Xibin and Jiang Jianheng (1987) that the thrust dies out beneath the basal Jurassic unconformity, and overlying Lower and Middle Jurassic strata show little lateral thickness variation across the Huoshaoshan and Shaqiuhe structures (Fig. 4B).

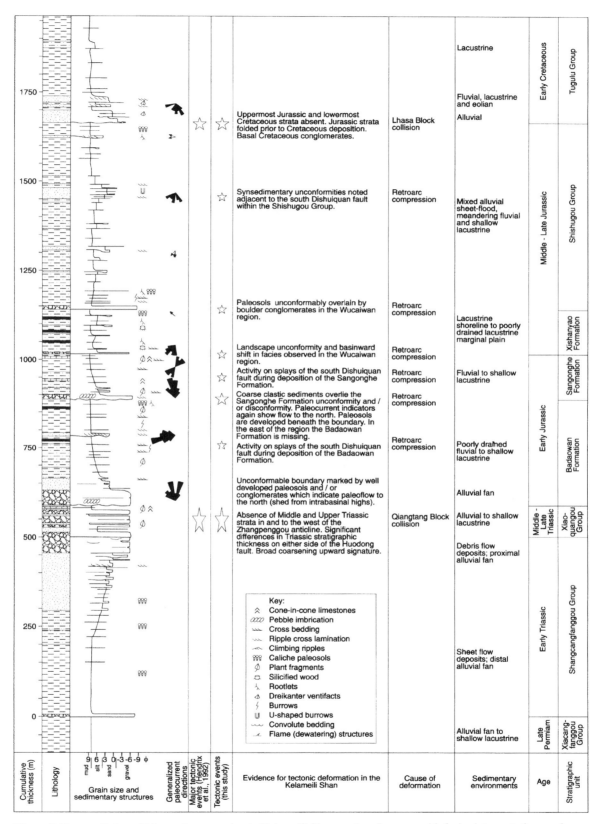

Figure 7. Composite log of Mesozoic succession in Kelameili Shan, eastern Junggar, with facies interpretations and summary of evidence for tectonic deformation in region and its likely causes. Compiled from CASP field work; additional stratigraphic data are from Compiling Team of Xinjiang Uygur Autonomous Region Regional Stratigraphic Tables (1981).

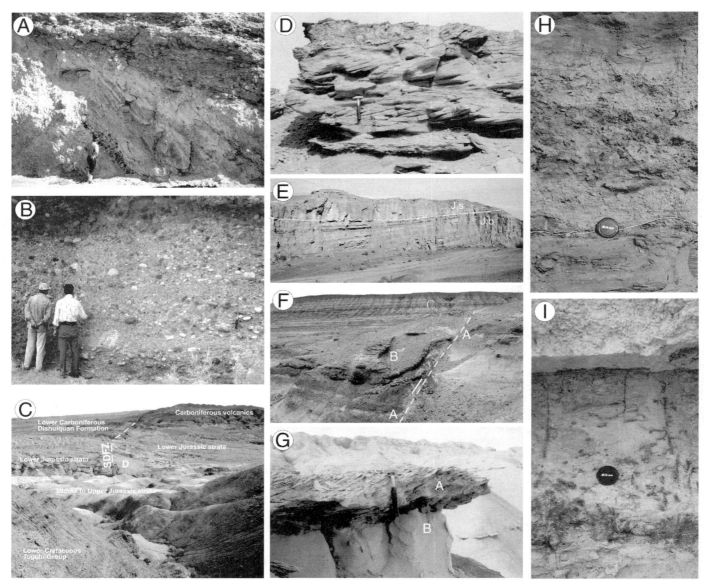

Figure 8. A: Unconformity (with angular discordance of ~34°) separating Lower Triassic Shangcangfanggou Group from overlying Lower Jurassic Badaowan Formation at 44°59.216′N, 89°07.575′E (Fig. 3, locality 12). B: Clast-supported, poorly sorted very large pebble to small cobble conglomerate within lower part of Lower Jurassic Badaowan Formation in Xidagou transect, at 44°57.091′N, 89°12.076′E (Fig. 3, locality 5). C: View from 44°15.189′N, 88°43.731′E (Figs. 3 and 6, locality 23), at Dishuiquan, looking east along trace of splay from south Dishuiquan fault zone (SDFZ), which thrusts Carboniferous volcanics northward over Carboniferous clastics of Dishuiquan Formation. The Carboniferous basement has been onlapped by Jurassic sediments, lower units of which (Lower Jurassic Badaowan and Sangonghe formations) have also been affected by reactivation on this fault (with opposite sense of dip-slip motion); downthrown block (D) is visible in center of view. Middle and Upper Jurassic and Cretaceous strata postdate fault movement. Onlap point of Lower Jurassic strata onto Carboniferous volcanics is ~1.75 km from viewer (see Fig. 6). D: Oil-stained trough cross-bedded fluvial sandstones from middle part of Badaowan Formation in Shaquihe structure at 45°03.771′N, 88°58.185′E (Figs. 3 and 6, locality 14). Paleoflow (illustrated in Fig. 7) was to west-southwest. Hammer is 28 cm long. E: Detail of subtle angular unconformity between Lower Jurassic Badaowan (J_1b) and Sangonghe (J_1s) formations in Dishuiquan transect near 45°15.299′N, 88°44.562′E, indicating uplift of Badaowan Formation to north (right in this view). See also Figure 9. Cliff is ~20 m high. F: Detail of east-west–orientated splay of south Dishuiquan fault shown in C, in Dishuiquan region west of 44°54.838′N, 89°10.842′E (Figs. 3 and 6, locality 21). This fault downthrows Lower Jurassic Sangonghe Formation to south (part of unit D in C), with offset of horizon A being ~5.6 m. Paleofault scarp controlled position of 3.6-m-thick Sangonghe Formation fluvial channel (B). Fault does not cut younger strata (note continuous nature of overlying sedimentary rocks in background [C]). G: Interbedded fluvial and eolian sandstone in Lower Cretaceous Tugulu Group in Dishuiquan transect (45°14.858′N, 88°42.592′E; Figs. 3 and 6, locality 24). Fluvial sandstone (A) contains 20-cm-high sets of trough cross-bedding to 210° and irregular scoured base with red mudstone rip-up clasts to 15 cm in diameter. These are interbedded with fine-grained, moderately to well-sorted eolian litharenite with hematitic clay rims, which contain traces of oil and have very good reservoir properties (B). Hammer is 28 cm long. H: Unconformity between Lower Jurassic Sangonghe Formation and Middle Jurassic Xishanyao Formation at 45°10.524′N, 88°58.064′E (Figs. 3 and 6, locality 16), west of Wucaiwan. Unconformity separates laminated lacustrine mudstone of Sangonghe Formation from 10-cm-thick, subangular to subrounded, small- to medium-pebble-grade fluvial conglomerate, composed largely of green volcanic tuff clasts at base of Xishanyao Formation. This represents abrupt basinward shift in facies and may be sequence boundary. Basal conglomerate grades into granule-grade litharenite containing irregular coal streaks and woody material before passing upward into carbonaceous fine-grained sandstone and siltstone at top of field of view. Lens cap is 52 mm in diameter. I: Paleosol and rootlet development beneath Lower Jurassic Sangonghe Formation unconformity in Dishuiquan transect. Paleosol includes 1–2-cm-diameter dusky red glaebules (10R3/2) within pale yellow matrix (2.5Y7/5). Lens cap is 52 mm in diameter.

Figure 9. Panorama showing Mesozoic succession in Dishuiquan region taken from 45°15.283'N, 88°44.636'E looking west-northwest. The panorama shows Carboniferous Dishuiquan Formation onlapped by Lower Jurassic Badaowan Formation via angular unconformity; Gilbert-type delta body shed from north is visible. Subtle angular unconformity at base of Sangonghe Formation that cuts down to north is marked (see Fig. 8, F and I). Variegated mudstone-dominated units of undifferentiated Middle to Upper Jurassic Xishanyao Formation and Shishugou Group overlie this, before being capped by poorly sorted pebble conglomerates at base of Lower Cretaceous Tugulu Group. Tugulu Group occurs above irregular angular unconformity that truncates southerly dipping underlying Jurassic sedimentary rocks progressively to north. Splay of south Dishuiquan fault cuts Lower Jurassic Badaowan and Sangonghe formations, but not younger strata (see also Fig. 8, C and F). The track at right margin of field of view (marked by star) is ~2m wide.

Petroleum exploration has revealed an unconformable relationship between Triassic and Jurassic strata on a number of other intrabasinal highs within the eastern part of the Junggar Basin, including the Beisantai and Santai rises (Li Xibin and Jiang Jianheng, 1987; Wang Mingming, 1992; Zhang Guojun, 1993). Middle to Upper Triassic strata have also been partially to completely removed from the Baijiahai Rise (Zhang Xuejun et al., 1999). Greene et al. (this volume) document uplift of the Bogda Shan at the southeastern side of the Junggar Basin during the ?Middle and Late Triassic.

JURASSIC

Jurassic strata to the south of the Kelameili Shan were examined in three major sections; these are, from east to west, Xidagou (localities 5–11), Wucaiwan (localities 15–19), and Dishuiquan (localities 21 and 22; Figs. 3 and 6). The basal Jurassic unit, the Badaowan Formation, was also examined in the Shaqiuhe structure, where these strata contain an exhumed hydrocarbon paleoreservoir. Four Jurassic stratigraphic units are recognized in the Kelameili Shan (Fig. 5), and these are described in turn.

Badaowan Formation

Lower boundary, distribution, and thickness. The Lower Jurassic Badaowan Formation is exposed to the west of 89°30′E along the southern flank of the Kelameili Shan (Fig. 3). At its easternmost limit it unconformably overlies the Middle to Upper Triassic Xianquangou Group, whereas on either flank of the Zhangpenggou anticline it unconformably overlies the Lower Triassic Shangcangfanggou Group (Fig. 8A). To the west of the Zhangpenggou anticline, the Badaowan Formation onlaps Devonian-Carboniferous basement (Fig. 9).

The Badaowan Formation was recorded to be as thick as 340 m in this study (Fig. 11), although the CTXUARRST (1981) reported thicknesses to 495 m. The formation is 469.5 m

thick in the Caican-2 well (Zhang Xuejun et al., 1999; located in Fig. 4A).

Sedimentology. Considerable variations in the thickness and character of the Badaowan Formation were observed in the study region (Fig. 11). These differences are related to the proximity of these deposits to north-south–trending structures within the basin, deposits at the flanks of the Shaqiuhe and Zhangpenggou structures being different from those farther north, close to the south Dishuiquan fault at Wucaiwan and Dishuiquan (Figs. 3 and 10).

The base of the Badaowan Formation in the Xidagou section and the core of the Shaqiuhe structure consists of clast-supported large pebble to small cobble conglomerate, which in the Shaqiuhe structure is at least 100 m thick (Figs. 8B and 11A). Clasts are composed of a mixture of basement lithologies, including quartzite, granodiorite, cleaved rhyolite, granite, graywacke, and chert. They are poorly to moderately sorted and subrounded to well rounded. Some clasts were reddened prior to deposition, suggesting that they underwent significant reworking at the Triassic-Jurassic boundary. The units are either massive or contain crude meter-scale subhorizontal bedding. A mean maximum clast size (MMCS) of 55 cm was recorded from the Xidagou section. Well-developed imbrication indicates paleoflow toward the north (Fig. 11, A and B). These deposits are interpreted to be alluvial in origin.

Both successions fine upward from ripple cross-bedded medium- and coarse-grained sandstone, through lenticular ripple cross-laminated sandstone sheets, to brown-yellow siltstone and pale gray mudstone with rare coal, plant fragments, and isolated pebble horizons. These are interpreted as poorly drained alluvial to shallow lacustrine facies.

Approximately halfway up the Xidagou section, at 44°56.630′N, 89°12.138′E (Fig. 3, locality 6, and Fig. 9A), 4 m of medium-grained, subangular to well-rounded, dark red (2.5YR4/6) litharenite pass upward into gray siltstone with medium-grained sandstone lenses. These contain low-angle planar cross-bedding and climbing ripple cross-lamination with

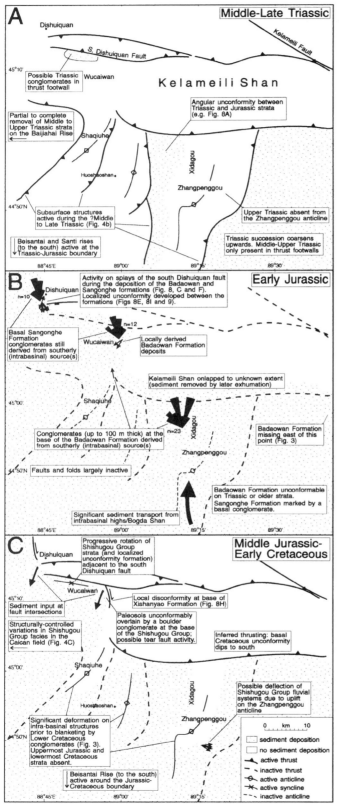

Figure 10. Schematic map representations of structural, stratigraphic, and sedimentological evidence for deformation in Middle-Late Triassic (A), Early Jurassic (B), and Middle Jurassic to Early Cretaceous (C) in Kelameili Shan region of eastern Junggar.

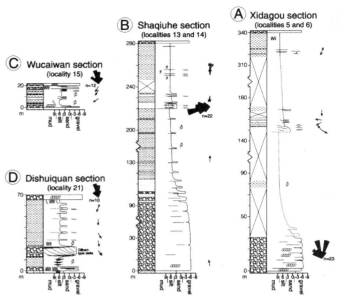

Figure 11. Detailed stratigraphic logs through Lower Jurassic Badaowan Formation in Kelameili Shan region. Locality numbers and approximate positions of logged sections are given in Figure 3; see Figure 7 for key.

migration directions toward 320° and 120°. This outcrop is interpreted as lacustrine shoreline sands, the low-angle planar cross-bedding and bidirectional ripples being indicative of the interaction between wave and current processes in a shoreface environment. These strata are overlain by medium- to fine-grained sandstone channels, with 30–50-cm-high sets of planar cross-bedding foresets indicating flow toward between 180° and 280°. The channels have ribbon geometries, are 1–5 m wide, and are isolated within fine-grained sedimentary rocks with scattered silicified tree remains. Wave ripple cross-lamination was not observed. Overall, this interval marks a shift from shallow-water lacustrine to fluvial facies in the Xidagou section.

Coarse clastic bodies are also developed within the middle to upper part of the Badaowan Formation in the Shaqiuhe structure, where at least 5 m of medium-grained feldspathic litharenite occurs (Fig. 8D). This sandstone interval is impregnated by degraded hydrocarbons (containing as much as 27% bitumen by volume) and displays trough and planar cross-bedding to 95 cm high. Paleocurrent indicators obtained from these structures are typically directed toward the west (Fig. 11B). The sandbodies are sheet-like in geometry, although individual cross-bedded units are lenticular parallel to paleoflow, and traceable for as much as 16 m. Foresets pass into bottomsets, which comprise horizontally laminated sandstone intervals that separate some cross-bedded horizons. Horizontally bedded ripple cross-laminated fine-grained sandstone caps some dune-scale trough cross-bedded intervals. Outcrops are interpreted as the deposits of a multichannel sandy bedload fluvial system. Laterally equivalent strata at 45°04.192′N, 88°58.670′E (Figs. 3 and 6, locality 13) comprise isolated ribbon bodies, some of which are also oil stained. These are likely to represent minor fluvial tributaries beyond the margins of the main channel system.

Above these coarse-grained intervals, the Badaowan Formation in both the Xidagou section and Shaqiuhe structure pass back into mudstone-dominated units with decimeter-thick medium- to fine-grained asymmetric ripple cross-laminated sandstone units that contain dwelling burrows (Fig. 11, A and B). Low-angle climbing bidirectional ripple trains are present, as are convolute lamination, dewatering structures, and rare coals. These deposits are interpreted to represent shallow to marginal lacustrine facies. Reddened horizons occur within the Xidagou section immediately beneath the base of the Sangonghe Formation.

In contrast to the Xidagou and Shaqiuhe sections, in the Wucaiwan transect, located close to the south Dishuiquan fault, the Badaowan Formation is only 20 m thick (Fig. 11C). The formation unconformably overlies Carboniferous dolerite and tuff and is marked by very coarse grained, poorly sorted sandstone with volcanic and igneous clasts, to 13 cm in diameter. These form part of a 2-m-thick basal coarse clastic unit that comprises a series of decimeter-scale fining-upward very coarse to fine-grained sandstone cycles. These are overlain by interbedded finely laminated carbonaceous mudstone and coal, brownish yellow silty mudstone and siltstone with plant debris, and a number of fining-upward lenticular medium-grained sandstone bodies that contain trough cross-bedding with foreset directions toward 235° and 210°. A nonpersistent, 2.5-m-thick, poorly sorted, large to very large pebble conglomerate containing clasts of lithic sandstone, quartz, and volcanic lithologies occurs near the top of the section. These deposits are probably alluvial to lacustrine in nature; the alluvial components are derived from the Kelameili Shan.

The base of the Badaowan Formation in the Dishuiquan transect, at 45°15.299′N, 88°44.562′E (Figs. 3 and 6, locality 21), consists of matrix-dominated, poorly sorted breccias that infill an irregular landscape topography (Figs. 9 and 11D). The breccias are dominated by clasts, with a MMCS of 27 cm, derived from the underlying Lower Carboniferous Dishuiquan Formation, and represent localized hillslope (colluvial) deposits. These sedimentary rocks also contain well-developed 2–3-m-thick laterite caliches.

The initial breccia horizon is overlain by 8–10 m of well-laminated, 7.5R4/3 weak red mudstone to fine-grained sandstone with randomly distributed <2 cm pebbles, carbonaceous partings and plant fragments. In the upper meter of this interval the strata coarsen upward slightly, and become yellow-brown before passing abruptly into a 10–12-m-thick breccia-conglomerate unit composed of subangular, poorly sorted medium to large pebbles, dominated by rhyolitic and tuffaceous lithologies, with a MMCS of 11 cm. Large-scale foresets, to 6 m high, are preserved within these deposits, and indicate a progradation direction toward 144°. Topset deposits comprise

fine- to medium-grained calcite-cemented volcanic arenite, and are capped by mottled 10R3/4 dusky red or massive 2.5Y5/2 grayish brown mudstone. The cross-bedded breccia-conglomerate unit is interpreted as a Gilbert-type delta body. The underlying strata were deposited under upper plane bed flow conditions and are interpreted to form toesets to the Gilbert-type delta that developed following the initial lacustrine flooding of the Jurassic landscape.

Several hundred meters to the north of the inferred Gilbert-type delta body, the same stratigraphic horizon is represented by a 2-m-thick medium-grained sandstone interval that includes 1.4-m-thick lenticular to ribbon-shaped channels, oriented north-south, with basal conglomerate clasts to 11 cm in diameter. These deposits are thought to represent the feeder fluvial system to the Gilbert-type delta body described here.

The Badaowan Formation in the Dishuiquan region is cut by splays from the south Dishuiquan fault zone, which extend approximately east-west through the Gilbert-type delta body and have a combined component of dip-slip displacement down to the south (Figs. 3, 6, and 9). To the north of the faults, beds overlying the Gilbert-type delta body consist of ~95% mudstone, which varies on a 1–2 m scale from light gray (5Y7/1), with plant fragments, to weak red (10R5/2). Thin, 20-cm-thick, small pebble to medium-grained sandstone horizons also occur, which include ripple-scale cross-bedding with migration directions toward between 150° and 180°. These deposits are interpreted as fluvial flood-plain to marginal lacustrine facies.

To the south of the fault splays, the upper part of the Badaowan Formation is dominated by massive to finely laminated 7.5R5/3 weak red mudstone and silty mudstone, and very fine- to medium-grained sandstone (Fig. 9). The proportion of sandstone to mudstone increases to ~50% at the top of the section (Figs. 8E and 9), in sharp contrast to the mud-prone equivalent section to the north of the fault traces. The sandstone occurs as either coarsening-upward units from underlying mudstone or 2–3-m-thick, sharply based intervals. These display convolute bedding and poorly defined wave ripple cross-lamination; some paleocurrent indicators demonstrate flow to the north. These deposits represent shallow lacustrine facies, the northerly ripple migration directions representing the direction of wave approach perpendicular to the east-west–oriented lake margin, and suggest a deepening of environments relative to those to the north of the fault traces.

Sedimentary evidence for deformation. Significant deformation occurred in the Late Triassic prior to the deposition of the Lower Jurassic Badaowan Formation (Figs. 3, 4B, and 10A). An angular unconformity is developed between Triassic and Jurassic strata; Triassic strata dip more steeply to the south, indicating uplift to the north in the Kelameili Shan (e.g., Fig. 8A). There was erosion or nondeposition of all or part of the Triassic succession across the Zhangpenggou, Huoshaoshan, and Shaqiuhe anticlines (Figs. 3, 4, and 10A). Basal Jurassic strata on the flanks of these structures are made up of coarse clastic sedimentary rocks that contain paleoflow indicators toward the north, despite the proximity (and activity) of the Kelameili Shan (Figs. 10B and 11, A and B). It is likely that these sediments were shed either from the Bodga Shan, or more likely from the remnant topography of intrabasinal structures, such as the Beisantai and Baijiahai rises or more localized structures that were active in the Late Triassic, prior to their onlap during the Jurassic.

Synsedimentary tectonic activity during the deposition of the Badaowan Formation is only seen in the Dishuiquan section, where southerly downthrowing splays of the south Dishuiquan fault have affected facies development (Fig. 10B). A transition from fluvial flood-plain and marginal lacustrine to shallow lacustrine conditions is interpreted, and an increase in sandstone proportion observed, southward across the fault splays in units of the Badaowan Formation developed above the Gilbert-type delta body. The juxtaposition of the Gilbert-type delta body and the fault splays might suggest that these faults were also responsible for creating the topographic break across which the Badaowan Formation Gilbert-type delta body developed.

Sangonghe Formation

Lower boundary, distribution, and thickness. East of 89°30'E, the Lower Jurassic Sangonghe Formation disconformably to unconformably overlies the Middle to Upper Triassic Xiaoquangou Group (Fig. 3). To the west of this, the Sangonghe Formation disconformably overlies the Badaowan Formation, except in the far west of the region (west of 88°47.6'E), where an unconformity is developed, the Sangonghe Formation dipping more gently toward the south to southwest (Figs. 8E and 9). The Sangonghe Formation in the Kelameili Shan region is reported to be between 50 and 253 m thick (CTXUARRST, 1981). According to Pu Renhai and Tang Zhonghua (1994), in the Cainan region the formation varies in thickness between 123 m in the Cai 8 and 24 wells and 194.5 m in the Sha 9 well (located in Fig. 4A). The following sedimentary descriptions are listed from east to west.

Sedimentology. Near Shuangjingzi at 44°45.760'N, 90°14.600'E (Fig. 3, locality 1), the base of the Jurassic succession is represented by conglomerate of the Sangonghe Formation, which is well rounded and typically large pebble to large cobble grade; the maximum clast size is 20 cm in diameter. Imbrication indicates paleoflow to 212°. Hendrix et al. (1992) also recorded paleocurrent indicators from the Sangonghe Formation in the eastern part of the Kelameili Shan; these showed a wide spread of paleocurrent values ranging from west to east-southeast, with flow to the south.

The base of the Lower Jurassic Sangonghe Formation in the Xidagou transect is marked by a 7-m-thick continuous sheet of well-rounded, but poorly sorted, large to very large pebble conglomerate. Clasts are dominated by acidic volcanics, and are commonly but not exclusively agate bearing. No paleocurrent indicators were seen. The conglomerate is overlain by 15 m of poorly exposed fine-grained strata, which includes gray mudstone.

Above this, a 7-m-high bluff at 44°55.761'N, 89°11.870'E (Fig. 3, locality 7), comprises 2 m of clean, pebbly coarse-grained massive sandstone overlain by trough cross-bedded coarse- to medium-grained litharenite, some of which is bitumen stained. Within the bluff, these pass upward into ripple and climbing ripple cross-laminated siltstone, overlain by laminated claystone with cone-in-cone limestone and fine-grained asymmetric, straight-crested ripple cross-laminated sandstone. Ripple cross-laminations indicate paleoflow both to and from 220°–230°. Adhesion worts, an irregular sandstone bedding-plane texture developed due to the adhesion of wind-blown sand across a damp sediment surface (Reineck, 1955; Collision, 1986), are also present. The upper 60 m of the Sangonghe Formation consist of gray silty mudstone and siltstone with symmetrical wave ripple cross-laminated fine- to very fine grained sandstone.

These facies are interpreted as a fluvial (basal conglomerate and overlying fine-grained strata), through lacustrine shoreline (massive to trough cross-bedded to climbing ripple cross-laminated sandstone with adhesion worts), to shallow lacustrine (mudstone with wave ripple cross-laminated sandstone) transition, indicating a progressive upward deepening of facies.

The base of the Sangonghe Formation in the Wucaiwan transect at 45°10.558'N, 88°59.345'E (Figs. 3 and 6, locality 15) is marked by a 2.5-m-thick, highly amalgamated coarse clastic horizon that fines upward from boulder-grade conglomerate, composed of granite, rhyolite, cleaved volcanics, ignimbrite, and sandstone clasts, to fine-grained cross-bedded litharenite. These deposits are interpreted as fluvial in origin and contain paleocurrent indicators to the north and north-northwest (Fig. 11C). The disconformity is underlain by a well-developed paleosol with vertically stacked calcareous glaebules to 1 cm in diameter and 4–5 cm high.

Overlying this coarse clastic unit, the lower part of the formation consists of noncalcareous variegated mudrock (which include dusky reds; 2.5YR3/4) with siderite nodules, light gray (10Y7/1) claystone, rare carbonaceous horizons, and minor cross-bedded channelized fine-grained litharenite, with crosssets between 15 and 50 cm high, indicating paleoflow toward the west-northwest. This part of the formation is 15–20 m thick and represents an occasionally poorly drained fluvial environment, dominated by the deposition of overbank mudstones. The mixed to suspended load of the fluvial system (with fines exceeding sands) suggests a meandering channel pattern. The paucity of organic material and the development of paleosols might suggest a semi-arid climate.

The uppermost part of the Sangonghe Formation within the Wucaiwan region is exposed in the hanging wall of a small north-south–trending thrust at 45°10.524'N, 88°58.064'E (Figs. 3 and 6, locality 16). This comprises gray, laminated mudrock and siltstone, which weathers light brown. Plant fragments, cone-in-cone limestone and symmetrical-ripple crosslaminated sandstone, indicating paleoflow to 060° and 240°, are developed. Shoaling-wave influenced, offshore transitional lacustrine facies below fair-weather wave base are interpreted, again indicating a broad basinward shift in facies through the deposition of the formation.

The Sangonghe Formation in the Dishuiquan transect overlies a subtle angular unconformity and onlaps the Badaowan Formation toward the north (toward the Kelameili fault zone; Figs. 8E and 9). This unconformity is associated with an underlying paleosol which is 1 m thick, with 1–2-cm-diameter dusky red glaebules (10R3/2) within a pale yellow matrix (2.5Y7/5), along with well-developed root traces (Fig. 8I). The base of the Sangonghe Formation is marked by a 10-m-thick stacked, multistorey channel body, which consists of medium-grained sandstone and lenses of pebble- to boulder-grade conglomerate containing clasts to 80 cm in diameter. Clasts are subangular to subrounded, moderately sorted, and are dominated by volcanic (particularly rhyolite) components, with granite, sandstone, graywacke, and rare quartzite. Imbrication and trough cross-bedding indicates paleoflow generally to the north-northwest (Fig. 11D).

A second, 6–7-m-thick unit of large pebble conglomerate is separated from the first by 6 m of dusky to weak red mudstone. Above this the succession is dominated by mudstone and sandy mudstone with paleosol horizonation, lesser fine- to medium-grained sandstone with paleocurrents to 155°, and cone-in-cone limestone. It is not possible to determine the boundary with the overlying Xishanyao Formation accurately at this locality (Fig. 9). These strata were probably deposited in a fluvial to marginal lacustrine environment.

Evidence for deformation. The basal influx of fluvial conglomerate, in places associated with paleosol development and/or an angular unconformity, provides evidence for the uplift responsible for the sedimentological definition of the Badaowan Formation–Sangonghe Formation boundary. The unconformity at Dishuiquan (Figs. 8E and 9) displays tilting toward the south, away from the Kelameili Shan. Paleocurrent data vary in this basal unit. Observations in the Wucaiwan and Dishuiquan transects indicate paleoflow unequivocally to the north to northwest (Fig. 11, C and D) and suggest the rejuvenation of source areas to the south (Fig. 10B); the basement-derived clasts indicate that these regions were yet to be onlapped completely by Jurassic sediments. In the Wucaiwan section, these lithologies may have been derived from the promontory of the Kelameili Shan immediately to its south (Figs. 3 and 6). This cannot be the case for the Dishuiquan section, and an intrabasinal high is assumed (Fig. 10B); it is interesting to note that the basal strata in this section were derived from the south and yet onlap the Badaowan Formation toward the north, implying a more subdued topography in this direction.

Synsedimentary fault activity is evident in the Sangonghe Formation in the Dishuiquan region (Fig. 8, C and F). Splays of the south Dishuiquan fault zone have offsets to 5.6 m, and influenced the positions of fluvial channels (Fig. 8F) and other facies. The fault splays strike at 055°–080° (Fig. 6), and are steeply dipping to subvertical. Fault planes are marked by scaley fabrics in fine-grained strata, and hematitic and local bitumen staining. None of the faults seem to propagate into overlying Jurassic strata (Fig. 8, C and F).

The transition toward finer grained basinal facies upward within the Sangonghe Formation is consistent with an inferred rise in lacustrine base level during the deposition of this unit reported by Qiu Dongzhou et al. (1994), Wang Longzhang (1994, 1995), and Shi Xuanyu et al. (1995).

Xishanyao Formation

Lower boundary, distribution, and thickness. Where present, the Middle Jurassic Xishanyao Formation is mapped as being conformable with the underlying Sangonghe Formation (RSTBGXUAR, 1965, 1966, 1980). This was observed to be the case in the field except at 45°10.524′N, 88°58.064′E (Figs. 3 and 6, locality 16), to the west of Wucaiwan, where a disconformity with ~1.5 m of relief was observed at the base of the formation. The formation is absent at outcrop in the region of Shuangjingzi, in the eastern part of the Kelameili Shan and to the south of Wucaiwan (Fig. 6).

The CTXUARRST (1981) described the formation as being up to 262 m thick, and being 118 m thick on the eastern limb of the Zhangpenggou anticline. It is 143 m thick in the Caican 2 well (Zhang Xuejun et al., 1999).

Sedimentology. The lower part of the Xishanyao Formation, at 44°55.536′N, 89°11.430′E (Fig. 3, locality 8) in the Xidagou transect, forms a distinctive white bleached appearing horizon, ~50 m thick. This is composed of three facies associations.

The lowermost facies association consists of fine- to medium-grained sandstone sheets, 60–200 cm thick, with primary current lineations and ripple trough cross-bedding. These occur within light gray mudstone to siltstone.

The second association includes a 3-m-thick medium-grained litharenite, in which massive to parallel-laminated facies with primary current lineations dominant, although low- to moderate-angle planar cross-bedding (< 9°), high-angle trough cross-bedding (18°), and sinuous, round-crested asymmetrical ripples are also developed. Sedimentary structures indicate paleocurrent directions typically toward 210° to 260°, although low-angle cross-bedding is also seen with a progradation direction of 120°. These units are extensive, uniform in nature, and contain no pebbles or evidence for scouring.

The boundary between the second and third facies associations is marked by large-scale trough cross-bedding in medium-grained sandstone, with troughs 1.5–2 m wide and 50 cm deep. Trough orientations indicate paleoflow to 310°. Above this, a continuous horizon of oil-stained, fine-grained volcanic litharenite and light gray mudstone occurs with current ripple cross-lamination indicating paleoflow toward 220°. The unit fines upward and is overlain by coaly material before the formation becomes reddened.

The lowermost facies association is interpreted to form a continuation of the offshore-transition lacustrine shoreline environment developed at the top of the Sangonghe Formation. Sandbody thickness increases within the upper part of this association, and some sandbody members may represent lower shoreface facies transitional into the second facies association, which is dominated by shoreface to foreshore facies. The predominant paleocurrent direction (220°) is thought to represent the offshore-backwash direction; lesser indices at high angles to this direction indicate longshore sediment transport. Coaly material in the upper facies association was deposited in a back-barrier lagoonal setting, such that overall a regressive trend is observed. The underlying trough cross-bedded units in the third association are therefore likely to have been deposited in a backshore position, possibly as eolian dunes.

The middle part of the Xishanyao Formation, around 44°55.063′N, 89°10.868′E (Fig. 3, locality 9), is composed of poorly exposed multicolored and fractured mudstone and siltstone effected by coal combustion. Rare, 30-cm-thick, medium-grained sandstone sheets with ripple cross-lamination indicating paleoflow toward 180° to 230° are developed. Many original textures and lithologies have been obliterated in this interval, making any environmental interpretations difficult. The fine-grained nature of the units and the original existence of coals, however, might indicate a low-gradient, vegetated, poorly drained lacustrine coastal plain or swamp environment.

The upper part of the Xishanyao Formation consists of pale gray mudstone and siltstone with minor coal and 30–50-cm-thick medium-grained, well-cemented sandstone sheets every several meters. Silicified tree remains and limonite concretions are also present. The top of the formation is marked by a laterally extensive, 2-m-thick trough cross-bedded fluvial sandstone channel, at 44°54.838′N, 89°10.842′E (Fig. 3, locality 10), with a paleocurrent direction of 220°. This is overlain by red and yellow-brown mudstone of the Middle to Upper Jurassic Shishugou Group.

The base of the Xishanyao Formation at 45°10.524′N, 88°58.064′E (Figs. 3 and 6, locality 16), west of Wucaiwan, is marked by a 10-cm-thick, subangular to subrounded, small- to medium-pebble–grade fluvial conglomerate, composed largely of green volcanic tuff clasts (Fig. 8H). This overlies laminated lacustrine mudstone of the Sangonghe Formation via an unconformity with ~1.5 m of relief seen over 20 m laterally. This boundary marks an abrupt basinward shift in facies and represents a candidate sequence boundary.

The Xishanyao Formation is at least 15 m thick at this locality, and is dominated by carbonaceous fine-grained sandstone, siltstone, and mudstone. Coaly horizons, 20–30-cm-thick, 30–50-cm-thick coarse-grained sublitharenite ribbon bodies, siderite nodules, rootlets, and silicified tree remains are also present. Paleocurrents are difficult to ascertain, but are generally toward 300°–320°. The unit generally fines upward. In the upper 6 m of the formation, background colors change from gray to yellow-brown and increasingly rubified paleosols, which include a 1.5-m-thick 2.5YR4/4 (dusky red) horizon, are developed.

The CTXUARRST (1981) reported a 1.3-m-thick conglomerate at the base of the Xishanyao Formation at the eastern limb of the Zhangpenggou anticline.

Within the Dishuiquan transect it was not possible to distinguish the Xishanyao Formation reliably from the underlying Sangonghe Formation or the overlying Shishugou Group (Fig. 9). The middle part of the succession is dominated by siltstone and mudstone, with rare fine- to very fine grained sandstone sheets. Olive-gray intervals occur in this part of the section and are speculatively correlated with some of the more poorly drained intervals in other sections.

Evidence for deformation. The Xishanyao Formation is marked regionally by a reduction in lacustrine conditions (Shi Xuanyu et al., 1995), and by a return to terrestrial environments in the Kelameili region, although the nature of this transition varies. In the region of Wucaiwan, an abrupt basinward shift in facies is recognized, fluvial conglomerate being incised into lacustrine mudstone. This evidence might be used to imply a marked drop in base level at the Sangonghe-Xishanyao formation boundary. In the Xidagou region, however, a regressive trend is observed with no facies dislocation and/or omission, and in the Dishuiquan region no change in facies was apparent.

Due to the variable expression of the Sangonghe-Xishanyao formation boundary, the localized incision in the Wucaiwan region is unlikely to be caused by a drop in lake level. Instead, the drop in base level is more readily explained by the differential uplift of a local structure, which was to later thrust the Sangonghe Formation over the Xishanyao Formation at this locality (Fig. 10C). Similarly, the change from poorly drained to paleosol-rich strata at the top of the formation might indicate further relative uplift on this structure prior to the unconformity development at the base of the overlying Shishugou Group. Pu Renhai et al. (1994a) defined the Xishanyao Formation in the east of the basin as a tectonically generated seismic sequence.

Shishugou Group

Lower boundary, distribution, and thickness. In the western part of the Kelameili Shan, the Middle and Upper Jurassic Shishugou Group conformably overlies the Xishanyao Formation, with the exception of the Wucaiwan region, where it is unconformable on either the Sangonghe or Xishanyao Formation (Figs. 3 and 6). To the east of 90°00′E, the group is either unconformable on the Sangonghe Formation, or disconformable on the Xishanyao Formation. According to the CTXUARRST (1981), the group is as much as 816 m thick, and is 597 m thick in the Shishugou section on the eastern limb of the Zhangpenggou anticline. It is 540 m thick in the Cai 24 well (Pu Renhai et al., 1994b).

Sedimentology. In the eastern part of the study area, McKnight et al. (1990) provided information on the Shishugou Group in their description of the Xinjiang petrified forest (the approximate position of which is marked by M in Fig. 3). They subdivided the group into three units. The lower unit is

180–380 m thick, unconformable on the Xishanyao Formation and older strata, and marked by a basal conglomerate overlain by a silty mudstone to sandstone in which the silicified tree remains were documented. Facies analysis indicated a shallow, wide, mixed load, meandering fluvial depositional environment, and a mild, possibly seasonal climate. Paleoflow was estimated to be to the south. The middle unit within the group was also reported to be unconformable and is dominated by purple-red silty shale, with common decimeter-thick, tabular beds of gray-green, fine- to medium-grained sandstone and minor limestone lenses. The silty shale is planar laminated and mudcracked, and the tabular sandstone displays abundant climbing ripples and, in places, trough to planar cross-bedding. The unit is 205–437 m thick and was interpreted to be the product of highly aggradational, ephemeral braided fluvial streams (Hendrix et al., 1992). The upper unit conformably overlies the middle unit and consists of 135–230 m of mudstone.

Scattered outcrops of the Shishugou Group were observed on the eastern side of the Zhangpenggou anticline at 44°49.947′N, 89°16.117′E (Fig. 3, locality 2). These comprise fluvial ribbon channel bodies, composed of medium-grained litharenite, within light gray mudstone with plant remains and silicified tree trunks. Cross-bedding indicates flow to the east-southeast.

The bulk of the Shishugou Group was not observed in detail in the Xidagou transect, but consists of siltstone and mudstone at its base that grade upward in color from weak red into brown-yellow. Very rare carbonate concretions are developed. Beneath the Cretaceous unconformity, however, at 44°55.196′N, 89°08.067′E (Fig. 3, locality 11), the group consists of poorly consolidated gray medium-grained lithic sandstone with possible low-angle trough cross-bedding to 290°, interbedded with lesser silty mudstone and mudstone. Qiu Dongzhou (1996, personal commun.) interpreted these strata as being shallow lacustrine in origin.

In the Wucaiwan region, at 45°11.229′N, 88°58.596′E (Figs. 3 and 6, locality 17), the base of the Shishugou Group unconformably overlies Carboniferous strata and is marked by a reddened, angular granite boulder–bearing breccia, with clasts to 2.2 m in diameter. This unit fines upward over 3–4 m, through poorly sorted, angular medium-pebble grade breccia, composed of green-gray metasediment, graywacke, rare volcanics and weathered granite into dusky red siltstone. It represents a very localized scree and/or alluvial fan deposit derived from the Kelameili Shan. The location of the fan was probably controlled by a north-south–trending tear fault immediately to the east, which had displaced the Carboniferous basement prior to the deposition of the Shishugou Group. A laterally nonpersistent weathered surface, including dusky red coloration (10R3/4) several meters deep, is developed below the unconformity surface.

The same stratigraphic relationship at the base of the Shishugou Group was observed at 45°10.524′N, 88°58.064′E (Figs. 3 and 6, locality 16), 1.6 km to the south-southwest, except that here the group overlies the Xishanyao Formation via a disconformity, below which paleosols are developed. Granite clasts at this locality are as much as 1.6 m in diameter. The Shishugou Group is mapped to directly overlie the Sangonghe Formation 2 km to the south of this (Fig. 6).

To the west of these basal units, the Shishugou Group crops out in a syncline in the footwall of the south Dishuiquan fault, which was examined between 45°12.178′N, 88°57.857′E and 45°11.958′N, 88°56.131′E (Figs. 3 and 6, localities 18 and 19). Panoramic views of the overlying stratigraphy show a series of color changes that may be traced along strike over 4–5 km. Lithologies are dominated by silty mudstone to fine-grained litharenite. Locally, mottled to reddened intervals with angular gravel stringers composed of basement lithologies, lenticular, laterally amalgamated pebbly granule-grade units with paleocurrent indicators to the southwest, and 3-m-thick ribbon channel bodies are developed. These are interpreted as alluvial ephemeral sheetflood and low-gradient suspended-load meandering fluvial facies. Gray intervals with symmetrical (wave) ripple cross-lamination and vertical dwelling burrows are interpreted as shallow lacustrine in nature; cross-lamination orientations indicate that wave approach directions were toward the north.

Immediately to the south of the south Dishuiquan fault, west of Wucaiwan, breccias interbedded within the Shishugou Group are interpreted to have been derived from the hanging wall of this structure in the Middle to Late Jurassic. Clast compositions vary along strike and it was possible in one instance to match phyllite in breccia in the Shishugou Group with phyllite in the hanging wall of the thrust. The breccia and the hanging-wall outcrop are offset by 2 km, suggesting 2 km of sinistral displacement on the fault since their deposition. This displacement is most likely to be Cenozoic, given that Cretaceous strata in the footwall of this thrust farther west are folded.

It is difficult to distinguish between the Xishanyao Formation and the Shishugou Group in the Dishuiquan transect. Observations from the upper part of the Jurassic succession, which is assumed to form part of the Shishugou Group, were made between 45° 14.610′N, 88°43.676′E and 45°13.984′N, 88°43.755′E (Figs. 3 and 6, locality 22). The succession comprises dusky red (10R3/4) to mottled light gray (5Y7/1.5) mudstone, siltstone, and sheets of fine- to medium-grained volcanic arenite, to 60 cm thick. Bioturbation (burrows and rootlets), weak paleosol horizonation, ripple cross-lamination, and cross-bedding (indicating flow toward 180°–218°) are developed. A single medium-grained wedge-shaped channel body, which was as deep as 6 m and 50 m wide was also observed. The unit is oriented east-west in cross section and contains trough cross-bedding that dips toward 260°. The channel fines upward and contains mottled rootlet horizons in its upper part. The channel fill is interpreted as the point-bar component of a meandering fluvial system. The entire group in this transect is interpreted as the product of a mixed to suspended load fluvial system, dominated by intermittently drained flood-plain facies.

Notable within all of the Shishugou Group is the sparsity of carbonaceous material even in poorly drained fluvial to shallow lacustrine facies. This suggests that terrestrial plant colonization was limited due to arid climatic conditions. A change to more arid climatic conditions in the Late Jurassic was noted by Hendrix et al. (1992) in southern Junggar, Turpan, and northern Tarim, and in western Mongolia by Sjostrom et al. (this volume).

Evidence for deformation. Evidence for deformation prior to or during the deposition of the Shishugou Group was observed at three localities.

In the Wucaiwan region, evidence for localized tectonic activity is provided by the unconformable contact between the Shishugou Group and the Xishanyao Formation, and the deposition of boulder conglomerate. These occur where north-south tear faults relayed displacement between strands of the south Dishuiquan fault zone, and allowed major alluvial systems to feed into the basin (Figs. 6 and 10C). The absence of the Xishanyao Formation beneath the unconformity to the south of Wucaiwan is also either the result of postdepositional uplift and erosion or onlap. Similar deformation is assumed in the Shuangjingzi region to account for the basal Shishugou Group unconformity and absence of the Xishanyao Formation.

At 45°13.440′N, 88°49.603′E (Figs. 3 and 6, locality 20), adjacent to the south Dishuiquan fault, three angular unconformities are preserved within the Middle-Upper Jurassic Shishugou Group; strata are progressively rotated toward the south. Cross-bedding within this group indicating paleoflow toward the north (i.e., toward the thrust) was observed at this locality. This implies that for at least part of the Middle-Late Jurassic there was little relief north of the present thrust fault (Fig. 10C).

East-southeast–directed paleocurrents on the eastern side of the Zhangpenggou anticline at 44°49.947′N, 89°16.117′E (Fig. 3, locality 2) are unusual in that most paleoflow indicators for the Shishugou Group, and other Jurassic formations, have a component of westerly flow, toward the basin center. Although the data should not be overinterpreted, it is suggested that these paleoflow directions might reflect the topographic influence of the Zhangpenggou anticline during deposition (these readings were taken from the eastern flank of the structure), due to active or earlier deformation on this structure (Fig. 10C).

Both McKnight et al. (1990) and Pu Renhai et al. (1994b) divided the Shishugou Group into three units, the latter correlating these units with the Toutunhe, Qigu, and Kalazha formations as defined in the south of the basin, although the basis of this correlation was not stated (Fig. 4C). McKnight et al. (1990) noted unconformities at the base of their first and second units. These surfaces are likely to be tectonically driven. Pu Renhai et al. (1994b) reported that the base of the Shishugou Group is generally conformable in the Cainan region to the west of the Kelameili Shan, except in the hanging wall of the south Dishuiquan fault and to the west of the Beisantai rise, where underlying units have been removed. Seismic reflectors in this unit are generally parallel and denote relatively stable basinwide subsidence.

The base of the second unit of the Shishugou Group ($J_{2-3}sh^2$; equivalent to the Qigu Formation) shows seismic onlap onto the Baijiahai rise, downlap in the region of the Cai 24 well (due to rapid sedimentation), and truncation on the Shaqiuhe structure, where the top of the first unit ($J_{2-3}sh^1$) was eroded. Northwest-southeast compression is interpreted to have uplifted northeast-southwest structures, such as Shaqiuhe and Baijiahai. Seismic reflectors are generally mounded to lenticular in this unit; these were interpreted to represent alluvial fan facies (Pu Renhai et al., 1994b). The second unit is missing from the Caican 2 well (Fig. 4C), and from the vicinity of the Zhang 2 well (Fig. 4B).

The third unit ($J_{2-3}sh^3$; equivalent to the Kalazha Formation) is missing from large areas including parts of the Baijiahai Rise (Fig. 4C); gentle uplift of positive structures is thought to have occurred toward the end of the Late Jurassic, although the restricted nature of this unit may well be due to truncation by the overlying Lower Cretaceous Tugulu Group.

Facies variations have been demonstrated in the Shishugou Group to the northwest of the Shaqiuhe structure by Pu Renhai et al. (1994b) (Fig. 4C). The succession in the Cai 24 well contains sandbodies between 2 and 10 m thick extending for as much as 500 m laterally; the proportion of sand decreases in wells to the west, where sandbodies are typically ~1 m thick and of limited lateral extent (Fig. 4C). The sandstone bodies were interpreted as distributary fluvial channels on the middle part of an alluvial fan, the apex of which is close to Dishuiquan, the Cai 24 well being interpreted to coincide with the location of a major channel unit. This channel would seem to be influenced by the location of both the Shaqiuhe and Baijiahai highs between which it flowed (Figs. 4 and 11). There is clearly also a distinct difference in the proportion of sandstones in the Cai 24 well and the Dishuiquan outcrops studied in our survey, highlighting the variability of facies away from sediment input points. Tuffs have been reported in the Toutunhe Formation by Zhou Jingcai et al. (1989) and in the Qigu Formation by Pu Renhai et al. (1994b); if correct, this would suggest that there was active volcanism regionally within the Middle and Late Jurassic.

CRETACEOUS

Tugulu Group

Lower boundary and thickness. The base Cretaceous unconformity is regionally developed and was examined at a number of locations in the Kelameili region (Fig. 9). In all places where it was examined, the base of the Lower Cretaceous Tugulu Group consists of poorly sorted, pebble-grade breccia; varied clast types (some of which have faceted surfaces) include angular reddened sandstone and/or graywacke, mudstone, volcanics, and tuff, and rare well-rounded quartzite and granite in an orange-brown sandstone matrix. Crude horizontal stratification, fining-upward trends, and scouring are observed.

According to Shen Yanbin and Mateer (1992), lowermost Cretaceous strata (equivalent to the Qingshuihe Formation in the Bodga Shan region) are missing from the Kelameili Shan region (Fig. 5). The Tugulu Group is 1206 m thick in the Caican 2 well (Zhang Xuejun et al., 1999).

Sedimentology. It was only in the Dishuiquan transect (Figs. 3 and 6, locality 24) that overlying units of the Lower Cretaceous Tugulu Group were observed. Here poorly sorted breccia grades into slightly better organized units in which channelization, trough cross-bedding, and primary current lineation are present. The breccia is poorly sorted and angular, contains dreikanter ventifact pebble lags, and is interbedded with medium-grained hematite-rimmed litharenite. Paleocurrent indicators show a high degree of variance, but are largely directed toward the south (represented in Fig. 7). These strata, which are ~40 m thick, probably represent ephemeral streamflood processes reworking stony desert pavements (reg) in a depositional setting resembling that of the present-day desert.

Overlying this, the section is dominated by thick-bedded units of moderately to well-sorted, fine-grained litharenite with hematitic clay rims. These units are massive, or contain parallel lamination with primary current lineations or large-scale trough cross-bedding, and display very good reservoir properties. Minor gravel lags with faceted pebbles are also present. These are interpreted as eolian deposits and are interbedded with thin (< 1 m) intervals of well-cemented trough cross-bedded sandstone with scoured bases and basal mudstone rip-up clasts to 15 cm in diameter (marked A in Figure 8G). Trough cross-sets are 20–30 cm high and indicate flow to 210°; these are interpreted to be alluvial in origin. Weak red laminated siltstone to mudstone horizons reported to contain bivalves, ostracods, charophytes, and plants (Shen Yanbin and Mateer, 1992), interbedded with climbing ripple cross-laminated sandstone (with paleoflow indicators to the south), are interpreted to be lacustrine in nature.

Interbedded fluvial and eolian strata dominate in the northern part of the section, whereas lacustrine strata increase in importance to the south where they pass into grayer (less oxidized) units. Overall, the section is interpreted to represent a lacustrine shoreline succession dominated by eolian deposits, which underwent intermittent alluvial reworking in the north, passing into deeper lacustrine conditions to the south. The section is reported to be 386-m-thick (Shen Yanbin and Mateer, 1992).

Evidence for deformation. Surface and subsurface observations indicate that a significant hiatus occurred prior to Cretaceous sedimentation. Nondeposition or erosion of uppermost Jurassic and lowermost Cretaceous strata are reported (Shen Yanbin and Mateer, 1992; Pu Renhai et al., 1994a). Outcrop patterns indicate the blanketing of structures that deform uppermost Jurassic strata in the Zhangpenggou and Shaqiuhe anticlines, and an east-west–orientated syncline within strata of the Shishugou Group in the footwall to the south Dishuiquan fault, close to Wucaiwan (Figs. 3, 6, and 10C). In general, pre-Cretaceous strata dip more steeply to the south than Cretaceous strata. Significant truncation of deformed pre-Cretaceous strata also occurs to the south within the Beisantai Rise (Wang Mingming, 1992).

DISCUSSION

Current models for the Mesozoic evolution of the Junggar Basin describe it as an intracontinental foreland basin (Watson et al., 1987; Hendrix et al., 1992) related to the collision of continental blocks with the evolving southern margin of Asia. Specific pulses of sedimentation in Junggar are correlated with specific collision events. These are the Late Triassic collision of the Qiangtang Block, the Late Jurassic collision of the Lhasa Block, and the Late Cretaceous collision of the Kohistan-Dras island arc (Fig. 1).

Subsurface data from the Junggar Basin indicate that compressional deformation was underway before Late Triassic time. This is best recorded along the northwestern basin margin, where extensive hydrocarbon exploration has been conducted in the Ke-Wu fault belt (Fig. 2). Marked variations in stratal thickness are present across thrusts; the strata affected range in age from the Early Triassic to the Late Jurassic (Zhang Jiyi, 1984; Xie Hong et al., 1988). Şengör (1990) noted that there was no corresponding Tethyan collision event at the southern margin of Asia in the Early or Middle Triassic, but that a major south-facing Andean-type margin was present along the southern side of the Tarim Block, within the Kunlun Shan. This arc system was apparently compressional (sensu Dewey, 1980), and therefore a plausible cause of compressional deformation across Central Asia to its north.

Late Triassic and Early Jurassic source-area rejuvenation in the Kelameili Shan region is demonstrated by the influx of conglomerate over this time. Our observation of a base Jurassic angular unconformity is the first evidence for coeval deformation to be reported from outcrops at the eastern margins of the Junggar Basin.

Evidence of deformation at later times in the Early Jurassic and in the Middle Jurassic is also present in the Kelameili Shan region (Figs. 5 and 10, B and C). There is synsedimentary faulting in the Badaowan and Sangonghe formations, an angular unconformity at the base of the Sangonghe Formation, and disconformities at the base of the Xishanyao Formation and Shishugou Group. The latter unconformity is locally overlain by a boulder conglomerate, with coarser clastic sediment preferentially entering the basin at the intersection of splays of the south Dishuiquan fault and north-south–trending tear faults. The absolute duration of this deformation is not known with any precision, but it appears that there was episodic tectonism in the Kelameili Shan and the rest of the Junggar area for tens of millions of years, rather than a single conglomeratic pulse near the Late Triassic–Early Jurassic boundary.

This episodic history of tectonism argues against a simple one-to-one correlation between an Early Jurassic sedimentary pulse in northwest China and collision of the Qiangtang Block with Asia. There are at least three possible explanations for this. First, the collision of Qiangtang could have been a protracted event, with considerable convergence following the initial impact. The Cenozoic India-Asia collision represents an excellent analogue for this possibility, with major deformation throughout Central Asia continuing for more than 50 m.y. after the initial impact of the Indian plate. Second, following the collision of the Qiangtang Block, a new convergent margin was established on its southern margin—now effectively the new Asian margin (Matte et al., 1996). If this arc was compressional, the resultant deformation could have affected the margins of the Junggar Basin in the same style as the Triassic deformation. Third, there may have been more than one continental block that collided with the southern margin of Asia in the Late Triassic to Early Jurassic interval. Mattern et al. (1996) suggested that, in addition to the Qiangtang Block, there may be a small "Uygur terrane" in the western Kunlun, which collided with Asia shortly before the main Qiangtang collision. Şengör et al. (1988) suggested that the Qiangtang Block was a composite continent, on the basis of the differing stratigraphy of its eastern and western portions. The age of the intervening suture, however, is not known with sufficient precision to gauge whether eastern and western Qiangtang amalgamated prior to their collision with the southern side of Asia, or whether they collided separately.

The Lhasa-Asia collision is a prime candidate for causing deformation across the Junggar Basin near the Jurassic-Cretaceous boundary. Within the Kelameili Shan, this is expressed by a regional angular unconformity at the base of the Lower Cretaceous Tugulu Group, omission of latest Jurassic–earliest Cretaceous strata, and the conglomerate pulse at the base of the Tugulu Group. We have no data to determine whether there was protracted deformation through the Cretaceous, in the manner of the Jurassic history outlined herein. Upper Cretaceous conglomerate progradation elsewhere in northwest China have been related to the Kohistan-Dras collision ca. 70–80 Ma (Hendrix et al., 1992). As with the Early Jurassic evolution, additional causes of deformation possibly include retroarc deformation north of the Andean-type margin at the southern edge of the Lhasa Block, or compression due to the docking of small continental fragments independent of the main Lhasa Block, for example the Risum Block of Matte et al. (1996). From their analysis of the sedimentary record of the Qaidam Basin, and a review of the timing of subsidence phases across northwest China, Ritts and Biffi (this volume) also argued against synchronous, regional, short-lived pulses of tectonism.

Cenozoic deformation and molasse sedimentation in all of the major mountain ranges and basins of northwest China are linked to compression arising from the India-Asia collision. Most Cenozoic deposition in the Tarim and Junggar Basins has occurred since the beginning of the Neogene ca. 23 Ma (e.g., Wang Qiuming, 1995; Zhang Guojun, 1993). This correlates well with estimates of the age of initial Cenozoic exhumation in

the Tian Shan (Hendrix et al., 1994; Sobel and Dumitru, 1997; Yin et al., 1998).

Because the initial collision of India and Asia is well determined as ca. 55 Ma (Searle et al., 1987), there is a gap of ~30 m.y. before the major effects of this collision appear in northwest China. This is consistent with deformation occurring in other areas of Central Asia, especially the Himalayas and Tibet, before propagating northward to affect the Tian Shan and adjacent areas. If the Mesozoic collisions described here behaved in a similar fashion to the India-Asia collision (each involved a Gondwanan continental block colliding with an active continental margin at the southern side of Asia [Hendrix et al., 1992; Matte et al., 1996], producing a southward-vergent collision), then it is possible that the timings of the collisions within Tibet are underestimated, and that in each case there was a time lag of 20–30 m.y. between the initial collision and the appearance of a pulse of clastic sediments in the Junggar Basin. This is difficult to prove, given the paucity of existing data on the Tibetan collisions, but is worthy of further investigation.

CONCLUSIONS

Field work from the Kelameili Shan, eastern Junggar, and subsurface data contained within the Chinese geological literature indicate that deformation affected sedimentary patterns throughout much of the Mesozoic history of the Junggar Basin. Evidence for this recurrent deformation comes in the form of angular and/or landscape unconformities, all of which are overlain by alluvial conglomerates and which, on occasion, are marked by mature paleosol horizons (Figs. 8, A, E, H, I, and 9), syn-sedimentary fault activity (Figs. 8, C and F), geological map- and subsurface-defined stratigraphic omissions and abrupt variations in sedimentary thickness (Figs. 3, 4, 5, 6, and 10), anomalous paleoflow directions (Figs. 10 and 11), and lateral facies variations (Fig. 4C). These data indicate that deformation occurred during the Middle to Late Triassic, at the Triassic-Jurassic boundary, at the boundary between and within the Lower Jurassic Badaowan and Sangonghe formations, at the boundary between the Lower Jurassic Sangonghe and Middle Jurassic Xishanyao formations, at the base of and within the Middle to Upper Jurassic Shishugou Group, and at the boundary between the Middle to Upper Jurassic Shishugou and Lower Cretaceous Tugulu groups (Figs. 7 and 10).

Contractile deformation is required, therefore, in addition to that produced by the collisions at the southern continental margin of Asia invoked by Hendrix et al. (1992) (Fig. 7): i.e., the Qiangtang Block in the Late Triassic, the Lhasa Block in the Late Jurassic, and the Kohistan-Dras island arc in the Late Cretaceous. Retroarc compression, north of each contemporary active continental margin at the southern side of Asia, is the most plausible mechanism. Prolonged collision of currently recognized tectonic units or the collision of additional poorly constrained units along Asia's southern margin are also possible causes. The time lag of ~30 m.y. between the initial collision of

India with Asia and syntectonic sedimentation within northwest China might suggest that current estimates of the timing of Mesozoic collisions at the southern side of Asia, based on an investigation of the sedimentary record of northwest China, are underestimates. Evidence for the timing of initial collision are likely to be lost or preserved along suture zones within the remote Tibetan plateau.

ACKNOWLEDGMENTS

We thank Qiu Dongzhou and Tang Zhanghua for advice during field work, and Christine Brouet-Menzies and Geng Quanru for help and logistical support. Critical reviews by David Eberth, Marc Hendrix, David Macdonald, and Bradley Ritts improved the manuscript, and we thank Marc Hendrix for his help and patience during the editorial process. Vladimir Merkouriev and Matt Hart drafted some of the figures. CASP China Basins Project is funded by a consortium of companies; we thank Agip, Amoco, Anderman-Smith, Arco, Chevron, Conoco, Deminex, Exxon, Japanese National Oil Corporation, Louisiana Land and Exploration, Mobil, Phillips, Texaco, and Unocal for their support. This is Cambridge University Earth Science contribution 5623.

REFERENCES CITED

Allen, M.B., Windley, B.F., and Zhang Chi, 1993, Paleozoic collisional tectonics and magmatism of the Chinese Tien Shan, Central Asia: Tectonophysics, v. 220, p. 89–115.

Allen, M.B., Şengör, A.M.C., and Natal'in, B.A., 1995, Junggar, Turfan, and Alakol basins as Late Permian sinistral pull-apart structures in the Altaid orogenic collage, central Asia: Geological Society of London Journal, v. 152, p. 327–338.

Biske, Y.S., 1995, Late Paleozoic collision of the Tarimskiy and Kirghiz-Kazakh paleocontinents: Geotectonics, v. 29, p. 26–34.

Bureau of Geology and Mineral Resources of Xinjiang Uygur Autonomous Region, 1993, Regional geology of Xinjiang Uygur Autonomous Region: Beijing, Geological Publishing House, Geological Memoirs, ser. 1, no. 32, 841 p. (in Chinese with brief English text).

Carroll, A.R., Graham, S.A., Hendrix, M.S., Ying, D., and Zhou, D., 1995, Late Paleozoic tectonic amalgamation of northwestern China: Sedimentary record of the northern Tarim, northwestern Turpan, and southern Junggar basins: Geological Society of America Bulletin, v. 107, p. 571–594.

Collision, J.D., 1986, Deserts, *in* Reading, H.G., ed., Sedimentary environments and facies (second edition): Oxford, Blackwell Scientific Publications, p. 95–112.

Compiling Team of the Xinjiang Uygur Autonomous Region Regional Stratigraphic Tables, 1981, Stratigraphic tables of North West China: Xinjiang Uygur Autonomous Region: Beijing, Geological Publishing House, 496 p. (in Chinese).

Dewey, J.F., 1980, Episodicity, sequence, and style at convergent plate boundaries, *in* Strangeway, D.W., ed., The continental crust and its mineral deposits: Geological Association of Canada Special Paper 20, p. 553–573.

Hendrix, M.S., Graham, S.A., Carroll, A.R., Sobel, E.R., McKnight, C.L., Schulein, B.J., and Wang Zuoxun, 1992, Sedimentary record and climatic implications of recurrent deformation in the Tian Shan: Evidence from Mesozoic strata of the north Tarim, south Junggar, and Turpan basins, northwest China: Geological Society of America Bulletin, v. 104, p. 53–79.

Hendrix, M.S., Dumitru, T.A., and Graham, S.A., 1994, Late Oligocene–early Miocene unroofing in the Chinese Tian Shan: An early effect of the India-Asia collision: Geology, v. 22, p. 487–490.

Li Xibin, and Jiang Jianheng, 1987, The survey of petroleum geology and the controlling factor for hydrocarbon distribution in the east part of the Junggar Basin: Oil and Gas Geology, v. 8, p. 99–107 (in Chinese with English abstract).

Liu Zhaosheng, 1993, Jurassic sporopollen assemblages from the Beishan coalfield, Qitai, Xinjiang: Acta Micropalaeontologica Sinica, v. 10, p. 13–36 (in Chinese with English summary).

Matte, P., Tapponnier, P., Arnaud, N., Bourjot, L., Avouac, J.P., Vidal, P., Liu Qing, Pan Yusheng, and Wang Yi, 1996, Tectonics of Western Tibet, between the Tarim and the Indus: Earth and Planetary Science Letters, v. 142, p. 311–330.

Mattern, F., Schneider, W., Li Yongan, and Li Xiangdong, 1996, A traverse through the western Kunlun (Xinjiang, China): Tentative geodynamic implications for the Paleozoic and Mesozoic: Geologische Rundschau, v. 85, p. 705–722.

McKnight, C.L., Graham, S.A., Carroll, A.R., Gan, Q., Dilcher, D.L., Zhao Min, and Hailiang Yun, 1990, Fluvial sedimentology of an Upper Jurassic petrified forest assemblage, Shishu Formation, Junggar Basin, Xinjiang, China: Palaeogeography, Palaeoclimatology, Palaeoecology, v. 79, p. 1–9.

Pu Renhai, and Tang Zhonghua, 1994, Preliminary research on investigating relative thickness of the sand body at the bottom of Sangonghe Formation using seismic amplitude information: Geophysical Prospecting for Petroleum, v. 33, p. 66–74 (in Chinese with English abstract).

Pu Renhai, Mei Zhichao, and Tang Zhonghua, 1994a, A preliminary discussion of Jurassic non-marine sequence stratigraphy, eastern Junggar Basin: Xinjiang Petroleum Geology, v. 15, p. 335–342 (in Chinese with English abstract).

Pu Renhai, Mei Zhichao, and Tang Zhonghua, 1994b, Division of seismic sequences and interpretation on sequence stratigraphy of Shishugou Group, Cainan Area: Earth Science, v. 19, p. 769–777 (in Chinese with English abstract).

Qin Subao, 1986, Analysis of Permian sedimentary formation and their oil and gas potential in Kelameili region: Oil and Gas Geology, v. 7, p. 281–287 (in Chinese with English abstract).

Qiu Dongzhou, Zhang Jiqing, Wang Xilin, Jiang Xinsheng, Zhao Yuguang, Wang Longzhang, Li Rong, Min Jikun, Chen Xinfa, Zhang Congzhen, Zeng Jun, and Zhao Yumei, 1994, Reservoir sedimentology, diagenesis and evaluation of Triassic and Jurassic systems in the northwestern margin of Junggar Basin: Chengdu, Chengdu University of Science Press, 122 p. (in Chinese with English abstract).

Regional Survey Team of the Bureau of Geology of the Xinjiang Uygur Autonomous Region, 1965, Geological Map L-45-XXX: Beijing, Ministry of Geology and Mineral Resources, scale 1:200 000.

Regional Survey Team of the Bureau of Geology of the Xinjiang Uygur Autonomous Region, 1966, Geological Map L-46-XXV: Beijing, Ministry of Geology and Mineral Resources, scale 1:200 000.

Regional Survey Team of the Bureau of Geology of the Xinjiang Uygur Autonomous Region, 1980, Geological Map L-46-XXXI: Beijing, Ministry of Geology and Mineral Resources, scale 1:200 000.

Reineck, H.E., 1955, Haftrippeln und Haftwarzen, Ablagerungsformen von Flugsand: Senckenbergiana Lethaea, v. 36, p. 347–357.

Searle, M.P., Windley, B.F., Coward, M.P., Cooper, D.J.W., Rex, A.J., Rex, D., Li Tingdong, Xiao Xuchang, Jan, M.Q., Thakur, V.C., and Kumar, S., 1987, The closing of Tethys and the tectonics of the Himalaya: Geological Society of America Bulletin, v. 98, p. 678–701.

Şengör, A.M.C., 1990, Plate tectonics and orogenic research after 25 years: A Tethyan perspective: Earth Science Review, v. 27, p. 1–201.

Şengör, A.M.C., and Natal'in, B.A., 1996, Paleotectonics of Asia: Fragments of a synthesis, *in* Yin, A., and Harrison, T.M., eds., The tectonic evolution of Asia: Cambridge, Cambridge University Press, p. 486–640.

Şengör, A.M.C., Altiner, D., Cin, A., Ustaomer, T., and Hsü, C.J., 1988, Origin and assembly of the Tethyside orogenic collage at the expense of Gondwana Land, *in* Audley-Charles, M.G., and Hallam, A., eds., Gondwana and Tethys: Geological Society Special Publication 37, p. 119–181.

Shen Yanbin, and Mateer, N.J., 1992, An outline of the Cretaceous System in northern Xinjiang, western China, *in* Mateer, N.J., and Chen Peiji, eds., Aspects of non-marine Cretaceous geology: Beijing, China Ocean Press, p. 49–77.

Shi Xuanyu, Yu Chunhui, Reyihanguli, Liu Shuhui, and Ha Yali, 1995, Relationship between Jurassic sedimentary system and its oil and gas in Junggar Basin: Xinjiang Petroleum Geology, v. 16, p. 233–241 (in Chinese with English abstract).

Sobel, E.R., and Dumitru, T.A., 1997, Thrusting and exhumation around the margins of the western Tarim Basin during the India-Asia collision: Journal of Geophysical Research, v. 102, p. 5043–5063.

Wang Longzhang, 1994, Approaches to the curves for the lake-level fluctuations in the Junggar Basin during Meso-Cenozoic time: Sedimentary Facies and Palaeogeography, v. 14, no. 6, p. 1–14 (in Chinese with English abstract).

Wang Longzhang, 1995, A study on Mesozoic-Cenozoic continental sequence stratigraphy and its application in Junggar Basin: Xinjiang Petroleum Geology, v. 16, p. 324–330 (in Chinese with English abstract).

Wang Mingming, 1992, Formation of oil-gas pools and distribution of oil-gas in the eastern part of Junggar Basin: Petroleum Exploration and Development, v. 19, no. 6, p. 1–8 (in Chinese with English abstract).

Wang Qiuming, ed., 1995, Xinjiang oil and gas province: Tarim, Turpan-Hami and other main basins: Petroleum Geology of China, v. 15, 660 p. (in Chinese).

Wang Yilin, 1996, Petroleum exploration results during the Eighth Five-Year Plan and exploration programme in the period of the Ninth Five-Year Plan: Xinjiang Petroleum Geology, v. 17, p. 1–14 (in Chinese with English abstract).

Watson, M.P., Hayward, A.B., Parkinson, D.N., and Zhang, Z.M., 1987, Plate tectonic history, basin development and petroleum source rock deposition onshore China: Marine and Petroleum Geology, v. 4, p. 205–225.

Windley, B.F., Allen, M.B., Zhang, C., Zhao, Z.Y., and Wang, G.R., 1990, Paleozoic accretion and Cenozoic redeformation of the Chinese Tienshan Range, central Asia: Geology, v. 18, p. 128–131.

Xie Hong, Zhao Bai, Lin Longdong, and You Qimei, 1988, Oil-bearing features along the Karamay Overthrust Belt, Northwestern Junggar Basin, China, *in* Wagner, H.C., et al., eds., Petroleum resources of China and related subjects: Houston, Circum-Pacific Council for Energy and Mineral Resources, Earth Science Series, v. 10, p. 387–402.

Yan Yugui, 1993, Fault-controlled hydrocarbon distribution in the Junggar Basin, NW China: Journal of Petroleum Geology, v. 16, p. 109–114.

Yin, A., Nie, S., Craig, P., Harrison, T.M., Ryerson, F.J., Qian Xianglin, and Yang Geng, 1998, Cenozoic tectonic evolution of the southern Tian Shan: Tectonics, v. 17, p. 1–27.

Zhang Guojun, ed., 1993, Xinjiang oil and gas province—1. Junggar Basin: Petroleum Geology of China, v. 15, 390 p. (in Chinese).

Zhang Jiyi, 1984, Petroleum accumulation in the northwest margin of Junggar Basin: Beijing, International Petroleum Geology Symposium, Abstracts, 16 p.

Zhang Xuejun, Li Peijun, and Li Gang, 1999, Structure evolution analyses of Cai 31 well in Baijiahai Arch: Xinjiang Petroleum Geology, v. 20, p. 473–475 (in Chinese).

Zhou Jingcai, Shi Xuanyu, Ma Xiaoyang, Ma Chuandong, and Song Zhigang, 1989, Study of Jurassic sedimentary facies, south margin of Junggar Basin: Xinjiang Petroleum Geology, v. 10, p. 39–54 (in Chinese).

Zonenshain, L.P., Kuzmin, M.I., and Natapov, L.M., 1990, Geology of the USSR: A plate tectonic synthesis: American Geophysical Union Geodynamics Series, v. 21, p. 1–242.

MANUSCRIPT ACCEPTED BY THE SOCIETY JUNE 5, 2000

Geological Society of America
Memoir 194
2001

Sedimentology and provenance of Mesozoic nonmarine strata in western Mongolia: A record of intracontinental deformation

Derek J. Sjostrom*[†]
Marc S. Hendrix[†]
Department of Geology, University of Montana, Missoula, Montana 59812, USA
Demchig Badamgarav[†]
Centre for Paleontology, Mongolian Academy of Sciences, Ulan Bator, Mongolia
Stephan A. Graham[†]
Department of Geological and Environmental Science, Stanford University, Stanford, California 94305-2115, USA
Bruce K. Nelson[†]
Department of Geological Sciences, University of Washington, Seattle, Washington 98195, USA

ABSTRACT

Western Mongolia is a structurally complicated and little-studied portion of the central Asia tectonic collage, yet contains well-exposed sequences of Mesozoic sedimentary strata that preserve an important record of ancient intraplate deformation. In order to document the record of Mesozoic sedimentary basin development and to provide a basis for interpreting the sedimentary record of intraplate deformation, we studied the sedimentology and provenance of Mesozoic strata at four locations in western Mongolia. Triassic sedimentary strata are missing across the study area, and Lower Jurassic strata unconformably overlie older rocks of various ages. Lower to Middle Jurassic basin fill in the northern portion of the study area is dominated by coarse, granite-bearing conglomerate and arkosic sandstone ($Qm_{35}F_{49}L_{16}$) with a volcanic-dominated lithic fraction ($Qp_{22}Lvm_{63}Lsm_{15}$). Paleocurrent indicators suggest southwest-directed sediment transport, reflecting uplift of an ancestral version of the Hanhöhiy Uul, a mountain range located north and east of the study area and composed mainly of Cambrian granite and associated volcanics. In the central portion of the study area, Lower to Middle Jurassic strata comprise an upward-fining kilometer-thick sequence from coarse matrix-supported fanglomerate to fine sandstone, shale, and coal suggestive of a mudload-dominated, meandering fluvial environment. Conglomerate clasts are dominantly basalt and granite, and sandstone is arkosic with higher QmFL%L and Qp-Lvm-Lsm%Lsm than equivalent sandstone to the north ($Qm_{41}F_{35}L_{24}$; $Qp_{24}Lvm_{41}Lsm_{35}$). Paleocurrent indicators suggest transport to the south-southwest. In the southernmost part of the study area, Lower through Upper Jurassic strata record a transition from mudload-dominated fluvial environments with abundant organic matter to fluvial flood-plain and/or mud flat with little organic matter and abundant calcisols. These strata are sharply overlain by

*Current address: Department of Earth Sciences, Dartmouth College, Hanover, New Hampshire 03755, USA.
[†]E-mails: Sjostrom, derek.sjostrom@dartmouth.edu; Hendrix, marc@ selway.umt.edu; Badamgarav, badam87@hotmail.com; Graham, graham@ pangea.stanford.edu; Nelson, bnelson@u. washington.edu.

Sjostrom, D.J., et al., 2001, Sedimentology and provenance of Mesozoic nonmarine strata in western Mongolia: A record of intracontinental deformation, *in* Hendrix, M.S., and Davis, G.A., eds., Paleozoic and Mesozoic tectonic evolution of central Asia: From continental assembly to intracontinental deformation: Boulder, Colorado, Geological Society of America Memoir 194, p. 361–388.

Upper Jurassic–Lower Cretaceous angular polymictic conglomerate that in turn is overlain by Lower Cretaceous lacustrine mudstone and siltstone. Mesozoic sandstone at the southernmost field location is lithic rich and contains higher QpLvmLsm%Lsm than elsewhere in the study area ($Qm_{31}F_{33}L_{36}$; $Qp_{10}Lvm_{50}Lsm_{40}$). Paleocurrent analysis suggests a fundamental reversal from south-directed paleoflow prior to deposition of coarse conglomerate to north-directed paleoflow during and after deposition of conglomerate. We interpret this reversal, the deposition of coarse conglomerate, and the lithic-rich sandstone compositions to reflect latest Jurassic uplift of the paleo-Altai Mountains to the south.

Sm-Nd isotopic analysis performed on sandstone samples from the central part of the study area (Jargalant locality) yield depleted mantle derivation ages of ca. 1 Ga and ε_{Nd} values of -3.8 to -4.8, suggesting significant contributions from old continental crust. Sm-Nd isotopic analysis on local basement rocks, as well as granite and basalt cobbles within Jurassic strata, yield depleted mantle derivation ages and an isochron age of ca. 600 Ma, with initial ε_{Nd} values similar to that expected of depleted mantle at that time, suggesting an origin from juvenile crust. We interpret these results to indicate that a significant fraction of sand-sized detritus was derived from Archean granitic basement forming the southern margin of the Siberian craton, ~250 km to the north.

We favor a compressional tectonic setting for Mesozoic sedimentary basins of western Mongolia, although a strike-slip setting for these basins is also possible. This interpretation is based on several lines of evidence: (1) Mesozoic volcanics that might be expected in an extensional setting are lacking from the study area; (2) Jurassic coal contains low vitrinite reflectance values suggestive of a low heat flow more typical of a flexural basin setting than an extensional setting; and (3) adjacent sedimentary basins in western China contain well-documented evidence of Mesozoic contractile deformation and bear many of the same sedimentologic attributes as basins in western Mongolia.

INTRODUCTION

The tectonic assembly of Asia has produced multiple episodes of significant Phanerozoic intracontinental deformation (e.g., Delville et al., this volume; Greene et al., this volume; Sobel et al., this volume), an important record of which is preserved in the fill of associated sedimentary basins (Carroll et al., 1995; Hendrix et al., 1992, 1995; Ritts and Biffi, this volume; Vincent and Allen, this volume). Situated in the midst of the central Asian tectonic collage in western Mongolia are several nonmarine sedimentary basins in the Valley of Lakes, a broad, sparsely populated region between the Altai Mountains and the Hangay Plateau (Pentilla, 1994) (Fig. 1A). Substantial thicknesses of Lower to Middle Jurassic through Lower Cretaceous strata are exposed locally in the Valley of Lakes, where they are in both fault and depositional contact with Vendian-Cambrian magmatic-arc–related igneous rocks (Fig. 1B; Shuvalov, 1968, 1969; Khosbayar, 1973; Devyatkin et al., 1975; Yanshin, 1989; Dobretsov et al., 1995).

During 1992 and 1995, we studied the major exposures of Mesozoic strata in the Valley of Lakes. Our work had two main objectives. First, we sought to provide baseline documentation of the sedimentology and provenance of Mesozoic strata in western Mongolia. Although several regional stratigraphic and paleontologic studies (Shuvalov, 1968, 1969; Khosbayar, 1973) and two largely interpretive works addressing the development of sedimentary basins in this region have been published

(Traynor and Sladen, 1995; Dobretsov et al., 1995), very little published work addresses the specific sedimentology and provenance of Mesozoic strata from this part of Asia.

Our second objective was to reconstruct Mesozoic sedimentary basin development in western Mongolia. Recent structural and related sedimentary studies in central Asia have documented the development of contractile, strike-slip, and extensional basins during Mesozoic time. Most workers agree that Mesozoic basins in northwestern China, west of our field area, were largely contractile in nature (e.g., Hendrix et al., 1992, 1995; Sobel, 1999; Greene, this volume; Ritts and Biffi, this volume; Vincent and Allen, this volume). In contrast, strong evidence for mixed contractile and extensional deformation and basin development has been described east of our study area in southeastern Mongolia and adjacent portions of north-central and northeastern China (Webb et al., 1999; Johnson et al., this volume; Darby et al., this volume; Davis et al., this volume). Other work has suggested that the structural heterogeneity of the central Asian crust (Şengör et al., 1993; Şengör and Natal'in, 1996) contributed directly to the development of strike-slip deformation and related basins for parts of central Asia south and west of our study area (Allen et al., 1995; Sobel, 1999; Vincent and Allen, 1999). It is clear that without accurate and detailed descriptions of Mesozoic strata from western Mongolia, the nature of sedimentary basin development in this key region will remain highly speculative, and even the broadest outline of Mesozoic tectonism across Asia will contain considerable uncertainty.

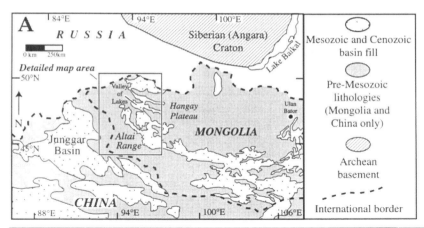

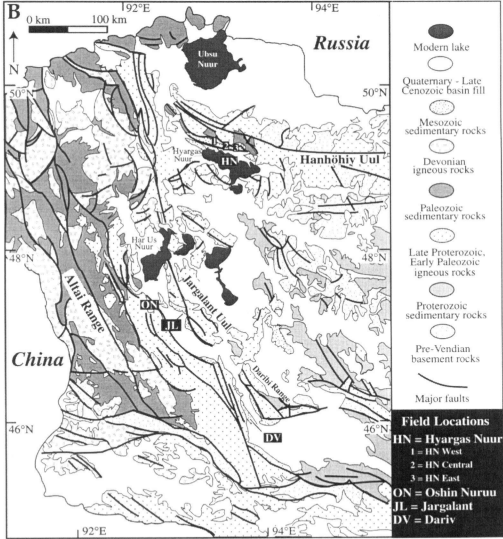

Figure 1. A: Regional map of western Mongolia and adjacent parts of northwestern China showing positions of major modern sedimentary basins and location of study area (highlighted in box). B: Simplified geologic map of western Mongolia, modified from Yanshin (1989) and Şengör and Natal'in (1996), and showing locations of four principal study localities described in this chapter.

METHODS

We chose four locations in western Mongolia for detailed study based on quality of exposure, accessibility, and relative completeness of Mesozoic section (Fig. 1). At each locality, we measured stratigraphic sections at a resolution of tens of m to document the overall Mesozoic sedimentary style (Fig. 2). We collected paleocurrent indicator measurements throughout each section and measured detailed sections locally at decimeter-scale resolution to provide a basis for paleoenvironmental interpretation (Fig. 3). We photodocumented important sedimentologic features (Fig. 4), and investigated lateral mesoscale facies changes by measuring multiple sections separated by several km along strike (Sjostrom, 1997).

Absolute dating of Mesozoic nonmarine strata in the study area is difficult due to a lack of interbedded, datable volcanic units and little reported magnetostratigraphic work. However, age control to the epoch level is possible due to extensive previous paleontologic work by Mongolian and Soviet Union geoscientists (Shuvalov, 1968, 1969; Khosbayar, 1973; Devyatkin et al., 1975) (Fig. 5). In this study we relied on this paleontologic work for age control.

Sandstone samples were collected from each study location for provenance analysis. All samples were point-counted using a modified Gazzi (1966)-Dickinson (1970) method (Table 1; Ingersoll et al., 1984). Typically, at least 500 counts were identified and tabulated per sample. Point-counting procedures and criteria for recognizing specific grains were described fully in Sjostrom (1997). Raw point-count data (Table 1) were recalculated into detrital modes following the methods of Ingersoll et al. (1984) (Table 2) and plotted on Qm-F-Lt, Qp-Lvm-Lsm, and Qm-P-K ternary diagrams (Fig. 6).

In order to complement provenance work derived from the petrographic analysis of sandstone samples, we performed Sm-Nd isotopic analysis on 10 whole-rock samples from the Jargalant location (Fig. 1), following the methods of Nelson et al. (1993) with slight modifications described in Sjostrom (1997). The ε_{Nd} values were calculated as deviations from CHUR in parts per 10^4, using present-day CHUR values of $^{143}Nd/^{144}Nd = 0.512683$ and $^{147}Sm/^{144}Nd = 0.1967$. The error on Sm and Nd concentration determinations is ~0.1% (Table 3). Model depleted mantle derivation ages (T_{DM}) were also calculated following the method of DePaolo et al. (1991) (Fig. 7).

SEDIMENTOLOGIC FRAMEWORK OF MESOZOIC STRATA

Hyargas Nuur study area

Observations. The Hyargas Nuur study area is located on the southern slope of Hanhöhiy Uul (Uul means mountains in Mongolian), west of the Hangay Plateau (Fig. 1B). At this locality, a complete section of Lower to Middle Jurassic rock is continuously exposed. We examined three separate sections,

each separated by about 10 km along structural strike (Fig.1B): Hyargas Nuur west (49°21′34″N, 93°7′12″E), Hyargas Nuur central (49°20′24″N, 93°16′41″E), and Hyargas Nuur east (49°21′17″N, 93°26′31″E). At each locality, the Jurassic section is bounded above and below by thrust faults that place the Jurassic section structurally above Neogene lacustrine deposits and structurally below Vendian-Cambrian igneous rocks (Yanshin, 1989).

Each of the three Hyargas Nuur sections is characterized by ~750 m of poorly organized, clast-supported beds of pebble to boulder conglomerate. Each section coarsens upward slightly (Fig. 2A); the maximum clast size at the base of each section is ~50 cm (average 5–10 cm by visual estimate) and at the top of each section is 60–70 cm (average 10–20 cm by visual estimate). Conglomerate beds typically contain coarse sandstone between pebble- and cobble-sized clasts (Fig. 4A). Individual conglomerate beds range in thickness from a few cms to several tens of m and can be traced for no more than several hundred m before pinching out or grading into a different lithology. Conglomerate intervals are typically uniform in grain size, although both upward-fining sequences with basal gravel lags and upward-coarsening sequences were observed locally (Fig. 3A). Most cobbles are subrounded (Fig. 4A), and imbrication is common. In contrast to the well-rounded, moderately sorted, imbricated conglomerate that makes up most of each section are several beds of poorly sorted, muddy, locally matrix-supported angular conglomerate (Figs. 3A and 4B). Pebbly coarse sandstone, along with minor mudstone, and siltstone are interbedded with conglomerate at each of the three Hyargas Nuur sections.

Traction structures and evidence of traction-related erosion are common within each of the Hyargas Nuur sections. Contacts between different lithologic units are typically scoured, and relief at the base of conglomerate and sandstone beds locally exceeds 1 m. Conglomeratic strata contain abundant low-angle depositional surfaces with amplitudes of 2–4 m. These surfaces can be traced laterally for tens of m and result from slight changes in grain size. Sandstone interbedded with conglomerate typically weathers in a flaggy fashion and contains planar lamination, meter-scale trough cross-beds with pebbles or cobbles mantling troughs, and local asymmetric ripples. Locally interbedded siltstone and mudstone is uniformly greenish-gray in color, fissile, and contains abundant wood fossils, including transported logs and leaf impressions. The orientation of imbricated clasts, trough axes, and ripple drift foresets were measured at each section (n = 348 for all three sections) and generally indicates paleoflow to the southwest (Fig. 2A).

Interpretations. The coarse grain size, laterally discontinuous facies, lack of thick fine-grained intervals, and presence of abundant tractional structures all suggest deposition in tractional, bedload-dominated streams. Matrix-supported angular conglomerate observed at the Hyargas Nuur west (7–8 m level; Fig. 3A) and Hyargas Nuureast (13–18 m level; Sjostrom, 1997) are interpreted as debris-flow deposits. Overall, this style of sedimentation is typical of the upper reaches of stream-dominated

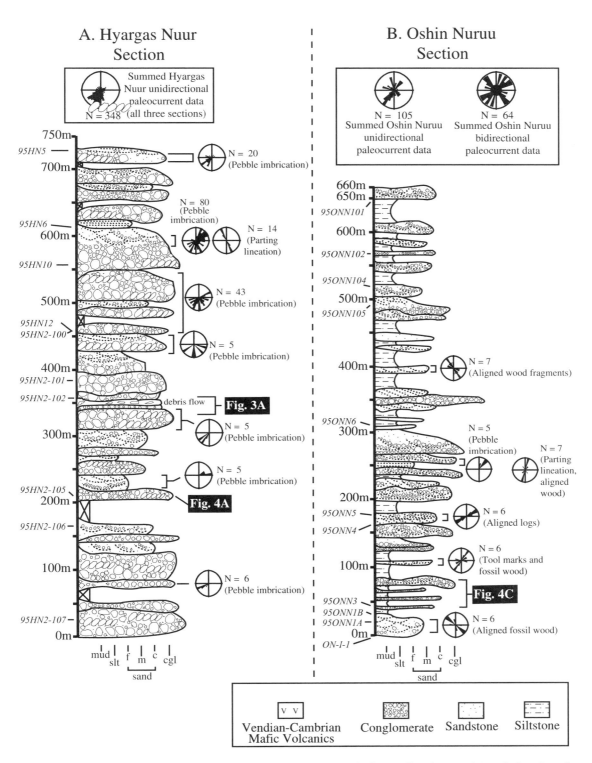

Figure 2. Stratigraphic sections, measured and restored paleocurrent indicator directions, and sample locations for Hyargas Nuur, Oshin Nuruu, Jargalant, and Dariv localities. Locations of sections are shown in Figure 1B. Detailed stratigraphic sections are shown in Figure 3. Note that scale of A and B is twice that of C and D. Mud—mudstone, slt—siltstone, f—fine sand, m—medium sand, c—coarse sand, cgl—conglomerate.

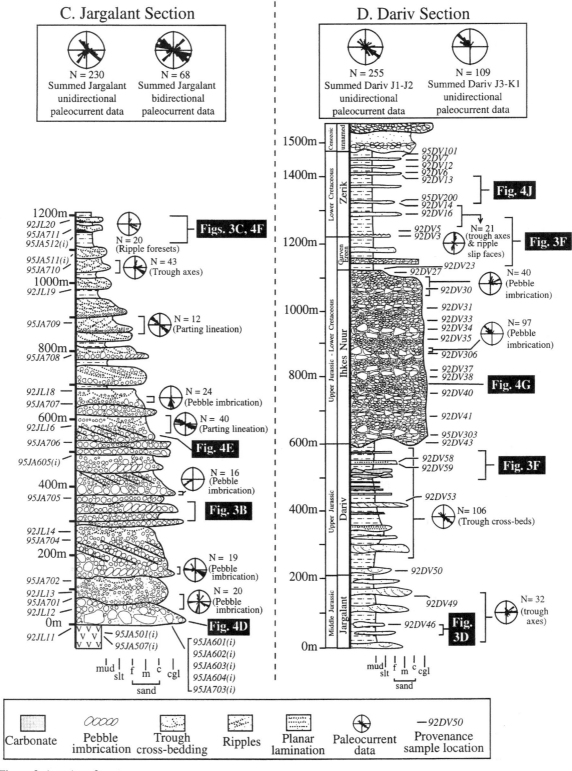

Figure 2. (*continued*)

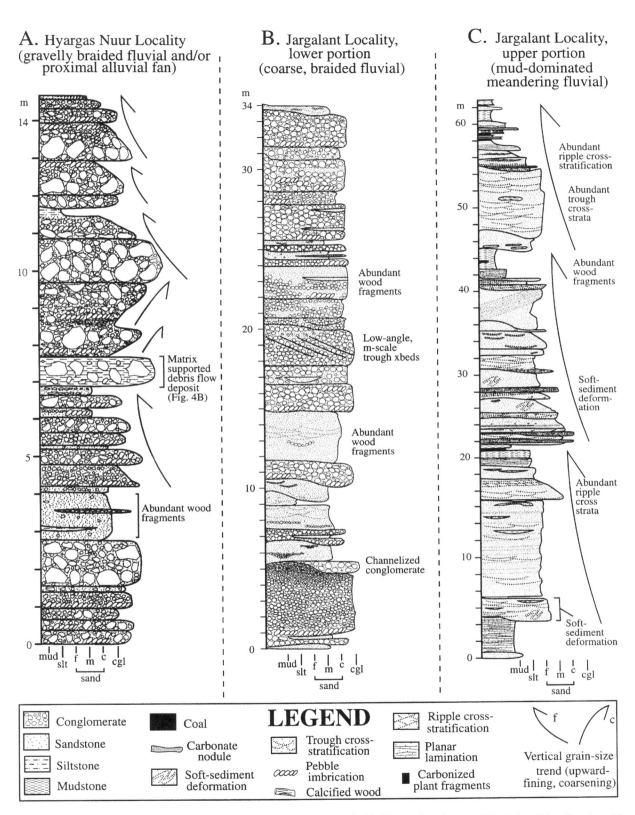

A. Hyargas Nuur Locality
(gravelly braided fluvial and/or
proximal alluvial fan)

m

14

10

5

0

Matrix
supported
debris flow
deposit
(Fig. 4B)

Abundant wood
fragments

mud slt f m c cgl
sand

B. Jargalant Locality,
lower portion
(coarse, braided fluvial)

m
34

30

20

10

0

Abundant
wood
fragments

Low-angle,
m-scale
trough xbeds

Abundant
wood
fragments

Channelized
conglomerate

mud slt f m c cgl
sand

C. Jargalant Locality,
upper portion
(mud-dominated
meandering fluvial)

m

60

50

40

30

20

10

0

Abundant
ripple cross-
stratification

Abundant
trough
cross-
strata

Abundant
wood
fragments

Soft-
sediment
deform-
ation

Abundant
ripple
cross
strata

Soft-
sediment
deformation

mud slt f m c cgl
sand

LEGEND

Conglomerate

Sandstone

Siltstone

Mudstone

Coal

Carbonate
nodule

Soft-sediment
deformation

Trough cross-
stratification

Pebble
imbrication

Calcified wood

Ripple cross-
stratification

Planar
lamination

Carbonized
plant fragments

f   c

Vertical grain-size
trend (upward-
fining, coarsening)

Figure 3. Detailed measured stratigraphic sections from Hyargas Nuur, Oshin Nuruu, Jargalant, and Dariv localities. Stratigraphic placement of these sections is shown in Figure 2. Mud—mudstone, slt—siltstone, f—fine sand, m—medium sand, c—coarse sand, cgl—conglomerate.

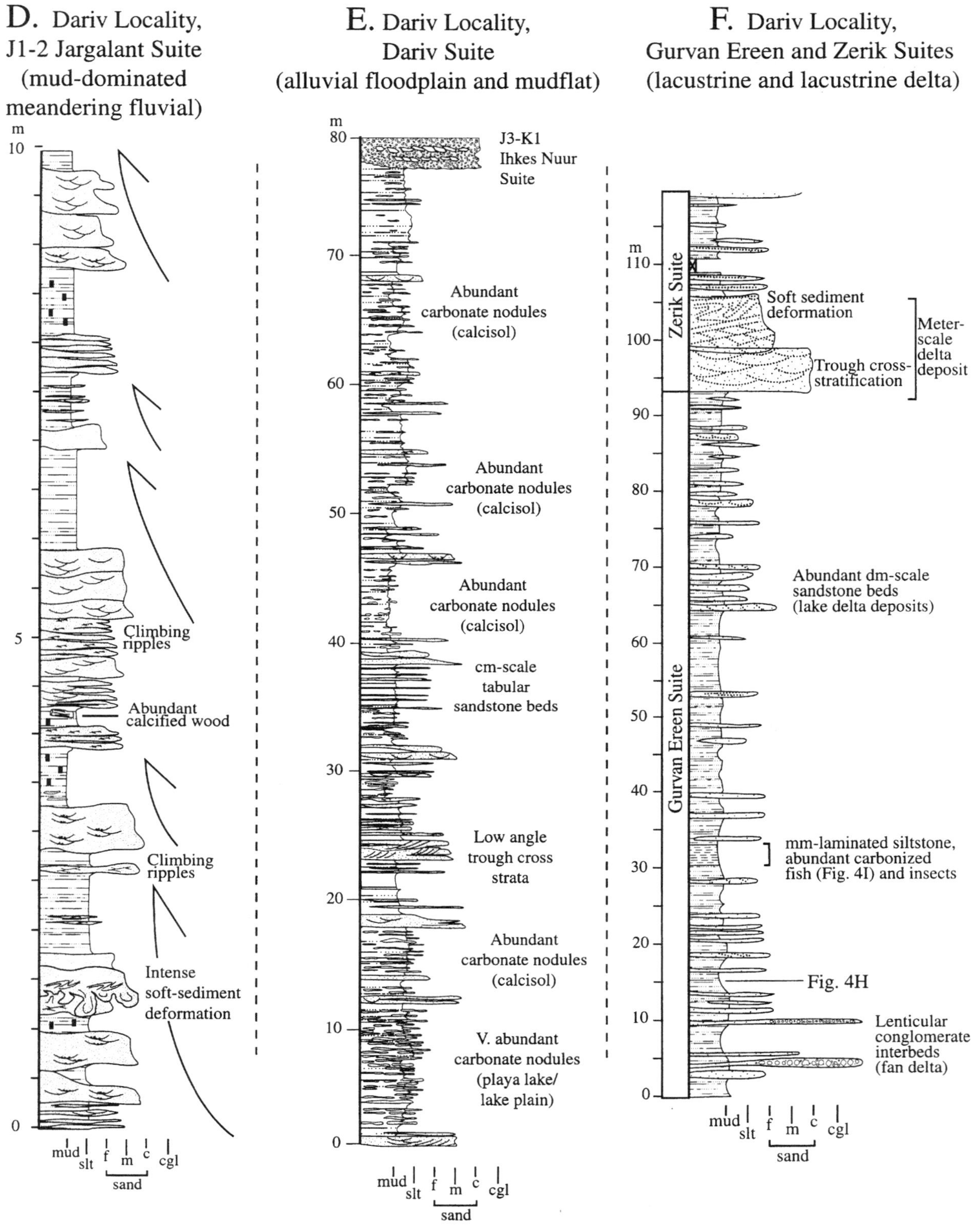

D. Dariv Locality,
J1-2 Jargalant Suite
(mud-dominated meandering fluvial)

m
10

Climbing ripples

Abundant calcified wood

Climbing ripples

5

Intense soft-sediment deformation

0

mud slt | f | m | c | cgl
sand

E. Dariv Locality,
Dariv Suite
(alluvial floodplain and mudflat)

m
80

J3-K1 Ihkes Nuur Suite

70

Abundant carbonate nodules (calcisol)

60

Abundant carbonate nodules (calcisol)

50

Abundant carbonate nodules (calcisol)

40

cm-scale tabular sandstone beds

30

Low angle trough cross strata

20

Abundant carbonate nodules (calcisol)

10

V. abundant carbonate nodules (playa lake/ lake plain)

0

mud slt | f | m | c | cgl
sand

F. Dariv Locality,
Gurvan Ereen and Zerik Suites
(lacustrine and lacustrine delta)

m
110

Zerik Suite

Soft sediment deformation

Trough cross-stratification

Meter-scale delta deposit

100

90

80

Gurvan Ereen Suite

Abundant dm-scale sandstone beds (lake delta deposits)

70

60

50

40

mm-laminated siltstone, abundant carbonized fish (Fig. 4I) and insects

30

20

Fig. 4H

10

Lenticular conglomerate interbeds (fan delta)

0

mud slt | f | m | c | cgl
sand

Figure 3. (*continued*)

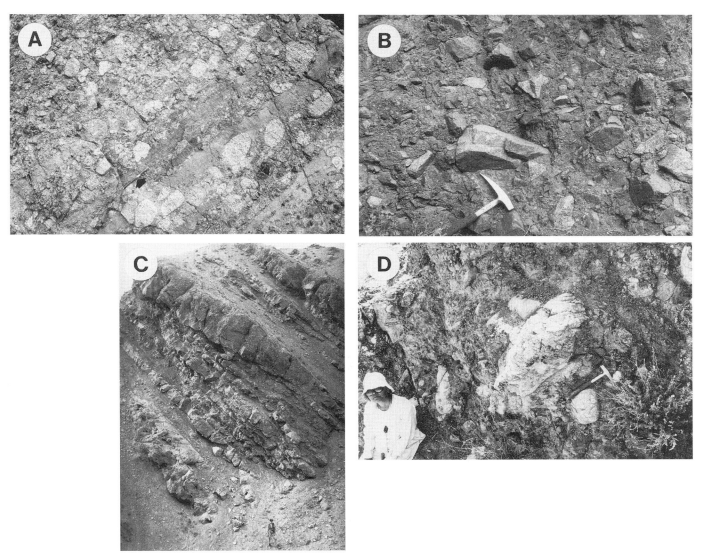

Figure 4. (this and facing page). A: Imbricated, well-rounded cobble conglomerate interbedded with sandstone, Hyargas Nuur locality. Clasts are predominantly granite with minor mafic volcanics. Stratigraphic position of photograph is shown in Figure 2A. Pen (highlighted with black arrow) for scale. B: Hyargas Nuur debris-flow deposit. Decimeter-scale angular cobbles of granite are supported by muddy matrix. Stratigraphic position of photograph is shown in Figure 3A. Hammer for scale. C: Channel deposits exposed at base of Oshin Nuruu location. Stratigraphic position of photograph is shown in Figure 2B. Resistant units are composed primarily of pebble conglomerate, interbedded with recessive siltstone and fine sandstone. Person for scale at bottom of photograph. D: Basal conglomerate exposed at Jargalant section. Boulders of basalt and granite are to 1 m in diameter and are subangular to subrounded, suggesting source-proximal deposition. E: Meter-scale tabular cross-stratification (highlighted with white dotted lines) and channelized sandstone (ch) from bedload-dominated fluvial facies exposed at Jargalant location. Cross-sets are composed of pebble to cobble conglomerate in coarse sandstone matrix. Stratigraphic "up" is to left. Stratigraphic position of photograph is shown in Figure 2C. Note person in lower right for scale. F: Three upward-fining sequences (ufs1, ufs2, ufs3) from mud-load–dominated meandering fluvial facies at Jargalant. Stratigraphic "up" is to left. Sequences have scoured base and fine upward from coarse sandstone to siltstone to organic-rich mudstone and coal. Note person (circled) for scale. Lateral accretion surfaces (highlighted in photograph) are present within sandy layers. See Figure 2C for stratigraphic position of photograph. G: Polymictic, poorly sorted, and poorly rounded conglomerate typical of Upper Jurassic–Lower Cretaceous Ihkes Nuur Suite. We interpret these deposits to be related to uplift of Altai Mountains adjacent to this location. Clast types found within this conglomerate include: sedimentary (s), metamorphic (m), felsic volcanic (fv), and mafic volcanic (mv). Stratigraphic position of photograph is shown in Figure 2D. H: Interbedded siltstone and shale typical of Lower Cretaceous Gurvan Ereen Suites, Dariv section. Note wave ripples (w) and lens cap (5 cm diameter) for scale. We infer these deposits to reflect deposition in shallow, oxic lake. Stratigraphic position of photograph is shown in Figure 3F. I: Fossil fish and pelecypods preserved in finely laminated siltstone of Lower Cretaceous Gurvan Ereen Suite, Dariv section. These deposits are interpreted to represent deposition in anoxic lake. Stratigraphic position of photograph is shown in Figure 3F. J: Typical outcrop of Zerik Suite, Dariv locality. The Zerik suite is predominantly siltstone and mudstone inferred to be deposited in an oxic lacustrine environment. Note person (circled) for scale. Stratigraphic position of photograph is shown in Figure 2D.

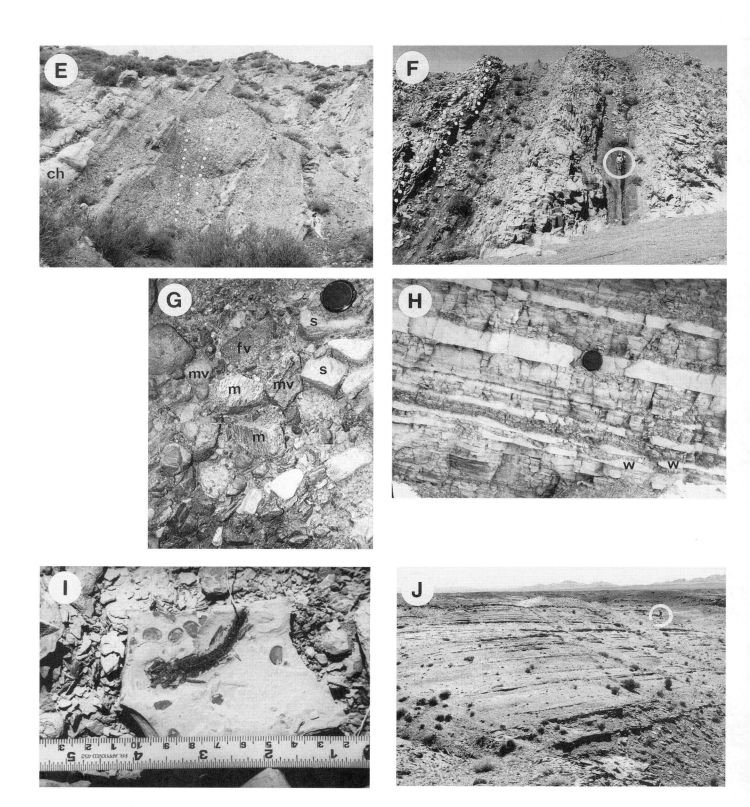

Figure 4. (*continued*)

Jargalant locality
Lower and Middle Jurassic strata
Flora (Devyatkin et al., 1975):
  Cladophlebis haiburnensis L. et H.
  Gingo digitata Brong.
  C. sibirica Heer.
  Pityophyllum longifolium Nath.
  Pityophyllim sp.
  Phoenicopsis sp.
  Ptherophyllum sp.
  Coniopteris burjensis Zal. (by Neiburg M.F.)
  Cladophlebis ex gr. williamsoni (Brong) by
    Wakhrameev V.A.

Insects (Rasnitsyn, 1985):
  Keleusticus acutus Ponomarenko, sp.nov.
  Dzeregia striata Ponomarenko, sp.nov.
  Taublattopsis costalis Vishniakova, sp.nov.
  Mesoblattina mongolica Vishniakova, sp.nov.
  maroblatta incompleta Vishniakova, sp.nov.

Molluscs (upper portion of the Jargalant section;
Devyatkin et al., 1975):
  Ferganoconcha cf. Estherineformis Tschern.
  F. sibirica Tschern.
  F. elongata Rag.
  F. tomiensis Rag.
  F. minor Martins.
  F. cf. Subcentralis Tschern.
  Ferganoconcha sp.
  Unio cf. Golavae Leb.
  U. porrectus Sow.
  U. jenissejensis Leb.
  Unio sp.
  Pseudocardinia sp.
  Sibireconcha jenissejensis Leb.

Oshin Nuruu locality
Flora (Devyatkin et al., 1975):
  Cladophlebis argulata (Heer) Rond

Molluscs (Devyatkin et al., 1975):
  Pseudocardinia cf. Londa Kol. (G.G. Martinson)

Insects (Rasnitsyn, 1985):
  Mesobaetis sibirica Brauer, Redtenbacher,
    Ganflbauer,1889
  Mesoneta utriculata Sinitshenkova, sp.nov.
  Siberiogenites rotundatus Sinitshenkova, sp.nov.
  Asianisca mongolica Yu.Popov, sp.nov.
  Timarchopsis mongolicus Ponomarenko, sp.nov.
  Chimaerocoleites gyrinides Ponomarenko, sp.nov.
  Dzeregia striata Ponomarenko, sp.nov.
  Orthophlebia mongolica Sukatsheva, sp.nov.
  Sibirioneura interrupta Pritykina, sp.nov.
  Hypsophlebia dubia Pritykina, sp.nov.
  Oshinia cellulata Pritykina, sp.nov.
  Karatawia mongolica Pritykina, sp.nov.
  Kobdoneura manca Pritykina, sp.nov.
  Sogdoblatta parvula Vishniakova, sp.nov.
  Taublatta grandis Vishniakova, sp.nov.
  Samaroblattula ashinensis Vishniakova, sp.nov.
  Mesoleuctroides latus Sinitshenkova, sp.nov.
  Trianguliperla orbiculata Sinitshenkova, sp.nov.
  Mongolohagla unicolor Zherichin, sp.nov.

Dariv locality
J3 Dariv Suite
Pelecypoda (Khosbayar, 1973):
  Tutuella globosa Kol.
  Tutuella sp.
  Arguniella sp.

J3-K1 Ihkes Nuur Suite
Phylopoda (Khosbayar, 1973):
  Pseudograpta andrewsi (Jones)
  P. murchisoniae (Jones)

Charophyta (Khosbayar, 1973):
  Aclistochara branconi Peck.

J3-K1 Gurvan Ereen Suite
Konhostraca (Devyatkin et al., 1975):
  Bairdestheria middendorfii Jones
  B. sinensis (Chi.)
  Liograpta (Kob. Et Kus.)
  Palaeoleptestheria wolchonini Nov.
  P. baicalia Kras.
  P. legiminiformis Kras
  P. gurvanerensis Step.
  Palaeoleptestheria sp.

Ostracoda (Devyatkin et al., 1975):
  Daurina mongolica sp.nov.
  Cypridea vitimensis Mand.
  C. Trita Lub., C. zagustaica Scoblo
  Mongolionella palmosa Mand.
  M. martini Neustr.
  Darwinula contracta Mand.
  Lycopterocypris sp.
  Rhinocypris potanini (Gal.)

Fish (Devyatkin et al., 1975):
  Stichopterus popovi sp.nov.
  Gurvanichthys mongolicus sp.nov.

Insects (Rasnitsyn, 1985):
  Ostracindusia biassica Suk.
  Corixonesta hosbajari sp.nov.
  Secrindusia translucens Suk.

Ostracoda (Devyatkin et al., 1975):
  Mongolianella palmosa Mand.
  Cypridea vitimensis Mand.
  C. trita Lub.
  C. zagustaica Scoblo.
  Darwinula contracta Mand.

Insects (Rasnitsyn, 1985):
  Baissocovixa Jacjewski J. Pop.

Dariv locality (cont.)
K1 Zerik Suite
Mollusca (Devyatkin et al., 1975):
  Unio tsentahozensis Jak.
  U. hangaensis Jak.
  U. cf. Hangaensis Jak.
  U. tudogoensis Jak.
  Leptesthes(?) bahatiriensis Jak.
  Limnocyrena anderssoni pusilla
    (Grab.)sub. sp. nov.
  L. zergensa Martins.
  L. anderssoni longa (Grab.)
    sub.sp.nov.
  Valvata subpiscinalis Martins.
  V. mongolica sp. nov.
  Probaicalia vitimensis Martins.
  P. gerassimovi (Reis.)
  Bithynia laaechioidea Martins.
  Guraulus sp.
  Konhostraca-Bairdestheria dahurica
    (Tschern.).

Ostracoda (Devyatkin et al., 1975):
  Darwinula ex gr. oblonga (Roemer)
  D. ex gr. contracta Mand.
  D. contracta Mand.
  Cypridea zacustaica Scoblo
  C. ex gr. trita Lub.
  C. tuberculisperga Gal.
  C. ex gr. vitimensis Mand.
  C. gurvanensis Sinitsa.
  C. prognata Lub.
  Cytheridae gen.nov.
  Rhinocypris ex gr.jurassica (Mart.)
  R. ex gr. barunbainensis (Lub.)
  R. ex gr. tugurigensis (Lub.)
  R. potanini (Gal.)
  Timiriasevia sp.

Insects (Rasnitsyn, 1985):
  Ostracindusia modesta Suk.
  O. biassica Suk.
  O. onusta Suk.
  O. conchifera Suk.
  Pelindusa ostracifera Suk.
  P. minae Suk.
  Cristocorixa gurvanica sp.nov.

Figure 5. Flora and fauna reported from the western Mongolian Mesozoic strata.

TABLE 1. RAW GRAIN COUNTS FOR MESOZOIC SANDSTONE OF WESTERN MONGOLIA

Sample	Sample age	Qm	Qp	Cht	K	P	Lv	Ls	Lm	unidL	bt	ms	chl	D	Cmt	Matr	Por	n
Hyargas Nuur East																		
95HN2-107	J1-2	125	13	0	122	126	38	3	4	7	6	2	10	13	1	30	0	500
95HN2-106	J1-2	130	16	0	100	160	8	2	9	6	1	5	12	23	0	25	3	500
95HN2-105	J1-2	126	23	0	73	131	21	4	8	4	1	2	9	29	14	51	4	500
95HN2-102	J1-2	130	13	0	95	158	26	8	3	9	15	3	6	10	0	24	0	500
95HN2-101	J1-2	142	14	0	84	140	44	7	9	4	4	1	9	25	0	15	2	500
95HN2-100	J1-2	150	9	0	70	132	48	11	6	10	6	6	4	14	3	31	0	500
95HN12	J1-2	149	10	0	77	135	63	1	3	7	5	0	2	20	7	21	0	500
95HN10	J1-2	131	10	6	20	187	114	10	0	0	3	2	1	6	6	4	0	500
95HN8	J1-2	166	2	0	101	129	62	8	2	2	7	1	0	9	0	9	2	500
95HN6	J1-2	73	1	1	39	61	43	6	3	3	5	0	1	5	0	8	1	250
95HN5	J1-2	169	24	0	17	101	106	3	4	4	10	1	5	11	0	45	0	500
Hyargas Nuur Central																		
95HNC104	J1-2	150	15	0	105	119	32	8	2	6	6	0	6	20	4	19	8	500
95HNC102	J1-2	149	8	0	83	141	7	1	3	5	18	14	9	28	0	31	3	500
95HN18	J1-2	174	7	0	116	130	29	3	0	2	1	5	6	9	0	13	5	500
95HN16	J1-2	161	5	0	73	194	15	0	5	3	8	6	5	13	0	8	4	500
Hyargas Nuur West																		
92HN10	J1-2	192	10	0	88	145	17	9	1	0	5	0	1	6	2	22	2	500
92HN13	J1-2	122	7	0	90	98	63	11	3	3	3	0	0	68	1	30	1	500
92HN14	J1-2	132	8	0	96	122	48	4	2	5	0	2	0	58	0	23	0	500
92HN15	J1-2	182	8	0	87	141	25	4	0	1	8	6	2	11	0	26	3	504
92HN16	J1-2	221	7	0	81	125	31	6	0	1	3	0	1	7	0	12	5	500
95HN1 DS	J1-2	173	34	0	79	137	21	0	1	2	3	5	0	5	0	27	11	498
92HN17	J1-2	158	9	0	79	144	59	4	13	6	1	2	1	13	0	23	0	512
92HN18	J1-2	199	8	2	58	130	48	4	0	1	11	0	1	2	0	25	1	490
Oshin Nuruu North																		
ON-I-1	J1-2	141	24	4	52	104	41	5	37	8	4	3	9	26	1	13	28	500
95ONN1A	J1-2	52	5	0	21	25	10	5	23	1	1	2	0	2	52	0	1	200
95ONN1B	J1-2	145	16	1	54	90	22	9	29	13	6	0	6	18	4	35	52	500
95ONN3	J1-2	180	12	0	87	66	25	6	50	12	3	1	2	8	32	11	5	500
95ONN4	J1-2	143	28	0	61	84	44	8	47	7	6	1	1	2	26	6	36	500
95ONN5	J1-2	76	13	1	26	41	10	5	39	0	0	0	3	1	2	13	20	250
95ONN6	J1-2	151	11	1	62	60	12	3	38	5	1	5	0	7	137	1	8	502
95ONN105	J1-2	149	40	2	59	87	17	0	41	4	8	7	5	9	40	13	19	500
95ONN104	J1-2	101	20	0	67	72	17	1	47	12	10	4	6	4	135	1	3	500
95ONN102	J1-2	174	30	0	46	87	14	3	47	6	9	6	4	7	47	6	14	500
95ONN101	J1-2	130	30	0	54	83	25	3	47	3	1	2	2	3	114	1	2	500
Oshin Nuruu South																		
95ONS202	J1-2	145	18	3	81	107	18	8	35	6	7	5	6	8	0	21	32	500
95ONS203	J1-2	156	27	0	114	9	48	4	5	1	4	10	3	2	104	13	0	500
95ONS205	J1-2	162	18	1	131	82	47	9	13	0	2	0	2	3	0	26	4	500
95ONS204	J1-2	175	24	0	114	32	64	14	24	1	4	2	5	5	0	11	25	500
95ONS206	J1-2	162	28	2	70	66	99	8	27	0	4	1	0	1	1	12	18	499
95ONS3	J1-2	137	27	0	51	91	8	3	45	3	8	7	6	10	77	4	23	500
95ONS5	J1-2	111	15	0	63	84	8	0	50	3	13	3	14	9	122	1	4	500
Jargalant																		
95JA701	J1-2	183	35	0	52	80	105	7	10	0	7	2	2	5	3	5	4	500
92JL11	J1-2	216	7	1	105	75	43	18	3	0	2	2	0	10	0	17	1	500
92JL12	J1-2	220	12	0	109	76	39	12	6	0	1	1	0	11	0	12	1	500
92JL13	J1-2	212	7	0	130	84	25	8	4	1	1	2	0	6	0	18	2	500
95JA702	J1-2	223	30	3	38	89	78	5	8	0	5	2	0	4	0	10	5	500
95JA703	J1-2	171	35	0	53	115	47	1	9	1	7	12	3	9	0	18	19	500
95JA704	J1-2	191	19	1	43	97	74	7	17	0	9	0	0	6	0	7	28	499

TABLE 1. RAW GRAIN COUNTS FOR MESOZOIC SANDSTONE OF WESTERN MONGOLIA (continued)

Sample	Sample age	Qm	Qp	Cht	K	P	Lv	Ls	Lm	unidL	bt	ms	chl	D	Cmt	Matr	Por	n
92JL14	J1-2	206	14	0	115	89	8	11	4	4	3	3	0	13	0	25	5	500
95JL705	J1-2	145	62	6	90	44	86	15	25	0	3	1	0	4	0	13	7	501
95JA706	J1-2	164	32	1	53	85	91	8	22	1	9	0	0	7	0	10	17	500
92JL16	J1-2	217	4	0	121	70	20	14	21	0	1	1	0	5	0	13	13	500
95JA707	J1-2	186	14	1	37	101	99	4	24	3	8	4	2	3	0	4	10	500
92JL18	J1-2	184	16	0	91	86	36	14	10	3	3	0	0	31	0	23	3	500
95JA708	J1-2	197	33	0	69	105	36	0	23	0	1	5	1	11	0	12	5	498
95JL709	J1-2	181	50	2	78	99	47	1	10	3	4	2	2	5	6	8	2	500
92JL19	J1-2	297	10	0	67	33	28	11	9	1	5	2	0	3	1	20	13	500
95JA710	J1-2	242	45	0	30	75	69	3	13	2	0	1	0	2	4	8	6	500
95 JA 511	J1-2	212	16	6	36	76	110	10	14	0	2	2	4	0	8	2	2	500
95JA711	J1-2	199	29	0	90	63	54	6	21	0	7	5	7	2	0	21	2	506
92JL20	J1-2	256	9	0	136	44	18	3	11	1	3	5	0	0	0	10	4	500
Dariv																		
Lower Cretaceous Zerik Formation																		
95DV3	K1	155	45	1	27	140	77	7	20	6	9	0	5	8	0	0	0	500
95DV5	K1	34	0	0	19	22	55	11	1	0	0	0	1	4	49	1	3	200
92DV16	K1	114	9	1	51	68	68	24	29	0	0	1	0	28	75	24	8	500
92DV14	K1	125	6	1	73	98	47	30	13	0	1	2	1	13	90	0	0	500
95DV200	K1	123	19	0	74	54	72	0	6	2	0	0	0	2	144	0	4	500
92DV13	K1	81	5	1	59	60	74	24	20	7	2	2	3	16	141	3	2	500
95DV6	K1	92	19	0	61	80	54	2	15	8	0	1	0	18	150	0	0	500
92DV12	K1	68	1	0	12	33	24	26	12	4	1	0	1	7	59	1	1	250
95DV7	K1	96	3	0	39	80	79	2	21	6	0	0	1	25	141	2	5	500
95DV101	K1	109	40	0	57	63	39	3	8	5	0	4	6	8	145	1	13	501
Lower Cretaceous Gurvan Ereen Formation																		
92DV27	K1	116	5	0	59	62	52	23	11	9	0	7	4	20	131	0	1	500
92DV23	K1	67	16	0	73	58	55	33	40	5	1	6	2	5	134	4	1	500
Upper Jurassic-Lower Cretaceous Ihkes Nuur Formation																		
92DV43	J3-K1	59	1	0	16	30	31	22	14	6	0	0	0	3	7	4	7	200
95DV303	J3-K1	128	7	0	29	108	36	3	26	7	1	2	2	17	124	2	8	500
92DV41	J3-K1	154	12	0	77	64	71	26	22	3	5	3	3	7	10	21	22	500
92DV40	J3-K1	64	14	0	31	31	42	27	21	3	1	1	1	3	7	9	25	280
92DV38	J3-K1	162	17	0	49	58	46	30	34	14	0	9	3	14	47	7	10	500
92DV37	J3-K1	44	3	0	13	16	44	14	12	7	0	1	1	13	6	5	21	200
95DV306	J3-K1	88	44	2	22	52	30	11	63	3	7	13	4	85	65	9	2	500
95DV35	J3-K1	66	5	0	22	16	30	54	12	8	0	2	0	6	25	1	3	250
92DV34	J3-K1	52	0	0	25	11	44	26	1	5	0	0	0	0	5	3	28	200
92DV33	J3-K1	59	10	0	47	33	128	45	10	7	0	3	1	3	153	0	1	500
92DV31	J3-K1	79	13	1	79	55	55	12	7	10	6	0	0	5	11	26	41	400
92DV30	J3-K1	82	8	0	58	62	67	31	25	6	0	4	3	7	145	1	1	500
Upper Jurassic Dariv Formation																		
92DV50	J3	176	5	0	32	62	38	22	35	2	0	9	0	30	83	0	6	500
92DV53	J3	106	7	0	33	45	24	24	9	0	0	0	4	5	3	5	29	294
92DV58	J3	127	6	0	43	77	43	12	43	13	3	8	0	10	93	12	10	500
92DV59	J3	87	10	0	77	71	30	61	27	10	1	2	2	9	67	34	12	500
Lower-Middle Jurassic Jargalant Formation																		
92DV46	J1-2	146	18	0	66	107	32	39	34	11	0	4	1	9	19	14	0	500
92DV49	J1-2	146	1	0	93	115	44	21	8	7	15	3	7	17	1	7	15	500

Note: Grain categories: Qm—monocrystalline quartz; Qp—polycrystalline quartz; cht—chert; K—potassium feldspar; P—plagioclase; Lv—lithic volcanic and metavolcanic grains; Ls—lithic sedimentary grains; Lm—lithic metamorphic grains; unidL—unidentified; lithic grains; bt—biotite; ms—muscovite; chl—chlorite; D—dense minerals; Cmt—cement; Matr—matrix; por—porosity; n—total number counts per sample.

TABLE 2. RECALCULATED DETRITAL MODES OF MESOZOIC SANDSTONE FROM WESTERN MONGOLIA

Sample	Sample age	QmFL%			QpLvmLsm%			QmPK%			Accessories	
		Qm	F	Lt	Qp	Lv	Lsm	Qm	P	K	M	D
Hyargas Nuur East												
95HN2-107	J1-2	28.5	56.6	14.8	22.4	65.5	12.1	33.5	33.8	32.7	3.8	2.8
95HN2-106	J1-2	30.2	60.3	9.5	45.7	22.9	31.4	33.3	41.0	25.6	3.8	4.9
95HN2-105	J1-2	32.3	52.3	15.4	41.1	37.5	21.4	38.2	39.7	22.1	2.8	6.7
95HN2-102	J1-2	29.4	57.2	13.3	26.0	52.0	22.0	33.9	41.3	24.8	5.0	2.1
95HN2-101	J1-2	32.0	50.5	17.6	18.9	59.5	21.6	38.8	38.3	23.0	2.9	5.2
95HN2-100	J1-2	34.4	46.3	19.3	12.2	64.9	23.0	42.6	37.5	19.9	3.4	3.0
95HN12	J1-2	33.5	47.6	18.9	13.0	81.8	5.2	41.3	37.4	21.3	1.5	4.2
95HN10	J1-2	27.4	43.3	29.3	11.4	81.4	7.1	38.8	55.3	5.9	1.2	1.2
95HN8	J1-2	35.2	48.7	16.1	2.7	83.8	13.5	41.9	32.6	25.5	1.6	1.8
95HN6	J1-2	31.7	43.5	24.8	3.7	79.6	16.7	42.2	35.3	22.5	2.5	2.1
95HN5	J1-2	39.5	27.6	32.9	17.5	77.4	5.1	58.9	35.2	5.9	3.5	2.4
Hyargas Nuur Central												
95HNC104	J1-2	34.3	51.3	14.4	26.3	56.1	17.5	40.1	31.8	28.1	2.6	4.3
95HNC102	J1-2	37.5	56.4	6.0	42.1	36.8	21.1	39.9	37.8	22.3	8.8	6.0
95HN18	J1-2	37.7	53.4	8.9	17.9	74.4	7.7	41.4	31.0	27.6	2.5	1.9
95HN16	J1-2	35.3	58.6	6.1	20.0	60.0	20.0	37.6	45.3	17.1	3.9	2.7
Hyargas Nuur West												
92HN10	J1-2	41.6	50.4	8.0	27.0	45.9	27.0	45.2	34.1	20.7	1.3	1.3
92HN13	J1-2	30.7	47.4	21.9	8.3	75.0	16.7	39.4	31.6	29.0	0.6	14.5
92HN14	J1-2	31.7	52.3	16.1	12.9	77.4	9.7	37.7	34.9	27.4	0.4	12.2
92HN15	J1-2	40.6	50.9	8.5	21.6	67.6	10.8	44.4	34.4	21.2	3.4	2.3
92HN16	J1-2	46.8	43.6	9.5	15.9	70.5	13.6	51.8	29.3	19.0	0.8	1.4
95HN1 DS	J1-2	38.7	48.3	13.0	60.7	37.5	1.8	44.5	35.2	20.3	1.7	1.1
92HN17	J1-2	33.5	47.2	19.3	10.6	69.4	20.0	41.5	37.8	20.7	0.8	2.7
92HN18	J1-2	44.2	41.8	14.0	16.1	77.4	6.5	51.4	33.6	15.0	2.6	0.4
Hyargas Nuur Mean (all sections)		35.1	49.4	15.6	21.5	63.2	15.3	41.7	36.7	21.6	2.7	3.8
Hyargas Nuur St. Dev. (all sections)		5.1	7.0	7.0	14.1	17.2	7.8	6.0	5.5	6.4	1.8	3.4
Oshin Nuruu North												
ON-I-1	J1-2	33.9	37.5	28.6	25.2	36.9	37.8	47.5	35.0	17.5	3.5	5.7
95ONN1A	J1-2	36.6	32.4	31.0	11.6	23.3	65.1	53.1	25.5	21.4	2.0	1.4
95ONN1B	J1-2	38.3	38.0	23.7	22.1	28.6	49.4	50.2	31.1	18.7	2.9	4.4
95ONN3	J1-2	41.1	34.9	24.0	12.9	26.9	60.2	54.1	19.8	26.1	1.3	1.8
95ONN4	J1-2	33.9	34.4	31.8	22.0	34.6	43.3	49.7	29.2	21.2	1.9	0.5
95ONN5	J1-2	36.0	31.8	32.2	20.6	14.7	64.7	53.1	28.7	18.2	1.4	0.5
95ONN6	J1-2	44.0	35.6	204	18.5	18.5	63.1	55.3	22.0	22.7	1.7	2.0
95ONN105	J1-2	37.3	36.6	26.1	42.0	17.0	41.0	50.5	29.5	20.0	4.7	2.1
95ONN104	J1-2	30.0	41.2	28.8	23.5	20.0	56.5	42.1	30.0	27.9	5.5	1.1
95ONN102	J1-2	42.8	32.7	24.6	31.9	14.9	53.2	56.7	28.3	15.0	4.4	1.6
95ONN101	J1-2	34.7	36.5	28.8	28.6	23.8	47.6	48.7	31.1	20.2	1.3	0.8
Oshin Nuruu South												
95ONS202	J1-2	34.4	44.7	20.9	25.6	22.0	52.4	43.5	32.1	24.3	4.0	1.8
95ONS203	J1-2	42.9	33.8	23.4	32.1	57.1	10.7	55.9	3.2	40.9	4.4	0.5
95ONS205	J1-2	35.0	46.0	19.0	21.6	53.4	25.0	43.2	21.9	34.9	0.9	0.6
95ONS204	J1-2	39.1	32.6	28.3	19.0	50.8	30.2	54.5	10.0	35.5	2.4	1.1
95ONS206	J1-2	35.1	29.4	35.5	18.3	60.4	21.3	54.4	22.1	23.5	1.1	0.2
95ONS3	J1-2	37.5	38.9	23.6	32.5	9.6	57.8	49.1	32.6	18.3	5.3	2.5
95ONS5	J1-2	33.2	44.0	22.8	20.5	11.0	68.5	43.0	32.6	24.4	8.0	2.4
Oshin Nuruu Mean (both sections)		37.0	36.7	26.3	23.8	29.1	47.1	50.2	25.8	23.9	3.2	1.7
Oshin Nuruu St. Dev. (both sections)		3.8	4.7	4.6	7.5	16.3	16.6	4.8	8.3	7.0	2.0	1.4
Jargalant												
95JA701	J1-2	38.8	28.0	33.3	22.3	66.9	10.8	58.1	25.4	16.5	2.3	1.0
92JL11	J1-2	46.2	38.5	15.4	11.1	59.7	29.2	54.5	18.9	26.5	0.8	2.1
92JL12	J1-2	46.4	39.0	14.6	17.4	56.5	26.1	54.3	18.8	26.9	0.4	2.3
92JL13	J1-2	45.0	45.4	9.6	15.9	56.8	27.3	49.8	19.7	30.5	0.6	1.3
95JA702	J1-2	47.0	26.8	26.2	26.6	62.9	10.5	63.7	25.4	10.9	1.4	0.8
95JA703	J1-2	39.6	38.9	21.5	38.0	51.1	10.9	50.4	33.9	15.6	4.8	1.9

Sample	Sample age	QmFL%			QpLvmLsm%			QmPK%			Accessories	
		Qm	F	Lt	Qp	Lv	Lsm	Qm	P	K	M	D
95JA704	J1-2	42.5	31.2	26.3	16.9	62.7	20.3	57.7	29.3	13.0	1.9	1.3
92JL14	J1-2	45.7	45.2	9.1	37.8	21.6	40.5	50.2	21.7	28.0	1.3	2.8
95JL705	J1-2	30.7	28.3	41.0	35.1	44.3	20.6	52.0	15.8	32.3	0.8	0.8
95JA706	J1-2	35.9	30.2	33.9	21.4	59.1	19.5	54.3	28.1	17.5	1.9	1.5
92JL16	J1-2	46.5	40.9	12.6	6.8	33.9	59.3	53.2	17.2	29.7	0.4	1.1
95JA707	J1-2	39.7	29.4	30.9	10.6	69.7	19.7	57.4	31.2	11.4	2.9	0.6
92JL18	J1-2	41.8	40.2	18.0	21.1	47.4	31.6	51.0	23.8	25.2	0.6	6.5
95JA708	J1-2	42.5	37.6	19.9	35.9	39.1	25.0	53.1	28.3	18.6	1.5	2.3
95JL709	J1-2	38.4	37.6	24.0	47.3	42.7	10.0	50.6	27.7	21.8	1.7	1.0
92JL19	J1-2	65.1	21.9	12.9	17.2	48.3	34.5	74.8	8.3	16.9	1.5	0.6
95JA710	J1-2	50.5	21.9	27.6	34.6	53.1	12.3	69.7	21.6	8.6	0.2	0.4
95 JA 511	J1-2	44.2	23.3	32.5	14.1	70.5	15.4	65.4	23.5	11.1	1.6	0.0
95JA711	J1-2	43.1	33.1	23.8	26.4	49.1	24.5	56.5	17.9	25.6	3.9	0.4
92JL20	J1-2	53.6	37.7	8.8	22.0	43.9	34.1	58.7	10.1	31.2	1.6	0.0
Jargalant Mean		44.2	33.8	22.1	23.9	52.0	24.1	56.8	22.3	20.9	1.6	1.4
Jargalant St. Dev.		7.1	7.3	9.4	11.0	12.4	12.2	6.9	6.7	7.8	1.2	1.4
Dariv		6.8	6.3	7.7	9.4	18.3	18.4	6.8	7.6	7.4	1.8	1.4
Lower Cretaceous Zerik Suite												
95DV3	K1	32.4	34.9	32.6	30.7	51.3	18.0	48.1	43.5	8.4	2.8	1.6
95DV5	K1	23.9	28.9	47.2	0.0	82.1	17.9	45.3	29.3	25.3	0.7	2.7
92DV16	K1	31.3	32.7	36.0	7.6	51.9	40.5	48.9	29.2	21.9	0.3	7.1
92DV14	K1	31.8	43.5	24.7	7.2	48.5	44.3	42.2	33.1	24.7	1.0	3.2
95DV200	K1	35.1	36.6	28.3	19.6	74.2	6.2	49.0	21.5	29.5	0.0	0.6
92DV13	K1	24.5	36.0	39.6	4.8	59.7	35.5	40.5	30.0	29.5	2.0	4.5
95DV6	K1	27.8	42.6	29.6	21.1	60.0	18.9	39.5	34.3	26.2	0.3	5.1
92DV12	K1	37.8	25.0	37.2	1.6	38.1	60.3	60.2	29.2	10.6	1.1	3.7
95DV7	K1	29.4	36.5	34.0	2.9	75.2	21.9	44.7	37.2	18.1	0.3	7.1
95DV101	K1	33.6	37.0	29.3	44.4	43.3	12.2	47.6	27.5	24.9	2.9	2.3
Lower Cretaceous Gurvan Ereen Suite												
92DV27	K1	34.4	35.9	29.7	5.5	57.1	37.4	48.9	26.2	24.9	3.0	5.4
92DV23	K1	19.3	37.8	42.9	11.1	38.2	50.7	33.8	29.3	36.9	2.5	1.4
Upper Jurassic-Lower Cretaceous Ihkes Nuur Suite												
92DV43	J3-K1	33.0	25.7	41.3	1.5	45.6	52.9	56.2	28.6	15.2	0.0	1.6
95DV303	J3-K1	37.2	39.8	23.0	9.7	50.0	40.3	48.3	40.8	10.9	1.4	4.6
92DV41	J3-K1	35.9	32.9	31.2	9.2	54.2	36.6	52.2	21.7	26.1	2.5	1.6
92DV40	J3-K1	27.5	26.6	45.9	13.5	40.4	46.2	50.8	24.6	24.6	1.3	1.3
92DV38	J3-K1	39.5	26.1	34.4	13.4	36.2	50.4	60.2	21.6	18.2	2.8	3.2
92DV37	J3-K1	28.8	19.0	52.3	4.1	60.3	35.6	60.3	21.9	17.8	1.2	7.7
95DV306	J3-K1	27.9	23.5	48.6	30.7	20.0	49.3	54.3	32.1	13.6	5.7	20.0
95DV35	J3-K1	31.0	17.8	51.2	5.0	29.7	65.3	63.5	15.4	21.2	0.9	2.7
92DV34	J3-K1	31.7	22.0	46.3	0.0	62.0	38.0	59.1	12.5	28.4	0.0	0.0
92DV33	J3-K1	17.4	23.6	59.0	5.2	66.3	28.5	42.4	23.7	33.8	1.2	0.9
92DV31	J3-K1	25.4	43.1	31.5	15.9	62.5	21.6	37.1	25.8	37.1	1.9	1.6
92DV30	J3-K1	24.2	35.4	40.4	6.1	51.1	42.7	40.6	30.7	28.7	2.0	2.0
Upper Jurassic Dariv Suite												
92DV50	J3	47.3	25.3	27.4	5.0	38.0	57.0	65.2	23.0	11.9	2.2	7.3
92DV53	J3	42.7	31.5	25.8	10.9	37.5	51.6	57.6	24.5	17.9	1.6	1.9
92DV58	J3	34.9	33.0	32.1	5.8	41.3	52.9	51.4	31.2	17.4	2.9	2.6
92DV59	J3	23.3	39.7	37.0	7.8	23.4	68.8	37.0	30.2	32.8	1.3	2.3
Lower-Middle Jurassic Jargalant Suite									###			
92DV46	J1-2	32.2	38.2	29.6	14.6	26.0	59.3	45.8	33.5	20.7	1.1	1.9
92DV49	J1-2	33.6	47.8	18.6	1.4	59.5	39.2	41.2	32.5	26.3	5.2	3.6
Dariv Mean (all suites)		31.2	32.6	36.2	10.5	49.5	40.0	49.1	28.2	22.8	1.7	3.7
Dariv St. Dev. (all suites)		6.5	7.7	9.7	10.2	15.4	16.2	8.4	6.7	7.7	1.4	3.7

Note: Abbreviations are: Qm—monocrytalline quartz, F—total feldspar, Lt—total lithics, Qp—polycrystalline quartz, Lv—lithic volcanic, Lsm—lithic sedimentary plus lithic metamorphic, P—plagioclase, K—potassium feldspar, M—percent mica, D—percent heavy (dense) minerals, St. Dev.—standard deviation.

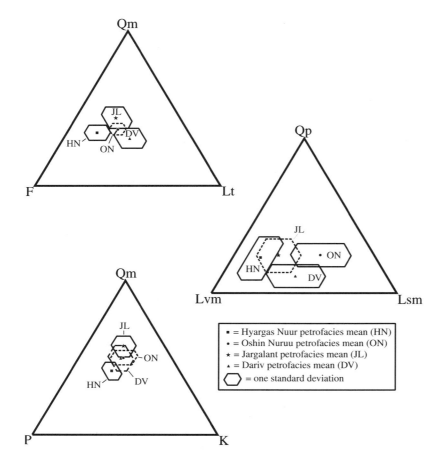

Figure 6. Ternary plots of compositions of Mesozoic sandstone samples from western Mongolia, showing mean composition and 1σ standard deviation for each petrofacies. Raw point-count data are shown in Table 2 and detrital modes are shown in Table 3. Qm = monocrystalline quartz, Qp = polycrystalline quartz, F = total feldspar, P = plagioclase, K = potassium feldspar, Lt = total lithic grains, Lvm = lithic volcanic and metavolcanic grains, Lsm = lithic sedimentary plus lithic metamorphic grains.

TABLE 3. WHOLE-ROCK Sm AND Nd ISOTOPIC DATA FROM JURASSIC SEDIMENTS AND VENDIAN–CAMBRIAN BASEMENT AND COBBLES

Sample	εNd(0)*	± 2 σ	^{143}Nd/^{144}Nd†	^{147}Sm/^{144}Nd	T(DM)Nd§	εNd (T)	Nd (ppm)	Sm (ppm)	Description
95 JA 501	6.32	0.16	0.512962±8	0.1853	0.68	7.18	8.5653	2.6237	Vendian-Cambrian mafic volcanic
95 JA 507	˙6.79	0.14	0.512986±7	0.1851	0.54	7.67	10.276	3.1439	Vendian-Cambrian mafic volcanic
95 JA 602	4.98	0.16	0.512894±8	0.1714	0.71	5.56	11.942	3.3832	Basalt cobble from basal conglomerate
95 JA 605	3.75	0.14	0.512830±7	0.1514	0.64	4.77	26.360	6.5962	Basalt cobble from braided sequence
95 JA 601	1.65	0.14	0.512723±7	0.1213	0.61	3.36	6.3489	1.2735	Granite cobble from basal conglomerate
95 JA 603	2.70	0.16	0.512776±8	0.1409	0.66	3.96	3.0602	0.7129	Granite cobble from basal conglomerate
95 JA 511	-3.75	0.14	0.512446±7	0.1278	1.17	-2.19	11.887	2.5110	Sandstone from meandering sequence
95 JA 512	-4.79	0.12	0.512392±6	0.1268	1.25	-3.21	30.634	6.4211	Siltstone from meandering sequence
95 JA 604	-4.72	0.12	0.512396±6	0.1164	1.11	-2.90	31.889	6.1389	Basal sandstone intraclast
95 JL 703	-4.55	0.12	0.512405±6	0.1179	1.11	-2.76	17.435	3.3970	Basal sandstone

*εNd values calculated relative to CHUR (DePaolo and Wasserburg, 1976) using present-day values of ^{143}Nd/^{144}Nd = 0.512638 and ^{147}Sm/^{144}Nd = 0.1967.

† Error in Nd isotope ratio represents within-run statistics, and is ± 2σ in the last significant digit. Sample reproducibility is ± 25 ppm at 2σ.

§ T(DM)Nd represents a single-stage depleted mantle model age (Ga) calculated after DePaolo et al. (1991).

Initial isotopic compositions are calculated for the age of bedrock and cobbles at 600 Ma and for the sandstone and siltstone at a deposition age of 180 Ma.

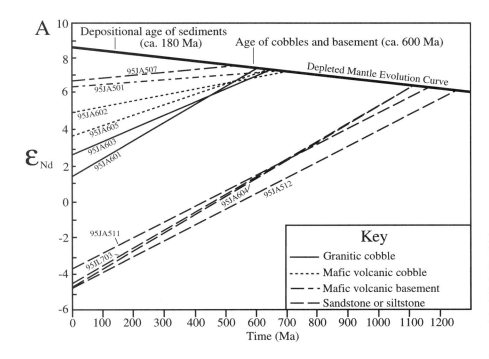

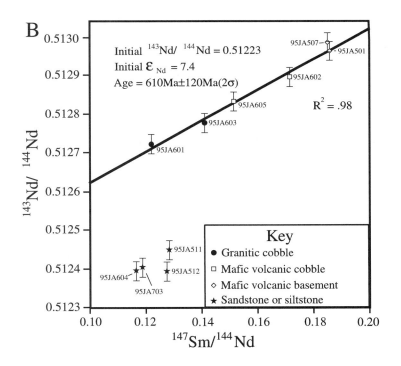

Figure 7. A: ε_{Nd} vs. time evolution for bedrock and cobbles (solid lines) and sandstone and siltstone (dashed lines) samples collected from Jargalant locality. Note distinctly different trends of two sample types, which suggests sampled bedrock was not only source for sandstone and siltstone samples. B: $^{143}Nd/^{144}Nd$ vs. $^{147}Sm/^{144}Nd$ for samples collected from Jargalant locality. Differences between clastic and cobble and/or basement signatures suggest that sandstone samples include component from older continental source. Isochron for basalt basement and basalt cobble samples defines age of 610 ± 120 Ma and initial ε_{Nd} of 7.4 for these samples (based on Model 1 Yorkfit). This age correlates with published dates for these lithologies. Similarity of calculated initial ε_{Nd} to ε_{Nd} inferred for 600 Ma depleted mantle suggests formation of basement and cobbles from juvenile crust.

alluvial fan environments (e.g., Gloppen and Steel, 1981; Nilsen, 1982; DeCelles et al., 1987; Blair and McPherson, 1992). We interpret the slight overall upward-coarsening trend, coarse grain size, and thickness of section to record syntectonic sedimentation and progradation of the alluvial system toward the southwest, as documented for analogous alluvial systems elsewhere (Wiltschko and Dorr, 1983; Burbank et al., 1988; Fillmore, 1991; Schwans, 1995).

Oshin Nuruu study location

Observations. Located on the southwest slope of Jargalant Uul, the Oshin Nuruu site consists of well-exposed Lower to Middle Jurassic strata (Fig. 1B). We measured two stratigraphic sections along strike at decimeter resolution: 645 m at Oshin Nuruu north (47°30′34.8″N, 92°22′28″E) (Fig. 2B); and 450 m at Oshin Nuruu south (47°29′02.2″N, 92°23′18″E) (Sjostrom, 1997). The base of each section is covered by modern alluvium and the top is unconformably overlain by Neogene lacustrine deposits (Devyatkin et al., 1975; Yanshin, 1989).

The Oshin Nuruu sections are dominated by sandstone with subordinate pebbly conglomerate and siltstone (Fig. 2B). The south section can be subdivided into a coarser grained lower portion containing mainly conglomerate and sandstone and a finer grained upper portion containing about 50% siltstone (Sjostrom, 1997). The north section is somewhat finer grained overall, but also contains more conglomerate in its lower part and more siltstone in its upper part. In each section, sandstone beds are typically several decimeters to several meters thick, locally pebbly, and medium to coarse grained. Individual beds of conglomerate and sandstone are laterally discontinuous and typically traceable for less than ~100 m (Fig. 4C). Most have erosional bases. Sandstone contains abundant trough cross-beds and planar lamination, with less common asymmetric ripples and soft-sediment deformation. Conglomerate in both sections is mostly clast supported, although several conglomerate beds near the base of the Oshin Nuruu south section are supported by a muddy matrix (Sjostrom, 1997). Clast diameter is typically <4 cm, and clasts are typically well sorted and moderately rounded. Conglomerate commonly contains trough or tabular cross-stratification with decimeter- to meter-scale relief (Fig. 2B). Siltstone is common at each of the Oshin Nuruu sections, particularly in the upper half of the north section, where it is fissile and contains abundant woody debris, including transported logs and leaf imprints, root casts, and decimeter-scale sandstone lenses. Paleocurrent trends measured on trough axes and oriented woody debris (n = 105) at Oshin Nuruu north are somewhat scattered, but appear to suggest transport to the southwest (Fig. 2B).

Interpretations. Strata exposed at the Oshin Nuruu south section suggest a change from deposition in a coarse bedload–dominated stream in the upper reaches of an alluvial systems to deposition in a shallow mixed-load stream system with muddy interfluves. This interpretation is based on the change from highly lenticular, coarse-grained strata with local matrix-supported debris flows at the base of the section (e.g., Gloppen and Steel, 1981; Nilsen, 1982; DeCelles et al., 1987; Blair and McPherson, 1992) to lenticular sandstone with meter-scale trough cross-beds and interstratified siltstone or mudstone in the upper part of the section (e.g., Smith, 1970, 1971, 1972; Cant, 1982; Collinson, 1986; Schmitt et al., 1991; Bromley, 1992).

Jargalant Study Area

Observations. The Jargalant study area is located along the southwest slope of Jargalant Uul, ~50 km southeast of Har Us Nuur (47°29′17″N, 92°34′26″E; Fig. 1). We measured a 1180 m section of Lower to Middle Jurassic strata (Fig. 2C), along with two decimeter-scale detailed sections (Fig. 3, B and C). At Jargalant, Jurassic strata depositionally overlie Vendian–Lower Cambrian mafic volcanics and/or metabasalts (Yanshin, 1989; Cunningham et al., 1996), although this contact appears to be locally shear modified. The upper part of the section is structurally disturbed and unconformably overlain by Neogene clastic rocks (Yanshin, 1989; Cunningham et al., 1996).

The Lower to Middle Jurassic deposits at Jargalant fine upward from boulder conglomerate to interbedded medium to coarse sandstone, siltstone, mudstone, and coal at the top of the section (Fig. 2C). Conglomerate is restricted mainly to the lower 640 m of the section, where it fines upward from a basal boulder conglomerate with clasts to 1.35 m in diameter (Fig. 4D) to well-sorted, clast-supported conglomerate with generally well-rounded clasts <20 cm in diameter. Conglomerate is locally imbricated and commonly contains horizontal stratification and low-angle tabular cross-stratification with relief of several meters (Fig. 4E). Individual conglomerate beds are lenticular over hundreds of meters and most fine upsection. Fossilized wood is abundant, particularly in association with thin, interbedded sandstone beds.

At approximately the 640 m level, abundant siltstone first appears and cobble conglomerate essentially disappears (Fig. 2C). By the 900 m level, the Jargalant section contains only minor pebbly conglomerate and is dominated by medium to coarse sandstone, siltstone, mudstone, and coal (Fig. 2C). Several upward-fining sequences involving all of these lithologies are present at the top of the section (Figs. 3C and 4F). Each upward-fining sequence contains a basal sandstone unit, typically ~10 m thick, with abundant meter-scale trough cross-bedding, soft-sediment deformation, asymmetric ripples, and coalified wood fragments. Sandstone grades upsection into blocky-weathering green and gray siltstone with abundant woody debris. Siltstone units are typically a few meters thick, locally contain thin tabular sandstone beds, and grade upsection into dark gray or brown mudstone which, in turn, grades into coal. Coal beds are as much as 30 cm thick and most grade back into mudstone before being truncated by the basal sandstone of the next sequence.

Imbrication and trough axis orientations (n = 163) in conglomerate and sandstone of the Jargalant section suggest flow dominantly to the southeast and locally to the southwest

(Fig. 2C). The orientation of parting lineation in sandstone and the elongation direction of most wood casts and coalified wood fragments (n = 68) is southeast-northwest, consistent with paleoflow toward the southeast.

Interpretations. Lower to Middle Jurassic deposits exposed at Jargalant suggest a series of transitional environments, from coarse-grained bedload-dominated fluvial deposition at the base of the section to a sandy bedload-dominated fluvial environment beginning at approximately the 640 m level, to a mudload-dominated fluvial facies above the 780 m level (Fig. 2C). The well-rounded shape of most clasts, the lenticularity of conglomerate beds, the presence of abundant large-scale tabular cross-stratification, and the locally abundant imbrication support the interpretation of a bedload-dominated fluvial environment. Sandy strata above the ~640 m level are inferred to be bedload-dominated fluvial strata based on the abundance of traction-transport structures, lenticular geometry of beds, and relative paucity of fine-grained strata. The upward-fining sequences above the ~780 m level are diagnostic of a mudload-dominated meandering fluvial environment (Gersib and McCabe, 1981; Cant, 1982; Davis, 1983; Smith, 1987). We infer the sharp, irregular contact at the base of each upward-fining sequence to reflect channel erosion and the basal layers of pebbly conglomerate to be channel lag deposits. Sandstone units likely record point bar migration, as suggested by the presence of gently dipping internal surfaces (e.g., Fig. 4F), the transition from trough cross-beds to ripples upsection, and the abundance of soft-sediment deformation (Sjostrom, 1997). We interpret the siltstone and coal capping each upward-fining unit as flood-plain deposits, and we interpret the decimeter-thick tabular sandstone layers within overbank siltstone as crevasse-splay deposits (Bridge, 1984; Dubiel, 1992).

Dariv study area

Observations. The Dariv locality is situated about 15 km southwest of the town of Dariv in the eastern foothills of the Altai Mountains (46°21'35"N, 94°06'35"E; Fig. 1B). There, ~1500 m of Lower to Middle Jurassic through Lower Cretaceous nonmarine strata are exposed (Fig. 2D). Devyatkin et al. (1975) subdivided Mesozoic strata at Dariv into five lithologically distinct suites: these include the Lower to Middle Jurassic Jargalant suite, the Upper Jurassic Dariv suite, the Upper Jurassic–Lower Cretaceous Ihkes Nuur suite, the Lower Cretaceous Gurvan Ereen suite, and the Lower Cretaceous Zerik suite (Khosbayar, 1973; Devyatkin et al., 1975; Graham et al., 1997) (Fig. 2D). Mesozoic strata at Dariv unconformably overlie Paleozoic volcanic rocks (Devyatkin et al., 1975) and are overlain by Cenozoic alluvial deposits. We measured a section at 10 m resolution through the entire Mesozoic sequence (Fig. 2D), and we measured several more detailed sections to better document the sedimentologic character of each of Devyatkin et al.'s (1975) suites (Sjostrom, 1997) (Fig. 3, D, E, and F).

The Jargalant suite is ~200 m thick at Dariv. Devyatkin et al. (1975) reported Lower and Middle Jurassic (pre-Callovian) ostracods and plant fossils from the Jargalant suite that provide the basis for its age assignment. At Dariv, the Jargalant suite consists of interbedded sandstone, siltstone, and mudstone (Fig. 3D). Most sandstone beds are <1 m thick, typically lenticular over several hundred meters, contain erosive bases, and fine upward to siltstone or sandy siltstone. Sandstone is trough cross-bedded with pebbles or granules mantling troughs, and many beds contain rippled tops and soft-sediment deformation. Interbedded siltstone and mudstone is green or gray and contains abundant carbonate concretions and calcified woody debris. The orientation of measured trough axes within the Jargalant suite suggests sediment transport to the east (Fig. 2D).

Gray-green strata of the Jargalant suite grade upsection into redbeds of the Dariv Suite, dated by Khosbayar (1973) as Upper Jurassic (Callovian and Oxfordian) on the basis of fresh-water pelecypods. Graham et al. (1997) reported the remains of a Late Jurassic sauropod (probably *Mamenchisaurus*) and abundant disarticulated dinosaur skeletal debris in redbeds of the Dariv suite. Overall, the Dariv suite fines upward from medium to coarse sandstone interbedded with sandy siltstone to interbedded red siltstone and mudstone (Figs. 2D and 3E). Sandstone beds at the base of the Dariv suite typically scour into underlying units and contain local pebble-mantled trough cross-beds and planar lamination (Fig. 2D). Trough axis orientations suggest transport to the southeast. Sandstone is interbedded with decimeter-scale pedoturbated red sandy siltstone and mudstone containing carbonate concretions and nodular calcisols that increase in abundance upsection (Graham et al., 1997). Two distinct, laterally continuous micritic horizons several meters thick are present near the top of the Dariv suite (Fig. 2D; 2–12 and 15–17 m marks, Fig. 3E).

Fine-grained redbeds of the uppermost Dariv suite are sharply overlain by angular-clast conglomerate of the Ihkes Nuur suite (Fig. 4G). Elsewhere, Khosbayar (1973) dated similar strata as latest Jurassic on the basis of phyllopods and charophytes. The Ihkes Nuur suite consists of ~600 m of interbedded conglomerate and coarse sandstone with local meter-scale boulders that attest to its proximal depositional position (Fig. 2D; Sjostrom, 1997). Bedding is crude and individual conglomerate beds are laterally traceable for a few tens of meters. Imbrication is abundant and indicates transport to the north-northwest. Locally interbedded pebbly sandstone beds are parallel laminated and/or trough cross-bedded.

Coarse-grained strata of the Ihkes Nuur suite are gradationally overlain by shale and siltstone of the Upper Jurassic–Lower Cretaceous Gurvan Ereen suite (Figs. 2D, 3F, and 4H). The Gurvan Ereen suite is ~200 m thick and contains fossil fish, phyllopods, ostracods, mollusks, insects, and a small ornithischian dinosaur that together provide the basis for its age assignment (Khosbayar, 1973; Devyatkin et al., 1975). Red siltstone and shale at the base of the Gurvan Ereen suite (Fig. 4H) is interbedded with an angular-clast conglomerate and sandstone characteristic of the Ihkes Nuur suite (Fig. 3F). Most of the Gurvan Ereen suite consists of fine-grained massive or laminated gray shale and siltstone. Symmetrically rippled, meter-scale sandstone and siltstone occur toward the top of the suite. One distinctively laminated shale occurs near the middle of the Gurvan Ereen suite and contains abundant fish, ostracod, and insect fossils (Figs. 3F and 4I).

The top of the Mesozoic section at Dariv consists of the Zerik suite, which Khosbayar (1973) and Devyatin et al. (1975) reported as Lower Cretaceous (Albian-Aptian) on the basis of the occurrence of fresh-water mollusks, phyllopods, and ostracods. The contact between the Gurvan Ereen suite and the Zerik suite is apparently conformable. The Zerik suite consists predominantly of light gray or yellow siltstone and mudstone with local decimeter- to meter-scale sandstone beds (Figs. 2D, 3F, and 4J). Sandstone contains pebble trains and lenses, meter-scale trough cross-stratification, planar lamination, convoluted bedding, asymmetric ripples, and local desiccation-cracked surfaces. Interbedded siltstone and mudstone is typically laminated or massive. Measured trough axes and ripple foresets from the Zerik suite suggest transport to the north-northeast (Fig. 2D).

Interpretations. Lower to Middle Jurassic through Lower Cretaceous strata exposed at Dariv represent a variety of depositional environments and contain a fundamental change in paleocurrent directions. We interpret strata of the Lower to Middle Jurassic Jargalant suite as moderate- to high-sinuosity mudload-dominated fluvial deposits. This interpretation is based on the presence of lenticular upward-fining sandstone sequences with sharp, erosive bases and common cross-stratification and convoluted bedding, the abundance of interbedded siltstone and shale that typically encases individual sandstone bodies, and the common occurrence of plant debris. The orientation of trough axes measured in the Jargalant suite suggests transport to the east.

Graham et al. (1997) interpreted overlying red sandstone, siltstone, and mudstone of the Upper Jurassic Dariv suite as recording deposition in a shallow, sandy-load dominated fluvial environment that fines upward to a flood-plain or mudflat depositional system. This interpretation was based on the dominance of lenticular, trough cross-bedded sandstone at the base of the suite and the abundance of sandy siltstone and mudstone with minor sandy interbeds and common calcisols near the top of the unit (Fig. 2D). We interpret the two meter-scale micritic carbonate units near the top of the Dariv suite as ephemeral playa lake deposits, and the numerous smaller calcareous nodules and concretions as calcisols. Both features are consistent with the interpretation of a broad alluvial plain.

Conglomerate in the Upper Jurassic–Lower Cretaceous Ihkes Nuur suite indicates an abrupt change in depositional style that likely was related to Mesozoic tectonism. Fine-grained mixed-load fluvial and flood-plain deposits of the upper Dariv suite are locally scoured by overlying coarse-grained bedload-dominated fluvial strata of the Ihkes Nuur suite, although an angular relationship between the two units was not observed. The

very coarse grain size, angularity of clasts (Fig. 4G), lateral discontinuity of individual beds, and abundant imbrication are typical of bedload-dominated fluvial deposits in proximal braided river systems or on the proximal parts of stream-dominated alluvial fans (Nilsen, 1982; DeCelles et al., 1987; Blair and McPherson, 1992; Ridgway and DeCelles, 1993).

Lower Cretaceous deposits of the Gurvan Ereen and Zerik suites mostly suggest a well-oxygenated lacustrine or playa depositional environment with small fringing delta complexes (Graham et al., 1997). Interstratified mud-rich angular-clast conglomerate and ostracod-bearing siltstone and shale at the Ihkes Nuur suite and Gurvan Ereen suite contact (Figs. 2D and 3F) suggest oxic lacustrine deposition adjacent to a topographic high. The presence of pastel colored massive or weakly laminated siltstone and mudstone with local desiccation cracks and a fresh-water fauna in most of the Gurvan Ereen suite indicates deposition in a well-oxygenated lacustrine or playa environment (Fig. 4I). The local occurrence of finely laminated shale with abundant fossils (Figs. 3F and 4J) suggests occasional development of a deeper water, locally anoxic lacustrine environment (e.g., Fouch and Dean, 1982). Interbedded sandstone and siltstone of the Zerik suite is interpreted to reflect small, fringing deltaic deposition, based on the occurrence of trough cross-bedding, symmetric ripples, channelized sandstone bodies, and local desiccation cracks (e.g., Coleman and Prior, 1982).

PROVENANCE ANALYSIS

Sandstone composition

In order to investigate sediment-source rock relations within the study area, we sampled sandstone at stratigraphic intervals of ~100 m from each study location for petrographic analysis. Sampled strata reflect deposition in alluvial, fluvial, and lacustrine systems fed from small tributaries or streams draining Mesozoic mountain ranges in central Asia (first- and second-order streams of Ingersoll et al., 1990). Hence, the sampling scale provided by the sandstone sample suite is probably too local to accurately predict plate tectonic setting (e.g., Dickinson and Suczek, 1979; Dickinson, 1985), but sandstone composition can be used to infer the Mesozoic distribution of detrital source rocks in the study area (e.g., Ingersoll et al., 1993). Along with qualitative observations of conglomerate clast compositions and measurements of paleocurrent indicator directions, sandstone compositional data provide a basis for our interpretations of provenance relations. Except for moderate compositional variation in sandstone from the Dariv locality, the composition of samples within each of the four localities is generally uniform (Sjostrom, 1997). Compositional variation between localities is significant, however, and forms the basis for discussion of sandstone composition in terms of four petrofacies (Fig. 6; Table 2). Each petrofacies consists of samples from one locality and is named for that locality.

Sandstone samples in the Hyargas Nuur petrofacies are characterized by lower QmFLt%Lt, higher QmFLt%F, and higher QmPK%P than other petrofacies (mean $Qm_{35}F_{49}Lt_{16}$, $Qm_{42}P_{37}K_{21}$) (Fig. 6; Table 2). The lithic grain fraction is dominated by Lvm and has lower modal percentages of Lsm than other petrofacies (mean $Qp_{22}Lvm_{63}Lsm_{15}$). The quartzofeldspathic and volcanic-rich sandstone composition is mirrored by the composition of associated conglomerate clasts, composed primarily of granite, andesite, and minor basalt. Percentages of dense minerals, including zircon, pyroxene, and amphibole are high (average 4%; maximum 14%). We interpret the composition of sandstone and conglomerate from Hyargas Nuur to reflect derivation from Cambrian granite, andesite, and mafic volcanic rock of Hanhöhiy Uul, located directly north of the Hyargas Nuur study location (Dergunov et al., 1983; Dergunov and Luvsandanzan, 1984; Yanshin, 1989) (Fig. 1B). This interpretation is supported by the dominance of southwest-directed paleocurrent indicators in the Hyargas Nuur section (Fig. 2A).

Sandstone of the Oshin Nuruu petrofacies is arkosic and lithic-rich (mean $Qm_{37}F_{37}Lt_{26}$) and has higher modal percentages of lithic metamorphic grains $Qp_{24}Lvm_{29}Lsm_{47}$ and mica grains (M = 3%) than any other petrofacies (Fig. 6; Table 2). The abundance of Lm and mica is consistent with observed abundant schist and gneiss pebble and cobble-sized clasts at the Oshin Nuuru locality. We interpret these metamorphic rock fragments and mica grains to have been derived principally from Late Proterozoic or early Paleozoic metamorphic rocks that crop out several km north of the section (Yanshin, 1989). Paleocurrent indicators at Oshin Nuruu are scattered but suggest transport dominantly to the south, consistent with derivation of metamorphic clasts from Vendian-Cambrian sources farther north. In addition to metamorphic clasts, granite and mafic volcanic pebbles and scarce cobbles occur in both Oshin Nuuru measured sections, but were generally minor, consistent with lower percentages of Lvm sandstone grains.

The Jargalant petrofacies is very similar to the Hyargas Nuur petrofacies, except that it is characterized by slightly higher percentages of Qm (mean $Qm_{44}F_{34}Lt_{22}$; $Qm_{57}P_{22}K_{21}$) and about half as much mica (mean = 1.6%). Cobbles and pebbles throughout the Jargalant section are mostly granitic. Some mafic volcanic clasts occur near the base of the section but decrease upsection, mirroring a decrease in QpLvmLsm%Lvm upsection (Sjostrom, 1997). The abundance of granite cobbles and pebbles in the Jargalant section likely reflects the erosion of proximal plutonic source areas. Paleocurrent measurements in the Jargalant section indicate flow dominantly to the southeast and southwest (Fig. 2C), so we interpret the source of coarse granitic detritus to be Middle to Upper Cambrian granite that, along with Riphian to Cambrian metabasalts, composes much of Jargalant Uul (Fig. 1, Yanshin, 1989). Domination of the lithic fraction by Lvm may have resulted from erosion of Cambrian volcanics proximal to Jargalant (e.g., Cambrian metabasalt upon which the Jargalant section rests unconformably) or from older metamorphic sources farther north, as

suggested by Sm-Nd isotopic data (see following). Similarly, the higher percentage of Qm at Jargalant, relative to other areas, may reflect compositional stabilization associated with greater transport distance from source areas north of the locality.

Sandstone from the Dariv locality is characterized by the most lithic-rich composition of all petrofacies (mean $Qm_{31}F_{33}Lt_{36}$) (Fig. 6). Lithic grains are dominated by subequal proportions of Lvm and Lsm grains and contain the lowest modal percentage of Qp of the four petrofacies (mean $Qp_{11}Lvm_{49}Lsm_{40}$). Like the Hyargas Nuur petrofacies, the Dariv petrofacies contains a high percentage of dense accessory minerals (mean D = 4%). Stratigraphic variation in sandstone composition is minor. Of the 30 samples collected from Lower to Middle Jurassic through Lower Cretaceous strata, standard deviations for each suite overlap on QmFLt, QpLvmLsm, and QmPK ternary diagrams (Sjostrom, 1997). The only notable stratigraphic change in sandstone composition is that samples from the upper part of the Ihkes Nuur suite (Upper Jurassic or Lower Cretaceous) and overlying strata have slightly higher QpLvmLsm%Lvm than underlying strata (Sjostrom, 1997). This stratigraphic change is coincident with a major reversal in paleocurrent indicator measurements; paleocurrents from the Jargalant, Dariv, and lower Ihkes Nuur suites trend mainly to the southeast, whereas paleocurrent indicators from the upper Ihkes, Gurvan Ereen, and Zerik suites trend to the north (Fig. 2D). Conglomerate clasts in the uppermost Jurassic or Lower Cretaceous Ihkes Nuur suite at Dariv are polymictic (Fig. 4G). Clast compositions include granite, mafic volcanics, and gneiss.

Both the paleocurrent reversal within the Ihkes Nuur suite (Fig. 2D) and the change in sandstone composition suggests a significant change in source area during latest Jurassic–Early Cretaceous time. According to the regional geologic map compiled by Yanshin (1989), Vendian-Cambrian greenstone, Riphian metamorphic rocks, and Cambrian granite are present a few km north of Dariv. South-trending paleocurrent indicators from the Jargalant, Dariv, and lower Ihkes Nuur suites suggest that these Riphean through Cambrian units were the likely sediment sources for these stratigraphic units. The change to north-trending paleocurrent indicators and more lithic volcanic-rich sandstone compositions within the Ihkes Nuur suite at Dariv suggests a shift in the basin facing direction and unroofing of the paleo-Altai Mountain Range several km southwest of the field location (Fig. 1B). According to Yanshin (1989), the Altai Mountains in the vicinity of Dariv consist mainly of Cambrian felsic volcanic rocks, Silurian and Devonian granite, and minor pre-Riphian(?) metamorphic rocks. We infer these units to be the principal source rocks for sandstone occurring above the lower Ihkes Nuur suite.

Sm-Nd isotope analyses

To complement our studies of sandstone composition, we conducted a limited pilot study of the Sm-Nd isotopic composition of 10 samples from the Jargalant locality. We chose the

Jargalant section (Fig. 1) for Sm-Nd isotopic analysis because at this location the Mesozoic sedimentary section is in depositional contact with the Vendian–Lower Cambrian crystalline basement rocks, allowing for direct isotopic comparison (e.g., Frost and Winston, 1987). In addition, Sm-Nd analysis of Jurassic sandstone samples yielded provenance information that could be directly compared with that from petrographic analysis of the same samples, and isotopic analysis could be performed on Jurassic siltstone that was too fine-grained for petrographic analysis. Our sample suite consisted of two samples of basement rock upon which the Jurassic sedimentary section unconformably rests, two basalt cobbles and two granite cobbles collected from within the Jurassic section, and four samples of Jurassic sandstone and siltstone (Fig. 2C).

It is interesting that basement rock and Jurassic conglomerate clasts yield ε_{Nd} values and depleted mantle derivation ages (T_{DM}) distinctly different from samples of Jurassic sandstone and siltstone (Fig. 7). All three sandstone samples and the siltstone sample have depleted mantle derivation ages between 1.1 and 1.3 Ga and negative $\varepsilon_{Nd}(0)$ values, ranging from −3.8 to −4.8 (Fig. 7A; Table 3). These data suggest derivation from a continental source with components at least as old as 1 Ga (e.g., DePaolo and Wasserburg, 1976; DePaolo, 1988; McLennan et al., 1993). In contrast, two granite cobbles from the Jurassic sedimentary section have (T_{DM}) ages of 610 Ma and 660 Ma and positive $\varepsilon_{Nd}(0)$ values of 1.7 and 2.7 (Fig. 7A; Table 3), suggesting derivation from a younger, perhaps local crust. The two mafic volcanic basement samples and two basaltic cobbles have $^{147}Sm/^{144}Nd$ ratios too similar to depleted mantle for meaningful model age interpretation; however, Sm/Nd data for the basalt cobble and basement samples define an isochron with an age of 610 Ma and an initial ε_{Nd} value of 7.4 (Fig. 7B), which agrees with published ages for the oceanic or subduction related basement of western Mongolia (Şengör et al., 1993; Fedorovskii et al., 1995; Brasier et al., 1996; Vhomentovsky and Gibsher, 1996). The granite cobbles also plot along this 610 Ma isochron and have depleted mantle model ages of 610 Ma and 660 Ma, suggesting that all of these igneous rocks formed from juvenile crust over a short time span.

It is unlikely that the difference in Sm-Nd compositions for sandstone and siltstone versus cobbles and basement samples are an artifact of the small number of cobble and basement samples because conglomerate clast compositions are uniform throughout the Jargalant section (granite dominant with subordinate mafic volcanic rocks) and we sampled each lithology at different places within the section (Fig. 2C). Rather, we suggest that sandstone and siltstone from the Jargalant locality were derived, at least in part, from a different, more distal source area than that of Jargalant cobbles and basement samples. Very few pre-Paleozoic potential source areas are mapped in western Mongolia close to the Jargalant section. Although Yanshin (1989) showed several occurrences of pre-Paleozoic granitic and metamorphic rocks throughout western Mongolia, recent U-Pb, Sm-Nd, and Rb-Sr isotopic results suggest that many of these rocks

are Paleozoic and hence unlikely to have produced the old mantle derivation ages of sandstone and siltstone samples at Jargalant (Fedorovskii et al., 1995). The closest likely source of dated, old, continental-affinity source rocks is ~250 km north of Jargalant, where a series of pre-2.5 Ga greenstones and metagranites crops out in the southern portion of the Siberian craton (Zorin et al., 1993a, 1993b; Fedorovskii et al., 1995). Mixing of detritus eroded from this ancient continental massif with locally derived detritus from Vendian-Cambrian juvenile crust could produce the observed ε_{Nd} values and (T_{DM}) ages for sandstone and siltstone of the Jargalant section. Similar mixed Sm-Nd isotopic compositions resulting from the erosion of multiple sources by well-integrated drainage systems are not uncommon and have been reported by Goldstien et al. (1984), Allègre and Rousseau (1984), Heller et al. (1985), Frost and Winston (1987), Nelson and DePaolo (1988), and Patchett et al. (1999).

Alternatively, sandstone and siltstone of the Jargalant locality may have been derived from erosion of Paleozoic plutonic rock in western Mongolia that incorporated older continental crustal material during its formation. This process would result in (T_{DM}) model ages older than the age of pluton crystallization and lower ε_{Nd} values than expected for depleted mantle-sourced plutonic rock (e.g., Farmer and DePaolo, 1983; Basu et al., 1990; Linn et al., 1991, 1992; Nelson et al., 1993). Plutonic rocks with inherited zircons suggestive of this mixing phenomenon have been reported >100 km southeast of the study area from the Daribi Range of western Mongolia and >500 km northeast of the study area from the Lake Baikal region of Russia (Fedorovskii et al., 1995; Fig. 1), but no evidence for crustal inheritance has been reported closer to the Jargalant locality.

SEDIMENTARY BASIN DEVELOPMENT AND INFERRED TECTONIC SETTING

Sedimentary facies data, sandstone compositions, paleocurrent information, and Sm-Nd isotopic analysis suggest that Lower to Middle Jurassic strata in western Mongolia were deposited in a semicontinuous, northwest-southeast–trending basin that drained axially to the southeast and probably contained several local basement uplifts (Fig. 8A). Lower to Middle Jurassic paleocurrent measurements across the study area suggest flow dominantly to the south. Coarse, synorogenic basin-fringing alluvial facies are preserved at Hyargas Nuur and Jargalant and possibly reflect Early Jurassic faulting along the flanks of adjacent, small mountain ranges. Sandstone and cobble clast compositions from each locality can be related directly to the erosion of local source rocks, but Sm-Nd isotopic compositions of sandstone and siltstone from Jargalant suggest derivation from continental crustal material from the southern margin of the Siberian craton, ~250 km to the north, supporting the interpretation of a large semicontinuous basin.

In contrast to coarse clastic Jurassic strata at Hyargas Nuur and Jargalant, Lower to Middle Jurassic clastic rocks at Oshin Nuruu and Dariv are finer grained. These strata are character-

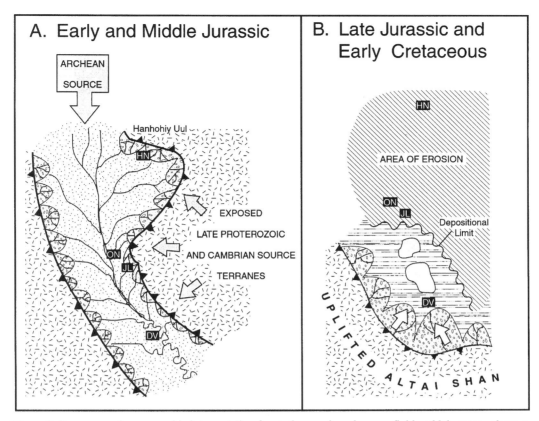

Figure 8. Summary paleogeographic interpretation for study area, based on our field and laboratory observations. A: During Early to Middle Jurassic time, single flexural basin extended across study area and drained largely to south. Early to Middle Jurassic deposition at Hyargas Nuur (HN) and lower portion of Jargalant (JL) section reflect basin-margin positions, whereas coeval deposition at Oshin Nuruu (ON) and Dariv (DV) suggests more basin-axial position. B: By latest Jurassic and earliest Cretaceous time, uplift of paleo-Altai Mountains south of study area produced reversal in paleocurrent directions and coarse clastic sedimentation at Dariv updip of small, oxic lake systems toward basin center. Upper Jurassic or Lower Cretaceous strata were neither encountered nor reported previously from Hyargas Nuur, Oshin Nuruu, or Jargalant localities.

ized by channelized sandstone with local soft-sediment deformation, abundant siltstone and shale, and common woody detritus, all of which suggest mudload-dominated fluvial deposition. Coal-bearing mudload-dominated meandering fluvial strata of Lower to Middle Jurassic age are also present at the top of the Jargalant section, suggesting a decrease in physiographic relief there by that time.

Although Upper Jurassic and Cretaceous strata are preserved only at Dariv, the abundance of redbeds and calcisol horizons in the Dariv suite suggest a change to more arid conditions (cf. Mack et al., 1993) by Late Jurassic time. Hendrix et al. (1992) reported a similar sedimentologic change in Middle to Upper Jurassic strata southwest of western Mongolia in the Junggar and Tarim basins of western China. Perhaps the most striking sedimentologic transition at Dariv, however, is the abrupt change from sandy siltstone and calcisol-bearing mudstone of the Dariv suite to boulder-bearing angular conglomerate of the Ihkes Nuur suite. Approximately coincident with this change is a major reversal of paleocurrent indicators from

southeast to northwest directed and a notable increase in the abundance of lithic volcanic clasts in sandstone of the Ihkes Nuur and younger stratigraphic units (Sjostrom, 1997). We interpret these changes in grain size and texture, paleocurrent directions, and sandstone composition to reflect uplift of the paleo-Altai Mountains, south of the Dariv locality, by latest Jurassic time (Fig. 8B). By Early Cretaceous time, development of an oxic lake system, evidenced by abundant preserved fish (Gurvan Eeren and Zerek suites), at Dariv suggests that the basin may have been more physiographically partitioned following uplift of the paleo-Altai range.

At present, the tectonic setting of Mesozoic sedimentary basins of western Mongolia is equivocal. No structural analyses that document Mesozoic deformation have been reported from the region, and sedimentary data permit interpretations that the basins may have been related to compression, extension, or strike-slip deformation involving basement rock. Coarse-grained, basement-derived strata are common in compressional basins (Heller et al., 1988; Fillmore, 1991; DeCelles and Giles,

1996), extensional basins (Miall, 1981), and strike-slip basins (Long, 1981; May et al., 1993; Ridgway and DeCelles, 1993; Ryang and Chough, 1997). Axial paleocurrent indicator directions are common both in underfilled foreland basins and extensional basins (Burbank et al., 1988; Jordan, 1995). At Jargalant and Oshin Nuuru, Lower to Middle Jurassic strata fine upward, an attribute described in foreland basin settings (Flemings and Jordan, 1990) and extensional basin settings (Chapin and Cather, 1994).

Although various tectonic settings are permitted by the currently available sedimentologic base, several lines of evidence suggest that a rift-related setting was unlikely during Mesozoic time. In contrast to Mesozoic rift basins of southeastern Mongolia (Johnson et el., this volume), basins of western Mongolia apparently lack interstratified volcanics. Thin, alkalic basalts were reported to be interbedded with Upper Jurassic strata in western Mongolia (Zorin et al., 1993a, 1993b; Shuvalov, 1968), but the exact location of these basalts, description of their contact relations, and absolute isotopic ages are not published, so it is impossible to evaluate these reports critically. In addition to the apparent lack of volcanics, we do not favor a rift-related setting because coal we sampled from the Jargalant locality has vitrinite reflectance values of 0.4%, suggesting a low post-Middle Jurassic heat flow more characteristic of flexural basins than rift-related basins (Sclater et al., 1980; Lysak, 1987; Şengör, 1995).

Modern strike-slip deformation is extensive throughout western Mongolia (Baljinnyam et al., 1993). Dobretsov et al. (1995) suggested that these modern structures are reactivated Mesozoic and older strike-slip faults. The role of strike-slip deformation in Mesozoic basins of western Mongolia is difficult to ascertain, but the apparent continuity of Lower to Middle Jurassic meandering fluvial facies across the central and southern portion of the study area suggests the development of a regional basin system that was at that time larger than most strike-slip basins. Strike-slip basins are generally relatively small features, with lateral dimensions of tens of kilometers (e.g., Long, 1981; May et al., 1993; Nilsen and Sylvester, 1995; Ryang and Chough, 1997). Sm-Nd isotopic provenance results from the Jargalant location suggest a possible source region as distant as southern Siberia. If this is the case, the basin in which these strata was deposited would have been hundreds of kilometers long, dimensions typically associated with foreland or extensional basins (Miall, 1981; Chapin and Cather, 1994; Jordan, 1995; Leeder, 1995; DeCelles and Giles, 1996), although a strike-slip basin setting cannot be ruled out by this alone. The relatively local occurrence of Upper Jurassic through Lower Cretaceous strata at Dariv is consistent with latest Jurassic (strike-slip?) partitioning of the more regional Lower to Middle Jurassic basin, although this possibility cannot be verified without additional subsurface data.

In our favored interpretation for the tectonic setting of the field area, Early and Middle Jurassic thrusting along faults bounding the western Mongolian basin(s) uplifted the paleo-Altai and paleo-Hangai Plateau relative to the basin (Fig. 8A).

This deformation produced an elongate flexural basin that drained to the southeast and was most analogous to a broken foreland basin (Dickinson and Snyder, 1978; Jordan et al., 1983). The apparent lack of Upper Jurassic and younger Mesozoic strata across much of the study area and the reversal in paleodispersal directions by Late Jurassic time at Dariv suggest uplift of the paleo-Altai Mountains south of the basin and the establishment of an internally drained lacustrine-dominated system (Fig. 8B).

The sedimentology and provenance of Mesozoic basins of western Mongolia are very similar to those of Mesozoic basins in adjacent northwest China (Hendrix et al., 1992; Hendrix, 2000); many workers have interpreted the basins in northwest China to be compressional and related to subduction and accretion along the southern margin of Asia (e.g., Zhang et al., 1984; Watson et al., 1987; Allen et al., 1991; Vincent and Allen, this volume). By analogy with Mesozoic basins of western China, Traynor and Sladen (1995) interpreted Lower to Middle Jurassic basins of western Mongolia to be compressional in nature. Although compressional Jurassic structures have yet to be identified in western Mongolia, contractile deformation of Jurassic age has been reported from the Junggar basin (Lee, 1985; Li and Jiang, 1987) and interpreted reflection seismic profiles from the Turpan-Hami basin show multiple episodes of contractile Mesozoic deformation (Greene et al., this volume).

CONCLUSIONS

On the basis of our field observations, provenance studies, and integration with regional geologic relations, we conclude the following.

1. Lower to Middle Jurassic strata exposed in western Mongolia record deposition in a variety of nonmarine siliciclastic depositional systems, including stream-dominated coarse alluvial fans, sandy bedload-dominated streams, mudload-dominated meandering streams, and ephemeral lakes. Coarse conglomerate at Hyargas Nuur likely reflects basin-fringing alluvial sedimentation, as does boulder conglomerate at the base of the Jargalant section. Sandy bedload-dominated fluvial to mudload-dominated fluvial deposition with south-directed paleocurrents at Oshin Nuuru, Jargalant, and Dariv likely reflects basin axial drainage. By latest Jurassic–earliest Cretaceous time, coarse, angular, poorly sorted conglomerate was shed into the Dariv locality. By Early Cretaceous time, the Dariv locality was the site of shallow lacustrine sedimentation. The lack of Late Jurassic and younger strata from localities in the central and northern portion of the study area suggests that they underwent erosion or nondeposition during this time.

2. Mesozoic sandstone of western Mongolia is lithic rich with varying proportions of specific grain types that provide the basis for four sandstone petrofacies. Sandstone of the Hyargas Nuur petrofacies is more quartzofeldspathic and has a higher QpLvmLsm%Lvm than other petrofacies (mean

$Qm_{35}F_{49}L_{16}$, $Qp_{22}Lvm_{63}Lsm_{15}$), reflecting erosion of Cambrian granite and volcanic rocks in the Hanhöhiy Uul north of the locality. Sandstone of the Oshin Nuruu petrofacies has higher QpLvmLsm%Lsm and %M than other petrofacies ($Qm_{37}F_{37}L_{26}$, $Qp_{24}Lvm_{29}Lsm_{47}$; M = 3%), reflecting erosion of Vendian-Cambrian metamorphic rocks exposed a few km north of the Oshin Nuuru locality. The Jargalant petrofacies is similar to the Hyargas Nuur petrofacies, but has higher QmFLt%Qm, possibly reflecting compositional stabilization due to greater transport distance. The Dariv petrofacies, which includes Lower Jurassic through Lower Cretaceous samples, is the most lithic rich of all petrofacies ($Qm_{31}F_{33}L_{36}$; $Qp_{10}Lvm_{50}Lsm_{40}$), reflecting the introduction of abundant lithic detritus from the uplifted ancestral Altai Shan south of Dariv by latest Jurassic time.

3. Limited Sm-Nd isotopic analyses suggest that conglomerate clasts and basement rock of the Jargalant Jurassic section were locally derived from and are associated with Vendian–Cambrian subduction or island-arc–affinity rocks. In contrast, Jurassic sandstone and siltstone in this section preserve the signal of input from older continental sources, possibly associated with the Siberian craton.

4. Although sedimentologic and provenance data do not directly define the Mesozoic tectonic setting of western Mongolia, they are consistent with the interpretation of a compressional basin that was broadly similar to a Laramide-style foreland basin. The paucity of Mesozoic volcanics, along with low vitrinite reflectance values for Jurassic coal, suggest a low heat flow uncharacteristic of rift-related basins. The apparent continuity of the basin during Early and Middle Jurassic time suggests that it was larger than most strike-slip basins. By latest Jurassic time, coarse alluvial strata was being deposited at Dariv and much of the remainder of the field area was apparently undergoing erosion. We infer the cause of this fundamental change in the sedimentary fill record to be the uplift of the paleo-Altai Shan to the south and associated strike-slip(?) partitioning of the basin by Late Jurassic time.

ACKNOWLEDGMENTS

This chapter summarizes the results of Sjostrom's Masters thesis, conducted under Hendrix at the University of Montana, Missoula. Support for this work was provided by National Science Foundation grants EAR-9315941 to Graham and EAR-9614555 to Hendrix and Graham. We are grateful for additional support from the American Association of Petroleum Geologists Student Research Fund and the Stanford-Mongolia Industrial Affiliates program, a consortium that has included BHP, Elf-Aquitaine, Exxon, Mobil, Nescor Energy, Occidental, Pecten, and Phillips. We thank E.R. Sobel, W. Kirschner, and B. Hendrix for field assistance. Work in western Mongolia would not have been possible without the generous logistical support from the Mongolian Academy of Sciences Centre for Paleontology.

We are especially grateful to R. Barsbold, T. Chimitsuren, and B. Denzen, and B. Ligden for their field support. We also thank Jerry Hinn for his assistance with the isotopic analyses, and M. Allen, M. Poage, T. Greene, S. Vincent, and R. Cox for their helpful reviews of an earlier version of this manuscript.

REFERENCES CITED

Allègre, C.J., and Rousseau, D., 1984, The growth of the continent through geological time studied by Nd isotope analysis of shales: Earth and Planetary Science Letters, v. 67, p. 19–34.

Allen, M.B., Windley, B.F., Chi, Z., Zhong-Yan, Z., and Guang-Rei, W., 1991, Basin evolution within and adjacent to the Tien Shan Range, NW China: Geological Society of London Journal, v. 148, p. 369–378.

Allen, M.B., Şengör, A.M.C., and Natal'in, B.A., 1995, Junggar, Turfan, and Alakol basins as Late Permian to ?Early Triassic extensional structures in a sinistral shear zone in the Altaid orogenic collage, central Asia: Geological Society of London Journal, v. 152, p. 327–338.

Baljinnyam, I., Bayasgalan, A., Borisov, B.A., Cisternas, A., Dem'yanovich, M.G., Ganbaatar, L., Kochetkov, V.M., Kurushin, R.A., Molnar, P., Philip, H., and Vashchilov, Y.Y., 1993, Ruptures of major earthquakes and active deformation in Mongolia and its surroundings: Geological Society of America Memoir 181, 62 p.

Basu, A.R., Sharma, M., and DeCelles, P.G., 1990, Nd, Sr-isotopic provenance and trace element geochemistry of Amazonian foreland basin fluvial sands, Bolivia and Peru: Implications for ensialic Andean orogeny: Earth and Planetary Science Letters, v. 100, p. 1–17.

Blair, T.C., and McPherson, J.G., 1992, The Trollheim alluvial fan and facies model revisited: Geological Society of America Bulletin, v. 104, p. 762–769.

Brasier, M.D., Dorjnamjaa, D., and Lindsay, J.F., 1996, The Neoproterozoic to Early Cambrian in southwest Mongolia: An introduction: Geological Magazine, v. 133, p. 365–369.

Bridge, J.S., 1984, Large scale facies sequences in alluvial overbank environment: Journal of Sedimentary Petrology, v. 54, p. 583–588.

Bromley, N.H., 1992, Variations in fluvial style as revealed by architectural elements, Kayenta Formation, Mesa Creek, Colorado, USA, *in* Miall, A.D., and Tyler, N., eds., The three-dimensional facies architecture of terriginous clastic sediments and its implications for hydrocarbon discovery and recovery: Society for Sedimentary Geology Concepts in Sedimentology and Paleontology, Series 3, p. 94–102.

Burbank, D.W., Beck, R.A., Raynolds, R.G.H., Hobbs, R., and Tahirkheli, R.A.K., 1988, Thrusting and gravel progradation in foreland basins: A test of post-thrusting gravel dispersal: Geology, v. 16, p. 1143–1146.

Cant, D.J., 1982, Fluvial facies models and their application, *in* Scholle, P.A., and Spearing, D., eds., Sandstone depositional environments: American Association of Petroleum Geologists Memoir 31, p. 115–137.

Carroll, A.R., Graham, S.A., Hendrix, M.S., Ying, D., and Zhou, D., 1995, Late Paleozoic tectonic amalgamation of northwest China: Sedimentary record of the northern Tarim, northwestern Turpan, and southwestern Junggar basins: Geological Society of America Bulletin, v. 107, p. 571–594.

Chapin, C.E., and Cather, S.M., 1994, Tectonic setting of the axial basins of the northern and central Rio Grande rift, *in* Keller, G.R., and Cather, S.M., eds., Basins of the Rio Grande Rift: Structure, stratigraphy, and tectonic setting: Geological Society of America Special Paper 291, p. 5–25

Coleman, J.M., and Prior, D.B., 1982, Deltaic environments of deposition, *in* Scholle, P.A., and Spearing, D., eds., Sandstone depositional environments: American Association of Petroleum Geologists Memoir 31, p. 139–178.

Collinson, J.D., 1986, Alluvial sediments, *in* Reading, H.G., ed., Sedimentary environments and facies: Oxford, Blackwell Scientific Publications, p. 20–62.

Cunningham, W.D., Windley, B.F., Dorjnamjaa, D., Badamgarov, G., and Saandar, M., 1996, A structural transect across the Mongolian Western Altai: Active transpressional mountain building in central Asia: Tectonics, v. 15, p. 142–156.

Davis, R.A., 1983, Depositional systems: London, Prentice-Hall, 669 p.

DeCelles, P.G., and Giles, K.A., 1996, Foreland basin systems: Basin Research, v. 8, p. 105–123.

DeCelles, P.G., Tolson, R.B., Graham, S.A., Smith, G.A., Ingersoll, R.V., White, J., Schmidt, C.J., Rice, R., Moxon, I., Lemke, L., Handschy, J.W., Follo, M.F., Edwards, D.P., Cavazza, W., Caldwell, M., and Bargar, E., 1987, Laramide thrust-generated alluvial fan sedimentation, Sphinx Conglomerate, southwestern Montana: American Association of Petroleum Geologists Bulletin, v. 71, p. 135–155.

DePaolo, D.J., 1988, Neodymium isotope geochemistry, an introduction, minerals and rocks: Berlin, Springer-Verlag, 187 p.

DePaolo, D.J., and Wasserburg, G.J., 1976, Nd isotopic variations and petrogenetic models: Geophysical Research Letters, v. 3, p. 249–252.

DePaolo, D.J., Linn, A.M., and Schubert, G., 1991, The continental crust age distribution: Methods of determining mantle separation ages from Sm-Nd isotopic data and application to the southwestern United States: Journal of Geophysical Research, v. 96, n. B2, p. 2071–2088.

Dergunov, A.B., and Luvsandanzan, B., 1984, Paleotectonic zones and nappe structures of western Mongolia: Geotectonics, v. 18, p. 215–225.

Dergunov, A.B., Luvsandanzan, B., Korobov, M.N., and Kheraskova, T.N., 1983, New data on the stratigraphy of the Vendian and Lower Cambrian deposits of the Khan-Khukhei (Haanhohiy) Range in western Mongolia: Russian Geology and Geophysics, v. 24, no. 3, p. 20–28.

Devyatkin, E.V., Martinson, G.G., Shuwalov, V.F., and Khosbayar, P., 1975, Stratigraphy of Mesozoic deposits of Mongolia, The Joint Soviet-Mongolian Scientific-Research Geological Expedition, Volume 13: Leningrad, Nauka, p. 25–49 (in Russian).

Dickinson, W.R., 1970, Interpreting detrital modes of graywacke and arkose: Journal of Sedimentary Petrology, v. 40, p. 695–707.

Dickinson, W.R., 1985, Interpreting provenance relations from detrital modes of sandstones, *in* Zuffa, G.G., ed., Provenance of arenites: NATO ASI Series C: Mathematical and Physical Sciences, 148, p. 333–361.

Dickinson, W.R., and Snyder, W.S., 1978, Plate tectonics of the Laramide orogeny, *in* Matthews, V., III, ed., Laramide folding associated with basement block faulting in the western United States: Geological Society of America Memoir 151, p. 355–366.

Dickinson, W.R., and Suczek, C.A., 1979, Plate tectonics and sandstone compositions: American Association of Petroleum Geologists Bulletin, v. 63, p. 2164–2182.

Dobretsov, N.L., Berzin, N.A., Buslov, M.M., and Ermikov, V.D., 1995, General aspects of the evolution of the Altai region and the interrelationships between its basement pattern and the neotectonic structural development: Russian Geology and Geophysics, v. 36, no. 10, p. 3–15.

Dubiel, R.F., 1992, Sedimentology and depositional history of the Upper Triassic Chinle Formation in the Unita, Piceance, and Eagle basins, northwestern Colorado and northeastern Utah: U.S. Geological Survey Bulletin 1787-W, 25 p.

Farmer, G.L., and DePaolo, D.J., 1983, Origin of Mesozoic and Tertiary granite in the western United States and implications for pre-Mesozoic crustal structure—I, Nd and Sr isotopic studies in the geocline of the northern Great Basin: Journal of Geophysical Research, v. 88, p. 3379–3401.

Fedorovskii, V.S., Khain, E.V., Vladimirov, A.G., Kargopolov, S.A., Gibsher, A.S., and Izokh, A.E., 1995, Tectonics, metamorphism, and magmatism of collisional zones of the central Asian Caledonides: Geotectonics, v. 29, p. 193–212.

Fillmore, R.P., 1991, Tectonic influence on sedimentation in the southern Sevier foreland, Iron Springs Formation (Upper Cretaceous), southwestern Utah, *in* Nations, J.D., and Eaton, J.G., eds., Stratigraphy, depositional environments, and sedimentary tectonics of the western margin, Cretaceous Interior Seaway: Geological Society of America Special Paper 260, p. 9–25.

Flemings, P.B., and Jordan, T.E., 1990, Stratigraphic modeling of foreland basins: Interpreting thrust deformation and lithosphere rheology: Geology, v. 18, p. 430–434.

Fouch, T.D., and Dean, W.E., 1982, Lacustrine and associated clastic depositional environments, *in* Scholle, P.A., and Spearing, D., eds., Sandstone depositional environments: American Association of Petroleum Geologists Memoir 31, p. 87–114.

Frost, C.D., and Winston, D., 1987, Nd isotopic systematics of coarse and fine grained sediments: Examples from the Middle Proterozoic Belt-Purcell Supergroup: Journal of Geology, v. 95, p. 309–327.

Gazzi, P., 1966, Le arenarie del flysch sopracretaceo dell' Appennino modense; correlaziani con il flysch di Monghidoro: Mineralogica e Petrografica Acta, v. 12, p. 69–97.

Gersib, G.A., and McCabe, P.J., 1981, Continental coal bearing sediments of the Port Hood Formation (Carboniferous), Cape Linzee, Nova Scotia, Canada, *in* Ethridge, F.G., and Flores, R.M., eds., Recent and ancient nonmarine depositional environments: Models for exploration: Society of Economic Paleontologists and Mineralogists Special Publication 31, p. 95–108.

Gloppen, T.G., and Steel, R.J., 1981, The deposits, internal structure and geometry in six alluvial fan-fan delta bodies (Devonian-Norway)—A study in the significance of bedding sequence in conglomerates, *in* Ethridge, F.G., and Flores, R.M., eds., Recent and ancient nonmarine depositional environments: Models for exploration: Tulsa, Society of Economic Paleontologists and Mineralogists Special Publication 31, p. 49–70.

Goldstien, S.L., O'Nions, R.K., and Hamilton, P.J., 1984, A Sm-Nd isotopic study of atmospheric dust and particulates from major river systems: Earth and Planetary Science Letters, v. 70, p. 221–236.

Graham, S.A., Hendrix, M.S., Barsbold, R., Badamgarav, D., Sjostrom, D.J., Kirschner, W., and McIntosh, J.S., 1997, Discovery and occurrence of the oldest known dinosaurs (Late Jurassic) in Mongolia: Palaios, v. 12, p. 292–297.

Heller, P.L., Peterman, Z.E., O'Neil, J.R., and Shafiqullah, M., 1985, Isotopic provenance of sandstones from the Eocene Tyee Formation, Oregon Coast Range: Geological Society of America Bulletin, v. 96, p. 770–780.

Heller, P.L., Angevine, C.L., Winslow, N.S., and Paola, C., 1988, Two-phase stratigraphic model of foreland-basin sequences: Geology, v. 16, p. 501–504.

Hendrix, M.S., 2000, Evolution of Mesozoic sandstone compositions, southern Junggar, northern Tarim, and western Turpan basins, northwest China: A detrital record of the ancestral Tian Shan: Journal of Sedimentary Research, v. 70, no. 3, p. 520–532.

Hendrix, M.S., Graham, S.A., Carroll, A.R., Sobel, E.R., McKnight, C.L., Schulein, B.J., and Wang, Z., 1992, Sedimentary record and climatic implications of recurrent deformation in the Tian Shan: Evidence from Mesozoic strata of the north Tarim, south Junggar, and Turpan basins, northwest China: Geological Society of America Bulletin, v. 104, p. 53–79.

Hendrix, M.S., Brassell, S.C., Carroll, A.R., and Graham, S.A., 1995, Sedimentology, organic geochemistry, and petroleum potential of Jurassic coal measures: Tarim, Junggar, and Turpan basins, northwest China: American Association of Petroleum Geologists Bulletin, v. 79, p. 929–959.

Ingersoll, R.V., Bullard, T.F., Ford, R.L., Grimm, J.P., Pickle, J.D., and Sares, S.W., 1984, The effect of grain size on detrital modes: A test of the Gazzi-Dickinson point-counting method: Journal of Sedimentary Petrology, v. 54, p. 103–116.

Ingersoll, R.V., Cavazza, W., Baldridge, W.S., Shafiqullah, M., 1990, Cenozoic sedimentation and paleotectonics of north-central New Mexico; implications for initiation and evolution of the Rio Grande Rift: Geological Society of America Bulletin, v. 102, p. 1280–1296.

Ingersoll, R.V., Kretchmer, A.G., and Valles, P.K., 1993, The effect of sampling scale on actualistic sandstone petrofacies: Sedimentology, v. 40, p. 937–953.

Jordan, T.E., 1995, Retroarc foreland basins, *in* Busby, C.J., and Ingersoll, R.V., eds., Tectonics of sedimentary basins: Cambridge, Massachusetts, Blackwell Science, p. 331–362.

Jordan, T.E., Isacks, B.L., Allmendinger, R.W., Brewer, J.A., Ramos, V.A., and Ando, C.J., 1983, Andean tectonics related to geometry of subducted Nazca plate: Geological Society of America Bulletin, v. 94, p. 341–361.

Khosbayar, P., 1973, New data on Upper Jurassic and Lower Cretaceous sediments of western Mongolia: Academy of Sciences, USSR, Doklady (Earth Sciences Section), v. 208, p. 115–116.

Lee, K.Y., 1985, Geology of the petroleum and coal deposits in the Junggar (Zhungear) basin, Xinjiang Uygur Zizhiqu, northwest China: U.S. Geological Survey Open-File Report 85-230, 53 p.

Leeder, M.R., 1995, Continental rifts and proto oceanic rift troughs, *in* Busby, C.J., and Ingersoll, R.V., eds., Tectonics of sedimentary basins: Cambridge, Massachusetts, Blackwell Science, p. 119–148.

Li, J., and Jiang, J., 1987, Survey of petroleum geology and the controlling factors for hydrocarbon distribution in the east part of the Junggar basin: Oil and Gas Geology, v. 8, p. 99–107.

Linn, A.M., DePaolo, D.J., and Ingersoll, R.V., 1991, Nd-Sr isotopic provenance analysis of Upper Cretaceous Great Valley fore-arc sandstones: Geology, v. 19, p. 803–806.

Linn, A.M., DePaolo, D.J., and Ingersoll, R.V., 1992, Nd-Sr isotopic, geochemical, and petrographic stratigraphy and paleotectonic analysis: Mesozoic Great Valley forearc sedimentary rock of California: Geological Society of America Bulletin, v. 104, p. 1264–1279.

Long, D.G.F., 1981, Strike-slip fault basins, Canadian Cordillera, *in* Miall, A.D., ed., Sedimentation and tectonics in alluvial basins: Geological Association of Canada Special Paper 23, p. 153–186.

Lysak, S.V., 1987, Terrestrial heat flow of continental rifts: Tectonophysics, v. 143, p. 31–41.

Mack, G.H., James, W.C., and Monger, H.C., 1993, Classification of paleosols: Geological Society of America Bulletin, v. 105, p. 129–136.

May, S.R., Ehman, K.D., Gray, G.G., and Crowell, J.C., 1993, A new angle on the tectonic evolution of the Ridge basin, a "strike-slip" basin in southern California: Geological Society of America Bulletin, v. 105, p. 1357–1372.

McLennan, S.M., Hemming, S., McDaniel, D.K., and Hanson, G.N., 1993, Geochemical approaches to sedimentation, provenance, and tectonics, *in* Johnsson, M.J., and Basu, A., eds., Processes controlling the composition of clastic sediments: Geological Society of America Special Paper 284, p. 21–40.

Miall, A.D., 1981, Alluvial sedimentary basins: Tectonic setting and basin architecture, *in* Miall, A., ed., Sedimentation and tectonics in alluvial basins: Geological Association of Canada Special Paper 23, p. 1–33.

Nelson, B.K., and DePaolo, D.J., 1988, Comparison of isotopic and petrographic provenance indicators in sediments from Tertiary continental basins of New Mexico: Journal of Sedimentary Petrology, v. 58, p. 348–357.

Nelson, B.K., Nelson, S.W., and Till, A.B., 1993, Nd- and Sr-isotope evidence for Proterozoic and Paleozoic crustal evolution in the Brooks Range, northern Alaska: Journal of Geology, v. 101, p. 435–450.

Nilsen, T.H., 1982, Alluvial fan deposits, *in* Scholle, P.A., and Spearing, D., eds., Sandstone depositional environments: American Association of Petroleum Geologists Memoir 31, p. 49–86.

Nilsen, T.H., and Sylvester, A.G., 1995, Strike-slip basins, *in* Busby, C.J., and Ingersoll, R.V., eds., Tectonics of sedimentary basins: Cambridge, Massachusetts, Blackwell Science, p. 425–457.

Patchett, P.J., Roth, M.A., Canale, B.S., de Freitas, T.A., Harrison, J.C., Embry, A.F., and Ross, G.M., 1999, Nd isotopes, geochemistry and constraints on sources of sediments in the Franklinian mobile belt, Arctic Canada: Geological Society of America Bulletin, v. 111, p. 578–589.

Pentilla, W.C., 1994, The recoverable oil and gas resources of Mongolia: Journal of Petroleum Geology, v. 17, p. 89–98.

Ridgway, K.D., and DeCelles, P.G., 1993, Steam-dominated alluvial fan and lacustrine depositional systems in Cenozoic strike-slip basins, Denali fault system, Yukon Territory, Canada: Sedimentology, v. 40, p. 645–666.

Ryang, W.H., and Chough, S.K., 1997, Sequential development of alluvial/lacustrine system: Southeastern Eumsung basin (Cretaceous), Korea: Journal of Sedimentary Research, v. 67, p. 274–285.

Schmitt, J.G., Jones, D.A., and Goldstrand, P.M., 1991, Braided stream deposition and provenance of the Upper Cretaceous–Paleocene(?) Canaan Peak Formation, Sevier foreland basin, southwestern Utah, *in* Nations, J.D., and Eaton, J.G., eds., Stratigraphy, depositional environments, and sedimentary tectonics of the western margin, Cretaceous Western Interior Seaway: Geological Society of America Special Paper 260, p. 27–45.

Schwans, P., 1995, Controls on sequence stacking and fluvial to shallow-marine architecture in a foreland basin, *in* Van Waggoner, J.C., and Bertram, G.T., eds., Sequence stratigraphy of foreland basin deposits: Outcrop and subsurface examples from the Cretaceous of North America: American Association of Petroleum Geologists Memoir 64, p. 55–102.

Sclater, J.G., Jaupart, C., and Galson, D., 1980, The heat flow through oceanic and continental crust and the heat loss of the Earth: Reviews of Geophysics and Space Physics, v. 18, p. 269–311.

Şengör, A.M.C., 1995, Sedimentation and tectonics of fossil rifts, *in* Busby, C.A., and Ingersoll, R.V., eds., Tectonics of sedimentary basins: Cambridge, Massachusetts, Blackwell Science, p. 117.

Şengör, A.M.C., and Natal'in, B.A., 1996, Paleotectonics of Asia: Fragments of a synthesis, *in* Yin, A., and Harrison, T.M., eds., The tectonic evolution of Asia: Cambridge, Cambridge University Press, p. 486–640.

Şengör, A.M.C., Natal'in, B.A., and Burtman, V.S., 1993, Evolution of the Altaid tectonic collage and Paleozoic crustal growth in Eurasia: Nature, v. 364, p. 299–307.

Shuvalov, V.F., 1968, More information about Upper Jurassic and Lower Cretaceous in the southeast Mongolian Altai: Academy of Sciences, USSR, Doklady (Earth Sciences Section), v. 197, p. 31–33.

Shuvalov, V.F., 1969, Continental redbeds of the Upper Jurassic of Mongolia: Academy of Sciences, USSR, Doklady (Earth Sciences Section), v. 189, p. 112–114.

Sjostrom, D.J., 1997, Lower-Middle Jurassic through Lower Cretaceous sedimentology, stratigraphy, and tectonics of western Mongolia [M.S. thesis]: Missoula, University of Montana, 189 p.

Smith, D.G., 1987, Meandering river point bar lithofacies models: Modern and ancient examples compared, *in* Ethridge, F.G., et al., eds., Recent developments in fluvial sedimentology: Society of Economic Paleontologists and Mineralogists Special Publication 39, p. 83–91.

Smith, N.D., 1970, The braided steam depositional environment: Comparison of the Platte River with some Silurian clastic rocks, north-central Appalachians: Geological Society of America Bulletin, v. 81, p. 2993–3014.

Smith, N.D., 1971, Transverse bars and braiding in the lower Platte River, Nebraska: Geological Society of America Bulletin, v. 82, p. 3407–3420.

Smith, N.D., 1972, Some sedimentological aspects of planar cross-stratification in a sandy braided river: Journal of Sedimentary Petrology, v. 42, p. 624–634.

Sobel, E.R., 1999, Basin analysis of the Jurassic–Lower Cretaceous southwest Tarim Basin, northwest China: Geological Society of America Bulletin, v. 111, p. 709–724.

Traynor, J.J., and Sladen, C., 1995, Tectonic and stratigraphic evolution of the Mongolian People's Republic and its influence on hydrocarbon geology and potential: Marine and Petroleum Geology, v. 12, p. 35–52.

Vhomentovsky, V.V., and Gibsher, A.S., 1996, The Neoproterozoic–Lower Cambrian in northern Govi-Altay, western Mongolia: Regional setting, lithostratigraphy and biostratigraphy: Geological Magazine, v. 133, p. 371–390.

Vincent, S.J., and Allen, M.B., 1999, Evolution of the Minle and Chaoshui Basins, China: Implications for Mesozoic strike-slip basin formation in central Asia: Geological Society of America Bulletin, v. 111, p. 725–742.

Watson, M.P., Hayward, A.B., Parkinson, D.N., and Zhang, Z.M., 1987, Plate tectonic history, basin development, and petroleum source rock deposition onshore China: Marine and Petroleum Geology, v. 4, p. 205–225.

Webb, L.E., Graham, S.A., Johnson, C.L., Badarch, G., and Hendrix, M.S., 1999, Occurrence, age, and implications of the Yagan–Onch Hayrhan metamorphic core complex, southern Mongolia: Geology, v. 27, p. 143–146.

Wiltschko, D.V., and Dorr, J.A., 1983, Timing and deformation in Overthrust Belt and foreland of Idaho, Wyoming, and Utah: American Association of Petroleum Geologists Bulletin, v. 67, p. 1304–1322.

Yanshin, A.L., 1989, Map of geological formations of the Mongolian People's Republic: Moscow, Academia Nauka USSR, scale 1:1 500 000.

Zhang, Z.M., Liou, J.G., and Coleman, R.G., 1984, An outline of the plate tectonics of China: Geological Society of America Bulletin, v. 95, p. 295–312.

Zorin, Y.A., Belichenko, V.G., Turutanov, E.K., Kozhenvnikov, V.M., Ruzhentsev, S.V., Dergunov, A.B., Filippova, I.B., Tomurtogoo, O., Arvisbaarar, N., Bayasgalan, T., Biambaa, C., and Khosbayar, P., 1993a, The south Siberia–central Mongolia transect: Tectonophysics, v. 225, p. 361–378.

Zorin, Y.Z., Belichenko, V.G., Turutanov, Y.K., Kozhevnikov, V.M., Ruzhentsev, S.V., Dergunov, A.B., Filippova, I.B., Tomurtogoo, O., Arvisbaator, N., Bayasgalan, T., Byamba, C., and Khosbayar, P., 1993b, The central Siberia-Mongolia transect: Geotectonics, v. 27, p. 103–117.

MANUSCRIPT ACCEPTED BY THE SOCIETY JUNE 5, 2000

Geological Society of America
Memoir 194
2001

Triassic synorogenic sedimentation in southern Mongolia: Early effects of intracontinental deformation

Marc S. Hendrix*
Mary A. Beck*
Department of Geology, University of Montana, Missoula, Montana 59812, USA
Gombosuren Badarch*
Institute of Geology and Mineral Resources, Mongolian Academy of Sciences, Ulaanbaatar, Mongolia 210351
Stephan A. Graham*
Department of Geological and Environmental Sciences, Stanford University, Stanford, California 94305-2115, USA

ABSTRACT

An important record of the earliest phases of intracontinental deformation in southern Mongolia is contained in widespread Upper Permian and Triassic sedimentary strata that crop out in the region. In this chapter, we provide documentation of stratigraphy, sedimentary style, sediment dispersal trends, and provenance relations from Upper Permian through lowermost Jurassic(?) strata at four separate localities in southern Mongolia: (1) Noyon Uul, a doubly-plunging syncline where we measured four separate sections through Permian, Triassic, and lowermost Jurassic(?) strata; (2) Tost Uul, a second doubly-plunging syncline where we measured one section in Permian, Triassic and lowermost Jurassic(?) strata; (3) Chonin Boom, a structurally disrupted series of Triassic strata located adjacent to the Tost fault; and (4) Toroyt, a well-exposed section of dominantly fine-grained strata in the western part of the study area.

Strata of the study area are subdivided into five sedimentary facies on the basis of geometry of coarse clastic units, observed sedimentary structures, and comparison with modern and ancient analogs. These include: (1) gravelly, braided-fluvial facies; (2) alluvial and lake-plain facies; (3) meandering fluvial facies; (4) lake delta and open-lacustrine facies; and (5) profundal lake facies. Sandstone compositions throughout the study area are strongly dominated by lithic volcanic detritus (mean composition $Qm_{18}F_{26}Lt_{56}$; $Qp_6Lvm_{88}Lsm_6$), reflecting the erosion principally of local Paleozoic arc source terranes. Silicified volcanic and chert duraclast lithologies dominate conglomeratic facies across the study area; there are subordinate clasts of pink alkali granite, syenite, diorite, and vein quartz. Clast counts from the Noyon syncline are internally consistent within each synclinal limb but differ between the two limbs, suggesting that each limb received sediment from a different source terrane. Paleocurrent indicators from the north limb of the Noyon syncline record transport dominantly to the south and west, whereas indicators from the south limb record transport dominantly to the north and west, consistent with the interpretation of a broad depositional trough during Triassic time that funneled sediment westward down its axis. Paleocurrent indicators at Tost Uul, located west of Noyon Uul along strike, are dominantly west-directed, consistent with the interpretation of an axial drainage system.

*E-mails: Hendrix, marc@selway.umt.edu; Beck, mbeck@geosc.psu.edu; Badarch, gbadarch@magicnet.mn; Graham, graham@pangea.stanford.edu.

Hendrix, M.S., et al., 2001, Triassic synorogenic sedimentation in southern Mongolia: Early effects of intracontinental deformation, *in* Hendrix, M.S., and Davis, G.A., eds., Paleozoic and Mesozoic tectonic evolution of central Asia: From continental assembly to intracontinental deformation: Boulder, Colorado, Geological Society of America Memoir 194, p. 389–412.

Restoration of 95–125 km of inferred post-depositional left-lateral strike-slip movement on the Tost fault, separating the Chonin Boom and Toroyt localities on the west from the Noyon Uul and Tost Uul localities on the east, brings the Chonin Boom section into subalignment with the Noyon Uul–Tost Uul synclinal pair and places the Toroyt locality farther to the northwest, obliquely down depositional dip. Combining our sedimentary environmental interpretations with this restored paleogeography, we infer that early intracontinental deformation in southern Mongolia produced a broad, east-west (present coordinates) depositional trough that funneled sediment through the Noyon Uul–Tost Uul–Chonin Boom localities to the Toroyt locality to the west. This trough was subsequently flooded in Late Triassic or earliest Jurassic(?) time by a regional lake. This history and style of sedimentation is broadly analogous to coeval Triassic–Jurassic sedimentation along tectonic strike in the Turpan-Hami basin of western China, where axial transport of coarse clastic sediments fed a similar regional lake system. Subsequent continued north-south shortening of southern Mongolia during Middle and Late Jurassic time folded and unroofed strata of the study area, as compression associated with thrusting in the Bei Shan intensified and propagated northward.

INTRODUCTION

Southern Mongolia is a complex collage of Paleozoic volcanic arc terranes, ophiolites, and associated marine strata that were assembled in piecemeal fashion along the active southern margin of paleo-Asia during late Paleozoic time (Lamb and Badarch, 1997; this volume). Following the amalgamation of these tectonic elements and the subsequent accretion of the north China block during Carboniferous–Permian time (Fig. 1A; Şengör et al., 1993), formerly active arc systems in southern Mongolia became extinct and the active southern continental margin of Asia shifted to the south and southeastern edge of the north China block (Zhang et al., 1984; Watson et al., 1987). At that time, southern Mongolia began a protracted period of intraplate deformation that continues today (e.g., Baljinnyam et al., 1993; Cunningham et al., 1995, 1996, 1997).

Built atop the accreted volcanic terranes of southern Mongolia are Mesozoic and Cenozoic sedimentary basins that have resulted from vertical tectonic movements associated with intraplate deformation (Fig. 1B). Because of the combined effects of structural dismemberment, thermal overprinting, and erosion that commonly affect orogenic belts, these sedimentary basins contain important information, not preserved elsewhere, regarding the post-assembly intraplate deformation that affected the region. In southern Mongolia, tectonic interpretations of intraplate deformation based on the analysis of sedimentary basins (e.g., Hendrix et al., 1996; Johnson et al., 1997) complement direct structural and geochronologic analyses of exposed Mesozoic orogenic systems (e.g., Webb et al., 1999; Dumitru and Hendrix, this volume; Johnson et al., this volume) and the documentation of volcanic and plutonic igneous rocks emplaced during intraplate deformation (Kovalenko et al., 1991). Despite the publication of these collective studies and their role in outlining Mesozoic tectonic history in the region, the early stages of intraplate deformation in southern Mongolia remain poorly known.

Key to understanding the history of intracontinental deformation in southern Mongolia are large, well-exposed outcrops of Triassic through lowermost Jurassic strata that appear prominently on

the 1:1 500 000 scale geologic map of Mongolia (Yanshin, 1989). In our initial work in this area (Hendrix et al., 1996) we reported sedimentologic and provenance results from the southern limb of the Noyon Uul syncline, where more than 1.6 km of Permian fine-grained fluvial sandstone and siltstone and more than 4 km of Triassic–lowermost Jurassic fluvial conglomerate, sandstone, siltstone, and laminated organic-rich lacustrine mudstone are exposed. Although the thickness of strata, facies assemblage, and sandstone compositional changes at this locality suggest the influence of vertical tectonic movements, it is critical that the remainder of the outcrop belt in this region be studied before a more complete picture can be assembled and specific hypotheses can be tested relating to the early effects of intracontinental deformation. For example, several isolated outcrops of lower Mesozoic strata occur across southern Mongolia, but at present it is uncertain whether these outcrops are erosional remnants of a single regionally extensive compressional basin (e.g., Hendrix et al., 1996), or whether they reflect deposition in smaller, discrete, possibly strike slip basins (e.g., May et al., 1993). Analysis of Triassic and lowermost Jurassic(?) nonmarine strata in southern Mongolia is also important because it will provide insight into the longer term tectonic history of eastern Asia. In particular, the relationships, if any, between Triassic–lowermost Jurassic(?) synorogenic sedimentation and recently postulated Jurassic(?) strike slip faulting in southern Mongolia (Lamb et al., 1999) are unresolved. Similarly, it is uncertain whether earlier phases of widespread extension during late Mesozoic time in southeastern Mongolia (e.g., Webb et al., 1999; Johnson et al., this volume) may have influenced the deposition of lower Mesozoic strata in the study area.

To address these problems, we undertook a regional study of Triassic sedimentary strata across southern Mongolia, visiting each major outcrop of Triassic sedimentary rocks depicted on the 1:1 500 000 scale geologic map of Mongolia (Yanshin, 1989) (Fig. 1C). Where permitted by exposure and access, we measured sedimentary sections at a resolution of ~10 m in order to establish a regional framework for understanding lower Mesozoic strata across the region, and we measured a number of detailed sections for paleoenvironmental interpretation. We measured the orientation of paleocurrent indicators to document the configuration of paleodispersal systems, and we collected sandstone and conglomerate clast samples for a regional petrologic analysis. Our studies were conducted at four principle localities (Fig. 1C): (1) Noyon Uul, a doubly-plunging syncline where ~4 km of Triassic and lowermost Jurassic strata are exposed (Fig. 1D); (2) Tost Uul, located west of Noyon Uul along strike and consisting of coeval strata of similar thickness, although not as well exposed; (3) Chonin Boom, an isolated series of hills consisting of outcropping Triassic strata; and (4) Toroyt, a series of well exposed Triassic fine-grained deposits located at the western end of the study area. In this chapter, we present our sedimentologic, paleocurrent, and petrologic data from each of these areas. We use our data to define the paleogeography of the region, and subsequently to develop an interpretive model for the early phases of intracontinental deformation in southern Mongolia.

METHODS

Field studies

Much of our field effort was concentrated around the flanks of the Noyon Uul syncline, where we studied four separate sections of Triassic through lowermost Jurassic strata (Fig. 1D). We concentrated our efforts in two canyons that cut transversely through the entire syncline, permitting easy access to well-exposed strata on both limbs: (1) Sain Sar Bulag Canyon, located to the west and named for a perennial water spring in the canyon, and (2) Goyot Canyon, located to the east and named after the Mongolian word for poplar tree. Herein, we present our observations for the north and south Sain Sar Bulag sections, located on the north and south limbs of the syncline, respectively, and the north and south Goyot Canyon sections, similarly located. Hendrix et al., (1996) reported reconnaissance sedimentologic and stratigraphic results from the lower part of the Sain Sar Bulag section, as well as data from a third transect across the syncline west of Sain Sar Bulag Canyon (their western section; Fig. 1D).

During the summer of 1997, we measured each of the stratigraphic sections at Noyon Uul and one section along the south flank of the Tost Uul syncline at a resolution of 10 m (Figs. 2 and 3). In addition, we measured smaller stratigraphic sections at submeter resolution and photographed important sedimentary structures (Fig. 4) to provide documentation of sedimentation style and a basis for paleoenvironmental interpretation. We collected paleocurrent measurements on trough and tabular crossbeds, bar sets, pebble imbrications, aligned wood fragments, parting lineations, and uncommon flute and gutter casts throughout each section. Restoration of paleocurrent data from tilted beds and treatment of planar data from trough cross-bed limbs followed the methods of DeCelles et al., (1983). The Triassic section at Chonin Boom was structurally dismembered and poorly exposed, permitting only qualitative observations

as to sedimentary style. At Toroyt, we measured several small sections at submeter resolution in order to provide a basis for environmental interpretation of Triassic strata at that locality (Fig. 5).

Provenance studies

At each locality we collected sandstone samples for compositional analysis, and we conducted clast composition counts from conglomerate units. We collected sandstone samples every 100–300 m of stratigraphic section at Noyon and Tost Uul and collected conglomerate clast samples every 400–500 m in conglomeratic parts of each section (Figs. 2 and 3). We also collected samples of sandstone and/or conglomerate at the Chonin Boom, Toroyt, Har Dell, and Gorwantes localities (Fig. 1C) to compare with compositional data from Noyon and Tost Uul. We grouped and counted conglomerate pebbles for provenance analysis on the basis of lithology. The number of individual conglomerate clasts identified per sample ranged from 108 to 359. We recognized and tabulated the following clast types: granite and/or granodiorite, syenite, diorite, volcanic porphyry with visible quartz, volcanic porphyry with no visible quartz but with visible K-spar, andesite, vein quartz, metasedimentary, including argillite and chert, agate, breccia, foliated metamorphic rock, and unknown. Because of the procedural difficulty distinguishing between silicified metasiltstone, argillite, and chert pebbles, we elected to combine these clast with metasedimentary clasts, recognizing that some of the chert clasts may be silicified tuff. Normalized conglomerate clast percentages were plotted on histograms.

Thin sections of sandstone samples collected for provenance analysis were stained for both potassium feldspar and plagioclase with cobaltinitrate and alizarin red stains, respectively (Laniz et al., 1964). All samples were point-counted using a modified Gazzi (1966)-Dickinson (1970) method (Table 1; Ingersoll et al., 1984). At least 500 individual grains were identified and tabulated per sample. Raw point-count data (Table 1) were recalculated into detrital modes following the methods of Ingersoll et al., (1984) (Table 2) and plotted on Qm-F-Lt, Qp-Lvm-Lsm, and Qm-P-K ternary diagrams.

AGE CONTROL

Of the four localities, only strata at Noyon Uul have been dated directly. Strata at Tost Uul, Chonin Boom, and Toroyt are correlated to strata exposed at Noyon Uul principally on the basis of similar lithostratigraphy. Locally, sedimentary strata at Noyon Uul unconformably overlie Carboniferous basalt and andesite porphyritic tuff (Zaitsev et al., 1973). In the eastern part of the field area, Upper Permian sedimentary strata overlie a suite of bimodal volcanics radiometrically dated as Early Permian (Anatol'eva, 1974; Yarmoliuk, 1978; Yarmulyuk and Tikhonov, 1982). Where we observed the base of the Permian sedimentary section, it was in fault contact with older, silicified volcanics.

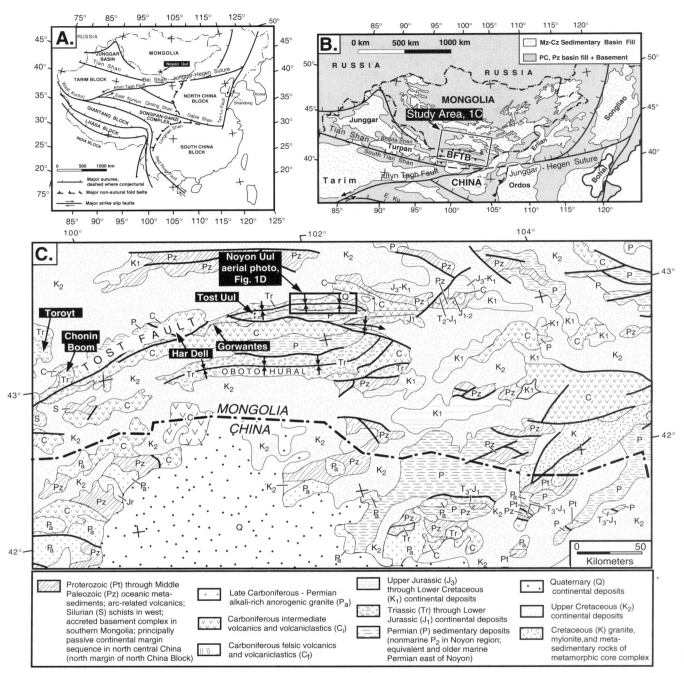

Figure 1. A: Simplified tectonic map of Asia, showing major features discussed in text. B: Map of major Mesozoic–Cenozoic sedimentary basins and structural features in central and eastern Asia. Study area in southern Mongolia (highlighted by box) is shown in detail in C. BFTB = Bei Shan fold and thrust belt. C: Geologic map of southern Mongolia and adjacent parts of north-central China showing major study locations discussed in this chapter. Note location of D shown by box. Simplified from Chen (1985), Yanshin (1989), Zhou et al. (1989), Bureau of Geology and Mineral Resources (1991), and Zheng et al. (1991). D (facing page): Aerial photograph mosaic of Noyon Uul syncline showing four transects discussed in this chapter, in addition to western transect of Hendrix et al. (1996). White dots are part of original referencing system for each photograph in mosaic and should be ignored.

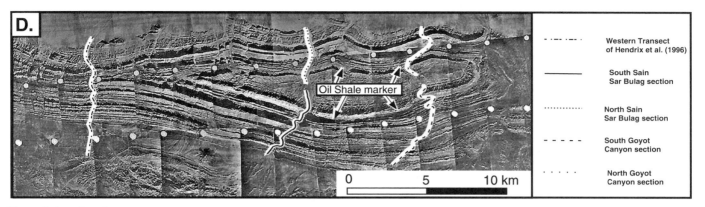

Figure 1. (*continued*)

Fine-grained strata below the first major conglomerate beds of the Noyon Uul syncline were dated by Zaitsev et al., (1973) as Late Permian based on floral assemblages, although Gubin and Sinitza (1993) suggested that the uppermost 300 m of fine-grained strata below the first major conglomerate ridge are Early Triassic in age, based on the discovery of the tetrapod *Lystrosaurus hedini*. Strata above the first major ridge of conglomerate at Noyon Uul have been dated using various plant assemblages, gastropods, insects, and crustacean fossils (Zaitsev et al., 1973; Badamgarav, 1985; Gubin and Sinitza, 1993). Using floral data, Zaitsev et al., (1973) determined a Middle and Late Triassic age and Gubin and Sinitza (1993) suggested a Ladinian–Carnian age for these strata. Zaitsev et al., (1973) also suggested that strata in the core of the syncline may be of Early Jurassic age, although they did not report any Early Jurassic index fossils. In this chapter, we adopt the position of these previous paleontologic studies and provisionally refer to fine-grained strata lower than 300 m below the basal conglomerate as Upper Permian and strata above this point as Triassic through lowermost Jurassic(?).

FACIES ANALYSIS

We recognize five distinct facies assemblages in Permian, Triassic, and Lower Jurassic(?) strata of the study area: (1) gravel-rich braided-fluvial deposits; (2) meandering fluvial and/or flood-plain facies; (3) lake delta facies; (4) shallow, well-oxygenated lake facies; and (5) profundal lacustrine facies. In the following text we summarize our descriptions of each of these facies and provide the basis for their interpretation.

Gravel-bearing braided-fluvial strata

Pebble to fine cobble conglomeratic strata are very common in each of the Noyon Uul transects, at Tost Uul, and at Chonin Boom (Figs. 2 and 3). Local outcrops of conglomeratic strata also were observed at the base of the Toroyt section. Individual beds of conglomerate range from 1 to 25 m thick and are typically lenticular on a scale of hundreds of meters. Most con-

glomeratic intervals have a scoured base and many consist of stacked lenses of pebbly conglomerate. Decimeter-scale interbedded lenses of massive, horizontally laminated, or trough cross-bedded sandstone are common. Conglomerate clasts are grain supported and are well rounded; muddy matrix-supported conglomerate beds were not observed. Sedimentary structures are abundant within the conglomerate and include imbrication (particularly at the base of conglomerate beds), planar stratification, and meter-scale trough and tabular cross-stratification. Of particular note are large tabular cross-beds with as much as 15 m of relief (Fig. 4A). Primary depositional dip on these cross-beds is typically ~20°–30°. Wood impressions are common in conglomeratic strata, especially on bedding surfaces of interstratified sandstone intervals.

We interpret gravel-bearing strata from the field area principally as representing coarse-grained deposition in a braided-fluvial environment. The lenticular shape of conglomeratic units probably reflects channel-filling events during low water and the common cut-and-fill patterns reflect the ephemeral nature of broad shallow channels within the stream system (Miall, 1992). We interpret the large-scale tabular cross-beds as downstream-accreting bar foresets (e.g., Siegenthaler and Huggenberger, 1993; Miall, 1996). Their high primary depositional dip angle suggests that they are not lateral accretion surfaces, as does the fact that most dip in the direction of transport recorded by other paleocurrent indicators. Although it is possible that some are Gilbert-delta foresets, we did not observe topset or bottomset strata, nor did we observe lake shoreline features on any of the sets, as Boorsma (1992) reported for Gilbert-deltaic strata from the Neogene of Spain.

Meandering fluvial and flood-plain facies

Strata interpreted to record meandering fluvial deposition were observed in all Noyon Uul sections and at Tost Uul (Figs. 2 and 3). This facies is characterized by interstratified sandstone, siltstone, and local shale (Fig. 4B). Most sandstone units >1 m thick are lenticular, sharp based, and upward fining. These sandstone beds commonly are trough cross-stratified at

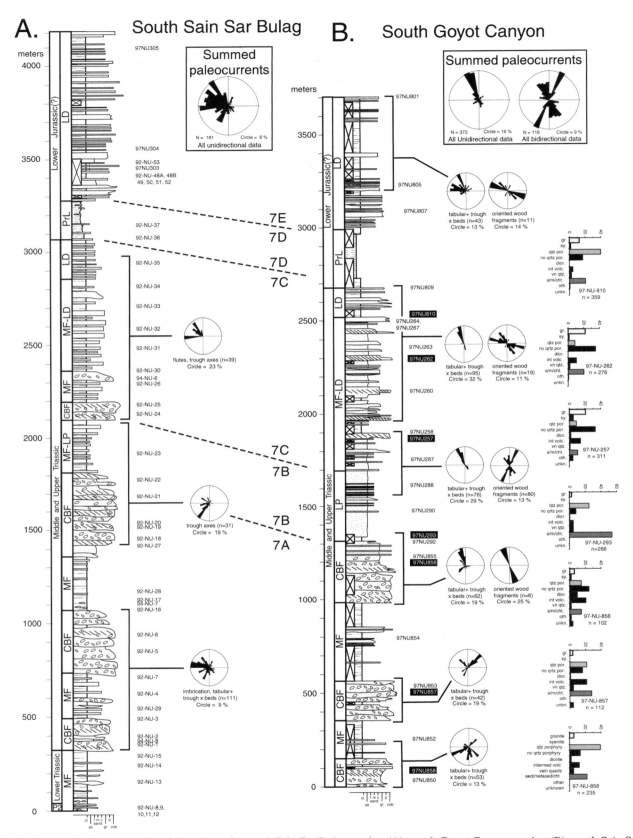

Figure 2. Simplified measured stratigraphic sections for south Sain Sar Bulag section (A); south Goyot Canyon section (B); north Sain Sar Bulag section (C); and north Goyot Canyon section (D), Noyon Uul syncline. Location of each section is shown in Figure 1D. Age, facies, and grain-size abbreviations are as follows: P2, Upper Permian; T1, Lower Triassic; CBF, coarse braided fluvial; MF, meandering fluvial; LP, lake plain; AP, alluvial plain; LD, lake delta; PrL, profundal lake facies; cl, clay; slt, silt; f, fine; m, medium; c, coarse; gr, granules; p, pebbles; cob, cobbles. Note stratigraphic placement of sandstone and conglomerate samples collected for provenance analysis. Rose diagrams to right of each

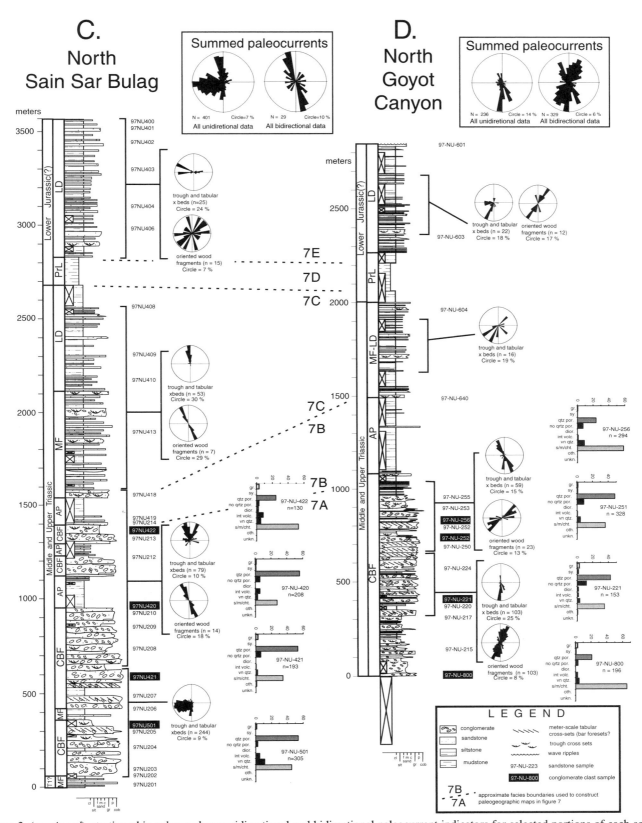

Figure 2. (*continued*) stratigraphic column show unidirectional and bidirectional paleocurrent indicators for selected portions of each section. Summed paleocurrent indicator data for each section are shown at top of each column. Conglomerate clast-count data are at far right side of each column, except for clast data from south Sain Sar Bulag, which were presented in Hendrix et al. (1996). Key to clast category abbreviations is shown in histogram for sample 97-NU-858, lower right corner of B.

Tost Uul Section

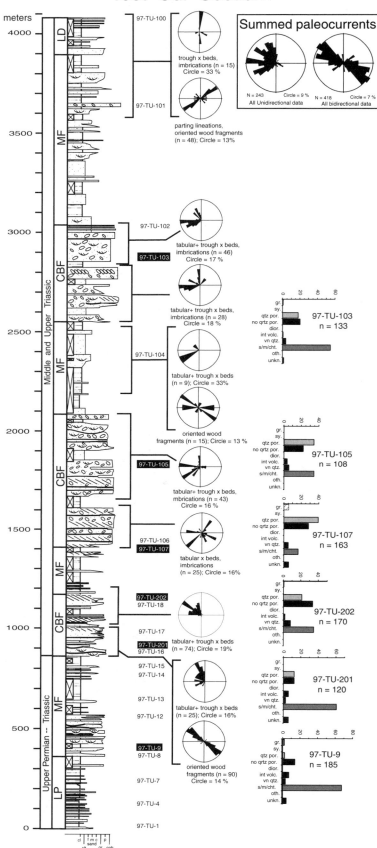

Figure 3. Simplified measured stratigraphic sections for Tost Uul stratigraphic section. Note stratigraphic placement of sandstone and conglomerate samples collected for provenance analysis. Rose diagrams to right of each stratigraphic column show unidirectional (left) and bidirectional (right) paleocurrent indicators for selected portions of each section. Summed paleocurrent indicator data for each section are also shown at top of each column. Symbols for lithology and sedimentary structures and all abbreviations are as for Figure 2.

Figure 4. A: Large tabular foresets in coarse pebble conglomerate, north Goyot Canyon section, Noyon Uul. Person for scale (circled). B: Outcrop style of meandering fluvial deposits, south Goyot Canyon, Noyon Uul. Scale bar in lower right portion of photograph is 10 m long. C: Slumped sandstone in lake-delta facies, South Goyot Canyon, Noyon Uul. Person for scale. (Fig. 4D–4G: see page 398.) D: Close-up of soft-sediment deformation, lake-delta facies, Goyot Canyon, Noyon Uul. Hammer for scale. E: Fresh-water bivalve coquina, north Sain Sar Bulag section, Noyon Uul. Finger pointing to individual mollusc. F: Outcrop style of Triassic–Lower Jurassic(?) strata at Toroyt interpreted to reflect deposition in regional lake. Scale bar in lower right is ~5 m. G: Millimeter-scale lamination, profundal lake facies, south Sain Sar Bulag section, Noyon Uul. Hammer head for scale.

the base, and pebble clasts and/or intraclasts commonly mantle cross-beds. Horizontal lamination and asymmetric ripple cross-lamination are common at the top of sandstone units. Some upward-fining sandstone packages contain internal bedding surfaces that dip <10°. In addition to these thicker sandstone units, this facies contains decimeter-scale tabular sandstone beds interstratified with siltstone or shale. These sandstone layers are usually sharp based, fine upward slightly, and commonly contain planar stratification and parting lineation in the lower portions and asymmetric climbing ripples in the upper portions.

Wood impressions and coaly organic matter are ubiquitous in this facies and are particularly notable on sandstone and siltstone bedding surfaces. Fresh-water molluscs are present locally. Much of this facies consists of gray-green siltstone with local coaly beds or red siltstone containing carbonate nodules. These lithologies are particularly common in Permian strata within the south Sain Sar Bulag and south Goyot Canyon sections at Noyon Uul and at Tost Uul.

Our interpretation of this facies as recording deposition in meandering fluvial environments is based on the lithologic

Figure 4. (*continued*)

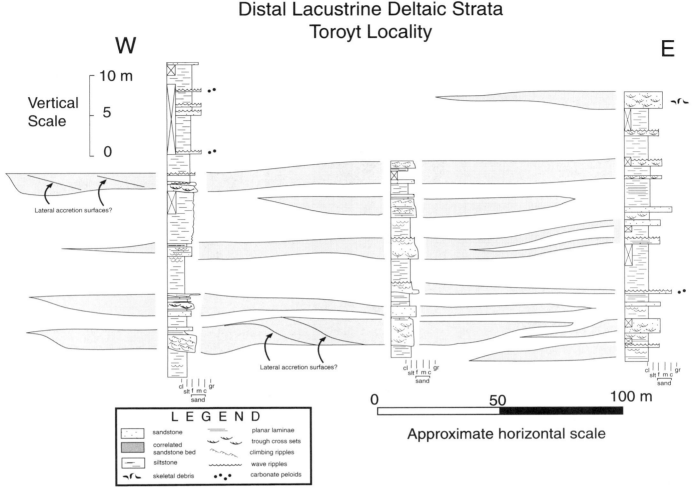

Figure 5. Measured stratigraphic sections and correlated sandstone bodies in canyon wall, Toroyt locality. This cross section was originally mapped on photomosaic. These strata are interpreted to have been deposited on distal end of lake delta. Most sandstone bodies are interpreted to represent distributary channel deposits. Grain-size abbreviations are explained in caption for Figure 2.

assemblage, the geometry and structure of sandstone bodies, and the presence of specific sedimentary structures. We interpret upward-fining lenticular sandstone units principally as representing lateral accretion on or within a point bar, as has been documented from numerous modern and ancient meandering fluvial systems (cf. Miall, 1996). Gently dipping internal bedding planes are interpreted as lateral accretion surfaces. We interpret decimeter-scale tabular sandstone beds interbedded with siltstone or shale and containing planar lamination and asymmetric climbing ripples as crevasse-splay deposits, and we interpret gray-green siltstone and shale containing local carbonate concretions and coaly horizons as overbank deposits. The abundance of wood impressions and coaly organic matter in this facies is consistent with the presence of bank-stabilizing plants (e.g., Smith, 1976) that may have contributed to channel sinuosity (Miall, 1996).

Alluvial and/or lake plain facies

Thick units of siltstone and shale are common in Triassic strata at Noyon Uul, where they form strike valleys hundreds of meters wide within the syncline. On the northern limb of the Noyon syncline, shale and siltstone facies range from brown to brick-red with local interbeds of gray-green siltstone and shale containing rare, centimeter-thick layers of carbonaceous mudstone and coal. Centimeter-scale tabular bedding is characteristic. Calcareous nodules and concretions are common, as are local, decimeter-scale interbeds of sandstone and uncommon pebble conglomerate. Thin sandstone and conglomerate layers are sharp based and appear to be lenticular over several hundred meters. On the south limb of the syncline, packages of siltstone and shale forming wide strike valleys are gray-green and somewhat coarser than equivalent strata on the northern limb of the syncline. Calcareous nodules are common.

TABLE 1. SANDSTONE POINT COUNT RAW DATA

Sample #	Age	Lat (°-'N)	Long (°-'E)	Qm	Qp	Cht	K	P	Lvm	Ls	Lc	Lm	Lu	bt	ms	chl	D	cem	matr	por	gran frags	rad cht	n
South Sain Sar Bulag Transect																							
97-NU-304	J1?	43-15	101-55	44	3	1	104	71	174	1	0	3	2	2	0	0	0	78	17	0	Y	0	500
97-NU-305	J1?	43-15	101-55	18	1	1	27	95	204	0	0	2	3	0	0	0	8	98	43	0	Y	0	500
South Goyot Canyon Transect																							
97-NU-850	Tr	43-13	102-01	166	21	1	75	82	122	1	0	6	2	4	0	0	3	15	2	0	Y	0	500
97-NU-852	Tr	43-13	102-01	143	20	17	0	47	161	4	0	1	3	0	0	0	0	72	32	0	Y	0	500
97-NU-853	Tr	43-13	102-01	60	11	3	39	57	163	1	0	3	14	1	0	0	0	148	0	0	N	0	500
97-NU-854	Tr	43-13	102-01	47	7	10	26	98	223	2	0	6	3	4	0	0	1	47	26	0	Y	0	500
97-NU-855	Tr	43-13	102-01	119	10	7	51	57	195	0	0	3	3	0	0	0	2	25	28	0	Y	0	500
97-NU-292	Tr	43-13	102-01	70	10	3	51	56	180	2	0	6	3	0	0	0	6	62	51	0	Y	0	500
97-NU-290	Tr	43-13	102-01	64	5	2	25	80	221	1	0	5	2	8	0	0	6	14	67	0	Y	0	500
97-NU-288	Tr	43-13	102-01	46	0	0	18	198	154	0	0	2	10	4	1	0	15	24	28	0	N	0	500
97-NU-287	Tr	43-13	102-01	41	4	1	0	156	171	0	0	5	4	6	1	1	16	57	37	0	Y	0	500
97-NU-258	Tr	43-13	102-01	26	5	3	31	80	229	2	0	14	4	3	0	0	4	75	24	0	Y	0	500
97-NU-260	Tr	43-13	102-01	55	11	2	48	56	199	0	0	5	2	3	0	0	13	46	60	0	Y	0	500
97-NU-263	Tr	43-14	102-01	37	7	0	58	136	155	1	0	8	4	4	0	0	8	52	30	0	Y	0	500
97-NU-264	Tr	43-14	102-01	116	4	0	43	93	194	4	0	10	6	1	0	0	5	14	0	10	N	0	500
97-NU-267	Tr	43-14	102-01	116	3	2	76	105	144	0	0	8	3	1	0	0	4	8	1	28	N	1	499
97-NU-809	J1?	43-14	102-01	93	1	0	65	126	158	0	0	6	3	4	0	0	13	27	4	0	N	0	500
97-NU-807	J1?	43-15	102-02	58	2	0	76	94	170	0	0	6	2	2	0	0	1	68	21	0	Y	0	500
97-NU-805	J1?	43-15	102-02	77	4	0	62	74	199	2	0	8	1	2	0	0	1	67	3	0	N	0	500
97-NU-801	J1?	43-15	102-02	111	2	1	36	102	178	0	0	6	3	4	0	0	4	26	0	27	N	0	500
North Sain Sar Bulag Transect																							
97-NU-201	Tr	43-17	101-56	12	7	35	1	44	361	0	0	1	2	0	0	0	10	9	18	0	Y	O	500
97-NU-202	Tr	43-17	101-56	9	8	8	0	12	188	28	0	15	6	0	0	0	10	202	3	11	N	O	500
97-NU-203	Tr	43-17	101-56	175	12	10	15	34	186	4	0	6	6	2	0	0	0	47	3	0	N	O	500
97-NU-204	Tr	43-17	101-56	150	12	8	0	36	208	2	0	2	3	0	0	0	1	68	10	0	Y	O	500
97-NU-205	Tr	43-17	101-56	115	13	2	54	28	167	0	0	10	3	1	0	1	12	94	0	0	N	O	500
97-NU-206	Tr	43-17	101-56	85	20	12	61	84	152	5	0	5	5	0	0	0	2	46	23	0	Y	O	500
97-NU-207	Tr	43-17	101-56	105	10	2	40	27	214	30	0	10	4	5	0	0	0	36	17	0	N	O	500
97-NU-208	Tr	43-17	101-56	95	11	0	13	24	227	0	0	8	4	4	0	0	10	41	63	0	N	O	500
97-NU-209	Tr	43-17	101-56	96	6	1	18	38	296	2	0	4	2	0	0	1	4	30	2	0	N	O	500
97-NU-210	Tr	43-17	101-56	94	11	12	64	20	183	3	0	1	0	0	0	0	4	25	83	0	Y	O	500
97-NU-212	Tr	43-17	101-56	85	9	2	23	84	255	1	0	1	0	0	0	0	18	16	5	1	N	O	500
97-NU-213	Tr	43-16	101-56	109	12	2	64	47	206	1	0	6	1	1	0	0	3	13	35	0	Y	O	500
97-NU-214	Tr	43-16	101-56	62	30	2	48	72	224	0	0	8	2	1	0	0	6	9	36	0	Y	O	500
97-NU-419	Tr	43-16	101-56	70	5	0	51	94	185	0	0	10	5	3	0	0	20	39	12	6	N	O	500
97-NU-418	Tr	43-16	101-56	59	11	4	55	60	160	0	0	3	3	2	0	0	6	73	64	0	Y	O	500
97-NU-413	Tr	43-16	101-56	21	3	0	0	111	237	0	0	0	3	1	0	0	18	63	43	0	Y	O	500
97-NU-410	Tr	43-16	101-56	80	0	1	22	172	123	1	0	7	10	0	0	0	37	15	28	4	N	O	500
97-NU-409	Tr	43-16	101-56	35	6	4	27	69	248	0	0	7	1	1	0	0	13	54	35	0	Y	O	500
97-NU-408	Tr	43-16	101-56	29	6	2	41	80	201	1	0	5	5	5	0	0	1	106	18	0	Y	O	500
97-NU-406	J1?	43-15	101-56	53	6	1	98	82	179	0	0	5	1	1	0	0	0	49	25	0	Y	O	500
97-NU-404	J1?	43-15	101-56	60	1	0	18	123	213	0	0	9	2	2	0	0	12	35	24	1	N	O	500
97-NU-403	J1?	43-15	101-56	98	1	0	42	131	190	0	0	6	1	0	0	0	7	10	7	7	N	O	500
97-NU-402	J1?	43-15	101-56	121	3	0	25	104	214	1	0	7	0	2	0	0	6	7	2	8	N	O	500
97-NU-401	J1?	43-15	101-56	97	5	0	42	129	164	1	0	8	3	3	0	0	14	7	15	12	N	O	500
97-NU-400	J1?	43-15	101-56	90	4	0	46	92	196	0	0	7	0	0	0	0	5	31	6	23	n	O	500

We interpret these thick packages of fine-grained strata to represent deposition at the distal fringes of an alluvial plain. We infer thin sandstone and pebble conglomerate layers to reflect periodic sheet-flood deposits and abundant carbonate nodules and concretions to reflect calcisol development (cf. Mack et al., 1993), although it is possible that some of the calcareous nodules may have been precipitated by groundwater. Although carbonaceous shale and thin coal layers are uncommon, they are consistent with a rise in the local water table and associated preservation of organic matter.

Lacustrine delta and shallow oxic lake facies

Strata interpreted to reflect deltaic deposition in well-oxygenated lakes were observed in each of the sections at Noyon Uul and at Toroyt. At Noyon, this facies is characterized by lenticular sandstone with soft-sediment deformation, commonly of large amplitude (Fig. 4, C and D). Deformed sandstone is interstratified with gray-green laminated siltstone and shale with abundant woody debris on bedding planes and local freshwater molluscs. Submeter-scale fine sandstone and silicified siltstone beds with abundant climbing ripples and local soft-sediment deformation are also present. Wave ripples cap these sandstone beds locally.

Although a gradation exists between meandering fluvial and lake-delta deposits at Noyon, several criteria support a deltaic interpretation for parts of the section. Meter-scale, soft-sediment–deformed sandstone units characteristic of this facies (Fig. 4, C and D) are inferred to be prograding lobe deposits, as Castle (1990) described from the Douglas Creek Member of the Green River Formation in Utah. Thinner, fine-grained sand-

TABLE 1. SANDSTONE POINT COUNT RAW DATA (continued)

Sample #	Age	Lat (°-'N)	Long (°-'E)	Qm	Qp	Cht	K	P	Lvm	Ls	Lc	Lm	Lu	bt	ms	chl	D	cem	matr	por	gran frags	rad cht	n
North Goyot Canyon Transect																							
97-NU-215	Tr	43-17	102-02	141	5	0	18	19	170	0	3	1	4	36	0	0	4	66	33	0	N	0	500
97-NU-217	Tr	43-17	102-02	118	13	10	41	24	166	11	0	5	3	1	0	0	2	93	13	0	Y	0	500
97-NU-220	Tr	43-17	102-02	123	13	12	25	44	169	6	0	1	5	2	0	0	2	98	0	0	Y	0	500
97-NU-224	Tr	43-17	102-02	71	9	5	69	50	215	3	0	3	3	5	0	0	1	13	53	0	Y	0	500
97-NU-250	Tr	43-17	102-02	93	20	7	45	47	213	0	0	3	4	6	0	0	3	22	37	0	Y	0	500
97-NU-252	Tr	43-17	102-02	53	9	11	42	39	272	1	0	0	2	1	0	0	11	19	40	0	Y	0	500
97-NU-253	Tr	43-17	102-02	71	18	1	30	65	199	3	0	7	1	1	0	0	2	34	68	0	Y	0	500
97-NU-255	Tr	43-17	102-02	60	13	1	57	85	206	3	0	7	2	1	0	0	2	53	10	0	Y	0	500
97-NU-640	Tr	43-16	102-01	44	11	4	33	66	197	1	0	10	4	7	0	0	19	24	80	0	Y	0	500
97-NU-604	Tr	43-16	102-01	72	11	3	30	79	207	0	0	1	4	0	0	0	20	59	14	0	Y	0	500
97-NU-603	J1?	43-16	102-01	50	1	1	77	81	162	0	0	3	2	0	0	0	12	90	21	0	Y	0	500
97-NU-601	J1?	43-15	102-01	42	3	0	44	105	177	0	0	0	5	2	0	0	7	56	59	0	Y	0	500
Tost Uul Transect																							
97-TU-1	P2(?)	43-14	101-33	28	8	1	0	81	289	4	0	4	22	1	0	0	0	48	14	0	N	0	500
97-TU-4	P2(?)	43-14	101-32	38	15	1	0	54	294	8	0	3	19	1	0	0	1	44	22	0	N	0	500
97-TU-7	P2(?)	43-14	101-32	23	3	1	0	70	294	10	0	1	26	0	0	0	1	46	25	0	N	0	500
97-TU-8	P2(?)	43-14	101-32	25	4	12	0	46	315	10	0	2	12	1	0	0	0	66	4	1	N	1	498
97-TU-12	P2(?)	43-14	101-32	8	1	10	0	36	315	12	0	0	22	0	0	0	0	92	2	0	N	2	498
97-TU-13	P2(?)	43-14	101-32	21	3	6	0	80	316	4	0	0	6	0	0	0	0	58	5	0	N	1	499
97-TU-14	P2(?)	43-14	101-32	16	8	12	0	67	284	32	0	1	13	0	0	1	1	54	11	0	N	0	500
97-TU-15	P2(?)	43-15	101-32	25	9	7	0	50	285	33	0	1	12	0	0	0	3	70	5	0	N	0	500
97-TU-16	P2(?)	43-15	101-32	44	3	9	0	80	260	18	0	1	17	0	0	0	5	51	9	0	N	3	497
97-TU-17	Tr	43-15	101-32	189	11	3	51	35	136	5	0	1	3	2	0	0	4	49	6	0	N	4	495
97-TU-18	Tr	43-15	101-32	166	19	5	49	28	115	12	0	1	8	3	0	0	5	82	7	0	N	0	500
97-TU-106	Tr	43-16	101-31	137	24	6	41	40	142	9	0	9	16	3	0	0	2	62	9	0	N	0	500
97-TU-104	Tr	43-16	101-31	85	14	5	41	102	150	13	0	4	10	2	0	0	4	52	17	0	N	1	499
97-TU-102	Tr	43-17	101-31	90	19	9	45	39	211	11	0	9	15	0	0	0	4	39	9	0	N	0	500
97-TU-101	Tr	43-17	101-31	104	19	3	60	73	141	9	0	23	9	3	0	0	0	45	10	0	N	1	499
97-TU-100	Tr	43-18	101-31	91	20	8	35	57	159	8	0	27	15	0	0	0	1	75	4	0	N	0	500
Chonin Boom Locality																							
97-CB-1B	Tr	43-06	99-46	69	14	3	17	80	172	10	0	37	14	1	0	0	9	56	9	9	N	0	500
97-CB-1C	Tr	43-06	99-46	57	15	0	18	75	171	15	0	62	14	2	0	1	5	55	9	1	N	0	500
97-CB-2A	Tr	43-06	99-47	65	12	6	46	62	170	8	0	55	12	4	0	1	1	50	8	0	N	0	500
97-CB-1A	Tr	43-06	99-46	78	2	3	22	64	199	16	0	32	12	0	0	0	16	42	0	14	N	0	500
Toroyt Locality																							
97-TO-15A	Tr?	43-13	99-30	98	4	0	28	76	253	6	1	9	0	2	0	0	5	12	2	4	N	0	500
97-TO-15B	Tr?	43-13	99-30	100	3	3	67	76	121	7	0	38	11	7	0	0	8	41	12	6	N	0	500
97-TO-15C	Tr?	43-13	99-30	100	12	0	78	73	133	3	0	24	14	1	0	0	8	41	10	3	N	0	500

Note: Abbreviations as follows: Lat—latitude; Long—longitude; Qm—monocrystalline quartz; Qp—polycrystalline quartz; cht—chert; K—potassium feldspar; P—plagioclase; Lvm—lithic volcanic and metavolcanic grains; Ls—lithic sedimentary grains; Lc—intrabasinal carbonate grains; Lm—lithic metamorphic grains; Lu—unidentified lithic grains; bt—biotite; ms—muscovite; chl—chlorite; D—dense minerals; cem—cement; matr—matrix; por—porosity; gran frags—granitic rock fragments; rad cht—radiolarian chert; n—total number counts per sample. P2—upper Permian; Tr—Triassic; J1—lower Jurassic.

stone with asymmetric climbing ripples and local soft-sediment deformation are interpreted as levee deposits, although it is also possible that they are interdistributary bay crevasse-splay deposits. The presence of wave ripples and the close stratigraphic association with a laminated lacustrine oil shale unit (see following) further support a lake-delta interpretation. The local occurrence of decimeter- to meter-thick beds of carbonate bivalve coquina likely reflects sublittoral winnowing and concentration of shelly debris (Fig. 4E). Based on observed sedimentary facies, we interpret lake deltas to be associated with an overfilled lake system (cf. Carroll and Bohacs, 1999).

At Toroyt, lake-delta facies are represented principally by quartzose siltstone and shale containing interbedded thin, tabular, medium to fine sandstone layers (Fig. 5). The siltstone is weakly tectonically cleaved and silica cemented. Several meter-scale lenticular sandstone lenses were observed with abundant trough cross-stratification and asymmetric rippling. Decimeter-scale, very well sorted fine-sandstone beds with symmetric ripples also were observed. Skeletal debris (fish?) and conchostracons are present on siltstone bedding planes. The uppermost exposed levels of strata at Toroyt contain several decimeter-thick beds of carbonate peloidal grainstone that are silica cemented and form a resistant topographic bench. Peloidal grainstone beds are well sorted, wave rippled, and commonly grade vertically into wave-rippled quartzose sandstone.

More open-lacustrine sedimentation at Toroyt is suggested by the presence of as much as 1 km of siltstone, fine sandstone, and interbedded carbonate peloidal grainstone (Fig. 4F). Medium-grained sandstone lenses that are several meters thick and contain trough cross-stratification and asymmetric rippling are interpreted as distributary mouth bar deposits, though it is possible that some may be subaqueous levee deposits (e.g., Castle, 1990). Sandstone at Toroyt is typically well sorted, consistent with a greater amount of transport there than at Noyon. In addition, the presence of abundant skeletal debris (fish?), conchostracons, and local carbonate grainstone suggests

TABLE 2. SANDSTONE DETRITAL MODES

Sample #	Age	Lat (°-'N)	Long (°-'E)	Qm	F	Lt	Qp	Lvm	Lsm	Qm	P	K	P/F	%M	%D
South Sain Sar Bulag Transect															
97-NU-304	J1?	43-15	101-55	10.9	43.4	45.7	2.2	95.6	2.2	20.1	32.4	47.5	0.4	0.5	0.0
97-NU-305	J1?	43-15	101-55	5.1	34.8	60.1	1.0	98.1	1.0	12.9	67.9	19.3	0.8	0.0	2.2
South Goyot Canyon Transect															
97-NU-850	Tr	43-13	102-01	34.9	33.0	32.1	14.6	80.8	4.6	51.4	25.4	23.2	0.5	0.8	0.6
97-NU-852	Tr	43-13	102-01	36.1	11.9	52.0	18.2	79.3	2.5	75.3	24.7	0.0	1.0	0.0	0.0
97-NU-853	Tr	43-13	102-01	17.1	27.4	55.6	7.7	90.1	2.2	38.5	36.5	25.0	0.6	0.3	0.0
97-NU-854	Tr	43-13	102-01	11.1	29.4	59.5	6.9	89.9	3.2	27.5	57.3	15.2	0.8	0.9	0.2
97-NU-855	Tr	43-13	102-01	26.7	24.3	49.0	7.9	90.7	1.4	52.4	25.1	22.5	0.5	0.0	0.4
97-NU-292	Tr	43-13	102-01	18.4	28.1	53.5	6.5	89.6	4.0	39.5	31.6	28.8	0.5	0.0	1.6
97-NU-290	Tr	43-13	102-01	15.8	25.9	58.3	3.0	94.4	2.6	37.9	47.3	14.8	0.8	1.9	1.4
97-NU-288	Tr	43-13	102-01	10.7	50.5	38.8	0.0	98.7	1.3	17.6	75.6	6.9	0.9	1.1	3.3
97-NU-287	Tr	43-13	102-01	10.7	40.8	48.4	2.8	94.5	2.8	20.8	79.2	0.0	1.0	2.0	3.9
97-NU-258	Tr	43-13	102-01	6.6	28.2	65.2	3.2	90.5	6.3	19.0	58.4	22.6	0.7	0.7	1.0
97-NU-260	Tr	43-13	102-01	14.6	27.5	57.9	6.0	91.7	2.3	34.6	35.2	30.2	0.5	0.8	3.3
97-NU-263	Tr	43-14	102-01	9.1	47.8	43.1	4.1	90.6	5.3	16.0	58.9	25.1	0.7	1.0	1.9
97-NU-264	Tr	43-14	102-01	24.7	28.9	46.4	1.9	91.5	6.6	46.0	36.9	17.1	0.7	0.2	1.1
97-NU-267	Tr	43-14	102-01	25.4	39.6	35.0	3.2	91.7	5.1	39.1	35.4	25.6	0.6	0.2	0.9
97-NU-809	J1?	43-14	102-01	20.6	42.3	37.2	0.6	95.8	3.6	32.7	44.4	22.9	0.7	0.9	2.8
97-NU-807	J1?	43-15	102-02	14.2	41.7	44.1	1.1	95.5	3.4	25.4	41.2	33.3	0.6	0.5	0.2
97-NU-805	J1?	43-15	102-02	18.0	31.9	50.1	1.9	93.4	4.7	36.2	34.7	29.1	0.5	0.5	0.2
97-NU-801	J1?	43-15	102-02	25.3	31.4	43.3	1.6	95.2	3.2	44.6	41.0	14.5	0.7	0.9	0.9
North Sain Sar Bulag Transect															
97-NU-201	Tr	43-17	101-56	2.6	9.7	87.7	10.4	89.4	0.2	21.1	77.2	1.8	1.0	0.0	2.1
97-NU-202	Tr	43-17	101-56	3.3	4.4	92.3	6.5	76.1	17.4	42.9	57.1	0.0	1.0	0.0	3.5
97-NU-203	Tr	43-17	101-56	39.1	10.9	50.0	10.1	85.3	4.6	78.1	15.2	6.7	0.7	0.4	0.0
97-NU-204	Tr	43-17	101-56	35.6	8.6	55.8	8.6	89.7	1.7	80.6	19.4	0.0	1.0	0.0	0.2
97-NU-205	Tr	43-17	101-56	29.3	20.9	49.7	7.8	87.0	5.2	58.4	14.2	27.4	0.3	0.5	3.0
97-NU-206	Tr	43-17	101-56	19.8	33.8	46.4	16.5	78.4	5.2	37.0	36.5	26.5	0.6	0.0	0.5
97-NU-207	Tr	43-17	101-56	23.8	15.2	61.1	4.5	80.5	15.0	61.0	15.7	23.3	0.4	1.1	0.0
97-NU-208	Tr	43-17	101-56	24.9	9.7	65.4	4.5	92.3	3.3	72.0	18.2	9.8	0.6	1.0	2.5
97-NU-209	Tr	43-17	101-56	20.7	12.1	67.2	2.3	95.8	1.9	63.2	25.0	11.8	0.7	0.2	0.9
97-NU-210	Tr	43-17	101-56	24.2	21.6	54.1	11.0	87.1	1.9	52.8	11.2	36.0	0.2	0.0	1.0
97-NU-212	Tr	43-17	101-56	18.5	23.3	58.3	4.1	95.5	0.4	44.3	43.8	12.0	0.8	0.0	3.8
97-NU-213	Tr	43-16	101-56	24.3	24.8	50.9	6.2	90.7	3.1	49.5	21.4	29.1	0.4	0.2	0.7
97-NU-214	Tr	43-16	101-56	13.8	26.8	59.4	12.1	84.8	3.0	34.1	39.6	26.4	0.6	0.2	1.3
97-NU-419	Tr	43-16	101-56	16.7	34.5	48.8	2.5	92.5	5.0	32.6	43.7	23.7	0.6	0.7	4.5
97-NU-418	Tr	43-16	101-56	16.6	32.4	51.0	8.4	89.9	1.7	33.9	34.5	31.6	0.5	0.6	1.7
97-NU-413	Tr	43-16	101-56	5.6	29.6	64.8	1.3	98.8	0.0	15.9	84.1	0.0	1.0	0.3	4.6
97-NU-410	Tr	43-16	101-56	19.2	46.6	34.1	0.8	93.2	6.1	29.2	62.8	8.0	0.9	0.0	8.2
97-NU-409	Tr	43-16	101-56	8.8	24.2	67.0	3.8	93.6	2.6	26.7	52.7	20.6	0.7	0.2	3.2
97-NU-408	Tr	43-16	101-56	7.8	32.7	59.5	3.7	93.5	2.8	19.3	53.3	27.3	0.7	1.3	0.3
97-NU-406	J1?	43-15	101-56	12.5	42.4	45.2	3.7	93.7	2.6	22.7	35.2	42.1	0.5	0.2	0.0
97-NU-404	J1?	43-15	101-56	14.1	33.1	52.8	0.4	95.5	4.0	29.9	61.2	9.0	0.9	0.5	2.7
97-NU-403	J1?	43-15	101-56	20.9	36.9	42.2	0.5	96.4	3.0	36.2	48.3	15.5	0.8	0.0	1.5
97-NU-402	J1?	43-15	101-56	25.5	27.2	47.4	1.3	95.1	3.6	48.4	41.6	10.0	0.8	0.4	1.2
97-NU-401	J1?	43-15	101-56	21.6	38.1	40.3	2.8	92.1	5.1	36.2	48.1	15.7	0.8	0.6	3.0
97-NU-400	J1?	43-15	101-56	20.7	31.7	47.6	1.9	94.7	3.4	39.5	40.4	20.2	0.7	0.0	1.1

more distal sedimentation with periods of reduced siliciclastic input.

Profundal lake strata

Hendrix et al. (1996) described ~200 m of orange- to tan-weathering, organic-rich siltstone and dolomitic oil shale in the Noyon Uul syncline and interpreted the unit as a poorly oxygenated lake deposit (underfilled lake of Carroll and Bohacs, 1999). This unit is prominent in each of the four measured sections and can easily be traced around the flanks of the syncline (Fig. 1D), so it is a convenient marker horizon. The deposit consists largely of laminated siltstone and shale (Fig. 4G). Soft-sediment deformation is common in the lower portion of the unit. Lamination is not as well developed in the upper portion of the deposit, and decimeter- to meter-scale dolomite concretions are more common there. Intraclast horizons, likely representing storm scour and deposition, are present locally. We did not observe a fauna in the laminated shale unit. Hendrix et al. (1996) reported total organic carbon (TOC) measurements from the oil shale ranging from 2.7% to 3.5% (average = 3.1%), although this range in TOC values is partly a function of outcrop weathering. These shales are thermally immature (average Ro = 0.46%; n = 5) and contain organic geochemical bio-

TABLE 2. SANDSTONE DETRITAL MODES (continued)

Sample #	Age	Lat (°-'N)	Long (°-'E)	Qm	F	Lt	Qp	Lvm	Lsm	Qm	P	K	P/F	%M	%D
North Goyot Canyon Transect															
97-NU-215	Tr	43-17	102-02	39.1	10.2	50.7	2.8	95.0	2.2	79.2	10.7	10.1	0.5	9.0	1.0
97-NU-217	Tr	43-17	102-02	30.2	16.6	53.2	11.2	81.0	7.8	64.5	13.1	22.4	0.4	0.3	0.5
97-NU-220	Tr	43-17	102-02	30.9	17.3	51.8	12.4	84.1	3.5	64.1	22.9	13.0	0.6	0.5	0.5
97-NU-224	Tr	43-17	102-02	16.6	27.8	55.6	6.0	91.5	2.6	37.4	26.3	36.3	0.4	1.2	0.2
97-NU-250	Tr	43-17	102-02	21.5	21.3	57.2	11.1	87.7	1.2	50.3	25.4	24.3	0.5	1.4	0.7
97-NU-252	Tr	43-17	102-02	12.4	18.9	68.8	6.8	92.8	0.3	39.6	29.1	31.3	0.5	0.2	2.5
97-NU-253	Tr	43-17	102-02	18.0	24.1	58.0	8.3	87.3	4.4	42.8	39.2	18.1	0.7	0.3	0.5
97-NU-255	Tr	43-17	102-02	13.8	32.7	53.5	6.1	89.6	4.3	29.7	42.1	28.2	0.6	0.2	0.5
97-NU-640	Tr	43-16	102-01	11.9	26.8	61.4	6.7	88.3	4.9	30.8	46.2	23.1	0.7	1.8	4.8
97-NU-604	Tr	43-16	102-01	17.7	26.8	55.5	6.3	93.2	0.5	39.8	43.6	16.6	0.7	0.0	4.7
97-NU-603	J1?	43-16	102-01	13.3	41.9	44.8	1.2	97.0	1.8	24.0	38.9	37.0	0.5	0.0	3.1
97-NU-601	J1?	43-15	102-01	11.2	39.6	49.2	1.7	98.3	0.0	22.0	55.0	23.0	0.7	0.5	1.8
Tost Uul Transect															
97-TU-1	P2(?)	43-14	101-33	6.4	18.5	75.1	2.9	94.4	2.6	25.7	74.3	0.0	1.0	0.2	0.0
97-TU-4	P2(?)	43-14	101-32	8.8	12.5	78.7	5.0	91.6	3.4	41.3	58.7	0.0	1.0	0.2	0.2
97-TU-7	P2(?)	43-14	101-32	5.4	16.4	78.3	1.3	95.1	3.6	24.7	75.3	0.0	1.0	0.0	0.2
97-TU-8	P2(?)	43-14	101-32	5.9	10.8	83.3	4.7	91.8	3.5	35.2	64.8	0.0	1.0	0.2	0.0
97-TU-19	P2(?)	43-14	101-32	2.0	8.9	89.1	3.3	93.2	3.6	18.2	81.8	0.0	1.0	0.0	0.0
97-TU-13	P2(?)	43-14	101-32	4.8	18.3	76.8	2.7	96.0	1.2	20.8	79.2	0.0	1.0	0.0	0.0
97-TU-14	P2(?)	43-14	101-32	3.7	15.5	80.8	5.9	84.3	9.8	19.3	80.7	0.0	1.0	0.2	0.2
97-TU-15	P2(?)	43-15	101-32	5.9	11.8	82.2	4.8	85.1	10.1	33.3	66.7	0.0	1.0	0.0	0.7
97-TU-16	P2(?)	43-15	101-32	10.2	18.5	71.3	4.1	89.3	6.5	35.5	64.5	0.0	1.0	0.0	1.1
97-TU-17	Tr	43-15	101-32	43.5	19.8	36.6	9.0	87.2	3.8	68.7	12.7	18.5	0.4	0.5	0.9
97-TU-18	Tr	43-15	101-32	41.2	19.1	39.7	15.8	75.7	8.6	68.3	11.5	20.2	0.4	0.7	1.2
97-TU-106	Tr	43-16	101-31	32.3	19.1	48.6	15.8	74.7	9.5	62.8	18.3	18.8	0.5	0.7	0.5
97-TU-104	Tr	43-16	101-31	20.0	33.7	46.2	10.2	80.6	9.1	37.3	44.7	18.0	0.7	0.5	0.9
97-TU-102	Tr	43-17	101-31	20.1	18.8	61.2	10.8	81.5	7.7	51.7	22.4	25.9	0.5	0.0	0.9
97-TU-101	Tr	43-17	101-31	23.6	30.2	46.3	11.3	72.3	16.4	43.9	30.8	25.3	0.5	0.7	0.0
97-TU-100	Tr	43-18	101-31	21.7	21.9	56.4	12.6	71.6	15.8	49.7	31.1	19.1	0.6	0.0	0.2
Chonin Boom Locality															
97-CB-1B	Tr	43-06	99-46	16.6	23.3	60.1	7.2	72.9	19.9	41.6	48.2	10.2	0.8	0.2	2.1
97-CB-1C	Tr	43-06	99-46	13.3	21.8	64.9	5.7	65.0	29.3	38.0	50.0	12.0	0.8	0.7	1.1
97-CB-2A	Tr	43-06	99-47	14.9	24.8	60.3	7.2	67.7	25.1	37.6	35.8	26.6	0.6	1.1	0.2
97-CB-1A	Tr	43-06	99-46	18.2	20.1	61.7	2.0	79.0	19.0	47.6	39.0	13.4	0.7	0.0	3.6
Toroyt Locality															
97-TO-15A	Tr?	43-13	99-30	20.6	21.9	57.5	1.5	92.7	5.9	48.5	37.6	13.9	0.7	0.4	1.0
97-TO-15B	Tr?	43-13	99-30	23.5	33.6	43.0	3.5	70.3	26.2	41.2	31.3	27.6	0.5	1.6	1.8
97-TO-15C	Tr?	43-13	99-30	22.9	34.6	42.6	7.0	77.3	15.7	39.8	29.1	31.1	0.5	0.2	1.8

Note: Abbreviations as follows: Qm—monocrytalline quartz; F—total feldspar; Lt—total lithics; Qp—polycrystalline quartz; Lvm—lithic volcanic and metavolcanic; Lsm—lithic sedimentary plus lithic metamorphic; P—plagioclase; K—potassium feldspar; P/F—ratio of potassium feldspar to total feldspar; %M—percent mica; %D—percent dense minerals. P2—upper Permian; Tr—Triassic; J1—lower Jurassic; Lat—latitude; Long—longitude.

marker compounds suggestive of anoxic and saline but not hypersaline conditions (Hendrix et al., 1996). We did not observe laminated organic-rich shale and siltstone at Tost Uul, probably because strata exposed there were stratigraphically lower than the level of the laminated oil shale at Noyon Uul, although it is also possible that the deposit at Noyon Uul did not extend as far west as Tost Uul. We also did not observe laminated, organic-rich siltstone and shale at either Chonin Boom or Toroyt.

ANALYSIS OF PALEOCURRENT INDICATORS

More than 1900 paleocurrent indicator measurements were collected in this study in order to provide a basis for paleogeographic reconstruction (Figs. 2 and 3). Paleocurrent indicators from Noyon and Tost Uul strongly suggest the presence of a broad, east-west–trending trough that opened to the west. Combined paleocurrent indicator measurements from the north Goyot Canyon suggest transport to the south-southwest. In contrast, combined measurements for the south Goyot Canyon section indicate flow to the north-northwest (Fig. 2). Combined paleocurrent indicators from the north Sain Sar Bulag section and from the south Sain Sar Bulag section are somewhat more scattered, but generally suggest transport to the west and north at each locality. Paleocurrent indicators at Tost Uul (Fig. 3) also suggest transport largely to the west, suggesting the presence of a connection to west-directed dispersal systems at Noyon Uul.

In addition to these first-order paleocurrent trends at Noyon and Tost Uul are several second-order trends of interest (Figs. 2 and 3). Bar set directions and trough cross-stratification in coarse, braided-fluvial facies at the base of the north Sain sar Bulag section suggest transport initially to the southwest but change upsection to north- and west-directed transport. Paleocurrent directions in the south Goyot Canyon section display considerable scatter initially but then indicate relatively uniform transport to the northwest farther upsection. At Tost, paleocurrent trends in coarse, braided-fluvial strata in the lower portion of the section are more consistent than paleocurrent trends in finer grained lake-delta strata in the upper portion of the section, probably because of the higher sinuosity associated with the latter.

PETROLOGY OF CONGLOMERATE AND SANDSTONE

Clast counts

Conglomerate clast compositions are remarkably similar across the study area (Figs. 2, 3, and 6A). All samples are dominated either by silicified volcanic porphyry clasts or metasedimentary duraclasts that include metasandstone, metasiltstone, argillite, and chert. Foliated metamorphic clasts are notably absent or are negligible (<1%) in all samples, as are carbonate and mafic igneous clasts.

Although clast populations across the study area are dominated by silicified volcanic porphyry and metasedimentary clasts, clast population data (Figs. 2, 3, and 6A) contain some clear trends. First, clast compositions from the south Goyot Canyon section are more variable than any other locality and include significant percentages of granite, syenite, diorite, and intermediate volcanic clasts. Of note, pink K-feldspar–bearing granite increases in abundance upsection from <5% to >20% (Fig. 2B). Hendrix et al., (1996) documented a similar trend along strike in the south Sain Sar Bulag section and interpreted it to reflect unroofing of structurally deeper plutonic levels over time. Although pink granite clasts are present locally across the rest of the study area, including the north limb of the Noyon Uul syncline, the upsection increase in granite clasts seems to be restricted to the south limb of the syncline. Second, clast populations in all four samples from the north Sain sar Bulag section are dominated by metasedimentary plus chert clasts and quartz-bearing silicified volcanic porphyry clasts, similar to clast populations in four samples collected from the north Goyot Canyon section (cf. Fig. 2, C and D). The distribution of clasts in these two sections is notably different than the distribution of clasts from the south side of the Noyon Uul syncline (Fig. 2B; Hendrix et al., 1996). These observations suggest that the two limbs of the syncline did not share source terranes, consistent with the observation of converging paleocurrent indicators on the two synclinal limbs. Third, the clast populations at Tost Uul (Fig. 3) change upsection from samples dominated by metasedimentary plus chert clasts to samples dominated by volcanic porphyry clasts. Samples from Tost Uul also contain a higher percentage of porphyritic volcanics with no visible quartz than all but three samples from Noyon Uul, suggesting some differences in the composition of source terranes for the two areas. Fourth, the relative percentages of quartz-bearing porphyritic clasts and non-quartz-bearing porphyritic clasts in samples from Chonin Boom, Gorwantes, Har Dell, and Toroyt are similar to those from the Tost Uul section, located immediately to the east (Figs. 3 and 6A).

Sandstone point-counts

Sandstone throughout the study area is lithic rich ($Qm_{18}F_{26}Lt_{56}$; Fig. 6B). In all samples, volcanic lithic fragments dominate the lithic grain population ($Qp_6Lvm_{88}Lsm_6$). We interpret the abundance of volcanic lithic fragments to reflect erosion principally of Paleozoic arc volcanic sequences that underpin most of southern Mongolia (Ruzhentsev and Pospelov, 1992; Lamb and Badarch, 1997, this volume). This interpretation is consistent with the high degree of silicification of volcanic pebbles, cobbles, and sand grains throughout the study area as well as the lack of reported Mesozoic volcanism in the area. Hence, the strongly volcanic signature of sandstone from the study area is relict and not a predictor of contemporaneous tectonic setting (cf. Mack, 1984; Graham et al., 1993). Minor percentages of foliated metamorphic grains and siliciclastic sedimentary grains are present throughout the study area, but they are most abundant in the upper portion of the Tost Uul section, at Chonin Boom, and at Toroyt, where they compose to 20% of the lithic fraction (Chonin Boom plus Toroyt mean $Qp_5Lvm_{75}Lsm_{20}$). Other source-diagnostic grain types that are present throughout the study area include granitic rock fragments, interpreted to be derived from Late Carboniferous and Permian alkali granite common across southern Mongolia (Fig. 1C), and radiolarian chert grains, inferred to be derived from Paleozoic chert of oceanic origin (Lamb and Badarch, 1997). Carbonate peloids, interpreted to be intrabasinal, are common in sandstone samples from Toroyt but are absent in sandstone collected elsewhere.

Analysis of sandstone collected at different stratigraphic levels at each of the Noyon Uul sections and at Tost Uul suggests several compositional trends through time (Hendrix et al., 1996; Beck, 1999). Perhaps most strikingly, sandstone collected below the basal conglomerate at Tost Uul is devoid of K-spar, in contrast to samples collected within and above the basal conglomerate that contain abundant K-spar (Fig. 6B). Hendrix et al., (1996) noted a similar trend in the south Sain Sar Bulag section and their western section and interpreted it to reflect erosional unroofing of K-spar–bearing granitic source rocks during and after deposition of the basal conglomerate. Of note, some sandstone samples from each of the Noyon Uul sections plot beyond the limit of detrital modes on Qm-P-K diagrams as proposed by Dickinson and Suczek (1979) (Fig. 6B). Almost certainly, the

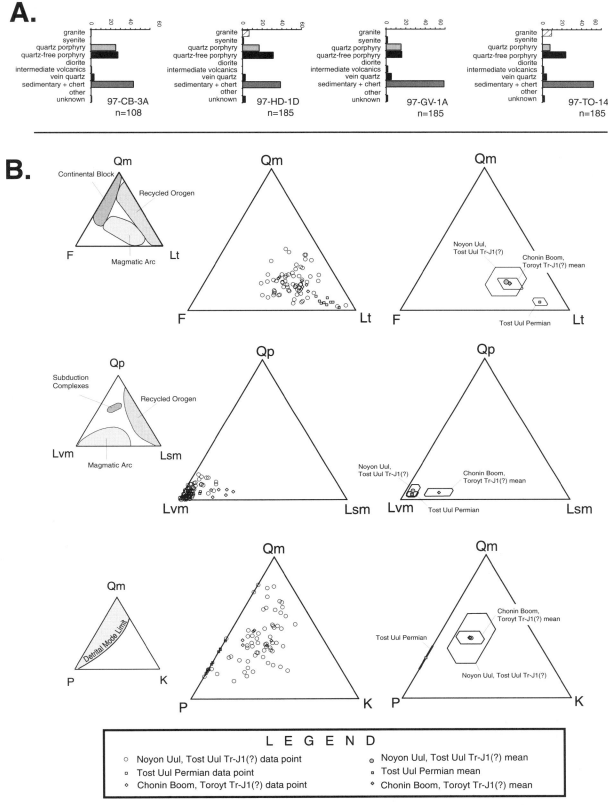

Figure 6. A: Conglomerate clast-count histograms for Chonin Boom (97-CB-3A), Har Dell (97-HD-1D), Gorwantes (97-GV-1A), and Toroyt (97-TO-14). See Figure 1C for sample locations. B: Modal sandstone compositional data, means, and standard deviations for study area plotted on QmFLt, QpLvmLsm, and QmPK ternary diagrams. Numeric data presented in Table 2. Qm, monocrystalline quartz; F, total feldpar; Lt, total lithic framework fraction; Qp, polycrystalline quartz; Lvm, lithic volcanic and metavolcanic grains; Lsm, lithic sedimentary plus metamorphic grains; P, plagioclase; K, potassium feldspar. Refer to Ingersoll et al. (1984) for methods used to calculate these detrital modes from raw data presented in Table 1. Ternary plots on left show tectonic petrofacies of Dickinson and Suczek (1979). Individual sandstone compositional data points are shown in middle column of ternary diagrams. Compositional means and standard deviations (2σ) are shown in ternary diagrams on right.

strongly K-spar–rich composition of these samples was produced by local erosion of Carboniferous–Permian K-spar–bearing alkali granite, combined with a relatively small scale drainage (first-order drainage of Ingersoll, 1990), in which K-spar–rich compositions were not diluted. Analogous combinations of K-spar–rich source terranes and local drainage systems were interpreted to have produced sandstone of similar composition in northwestern China (Graham et al., 1993; Carroll et al., 1995) and the Rio Grande rift (Ingersoll and Cavazza, 1991).

Relative to the strongly volcanic-dominated lithic compositions at Noyon Uul, the occurrence of lithic metamorphic grains in the upper part of the Tost Uul section and in samples from Chonin Boom is somewhat anomalous (Fig. 6B). We interpret the influx of foliated metamorphic grains to represent the erosion of Silurian schist. Currently, Silurian schist is exposed southeast of the Chonin Boom locality in the western portion of the study area, but not the eastern portion of the study area (Fig. 1C). Most likely, Silurian schist exposures were more widespread during Triassic time, prior to the deposition of widespread Upper Cretaceous continental strata.

FACIES RECONSTRUCTIONS

Integrated analyses of sedimentologic, stratigraphic, paleocurrent, and provenance data for the field area suggest that Triassic and lowermost Jurassic(?) strata preserve a changing record of nonmarine depositional environments in an east-west–trending, actively deforming orogen (Fig. 7). Although preserved only at Tost Uul, south Sain Sar Bulag, and south Goyot Canyon, the transition between underlying, fine-grained strata of probable Late Permian age and overlying coarser Triassic strata represents a significant change in paleogeography in the study area. Not only are the coarse-grained, braided-fluvial, conglomeratic facies very unlike finer grained, meandering fluvial Permian strata, but the first appearance of K-spar in Triassic braided-fluvial strata suggests renewed downcutting and erosion of granitic source terranes beginning in Triassic time. Triassic strata of the Noyon and Tost synclines are mapped as being in depositional contact with Upper Permian strata in the east and Carboniferous strata in the west (Yanshin, 1989), consistent with the presence of a significant unconformity at the base of the Triassic strata.

The lower part of the Triassic section at Noyon Uul and Tost Uul is interpreted to record a large, high-discharge braided-fluvial system that drained source terranes to the north and south and was funneled westward into a broad trough (Fig. 7A). Paleocurrents from braided-fluvial strata in the north and south limbs of the Noyon syncline converge, and there are clear differences in conglomerate clast populations between the limbs of the syncline, suggesting the presence of two different dispersal systems draining different source terranes. Consistent with this interpretation is the preservation of thick deposits of brick-red shale and siltstone on the north limb of the Noyon Uul syncline, but not the south limb. We interpret these redbeds, which

contain abundant calcisols, to represent deposition in a seasonally arid alluvial plain that drained to the south, toward more high-discharge north-draining meandering fluvial dispersal systems with associated shallower water tables and no preserved redbeds and calcisols (Fig. 7B). The presence of abundant west-directed paleocurrent indicators in the south Sain Sar Bulag, north Sain sar Bulag, and Tost Uul sections suggests that dispersal systems converged into a longitudinal drainage system that flowed westward (Fig. 7, A and B). This inference is supported by qualitative observations of meter-scale bar sets indicating west-directed paleoflow in conglomerate at Chonin Boom.

A fundamental change from sedimentation in braided-fluvial and alluvial-plain environments to sedimentation in meandering-fluvial and lacustrine systems is represented across the study area. Paleocurrent trends and facies patterns in meandering-fluvial and lacustrine-deltaic strata suggest that dispersal systems converged into a broad, east-west–trending trough that was at times occupied by a shallow lake system (Fig. 7C). In each of the Noyon Uul transects, meandering-fluvial strata are overlain by lacustrine-deltaic strata, which are in turn overlain by 200 m of organic-rich, laminated shale and siltstone interpreted to represent a deep-water, anoxic lake deposit (Figs. 2 and 7D). Following deposition of deep-water anoxic lake deposits at Noyon Uul, the remainder of each Noyon section consists of shallow, well-oxygenated lake deposits and delta distributary sand complexes interbedded with meandering fluvial deposits (Fig. 7E).

The development of a long-lived and regional lake system during Late Triassic time is suggested by more than 2 km of open-lacustrine, lacustrine-deltaic, and associated meandering fluvial strata in the upper portion of the Noyon Uul sections and at least 1 km of similar strata at Tost Uul. In addition, the top of the exposed section at Toroyt consists of as much as 1 km of fine-grained strata with remarkably tabular bedding (Fig. 4F), suggestive of a long-lived lacustrine system. The occurrence of these fine-grained deposits at Toroyt atop coarse-grained braided-fluvial strata at the base of the section is similar to the overall upward-fining character of strata measured at Tost and Noyon Uul (Figs. 2 and 3). In the absence of direct age control, strata at Toroyt are tentatively assigned a Late Triassic–earliest Jurassic age by local Mongolian geologic mapping teams (G. Badarch, 1998, personal commun.), based on a broad lithostratigraphic correlation with lake deposits at Noyon. Recognizing the uncertainties in age control and the correlation between fine-grained deposits across the study area, we provisionally suggest that the thick, fine-grained lake deposits at Toroyt are equivalent to meandering fluvial deposits and associated lacustrine strata in the upper portion of the Tost and Noyon Uul sections (Fig. 7, C–E). This inference is supported not only by the regional interpretation of depositional systems based on our sedimentologic studies, but by the similarity in conglomerate clast populations and sandstone point-count data from across the study area (Figs. 2, 3, and 6; Tables 1 and 2).

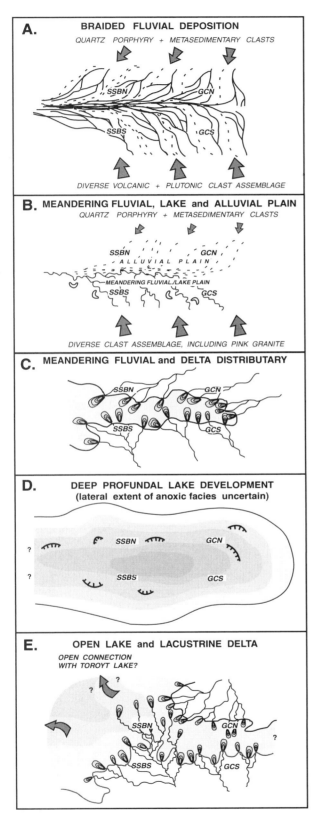

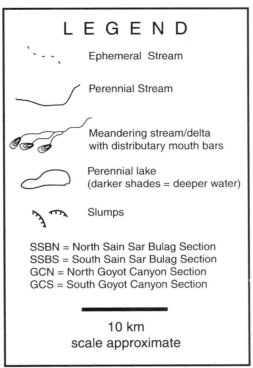

Figure 7. Interpreted paleogeographic evolution of Noyon Uul region during Triassic–lowermost Jurassic(?) time, based on sedimentary facies, measured paleocurrent indicators, and conglomerate clast petrology for study area. A: Braided fluvial deposition corresponding to conglomeratic intervals in basal 1–1.5 km of each section. Paleocurrent indicators that converge toward modern synclinal axis and different clast compositions on each limb of syncline strongly suggest that Noyon region was east-west depositional trough that received sediment from different source areas to north and south and funneled it to west. B: Fluvial and alluvial-plain deposition during which fine-grained redbeds were deposited on north limb of depositional trough and somewhat coarser sediment was deposited in meandering rivers on south limb of depositional trough. C: Meandering fluvial and delta distributary deposition representing early stages of flooding of Noyon Uul syncline by lake waters. We are uncertain whether change to meandering fluvial and lacustrine deposition occurred as function of climate change or ponding of lake waters in syndeforming Noyon trough. D: Profundal lake development corresponding to deposition of organic-rich laminated oil shale in Noyon syncline. E: Open lake and lacustrine deltaic deposition. North-directed paleocurrent indicators in north Sain Sar Bulag section suggest that lake may have opened to northwest and was no longer confined to Noyon Uul depositional trough.

TRIASSIC SEDIMENTATION IN THE CONTEXT OF ASIAN TECTONICS

Numerous attributes of lower Mesozoic deposits in the study area suggest that they are synorogenic. The thick nature of nonmarine strata at Noyon, Tost, and Toroyt suggests the influence of active, tectonically driven subsidence. Locally thick, coarse clastic deposits across the study area indicate the presence of significant physiographic relief and the active erosion of source terranes. Significant changes in paleocurrent dispersal patterns indicate an evolving topography consistent with active tectonism. Different conglomerate clast populations on the north and south limbs of the Noyon Uul syncline, combined with paleocurrent directions that converge toward the synclinal axis, suggest the presence of a broad depositional trough that preceded the formation of the Noyon syncline and likely in part controlled the location of syncline formation (cf. Dumitru and Hendrix, this volume). A structurally similar, east-west–trending syncline of Triassic strata is present immediately to the south of the Noyon–Tost Uul paired synclines at Oboto Hural (Fig. 1C). Both the Noyon Uul–Tost Uul synclinal pair and the Oboto Hural syncline are overlapped by gently tilted but otherwise undeformed Upper Cretaceous strata (Fig. 1C), indicating that neither is a Cenozoic feature (cf. Hendrix et al., 1996). Although the Oboto Hural syncline is largely covered and we did not study it in detail, its presence and the fact that it is overlapped by relatively undeformed Upper Cretaceous strata suggest that Triassic compression produced at least two actively deforming depositional troughs that were subsequently folded and partially eroded (Fig. 8). A third, east-plunging syncline involving Triassic strata and overlapped by Upper Cretaceous deposits is shown on the 1:1 500 000-scale geologic map by Yanshin (1989) ~30 km southeast of the Noyon Uul syncline (Fig. 1C)

Triassic synorogenic sedimentation in southern Mongolia is almost certainly related to contractile structural deformation that occurred during Triassic tectonic amalgamation of central and southern Asia, but the specific configuration of Triassic structures and their relationship to evolving sedimentary basins remains uncertain. During Triassic time, the combined South China block–Qinling–Kunlun Shan collided with the southern margin of Asia (Fig. 1A; Şengör et al., 1993; Yin and Nie, 1996; Zhou and Graham, 1996). This collision, or series of collisions, likely produced far-field deformation that included the development of a series of compressional sedimentary basins (collisional successor basins of Graham et al., 1993). An analogous modern scenario is present in the profound deformation of the Tian Shan, located more than 1000 km from the India-Asia suture zone, and the immense aprons of upper Cenozoic sediment shed into basins adjacent to the Tian Shan in western China (Zhang, 1981). Ancient analogs have been recognized or interpreted from other sedimentary basins of central Asia (Hendrix et al., 1992; Ritts and Biffi, this volume; Sjostrom et al, this volume; Vincent and Allen, this volume).

We interpret that, as contractile deformation intensified during Jurassic time, the Noyon Uul–Tost Uul depositional trough was folded to form the synclinal pair. As shown on the 1:1 500 000 geologic map of Mongolia (Yanshin, 1989), Triassic–Lower Jurassic strata at Noyon and Tost Uul are locally overlapped by less-deformed Upper Jurassic and Lower Cretaceous strata and regionally overlapped by gently tilted Upper Cretaceous strata. On the basis of paleontologic and stratigraphic evidence, Shuvalov (1968, 1969) concluded that a locally angular, regional unconformity exists between Middle and Upper Jurassic strata across much of southern Mongolia. Hendrix et al., (1996) interpreted these collective relations to suggest that Triassic and lowermost Jurassic(?) strata at Noyon and Tost Uul were deformed during late Middle Jurassic time due to shortening associated with the distal portion of the active Bei Shan thrust system (Zheng et al., 1996; Fig. 1B). Detrital apatite fission-track age results from Triassic strata across the study area (Dumitru and Hendrix, this volume) support the interpretation that coarse clastic Triassic strata were erosionally unroofed and rapidly cooled during Middle Jurassic time.

Although compressional deformation likely played an important role in the development of lower Mesozoic sedimentary basins in southern Mongolia, the role of strike-slip deformation in localizing deposition is less certain. Hendrix et al., (1996) argued against a strike slip origin for Triassic strata of southern Mongolia, citing the larger size of the inferred Triassic basin and the greater continuity of facies in the Noyon Uul syncline, relative to most strike slip basins. The results presented in this chapter appear to be consistent with this interpretation, at least to the extent that each of the four study locations contain strata of similar sedimentary style; the rapid facies transitions characteristic of strike-slip basins (e.g., Nilsen and Sylvester, 1995) do not appear to be present. Rather, we are able to correlate individual stratigraphic units (e.g., anoxic lacustrine facies) for tens of kilometers around the Noyon syncline. Likewise, the longevity of early Mesozoic deposition in southern Mongolian basins (~50 m.y.) is inconsistent with a strike-slip origin (Angevine et al., 1990), although it is conceivable that strike slip could have influenced deposition during part of the time represented by preserved strata.

To date, there is no evidence that lower Mesozoic basins of southern Mongolia were extension related. Volcanics associated with late Mesozoic extension (Kovalenko et al., 1991; Keller and Hendrix, 1997) postdate the deposition of strata discussed in this chapter and do not occur in the field area. Similarly, late Mesozoic extensional structures that have been recognized east of the study area (Webb et al., 1999; Johnson et al., this volume) have not been recognized as far west as Noyon Uul and postdate the compressional folding of the Noyon Uul syncline (Dumitru and Hendrix, this volume). Moreover, Zheng et al. (1991) reported that a major extensional detachment, located east of Noyon Uul and inferred to be Early Cretaceous, crosscuts one of the major thrust faults associated with the contractile defor-

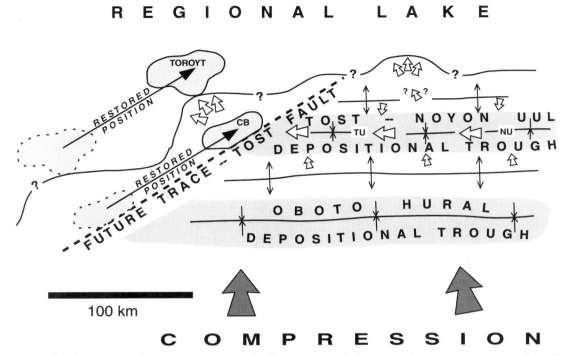

Figure 8. Inferred regional Triassic paleogeography for study area. Sedimentary facies, paleocurrent trends, and provenance relationships suggest that modern Noyon Uul–Tost Uul paired synclines were single, east-west (present coordinates) depositional trough into which sediment was shed and funneled westward in axial drainage system. Restoration of 95–125 km of left-lateral strike slip on Tost Uul fault (Lamb et al., 1999) places Chonin Boom locality in approximate alignment with Tost–Noyon Uul depositional trough, as supported by our observations of similar facies at three localities. Notably finer grain size and sedimentary facies at Toroyt suggest that it is in more distal lacustrine position and may reflect presence of regional lake system west and north of Noyon–Tost Uul region. We infer formation of Noyon Uul–Tost Uul depositional trough and Oboto Hural depositional trough farther south to have resulted from regional north-south compression associated with intracontinental deformation in region. Large dark gray arrows indicate vergence of Triassic contractile deformation inferred to have produced Oboto Hural and Noyon Uul–Tost Uul depositional troughs. Smaller white arrows represent inferred direction of sediment transport during deformation. Small black arrows represent approximate restored position of Toroyt and Chonin Boom localities. Open arrows denote structural trends associated with syndeforming anticlines and synclines.

mation, suggesting that Middle Jurassic contractile deformation gave way to more regional extension during Cretaceous time.

PALEOGEOGRAPHIC IMPLICATIONS OF POST-TRIASSIC DEFORMATION

Although our data do not suggest that strike-slip faulting played a major role in localizing deposition of Triassic strata in southern Mongolia, Triassic paleogeographic reconstructions of southern Mongolia must take into account recent interpretations of significant Jurassic left-lateral strike slip in the area. Lamb et al. (1999) described a series of northeast-trending, left-lateral strike-slip faults that offset a variety of Paleozoic marker beds along the China-Mongolia border. On the basis of the offset of three Paleozoic marker horizons, Lamb et al. (1999) interpreted 95–125 km of sinistral movement on the Tost fault, which extends directly through the present study area and separates the

Tost and Noyon Uul localities from Chonin Boom and Toroyt farther west (Fig. 1B). Although Lamb et al. (1999) were able to determine the maximum age of deformation on the Tost fault only as Early Permian, their analysis showed the maximum age of related left-lateral strike-slip faults in southern Mongolia to be Late Permian. Regional overlap of the Tost fault by Upper Cretaceous strata provides the minimum age constraint for the timing of displacement. Lamb et al. (1999) favored a late Middle Jurassic timing for sinistral faults in southern Mongolia, based on an assumed association between strike-slip faulting and large-magnitude north-verging Jurassic thrusting south of the area (Zheng et al., 1991, 1996; Zou et al., 1992). Direct evidence of Jurassic movement on the Tost fault was described by Dumitru and Hendrix (this volume), who estimated 3–4 km of unroofing in the vicinity of the fault between 200 and 160 Ma (Early to Middle Jurassic), based on the analysis of detrital apatite fission-track data. Assuming that the unroofing documented by apatite annealing is related to the sinistral offset of post-Early Permian marker beds (Lamb et al., 1999), lacustrine deposits at Toroyt and strata at Chonin Boom have been displaced to the southwest 95–125 km relative to the deposits at Tost Uul and Noyon Uul.

Restoring the Toroyt and Chonin Boom localities to their original position places the Chonin Boom locality directly west of the Noyon Uul–Tost Uul paired synclines and places the Toroyt locality northwest of the western end of the Tost Uul syncline (Fig. 8). This paleogeographic reconstruction suggests that clastic deposits at Chonin Boom were the downdip continuation of deposits at Tost Uul and Noyon Uul. Deposition of lacustrine strata at Toroyt likely did not occur directly downdip from the Noyon–Tost–Chonin Boom system, but rather somewhat to the north, obliquely downdip.

This paleogeographic interpretation suggests a sedimentary basin system analogous to early Mesozoic lakes in the Turpan-Hami basin of western China, where Greene et al. (this volume) documented Jurassic longitudinal drainage between the ancestral Bogda Shan and south Tian Shan (Fig. 1B). Although our paleogeographic reconstruction of Triassic–lowermost Jurassic(?) depositional systems from southern Mongolia is only approximate, it suggests that an axial drainage system during Early Triassic time flowed to the west in a broad depositional trough that connected the Noyon Uul, Tost Uul, and Chonin Boom localities, analogous to Jurassic axial flow in the Turpan basin. The regional lake system represented by thick, fine-grained deposits with interbedded carbonate strata at Toroyt appears to have been located obliquely down depositional dip to the west-northwest (Fig. 8). Hendrix et al. (1996, 1998) speculated that Triassic deposits in southern Mongolia may reflect the early stages of a regional foreland basin system that extended across southern Mongolia and perhaps west into the Junggar and Turpan basins (Fig. 1). Although our data set does not demonstrate a connection between the southern Mongolian Triassic lake system and coeval lake systems of western China, it is consistent with the development of a regional Triassic–lowermost Jurassic(?) lake system in southwestern Mongolia that was located along tectonic strike from coeval, regional lake systems in western China.

CONCLUSIONS

On the basis of our field and laboratory studies of Triassic strata in southern Mongolia, we conclude the following.

1. Thick Triassic through Lower Jurassic(?) synorogenic fluvial and lacustrine strata are preserved in southern Mongolia and include the following facies: (1) gravelly braided-fluvial facies, occurring in the lower portion of the Triassic section and consisting mainly of stacked lenses of well-rounded grain-supported conglomerate with large trough and tabular cross-bedding; (2) alluvial-plain facies, characterized by maroon to brick-red shale and siltstone with local carbonate nodules and horizons interpreted to be calcisols; (3) meandering fluvial facies, dominated by abundant gray-green overbank siltstone but also containing isolated upward-fining channel sandstone and tabular fine sandstone interpreted as crevasse-splay deposits; (4) lacustrine delta and open lacustrine facies, occurring mainly in the upper portion of the Triassic section and dominated by thick, soft-sediment-deformed distributary mouth bar sandstone interbedded with decimeter- to meter-scale siltstone and fine sandstone with local climbing ripples, wave ripples, and mollusc coquinas; and (5) profundal lake facies marked by organic-rich, millimeter-laminated shale and siltstone with local dolomite concretions and no observed faunal remains.

2. Paleocurrent analysis of Triassic strata at Noyon and Tost Uul is consistent with the interpretation of an actively deforming, broad depositional trough that funneled sediments shed into it to the west. Paleocurrent indicator directions on both limbs of the present-day eastern portion of the Noyon Uul syncline converge toward the central area of the outcrop belt, whereas indicator directions farther west along strike on both limbs converge obliquely with a significant component of transport to the west. Farther west at Tost Uul, paleocurrent indicator directions trend uniformly to the west.

3. Conglomerate clasts across the study area are dominated by porphyritic volcanics with felsic to intermediate compositions and metasedimentary clasts including metasandstone, metasiltstone, and metachert. Clasts of granite occur in increased abundance in younger strata on the south limb of the Noyon Uul syncline, but are spotty elsewhere in the field area. Clast counts from the northern and southern limbs of the Noyon Uul syncline are markedly similar within each limb, but significantly different between the two limbs, consistent with the interpretation of a depositional trough receiving sediment from different source terranes to the north and south. Conglomerate clast compositions farther west at Tost Uul, Chonin Boom, and Toroyt are broadly similar in that they are dominated by volcanic and sedimentary duraclasts with local occurrence of granite, but differ slightly in the relative proportion of specific grain types, suggesting the addition of local erosional detritus to these areas.

4. Sandstone compositions across the study area are dominated by lithic volcanic grains ($Qm_{18}F_{26}Lt_{56}$; $Qp_6Lvm_{88}Lsm_6$), inferred to reflect the erosion of Paleozoic arc-related basement rock underpinning the study area. Lithic sedimentary and foliated metamorphic grains, inferred to reflect erosion of Silurian schist in the western part of the study area, compose as much as 20% of the lithic fraction from the upper portion of the Tost Uul section, from Chonin Boom, and from Toroyt. Toroyt sandstone also contains abundant carbonate peloids inferred to be intrabasinal. There are few vertical trends in grain composition at Noyon or Tost Uul, except that the first appearance of K-spar at Noyon and Tost Uul occurs in the basal Triassic conglomerate, suggesting the unroofing of Carboniferous to Permian K-spar-bearing granite by that time.

5. Restoration of 95–125 km of Jurassic left-lateral slip on the Tost fault, as proposed by Lamb et al., (1999), brings the Chonin Boom locality into subalignment with the Noyon and Tost Uul localities and places the Toroyt locality to the northwest of the other three locations. This paleogeographic reconstruction suggests that the Noyon, Tost, and Chonin Boom localities were part of a continuous, axial drainage system that occupied a depositional low between two broad east-west–trending upwarps. It also suggests that the more distal lake environment present at Toroyt was located obliquely across strike to the west-northwest of this axial drainage system. This scenario is consistent with earlier interpretations of a more regional foreland basin to the north and west that may have been connected to an analogous foreland system to the west in China.

ACKNOWLEDGMENTS

The research was supported by National Science Foundation grant EAR-9614555 to Hendrix and Graham. We gratefully acknowledge the technical support of the Mongolian Academy of Sciences Geological Institute and, in particular, support from R. Barsbold and D. Badamgarav. We thank C. Johnson, R. Lenegan, D. Sjostrom, and L. Webb for outstanding field assistance, and we are grateful for the logistical field support provided by A. Chimitsuren and B. Ligden. Critical reviews that greatly improved earlier versions of this manuscript were provided by A.R. Carroll, R. Cole, and G.H. Mack.

REFERENCES CITED

Anatol'eva, A.I., 1974, The structure and composition of the Permian red volcanogenic-sedimentary complex in southern Mongolia: Soviet Geology and Geophysics, v. 15, no. 10, p. 25–33.

Angevine, C.L., Heller, P.L., and Paola, C., 1990, Quantitative sedimentary basin modeling: American Association of Petroleum Geologists Continuing Education Course Note Series, v. 32, 132 p.

Badamgarav, D., 1985, Crustacean records; kazacharthrids from the Triassic of Mongolia: Paleontological Journal, v. 19, p. 135–138.

Baljinnyam, I., Bayasgalan, A., Borisov, B.A., Cisternas, A., Dem'yanovich, M.G., Ganbaatar, L., Kochetkov, V.M., Kurushin, R.A., Molnar, P., Herve, P., and Vashchilov, Y.Y., 1993, Ruptures of major earthquakes and active deformation in Mongolia and its surroundings: Geological Society of America Memoir 181, 62 p.

Beck, M.A., 1999, Sedimentologic and tectonic characterization of Mesozoic strata, Noyon Uul syncline, southern Mongolia [M.S. thesis]: Missoula, University of Montana, 149 p.

Boorsma, L.J., 1992, Syn-tectonic sedimentation in a Neogene strike slip basin containing a stacked Gilbert-type delta (SE Spain): Sedimentary Geology, v. 81, p. 105–123.

Bureau of Geology and Mineral Resources, 1991, Regional geology of Nei Mongol (Inner Mongolia) Autonomous Region: Geological Memoirs of the People's Republic of China Ministry of Geology and Mineral Resources, Volume 1: Beijing, Geological Publishing House, 725 p.

Carroll, A.R., and Bohacs, K.M., 1999, Stratigraphic classification of ancient lakes: Balancing tectonic and climatic controls: Geology, v. 27, p. 99–102.

Carroll, A.R., Graham, S.A., Hendrix, M.S., Ying, D., and Zhou, D., 1995, Late Paleozoic tectonic amalgamation of northwestern China: Sedimentary record of the northern Tarim, northwestern Turpan, and southern Junggar basins: Geological Society of America Bulletin, v. 107, p. 571–594.

Castle, J.W., 1990, Sedimentation in Eocene Lake Uinta (Lower Green River Formation), northeastern Uinta basin, Utah, in Katz, B.J., ed., Lacustrine basin exploration, case studies and modern analogs: American Association of Petroleum Geologists Memoir 50, p. 243–263.

Chen, Z., 1985, Geologic map of Xinjiang Uygur Autonomous Region, China: Beijing, Geological Publishing House, 1:2 000 000 scale.

Cunningham, W.D., Windley, B.F., Dorjnamjaa, D., Badamgarav, G., and Saandar, M., 1995, A structural transect across the Mongolian Altai: Active transpressional mountain building in central Asia: Tectonics, v. 15, p. 142–156.

Cunningham, W.D., Windley, B.F., Dorjnamjaa, D., Badamgarav, G., and Saandar, M., 1996, Late Cenozoic transpression in southwestern Mongolia and the Gobi Altai–Tien Shan connection: Earth and Planetary Science Letters, v. 140, p. 67–81.

Cunningham, W.D., Windley, B.F., Owen, L.A., Barry, T., Dorjnamjaa, D., and Badamgarav, G., 1997, Geometry and style of partitioned deformation within a late Cenozoic transpressional zone in the eastern Gobi Altai Mountains, Mongolia: Tectonophysics, v. 277, p. 285–306.

DeCelles, P.G., Langford, R.P., and Schwartz, R.K, 1983, Two new methods of paleocurrent determination from trough cross-stratification: Journal of Sedimentary Petrology, v. 53, p. 629–642.

Dickinson, W.R., 1970, Interpreting detrital modes of graywacke and arkose: Journal of Sedimentary Petrology, v. 40, p. 695–707.

Dickinson, W.R., and Suczek, C.A., 1979, Plate tectonics and sandstone compositions: American Association of Petroleum Geologists Bulletin, v. 63, p. 2164–2182.

Gazzi, P., 1966, Le arenarie del flysch sopracretaceo dell'Appennino modense; correlaziani con il flysch di Monghidoro: Mineralogica e Petrografica Acta, v. 12, p. 69–97.

Graham, S.A., Hendrix, M.S., Wang, L.B., and Carroll, A.R., 1993, Collisional successor basins of western China: Impact of tectonic inheritance on sand composition: Geological Society of America Bulletin, v. 105, p. 323–344.

Gubin, Y.M., and Sinitza, S.M., 1993, Triassic terrestrial tetrapods of Mongolia and the geological structure of the Sain-Sar-Bulak locality, in Lucas, S.G., and Morales, M., eds., The nonmarine Triassic: New Mexico Museum of Natural History and Science Bulletin no. 3, p. 169–170.

Hendrix, M.S., Graham, S.A., Carroll, A.R., Sobel, E.R., McKnight, C.L., Schulein, B.J., and Wang, Z., 1992, Sedimentary record and climatic implications of recurrent deformation in the Tian Shan: Evidence from Mesozoic strata of north Tarim, south Junggar, and Turpan basins, northwest China: Geological Society of America Bulletin, v. 104, p. 53–79.

Hendrix, M.S., Graham, S.A., Amory, J.Y., and Badarch, G., 1996, Noyon Uul (King Mountain) syncline, southern Mongolia: Early Mesozoic sedimentary record of the tectonic amalgamation of central Asia: Geological Society of America Bulletin, v. 108, p. 1256–1274.

Hendrix, M.S., Beck, M.A., Lenegan, R., Graham, S.A., Johnson, C.J., Webb, L., and Sjostrom, D.J., 1998, Early Mesozoic development of a regional lake system in southern Mongolia: American Association of Petroleum Geologists Abstracts with Programs (CD-Rom).

Ingersoll, R.V., 1990, Actualistic sandstone petrofacies: Discriminating modern and ancient source rocks: Geology, v. 18, p. 733–736.

Ingersoll, R.V., and Cavazza, W., 1991, Reconstruction of Oligo-Miocene volcaniclastic dispersal patterns in north-central New Mexico using sandstone petrofacies, in Fisher, R.V., and Smith, G.A., eds., Sedimentation in volcanic settings: SEPM (Society for Sedimentary Geology) Special Publication 45, p. 227–236.

Ingersoll, R.V., Bullard, T.F., Ford, R.L., Grimm, F.P., Pickle, J.D., and Sares, S., 1984, The effect of grain size on detrital modes: A test of the Gazzi-Dickinson point-counting method: Journal of Sedimentary Petrology, v. 54, p. 103–116.

Johnson, C.L., Graham, S.A., Hendrix, M.S., and Badarch, G., 1997, Sedimentary record of Jurassic-Cretaceous rifting, southeastern Mongolia: Implications for the Mesozoic tectonic evolution of central Asia: Geological Society of America Abstracts with Programs, v. 29, no. 6, p. A-228.

Keller, A.M., and Hendrix, M.S., 1997, Paleoclimatologic analysis of a Late Jurassic petrified forest, southeastern Mongolia: Palaios, v. 12, p. 282–291.

Kovalenko, V.I., Dergunov, A.B., Barsbold, R., Luwsandanzan, B., Knipper, A.L., Janshin, A.L., Zaitsev, N.S., and Gerbova, V.G., eds., 1991, Volcanoplutonic associations of central Mongolia: Nauka Press, The Joint Soviet-Mongolian Scientific Research Geological Expedition Transactions, v. 50, 229 p.

Lamb, M.A., and Badarch, G., 1997, Paleozoic sedimentary basins and volcanic-arc systems of southern Mongolia: New stratigraphic and sedimentologic constraints: International Geology Review, v. 39, p. 542–576.

Lamb, M.A., Hanson, A.D., Graham, S.A., Badarch, G., and Webb, L.E., 1999, Left-lateral sense offset of upper Proterozoic to Paleozoic features across the Gobi Onon, Tost, and Zuunbayan faults in southern Mongolia and implications for other central Asian faults: Earth and Planetary Science Letters, v. 173, p. 183–194.

Laniz, R.V., Stevens, R.E., and Norman, M.B., 1964, Staining of plagioclase feldspar and other minerals with F.D. and C. Red No. 2: U.S. Geological Survey Professional Paper 501-B, p. B152–B153.

Mack, G.H., 1984, Exceptions to the relationship between plate tectonics and sandstone composition: Journal of Sedimentary Petrology, v. 54, p. 212–220.

Mack, G.H., James, W.C., and Monger, H.C., 1993, Classification of paleosols: Geological Society of America Bulletin, v. 105, p. 129–136.

May, S.R., Ehman, K.D., Gray, G.G., and Crowell, J.C., 1993, A new angle on the tectonic evolution of the Ridge basin, a "strike slip" basin in southern California: Geological Society of America Bulletin, v. 105, p. 1357–1372.

Miall, A.D., 1992, Alluvial models, *in* Walker, R.G., and James, N.P., eds., Facies models, response to sea-level change: Ontario, Geological Association of Canada, p. 119–142.

Miall, A.D., 1996, The geology of fluvial deposits: Sedimentary facies, basin analysis, and petroleum geology: New York, Springer-Verlag, 582 p.

Nilsen, T.H., and Sylvester, A.G., 1995, Strike-slip basins, *in* Busby, C.J., and Ingersoll, R.V., eds., Tectonics of sedimentary basins: Cambridge, Blackwell Science, p. 425–457.

Ruzhentsev, S.V., and Pospelov, I.I., 1992, The South Mongolian Variscan fold system: Geotectonics, v. 26, p. 383–395.

Şengör, A.M.C., Burke, K., and Natal'in, B.A., 1993, Asia: A continent built and assembled over the past 500 million years: Geological Society of America Short Course Notes, 262 p.

Shuvalov, V.F., 1968, More information about Upper Jurassic and Lower Cretaceous in the southeast Mongolian Altai: Academy of Sciences, USSR, Doklady (Earth Sciences Section), v. 197, p. 31–33.

Shuvalov, V.F., 1969, Continental redbeds of the Upper Jurassic of Mongolia: Academy of Sciences, USSR, Doklady (Earth Sciences Section), v. 189, p. 112–114.

Siegenthaler, C., and Huggenberger, P., 1993, Pleistocene Rhine gravel: Deposits of a braided river system with dominant pool preservation, *in*

Best, J.L., and Bristow, C.S., eds., Braided rivers: Geological Society [London] Special Publication 75, p. 147–162.

Smith, D.G., 1976, Effect of vegetation on lateral migration of anastomosed channels of a glacial meltwater river: Geological Society of America Bulletin, v. 87, p. 857–860.

Watson, M.P., Hayward, A.B., Parkinson, D.N., and Zhang, Z.M., 1987, Plate tectonic history, basin development, and petroleum source rock deposition onshore China: Marine and Petroleum Geology, v. 4, p. 205–225.

Webb, L.E., Graham, S.A., Johnson, C.L., Badarch, G., and Hendrix, M.S., 1999, Occurrence, age, and implications of the Yagan–Onch Hayrhan metamorphic core complex, southern Mongolia: Geology, v. 27, p. 143–146.

Yanshin, A.L., 1989, Map of geological formations of the Mongolian People's Republic: Moscow, Academia Nauka USSR, scale 1:1 500 000.

Yarmoliuk, V.V., 1978, Upper Paleozoic volcanic association and structural-petrological features of their formation: Moscow, Nauka Press, Transactions of Joint Soviet-Mongolian Scientific Research Geologic Expedition, 138 p.

Yarmolyuk, V.V., and Tikhonov, V.I., 1982, Late Paleozoic magmatism and fault tectonics in the Trans-Altai Gobi, Mongolia: Geotectonics, v. 16, p. 123–130.

Yin, A., and Nie, S., 1996, A Phanerozoic palinspastic reconstruction of China and its neighboring regions, *in* Yin, A., and Harrison, M.T., eds., Tectonic evolution of Asia: New York, Cambridge University Press, p. 442–485.

Zaitsev, N.S., Mossakovsky, A.A., and Shishkin, M.A., 1973, Type section of upper Paleozoic and Triassic with first remains of Labyrintodonts, Southern Mongolia: Izvestia Akademii Nauk USSR, v. 7, p. 133–144.

Zhang, X., 1981, Regional stratigraphic chart of northwestern China: Beijing, China, Geological Publishing House, 496 p.

Zhang, Z.M., Liou, J.G., and Coleman, R.G., 1984, An outline of the plate tectonics of China: Geological Society of America Bulletin, v. 95, p. 295–312.

Zheng, Y.D., Wang, S.Z., and Wang, Y.F., 1991, An enormous thrust nappe and extensional metamorphic core complex newly discovered in Sino-Mongolian boundary area: Science in China, v. 34, p. 1145–1154.

Zheng, Y., Zhang, Q., Wang, Y., Liu, R., Wang, S.G., Zuo, G., Wang, S.Z., Lkaasuren, B., Badarch, G., and Badamgarav, Z., 1996, Great Jurassic thrust sheets in Beishan (North Mountains); Gobi areas of China and southern Mongolia: Journal of Structural Geology, v. 18, p. 1111–1126.

Zhou, D., and Graham, S.A., 1996, Songpan-Ganzi complex of the west Qinling Shan as a Triassic remnant-ocean basin, *in* Yin, A., and Harrison, M., eds., Tectonic evolution of Asia: Cambridge, Cambridge University Press, p. 281–299.

Zhou, Z., Zhao, R., Mao, J., Lun, Z., and Zhu, Y., 1989, Geologic map of Gansu Province, China: Beijing, Geological Publishing House, scale 1:1 500 000.

Zou, G., Feng, Y., Liu, C., and Zheng, Y., 1992, A new discovery of early Yanshanian strike slip compressional nappe zones on middle-southern segment of Beishan Mts., Gansu: Scientia Geologica Sinica, v. 10, p. 309–316.

MANUSCRIPT ACCEPTED BY THE SOCIETY JUNE 5, 2000

Printed in the U.S.A.

Geological Society of America
Memoir 194
2001

Sedimentary and structural records of late Mesozoic high-strain extension and strain partitioning, East Gobi basin, southern Mongolia

Cari L. Johnson
Laura E. Webb*
Stephan A. Graham
Department of Geological and Environmental Sciences, Stanford University, Stanford, California 94305-2155, USA
Marc S. Hendrix
Department of Geology, University of Montana, Missoula, Montana 59812, USA
Gombosuren Badarch
Institute of Geology and Mineral Resources, Mongolian Academy of Sciences, P.O. Box 118, 63 Peace Avenue, Ulaanbaatar, Mongolia 210351

ABSTRACT

Contrasting styles of late Mesozoic sedimentation in the East Gobi basin of southern Mongolia in part reflect varying rates and structural modes of local extension. High-strain extension was associated with formation of an Early Cretaceous metamorphic core complex in the Onch Hayrhan area. Recent mapping illustrates newly recognized structural features of this core complex, including ductile structures that indicate subhorizontal south-southeast–directed extension ca. 126 Ma. A regionally extensive, east-west–trending detachment fault is locally domed, as defined by the rollover of foliation planes from north to south dipping. This detachment fault terminates to the east against a north-south–trending dextral strike slip transfer zone. Lower Cretaceous strata associated with this transfer zone are generally synextensional breccias and alluvial-fan deposits consisting of unsorted boulder conglomerate, interbedded debris-flow units, and rock-fall deposits that represent rapidly deposited components of proximal, syntectonic depositional systems. In contrast, several hundred kilometers to the northeast of this core complex late Mesozoic extension was characterized by more typical synrift structural geometries (e.g., high-angle normal faults bounding half-graben subbasins). Comparatively mature depositional systems represented by dominantly fluvial to lacustrine environments characterized the Early Cretaceous record in this part of the basin.

The combined structural and stratigraphic studies we present in this chapter demonstrate that the East Gobi basin was segmented into regions of high- and low-strain extension, and that transfer zones accommodated these variations. In southern Mongolia and other parts of central Asia, late Mesozoic metamorphic core complex formation is typically associated with a rapid transition from compressional to extensional tectonics, particularly along orogenic belts associated with convergent terrane and plate margins.

*Current address: Department of Earth Sciences, 204 Heroy Geology Laboratory, Syracuse University, Syracuse, New York 13244-1070, USA

Johnson, C.L., et al., 2001, Sedimentary and structural records of late Mesozoic high-strain extension and strain partitioning, East Gobi basin, southern Mongolia, *in* Hendrix, M.S., and Davis, G.A., eds., Paleozoic and Mesozoic tectonic evolution of central Asia: From continental assembly to intracontinental deformation: Boulder, Colorado, Geological Society of America Memoir 194, p. 413–433.

This implies that gravitational collapse along tectonic boundaries may have played an important role in the localization of high-strain extension throughout the region.

INTRODUCTION

Mesozoic sedimentary basins of eastern China have long been recognized as a group of intracontinental rift basins (Watson et al., 1987). They are distinct from many of the basins of western China that are associated with Mesozoic compression (collisional successor basins of Graham et al., 1993; Fig. 1). Extension commenced in the Late Jurassic to Early Cretaceous in north-central China, (e.g., Erlian Basin; Li Chansong et al., 1997), younging eastward to the Tertiary offshore basins (e.g., Bohai Bay; Watson et al., 1987). This extensional regime has also been recognized in the Tamsag and East Gobi basins of eastern Mongolia, which are contiguous with Hailar and Erlian basins, respectively (Shuvalov, 1975; Traynor and Sladen, 1995; Fig. 1).

Many important aspects of these rift basins remain poorly documented, including the magnitude, style, and kinematics of extension, as well as details of sedimentary fill patterns. Furthermore, sedimentary studies of rift basins in central Asia have raised several important regional questions, including what factors may have driven extension, and how these basins may relate to documented early Mesozoic compression (Graham et al., 1996; Hendrix et al., 1996; Davis et al., 1998b; Dumitru and Hendrix, this volume).

In order to better understand late Mesozoic basin evolution, we focused on mapping the structure and the sedimentary and volcanic basin fill of the East Gobi basin of southern Mongolia. Previously, we documented the existence of the Cretaceous Yagan–Onch Hayrhan metamorphic core complex at the southwestern edge of the basin, along the China-Mongolia border (Webb et al., 1999; Fig. 2). Our regional studies indicate that two fundamentally different styles or end members of extension occurred in the East Gobi basin during Early Cretaceous time: (1) high-strain detachment faulting in the southwest, and (2) rifting and half-graben formation in the northeast. The differences in strain may be accommodated by transfer zones similar to the strike slip shear zone at Onch Hayrhan, as well as possibly unrecognized accommodation zones in other parts of the basin. In this chapter, we present new data from the Yagan–Onch Hayrhan core complex, briefly review the sedimentary record of rifting in the East Gobi basin, and consider the differences in extensional style within the context of basin-scale and regional-scale strain partitioning.

Geology of the East Gobi basin

The East Gobi basin is a northeast-southwest–oriented elongate basin that is divided into a series of fault-bounded sub-basins, including most notably the petroliferous Unegt and Zuunbayan subbasins (Fig. 2). Northwest-southeast–trending faults and sediment-covered corridors interrupt the northeast-southwest strike continuity of rift structures and further partition Mesozoic sedimentary depocenters. Along its southern margin, the East Gobi basin is separated from the Erlian basin in China

by an uplifted block of Precambrian–Paleozoic basement rocks (the Toto Shan block of Zonenshain et al., 1971) and by the Zuunbayan fault (Fig. 2; Suvorov, 1982).

Although little is known about the Zuunbayan fault, late Cenozoic strike slip fault activity is well documented in several areas of western and central Mongolia (Cunningham et al., 1997; Baljinnyam et al., 1993). Recent models of the Mesozoic–Cenozoic tectonic evolution of Mongolia have featured differing interpretations of the timing of initial fault activity (cf. Lamb et al., 1999; Yue and Liou, 1999). Proprietary seismic reflection data confirm a component of dip-slip normal faulting along the Zuunbayan fault zone during the Early Cretaceous, although strike slip faulting may have also played an important role in Mesozoic basin formation.

A series of Late Jurassic normal faults, reactivated as reverse faults during mid-Cretaceous compression, generally define the northern margin of the East Gobi basin (Suvorov, 1982; Traynor and Sladen, 1995), although detailed mapping of this margin is not currently available. To the south and east, the East Gobi basin extends into China and is generally contiguous with the northern edge of the Erlian basin (Fig. 2). To the southwest, it becomes a region of high-strain extension typified by the Yagan–Onch Hayrhan core complex (Zheng and Zhang, 1993).

Rifting in the East Gobi basin began in Late Jurassic time; as much as 2–3 km of nonmarine synrift sediment and volcanic flows fill asymmetric half-grabens (Fig. 3A; Graham et al., 1996). Widespread mid-Cretaceous uplift, possibly related to transpression, inverted the synrift sequence along the basin margins. Typically flat-lying Upper Cretaceous strata unconformably overlie tilted synrift strata (Fig. 3B; Traynor and Sladen, 1995), signaling the end of Mesozoic tectonism (Shuvalov, 1968).

MESOZOIC HIGH-STRAIN EXTENSION IN THE ONCH HAYRHAN REGION

Structure of the Yagan–Onch Hayrhan metamorphic core complex

The Yagan–Onch Hayrhan metamorphic core complex (Fig. 2) was first recognized in China by Zheng et al. (1991), and its continuation into southern Mongolia was documented by Webb et al. (1997, 1999). Its classic core complex components grade structurally upward from a plutonic and gneissic core, through an ~1-km-thick mylonite zone, to a topographically well-expressed detachment fault with a 5–10-m-thick chloritic breccia zone (Webb et al., 1999). Kinematic indicators from ductile and brittle structures indicate south-southeast–directed subhorizontal extension, consistent with the overall trend of the East Gobi basin. Synkinematic biotite yielded $^{40}Ar/^{39}Ar$ ages ranging from 129 to 126 Ma (Webb et al., 1999). This Early Cretaceous age of deformation closely corresponds with the age of

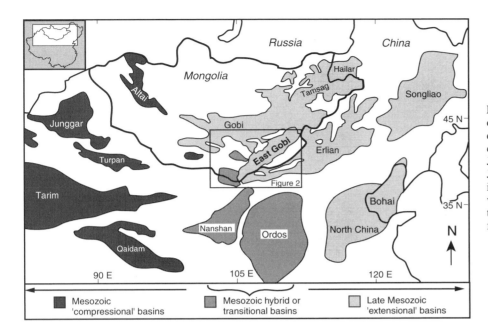

Figure 1. Mesozoic sedimentary basins of central Asia. Basins are generally divided into western basins formed by contractional tectonics during Triassic–Jurassic, eastern basins formed during Jurassic–Cretaceous intracontinental rifting, and hybrid or transitional basins with multiphase, transtensional and/or transpressional tectonic histories (modified from Watson et al., 1987).

volcanic tuffs and basalt sequences preserved in the basin (Shuvalov, 1968; Keller and Hendrix, 1997; Johnson et al., 1997a).

Following the initial results of Webb et al., 1999, we completed more detailed mapping on 1:100 000-scale topographic maps and 1:20 000-scale aerial photographs. Geologic transects included mapping, measurement, and documentation of foliations, lineations, faults and/or shear zones and folds, and their relative ages and types (Figs. 4 and 5). Shear sense was determined from *schistositi-cisaillement* fabrics, shear bands, and offset features. In domains of homogeneous deformation, shear sense was assessed from rotated and/or asymmetric features such as sigma and delta clasts. Mesoscopic fault-slip data were collected to understand fault arrays. The orientation and sense of slip on faults was used to calculate principal stress orientations and shape factors, qualitatively describing the stress geometry (Fig. 6). See Angelier (1994) and Passchier and Trouw

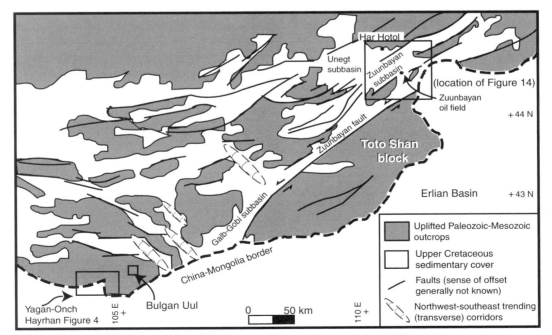

Figure 2. Generalized structure of East Gobi basin. Shaded areas represent uplifted regions with outcrops of Paleozoic–Lower Cretaceous strata separating fault-bounded subbasins. Overall northeast-southwest–oriented structural grain of basin is further subdivided by northwest-southeast–trending sediment-covered corridors.

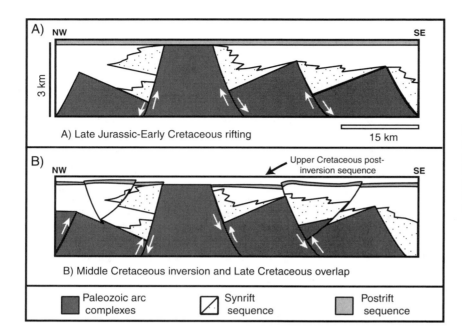

Figure 3. Schematic evolutionary cross sections of northeastern part of East Gobi basin near Zuunbayan (Fig. 2). Modified from Traynor and Sladen (1995). Cross sections are shown for rift development stage (A) and postrift inversion and overlap stage (B).

(1996) for comprehensive summaries and critical discussions of these structural methods.

We now present new field and structural data from the Yagan–Onch Hayrhan core complex (Figs. 4–6) concerning: (1) older ductile features that are overprinted by structures associated with south-southeast–directed extension and intruded by inferred Early Cretaceous plutons; (2) the northward rollover of the south dipping detachment into north-dipping foliation; and (3) a dextral transfer fault that forms the eastern boundary of the core complex and includes a series of high-angle brittle normal fault splays branching eastward.

Pre-core complex structures. The northern zone of the Onch Hayrhan core complex comprises a northeast-trending swath of metasedimentary rocks including interbedded stretched-pebble conglomerate, metasandstone, slate, and limestone (Fig. 4). The level of metamorphism decreases to the northeast from chloritic greenschist facies in the core of the complex to a relatively undeformed, unmetamorphosed sedimentary protolith section. This sequence of interbedded carbonate, conglomerate, sandstone, and shale is strikingly similar to Lower Permian rocks exposed at Bulgan Uul, ~30 km northeast of Onch Hayrhan Mountain (Ruzhentsev et al., 1989; Pavlova et al., 1991; Fig. 2). Both sections are distinctly different from outcrops of Carboniferous–Devonian arc-related marine sedimentary rocks in the area (Lamb and Badarch, 1997), and they bear no resemblance to the poorly lithified, proximal, synextensional conglomerate units of the Lower Cretaceous.

Foliations in the metasedimentary rocks typically dip steeply to the southeast (Fig. 5), although localities with consistent dips of almost every orientation were observed (Fig. 6, group i). Stretching lineations plunge moderately to subhorizontally southwest or northeast. In the higher grade and more strongly deformed sec-

tions of these rocks, shear sense is nearly universally top-to-the-southwest. However, two localities with top-to-the-northeast shear sense (Fig. 6, group i, plots I and J) were observed. Southward toward the detachment, tension gashes indicating northwest-southeast subhorizontal extension become increasingly abundant and crosscut the fabrics related to top-to-the-southwest shearing (Fig. 6, group ii). These metasedimentary rocks are also intruded by undeformed dikes and plutons of inferred Early Cretaceous age, and are in normal fault contact with Lower Cretaceous sedimentary rocks (Figs. 4, 5, and 7; Webb et al., 1999).

Thus, the top-to-the-southwest shear-sense indicators appear to be overprinted by younger shear indicators associated with Early Cretaceous northwest-southeast detachment faulting. Furthermore, top-to-the-southwest structures observed in the northern part of the Onch Hayrhan locality underlie Precambrian(?) carbonate and quartzite klippen (Figs. 4 and 8). These klippen are inferred to be the remnants of a large fold and thrust belt emplaced during Jurassic time (Zheng et al., 1991, 1996; Webb et al., 1999). We propose that the age of the southwest-oriented shear deformation in the metasedimentary units is bracketed between Permian (age of protolith sedimentary section) and Early Cretaceous (age of crosscutting features), and may be related to contractional deformation prior to core complex formation.

Domal structure. In the high-grade region of the core complex south of the metasedimentary section described here, foliations in schist and gneiss dip gently to the north-northeast, and stretching lineations plunge moderately to subhorizontally to the northwest (Figs. 4 and 6, group iii, plots A–D). These fabrics gradually rollover within a zone of ~1 km into southwest- and southeast-dipping foliations and southeast-plunging lineations associated with the detachment fault along the southern boundary of the core complex (Figs. 5 and 6, group iii, plots E

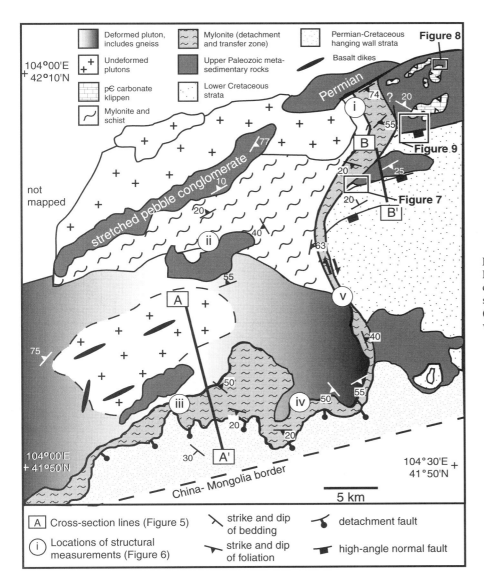

Figure 4. Outcrop structure of Onch Hayrhan region, including locations of cross sections (Fig. 5), structural measurements (Fig. 6), field photographs (Figs. 7 and 8), and sedimentary observations (Fig. 9).

and F, and group iv). Shear sense is uniformly top-to-the-southeast. This domal feature dies out eastward along the detachment, which has only south dipping foliation as it approaches the transfer fault (Figs. 4 and 5).

Transfer zone and brittle structures. From where it first crosses the border from China into Mongolia, the south dipping detachment surface can be traced eastward along strike for ~30 km (Fig. 4). At the point where the domal feature dies out, the detachment changes abruptly into a north-south–trending dextral shear zone. The shear zone is ~100 m wide, and separates mylonitic rocks of the core complex from a Lower Cretaceous sedimentary section to the east (Fig. 7). The transfer fault can be traced north from the detachment for ~15 km. Along its length, a series of brittle normal faults splay off to the east until the transfer zone dies out to the north (Figs. 4 and 7). Principal stress directions computed for the fault-slip data at localities within the transfer zone indicate that deformation was associated with subhorizontal south-southeast extension (Fig. 6, group v), and is identical to the extension direction indicated by stretching lineations in the metamorphic footwall.

Following the terminology of Faulds and Varga (1998), this feature can be classified as a dextral synthetic transfer, similar in geometry to the Las Vegas Valley shear zone of the western United States (Duebendorfer and Wallin, 1991). This dextral transfer zone is a key feature, in that it is among the structures that accommodate the transition from the high-strain extensional province of the Yagan–Onch Hayrhan core to a lower strain extensional province of the East Gobi basin.

Discussion. These new observations strengthen the interpretation of the Onch Hayrhan area as a metamorphic core complex. This region shares many similarities with classic examples from the western United States (Coney, 1980), including a domal, gently dipping detachment zone that juxtaposes ductile fabrics in mid-crustal rocks of the lower plate with relatively

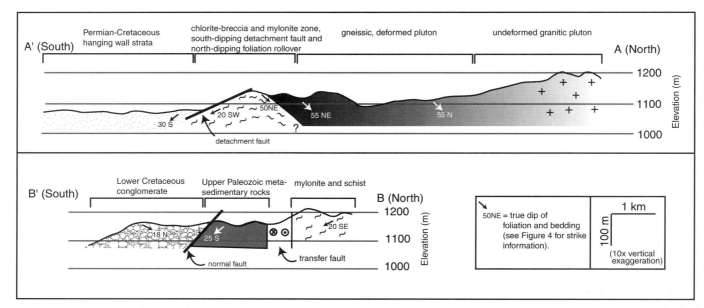

Figure 5. Cross-sections A'-A and B'-B (see Fig. 4 for location). Note domal structure of mylonitic foliation along detachment fault (A'-A), and high-angle normal fault that downdrops synextensional Lower Cretaceous conglomerate against Paleozoic basement rocks (B'-B). Detailed structural data are shown in Figure 6.

unmetamorphosed upper plate rocks, and termination of the detachment fault against a transfer zone oriented obliquely to the extension direction. The Early Cretaceous age of extension (Webb et al., 1999) implies that high-strain extension played a major role in controlling basin structure in parts of the Cretaceous East Gobi basin.

In addition, most of the crystalline rocks associated with this core complex were originally mapped as Precambrian (Yanshin, 1989) and were thought to represent part of the South Gobi microcontinent (Şengör et al., 1993). Although we cannot yet entirely discount the possibility of a Precambrian protolith in southern Mongolia, our field work indicates that the main protoliths at Onch Hayrhan are likely to be upper Paleozoic metasedimentary rocks. We currently are testing this hypothesis by sensitive high–resolution ion microprobe (SHRIMP) analyses of zircons in order to establish protolith ages of metamorphic rocks from Onch Hayrhan and other East Gobi basin localities. Preliminary results indicate that Onch Hayrhan protoliths are younger than Precambrian (T. Cope, 1999, personal commun.). This work further supports the idea that the crust of southern Mongolia is exclusively a collage of Paleozoic arc complexes accreted during the middle to late Paleozoic (Lamb and Badarch, 1997).

Field observations also suggest the existence of an older, overprinted deformation represented by top-to-the-southwest features found north of the detachment fault. In the northeast, these features underlie large carbonate and quartzite klippen (Fig. 8). Similar Triassic–Jurassic thrust-sheet remnants have been reported along the Chinese side of the border (Zheng

et al., 1996). Other evidence of early Mesozoic compression includes inferred Permian–Jurassic foreland basin strata northwest of Onch Hayrhan at Noyon Uul (Hendrix et al., 1996, this volume; Dumitru and Hendrix, this volume), and concurrent thrust faulting in Inner Mongolia (Davis et al., 1998c; Darby et al., this volume). Thus, we propose that the overprinted top-to-the-southwest structures may record a compressional precursor to extension.

Sedimentary record of high-strain extension

Lower Cretaceous sedimentary rocks associated with the Onch Hayrhan core complex crop out east of the transfer zone in a series of discrete subbasins bounded by high-angle normal faults that offset upper Paleozoic (Permian) metasedimentary rocks (Figs. 7 and 9). Although no volcanic layers have yet been found within the sedimentary section, the remarkably coarse grained, proximal nature of the deposits strongly suggests a synextensional origin. In addition, the conglomerate contains clasts of upper Paleozoic sedimentary rocks of the hanging wall, as well as granite, gneiss, and mylonite lithologies found within the footwall of the core complex. The north-dipping conglomerate is overlapped by relatively flat lying Upper Cretaceous clastic rocks that form the postrift sedimentary cover throughout southern Mongolia. Thus, this sedimentary sequence postdates core complex metamorphism and predates the overlap sequence, which generally is early Cenomanian at its base in southern Mongolia (Jerzykiewicz and Russell, 1991). Based on its syntectonic character, we infer that the Lower Cretaceous

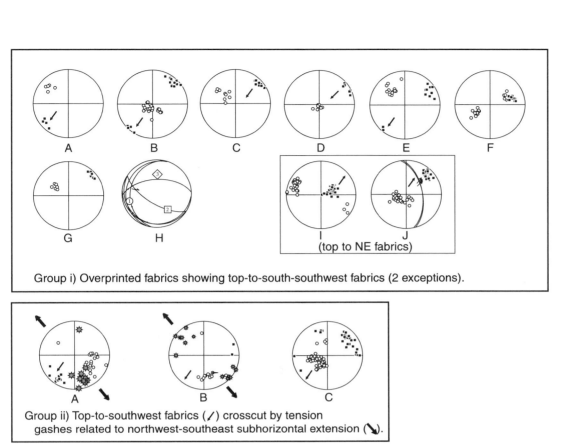

Group i) Overprinted fabrics showing top-to-south-southwest fabrics (2 exceptions).

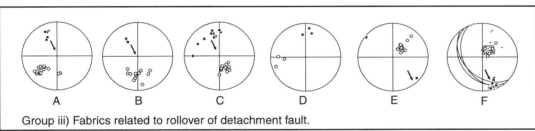

Group ii) Top-to-southwest fabrics (✓) crosscut by tension gashes related to northwest-southeast subhorizontal extension (↘).

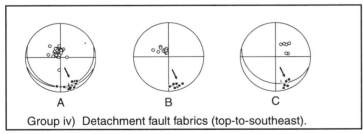

Group iii) Fabrics related to rollover of detachment fault.

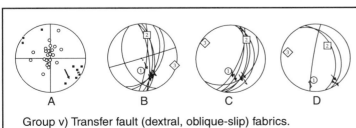

Group iv) Detachment fault fabrics (top-to-southeast).

Group v) Transfer fault (dextral, oblique-slip) fabrics.

○ Pole to foliation

■ Stretching lineation

✳ Pole to vein

▼ Pole to dike

↙ Shear-sense indicator (older, overprinted fabrics)

↖ Shear-sense indicator (younger, subhorizontal crosscutting fabrics)

Ⓦ Shear band (↓) and fault (↓) slip direction

◇ Principal stress direction

Figure 6. Stereonets of kinematic indicators from brittle and ductile fabrics, locations (i–v) in Figure 4. Shear sense in older fabrics (group i) is generally top to southwest, and is overprinted by top to southeast fabrics related to detachment faulting (groups ii to iv), and by oblique dextral slip indicators near transfer fault (group v).

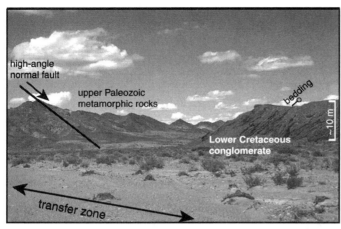

Figure 7. Photograph looking east across north–south–trending transfer zone. Transfer fault is exposed just west of frame of view, and extends northward, parallel to drainage shown in picture (see Fig. 4 for location). East-west–trending high-angle normal fault downdrops Lower Cretaceous conglomerate sequence against upper Paleozoic metamorphic rocks.

sequence formed during and/or immediately following high-strain extension ca. 126 Ma.

A complete measured stratigraphic section is not possible due to the formation of separate half-graben depocenters, the possibility of offlapping stratigraphic shingling (cf. Crowell, 1974), and intermittent exposures. Although not continuous, exposure is superb locally, and we chose to measure a series of smaller sections at decimeter resolution in order to typify depositional style and detrital composition (Figs. 9 and 10). Based on structural and facies relations, we are able to place these measured sections in their relative stratigraphic positions in Figure 10. The section is disrupted by a covered interval and a minor antithetic normal fault between sections C and D (Fig. 10); however, the reconstructed composite section provides a sense of minimum thickness of the basin fill, and permits inferences

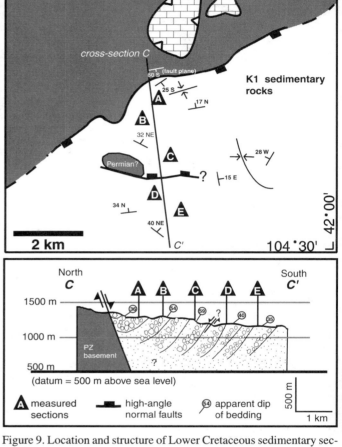

Figure 9. Location and structure of Lower Cretaceous sedimentary section. Locations of detailed measured sections (Fig. 10) are shown by black triangles. Black strike and dip symbols are bedding orientations. See Figure 4 for location.

Figure 8. Photograph of large klippe of Precambrian carbonate overlying Permian and upper Paleozoic schist across flat thrust-fault contact. See Figure 4 for location.

about the evolution of the extensional system and Early Cretaceous landscape.

Lower Cretaceous facies. Lower Cretaceous strata are exposed east of the transfer zone, adjacent to and rotated toward high-angle normal faults that offset Paleozoic basement (Figs. 7 and 9). Bedding dip angles increase slightly in the older units, indicating progressive fault activity and bed rotation during deposition (Fig. 9: cross section and measured sections A–C). The sedimentary sequence mainly consists of coarse-grained (clasts commonly 0.25 to >1 m diameter), poorly sorted, cobble to boulder conglomerate that indicates proximal, rapid, synextensional deposition. Boulders of Paleozoic crinoidal limestone testify to the proximal nature of these deposits, because meter-scale boulders of coarsely crystalline carbonate would not have

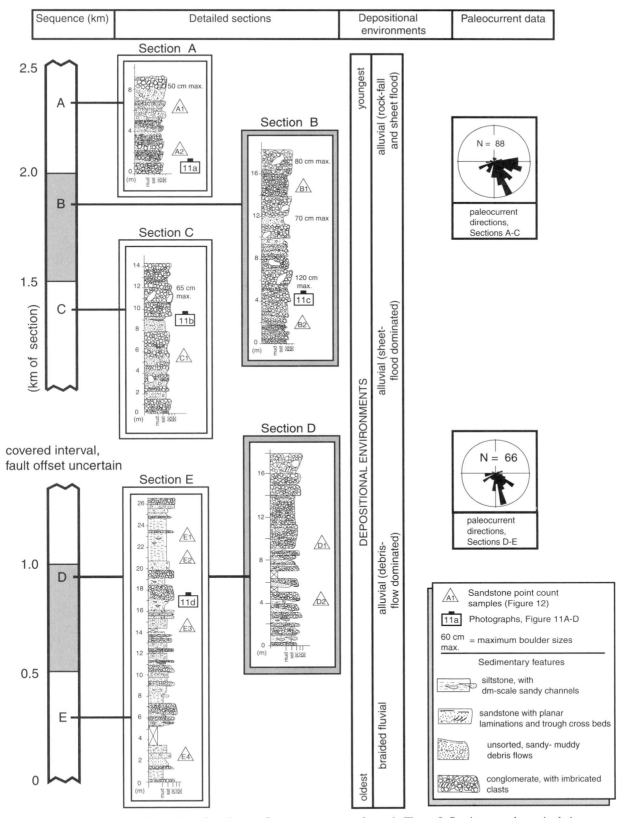

Figure 10. Detailed stratigraphic sections from Lower Cretaceous transect shown in Figure 9. Sections are shown in their approximate stratigraphic positions relative to overall sequence (left column), although thickness missing between sections C and D is not certain. All paleocurrent data are from pebble imbrication measurements. Vertical scales are same for all sections. Grain-size scale: mud, mudstone/siltstone; sst, sandstone; pc, pebble conglomerate; cc, cobble conglomerate; bc, boulder conglomerate.

survived transport over distances of more than a few kilometers. Paleocurrent directions from imbricated clasts are directed toward the south-southeast throughout the section (Fig. 10), indicating drainage from the footwall blocks of half-grabens within the Paleozoic basement.

Sedimentary facies are characteristic of alluvial fan deposits, and can be divided into three general categories following the terminology of Blair and McPherson (1994). These groups are defined as rock-fall units, debris flows, and sheet-flood deposits, in order of increasing abundance in the section.

Rock-fall units are represented by unsorted, clast-supported, extremely coarse deposits of angular boulders with very little matrix (Fig. 11A). These preserved talus cones are ~10 m in lateral width (i.e., cone diameter). They are rare in the section, and likely are formed due to slope failure along steep fault scarps (Beratan, 1998). Larger rock slides and rock avalanches were not observed, though they may have been transported further into the basin, thus bypassing the proximal fan deposits.

Debris-flow beds are present as matrix-rich, internally unorganized planar beds up to 2 m thick. Matrix ranges from sandy to clayey, and beds are unsorted, with floating, cobble-sized angular clasts (Fig. 11B). Debris-flow units are locally erosive at their bases, and also show reworking of the upper parts of the underlying bed. These are interpreted as cohesive, hyperconcentrated, and high-density flows formed by slope failure in water-saturated colluvium. The debris flows are most common in the upper parts of the sequence (Fig. 9, sections A–C), where they are interbedded with clast-supported conglomerate beds interpreted as sheetflood units (Blair and McPherson, 1994).

Sheetflood deposits constitute most of the volume of the Lower Cretaceous strata. Bed thicknesses range from 0.5 m to >2 m, and include interbedded coarse conglomerate and sandier units. Most typically, conglomerate beds display crude to well-defined stratification, erosional channelization, and clast imbri-

cation, indicating some degree of water working and bed-scale traction structures (Fig. 11C). Smaller scale structures (e.g., ripples, cross-beds) are rare, but in overall finer grained intervals, conglomerate with better rounded pebbles occurs in thin channel-form lenses encased in plane- to cross-laminated sandstone. Grain-supported conglomerate beds display both upward-coarsening and upward-fining trends, and are frequently interbedded with matrix-supported unstratified to poorly stratified pebbly mudstone and sandstone.

In addition to the alluvial facies described here, the lower part of the section (typified by detailed section E, Fig. 10) displays finer grained, pervasively channelized, imbricated, and rounded pebble to cobble conglomerates (Fig. 11D). The section also contains fine-grained, mottled, oxidized units interpreted as possible overbank or bar-top deposits and paleosols. In contrast to the rest of the sequence, which is dominated by alluvial processes and flash-flood, episodic sedimentation, these facies are interpreted as representing a seasonal braided stream environment.

Conglomerate and sandstone provenance. We collected compositional data from the sandstone and conglomerate fractions of the Lower Cretaceous section. Conglomerate provenance data were collected in the field by tabulating clast composition counts at four positions in the stratigraphic column (Fig. 12). We counted 100 clasts in each of the conglomerate samples, and in order to minimize the uncertainty associated with field identification of lithology, only four major clast types were tabulated (general sedimentary [sandstone, mudstone]; quartzite-carbonate; metamorphic; volcanic). A greater number of sandstone samples were collected and point-counted using the Gazzi-Dickinson method following the procedures of Ingersoll et al. (1984) (Figs. 12 and 13; Table 1).

Two main petrofacies can be recognized from the conglomerates (Fig. 12), a volcanic-sedimentary petrofacies, and a mixed volcanic-sedimentary-metamorphic petrofacies. The lat-

TABLE 1. RAW POINT-COUNT DATA FROM LOWER CRETACEOUS SANDSTONES OF THE ONCH HAYRHAN REGION

Sample	Latitude (N)	Longitude (E)	Qm	Qp	Cht	K	P	Lvm	Lslt	Lm	unid L	bt	ms	chl	heav	cmt	matr
A1	42.05' 54.2"	104.25' 48.3"	115	48	4	17	58	140	14	0	1	5			1	19	67
A2	42.05' 54.2"	104.25' 48.3"	193	66	6	5	78	108	13	4	2	3	6			4	16
B1	42.05' 07.8"	104.26' 05.5"	57	33	48	0	15	63	222	17	11	3				3	28
B2	42.05' 07.8"	104.26' 05.5"	75	29	21	1	21	63	·128	46						10	102
C1	42.05' 46.7"	104.26' 09.9"	96	48	20	13	28	41	16	18	19				4	75	132
D1	42.05' 05.5"	104.26' 23.9"	105	51	9	4	29	104	49	40					3	100	2
D2	42.05' 05.5"	104.26' 23.9"	73	71	27	1	9	18	29	126					3	48	93
E1	42.04' 38.9"	104.26' 26.4"	92	23	5	52	42	104	93	37		4					48
E2	42.04' 38.9"	104.26' 26.4"	115	43	8	48	48	87	63	31				5		15	35
E3	42.04' 38.9"	104.26' 26.4"	84	47	73	53	45	32	73	2		5			2	71	9
E4	42.04' 38.9"	104.26' 26.4"	60	69	111	8	9	93	44	26		5			4	74	12

Note: Point counts were performed using the Gazzi-Dickison method (Ingersoll et al., 1984), with 500 counts per slide. Qm—monocrystalline quartz, Qp—polycrystalline quartz, Cht—chert, K—potassium feldspar, P—plagioclase, Lvm—volcanic and metavolcanic rock fragments, Lslt—sedimentary rock fragments, Lm—metamorphic rock fragments, unid L—undifferentiated rock fragments, bt—biotite, ms—muscovite, chl—chlorite, heav—dense minerals, cmt—cement, matr—matrix. See Figures 4 and 9 for sample locations.

Figure 11. Outcrop photos of typical sedimentary styles of Lower Cretaceous section at Onch Hayrhan. Photos are keyed to detailed sections of Figure 10. A: Unsorted, clast-supported, matrix-free boulder bed, interpreted as talus slope deposit. B: Matrix-supported debris-flow deposit. Field book for scale (12 × 19 cm). C: Conglomerate and sandstone sequence; note large outsized boulders and crude bedding. Person for scale. D: Lenses of moderately rounded pebble conglomerate interstratified with plane- to cross-laminated sandstone and siltstone, interpreted as braided fluvial deposit. Hammer for scale.

ter is especially important, because some of the metamorphic clasts are recognizable as mylonite similar to that exposed in the footwall of the Yagan–Onch Hayrhan core complex (Fig. 14). Other distinctive lithologies include carbonate and quartzite clasts derived from the Permian and older section. Volcanic clasts present in conglomerate and present to abundant in sandstone (Fig. 12) appear to be mostly Paleozoic, rather than Mesozoic and synextensional, based on fabric and degree of alteration (paleovolcanic, rather than neovolcanic in the usage of Zuffa, 1980).

Clast counts, although limited in number, generally show lower percentages of high-grade metamorphic clasts in the younger parts of the stratigraphic section, including the disap-

pearance of mylonitic rock fragments somewhere between sections C and B (Fig. 12). Sandstone point counts from thin sections demonstrate the lithic-rich nature of these samples (Fig. 13). The point counts indicate a comparatively weaker evolutionary trend (Fig. 12), probably reflecting greater dispersal of sand in the environment as well as the difficulty in distinguishing between different types of metamorphic lithologies in medium-grained sand fractions (250–500 µm).

Discussion. The Lower Cretaceous sedimentary section associated with the metamorphic core complex in the Onch Hayrhan area strongly suggests synsedimentary tectonism and structural localization of sediment accumulation. The overall coarse-grained nature of the sequence (Fig. 10), its proximity to

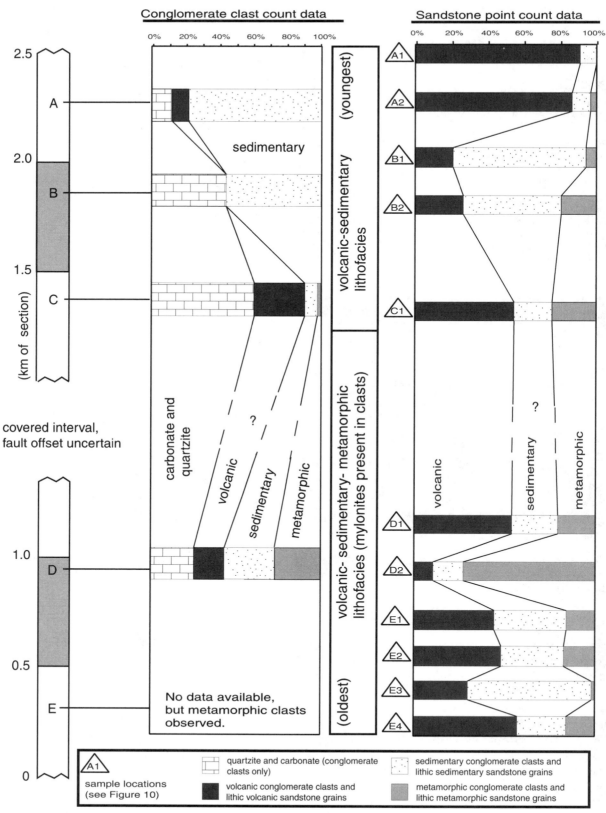

Figure 12. Conglomerate and sandstone clast composition data plotted by stratigraphic position. Sandstone point-count data are recorded in Table 1.

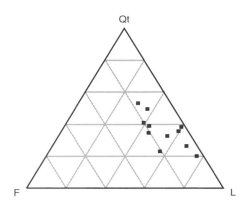

Figure 13. Ternary plot of total framework grains (QtFL) of Lower Cretaceous sandstone. No stratigraphic trends from youngest to oldest samples are evident in QtFL data. See Table 1 for sandstone point-count data.

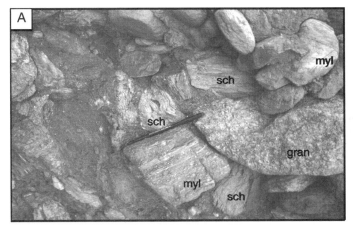

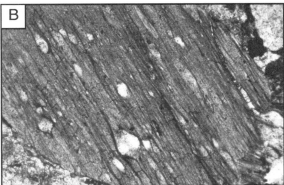

Figure 14. A: Photographs of metamorphic detritus from Lower Cretaceous sequence. Boulder bed from metamorphic petrofacies (lowermost conglomerate sample of Fig. 12). Clasts are predominantly mylonite (myl), schist (sch), and granite (gran). Pen for scale. B: Photomicrograph of coarse sand grain of metamorphic lithic rock, from sample E-1 (Fig. 12). Field of view is ~1 mm across.

probably coeval faults and the transfer zone (Fig. 4), and progressively rotated beds in a fanning-dip pattern (Fig. 9) indicate structurally induced, proximal deposition. Sedimentary units are inferred to be sheetflood dominated (type II alluvial fans of Blair and McPherson, 1994), produced by rapid uplift and intense mechanical erosion of the source area.

General trends in the evolution of depositional systems at Onch Hayrhan are evident from the oldest to youngest stratigraphic sequences (Fig. 10). As shown in section E (Fig. 10), the older parts of the synextensional sequence reflect finer grained, more distal, braided fluvial deposition. Older units grade up-section into extremely coarse, proximal, unsorted alluvial deposits (Figs. 11, A and B). We interpret this shift as recording progradation of an alluvial fan where a more distal braid plain sequence shifts vertically into increasingly proximal middle to upper fan deposits.

The limited size of the provenance data set precludes any firm conclusions regarding evolution of the sediment source terrane. Particularly in the case of proximal sedimentary sequences, conglomerate and sandstone provenance is strongly dependent on source lithologies that crop out in small, localized drainages that may not accurately represent evolution of the system overall (see first-order petrologic models of Ingersoll, 1990). Thus, evolution of the sequence toward increasingly localized drainage of a specific source area in which high-grade metamorphic rocks were not exposed may explain the lack of mylonitic conglomerate pebbles in the younger parts of the section. Metamorphic sand grains that persist throughout the sequence (Fig. 12) probably represent more integrated drainages in the source area, as well as the greater dispersion of sand-sized detritus across the basin.

Mesozoic climate was also an important factor influencing the formation of synextensional sedimentary facies at Onch Hayrhan. Several lines of evidence indicate that southern Mongolia had a relatively wet climate during the Jurassic–Early Cretaceous. Tree-ring analyses from a Late Jurassic petrified forest in the East Gobi basin indicate strong seasonal fluctuations in water supply and possibly a monsoonal climate (Keller and Hendrix, 1997). Jurassic–Cretaceous sedimentary sequences in other parts of the East Gobi basin support the presence of widespread stratified lakes, coal-forming swamps, and soil formation (discussed in the following; Figs. 15–17). Thus, the Cretaceous sedimentary deposits at Onch Hayrhan differ from many of the classic detachment-related basins documented in the western United States because they probably did not form in a dominantly arid environment (Friedmann and Burbank, 1995; exceptions include Muddy Creek half graben, southwest Montana; Janecke et al., 1999). In addition to structural activity in the source area, seasonal changes in water supply likely influenced the preserved sequence at Onch Hayrhan. For example, seasons of high rainfall may have triggered more of the debris-flow units, whereas dry-season sedimentation was probably characterized by rock falls and sheetfloods.

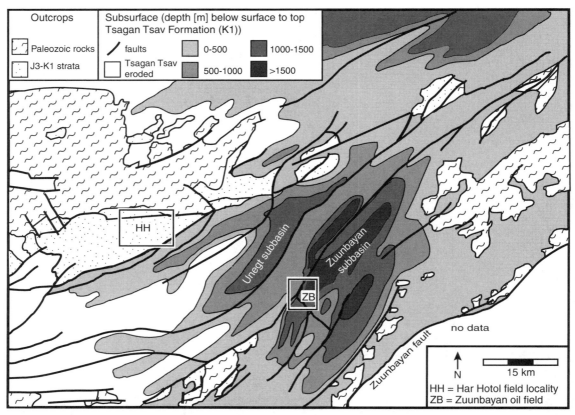

Figure 15. Depth-structure map on Tsagan Tsav (Lower Cretaceous) horizon of Zuunbayan and Unegt subbasins, north-eastern East Gobi basin (Shirakov and Kopytchenko, 1983). Shaded contours represent meters below present-day surface to top of Tsagan Tsav Formation, as compiled from seismic, magnetic, and well-log data (Shirakov and Kopytchenko, 1983). See Figure 2 for location.

Unfortunately, we currently lack geochronologic data sufficient to infer sedimentation or unroofing rates. A paucity of ash or other interstratified datable materials in this unfossiliferous section probably renders calculations of rates impossible, but linked thermochronologic study of metamorphic clasts and the metamorphic core complex may eventually shed some light on the timing of structural and/or erosional unroofing and sedimentation rates.

REGIONAL IMPLICATIONS OF THE YAGAN–ONCH HAYRHAN CORE COMPLEX

Strain partitioning and sedimentary basin development in southern Mongolia

The Onch Hayrhan core complex locality represents the high-strain end member of the Mesozoic extensional regime in southern Mongolia. In contrast, the low-strain extensional end member is represented by intracontinental rifting in the area around the Zuunbayan oil field in the northeastern part of the basin (Figs. 2 and 15). Rifting in this part of the basin resulted in the formation of a series of subbasins bounded by high-angle

normal faults, as demonstrated by field mapping (Graham et al., 1996; Johnson et al., 1997b), and unpublished reflection seismic data gathered by the petroleum industry. The presence of a detachment fault at depth below these half-grabens cannot be completely discounted; however, proprietary seismic reflection data show no evidence of a regional detachment within ~4 km of the surface.

In addition to the two contrasting structural styles of Mesozoic extension in the East Gobi basin, sedimentary successions associated with each of the structural settings are equally distinctive. We have studied the sedimentary sequences of rifted portions of the East Gobi basin in some detail (Graham et al., 1996; Johnson et al., 1997a), and we briefly summarize that work for comparison with the sedimentary sequence associated with the high-strain extensional regime described at Onch Hayrhan.

The most complete synrift stratigraphic section exposed in the East Gobi basin is located along the northeastern margin of the basin at Har Hotol, near the Zuunbayan oil field (Johnson et al., 1997a; Fig. 15). At this and nearby localities, >2 km of synextensional strata are exposed (Fig. 16). The sequence accumulated between ca. 155–131 Ma. (Late Jurassic–Early Cretaceous) (Graham et al., 1996) and dominantly reflects mature

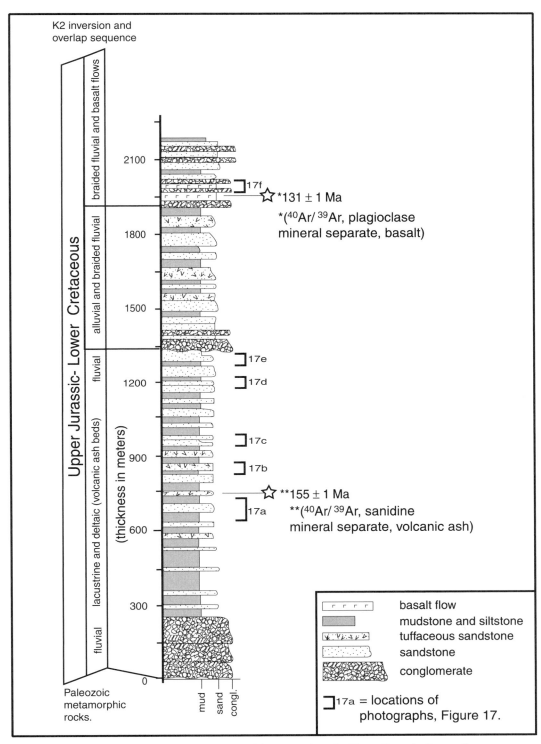

Figure 16. Stratigraphic column and environmental interpretations for Upper Jurassic–Lower Cretaceous syn-rift strata of Har Hotol area. See Figure 15 for location.

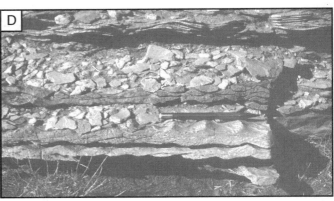

Figure 17. Outcrop photos of typical lithofacies of Har Hotol Upper Jurassic–Lower Cretaceous section. Stratigraphic positions of outcrops are noted in Figure 16. A: Lenticular channel sandstone beds over prodelta mudstones; person at base of cliff for scale. B: Slumping and soft-sediment deformation in fluvial-deltaic sandstone sequence. Field book for scale (12 × 19 cm); C: Upward-thickening sandstone sequence, interpreted as upward shoaling along prograding deltaic lacustrine margin; height of outcrop is ~20 m. D: Lacustrine facies distal to prograding delta margins, including centimeter-scale turbidite layers and symmetric wave ripples; pencil for scale. E: Calcisol development marked by white carbonate nodules within fluvial-alluvial redbed mudstone; hammer for scale. F: Interstratified conglomerate (labeled c) and basalt (labeled b) near top of Har Hotol sequence.

deltaic-fluvial-lacustrine depositional environments and numerous volcanic-ash–dispersal events (Fig. 16).

The basement rocks in this part of the basin are mainly metamorphosed Devonian–Carboniferous volcanic arc sequences (Lamb and Badarch, 1997). Prerift sedimentary rocks are Lower to Middle Jurassic conglomerate, sandstone, and carbonaceous mudstone that unconformably overlie the Paleozoic sequence. The initiation of extension is marked in the synrift stratigraphic sequence by a distinctive ~250-m-thick fluvial conglomerate composed of rounded, imbricated, and mainly basement-derived clasts (0–275 m mark in Fig. 16). This basal conglomerate fines upward abruptly into a siltstone and sandstone unit.

Excellent three-dimensional exposures of this fine-grained unit reveal low aspect-ratio, pervasively slumped, channelized sandstone units capping successive upward-coarsening sequences (Figs. 16 and 17, A–C). We interpret the sequence as a series of prograding lacustrine deltas, which grade laterally into a distal lacustrine sequence. The latter exhibits wave ripples and thin turbidite layers (Fig. 17D). A return to alluvial-fluvial sedimentation occurs in the middle part of the sequence (~1350 m, Fig. 16), with a conglomerate-sandstone redbed section that includes subaerial exposure surfaces and paleosols (Fig. 17E). Fluvially resedimented volcanic ash beds are present throughout much of the upper half of the section, which is capped by multiple interbedded basalt flows, ash layers, and sandstone and conglomerate of braided-fluvial origin (Fig. 17F). A period of structural inversion during mid-Cretaceous time resulted in tilting and erosion of synrift strata, which are overlain by relatively flat lying Upper Cretaceous strata (Fig. 3B).

These stratigraphic studies, combined with our examination of subsurface structure from proprietary seismic lines, indicate that extension along the northeast portion of the East Gobi basin was characterized by high-angle normal faulting and formation of half-graben subbasins. The sedimentary units at Har Hotol compose a typical rift sequence, including multiple ash layers and rapid lateral facies changes from basin-margin alluvial fans to distal, isolated lakes (cf. Friedmann and Burbank, 1995). Sedimentation was dominantly fluvial-lacustrine; limited pulses of conglomerate deposition probably represent initiation and reactivation of normal faults prior to basin inversion. These conglomerate units consist entirely of clasts derived from Paleozoic basement rocks and display tractive structures indicating water working in a braided fluvial depositional system.

Although most of the Jurassic–Cretaceous strata in the East Gobi basin reflect relatively mature, stable fluvial-lacustrine environments in conventional rift systems, as typified by the Zuunbayan–Har Hotol area (Figs. 16 and 17), the southern part of the basin around Onch Hayrhan was subject to large-magnitude, high-strain extension resulting in proximal, rapidly deposited alluvial sequences. Proximal alluvial fan deposits are found in a variety of tectonic settings, and coarse-grained clastic facies are common in classic intracontinental rifts (Gawthorpe et al., 1994; Soreghan et al., 1999), so the presence

of these sequences does not specifically require a detachment fault setting. Conversely, fine-grained lacustrine and playa lake deposits have been described in supradetachment basins in the United States (Beratan, 1991; Janecke et al., 1999). However, the distinctive and highly metamorphic provenance of the Onch Hayrhan synextensional deposits, the limited geographic extent of the subbasins, and the overall thickness of the typically coarse, poorly organized, aggradational conglomerate all indicate dramatic uplift of the source area and rapid creation of accommodation space. Combined with our structural data, these deposits appear to represent an extensional setting unique to the southern part of the East Gobi basin.

Thus, the East Gobi basin was segmented into provinces characterized by differences in rate and magnitude of extension, as has been described for the Basin and Range province of the western United States (Faulds and Varga, 1998). The record of strain variation is reflected in the synextensional sedimentary deposits of the basin (cf. Figs. 10 and 16), demonstrating the considerable utility of stratigraphic studies in unraveling this complex extensional system. Strain variations probably were facilitated by transverse transfer zones such as the one we describe in the Onch Hayrhan area (Figs. 4 and 7). In addition, accommodation zones of linked en echelon faults may have helped absorb regional strain variation, a process observed in both the North American Basin and Range (Faulds and Varga, 1998), and the East African Rift (e.g., Ebinger et al., 1987). Although accommodation zones have not specifically been recognized in the East Gobi basin, they may be represented in the subsurface by northwest-southeast–trending (transverse) corridors now covered by the Upper Cretaceous overlap sequence (Fig. 2).

Relation of Mesozoic extension to earlier contractile orogeny

Possible driving mechanisms for Mesozoic–Cenozoic rift-basin formation in east-central Asia include backarc extension west of the Asia-Pacific Cretaceous convergent margin (Watson et al., 1987). In addition, oroclinal closure of the Mongol-Okhotsk ocean (Fig. 18; Zonenshain et al., 1990) may have induced transtension, although the magnitude and timing of this event are poorly documented. We also note that in southernmost Mongolia extension closely followed two important compressional events: (1) closure of the Junggar-Hegen ocean between the North China block and the southern Mongolian arcs at the end of the Permian (Amory et al., 1994; Amory, 1996), and (2) emplacement of north-vergent thrust sheets (Fig. 8) during the Jurassic (Hendrix et al., 1996; Zheng et al., 1996). Recognition of the rapid transition from compressional to extensional tectonics during the middle Mesozoic introduces another possible factor driving regional extension in southern Mongolia: gravitational collapse (Graham et al., 1996).

Mesozoic metamorphic core complexes similar to Yagan–Onch Hayrhan were described by Davis et al. (1996) in the Yinshan-Yanshan belt, by van der Beek et al. (1996) in the Baikal region of southern Siberia (Fig. 18), and by Hacker et al.

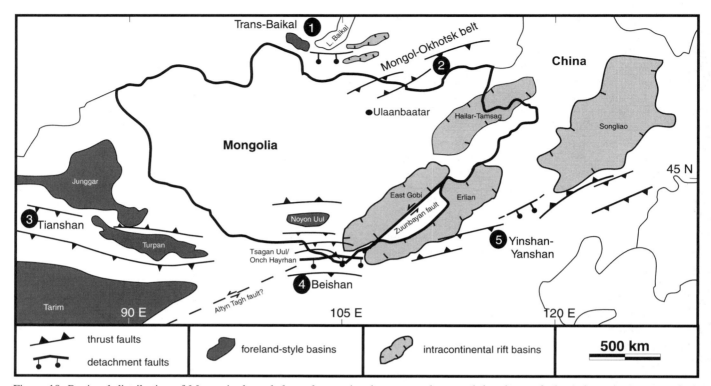

Figure 18. Regional distribution of Mesozoic thrust belts and extensional core complexes and detachment faults (schematic representation). 1, Trans-Baikal region. Late Triassic–Jurassic rift basin initiation, Middle Jurassic inversion, Late Jurassic development of west Baikal foredeep, and Early Cretaceous (140–120 Ma) core complex denudation (van der Beek et al., 1996; cf. Delvaux et al., 1995); 2, Mongol-Okhotsk fold belt. Diachronous west-east closure of Mongol-Okhotsk ocean from Permian to Cretaceous(?) (Zonenshain et al., 1990). 3, Tian Shan. Thrusting and foreland-style basin development, early Mesozoic–Cenozoic (Hendrix et al., 1992). 4, Beishan region. Late Permian–Middle Jurassic(?) thrusting, followed by Middle-Late Jurassic extension and core complex formation (Zheng et al., 1991). 5, Yinshan-Yanshan belt. Jurassic–Cretaceous thrusting and Cretaceous extensional detachment faulting (Davis et al., 1996, 1998a).

(1995) in the Dabie Shan. The association of all of these regions of detachment faulting with preceding episodes of significant crustal shortening suggests a common denominator. Gravitational collapse of overthickened crust underpinning contractile orogenic mountain belts may initiate detachment faulting, as has been suggested for many core complexes and related collapse basins worldwide (Burchfiel et al., 1992; Constenius, 1996). This mechanism also may have been an important factor in the regional segmentation of high- and low-strain extensional provinces of Mesozoic east-central Asia.

The late Mesozoic central Asian extensional province is superposed on an accretionary collage of diverse terranes (Zhang et al., 1984), and thus it provides an interesting comparison to the well-studied U.S. Basin and Range province (Fig. 19). Both examples show similarities in terms of the close timing between periods of shortening and extension, and the formation of both metamorphic core complexes and rift provinces. However, metamorphic core complexes of the western United States formed on long-stabilized Precambrian crust of the North American continent that heated and thickened during Mesozoic arc and/or retroarc tectonism (Graham, 1996). In contrast, highly extended regions in Asia such as the Onch Hayrhan area formed primarily along plate and microplate suture zones, on crust pre-

viously thickened during plate collision. Thus, the distribution of these core complexes appears to be strongly controlled by the locations of collisional orogens formed during the tectonic amalgamation of the Asian continent.

CONCLUSIONS

Structural and sedimentary data from the Yagan–Onch Hayrhan metamorphic core complex and from Har Hotol in the East Gobi basin indicate that both high- and low-strain extensional regimes were active during the Early Cretaceous in southern Mongolia. As a result, both a metamorphic core complex and a classic intracontinental rift system are preserved, along with their associated and contemporaneous sedimentary deposits. A transfer fault was partly responsible for locally accommodating strain along the surface termination of the main detachment fault at the Onch Hayrhan metamorphic core complex. Other, unmapped accommodation zones probably helped to partition the high- and low-strain provinces of the East Gobi basin.

Sedimentary sequences at either end of the East Gobi basin also reflect distinct structural settings. Alluvial conglomerate of the Lower Cretaceous Onch Hayrhan section is indicative of rapid, proximal sedimentation driven by large-magnitude ex-

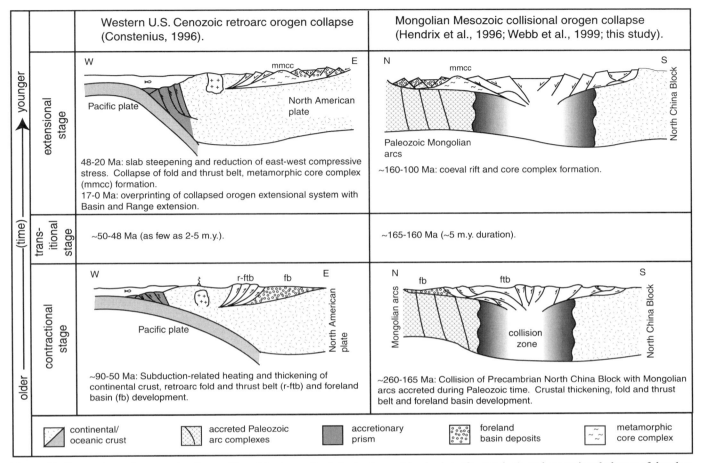

Figure 19. Comparison of contractional, transitional (between final shortening and beginning of extension), and extensional phases of development in Cenozoic of western United States (left column) and Mesozoic of Mongolia (right column). Cross sections are schematic representations only, modeled after Dickinson (1976) and Muñoz (1992).

tension in the area. By comparison, rift sedimentation in the half-graben subbasins near the Zuunbayan oil field was dominantly fluvial-lacustrine and suggestive of longer lived and relatively stable depositional systems.

Regionally, the East Gobi basin is one of several extensional provinces in central Asia that are recognized to have formed immediately following contractional tectonics. The presence of core complexes in many of these areas supports the idea that gravitational collapse of overthickened crust that formed along collisional terrane boundaries may have helped focus regions of high-strain extension during the Mesozoic.

ACKNOWLEDGMENTS

Funding for this project was provided by Nescor Energy and Roc Oil Company, Limited, and by National Science Foundation grants EAR-9708207 and EAR-961455 to S. Graham and M. Hendrix, respectively. C. Johnson was supported by a Gabilan Stanford Graduate Fellowship. We thank Julio Freidman and Kenneth Ridgway, whose comments significantly enhanced the quality of the final manuscript. We also thank A. Chimitsuren and our colleagues at the Mongolian Academy of Sciences, Institute of Geology and Mineral Resources, and the Mongolian Paleontological Institute for their scientific and logistic support.

REFERENCES CITED

Amory, J., 1996, Permian sedimentation and tectonics of southern Mongolia [M.S. thesis]: Stanford, California, Stanford University, 183 p.

Amory, J.Y., Hendrix, M.S., Lamb, M., Keller, A.M., Badarch, G., and Tomurtogoo, O., 1994, Permian sedimentation and tectonics of southern Mongolia: Implications for a time-transgressive collision with north China: Geological Society of America Abstracts with Programs, v. 26, no. 7, p. A-242.

Angelier, J., 1994, Fault slip analysis and palaeostress reconstruction, *in* Hancock, P.L., ed., Continental deformation: Tarrytown, New York, Pergamon Press, p. 53–100.

Baljinnyam, I., Bayasgalan, A., Borisov, B.A., Cisternas, A., Dem'yanovich, M.G., Ganbaatar, L., Kochetkov, V.M., Kurushin, R.A., Molnar, P., Philip, H., and Vaschilov, Y.Y., 1993, Ruptures of major earthquakes and active deformation in Mongolia and its surroundings: Geological Society of America Memoir 181, 62 p.

Beratan, K., 1991, Miocene synextension sedimentation patterns, Whipple Mountains, southeastern California: Implications for the geometry of the Whipple detachment system: Journal of Geophysical Research, v. 96, p. 12425–12442.

Beratan, K., 1998, Structural control of rock-avalanche deposition in the Colorado River extensional corridor, southeastern California–western Arizona, *in* Faulds, J.E., and Stewart, J.H., eds., Accommodation zones and transfer zones: The regional segmentation of the Basin and Range Province: Geological Society of America Special Paper 323, p. 115–125.

Blair, T.C., and McPherson, J.G., 1994, Alluvial fans and their natural distinction from rivers based on morphology, hydraulic processes, sedimentary processes, and facies assemblages: Journal of Sedimentary Research, v. A64, p. 450–489.

Burchfiel, B.C., Chen, Z., Hodges, K., Liu, Y., Royden, L., Deng, C., and Xu, J., 1992, The South Tibetan detachment system, Himalayan Orogen; extension contemporaneous with and parallel to shortening in a collisional mountain belt: Geological Society of America Special Paper 269, 41 p.

Coney, P.J., 1980, Cordilleran metamorphic core complexes; an overview, *in* Crittenden, M.D., Jr., et al., eds., Cordilleran metamorphic core complexes: Geological Society of America Memoir 153, p. 7–31.

Constenius, K.N., 1996, Late Paleogene extensional collapse of the Cordilleran foreland fold and thrust belt: Geological Society of America Bulletin, v. 108, p. 20–39.

Crowell, J.C., 1974, Origin of late Cenozoic basins in southern California, *in* Dickinson, W.R., ed., Tectonics and sedimentation: Society of Economic Paleontologists and Mineralogists Special Publication 22, p. 190–204.

Cunningham, W.D., Windley, B.F., Owen, L.A., Barry, T., Dorjnamjaa, D., and Badamgarav, J., 1997, Geometry and style of partitioned deformation within a late Cenozoic transpressional zone in the eastern Gobi Altai Mountains, Mongolia: Tectonophysics, v. 277, p. 285–306.

Davis, G., Xianglin, Q., Zheng, Y., Tong, H., Yu, H., Wang, C., Gehrels, G., Shafiquallah, M., and Fryxell, J., 1996, Mesozoic deformation and plutonism in the Yunmeng Shan: A metamorphic core complex north of Beijing, China, *in* Yin, A., and Harrison, M., eds., The Tectonic Evolution of Asia: Cambridge, Cambridge University Press, p. 253–280.

Davis, G.A., Cong, W., Zheng, Y., Zhang, J., Zhang, C., and Gehrels, G.E., 1998a, The enigmatic Yinshan fold-and-thrust belt of northern China; new views on its intraplate contractional styles: Geology, v. 26, p. 43–46.

Davis, G.A., Yadong, Z., Cong, W., Darby, B.J., Changhou, Z., and Gehrels, G., 1998b, Geometry and geochronology of Yanshan belt tectonics: Collected works of international symposium on geological science: Beijing, China, Peking University, p. 275–292.

Davis, G.A., Zheng Yadong, Wang Cong, Darby, B.J., and Hua, Yonggang, 1998c, Geologic introduction and field guide to the Daqing Shan thrust, Daqing Shan, Nei Mongol, China: Hohhot, China, Nei Mongol Bureau of Geology and Mineral Resources, Yinshan-Yanshan major thrust and nappe structures field conference, May 8–11, Hohhot, Nei Mongol, China, 23 p.

Delvaux, D., Moeys, R., Stapel, G., Melnikov, A., and Ermikov, V.Y.V., 1995, Palaeostress reconstructions and geodynamics of the Baikal region, Central Asia; Part I, Palaeozoic and Mesozoic pre-rift evolution: Tectonophysics, v. 252, p. 61–101.

Dickinson, W.R., 1976, Plate tectonic evolution of sedimentary basins, *in* Dickinson, W.R., and Yarborough, H., eds., Plate tectonics and hydrocarbon accumulation: American Association of Petroleum Geologists Continuing Education Course Note Series 1, p. 1–62.

Duebendorfer, E.M., and Wallin, E.T., 1991, Basin development and syntectonic sedimentation associated with kinematically coupled strike-slip and detachment faulting, southern Nevada: Geology, v. 19, p. 87–90.

Ebinger, C.J., Rosendahl, B.R., and Reynolds, D.J., 1987, Tectonic model of the Malawi rift, Africa: Tectonophysics, v. 141, p. 215–235.

Faulds, J.E., and Varga, R.J., 1998, The role of accommodation zones and transfer zones in the regional segmentation of extended terranes, *in* Faulds, J.E., and Stewart, J.H., eds., Accommodation zones and transfer zones: The regional segmentation of the Basin and Range Province: Geological Society of America Special Paper 323, p. 1–45.

Friedmann, J.S., and Burbank, D.W., 1995, Rift basins and supradetachment basins: Intracontinental extensional end members: Basin Research, v. 7, p. 109–127.

Gawthorpe, R.L., Fraser, A.J., and Collier, R.E., 1994, Sequence stratigraphy in active extensional basins: Implications for the interpretation of ancient basin-fills: Marine and Petroleum Geology, v. 11, p. 642–658.

Graham, S.A., 1996, Controls on intracontinental deformation in central Asia: Geological Society of America Abstracts with Programs, v. 28, no. 7, p. A-112.

Graham, S.A., Hendrix, M.S., Wang, L.B., and Carroll, A.R., 1993, Collisional successor basins of western China: Impact of tectonics inheritance of sand composition: Geological Society of America Bulletin, v. 105, p. 323–344.

Graham, S.A., Hendrix, M.S., Badarch, G., and Badamgarav, D., 1996, Sedimentary record of transition from contractile to extensional tectonism, Mesozoic, southern Mongolia: Geological Society of America Abstracts with Programs, v. 28, no. 7, p. A-68.

Hacker, B., Ratschbacher, L., and Webb, L., 1995, What brought them up? Exhumation of the Dabie Shan ultrahigh-pressure rocks: Geology, v. 23, p. 743–746.

Hendrix, M.S., Graham, S.A., Carroll, A.R., Sobel, E., McKnight, C.L., Schulein, B.S., and Wang, Z., 1992, Sedimentary record and climatic implications of recurrent deformation in the Tian Shan: Evidence from Mesozoic strata of the north Tarim, South Junggar, and Turpan basins, northwest China: Geological Society of America Bulletin, v. 104, p. 53–79.

Hendrix, M.S., Graham, S.A., Amory, J.Y., and Badarch, G., 1996, Noyon Uul syncline, southern Mongolia; lower Mesozoic sedimentary record of the tectonic amalgamation of Central Asia: Geological Society of America Bulletin, v. 108, p. 1256–1274.

Ingersoll, R.V., 1990, Actualistic sandstone petrofacies; discriminating modern and ancient source rocks: Geology, v. 18, p. 733–736.

Ingersoll, R.V., Bullard, T.F., Ford, R.L., Grimm, J.P., Pickle, J.D., and Sares, S.W., 1984, The effect of grain size on detrital modes: A test of the Gazzi-Dickinson point-counting method: Journal of Sedimentary Petrology, v. 54, p. 103–116.

Janecke, S., McIntosh, W., and Good, S., 1999, Testing models of rift basins: Structure and stratigraphy of an Eocene-Oligocene supradetachment basin, Muddy Creek half graben, south-west Montana: Basin Research, v. 11, p. 143–165.

Jerzykiewicz, T., and Russell, D.A., 1991, Late Mesozoic stratigraphy and vertebrates of the Gobi Basin: Cretaceous Research, v. 12, p. 345–377.

Johnson, C., Graham, S., Webb, L., Hendrix, M., Badarch, G., Sjostrom, D., Beck, M., and Lenegan, R., 1997a, Sedimentary response to Late Mesozoic extension, southern Mongolia: Eos (Transactions, American Geophysical Union), v. 78, p. F175.

Johnson, C.L., Graham, S.A., Hendrix, M.S., and Badarch, G., 1997b, Sedimentary record of Jurassic-Cretaceous rifting, southeastern Mongolia: Implications for the Mesozoic tectonic evolution of central Asia: Geological Society of America Abstracts with Programs, v. 29, no. 6, p. A-228.

Keller, A.M., and Hendrix, M.S., 1997, Paleoclimatologic analysis of a Late Jurassic petrified forest, southeastern Mongolia: Palaios, v. 12, p. 282–291.

Lamb, M.A., and Badarch, G., 1997, Paleozoic sedimentary basins and volcanic-arc systems of southern Mongolia; new stratigraphic and sedimentologic constraints: International Geology Review, v. 39, p. 542–576.

Lamb, M.A., Hanson, A.D., Graham, S.A., Badarch, G., and Webb, L.E., 1999, Left-lateral sense offset of Upper Proterozoic to Paleozoic features across the Gobi Onon, Tost, and Zuunbayan faults in southern Mongolia and implications for other central Asian faults: Earth and Planetary Science Letters, v. 173, p. 183–194.

Lin Chansong, Li Sitian, Wan Yongxian, Ren Jangye, and Zhang Yanmei, 1997, Depositional systems, sequence stratigraphy and basin filling evolution of Erlian fault lacustrine basin, northeast China, *in* Lia Baojun and Li Sitian, eds., Basin analysis, global sedimentology, geology, and sedimentology, Proceedings of the 30th International Geological Congress: VSP, Utrecht, The Netherlands, p. 163–175.

Muñoz, J.A., 1992, Evolution of a continental collision belt: ECORS-Pyrenees crustal balanced cross-section, *in* McClay, K.R., ed., Thrust tectonics: London, Chapman and Hall, 447 p.

Passchier, C.W., and Trouw, R.A.J., 1996, Microtectonics: New York, Springer, 289 p.

Pavlova, E.E., Manankov, I.N., Morozova, I.P., Solovjeva, M.N., Suetenko, O.D., and Bogoslovskaya, M.F., 1991, Permian invertebrates of southern Mongolia: Joint Soviet-Mongolian Paleontological Expedition Transactions, v. 40, 173 p.

Ruzhentsev, S.V., Pospelov, I.I., and Badarch, G., 1989, Tectonics of the Mongolian Indosinides: Geotectonics, v. 23, p. 476–486.

Şengör, A.M.C., Natal'in, B.A., and Burtman, V.S., 1993, Evolution of the Altaid tectonic collage and Palaeozoic crustal growth in Eurasia: Nature, v. 364, p. 299–307.

Shirokov, V.Y., and Kopytchenko, V.N., 1983, Schematic geologic-structure map of the Unegt and Zuunbayan basins, *in* Analysis and summary of geological and geophysical materials on possibility of oil and gas-bearing basins of Mongolia, Volume 1, book 1: Moscow, All-Union Export and Import Corporation, 291 p.

Shuvalov, V.F., 1968, More information about the Upper Jurassic and Lower Cretaceous in the southeastern Mongolia Altai: Academy of Sciences, USSR, Doklady, Earth Sciences Section, v. 179, p. 31–33.

Shuvalov, V.F., 1975, Stratigraphy of Mesozoic deposits of central Mongolia, *in* Zaitsev, N.S., Luwsandansan, B., Martinson, G.G., Menner, V.V., Pavlova, T.G., Peive, A.V., Timfeev, P.P., Tumortogoo, O., and Yanshin, A.L., Stratigraphy of Mesozoic deposits of Mongolia: Transactions of the joint Soviet-Mongolia scientific research geological expedition, v. 13, p. 50–112.

Soreghan, M.J., Scholz, C.A., and Wells, J.T., 1999, Coarse-grained, deep-water sedimentation along a border fault margin of Lake Malawi, Africa: Seismic stratigraphic analysis: Journal of Sedimentary Research, v. 69, p. 832–846.

Suvorov, A.I., 1982, Strukturnyy plan i razlomy territorii Mongolii: Izvestiya Akademii Nauk SSSR, Seriya Geologicheskaya, v. 1982, p. 122–136.

Traynor, J.J., and Sladen, C., 1995, Tectonic and stratigraphic evolution of the Mongolian People's Republic and its influence on hydrocarbon geology and potential: Marine and Petroleum Geology, v. 12, p. 35–52.

van der Beek, P., Delvaux, D., Andriessen, P., and Levi, K., 1996, Early Cretaceous denudation related to convergent tectonics in the Baikal region, SE Siberia: Geological Society of London Journal, v. 153, p. 515–523.

Watson, M.P., Hayward, A.B., Parkinson, D.N., and Zhang, Z.M., 1987, Plate tectonic history, basin development and petroleum source rock deposition onshore China: Marine and Petroleum Geology, v. 4, p. 205–225.

Webb, L., Graham, S., Badarch, G., Johnson, C., Hendrix, M., Beck, M., Sjostrom, D., and Lenegan, R., 1997, Characteristics and implications of the Onch Hayrhan metamorphic core complex of southern Mongolia: Eos (Transactions, American Geophysical Union), v. 78, p. F174.

Webb, L.E., Graham, S.A., Johnson, C.L., Badarch, G., and Hendrix, M.S., 1999, Occurrence, age, and implications of the Yagan-Onch Hayrhan metamorphic core complex, southern Mongolia: Geology, v. 27, no. 2, p. 143.

Yanshin, A.L., 1989, Map of geological formations of the Mongolian People's Republic: Moscow, Academia Nauka USSR, scale 1:1 500 000.

Yue, Y., and Liou, J.G., 1999, Two stage evolution model for the Altyn Tagh fault, China: Geology, v. 27, p. 227–230.

Zhang, Z., Liou, J.G., and Coleman, R., 1984, An outline of the plate tectonics of China: Geological Society of America Bulletin, v. 95, p. 295–312.

Zheng, Y., and Zhang, Q., 1993, The Yagan metamorphic core complex and extensional detachment fault in Inner Mongolia: Acta Geologica Sinica, v. 67, p. 301–309.

Zheng, Y., Wang, S., and Wang, Y., 1991, An enormous thrust nappe and extensional metamorphic core complex in Sino-Mongolian boundary area: Science in China, ser. B, v. 34, p. 1145–1152.

Zheng, Y., Zhang, Q., Wang, Y., Liu, R., Wang, S.G., Zuo, G., Wang, S.Z., Lkaasuren, B., Badarch, G., and Badamgarav, Z., 1996, Great Jurassic thrust sheets in Beishan (North Mountains); Gobi areas of China and southern Mongolia: Journal of Structural Geology, v. 18, p. 1111–1126.

Zonenshain, L.P., Markova, N.G., and Nagibina, M.S., 1971, Relationship between the Paleozoic and Mesozoic structures of Mongolia: Geotectonics, v. 4, p. 229–233.

Zonenshain, L., Kuzmin, M., Natapov, L., and Page, B., 1990, Geology of the USSR: A plate tectonic synthesis: American Geophysical Union Geodynamics Series, v. 21, 242 p.

Zuffa, G.G., 1980, Hybrid arenites: Their composition and classification: Journal of Sedimentary Petrology, v. 50, p. 21–29.

MANUSCRIPT ACCEPTED BY THE SOCIETY JUNE 5, 2000

Index

[Italic page numbers indicate major references]